Gmelin Handbook of Inorganic Chemistry

8th Edition

Organometallic Compounds in the Gmelin Handbook

The following listing indicates in which volumes these compounds are discussed or are referred to:

Ag	Silber B 5 (1975)
Au	Organogold Compounds (1980)
Bi	Bismut-Organische Verbindungen (1977)
Co	Kobalt-Organische Verbindungen 1 (1973), 2 (1973), Kobalt Erg.-Bd. A (1961), B 1 (1963), B 2 (1964)
Cr	Chrom-Organische Verbindungen (1971)
Cu	Organocopper Compounds 1 (1985), 2 (1983)
Fe	Eisen-Organische Verbindungen A 1 (1974), A 2 (1977), A 3 (1978), A 4 (1980), A 5 (1981), A 6 (1977), A 7 (1980), Organoiron Compounds A 8 (1986) present volume, Eisen-Organische Verbindungen B 1 (partly in English; 1976), B 2 (1978), B 3 (partly in English; 1979), B 4 (1978), B 5 (1978), Organoiron Compounds B 6 (1981), B 7 (1981), B 8 to B 10 (1985), B 11 (1983), B 12 (1984), Eisen-Organische Verbindungen C 1 (1979), C 2 (1979), Organoiron Compounds C 3 (1980), C 4 (1981), C 5 (1981), C 7 (1986), and Eisen B (1929–1932)
Hf	Organohafnium Compounds (1973)
Nb	Niob B 4 (1973)
Ni	Nickel-Organische Verbindungen 1 (1975), 2 (1974), Register (1975), Nickel B 3 (1966), and C 1 (1968), C 2 (1969)
Np, Pu	Transurane C (partly in English; 1972)
Pt	Platin C (1939) and D (1957)
Ru	Ruthenium Erg.-Bd. (1970)
Sb	Organoantimony Compounds 1 (1981), 2 (1981), 3 (1982)
Sc, Y, La to Lu	D 6 (1983)
Sn	Zinn-Organische Verbindungen 1 (1975), 2 (1975), 3 (1976), 4 (1976), 5 (1978), 6 (1979), Organotin Compounds 7 (1980), 8 (1981), 9 (1982), 10 (1983), 11 (1984), 12 (1985)
Ta	Tantal B 2 (1971)
Ti	Titan-Organische Verbindungen 1 (1977), 2 (1980), 3 (1984), 4 and Register (1984)
U	Uranium Suppl. Vol. E 2 (1980)
V	Vanadium-Organische Verbindungen (1971), Vanadium B (1967)
Zr	Organozirconium Compounds (1973)

Gmelin Handbook of Inorganic Chemistry

8th Edition

Gmelin Handbuch der Anorganischen Chemie

Achte, völlig neu bearbeitete Auflage

Prepared and issued by

Gmelin-Institut für Anorganische Chemie
der Max-Planck-Gesellschaft
zur Förderung der Wissenschaften

Director: Ekkehard Fluck

Founded by	Leopold Gmelin
8th Edition	8th Edition begun under the auspices of the Deutsche Chemische Gesellschaft by R. J. Meyer
Continued by	E. H. E. Pietsch and A. Kotowski, and by Margot Becke-Goehring

Springer-Verlag Berlin · Heidelberg · New York · Tokyo 1986

Gmelin Handbook of Inorganic Chemistry

8th Edition

Fe Organoiron Compounds

Part A

Ferrocene 8

Mononuclear Disubstituted Ferrocene Derivatives with C-, H-, and O-Containing Substituents (continued from "Ferrocene" 7)

With 14 illustrations

AUTHOR Marianne Drößmar-Wolf

EDITOR Adolf Slawisch

FORMULA INDEX Edgar Rudolph

CHIEF EDITOR Adolf Slawisch

Springer-Verlag Berlin · Heidelberg · New York · Tokyo 1986

LITERATURE CLOSING DATE: 1984
IN SOME CASES MORE RECENT DATA HAVE BEEN CONSIDERED

Library of Congress Catalog Card Number: Agr 25-1383

ISBN 3-540-93527-4 Springer-Verlag, Berlin · Heidelberg · New York · Tokyo
ISBN 0-387-93527-4 Springer-Verlag, New York · Heidelberg · Berlin · Tokyo

Typesetting, printing, and bookbinding: LN-Druck Lübeck

Preface

"Organoiron Compounds" A, Ferrocene 8 systematically covers the literature through the end of 1984 and even includes some references published more recently. A formula index for the volumes "Organoiron Compounds" A, Ferrocene 7 and 8 provides ready access to the compounds covered.

This volume continues the description of ferrocenes $FeC_{10}H_8R^1R^2$ in which at least R^1 contains C, H, and O. The description of these type of compounds started in "Organoiron Compounds" A, Ferrocene 7 (with alcohols and phenols, their esters, ethers and acetones, and aldehydes).

Series A so far comprises volumes A 1 to A 8, and it has been surveyed in the preface to A 7 (1980).

The data in tables are given in abbreviated form without dimensions; for dimensions, explanations, and further abbreviations used, see p. VIII (next page). Additional remarks, if necessary, are given in the headings of the tables.

Frankfurt
November 1985

Adolf Slawisch

Remarks on Abbreviations and Dimensions

Many compounds in this volume are presented in tables in which numerous abbreviations are used, and the dimensions are omitted for the sake of conciseness. This necessitates the following clarification:

Temperatures are given in °C, otherwise K stands for Kelvin. Abbreviations used with temperatures are m.p. for melting point, b.p. for boiling point, dec. for decomposition, and subl. for sublimation. The solvent in parentheses following the m.p. is the solvent from which the compound is recrystallized.

NMR represents **nuclear magnetic resonance**. Chemical shifts are given as δ values in ppm; reference substances and signs for δ are shown in the scheme below:

		increasing field →	
		$\delta = 0$ for	
^1H	+	$Si(CH_3)_4$	−
^{11}B	+	$BF_3 \cdot O(C_2H_5)_2$	−
^{13}C	+	$Si(CH_3)_4$	−
^{19}F	−	$CFCl_3$	+
^{31}P	+	H_3PO_4	−

Coupling constants in Hz are given as $^nJ(A, B)$ and refer to labelled structural formulas, n is number of bonds between the coupled nuclei.

Multiplicities of the signals are abbreviated as s, d, t, q (singlet to quartet), quint, sext, sept, oct (quintet to octet), and m (multiplet); terms like dd (double doublet) and dq (double quartet) are also used. Assignments referring to labelled structural formulas are given in the form C-4, H-3,5.

For **Mössbauer spectra** both the isomer shift δ and the quadrupole splitting Δ are given in mm/s; the reference substances for δ are indicated after the numerical value, e.g. $\delta = 0.31$ (^{57}Co/Cu).

Optical spectra are labelled as IR (infrared) and UV (electronic spectrum including the visible region): IR bands are given in cm^{-1}; the assigned bands are usually labelled with the symbols ν for stretching vibration and δ and γ for deformation vibrations in-plane and out-of-plane, respectively. The indices "s" and "a" mean symmetrical and asymmetrical. Intensities occur in parentheses in the common qualitative terms (s, m, w, vs, br., etc.). The UV absorption maxima, λ_{max}, are given in nm followed by the extinction coefficient ε ($L \cdot cm^{-1} \cdot mol^{-1}$) or log ε in parentheses; sh means shoulder.

Solvents used and the temperatures (in °C or K) are given in parentheses immediately after the spectral symbol. Common solvents are given by their formula or name except THF and DMF, which represents tetrahydrofuran and $(CH_3)_2NCHO$, respectively. For solvent mixtures are given volume ratios.

For the **mass spectrum** a parenthetical (CI) means chemical ionization. In all other cases the ions are generated by electron impact. $[M]^+$ denotes the molecular ion.

Further abbreviations used are n_D^t for refractive index for the sodium line, 589 nm, at t in °C; $[\alpha]_\lambda^t$ and $[M]_\lambda^t$ for specific and molar rotation for light of wavelength λ in nm at t in °C; c for concentration, when no unit is stated either g/L or grams per 100 mL can be assumed; mol. wt for molecular weight; (meas.) for measured and (calc.) for calculated values; ESR for electron spin resonance; hfs for hyperfine structure.

Table of Contents

Mononuclear Disubstituted Ferrocene Derivatives with C-, H-, and O-Containing Substituents

Ferrocene 8

Mononuclear Disubstituted Ferrocene Derivatives with C-, H-, and O-Containing Substituents (continued from "Ferrocene" 7)

General References, in addition to those in

"Eisen-Organische Verbindungen" A Ferrocen 1, 1974, pp. 2/5,
"Eisen-Organische Verbindungen" A Ferrocen 2, 1977, p. 1,
"Eisen-Organische Verbindungen" A Ferrocen 4, 1980, p. 1,
"Eisen-Organische Verbindungen" A Ferrocen 5, 1981, p. 1,
"Eisen-Organische Verbindungen" A Ferrocen 6, 1977, p. 1,
"Eisen-Organische Verbindungen" A Ferrocen 7, 1980, pp. 1/2, are:

D. Astruc, Organo-Iron Complexes of Aromatic Compounds. Applications in Synthesis, Tetrahedron **39** [1983] 4027/95.

P. N. Gaponik, A. I. Lesnikovich, Yu. G. Orlik, Heterometallic Ferrocene-Based Transition Metal Compounds, Usp. Khim. **52** [1983] 294/320; Russ. Chem. Rev. **52** [1983] 168/83.

E. G. Perevalova, M. D. Reshetova, K. I. Grandberg, Metody Elementoorganicheskoi Khimii: Zhelezoorganicheskie Soedineniya: Ferrotsen [Methods of Organometallic Chemistry: Organoiron Compounds: Ferrocene], Nauka, Moscow 1983, pp. 1/544.

E. R. Milaeva, A. Z. Rubezhov, A. I. Prokof'ev, O. Yu. Okhlobystin, Unpaired Electron in Transition Metal Complexes, Usp. Khim. **51** [1982] 1638/73; Russ. Chem. Rev. **51** [1982] 942/60.

A. N. Nesmeyanov et al., Ferrocene and Related Compounds, Nauka, Moscow 1982, pp. 1/439.

I. Omae, Organometallic Intramolecular-Coordination Compounds Containing a Cyclopentadienyl Donor Ligand, Coord. Chem. Rev. **42** [1982] 31/54.

S. P. Solodovnikov, Investigation of Paramagnetic Metallocenes and Diarene Complexes of Transition Metals by Magnetic Resonance Methods, Usp. Khim. **51** [1982] 1674/97; Russ. Chem. Rev. **51** [1982] 961/74.

W. R. Cullen, J. D. Woollins, Ferrocene-Containing Metal Complexes, Coord. Chem. Rev. **39** [1981] 1/30.

J. C. Green, Gas Phase Photoelectron Spectra of d- and f-Block Organometallic Compounds, Struct. Bonding [Berlin] **43** [1981] 37/112.

G. B. Shul'pin, Diastereotopy in Transition Metal Complexes, Usp. Khim. **49** [1980] 1214/33; Russ. Chem. Rev. **49** [1980] 645/54.

R. E. Bozak, Photochemistry in the Metallocenes, Advan. Photochem. **8** [1971] 227/44.

Annual surveys are published in J. Organometal. Chem. and in Organometal. Chem.; see also Advan. Organometal. Chem. and Organometal. Chem. Rev.

5.2.3.5 Ketones

5.2.3.5.1 1,1′-Compounds

5.2.3.5.1.1 1,1′-Diacetylferrocene $Fe(C_5H_4-CO-CH_3)_2$

5.2.3.5.1.1.1 Formation and Preparation

From Ferrocene by the Friedel-Crafts Method. For general information regarding this method see "Ferrocene" 1, 1974, p. 142 ff. and "Ferrocene" 2, 1977, p. 181 ff. The industrial preparation is effected with $CH_3COCl/AlCl_3$ [34]. In the laboratory the acetylation is conducted predominantly by stepwise addition of ferrocene in CH_2Cl_2 to a mixture of $CH_3COCl/AlCl_3$ (mole ratio 1/4/3 [76]; 1/(2.2 to 3)/(2.2 to 3) [2, p. 107], [3, 4, 8, 14, 24, 29, 30], [35, p. 143], [37, 58]; 1/(2 to 2.5)/(2 to 5) [10]; 1/2.8/0.25 [12, 33]) in CH_2Cl_2 with a reaction time of $\leqq$2 h at the boiling point [3, 4, 10, 14, 76], 30 to 35°C [30], or (better) room temperature [2, p. 107], [8, 12, 24, 29, 33], [35, p. 143], [37, 58]. The latter reduces the formation of tarry by-products. According to [3] no tarry products form even at the boiling point. CH_2Cl_2 is a favorable solvent for the preparation of diacylation products [3], which has the advantage, for instance with respect to CS_2, of giving higher yields and of being easier and safer to handle [10]. According to [8, 14] the reaction product contains no $C_5H_5FeC_5H_4-CO-CH_3$. The yields of $Fe(C_5H_4-CO-CH_3)_2$ are between 58 and 86% [2, p. 107], [3, 4, 10, 12, 14, 24, 29, 33], [35, p. 143], [58, 76, 77]. The method is also appropriate for carrying out in semi-industrial quantities [30]. Adding the ferrocene solution dropwise at room temperature over a period of 7 h to the CH_2Cl_2 solution of $CH_3COCl/AlCl_3$ (mole ratio 1/12/11) and holding 4 h at room temperature before further processing gives a reaction product in 93% yield that consists of ~88.9% $Fe(C_5H_4-CO-CH_3)_2$, ~8.9% $C_5H_5FeC_5H_3(CO-CH_3)_2$-1,2, and ~2.8% $C_5H_5FeC_5H_4-CO-CH_3$ [45].

The reaction of ferrocene with $CH_3COCl/AlCl_3$ in CS_2 proceeds smoothly [1]. The yield amounts on the average to 50%. Slow addition of a mixture of ferrocene in CS_2 and CH_3COCl to a suspension of $AlCl_3$ in CS_2 (mole ratio 1/(2 to 2.2)/(2.5 to 3)) and subsequent 90 min boiling gives a yield of ca. 60% [7, 10, 37], 50% [17]. This can according to [20, 26] be increased to 73.5% if after 4 h boiling one lets the reaction mixture stand overnight before processing. On addition of ferrocene to $CH_3COCl/AlCl_3$ in CS_2 (mole ratio 1/2/2) the yield after 5 h reaction at room temperature is 67% [2, p. 111]. Conducting the reaction with a mole ratio of 1/3.5/3 with 12 h boiling gives a yield of ~33 % [68]. A mole ratio of 1/3/4 with weaker boiling gives a yield of 24%. This increases to 51% on heating 45 parts by weight of $AlCl_3$ and 37 parts by volume of CH_3COCl in CS_2 slowly to boiling and reacting the mixture with 27 parts by weight of ferrocene up to the formation of tarry products; subsequently ferrocene and $AlCl_3$ are added alternately up to a total of 54 parts by weight of ferrocene and 90 parts by weight of $AlCl_3$, and finally once more 37 parts by volume of CH_3COCl [3]. Addition of $AlCl_3$ in CS_2 slowly to CH_3COCl and ferrocene in CS_2 and stirring overnight is recommended by [21, 22]; the yield is 17.5% [21].

Addition of $CH_3COCl/AlCl_3$ to ferrocene (mole ratio >3/>3/1) in $CHCl_3$ at 0°C gives after 1 h reaction $Fe(C_5H_4-CO-CH_3)_2$ in 73% yield, with $C_5H_5FeC_5H_4-CO-CH_3$ in 3% yield and a 6% recovery of the starting ferrocene. If the ferrocene before addition of the $CH_3COCl/AlCl_3$ complex (mole ratio 1/<1/<1) is in addition mixed with the equivalent quantity of $AlCl_3$, then the yields obtained after a reaction time of 1.5 h are 24.5 and 8.8%, respectively; 56% of the ferrocene is recovered [9]. The reaction in ClH_2CCH_2Cl at 40°C for 1.75 h yields in addition to a tarry mass only small quantities of $Fe(C_5H_4-CO-CH_3)_2$ [3]. None was obtained in $C_6H_5NO_2$ or petroleum ether at a water bath temperature of 60°C [10].

The acetylation with $(CH_3CO)_2O/AlCl_3$ produces according to [10] less tarry by-products than that with CH_3COCl. Conditions are optimal when $(CH_3CO)_2O/AlCl_3$ in CS_2 reacts at the boiling point with ferrocene (2/5/1 mole ratio) in CS_2 and the hydrolysis is conducted after a reaction time of 2 h only with ice, without HCl addition (74% yield) [10]. The yield is 46% when $AlCl_3$ and ferrocene in HCl_2CCCl_2H are reacted for 1 h at 0°C, then 1 h at room temperature with $(CH_3CO)_2O$ (mole ratio >5/1/<3) in HCl_2CCCl_2H [11]. A raw product is obtained after 1 h in CH_2Cl_2 in the presence of an excess of BF_3 as catalyst at 0°C, that consists of ~86 wt% $Fe(C_5H_4-CO-CH_3)_2$, 9 wt% $C_5H_5FeC_5H_4-CO-CH_3$, and 4 wt% $C_5H_5FeC_5H_3(CO-CH_3)_2$-1,2. It is expedient to remove the by-products and unreacted ferrocene from the raw product by extraction with petroleum ether, which dissolves only a small part of the $Fe(C_5H_4-CO-CH_3)_2$. It is important to use an HF-free BF_3, since even low HF concentrations hinder the formation of the diacetyl compound [36]. Also, with 85% H_3PO_4 as the catalyst the isomeric diacetyl compounds and $C_5H_5FeC_5H_4-CO-CH_3$ were obtained [27]. Warming the reaction mixture [49, 51] and lengthening the reaction time increase the yield of $Fe(C_5H_4-CO-CH_3)_2$ at the expense of the $C_5H_5FeC_5H_4-CO-CH_3$ [49]. The course of the reaction can be followed by thin layer chromatography on silica gel in benzene or CH_2Cl_2 with 30/1 C_6H_6/C_2H_5OH as developing solvent [27] or by high pressure liquid chromatography on SiO_2 with 10/1 ether/CH_3OH as mobile phase [49].

From Ferrocene Derivatives. Acetylation of $C_5H_5FeC_5H_4-CO-CH_3$ with $CH_3COCl/AlCl_3$ (1/1 mole ratio) by the Friedel-Crafts method in $CHCl_3$ requires the addition of solid $AlCl_3$. The yield after 2 h reaction then is 85% with 10% recovery of the unreacted $C_5H_5FeC_5H_4-CO-CH_3$. Without addition of solid $AlCl_3$ the yield is only 1% [9]. In CS_2 the yield is 89% with a 1/1/2 mole ratio of $C_5H_5FeC_5H_4-CO-CH_3/CH_3COCl/AlCl_3$ after a subsequent 30 min refluxing [6]. Oxidation of $Fe(C_5H_4-CHOH-CH_3)_2$ or $CH_3-CO-C_5H_4FeC_5H_4-CHOH-CH_3$ with $[(C_6H_5)_3SiO]_2CrO_2$ at 50°C in C_6H_6 in the dark for 24 h gives yields of 68 and 78%, respectively [59]. Hydrolysis of I with KOH/H_2O gives a yield of ~85% [38]. The compound is formed from $C_2H_5-C_5H_4FeC_5H_4-CO-CH_3$ by oxidation with MnO_2 in $CHCl_3$ at the boiling point (17 d reaction time) [46].

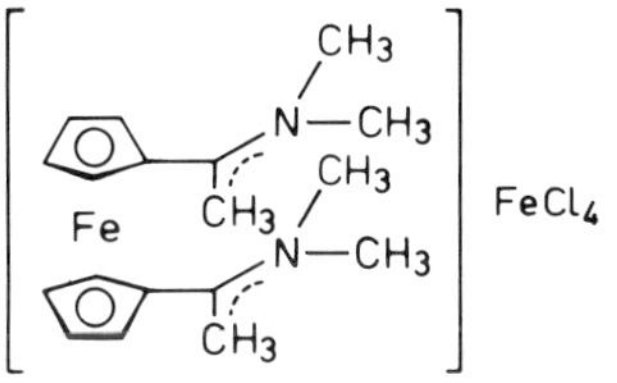

I

$Fe(C_5H_4-CO-CH_3)_2$ is also formed by the acetylation of $C_5H_5FeC_5H_4-I$ with a slight excess of $CH_3COCl/AlCl_3$ in CH_2Cl_2 at room temperature (yield ~15%) along with $C_5H_5FeC_5H_4-CO-CH_3$ (<1%) [18] or of $C_5H_5FeC_5H_4-Cl$ at 0°C to the boiling point (total reaction time 3.5 h; 6% yield) along with $CH_3-CO-C_5H_4FeC_5H_3(CO-CH_3-1)Cl$-3 (yield <9%) and $CH_3-CO-C_5H_4FeC_5H_4-Cl$ (5%) [53] as well as by oxidation of $Fe(C_5H_4-C_2H_5)_2$ with MnO_2 in boiling methylcyclohexane (yield 1.4%) along with $CH_3-CO-C_5H_4FeC_5H_4-C_2H_5$ (19%); the recovery of $Fe(C_5H_4-C_2H_5)_2$ was 44% [35, p. 144]. The reaction of $[FeC_{10}H_{10}]FeX_4$ (X = Cl, Br) with excess $(CH_3CO)_2O$ at the boiling point gives a 12% yield along with 4% of $C_5H_5FeC_5H_4-CO-CH_3$; the reaction is accompanied by considerable resin formation. In the presence of $AlCl_3$ only traces of the compound form. $[FeC_{10}H_{10}]BF_4$ gives with boiling $(CH_3CO)_2O$ the $Fe(C_5H_4-CO-CH_3)_2$ in 1% yield along with $C_5H_5FeC_5H_4-CO-CH_3$ (28%) and $FeC_{10}H_{10}$ (34%). In boiling $C_6H_5NO_2$ the di- and monoacetyl compounds are obtained in yields of 6 and 10%, respectively [32]. If $C_5H_5FeC_5H_4-CO-CH_3$ in isopropanol/hydrochloric acid is irradiated and the oil thus formed is treated with KOH and $FeCl_2$, then the compound forms in an ~5% yield [39]. It forms also on standing of $Fe(C_5H_4-C(Cl)=CH_2)_2$ in an acidic medium [23] and in traces during the autoxidation of $Fe(C_5H_4-C(CH_3)=CH_2)_2$ in CCl_4 on introducing O_2 in the dark in the presence of WCl_6 as a catalyst. Under basic conditions in CH_2Cl_2 or CH_2Cl_2/CH_3OH oxygenation occurs on the other hand only after photosensitization by addition of sodium copper chlorophyllin and irradiation with light of an Na vapor lamp [32]; cf. "Ferrocene" 7, 1980, p. 87.

References on pp. 5/7

From Acetylcyclopentadienide. $[C_5H_4{-}CO{-}CH_3]^-$ obtained from C_5H_6, Na, and CH_3COCl (molar ratio 1/2/1) in $CH_3OCH_2CH_2OCH_3$ [5] or tetrahydrofuran [15] at −70 to −50°C [5] and room temperature [15], respectively, has been reacted with $FeCl_2$ at room temperature in the corresponding solvent [5, 15] to give a 2% yield [15]. Isolated $Na[C_5H_4{-}CO{-}CH_3]$ gives a yield of 30% [72].

Deuterated and ^{59}Fe-Labelled Products. $Fe(C_5H_4{-}CO{-}CD_3)_2$ is gotten by acetylation of ferrocene with CD_3COCl [19, 64]. $^{59}Fe(C_5H_4{-}CO{-}CH_3)_2$ is obtained in a radiochemical yield of 15% on 2 h heating of $Fe(C_5H_4{-}CO{-}CH_3)_2$ with $^{59}FeCl_3$ in a 328/1 mole ratio in a sealed tube to 180°C and subsequent chromatographic processing [71]. On 60 min heating of $C_5H_5{}^{59}FeC_5H_4{-}CO{-}CH_3$ with $C_5H_5FeC_5H_4{-}CO_2CH_3$ in a 1/1 mole ratio in the presence of catalytic quantities of $FeCl_3$ in a sealed tube it forms in 3.8% yield in addition to 0.8% $^{59}Fe(C_5H_4{-}CO_2CH_3)_2$. Identification and purification take place by thin layer chromatography on SiO_2 in 5/1 $C_6H_6/CH_3CO_2C_2H_5$ [50].

Purification

The compound is purified by recrystallization from H_2O [3, 4, 57], 3/1 H_2O/acetone [30], isopropanol [24], petroleum ether [10, 66], hexane [61], cyclohexane [29], benzene [7, 36], benzene/hexane [20], CCl_4 [66], and/or sublimation in vacuum [56, 58, 61] at 120°C [58].

Separation

The separation from the most frequently occurring accompanying compounds from the Friedel-Crafts acetylation, ferrocene and $C_5H_5FeC_5H_4{-}CO{-}CH_3$, can be carried out by preparative dry-column chromatography on Al_2O_3 (activity III) with CH_2Cl_2 as developing solvent [41], by size-exclusion chromatography with pyridine as the mobile phase [55], or by high pressure [47, 49] and high performance liquid chromatography [54] with use of an inductively coupled Ar plasma emission source detector [54] with SiO_2 [47, 49, 54] as the stationary phase and toluene [54], 10/1 ether/CH_3OH [49], 30/1 ether/CH_3OH, 5/4 cyclohexane/ether, or 1/1 hexane/$CHCl_3$ [47] as the mobile phase. Good results are achieved by gas chromatography on Chromosorb W with 1% silicone rubber SE30 or 2.5% Apiezon-L at 200°C [60], see also [42]. The solution heat in silicone rubber SE30 has been calculated to be $\Delta H_s = 15.4$ kcal/mol (64.5 kJ/mol) from the dependence measured by Ayers et al. [25] of the retention times on the column temperature in the range from 125 to 200°C [62]. The retention sequence is ferrocene, $C_5H_5FeC_5H_4{-}CO{-}CH_3$, $Fe(C_5H_4{-}CO{-}CH_3)_2$ [47]. This agrees with the sequence of the R_f values ($Fe(C_5H_4{-}CO{-}CH_3)_2$ having the smallest value), as they hold for the thin layer chromatographic separation on Al_2O_3 [52] or silica gel [16, 27, 65, 74, 75] with C_6H_6/C_2H_5OH [16, 27, 52], 30/1 light benzine/C_2H_5OH, 30/4 light benzine/acetone, 30/2/2 toluene/ light benzine/acetone [65] or 15/1/3 toluene/petroleum ether/acetone [74, 75] as eluent.

If the compound is present in a distinct excess, it is appropriate to remove ferrocene, $C_5H_5FeC_5H_4{-}CO{-}CH_3$, and $C_5H_5FeC_5H_3(CO{-}CH_3)_2$-1,2 by extraction with petroleum ether from the raw product, during which process only a little $Fe(C_5H_4{-}CO{-}CH_3)_2$ goes into the extract [36].

The separation from mixtures with ferrocene, $Fe(C_5H_4{-}CH_2OH)_2$, and $Fe(C_5H_4{-}CO_2CH_3)_2$ can take place by gas chromatography with 2.5% Apiezon-L on Chromosorb W at 200°C [60] or thin layer chromatography on SiO_2 with toluene/light benzine/acetone (30/2/2) as the eluent [65].

$Fe(C_5H_4{-}CO{-}CH_3)_2$ can also be separated from ferrocene by column chromatography in cyclohexane on a gel of polystyrene crosslinked with 2% divinylbenzene [43] or in a chemical way. For such chemical separation the solution of the mixture in C_6H_6 is shaken 15 h with solid thiourea which is moistened with CH_3OH, whereupon the solid adduct $FeC_{10}H_{10} \cdot nSC(NH_2)_2$ separates while $Fe(C_5H_4{-}CO{-}CH_3)_2$ remains in solution, from which it, to be sure with contamination by ferrocene, can be isolated [44, 70]. $Fe(C_5H_4{-}CO{-}CH_3)_2$ is separated by thin layer

References on pp. 5/7

chromatography on Al_2O_3 from $C_5H_5FeC_5H_4$–CO–CH_3 and $C_5H_5FeC_5H_3(CO–CH_3)_2$-1,2 with 60/1 ether/$H_2O$ [45]. For this and the separation from the 1,2-diacetyl compound by extraction with petroleum ether of the raw product, see [36] and above. For separation of the three positional isomers of diacetylferrocene gas-liquid chromatography on polar phases in the temperature range 100 to 200°C is unsuitable [28]. The compound is separated by column chromatography in C_6H_6 on SiO_2 from C_2H_5–$C_5H_4FeC_5H_4$–CO–CH_3 from which the monoacetyl compound is eluted first with benzene containing 5% $CH_3CO_2C_2H_5$ and then the $Fe(C_5H_4–CO–CH_3)_2$ with benzene containing 20% $CH_3CO_2C_2H_5$ [35]. The separation from $Fe(C_5H_4–CO_2CH_3)_2$ takes place by thin layer chromatography on SiO_2 with 5/1 $C_6H_6/CH_3CO_2C_2H_5$ [50] and from II by size exclusion chromatography [55].

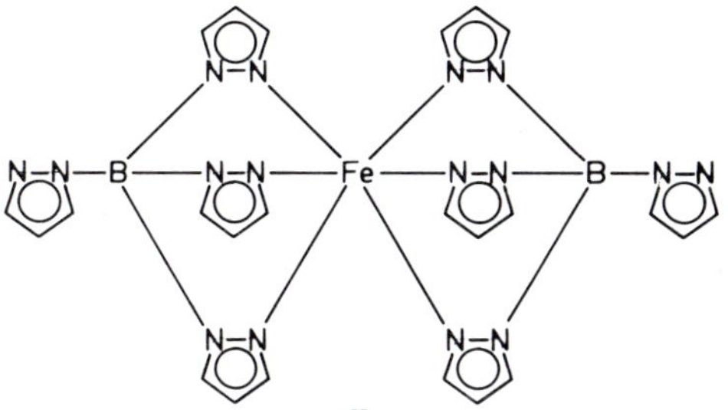

II

The R_f values for the thin layer chromatography on microcrystalline cellulose are 0.33 in 30/1 ether/CH_3OH, 0.66 in pentane, and 0.79 in CH_3OH [48]; on Al_2O_3 the R_f are 0.0375 in benzene, 0.131 in 50/1 benzene/ethanol [31], 0.19 in 60/1 ether/H_2O [45], 0.345 in 30/1 benzene/ethanol, and 0.665 in 15/1 benzene/ethanol [31].

Analysis

For the identification of polycrystalline samples by means of X-ray diffraction patterns see [13]. The high oxidation potential of $Fe(C_5H_4–CO–CH_3)_2$ prevents its determination by potentiometric titration with $Cu[ClO_4]_2$ in $CH_3CN/HClO_4/H_2O$ [63] or with $K_2Cr_2O_7$ in CH_3CO_2H/H_2O/acetone; beforehand it must be reduced back to $Fe(C_5H_4–C_2H_5)_2$ with $LiAlH_4/AlCl_3$ [67]. For a nondestructive determination of the Fe content by X-ray fluorescence analysis see [40]. After decomposition by wet digestion of the compound with 4/1 $HClO_4/HNO_3$ with addition of H_2O_2 the Fe content is determined photometrically by back titration of the ethylene-diaminetetraacetic acid excess with Bi^{3+} with use of 4-(2-pyridylazo)resorcinol as indicator [54].

For identification of $Fe(C_5H_4–CO–CH_3)_2$ in $C_5H_5FeC_5H_4$–CO–CH_3 by oscillography in $C_2H_5OH/HClO_4$ see [69].

References:

[1] R. B. Woodward, M. Rosenblum, M. C. Whiting (J. Am. Chem. Soc. **74** [1952] 3458/9). – [2] M. Rosenblum (Diss. Harvard Univ. 1953). – [3] N. V. De Bataafsche Petroleum Maatschappij (Neth. 96227 [1953/60]; C.A. **57** [1962] 13808). – [4] N. V. De Bataafsche Petroleum Maatschappij, A. C. Nixon, T. P. Rudy (Brit. 763550 [1954/56]; C.A. **1957** 10588; Ger. 1003727 [1954/57]). – [5] Union Carbide Corp., M. A. Lynch, J. C. Brantley (Brit. 785760 [1954/57]; C.A. **1958** 11126).

[6] A. N. Nesmeyanov, N. A. Volkenau (Dokl. Akad. Nauk SSSR **111** [1956] 605/8; Proc. Acad. Sci. USSR Chem. Sect. **106/111** [1956] 689/91). – [7] R. Riemschneider, D. Helm (Chem. Ber. **89** [1956] 155/61). – [8] M. Vogel, M. Rausch, H. Rosenberg (J. Org. Chem. **22** [1957] 1016/8). – [9]

G. D. Broadhead, J. M. Osgerby, P. L. Pauson (J. Chem. Soc. **1958** 650/6). – [10] W. Göller (Diss. Stuttgart T.H. 1958).

[11] K. L. Motz (Diss. Univ. Illinois 1958, pp. 1/130; Diss. Abstr. **19** [1958] 958). – [12] M. Rosenblum, R. B. Woodward (J. Am. Chem. Soc. **80** [1958] 5443/9). – [13] W. L. Baun (Anal. Chem. **31** [1959] 1308/11). – [14] M. R. Rausch, E. O. Fischer, H. Grubert (J. Am. Chem. Soc. **82** [1960] 76/82). – [15] J. M. Osgerby, P. L. Pauson (J. Chem. Soc. **1961** 4604/9).

[16] K. Schlögl, H. Pelousek, A. Mohar (Monatsh. Chem. **92** [1961] 533/41). – [17] M. Okawara, Y. Takemoto, H. Kitaoka, E. Haruki, E. Imoto (Kogyo Kagaku Zasshi **65** [1962] 685/8 from C.A. **58** [1963] 577). – [18] D. W. Hall, J. H. Richards (J. Org. Chem. **28** [1963] 1549/54). – [19] L. N. Mulay, A. Attalla (J. Am. Chem. Soc. **85** [1963] 702/6). – [20] K. Yamakawa, H. Ochi, K. Arakawa (Chem. Pharm. Bull. [Tokyo] **11** [1963] 905/11).

[21] Kodak-Pathé, J.-P. C. G. Dubosc (Fr. 1361470 [1963/64]; C.A. **63** [1965] 13461). – [22] Eastman Kodak Co., J.-P. C. G. Dubosc (U.S. 3335008 [1964/67]). – [23] K. Schlögl, W. Steyrer (Monatsh. Chem. **96** [1965] 1520/35). – [24] Colgate-Palmolive Co., J. E. Robertson (U.S. 3377248 [1965/68]; C.A. **69** [1968] No. 96878). – [25] O. E. Ayers, T. G. Smith, J. D. Burnett, B. W. Ponder (Anal. Chem. **38** [1966] 1606/7).

[26] K. Ban, T. Saegusa, J. Furukawa (Kogyo Kagaku Zasshi **69** [1966] 148/51, A9). – [27] J. E. Herz (J. Chem. Educ. **43** [1966] 599). – [28] A. N. Nesmeyanov [Nessmejanow], L. P. Yur'eva [Jurjewa], N. S. Kotchetkova [Kotschetkowa], S. V. Vitt [Witt] (Izv. Akad. Nauk SSSR Ser. Khim. **1966** 560/2; Bull. Acad. Sci. USSR Div. Chem. Sci. **1966** 530/2; Abhandl. Deut. Akad. Wiss. Berlin Kl. Chem. Geol. Biol. **1966** E265/E268). – [29] P. Petrovitch (Double Liaison **1966** No. 133, pp. 1093/109). – [30] A. N. Nesmeyanov, N. S. Kochetkova, V. J. Malakhov, V. D. Vil'chevskaya, R. K. Dybovskii, N. P. Palitsyn, E. M. Piskunova (U.S.S.R. 201428 [1966/67]; C.A. **69** [1968] No. 19312).

[31] N. Ya. Mar'yashkin, T. P. Vishnyakova, A. A. Koridze (Tr. Mosk. Inst. Neftekhim. Gaz. Prom. No. 72 [1967] 67/70; C.A. **69** [1968] No. 52260). – [32] A. N. Nesmeyanov, E. G. Perevalova, L. N. Yur'eva (Izv. Akad. Nauk SSSR Ser. Khim. **1967** 694/6; Bull. Acad. Sci. USSR Div. Chem. Sci. **1967** 675/6). – [33] D. E. Bublitz, K. L. Rinehart (Org. React. **17** [1969] 1/154, 79). – [34] P. Pratter (private communication from [33, p. 27]). – [35] F. M. Vigo (Diss. Univ. Illinois 1969; Diss. Abstr. Intern. B **31** [1970] 598/9).

[36] P. Carty, M. F. A. Dove (J. Organometal. Chem. **21** [1970] 195/201). – [37] E. B. Sokolova, M. P. Shebanova, L. A. Datskevich (Sin. Fiz. Khim. Polim. No. 6 [1970] 69/72; C.A. **73** [1970] No. 131397). – [38] U. Mueller-Westerhoff (Tetrahedron Letters **1972** 4639/42). – [39] R. E. Bozak, H. Javaheripour (Chem. Ind. **1973** 696/7). – [40] N. E. Gel'man, O. L. Lependina, E. A. Bozhevol'nov, K. I. Nikolaeva (Zh. Analit. Khim. **28** [1973] 1231/3; J. Anal. Chem. [USSR] **28** [1973] 1099/101).

[41] J. C. Gilbert, S. A. Monti (J. Chem. Educ. **50** [1973] 369/70). – [42] G. Guiochon, C. Pommier (Gas Chromatography in Inorganics and Organometallics, Ann Arbor, Mich., 1973, pp. 192/5). – [43] R. Einarsson, M. Zeppezauer (J. Organometal. Chem. **71** [1974] 277/82). – [44] A. N. Nesmeyanov, M. I. Rybinskaya, G. B. Shul'pin (Ger. Offen. 2438427 [1974/76]; C.A. **85** [1976] No. 5887). – [45] A. N. Nesmeyanov, E. V. Leonova, N. S. Kochetkova, A. I. Malkova (J. Organometal. Chem. **96** [1975] 271/4).

[46] A. N. Nesmeyanov, E. V. Leonova, N. S. Kochetkova, A. I. Malkova, A. G. Makarovskaya (J. Organometal. Chem. **96** [1975] 275/8). – [47] D. T. Haworth, T. Liu (J. Chromatog. Sci. **14** [1976] 519/20). – [48] D. T. Haworth, T.-C. Liu (Sep. Sci. **11** [1976] 329/34). – [49] D. T. Haworth, T. Liu (J. Chem. Educ. **53** [1976] 730/1). – [50] E. A. Stadlbauer, E. Nipper, M. Wenzel (J. Labelled Compounds Radiopharm. **13** [1977] 491/508).

[51] Bayer A.-G., V. Mues, W. Behrenz (Ger. Offen. 2711546 [1977/78]; C.A. **90** [1979] No. 1705). – [52] M. Hisatome, T. Hashiyama, K. Yamakawa (Tetrahedron Letters **1978** 3759/62). – [53] Š. Toma, M. Paričiova, E. Solčaniová (Acta Fac. Rerum Nat. Univ. Comenianae Chim. No. 26 [1978] 61/71). – [54] C. H. Gast, J. C. Kraak, H. Poppe, F. J. M. J. Maessen (J. Chromatog. **185** [1979] 549/61). – [55] D. W. Hausler, L. T. Taylor (Anal. Chem. **53** [1981] 1227/31).

[56] M. Pelino, M. Tomassetti, V. Piacente, G. d'Ascenzo (Thermochim. Acta **44** [1981] 89/99). – [57] L. Tătaru, I. Mazilu, M. Vâtă, T. Lixandru, C. Simionescu (J. Organometal. Chem. **214** [1981] 107/14). – [58] M. Tomassetti, R. Curini, G. d'Ascenzo, G. Ortaggi (Thermochim. Acta **48** [1981] 333/41). – [59] J. Holeček, K. Handlíř, J. Klikorka, M. Nádvorník (J. Prakt. Chem. **324** [1982] 345/8). – [60] K. Tanikawa, K. Arakawa (Chem. Pharm. Bull. [Tokyo] **13** [1965] 926/31).

[61] A. M. Tarr, D. M. Wiles (Can. J. Chem. **46** [1968] 2725/31). – [62] C. Pommier, G. Guiochon (Chromatographia **2** [1969] 346/9). – [63] P. F. Quirk, B. Kratochvil (Anal. Chem. **42** [1970] 535/6). – [64] I. J. Spilners, J. G. Larson (Org. Mass. Spectrom. **3** [1970] 915/24). – [65] K. Tanikawa, K. Arakawa (Bunseki Kagaku **20** [1971] 278/81).

[66] Ya. M. Paushkin, A. M. Shevchik (Vestsi Akad. Navuk Belarusk. SSR Ser. Khim. Navuk **1973** No. 3, pp. 95/8; C.A. **79** [1973] No. 53505). – [67] M. Peterlik, K. Schlögl (Z. Anal. Chem. **195** [1963] 113/7). – [68] Union Oil Co. of California, C. Rai (U.S. 3461287 [1963/69], C.A. **71** [1969] No. 81541). – [69] J. Tirouflet, R. Dabard, E. Laviron (Bull. Soc. Chim. France **1963** 1655/60). – [70] Institute of Heteroorganic Compounds Academy of Sciences USSR (Japan. Kokai 76-32544 [1974/76] from C.A. **85** [1976] No. 143306).

[71] D. Langheim, M. Wenzel, E. Nipper (Chem. Ber. **108** [1975] 146/54). – [72] W. P. Hart (Diss. Univ. Massachusetts 1981 from [73]). – [73] D. W. Macomber, W. P. Hart, M. D. Rausch (Advan. Organometal. Chem. **21** [1982] 1/55, 8). – [74] X.-J. Yang, Z.-Z. Xie, F.-S. Ni, S.-J. Shi (Gaodeng Xuexiao Huaxue Xuebao **1** [1980] 48/60 from C.A. **94** [1981] No. 219071). – [75] X. Yang, Z. Xie, F. Ni, S. Shi (Fundam. Res. Organometal. Chem. Proc. China-Japan-US Trilateral Semin. Organometal. Chem., Beijing 1980 [1982], pp. 765/9).

[76] R. Yin, X. Zehn (Lan-Chou Ta Hsueh Hsueh Pao Tzu Jan K'o Hsueh Pan **19** [1983] 41/5 from C.A. **100** [1984] No. 103578). – [77] K. Gonsalves, L. Zhan-ru, M. D. Rausch (J. Am. Chem. Soc. **106** [1984] 3862/3).

5.2.3.5.1.1.2 Physical and Structural Properties

1,1'-Diacetylferrocene forms red [1, 5, 7, 11, 13, 22, 30, 52], red-orange [97, 98] to orange [14, 55] crystals which according to [7, 13, 30, 55, 97, 98] are needle-like fibers and were obtained by [52] in the form of red prisms and by [2, 15] as large ruby red rectangular platelets.

Thermal Properties

Melting point in °C (solvent for recrystallization): 119 to 120 (petroleum ether or CCl_4) [71], 122 to 124 (C_6H_6/hexane) [30, 38], 123 to 124 (H_2O/acetone 3/1) [43], 124 [32], 124 to 125 (H_2O) [97, 98], 124 to 126 [4], 125 to 126 (H_2O) [3], 125 to 127 [52, 85], 126 to 126.5 [57], 126 to 127 (isopropanol) [37], (petroleum ether) [13], 126 to 127.5 [74], 126.5 to 127 [14], 127 [48], 127.0 to 127.5 [11, 44], 127 to 128 (C_6H_6/hexane) [89], 127.5 to 128.5 [2, p. 111], [15], 128 [54, 94], 128 to 129 [73], 129 [42], 130 to 131 [5, 6, 16, 18, 22, 60, 63, 103], 130.5 ± 0.5 [96, 100].

Vapor Pressure. Measurements by the Knudsen effusion method and by the torsion method between 360 and 400 K yield for the solid compound the vapor pressure equation (P in kPa, T in K) [96]:

$$\log P = (8.82 \pm 0.11) - (4289 \pm 44)/T$$

References on pp. 18/20

Molar heat capacity C_p in $cal \cdot mol^{-1} \cdot K^{-1}$. The following table presents for the temperatures (±0.5°C) values (±0.5 $cal \cdot mol^{-1} \cdot K^{-1}$) for $C_{p,\,meas}$ measured by differential scanning calorimetry (DSC) and for $C_{p,\,calc}$ calculated by the equation $C_{p,\,calc} = 63.22 + 0.247\,t$ for the solid substance [100]:

t in °C	20	30	40	50	60	70	80	90	100	110	120
$C_{p,\,meas}$	68.4	70.1	72.9	76.0	78.2	80.1	83.3	85.1	88.2	90.1	92.8
$C_{p,\,calc}$	68.2	70.6	73.1	75.5	78.0	80.5	82.9	85.4	87.9	90.3	92.8

Enthalpy of sublimation at 298 K $\Delta H^\circ_{subl} = 91.9 \pm 2.5$ kJ/mol [96], calculated from the vapor pressure values and the enthalpy functions for the gas phase by [91] (see below) and for the solid phase by [99].

Thermodynamic Functions. Statistical mechanical calculations for the gas phase based on the published molecular data with consideration only of the electronic ground state, assumed to be $^1\Sigma$, gave as functions of the temperature the following values for $(H^\circ_T - H^\circ_0)$ in ±10 cal/mol, S°_T in ±0.25 $cal \cdot mol^{-1} \cdot K^{-1}$, and $-(G^\circ_T - H^\circ_0)/T$ in ±0.25 $cal \cdot mol^{-1} \cdot K^{-1}$ [91]:

T in K	298.15	300	325	350	375	400	425	450
$(H^\circ_T - H^\circ_0)$	7324	7432	8848	10206	11496	12706	13823	14850
S°_T	96.52	96.87	101.33	105.27	108.77	111.89	114.66	117.14
$-(G^\circ_T - H^\circ_0)/T$	71.95	72.09	74.10	76.11	78.115	80.12	82.135	84.14

From these data, the sublimation enthalpy, and the C_p data values have been calculated for the solid compound for $-(G^\circ_T - H^\circ_{298})/T$ (in $J \cdot mol^{-1} \cdot K^{-1}$ for the first set of values and $cal \cdot mol^{-1} \cdot K^{-1}$ for the second) [96]:

T in K	298.15	300	320	340	360	380	400
$-(G^\circ_T - H^\circ_{298})/T$	240.50	240.71	243.51	245.98	249.24	252.92	256.90
	57.442	57.493	58.161	58.751	59.53	60.41	61.36

Dipole Moment

Measurements of the dielectric constants in benzene yield the following values for the dipole moment μ:

t in °C	20	30	45
μ in D	4.17	4.23	4.25
Ref.	[58]	[5, 58]	[58]

Regarding the implications for the molecular structure, see p. 17.

Molar Refraction

The refraction is 65.2 cm^3/mol for the Na D line [5].

Nuclear Resonance Spectra

For spectra in protonaceous solvents see p. 52.

^{1}H NMR Spectrum. The following shifts, δ in ppm, and coupling constants, J in Hz, have been measured at room temperature (t = triplet):

solvent	H-2,5 of C_5H_4	H-3,4 of C_5H_4	CH_3	Ref.
$CHCl_2F$	4.748t	4.475t	2.310	[82]
$CDCl_3$	4.77t (J=1.91)	4.51t (J=1.91)	—	[29]
CCl_4	4.62t	4.35t	—	[34, 39]

The shift of the CH_3 protons $\Delta\delta_{CH_3}$ in Hz (referred to that for $C_5H_5FeC_5H_4$–CO–CH_3 as zero) does not obey the equation valid for the compounds CH_3–CO–$C_5H_4FeC_5H_4$–R (R = C_2H_5, Cl, CN), $\Delta\delta_{CH_3} = 4.4\ \sigma_p + 0.1$, which connects $\Delta\delta_{CH_3}$ with the Hammett constants σ_p of the substituents R [56].

The line widths ΔH (measured from peak to peak of the first derivative curve) and the second moments ΔH_2^2 calculated from them are for the solid ^{1}H NMR spectra as a function of the temperature:

T in K	4.2	77	194	298
ΔH in Oe	—	8.0	7.0	7.0
ΔH_2^2 in Oe^2	13.8±2.0	9.6	9.6	9.3
Ref.	[70]	[16, 18]	[16, 18]	[16, 18]

ΔH_2^2 is plotted in a figure in [70] from 4.2 to 300 K, and ΔH and ΔH_2^2 for 140 to 298 K for $Fe(C_5H_4$–CO–$CD_3)_2$ in figures in [25, 28]. Values derived from these data for the reorientation process in the molecule are covered in the section on molecular properties.

^{13}**C NMR spectra** at room temperature (δ in ppm):

solvent	C-1	C-2,5 (cyclopentadienyl ring)	C-3,4	C-6 (CO)	C-7 (CH_3)	Ref.
C_6D_6	81.7	69.9	73.0	—	—	[79]
$ClCH_2CH_2Cl$	80.1	70.4	73.1	200.4	—	[86]
$CHCl_2F/CD_3COCD_3$ (80/20)	81.7	71.6	74.1	201.2	27.7	[82]
$CHCl_3$	80.1	70.4	73.1	200.4	27.1	[78, 79]
$CDCl_3$	80.9	71.0	73.5	200.6	27.5	[102]

^{13}C,^{1}H coupling constants, J in Hz, are: for C-2,5, ^{1}J=179.2, ^{2}J=6.4, ^{3}J=6.4; for C-3,4, ^{1}J=177.9; for CH_3, J=127.5 [102]. Coupling constants measured in CH_2Cl_2: J(^{13}C, ^{57}Fe) are 4.39 for C-2,5, 4.45 for C-3,4, 4.88 for C-1, all in Hz [86]. Values derived for the rotation of the CH_3CO group from the ^{13}C DNMR spectrum as a function of the temperature are covered under molecular properties.

^{57}**Fe NMR Spectrum.** Shifts δ referred to $FeC_{10}H_{10}$ as an internal standard are in CH_2Cl_2 425.6 [86], and in $CDCl_3$ 437.2±2.5 [95], 436.9 [107] ppm.

^{57}Fe Mössbauer Spectrum

Pure Compound. The following table shows the isomer shift δ (vs. the standard substances listed in the following column), the quadrupole splitting Δ, the relative probability f′ of a recoil-free transition (referred to that for ferrocene of 1 at 300 K), and the intensity ratio I_π/I_σ for the quadrupole doublet peaks:

T in K	δ in mm/s	standard substance	Δ in mm/s	f′	I_π/I_σ	Ref.
78	0.30 ± 0.02	^{57}Co/Cu	2.15 ± 0.05	—	—	[33]
80	0.49 ± 0.05	^{57}Co/stainless steel	2.27 ± 0.05	7.0 ± 0.7	0.85 ± 0.08	[20]
80	0.54 ± 0.01	natural iron	2.14 ± 0.05	—	—	[106]
80	0.54 ± 0.01	natural iron	2.15 ± 0.01	—	—	[108]
300	0.43 ± 0.05	^{57}Co/stainless steel	2.2 ± 0.05	2.8 ± 0.3	1.12 ± 0.1	[20]

See also [76]. The spectra show at 95 and 293 K the standard quadrupole doublet [88]. Mössbauer transmission spectra at 80 and 298 K for a powdered sample mixed with 50 wt% Li_2CO_3 are shown in figures in the paper. The equality of the intensities of the two quadrupole peaks indicates the absence of significant intramolecular vibrational anisotropy [83]. The dependence on temperature of ln f_T/f_0, where f_T and f_0 are the recoil-free fractions at the temperatures T and 0 K, respectively, is linear (for T = 100 to 300 K, see figure in the paper) [94]. The characteristic Debye temperature for the polycrystalline sample has been derived from the linear temperature dependence between 280 and 380 K under the assumption of the applicability of the Debye model, as $\Theta_D = 80 \pm 6$ K. The square of the average thermal vibrational amplitude of Fe, $\langle \bar{r}^2 \rangle$, at 300 K is $(3.7 \pm 0.6) \times 10^{-18}$ cm^2 [90].

Mössbauer Studies on Liquid Crystals. Since 1,1′-diacetylferrocene possesses a pronounced molecular axis, forms in the solid state molecular crystals, does not react chemically with organic compounds, and shows via the ^{57}Fe a large quadrupole splitting in liquid crystal systems, $Fe(C_5H_4{-}CO{-}CH_3)_2$ is suitable as a solute for conducting Mössbauer studies which give both information concerning its degree of ordering in the liquid crystal matrix and also concerning the liquid crystal phases themselves, see for example [59, 72, 83].

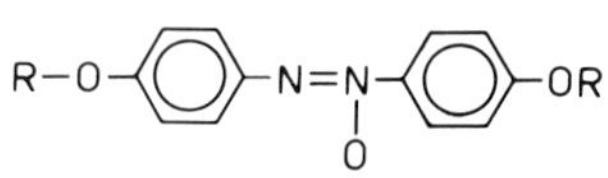

I

a: R = n-C_7H_{15} (HOAOB)
b: R = n-C_8H_{17} (OOAOB)

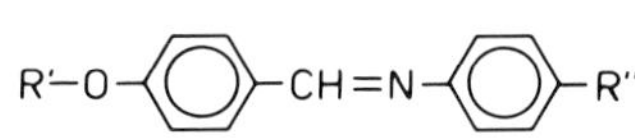

II

a: R′ = CH_3; R″ = C_4H_9 (MBBA)
b: R′ = C_4H_9; R″ = C_2H_5 (BBEA)
c: R′ = n-C_4H_9; R″ = n-C_8H_{17} (BBOA)
d: R′ = n-C_5H_{11}; R″ = C_3H_7 (PBPA)
e: R′ = n-C_6H_{13}; R″ = CH_3 (HBT)
f: R′ = n-C_6H_{13}; R″ = n-C_3H_7 (HBPA)

HOAOB (Ia) forms liquid crystals with a nematic and a smectic C phase. Its solutions with $Fe(C_5H_4{-}CO{-}CH_3)_2$ (0.25 to 10 wt%; enriched with ^{57}Fe so that its content in the solution is constant) then show a Mössbauer effect only when the ferrocene compound is present in the smectic phase as a solid component, i.e., as a suspension of small crystallites [67, 80]. Studies between 30 and 90°C of solutions with 5 wt% give according to [68] γ-resonance both in the smectic and the nematic phase. The latter is perhaps possibly due to the preservation of smectic structure fragments, so-called cybotactic groups, in the transition from the smectic to the nematic phase. Over the total range, the isomer shift δ = 0.65 mm/s (referred to $Na_2[Fe(CN)_5NO] \cdot 2H_2O$) and the quadrupole splitting Δ = 2.15 mm/s, values which correspond to those of the pure compound. No separated phases of $Fe(C_5H_4{-}CO{-}CH_3)_2$ were observed [67, 68]. The linear temperature dependence of ln f′ between 295 K and the transition temperature from the solid to the smectic phase yields $\Theta_D = 77 \pm 6$ K under the assumption of the applicability of the Debye model. At 300 K, $\langle \bar{r}^2 \rangle = (2.7 \pm 0.4) \times 10^{-18}$ cm^2 [90].

If the liquid crystals are oriented by cooling of the solution from the isotropic liquid to the smectic phase in a magnetic field of 4 or 9 kOe, a systematic change which depends on the strength of the orienting field can be observed for A_π/A_σ, the ratio of the area under the two peaks of the quadrupole doublet, as Θ (the angle between the magnetic field and γ ray) varies from 0° to 90°. The dependence of the recoil-free fraction on Θ is ascribed to the anisotropy of the liquid crystal lattice forces [59, 80]. If a 2 kOe magnetic field is used, no variation of A_π/A_σ with Θ can be determined within the error limits. In contrast, a strong dependence on Θ is reported for $\alpha f'$ (total area under the resonance spectrum); the absolute probability f of the Mössbauer effect at room temperature is 0.22 ± 0.01 for $\Theta = 0°$ and 0.11 ± 0.01 for $\Theta = 90°$; $f = 0.17 \pm 0.01$ for the isotropic melt pulverized after cooling to room temperature [62, 68]. From the experimental values for f, the squares of the average vibrational amplitudes of Fe perpendicular or parallel to the smectic layer are $\langle \bar{r}_\perp^2 \rangle = (0.08 \pm 0.04) \times 10^{-18}$ and $\langle \bar{r}_\parallel^2 \rangle = (0.7 \pm 0.2) \times 10^{-18}$ cm², respectively [61]. Under the assumption of a "cone of planes" structure of the oriented smectic liquid crystal phase $\langle \bar{r}_\perp^2 \rangle = (0.1 \pm 0.05) \times 10^{-17}$ and $\langle \bar{r}_\parallel^2 \rangle = (0.5 \pm 0.2) \times 10^{-17}$ cm². The angle between the magnetic field and the smectic layer is $(40 \pm 15)°$ [68].

Like HOAOB the compound OOAOB (Ib) also forms a smectic and a nematic mesophase. γ-Resonance was observed in both the smectic and the nematic phases in solutions with 5 wt% $Fe(C_5H_4{-}CO{-}CH_3)_2$ studied between 25 and 100°C; no separate phases of the ferrocene derivative separated. Over the total temperature range the isomer shift and the quadrupole splitting correspond to the values for the pure compound [67, 68]. The linear temperature dependence of ln f′ between 300 and >340 K (solid crystal phase up to the transition temperature to the smectic liquid crystal phase) gives under the assumption of the validity of the Debye model $\Theta_D = 80 \pm 6$ K. At 300 K, $\langle \bar{r}^2 \rangle = (2.2 \pm 0.4) \times 10^{-18}$ cm² [90].

BBEA (IIb) and PBPA (IId) form liquid crystals with a smectic H phase, in which $Fe(C_5H_4{-}CO{-}CH_3)_2$ (studies on solutions with 0.25 to 10 wt%) only exhibits a Mössbauer effect if it is present not in solution but as a suspension of small crystallites. The crystallites are small enough to experience a certain orientation by the liquid crystals, as shown by measurements on samples cooled in a magnetic field from the isotropic liquid to the smectic phase [80].

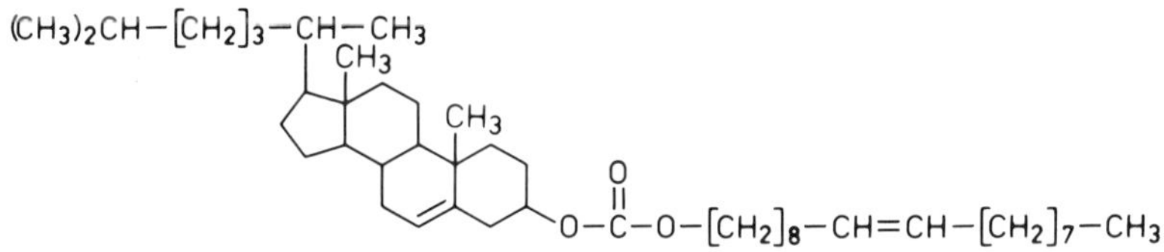

III (COC)

For solid glassy phases of smectic liquid crystals which are obtained by cooling of COC (III) containing 3 wt% $Fe(C_5H_4{-}CO{-}CH_3)_2$, the resonance absorption is very small at 95 K, at 293 K practically nonexistent. If the isotropic solution is cooled in an 18 kOe magnetic field, it forms at ~280 K an oriented smectic structure of the host phase, which leads to a partial orientation of the diacetylferrocene. In the solid glassy phase, the probe molecules retain the partial alignment and give rise to enhanced resonance absorption at 95 K around >200% for $\Theta = 0°$ and 90°; $I_\pi/I_\sigma \neq 1$. With increasing temperature the resonance absorption falls off rapidly and reaches at 134 K approximately the value of the unoriented sample. The rapid drop with increasing temperature indicates that the main contribution to this absorption is coming from temperature-dependent intermolecular contributions [88].

The differential thermal analysis of MBBA (IIa) with 2.0 wt% $Fe(C_5H_4{-}CO{-}CH_3)_2$ shows between ~292.5 and 314 K a liquid crystal phase with a nematic structure and a glass transition temperature at $T_G \approx 183$ K. From Mössbauer studies from 99 to 245 K in a 9 kOe magnetic field

References on pp. 18/20

on a sample cooled from room temperature (nematic phase) to the temperature of liquid nitrogen, the isomer shift $\delta = +0.24$ mm/s (referred to $^{57}Co/Pd$) and the quadrupole splitting $\Delta = 2.17$ mm/s correspond to those of the pure compound, so that a reaction here also is excluded between the guest and the host materials. In the glass phase, at lower temperatures $I_\pi > I_\sigma$ for $\Theta = 0°$ and $I_\pi < I_\sigma$ for $\Theta = 90°$. With rising temperature at first there is hardly any change evident in the line width, although the anisotropy decreases. Nevertheless at about 175 K a broadening begins for the Mössbauer lines (stronger for $\Theta = 0°$ than for $\Theta = 90°$) and after ~210 K the Mössbauer effect vanishes; only at temperatures above approximately 220 K with a line width which corresponds to the pure $Fe(C_5H_4–CO–CH_3)_2$ powder, does it again appear. This is evidently due to the fact that the nematic liquid crystal glass first converts into the supercooled liquid, which then recrystallizes as the supercooled liquid crystal. The anisotropy of the line broadening is due to the fact that diffusion of the ferrocene derivative takes place easier parallel to the MBBA molecular direction than perpendicular to it; the ratio of the diffusion coefficients $D_{\parallel}/D_{\perp} \approx 2.12$ at 210 K [101]. In a 0.20% solution (enriched to 50% in ^{57}Fe) after orientation in a 9 kOe magnetic field, Mössbauer data obtained in the range 100 to 300 K reveal from the temperature dependence of ln f a glass transition temperature region of 176 to 186 K. The line broadening observed in the range 230 to 240 K and decrease of the quadrupole splitting are explained in terms of a stochastic line-shape model for torsional or rotational molecular motions of diacetylferrocene molecules in the cold liquid. Independent of the temperature and orientation $A_\pi/A_\sigma \approx 1$, which excludes an orientation of the $Fe(C_5H_4–CO–CH_3)_2$ probe by the MBBA molecules [87].

A eutectic mixture of cyanobiphenylene and terphenylene with the trade name E8 forms like MBBA a supercooled nematic liquid crystal phase; for the solution containing 2.0 wt% $Fe(C_5H_4–CO–CH_3)_2$ the glass transition temperature is located at 174 K, according to DTA measurements. Corresponding studies on the samples oriented in a 10 kOe magnetic field give the same results as with MBBA. The line broadening begins at about 170 K, the Mössbauer effect vanishes at ~200 and reappears at ~210 K. At 193 K $D_{\parallel}/D_{\perp} \approx 2.26$ [101].

Investigations on supercooled glasses oriented in a 9 kOe magnetic field, with the smectic B liquid crystal structure, of BBOA (IIc) with 0.2% $Fe(C_5H_4–CO–CH_3)_2$ (enriched to 50% ^{57}Fe) give from the temperature dependence of ln f an orientation-dependent Debye temperature (64 K for $\Theta = 0°$ and 49 K for $\Theta = 90°$), which implies a temperature-dependent lattice contribution to the nuclear vibration ellipsoid, and a glass transition region of 175 to 186 K. $A_\pi/A_\sigma \approx 1$ independent of temperature and Θ [87].

The EFG (electric field gradient) tensor on the Fe nucleus is positive for supercooled liquid crystals oriented in a 10 kOe magnetic field, of the monotropic smectic B phase of HBT (IIe) with 8 wt% $Fe(C_5H_4–CO–CH_3)_2$, from the intensity ratio I_π/I_σ as a function of the temperature between 80 and 360 K at $\Theta = 0°$ and 90° as well as a function of Θ from 0° to 90° at 117 K. I_π/I_σ for $\Theta = 0°$ varies between 0.87 and 1.0, for $\Theta = 90°$ between 1.49 and 1.0. From this it is apparent that the liquid crystal molecules indeed exert a certain orienting influence on the $Fe(C_5H_4–CO–CH_3)_2$ molecules but they do not perfectly align them. The plots of f vs. temperature between 100 and 330 K for $\Theta = 0°$ and 90° show a sharp drop at the glass transition region around 158 K. The f values are higher for $\Theta = 0°$ than for 90°, accordingly the square of the average vibrational amplitude along the longer molecular axis is smaller than that perpendicular to it. No separate phases of the ferrocene derivative had separated out [83].

A constant resonance line width is observed between 100 and 233 K for $\Theta = 0°$ and 90°, from 233 to 237 K an increasing diffusive broadening, on supercooled glasses of the smectic H phase of HBPA (IIf) with 0.2 wt% $Fe(C_5H_4–CO–CH_3)_2$ (enriched to 50% ^{57}Fe), which were obtained by cooling of the isotropic solution in a 9 kOe magnetic field through the nematic phase. This is for $\Theta = 90°$ larger than for $\Theta = 0°$, and has been attributed by Detjen, Uhrich [75] to onset of translational diffusion of the guest molecules, which at 237 K have a diffusion rate for $\Theta = 90°$ approximately fourfold that for $\Theta = 0°$. The glass transition temperature for HBPA, $T_G \approx 233$ K, is derived from the temperature of beginning of diffusion [75]. In contrast to [75] LaPrice [87] refers the line broadening and decreasing quadrupole splitting to rotational or torsional motions of the probe molecules. The temperature dependence of ln f for the smectic H phase is linear only for 100 to 150 K. Above 150 K it possesses no Debye character; f decreases strongly and approaches zero in the region of the line broadening. From the slope of the line the Debye temperature is 66 K. The same value is also derived for HBPA in the solid state, from the linear temperature dependence of ln f between 100 and 300 K for $Fe(C_5H_4{-}CO{-}CH_3)_2$/HBPA in the solid state [65, 72]. $A_\pi/A_\sigma \approx 1$ independent of temperature and Θ [65, 87].

Solutions of 4-n-$C_7H_{15}O{-}C_6H_4{-}C(O)SC_6H_4{-}C_5H_{11}$-n-4′ ($\bar{7}$S5) containing 0.2 wt% $Fe(C_5H_4{-}CO{-}CH_3)_2$ (^{57}Fe enriched) show in the crystalline phase in the temperature range from 110 to 300 K a Debye-like behavior with a linear plot for ln f vs. T, Debye temperature of the host (65 ± 3) K [109].

IR and Raman Spectra

The IR spectrum is shown from 667 to 5000 cm^{-1} in a figure in [42]

In the Solid State. In Nujol (see figure in [14, p. 117]) the observed frequencies in cm^{-1}, their relative intensities (vw = very weak, w = weak, ms = moderately strong, vs = very strong, sh = shoulder), and their assignments (indices: a = asymmetric, s = symmetric) are from [41, 104]: 445 ms, 460 ms, sh, 486 s (ν_a Fe-ring), 506 s (asym. ring tilt), 546 s, 559 sh, 598 w, 625 s (δC=O), 670 ms, 825 s, 845 vs, 894 s, 918 w, 973 ms (CH_3, rocking), 1022 s, 1035 s, 1120 s ($\nu C{-}CH_3$), 1288 vs ($\nu C_{ring}{-}C_{acetyl}$), 1367 s ($\delta CH_3$), 1437 w ($\delta_a CH_3$), 1676 vs ($\nu$C=O). In mineral oil: 830 [21], 1115 ($\nu_s C{-}CH_3$) [92], 1280 ($\nu_a C_{ring}{-}C_{acetyl}$) [92], 1665 ($\nu$C=O) [21, 92]. In KBr: 1115 ($\nu C{-}CH_3$) [19, 30], 1658 [19], 1660 [30] (νC=O).

In Solution. The following frequencies in cm^{-1}, half widths in cm^{-1}, and integrated intensities in $M^{-1} \cdot cm^{-2}$ have been reported for the ν(C=O) bands as a function of the solvent:

solvent	frequency	half-width	integrated intensity	Ref.
heptane	1687	12.0	3.70×10^4	[26]
CCl_4	1672.1	—	—	[50]
CCl_4	1677	13.0	5.65×10^4	[26]
CCl_4	1683	—	—	[48]
CCl_4/CH_3OH (95/5)	1683	—	—	[48]
dioxane	1680	13.0	5.40×10^4	[26]
CS_2	1683	—	—	[104]

In $CHCl_3$ (see figure in [2, p. 153]) bands are located at 893.6, 1117.3 [15], 3096 to 3112 (νC–H from C_5H_4; integral intensity 3000 $M^{-1} \cdot cm^{-2}$), all in cm^{-1}. The intensity of this last band, but not its location, is affected substantially by the substituents on the C-1; a relationship exists between the intensity and the Hammett constants σ_p of the substituents [27]. Additional bands in CCl_4 are located at 482, 517, 534, 619, 840, 894, 960 (CH_3, rocking), 1020, 1035, 1054, 1115 ($\nu C{-}CH_3$), 1185, 1216, 1276 [104] (1279 [48]) ($\nu C_{ring}{-}C_{acetyl}$), 1358 ($\delta_s CH_3$), 1375 ($\delta CH_3$), 1399, 1428 ($\delta_a CH_3$), 1484 sh [104], all in cm^{-1}. Like the ν(C=O) band, the band at 1279 cm^{-1} in CCl_4 is scarcely shifted by addition of 5% CH_3OH to the solvent (to 1278 cm^{-1}), but is substantially broadened, which indicates an interaction of the CO group with the CH_3OH [48]. Additional

 References on pp. 18/20

bands in CS_2: 824, 891, 956 (CH_3, rocking), 1018, 1034, 1054, 1113 ($\nu C–CH_3$), 1211, 1273 ($\nu C_{ring}–C_{acetyl}$), 1355 ($\delta_s CH_3$), 1375 (δCH_3) [104], in cm^{-1}.

In the Gaseous State. The following table presents for 400 K the bands (wave numbers in cm^{-1}) and their assignments for the IR and Raman spectra, the latter excited with an He/Ne laser. The description of the ring vibrations was based on local C_{2v} symmetry [91].

IR	ring vibrations:	—	145	205	222	—
Raman		41*)	145	204	221	290
assignment		torsion	ring tilt	δ(ring-$COCH_3$)	δ(C–H)	γ(ring-$COCH_3$)

IR	298	311	320	371	470	—
Raman	298	310	322	370	471	567
assignment	ν_s(Fe-ring)	—	—	—	ν_a(Fe-ring)	γ(C–C)

IR	777	788	850	887	930	987
Raman	777	320	852	888	932	989
assignment	γ(C–H)	γ(C–H)	δ(C–C)	—	γ(C–H)	—

IR	1000	1045	1050	1260*)	1410	1445
Raman	1001	1044	1050	1260*)	1410	1444
assignment	ν(ring-$COCH_3$)	breathing	δ(C–H)	δ(C–H)	—	ν(C–C)

IR	1533	1620	3070	3090	3110	3120
Raman	1535	1620	3068	3090	3110	3120
assignment	—	ν(C–C)	ν(C–H)	—	ν(C–H)	ν(C–H)

IR	CH_3 internal modes:	1018	1377	1455	1466
Raman		1016	1377	1456	1465
assignment		ϱ(CH_3)	δ_s(CH_3)	—	δ_a(CH_3)

IR	CH_3 internal modes:	2907	2935	2955
Raman		2906	2933	2955
assignment		ν_s(C–H)	ν_a(C–H)	—

IR	CH_3–CO vibrations:	355	509	520	645	666
Raman		356	509	520	646	665
assignment		—	γ(C=O)	γ(CH_3)	δ(C–O)	δ(CH_3)

IR	CH_3–CO vibrations:	718	1220	1672	1687
Raman		716	1220	1673	1688
assignment		—	ν(C–CH_3)	—	ν(C=O)

*) Observed at an N_2 pressure of 5 atm
(δ = deformation in the plane; γ = deformation out of plane; ϱ = rocking).

Electronic Absorption Spectra

The absorption maxima λ_{max} in nm and the associated extinction coefficients ε in $M^{-1} \cdot cm^{-1}$ (log ε when in parentheses) at room temperature for various solvents plus their assignments according to [51] are shown in Table 1 on p. 16. Data regarding the oscillator strengths can be found in [35, 44, 51], and figures illustrating the spectrum in [17] for the 210 to 500 nm range and in [21] for 400 to 600 nm. For assignment of the long wavelength band to the symmetry-forbidden d-d transition at the Fe, see also [47, 77]; these agree with the hypothesis of Scott, Becker [24]. The effect of the substituents on the location and intensity of the bands is additive [40], the shifts of the bands versus those of ferrocene are bathochromic [77]. The high oscillator strength of the shortwave band at 196 nm confirms its assignment to a $\pi \rightarrow \pi^*$ transition in the ring [35].

The band at 462 nm in C_6H_6 is hypsochromically shifted to 384 nm by addition of CBr_4 due to the formation of the intermolecular charge transfer complex $Fe(C_5H_4{-}CO{-}CH_3)_2 \cdot CBr_4$ [93]. Intermolecular charge transfer bands cannot be observed in either CCl_4 or $CHBr_3$. Since these are to be expected for CCl_4 between 286 and 294 nm and for $CHBr_3$ between 312 and 323 nm, they are possibly overlapped by the intense $\pi \rightarrow \pi^*$ and $n \rightarrow \pi^*$ bands [36].

Magnetic Circular Dichroism

In the region of the d-d transition an S-type MCD band is observed with peaks in cyclohexane at $\lambda_1 = 505$ and $\lambda_2 = 435$ nm ($\lambda_{\varepsilon=0} = 457$ nm), partially overlapped by a Gaussian MCD curve with $\lambda_{max} = 360$ nm [64]. In C_6H_6 the peaks are at $\lambda_1 = 495$ and $\lambda_2 = 440$ nm (A/D ≈ -1.23) [49].

Photoelectron Spectrum

The ionization energies in the gas phase are (in eV) 7.06 (Fe d; $\rightarrow {}^2E_{2g}$), 7.36 (Fe d; $\rightarrow {}^2A_{1g}$), 8.97 (from e_{1u}; ligand π), 9.61 (from e_{1g}; ligand π) [69]. The first adiabatic ionization potentials calculated from the photoelectron spectrum are, in eV, 7.08 ± 0.05 [53], 7.05 ± 0.05 [81]; the vertical value is 7.41 ± 0.05 [81].

Crystal Structure

$Fe(C_5H_4{-}CO{-}CH_3)_2$ crystallizes in a monoclinic structure of space group $P2_1/c{-}C^5_{2h}$ (No. 15) with $a = 5.898 \pm 0.02$, $b = 13.036 \pm 0.005$, $c = 14.962 \pm 0.006$ Å, $\beta = 90.68° \pm 0.04°$ [55], or $a = 5.90 \pm 0.03$, $b = 13.03 \pm 0.06$, $c = 14.89 \pm 0.07$ Å, $\beta = 90° \pm 1°$ [8 to 10]. The density calculated with $Z = 4$ molecules per unit cell is 1.559 ± 0.002 g/cm^3; the density measured by flotation is 1.54 ± 0.01 g/cm^3 [55].

Molecular Structure

Fig. 1, p. 17, gives the molecular structure (see a) with bond distances in Å and angles (see b). The configuration is nearly eclipsed with an average dihedral angle of 4°41′. The cyclopentadienyl rings are planar and nearly parallel (angle between the two planes of 1°9′), their dimensions and the Fe-C_{ring} distances are normal. The acetyl groups are in the positions 1 and 3′. They are twisted slightly (around 11°1′ for C6–C7–O1 and 8°48′ for C6′–C7′–O1′) and bent out of the plane of the respective rings. The two rotations are nevertheless different, so that in one case the O-1, in the other the C-7′, is closer to the opposite cyclopentadienyl ring. Presumably this difference in rotation permits a more favorable packing in the crystalline state [55], see also [76].

References on pp. 18/20

Table 1
Electronic Absorption Maxima λ_{max} in nm for $Fe(C_5H_4\text{–}CO\text{–}CH_3)_2$.

solvent	$A_{2g}(e_{2g} \rightarrow e_{2g}^*)$ λ_{max}	ε	$A_{1u}(e_{2g} \rightarrow e_{2u})$ λ_{max}	ε	charge transfer to $COCH_3$ λ_{max}	ε	$A_{2u}(e_{2g} \rightarrow e_{2u})$ λ_{max}	ε	$E_{1u}(e_{2g} \rightarrow e_{2u})$ λ_{max}	ε	$E_{1u}(e_{2g} \rightarrow e_{2u})$ λ_{max}	ε	Ref.
n-hexane	456.5	353	366	423	321.5	1403	—		—		—		[36, 44]
n-heptane	440	—	—	—	320	—	262	—	225	—	—	—	[17]
	455	320	—	—	—	—	—	—	—	—	—	—	[40]
isooctane	455	350	—	—	318	1600	260	10400	224	26000	196	25000	[35, 51]
	455	340	—	—	318	1580	—	—	—	—	—	—	[31]
	456	305	—	—	320	1370	258	9500	224	24000	—	—	[48]
cyclohexane	463	417	—	—	—	—	—	—	—	—	—	—	[64]
C_6H_6	455	415	—	—	—	—	—	—	—	—	—	—	[49]
	460	411	—	—	324	1645	—	—	—	—	—	—	[66, 105]
	462	—	—	—	—	—	—	—	—	—	—	—	[93]
$CHCl_3$	470	(2.64)	—	—	334	(3.15)	—	—	—	—	—	—	[2, p. 161]
	463.0	540	375.0	713	332.0	1808	—	—	—	—	—	—	[44]
CCl_4	459.5	460	371.0	603	326.5	1704	—	—	—	—	—	—	[36, 44]
C_2H_5Br	461.0	448	366.5	673	324.5	1632	—	—	—	—	—	—	[44]
$CHBr_3$	—	—	373	705	336.7	1820	—	—	—	—	—	—	[36]
C_2H_5I	459.5	460	353.5	934	—	—	—	—	—	—	—	—	[44]
H_2O	476	650	—	—	338	1865	—	—	—	—	—	—	[66]
C_2H_5OH	461	530	—	—	—	—	263	12000	—	—	—	—	[30]
	463.0	560	378.0	660	331	1940	—	—	—	—	—	—	[44]
	464	510	—	—	332	1755	—	—	—	—	—	—	[45]
	—	—	—	—	333	(3.25)	262	(4.10)	225	(4.32)	—	—	[46]
C_2H_5OH/H_2O (96/4)	460	—	—	—	—	—	—	—	—	—	—	—	[21, 77]
	—	—	—	—	—	—	262	14620	223	24000	—	—	[23]
$(C_2H_5)_2O$	456.5	433	369.5	520	322.5	1623	—	—	—	—	—	—	[44]
CH_3CN	455	—	—	—	—	—	—	—	—	—	—	—	[47]
	462.0	740	—	—	—	—	—	—	—	—	—	—	[44]

References on pp. 18/20

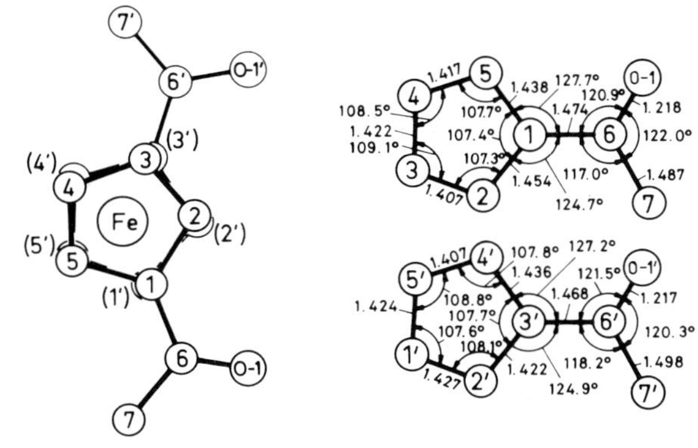

a b

Fig. 1. Molecular structure of 1,1'-diacetylferrocene from [55].

The correlation time τ_c and, by means of the equation $\tau_c = \tau_0 \exp(E_A/RT)$, the activation energy E_A have been calculated for the reorientation process of the molecule in the crystalline state from the temperature dependence of the line width in the solid state ^{1}H NMR spectrum and second moments (see p. 9) [18, 25]. According to [16] a reorientation process begins above 77 K; the molecular motion, however, does not consist of a free rotation of the cyclopentadienyl rings. From measurements on $Fe(C_5H_4{-}CO{-}CD_3)_2$ between 140 and 298 K the activation energy for reorientation of the rings amounts to $E_A = 5.0$ kcal/mol (20.9 kJ/mol); $\tau_0 = 6.0 \times 10^{-15}$ s [28]. Studies of the temperature dependence of the second moment between 4.2 and 300 K on $Fe(C_5H_4{-}CO{-}CH_3)_2$ indicate a beginning of the free rotation of the rings above 300 K and an activation energy $E_A > 11.0$ kcal/mol (46 kJ/mol). The free rotation of the CH_3 groups begins at 40 to 50 K; $E_A = 1.5$ kcal/mol (6.3 kJ/mol); $\tau_0 = 1.9 \times 10^{-14}$ s [70].

For the compound dissolved in C_6H_6 Richmond, Freiser [5] conclude from the agreement of the experimentally determined values of the dipole moment (see p. 8) with those calculated on the assumption of free rotation that there is free rotation of the rings. According to Dyatkina [12] this conclusion is not yet unequivocal. Studies of the conformation indicate from the correlation of the dipole moments of $Fe(C_5H_4{-}CO{-}CH_3)_2$ and IV that the free molecule dissolved in C_6H_6 prefers to exist in the racemic form V, where the rotation of the rings is nearly free, and less often in the rotationally isomeric meso form VI with slightly more restricted rotation [58]. The dynamic ^{13}C NMR spectrum in $CHCl_2F/CD_3COCD_3$ (80/20) gave for the rotation of the CH_3CO groups around the C(1)–C(6) bond the activation parameter $\Delta G^{\neq} = 7.8 \pm 0.2$ kcal/mol (32.66 kJ/mol) and the rate constant $k = 117\ s^{-1}$ for $T_c = 163$ K, the coalescence temperature of the signals of C-2 and C-5. The rotational barrier is around 0.7 kcal/mol (2.9 kJ/mol) less than that for $C_5H_5FeC_5H_4{-}CO{-}CH_3$, which is attributed to a true interannular effect, in which one CH_3CO group inductively diminishes the interaction of the other ring with its CH_3CO group [82], see also [84].

For the molecule in the gaseous state the rotational barrier for the cyclopentadienyl rings has been derived as 1.0 ± 0.1 kcal/mol (4.2 kJ/mol), from the Raman-active torsional fundamental at 41 cm^{-1} (see p. 14) [91].

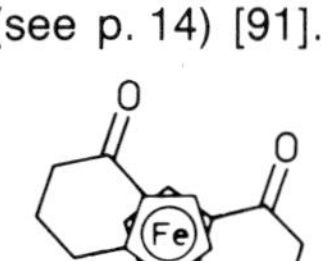

IV

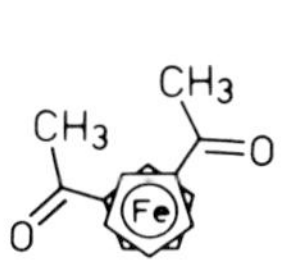

V

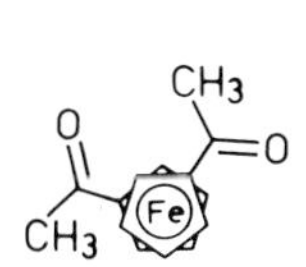

VI

References:

[1] R. B. Woodward, M. Rosenblum, M. C. Whiting (J. Am. Chem. Soc. **74** [1952] 3458/9). – [2] M. Rosenblum (Diss. Harvard Univ. 1953). – [3] N. V. De Bataafsche Petroleum Maatschappij, A. C. Nixon, T. P. Rudy (Brit. 763550 [1954/56]; C.A. **1957** 10588). – [4] Union Carbide Corp., M. A. Lynch, J. C. Brantley (Brit. 785760 [1957]; C.A. **1958** 11126). – [5] H. H. Richmond, H. Freiser (J. Am. Chem. Soc. **77** [1955] 2022/3).

[6] A. N. Nesmeyanov, N. A. Volkenau (Dokl. Akad. Nauk SSSR **111** [1956] 605/8; Proc. Acad. Sci. USSR Chem. Sect. **106/111** [1956] 689/91). – [7] R. Riemschneider, D. Helm (Chem. Ber. **89** [1956] 155/61). – [8] Yu. T. Struchkov [G. T. Stručkov] (Dokl. Akad. Nauk SSSR **110** [1956] 67/70). – [9] Yu. T. Struchkov [G. T. Stručkov], T. L. Khotsyanova [T. L. Khocjanova] (Kristallografiya **2** [1957] 382/3; Soviet Phys.-Cryst. **2** [1957] 376/7). – [10] G. T. Stručkov, T. L. Khocjanova (Acta Cryst. **10** [1957] 796/7).

[11] M. Vogel, M. Rausch, H. Rosenberg (J. Org. Chem. **22** [1957] 1016/8). – [12] M. E. Dyatkina (Usp. Khim. **27** [1958] 57/93, 62). – [13] W. Göller (Diss. Stuttgart T.H. 1958). – [14] K. L. Motz (Diss. Univ. Illinois 1958; Diss. Abstr. **19** [1958] 958). – [15] M. Rosenblum, R. B. Woodward (J. Am. Chem. Soc. **80** [1958] 5443/9).

[16] N. Ockman (AD-229397 [1959] 1/62, 29). – [17] T. N. Bolotnikova (Fiz. Probl. Spektrosk. Mater. 13th Soveshch., Leningrad 1960 [1962], Vol. 1, pp. 254/8; C.A. **59** [1963] 8263). – [18] L. N. Mulay, E. G. Rochow, E. O. Stejskal, N. E. Weliky (J. Inorg. Nucl. Chem. **16** [1960] 23/8). – [19] M. R. Rausch, E. O. Fischer, H. Grubert (J. Am. Chem. Soc. **82** [1962] 76/82). – [20] R. A. Stukan, S. P. Gubin, A. N. Nesmeyanov, V. I. Gol'danskii, E. F. Makarov (Teor. Eksperim. Khim. **2** [1966] 805/11; Theor. Exptl. Chem. [USSR] **2** [1966] 581/4).

[21] M. Furdík, Š. Toma, J. Suchý, P. Elecko (Chem. Zvesti **15** [1961] 45/62). – [22] M. Okawara, Y. Takemoto, H. Kitaoka, E. Haruki, E. Imoto (Kogyo Kagaku Zasshi **65** [1962] 685/8 from C.A. **58** [1963] 577). – [23] K. Schlögl, M. Peterlik, H. Seiler (Monatsh. Chem. **93** [1962] 1309/27). – [24] D. R. Scott, R. S. Becker (J. Chem. Phys. **35** [1961] 516/21). – [25] A. Attalla (Diss. Univ. Cincinatti 1962 from Diss. Abstr. **24** [1963] 1843/4).

[26] G. G. Dvoryantseva, Yu. N. Sheinker (Izv. Akad. Nauk SSSR Otd. Khim. Nauk **1963** 924/7; Bull. Acad. Sci. USSR Div. Chem. Sci. **1963** 832/5). – [27] G. G. Dvoryantseva, M. I. Struchkova, Yu. N. Sheinker (Dokl. Akad. Nauk SSSR **152** [1963] 617/20; Dokl. Chem. Proc. Acad. Sci. USSR **148/153** [1963] 740/3). – [28] L. N. Mulay, A. Attalla (J. Am. Chem. Soc. **85** [1963] 702/6). – [29] M. D. Rausch, V. Mark (J. Org. Chem. **28** [1963] 3225/8). – [30] K. Yamakawa, H. Ochi, K. Arakawa (Chem. Pharm. Bull. [Tokyo] **11** [1963] 905/11).

[31] B. M. Yavorskii, N. S. Kochetkova, G. B. Zaslavskaya, A. N. Nesmeyanov (Dokl. Akad. Nauk SSSR **149** [1963] 111/3; Dokl. Chem. Proc. Acad. Sci. USSR **148/153** [1963] 202/4). – [32] Kodak-Pathé, J.-P. C. G. Dubosc (Fr. 1361470 [1963/64]; C.A. **63** [1965] 13461). – [33] A. V. Lesikar (J. Chem. Phys. **40** [1964] 2746/7). – [34] G. G. Dvoryantseva, S. L. Portnova, K. I. Grandberg, S. P. Gubin, Yu. N. Sheinker (Dokl. Akad. Nauk SSSR **160** [1965] 1075/8; Dokl. Chem. Proc. Acad. Sci. USSR **160/165** [1965] 155/8). – [35] A. N. Nesmeyanov, B. M. Yavorskii, G. B. Zaslavskaya, N. S. Kochetkova (Dokl. Akad. Nauk SSSR **160** [1965] 837/40; Dokl. Chem. Proc. Acad. Sci. USSR **160/165** [1965] 131/4).

[36] I. Pavlík, J. Klikorka (Z. Chem. [Leipzig] **5** [1965] 345/6). – [37] Colgate-Palmolive Co., J. E. Robertson (U.S. 3377248 [1965/68]; C.A. **69** [1968] No. 96878). – [38] O. E. Ayers, T. G. Smith, J. D. Burnett, B. W. Ponder (Anal. Chem. **38** [1966] 1606/7). – [39] G. G. Dvoryantseva, S. L. Portnova, Yu. N. Sheinker, L. P. Yur'eva, A. N. Nesmeyanov (Dokl. Akad. Nauk SSSR **169** [1966] 1083/6; Dokl. Chem. Proc. Acad. Sci. USSR **166/171** [1966] 773/5). – [40] K. I. Grandberg, S. P. Gubin (Izv. Akad. Nauk SSSR Ser. Khim. **1966** 551/4; Bull. Acad. Sci. USSR Div. Chem. Sci. **1966** 519/21).

[41] I. Pavlík, J. Klikorka, K. Handliř, P. Sedlmayer (Z. Chem. [Leipzig] **6** [1966] 427/8). – [42] P. Petrovitch (Double Liaison **1966** No. 133, pp. 1093/109). – [43] A. N. Nesmeyanov, N. S. Kochetkova, V. I. Malakhon, V. D. Vil'chevskaya, R. K. Dybovskii, N. P. Palitsyn, E. M. Piskunova (U.S.S.R. 201428 [1966/67]; C.A. **69** [1968] No. 19312). – [44] I. Pavlík, R. Foniok, J. Grégr (Sb. Ved. Praci Vys. Sk. Chem. Technol. Pardubice **15** [1967] 25/78; C.A. **68** [1968] No. 82728). – [45] T. H. Barr, W. E. Watts (J. Organometal. Chem. **15** [1968] 177/85).

[46] T. H. Barr, W. E. Watts (Tetrahedron **24** [1968] 3219/35). – [47] H. Hennig, O. Gürtler (J. Organometal. Chem. **11** [1968] 307/16). – [48] A. M. Tarr, D. M. Wiles (Can. J. Chem. **46** [1968] 2725/31). – [49] H. Falk (Monatsh. Chem. **100** [1969] 411/8). – [50] W. F. Little, H. T. Busby (4th Intern. Conf. Organometal. Chem., Bristol 1969, p. G2).

[51] A. N. Nesmeyanov, G. B. Zaslavskaya, B. M. Yavorskii, N. S. Kochetkova, N. P. Gambaryan (Opt. Spektroskopiya **26** [1969] 945/9; Opt. Spectrosc. [USSR] **26** [1969] 512/5). – [52] F. M. Vigo (Diss. Univ. Illinois 1969; Diss. Abstr. Intern. B **31** [1970] 598/9). – [53] F. I. Vilesov (Usp. Fotoniki **1** [1969] 5/45, 27). – [54] P. Carty, M. F. A. Dove (J. Organometal. Chem. **21** [1970] 195/201). – [55] G. J. Palenik (Inorg. Chem. **9** [1970] 2424/30).

[56] M. Sato, M. Koga, I. Motoyama, K. Hata (Bull. Chem. Soc. Japan **43** [1970] 1142/7). – [57] E. B. Sokolova, M. P. Shebanova, L. A. Datskevich (Sin. Fiz. Khim. Polim. No. 6 [1970] 69/72). – [58] S. Sorriso, H. Lumbroso (Tetrahedron **26** [1970] 2695/701, 4674). – [59] D. L. Uhrich, J. M. Wilson, W. A. Resch (Phys. Rev. Letters **24** [1970] 355/9). – [60] K. Tanikawa, K. Arakawa (Bunseki Kagaku **20** [1971] 278/81).

[61] V. I. Gol'danskii, O. P. Kevdin, N. K. Kivrina, E. F. Makarov, V. Ya. Rochev, R. A. Stukan (Zh. Eksperim. Teor. Fiz. **63** [1972] 2323/8; Soviet Phys.-JEPT **36** [1973] 1226/8). – [62] V. I. Gol'danskii, O. P. Kevdin, N. K. Kivrina, E. F. Makarov, V. Ya. Rochev, R. A. Stukan (Sb. 2nd Dokl. Vses. Nauchn. Konf. Zhidk. Krist. Simp. Ikh Prakt. Primen., Ivanovo, USSR, 1972 [1973], pp. 63/7; C.A. **81** [1974] No. 127882). – [63] H. Imai (Bull. Soc. Chem. Japan **45** [1972] 1264/5). – [64] D. Nielson, D. Boone, H. Eyring (J. Phys. Chem. **76** [1972] 511/5). – [65] R. E. Detjen, D. L. Uhrich, C. F. Shelley (Phys. Letters A **42** [1972/73] 522/4).

[66] G. Cerichelli, B. Floris, G. Illuminati, G. Ortaggi (Gazz. Chim. Ital. **103** [1973] 911/22). – [67] V. I. Gol'danskii, O. P. Kevdin, N. K. Kivrina, V. Ya. Rochev, E. F. Makarov, R. A. Stukan, I. G. Chistyakov, L. S. Shabyshev (Sb. 2nd Dokl. Vses. Nauchn. Konf. Zhidk. Krist. Simp. Ikh Prakt. Primen., Ivanovo, USSR, 1972 [1973], pp. 56/62; C.A. **82** [1975] No. 10161). – [68] V. I. Gol'danskii, O. P. Kevdin, N. K. Kivrina, V. Ya. Rochev, R. A. Stukan, I. G. Chistyakov, L. S. Shabishev (Mol. Cryst. Liquid Cryst. **24** [1973] 239/47). – [69] S. E. Jackson (Diss. Univ. Oxford 1973 from J. C. Green, Struct. Bonding [Berlin] **43** [1981] 37/112, 49). – [70] M. K. Makova, E. V. Leonova, Yu. S. Karimov, N. S. Kochetkova (J. Organometal. Chem. **55** [1973] 185/90).

[71] Ya. M. Paushkin, A. M. Shevchik (Vestsi Akad. Navuk Belarusk. SSR Ser. Khim. Navuk **1973** No. 3, pp. 95/8; C.A. **79** [1973] No. 53505). – [72] D. L. Uhrich, R. E. Detjen, J. M. Wilson (Mößbauer Eff. Methodol. **8** [1973] 175/90). – [73] A. I. Volozhin, L. Yu. Anishchenko, N. R. Prokopchuk, Ya. M. Paushkin, E. M. Aizenshtein, L. P. Repina, K. A. Bogdanov, V. A. Levdanskii, P. N. Zernov (U.S.S.R. 484752 [1973/79]). – [74] K. Yamakawa, M. Hisatome (J. Organometal. Chem. **52** [1973] 407/24). – [75] R. E. Detjen, D. L. Uhrich (Mößbauer Eff. Methodol. **9** [1974] 113/25).

[76] M. L. Good, J. Buttone, D. Foyt (Ann. N.Y. Acad. Sci. **239** [1974] 193/207). – [77] A. N. Nesmeyanov, G. B. Shul'pin, P. V. Petrovskii, V. I. Robas, M. I. Rybinskaya (Dokl. Akad. Nauk SSSR **215** [1974] 865/8; Dokl. Chem. Proc. Acad. Sci. USSR **214/219** [1974] 211/5). – [78] A. N. Nesmeyanov, G. B. Shul'pin, M. I. Rybinskaya (Dokl. Akad. Nauk SSSR **218** [1974] 1107/10; Dokl. Chem. Proc. Acad. Sci. USSR **214/219** [1974] 705/7). – [79] A. N. Nesmeyanov, G. B. Shul'pin, M. I. Rybinskaya, P. V. Petrovskii (Dokl. Akad. Nauk SSSR **215** [1974] 599/602; Dokl. Chem.

Proc. Acad. Sci. USSR **214/219** [1974] 188/91). – [80] J. M. Wilson, D. L. Uhrich (Mol. Cryst. Liquid Cryst. **25** [1974] 113/21).

[81] A. J. Ashe, E. Meyers, P. Shu, T. V. Lehmann (J. Am. Chem. Soc. **97** [1975] 6865/6). – [82] J. Sandström, J. Seita (J. Organometal. Chem. **108** [1976] 371/80). – [83] V. G. Bhide, M. C. Kandpal, S. Chandra (Solid State Commun. **23** [1977] 459/62). – [84] B. E. Mann (Progr. Nucl. Magn. Reson. Spectrosc. **11** [1977] 95/114). – [85] Bayer A.-G., W. Behrenz (Ger. Offen. 2711546 [1977/78]; C.A. **90** [1979] No. 1705).

[86] A. A. Koridze, N. M. Astakhova, P. V. Petrovskii, A. I. Lutsenko (Dokl. Akad. Nauk SSSR **242** [1978] 117/20; Dokl. Chem. Proc. Acad. Sci. USSR **238/243** [1978] 416/8). – [87] W. J. LaPrice (Diss. Kent State Univ. 1978 from Diss. Abstr. Intern. B **39** [1979] 4431). – [88] R. P. Sharma, M. B. Kurup, B. V. Thosar (Proc. Nucl. Phys. Solid State Phys. Symp. C **21** [1978] 414/6). – [89] P. Canonne, G. Foscolos, R. Harder (J. Organometal. Chem. **178** [1979] 331/42). – [90] V. Ya. Rochev, N. K. Kivrina, O. P. Kevdin (Acta Phys. Polon. A **56** [1979] 137/42).

[91] L. Bencivenni, D. Ferro, M. Pelino, R. Teghil (J. Indian. Chem. Soc. **57** [1980] 1062/5). – [92] L. M. Dyagileva, O. S. Morozov, N. N. Vyshinskii, V. G. Tsvetkov, Yu. A. Aleksandrov (Zh. Obshch. Khim. **50** [1980] 1859/62; J. Gen. Chem. [USSR] **50** [1980] 1516/8). – [93] A. N. Nesmeyanov, A. V. Vannikov, V. A. Zver'kov, L. V. Balabanova, M.-G. A. Shvekhgeimer, N. S. Kochetkova (Dokl. Akad. Nauk SSSR **252** [1980] 143/6; Dokl. Phys. Chem. Proc. Acad. Sci. USSR **250/255** [1980] 347/50). – [94] K. Sato, M. Katada, H. Sano (Chem. Letters **1980** 217/20). – [95] E. Haslinger, W. Robien, K. Schlögl, W. Weissensteiner (J. Organometal. Chem. **218** [1981] C11/C14).

[96] M. Pelino, M. Tomassetti, V. Piacente, G. d'Ascenzo (Thermochim. Acta **44** [1981] 89/99). – [97] L. Tătaru, I. Mazilu, M. Vâtă, T. Lixandru, C. Simionescu (J. Organometal. Chem. **214** [1981] 107/14). – [98] L. Tătaru, M. Vâţă, I. Mazilu, T. Lixandru, C. Simionescu (Angew. Makromol. Chem. **101** [1981] 19/27). – [99] M. Tomassetti, G. d'Ascenzo (private communication from [96]). – [100] M. Tomassetti, R. Curini, G. d'Ascenzo, G. Ortaggi (Thermochim. Acta **48** [1981] 333/41).

[101] M. C. Kandpal, V. G. Bhide (Physica B + C **112** [1982] 57/66). – [102] S. Braun, T. S. Abram, W. E. Watts (J. Organometal. Chem. **97** [1975] 429/41). – [103] Union Oil Co. of California, C. Rai (U.S. 3461287 [1963/69]; C.A. **71** [1969] No. 81541). – [104] I. Pavlík, K. Handlíř (Collection Czech. Chem. Commun. **31** [1966] 1958/74). – [105] G. Cerichelli, B. Floris, G. Illuminati, G. Ortaggi (Ann. Chim. [Rome] **64** [1974] 119/23).

[106] G. Neshvad, R. M. G. Roberts, J. Silver (J. Organometal. Chem. **236** [1982] 349/58). – [107] E. Haslinger, K. Koci, W. Robien, K. Schlögl (Monatsh. Chem. **114** [1983] 495/9). – [108] M. C. Clemance, R. M. G. Roberts, J. Silver (J. Organometal. Chem. **247** [1983] 219/22). – [109] D. G. Todoroff, R. Marande, D. Boyd, D. L. Uhrich (Mol. Cryst. Liquid Cryst. **95** [1983] 367/71).

5.2.3.5.1.1.3 Solubility

$Fe(C_5H_4–CO–CH_3)_2$ forms red-orange solutions in organic solvents [1]. It is very soluble in $CH_3CO_2C_2H_5$ [1, 4], $CHCl_3$ [1, 4, 6], CH_2Cl_2 [6], or CCl_4 [1, 4], relatively well soluble in C_6H_6 [2, 6], CH_3OH [3], or C_2H_5OH [2], moderately soluble in cyclohexane [1, 4], diethyl ether [1, 2, 4, 6, 7], or CS_2 [2], and slightly soluble in petroleum ether [6]. In aliphatic hydrocarbons it dissolves more poorly than does 1,2-diacetylferrocene [5].

The solubility in H_2O is slight [1, 4, 7]. The aqueous solution is not very stable; it decomposes easily on evaporation with deposition of iron hydroxide, even at room temperature [2].

References on p. 21

Strong mineral acids [6] form intensely violet colored solutions (see p. 31), from which the compound can be extracted with $CHCl_3$ after dilution, unchanged [1, 4]. The solution in dilute sulfuric acid is not very stable [7].

References:

[1] M. Rosenblum (Diss. Harvard Univ. 1953). – [2] R. Riemschneider, D. Helm (Chem. Ber. **89** [1956] 155/61). – [3] K. Yamakawa, H. Ochi, K. Arakawa (Chem. Pharm. Bull. [Tokyo] **11** [1963] 905/11). – [4] M. Rosenblum, R. B. Woodward (J. Am. Chem. Soc. **80** [1958] 5443/9). – [5] P. Carty, M. F. A. Dove (J. Organometal. Chem. **21** [1970] 195/201).

[6] P. Carty, M. F. A. Dove (J. Organometal. Chem. **28** [1971] 125/32). – [7] J. Holeček, K. Handlíř, J. Klikorka (Collection Czech. Chem. Commun. **44** [1979] 1060/9).

5.2.3.5.1.1.4 Electrochemical Behavior

The following potentials have been measured with a Pt electrode at room temperature for the reversible oxidation to the ferrocenium ion according to $Fe(C_5H_4\text{-}CO\text{-}CH_3)_2 - e^- \rightleftharpoons [Fe(C_5H_4\text{-}CO\text{-}CH_3)_2]^+$:

solvent; supporting electrolyte	measurement method	potential in V	reference electrode	Ref.
CH_3CN; 0.2M $LiClO_4$	chrono-potentiometry	$E_{\tau/4} = 0.796 \pm 0.04$	SCE*)	[1, 2]
CH_3CN; 0.2M $LiClO_4$	chrono-potentiometry	$E_{\tau/4} = 0.799$	SCE	[6]
CH_3CN; 0.2M $NaClO_4$	chrono-potentiometry	$E_{\tau/4} = 0.551$	$Ag/AgClO_4$**)	[8]
CH_3CN; 0.1M $[(n\text{-}C_4H_9)_4N]ClO_4$	cyclic voltammetry	$E_{1/2} = 0.902$	saturated Ag/AgCl	[14]
CH_3CN; 0.1M $[(n\text{-}C_4H_9)_4N]ClO_4$	cyclic voltammetry	$E_{1/2} = 0.85$	SCE	[18]
CH_3CN; 0.1M $[(n\text{-}C_4H_9)_4N]ClO_4$	cyclic voltammetry	$E^\circ = 0.83$	SCE	[11]
CH_3CN; 0.1M $NaClO_4$	polarography	$E_{1/2} = 0.790$	SCE	[9]
C_2H_5OH/H_2O (50/50); 0.5M $HClO_4$	polarography	$E_{1/2} = 0.84$	SCE	[4]

*) Saturated aqueous calomel electrode. – **) Ag/0.01M $AgNO_3$ + 0.2M $[(C_2H_5)_4N]ClO_4$ in CH_3CN.

Due to its height the redox potential cannot be determined by potentiometric titration with $K_2Cr_2O_7$ in $CH_3CO_2H/HClO_4$ [7] or with $Cu(ClO_4)_2$ in $CH_3CN/HClO_4/H_2O$ [17]. Regarding the effects of substituents on the redox potential of ferrocene derivatives, see "Ferrocene" 1, 1974, pp. 89/90, as well as [2 to 6, 8]. The contribution of the CO groups is higher the better the coplanarity and thus the conjugation with the cyclopentadienyl ring [14]. On the n-type semiconductor electrodes MoS_2 or WS_2 in CH_3CN/0.1 M $[(n\text{-}C_4H_9)_4N]ClO_4$ there are in the dark

 References on p. 22

no oxidation sites; on illumination at 632.8 nm with an He/Ne laser the anode potential rises at n-MoS_2 to 0.59 [11], at n-WS_2 to 0.22 [18] V (against SCE). Investigations on the p-type WS_2 semiconductor electrode in the same solvent with the same electrolyte give for the locations of the cyclic voltammetric waves no differences versus those of Pt; the reduction of the $[Fe(C_5H_4–CO–CH_3)_2]^+$ ions required, nevertheless, irradiation of the p-WS_2 with an He/Ne laser [15].

$Fe(C_5H_4–CO–CH_3)_2$ does not reduce on n-type Si photoelectrodes derivatized with I (R′ = R″ = CH_3, Cl) in C_2H_5OH/0.1 M $[(n\text{-}C_4H_9)_4N]ClO_4$, surface-confined ferrocenium ions photogenerated by an He/Ne laser, at an applied potential of $-<0.2$ V (against SCE) [12].

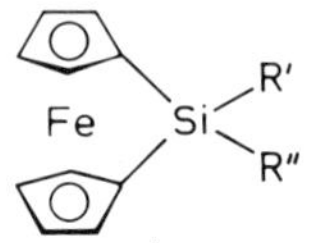

I

Electrochemical reduction forms in the successive one electron transitions the radical anion $[Fe(C_5H_4–CO–CH_3)_2]^-$ (see below) and the dianion $\mathbf{[Fe(C_5H_4–CO–CH_3)_2]^{2-}}$, whose half-wave potentials measured polarographically in $(CH_3)_2NCHO$/0.1 M $[(C_4H_9)_4N]PF_6$ amount to $E_{1/2} = -1.95$ and -2.65 V, respectively (against SCE). If the compound is electrolyzed in this solution or tetrahydrofuran/$[(C_4H_9)_4N]PF_6$ at a controlled potential ($->1.95$ or $->2.65$ V), then metallic iron and $[CH_3–CO–C_5H_4]^-$ are formed with use of 2 F according to the equation

$$Fe(C_5H_4–CO–CH_3)_2 + 2e^- \xrightarrow[\text{DMF or THF}]{[(C_4H_9)_4N]PF_6} 2[(C_4H_9)_4N][C_5H_4COCH_3] + Fe + 2PF_6^- \quad [10, 16].$$

If the electrolysis is conducted at low temperatures (0 to -25°C) in tetrahydrofuran and an inert atmosphere at the level of the first step, then the stable green $\mathbf{[Fe(C_5H_4–CO–CH_3)_2]^-}$ forms in the reversible reaction; at -25°C the reaction is nearly quantitative. Electrolysis at 10°C in the presence of CO leads in a two-electron-transfer process with CO incorporation to $[CH_3–CO–C_5H_4]^-$ and $[CH_3–CO–C_5H_4Fe(CO)_2]^-$ [13].

References:

[1] D. E. Bublitz, G. Hoh, T. Kuwana (Chem. Ind. [London] **1959** 635/6). – [2] T. Kuwana, D. E. Bublitz, G. Hoh (J. Am. Chem. Soc. **82** [1960] 5811/7). – [3] J. Komenda, J. Tirouflet (Compt. Rend. **254** [1962] 3093/5). – [4] J. Tirouflet, E. Laviron, R. Dabard, J. Komenda (Bull. Soc. Chim. France **1963** 857/62). – [5] J. Komenda (Chem. Zvesti **18** [1964] 378/84).

[6] W. F. Little, C. N. Reilley, J. D. Johnson, A. P. Sanders (J. Am. Chem. Soc. **86** [1964] 1382/6). – [7] E. G. Perevalova, S. P. Gubin, S. A. Smirnova, A. N. Nesmeyanov (Dokl. Akad. Nauk SSSR **155** [1964] 857/60; Dokl. Chem. Proc. Acad. Sci. USSR **154/159** [1964] 328/31). – [8] D. W. Hall, C. D. Russell (J. Am. Chem. Soc. **89** [1967] 2316/22). – [9] H. Hennig, O. Gürtler (J. Organometal. Chem. **11** [1968] 307/16). – [10] N. El Murr, A. Chaloyard, E. Laviron (Nouv. J. Chim. **2** [1978] 15/7).

[11] L. F. Schneemeyer, M. S. Wrighton (J. Am. Chem. Soc. **101** [1979] 6496/500). – [12] N. S. Lewis, A. B. Bocarsly, M. S. Wrighton (J. Phys. Chem. **84** [1980] 2033/43). – [13] N. El Murr, A. Chaloyard (J. Organometal. Chem. **193** [1980] C60/C62). – [14] E. Fujita, B. Gordon, M. Hillman, A. G. Nagy (J. Organometal. Chem. **218** [1981] 105/14). – [15] J. A. Baglio, G. S. Calabrese, D. J. Harrison, E. Kamieniecki, A. J. Ricco, M. S. Wrighton, G. D. Zoski (J. Am. Chem. Soc. **105** [1983] 2246/56).

[16] A. Chaloyard, N. El Murr (Inorg. Chem. **19** [1980] 3217/20). – [17] P. F. Quirk, B. Kratochvil (Anal. Chem. **42** [1970] 535/6). – [18] J. A. Baglio, G. S. Calabrese, E. Kamieniecki, R. Kershaw, C. P. Kubiak, A. J. Ricco, A. Wold, M. S. Wrighton, G. D. Zoski (J. Electrochem. Soc. **129** [1982] 1461/72).

5.2.3.5.1.1.5 Chemical Behavior

5.2.3.5.1.1.5.1 Reactions with Cleavage of the Iron-Cyclopentadienyl Bond

In the following only the reactions are described which proceed primarily with destruction of the ferrocene molecule. Reactions which lead secondarily to decomposition are discussed in 5.2.3.5.1.1.5.3, p. 28, under "Reduction" and in 5.2.3.5.1.1.5.4, p. 30, under "Substitution Reactions on the Cyclopentadienyl Rings" as well as on p. 33.

Thermal Decomposition

On heating in air at 10 K/min the thermal decomposition takes place in two steps. In the range 155 to 262°C (maximum rate constant at 230°C) the cleavage of the CH_3CO groups occurs in a reaction of the 0.20 order; the activation energy E_A is 16.1 kcal/mol (67.4 kJ/mol), the weight loss 22.85%. Between 262 and 450°C (maximum rate constant about 330°C), a reaction of order 2.10 occurs with the decomposition of the ferrocene residue; $E_A = 25.5$ kcal/mol (106.8 kJ/mol); weight loss 52.85% [28, 29].

In the gas phase in a closed system the compound decomposes between 480 and 540°C at an initial pressure of 72 to 125 Torr corresponding to a first order reaction; the temperature dependence of the rate constant gives $E_A = 48.3$ kcal/mol (202.2 kJ/mol) [12, 13]. In addition to Fe and C, gaseous pyrolysis products form comprised predominantly of CO (45.5 mol% of the gaseous product at 520°C) plus CH_4 (28.1%), H_2 (13.7%), C_2H_4 (6.8%), and C_2H_6 (0.7%) as well as higher hydrocarbons. If the ratio of the surface (S) to the volume (V) of the glass reaction vessel is raised from 1.5 to the range of 10 to 25 cm^{-1}, then the CO component in the gaseous reaction products increases slightly, while the proportions of H_2 and of the hydrocarbons drop somewhat [11, 17, 26]. At 550°C and S/V = 1.5 cm^{-1} the rate constant of $k = 4.47 \times 10^{-3}$ s^{-1} is 447 times larger than that of ferrocene under the same conditions [26, 27], see also [11]. The reaction takes place by a radical mechanism [12], which also is confirmed by the reduction of the reaction rate on addition of cyclopentadiene. The decrease of the reaction rate with increasing C_5H_6 concentration reaches a limiting value, where further concentration increase does not cause any further rate decrease [13].

The compound absorbed on silica gel forms on heating in an inert atmosphere at 350°C Fe^{III} silicate, at 550°C Fe^{II} silicate, and at 800°C Fe_2SiO_4 [31].

Mass Spectrum

On fragmentation by electron impact the molecular ion (m/e 270) possesses the highest relative intensity (set equal to 100%). It decomposes primarily according to the scheme in I either via m/e 255 → 199, which has been proven by metastable peaks at m/e 240.8 and 155.3, or via m/e 227 → 199, indicated by metastable peaks at m/e 190.8 and 174.4. The fragment occurring at m/e 163 $[FeC_5H_4{-}CO{-}CH_3]^+$ (relative intensity 19%) could be formed from the molecular ion by splitting off of $\dot{C}_5H_4COCH_3$ or from m/e 227 by cleaving off C_5H_4 [14]. If one starts with the absolute intensity (total intensity of all peaks is 100%), then the intensity of the molecular ion is 29%, that of $[C_5H_5Fe]^+ + [CH_3COC_5H_4Fe]^+$ 8.1%, and that of Fe-free ions 7.5%. On fragmentation in an arc discharge (0.2 to 0.3 A; discharge voltage 100 to 300 V) in contrast the metal-free ions make up with 32.5% the largest share; the intensity shares of the molecular ion and of $[C_5H_5Fe]^+ + [CH_3COC_5H_4Fe]^+$ amount to 10.5 and 6.2%, respectively [10].

References on pp. 25/6

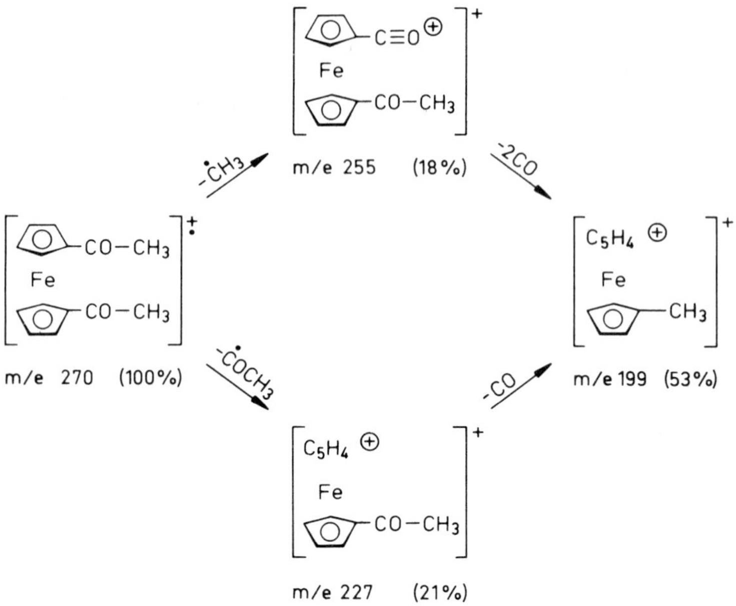

I from [14]

Photochemical Reactions

$Fe(C_5H_4{-}CO{-}CH_3)_2$ is photolabile [1]. When a CH_3OH solution sits in daylight for a long time, the color is bleached and finally a precipitate separates [8]. In isopropanol/hydrochloric acid a photochemical reaction takes place with heterolytic cleavage of the Fe-ring bond and formation of acetylpentadiene [16]. In contrast to its 1,2-isomer, 1,1′-diacetylferrocene does not decompose in an He/Ne laser beam (6228 Å) [9].

Other Decomposition Reactions

Oxidation with $Ce(SO_4)_2$ in H_2SO_4/H_2O leads to complete decomposition of the compound according to $Fe(C_5H_4{-}CO{-}CH_3)_2 + 3Ce^{IV} \rightarrow Fe^{3+} + 3Ce^{III} + 2C_5H_4{-}CO{-}CH_3$. This equation is only valid when the oxidation is so conducted that at no time is an excess of Ce^{IV} present. The first reaction step represents the complete one-electron oxidation to $[Fe(C_5H_4{-}CO{-}CH_3)_2]^+$, which decomposes into Fe^{3+} and $[C_5H_4{-}CO{-}CH_3]^-$. The latter acts versus the ferrocenium ion as a one-electron reducing agent, versus Ce^{IV} as a polyelectron reductant. This explains the superstoichiometric consumption of Ce^{IV} (up to 10 mol per 1 mol $Fe(C_5H_4{-}CO{-}CH_3)_2$) in the presence of excess Ce^{IV} [30].

The thermal treatment of mixtures of $Fe(C_5H_4{-}CO{-}CH_3)_2$ and $^{103}RuCl_3$ in a sealed ampule at 180°C yields after column chromatography on Al_2O_3 of the C_2H_5OH solution of the products with C_2H_5OH an eluate of almost pure $^{103}Ru(C_5H_4{-}CO{-}CH_3)_2/Fe(C_5H_4{-}CO{-}CH_3)_2$ [21].

Ligand exchange takes place to form $[Fe(C_6H_{6-n}(CH_3)_n)_2]^{2+}$ in 60 to 80% yield after 5 h reaction with $C_6H_{6-n}(CH_3)_n$ (^6L; $1 \leqq n \leqq 6$), in the absence of solvents when ^6L is liquid, otherwise in nonane or decalin at 110 to 200°C, in each case in the presence of $AlCl_3$ (at least 2, better 4 mol per 1 mol $Fe(C_5H_4{-}CO{-}CH_3)_2$) [15, 23]. Quite generally the formation of complexes by $Fe(C_5H_4{-}CO{-}R)_2$ with $AlCl_3$ takes place on neither the Fe nor the C_5H_4 rings, according to a more recent evaluation (cf. "Ferrocene" 7, 1980, p. 44), but on the two CO groups where it leads to the complete splitting of the molecule with formation of $[Fe^{6}L_2]^{2+}$, however, only at elevated temperatures. At lower temperatures (about 100°C), the 1,1′-diacylferrocenes are practically inert with respect to ligand exchange [19, 32, 33]; see also [20, 35]. At temperatures above 130°C loss of the CH_3 group and isomerization occur to some extent for $Fe(C_5H_4{-}CO{-}CH_3)_2$ [33]. These results supersede the investigations of Nesmeyanov et al. [5 to 7], who on reaction of $Fe(C_5H_4{-}CO{-}CH_3)_2$ in boiling mesitylene in the presence of $AlCl_3$ (mole ratio $Fe(C_5H_4{-}CO{-}CH_3)_2/AlCl_3 = 1/4$) after 4 to 5 h believed they had obtained monocations formed by single ligand exchange. The dication is isolable as the PF_6 or $B(C_6H_5)_4$ salt [15]. No exchange is observed with C_6H_6 boiling [19] even after 24 h in the presence of a large $AlCl_3$ excess [23, 24, 26], likewise not with coronene in boiling methylcyclohexane in the presence of $AlCl_3/Al$ [25].

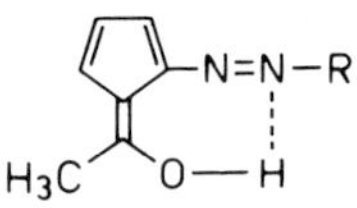

I

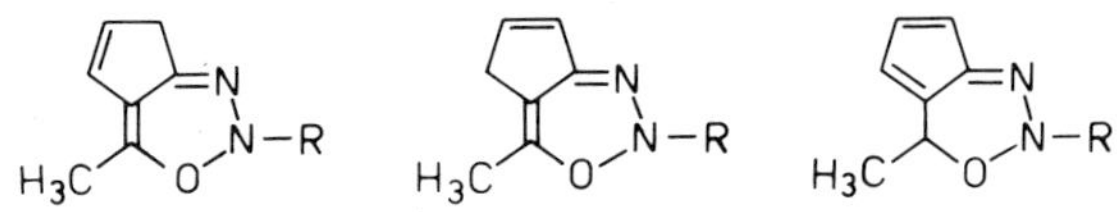

a b c

II

1,1′-Diacetylferrocene behaves differently than ferrocene with respect to diazonium salts. Thus, it reacts with $[C_6H_5N_2]Cl$ in glacial acetic acid at room temperature not with arylation but with formation of the Fe-free fulvene derivative I ($R = C_6H_5$) [4]. Nesmeyanov et al. [2, 3] early identified the disruption of the ferrocene nucleus but hypothesized as the initial product a 1,2,3-oxadiazine derivative of the structure II (most likely IIa) [3] or III [2]. The reaction with $[4\text{-}O_2N{-}C_6H_4N_2]Cl$ forms I ($R = 4\text{-}O_2N{-}C_6H_4$) and in small quantities the 4-nitrophenyl derivative IV along with considerable quantities of polymers insoluble in C_6H_6 [22]; III ($R = 4\text{-}O_2N{-}C_6H_4$) was assumed by [2] to be the reaction product. The reaction with $[4\text{-}CH_3{-}C_6H_4N_2]Cl$ proceeds slowly according to [2] to form III ($R = 4\text{-}CH_3{-}C_6H_4$). I ($R = 4\text{-}(C_3H_7)_2N{-}C_6H_4$) is also obtained by coupling under UV irradiation of $Fe(C_5H_4{-}CO{-}CH_3)_2$ with $[4\text{-}(C_3H_7)_2N\text{-}C_6H_4N_2]BF_4$ at 30 to 40°C in $HOCH_2CH_2OH$ [18].

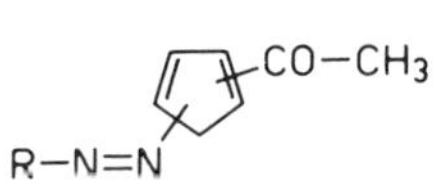

III

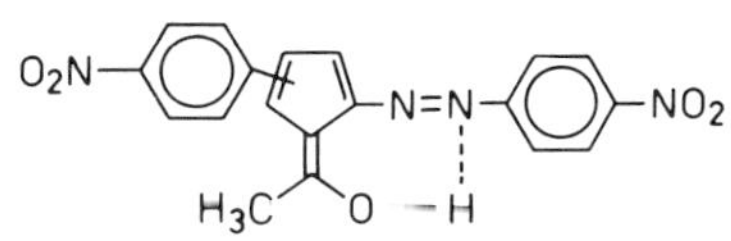

IV

References:

[1] R. B. Woodward, M. Rosenblum, M. C. Whiting (J. Am. Chem. Soc. **74** [1952] 3458/9). – [2] A. N. Nesmeyanov, E. G. Perevalova, R. V. Golovnya, N. A. Simukova, O. V. Starovsky (Izv. Akad. Nauk SSSR Otd. Khim. Nauk **1957** 638/40; Bull. Acad. Sci. USSR Div. Chem. Sci. **1957** 647/9). – [3] A. N. Nesmeyanov, E. G. Perevalova, N. A. Simukova, Yu. N. Sheinker, M. D. Reshetova (Dokl. Akad. Nauk SSSR **133** [1960] 851/4; Proc. Acad. Sci. USSR Chem. Sect. **130/135** [1960] 879/82). – [4] R. E. Bozak, K. L. Rinehart (J. Am. Chem. Soc. **84** [1962] 1589/93). – [5] A. N. Nesmeyanov, N. A. Vol'kenau, I. N. Bolesova (Tetrahedron Letters **1963** 1725/9).

[6] A. N. Nesmeyanov, N. A. Vol'kenau, I. N. Bolesova (Dokl. Akad. Nauk SSSR **149** [1963] 615/8; Dokl. Chem. Proc. Acad. Sci. USSR **148/153** [1963] 267/70). – [7] A. N. Nesmeyanov, N. A. Vol'kenau, L. S. Shilovtseva (Dokl. Akad. Nauk SSSR **160** [1965] 1327/30; Dokl. Chem. Proc. Acad. Sci. USSR **160/165** [1965] 203/6). – [8] A. M. Tarr, D. M. Wiles (Can. J. Chem. **46** [1968] 2725/31). – [9] P. Carty, M. F. A. Dove (J. Organometal. Chem. **21** [1970] 195/201). – [10] N. V. Zakurin, S. P. Gubin, V. P. Bochin (J. Organometal. Chem. **23** [1970] 535/44).

[11] L. M. Dyagileva, B. Y. Andreyev, G. I. Feklisov (5th Intern. Conf. Organometal. Chem., Moscow 1971, Vol. II, Abstr. pp. 387/8). – [12] L. M. Dyagileva, B. Ya. Andreev, G. I. Feklisov

(Dokl. Akad. Nauk SSSR **200** [1971] 1110/1; Dokl. Chem. Proc. Acad. Sci. USSR **196/201** [1971] 836/7). – [13] L. M. Dyagileva, B. Ya. Andreev (Tr. Khim. Khim. Tekhnol. **1972** 70/2). – [14] H. Imai (Bull. Chem. Soc. Japan **45** [1972] 1264/5). – [15] D. Astruc (Tetrahedron Letters **1973** 3437/9).

[16] R. E. Bozak, H. Javaheripour (Chem. Ind. [London] **1973** 696/7). – [17] L. M. Dyagileva (Tr. Khim. Khim. Tekhnol. **1974** No. 4, pp. 108/9). – [18] Ricoh Co., Ltd., K. Kojima, T. Goto (Japan. Kokai 75-149721 [1974/75] from C.A. **84** [1976] No. 121416). – [19] D. Astruc, R. Dabard (7th Intern. Conf. Organometal. Chem., Venice 1975, Abstr. No. 209). – [20] D. Astruc, R. Dabard (Bull. Soc. Chim. France **1975** 2571/4).

[21] D. Langheim, M. Wenzel, E. Nipper (Chem. Ber. **108** [1975] 146/54). – [22] Š. Toma, E. Solčaniová, A. Maholanyiová, J. Leško (Collection Czech. Chem. Commun. **40** [1975] 1629/40). – [23] D. Astruc, R. Dabard (Tetrahedron **32** [1976] 245/9). – [24] D. Astruc, R. Dabard (J. Organometal. Chem. **111** [1976] 339/47). – [25] G. Schmitt, W. Keim, J. Fleischhauer, U. Walbergs (J. Organometal. Chem. **152** [1978] 315/27).

[26] L. M. Dyagileva, E. I. Tsyganova, Yu. A. Aleksandrov (Zh. Obshch. Khim. **50** [1980] 2257/62; J. Gen. Chem. [USSR] **50** [1980] 1824/8). – [27] L. M. Dyagileva, E. I. Tsyganova, Yu. A. Aleksandrov (Zh. Obshch. Khim. **51** [1981] 1993/7; J. Gen. Chem. [USSR] **51** [1981] 1715/8). – [28] L. Tătaru, M. Dărânga, M. Vâţă, T. Lixandru, C. Simionescu (J. Therm. Anal. **22** [1981] 259/68). – [29] L. Tătaru, I. Mazilu, M. Vâtă, T. Lixandru, C. Simionescu (J. Organometal. Chem. **214** [1981] 107/14). – [30] J. Holeček, K. Handlíř, J. Klikorka (Collection Czech. Chem. Commun. **44** [1979] 1060/9).

[31] C. F. Cook, P. R. Gray, H. M. Barton (ORNL-IIC-10-Pt. 1 [1967] 130/41; C.A. **69** [1968] No. 46289). – [32] R. G. Sutherland (J. Organometal. Chem. Libr. **3** [1977] 311/41, 314). – [33] D. Astruc (Diss. Univ. Rennes from [34]). – [34] D. Astruc (Tetrahedron **39** [1983] 4027/95, 4036). – [35] E. E. Roman, D. Astruc (Nouv. J. Chim. **1** [1977] 183/5).

5.2.3.5.1.1.5.2 Oxidation

Electrochemical oxidation is covered in Section 5.2.3.5.1.1.4, p. 21. For oxidation by Ce^{IV} see p. 24.

To the Ferrocenium Ion $[Fe(C_5H_4-CO-CH_3)_2]^+$

The rate constant $k = 1.84 \times 10^4\ M^{-1} \cdot s^{-1}$ has been determined for the fast homogeneous electron transfer in CH_3CN from $Fe(C_5H_4-CO-CH_3)_2$ to the radical cation $[(4-CH_3-O-C_6H_4)_3N]^{+\cdot}$ by means of computer controlled internal reflection spectroelectrochemical measurements; this is 1.7×10^{-5} times that of ferrocene and is in good agreement with the value calculated by Marcus' theory [16, 21].

Rate constant and radical yield for the electron transfer reaction of $Fe(C_5H_4-CO-CH_3)_2$ (^1D) to methylene blue in the excited triplet state ($^3\overset{*}{A}{}^+$) in CH_3CN at 23°C are $k = 5.1 \times 10^9\ M^{-1} \cdot s^{-1}$ and $\Phi_r < 0.015$; the Gibbs free energy $\Delta G = 0.26$ eV. The following mechanism has been proposed for the process:

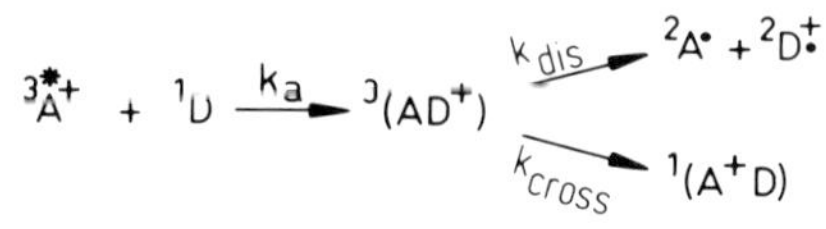

The triplet exciplex $^3(AD^+)$ formed with a diffusion-controlled rate constant k_a either dissociates with the rate constant k_{dis} into the free radicals or undergoes intersystem crossing (k_{cross}) to the singlet ground state. The uncommonly low value for Φ_r is attributed to Fe, which by increasing the spin-orbit coupling in $^3(AD^+)$ leads to enhanced intersystem crossing rates [23].

Uranyl cation luminescence is quenched by 1,1′-diacetylferrocene. In acetone at 25°C the quenching constant $k_Q = (1.00 \pm 0.05) \times 10^9\ M^{-1} \cdot s^{-1}$; the change of the free energy calculated for the one-electron transfer from $Fe(C_5H_4{-}CO{-}CH_3)_2$ to $(UO_2^{2+})^*$ is $\Delta G_Q = -1.80$ V (vs. SCE), so that the process is thermodynamically possible [20].

There is no formation of the cation with H_2SO_4 [13], $K_2Cr_2O_7$ in CH_3CO_2H/H_2O/acetone [5], Fe^{3+} [24], Cu^{2+} in $CH_3CN/HClO_4$ [26], or Ag^+ [25]. When $Fe(C_5H_4{-}CO{-}CH_3)_2$ in acetone is reacted with I_2, a shift toward lower fields of the cyclopentadienyl proton signals for the paramagnetic cation can be observed only for a short time, then the signal vanishes due to decomposition of the cation [9]. After oxidation with KIO_3, $[NH_4]_2[Ce(NO_3)_6]$, or $AgClO_4$ in the presence of several drops of H_2SO_4 in acetone, an ESR spectrum ($g_{\parallel} = 3.90$, $g_{\perp} = 1.65$) is obtained which resembles that of α-substituted 1,1′-dialkylferrocenium ions. Possibly the acetyl is converted into the hydroxyl group [13].

To $Fe(C_5H_4{-}CO{-}CH_2OH)_2$ and $Fe(C_5H_4{-}CO{-}CHO)_2$

Hydroxylation in the position α to the keto group takes place with $C_6H_5I(O_2CCH_3)_2$ in CH_3OH/KOH in a highly selective reaction without disturbance at the Fe or further oxidation. After hydrolysis with acetone/4-$CH_3C_6H_4SO_3H$ the yield is 36% [22]. SeO_2 oxidizes the diacetylferrocene in dioxane/H_2O at the boiling point to $Fe(C_5H_4{-}CO{-}CHO)_2 \cdot 2H_2O$ [19].

To $Fe(C_5H_4{-}CO_2H)_2$ and $HO_2C{-}C_5H_4FeC_5H_4{-}CO{-}CH_3$

Oxidation with O_2 in $OP(N(CH_3)_2)_3/KOC(CH_3)_3$ at 80°C yields besides $Fe(C_5H_4{-}CO_2H)_2$ (17% yield) the products $C_5H_5FeC_5H_4{-}CO_2H$ (34%) and ferrocene (26%); the low fraction of the dicarboxylic acid is attributed to the cleavage of the diketone occurring in the presence of strong bases [17]. The reaction goes practically quantitatively with NaOCl [4, 6, 8, 11, 18, 27] or KOCl [10, 12, 14] in H_2O [4, 6, 11, 12] or CH_3OH/H_2O [6, 8] at 45 to 50°C [6, 11, 12] or in $O(CH_2CH_2OCH_3)_2/H_2O$ at 20 to 25°C [18], with NaOBr in H_2O at 100°C [2, 7] or in dioxane/H_2O below 5°C [15] as well as with hypoiodite [1, 3].

I_2 in pryridine at room temperature forms $HO_2C{-}C_5H_4FeC_5H_4{-}CO{-}CH_3$, that can be further oxidized at temperatures above 100°C [6]. According to [2] the dicarboxylic acid forms from $Fe(C_5H_4{-}CO{-}CH_3)_2$ as low as 95°C, albeit only in 7.5% yield.

References:

[1] R. B. Woodward, M. Rosenblum, M. C. Whiting (J. Am. Chem. Soc. **74** [1952] 3458/9). – [2] W. Göller (Diss. Stuttgart T.H. 1958). – [3] K. L. Motz (Diss. Univ. Illinois 1958; Diss. Abstr. **19** [1958] 958). – [4] F. W. Knobloch, W. H. Rauscher (J. Polym. Sci. **54** [1961] 651/6). – [5] M. Peterlik, K. Schlögl (Z. Anal. Chem. **195** [1963] 113/7).

[6] K. Yamakawa, H. Ochi, K. Arakawa (Chem. Pharm. Bull. [Tokyo] **11** [1963] 905/11). – [7] Union Oil Co. of California, C. Rai (U.S. 3461287 [1963/69]; C.A. **71** [1969] No. 81541). – [8] Colgate-Palmolive Co., J. T. Suh (U.S. 3382267 [1965/68]; C.A. **69** [1968] No. 59378). – [9] H. P. Fritz, H. J. Keller, K. E. Schwarzhans (J. Organometal. Chem. **6** [1966] 652/9). – [10] I. A. Golubeva, T. P. Vishnyakova (Neftekhimiya i Pererabotka Nefti i Gaza: Tematicheskie i Analiticheskie Obsary TSNITT Eneftekhimiy, Moscow 1966, pp. 36/42).

[11] P. Petrovitch (Double Liaison **1966** 1093/109). – [12] I. A. Golubeva, T. P. Vishnyakova, N. N. Bul'on (Tr. Mosk. Inst. Neftekhim. Gaz. Prom. No. 72 [1967] 61/7). – [13] R. Prins, A. R. Korswagen (J. Organometal. Chem. **25** [1970] C74/C76). – [14] E. B. Sokolova, M. P. Shebanova, L. A. Datskevich (Sin. Fiz. Khim. Polim. **1970** No. 6, pp. 69/72). – [15] A. Sonoda, I. Moritani (J. Organometal. Chem. **26** [1971] 133/40).

[16] N. Winograd, T. Kuwana (J. Am. Chem. Soc. **93** [1971] 4343/50). – [17] G. Schmitt, S. Özman (J. Org. Chem. **41** [1976] 3331/2). – [18] V. P. Tverdokhlebov, I. V. Tselinskii, B. V. Gidaspov, G. Yu. Chikisheva (Zh. Org. Khim. **12** [1976] 2335/40; J. Org. Chem. [USSR] **12** [1976] 2268/72). – [19] M. Vernois, H. L. Williams (J. Appl. Polym. Sci. **23** [1979] 1601/6). – [20] O. Traverso, R. Rossi, L. Magon, A. Cinquantini, T. J. Kemp (J. Chem. Soc. Dalton Trans. **1978** 569/72).

[21] J. C. Holland (Diss. Purdue Univ. 1979 from Diss. Abstr. Intern. B **40** [1979] 2637). – [22] R. M. Moriarty, H. Hu, S. C. Gupta (Tetrahedron Letters **22** [1981] 1283/6). – [23] S. Tamura, K. Kokuchi, H. Kokubun, A. Weller (Z. Physik. Chem. [N. F.] **121** [1980] 165/72). – [24] M. Rosenblum (Diss. Harvard Univ. 1953). – [25] P. Carty, M. F. A. Dove (J. Organometal. Chem. **28** [1971] 125/32).

[26] P. F. Quirk, B. Kratochvil (Anal. Chem. **42** [1970] 535/6). – [27] K. Gonsalves, L. Zhan-ru, M. D. Rausch (J. Am. Chem. Soc. **106** [1984] 3862/3).

5.2.3.5.1.1.5.3 Reduction

For comparison with the subject matter here see also the material for the nucleophilic addition to the CO group in Section 5.2.3.5.1.1.5.6.1, p. 33; for electrochemical reduction see Section 5.2.3.5.1.1.4, p. 21.

Hydrogenolysis at elevated temperatures and pressures produces $Fe(C_5H_4{-}C_2H_5)_2$ and H_2O. Suitable catalysts and solvents include copper chromite in absolute ethanol [1], Raney nickel in diethyl ether [2] or tetrahydrofuran [22], Pt/C in methanol [4], or Pt in glacial acetic acid [6]. At 1 atm H_2 pressure hydrogenolysis succeeds over Ni-Re in $(CH_3)_2CHOH$ at 70°C; the yield is 85 to 90%, the purity 67 to 70% [35]. An investigation of the catalytic effectiveness of Ni-Al + M, M = Ti, Fe-Ti, Cr-Ti, or Cu, gave for Ni-Al and Ni-Al-Ti the greatest activity [36]. Hydrogenolysis in 10% $HClO_4$ in CH_3CO_2H at 1 atm H_2 and not more than 55°C with 10% Pd/C as catalyst yields $Fe(C_5H_4{-}C_2H_5)_2$ in a ratio $Fe(C_5H_4{-}CO{-}CH_3)_2$/catalyst of 1/3, and $CH_3{-}CO{-}C_5H_4FeC_5H_4{-}C_2H_5$ in a ratio of 1/1 [37]. If the hydrogenation is conducted at room temperature and pressure over PtO_2 in CH_3CO_2H, then $CH_3{-}CO{-}C_5H_4FeC_5H_4{-}C_2H_5$ is also obtained in addition to $Fe(C_5H_4{-}C_2H_5)_2$ (yields 30 and 26%, respectively) [20]. On conducting the reaction over Raney nickel at a working pressure of 240 atm H_2 and 345°C in C_2H_5OH decomposition of the molecule often occurs, so that ethylcyclopentane (yield 53%) forms along with $Fe(C_5H_4{-}C_2H_5)_2$ (yield 15%) [3].

Reduction to $Fe(C_5H_4{-}C_2H_5)_2$ succeeds also following Clemmensen's method with amalgamated zinc and concentrated HCl in glacial acetic acid [2, 12, 15]. Reaction with H_2O in boiling dioxane over Ni deposited on Zn dust gave after 12 h $CH_3{-}CO{-}C_5H_4FeC_5H_4{-}C_2H_5$ (40%) and traces of $CH_3{-}CO{-}C_5H_4FeC_5H_4{-}CHOH{-}CH_3$ along with $Fe(C_5H_4{-}C_2H_5)_2$ (yield 10%); more than 40% of the initial diacetylferrocene was recovered [18].

The reaction with $LiAlH_4/AlCl_3$ in tetrahydrofuran proceeds smoothly to form $Fe(C_5H_4{-}C_2H_5)_2$ [8]; likewise with $NaBH_4/AlCl_3$ in $CH_3OCH_2CH_2OCH_3$ [28] or with $NaBH_4$ in the same solvent with introduction of a weak stream of BF_3 [27]. If concentrated H_2SO_4 in small amounts is added to a mixture of $Fe(C_5H_4{-}CO{-}CH_3)_2$ and $NaBH_4$ in dioxane, in about 20 min a mixture forms of

$Fe(C_5H_4{-}C_2H_5)_2$ and $CH_3{-}CO{-}C_5H_4FeC_5H_4{-}C_2H_5$ (yields 27 and 32%, respectively) [31]. Reduction of $Fe(C_5H_4{-}CO{-}CH_3)_2$ with $LiAlD_4/AlCl_3$ forms $Fe(C_5H_4{-}CD_2{-}CH_3)_2$, analogous reduction of $Fe(C_5H_4{-}CO{-}CD_3)_2$ forms $Fe(C_5H_4{-}C_2D_5)_2$ [21].

Hydrogenation with $(C_2H_5)_3SiH$ in CF_3CO_2H yields $Fe(C_5H_4{-}C_2H_5)_2$ only at elevated temperatures and on use of excess silane; otherwise it forms $CH_3{-}CO{-}C_5H_4FeC_5H_4{-}C_2H_5$, whose C_2H_5 group hinders the hydrogenation of the second keto group [26]. The reaction of 1,1′-diacetylferrocene with R_3SnH ($R = C_4H_9$, $CH_2C_6H_5$, C_6H_5) in the presence of CH_3OH (mole ratio 1/2/4) in C_6H_6 produces $CH_3{-}CO{-}C_5H_4FeC_5H_4{-}C_2H_5$ and $Fe(C_5H_4{-}C_2H_5)_2$, with excess reductant in an extended reaction time ultimately only the latter. Instead of R_3SnH one can also employ $(C_6H_5)_2SnH_2$, whose instability, however, limits its applicability [19, 23].

$Fe(C_5H_4{-}CHOH{-}CH_3)_2$ is recovered as a 1/1 mixture of the meso and the racemic form [24] (cf. "Ferrocene" 7, 1980, pp. 167, 174/5) in the reduction with $LiAlH_4$ [34] in absolute diethyl ether [5, 7, 10, 15, 30] or tetrahydrofuran [32, 33]. $LiAlD_4$ and subsequent hydrolysis with H_2O forms $Fe(C_5H_4{-}CDOH{-}CH_3)_2$ [17], similarly $LiAlD_4$ and D_2O give the $Fe(C_5H_4{-}CDOD{-}CH_3)_2$ [21]. $NaBH_4/Fe(C_5H_4{-}CO{-}CH_3)_2$ in the mole ratio 2/1 in isopropanol forms $Fe(C_5H_4{-}CHOH{-}CH_3)_2$ [9, 11, 13, 14], in the mole ratio 1/1 in CH_3OH the $CH_3{-}CO{-}C_5H_4FeC_5H_4{-}CHOH{-}CH_3$ compound [16, 25].

No reduction takes place with fresh baker's yeast [29].

References:

[1] N. V. De Bataafsche Petroleum Maatschappij, A. C. Nixon, T. P. Rudy (Brit. 763550 [1954/56]; C.A. **1957** 10588; Ger. 1003727 [1954/57]). – [2] A. N. Nesmeyanov, N. A. Vol'kenau (Dokl. Akad. Nauk SSSR **107** [1956] 262/4). – [3] A. N. Nesmeyanov, E. G. Perevalova, R. V. Golovnya, T. V. Nikitina, N. A. Simukova (Izv. Akad. Nauk SSSR Otd. Khim. Nauk **1956** 739/41; Bull. Acad. Sci. USSR Div. Chem. Sci. **1956** 749/51). – [4] M. Vogel, M. Rausch, H. Rosenberg (J. Org. Chem. **22** [1957] 1016/8). – [5] W. Göller (Diss. Stuttgart T.H. 1958).

[6] K. L. Motz (Diss. Univ. Illinois 1958; Diss. Abstr. **19** [1958] 958). – [7] K. Arakawa, K. Yamakawa (Kagaku No Ryoiki **14** [1960] 632/44) – [8] K. Schlögl, A. Mohar, M. Peterlik (Monatsh. Chem. **92** [1961] 921/6). – [9] E. C. Winslow, E. W. Brewster (J. Org. Chem. **26** [1961] 2982). – [10] K. Yamakawa, H. Ochi, K. Arakawa (Chem. Pharm. Bull. [Tokyo] **11** [1963] 905/11).

[11] Colgate-Palmolive Co., J. E. Robertson (U.S. 3377248 [1965/68]; C.A. **69** [1968] No. 96878). – [12] G. Tainturier, J. Tirouflet (Bull. Soc. Chim. France **1966** 600/4). – [13] N. C. Fawcett, D. R. Anderson, P. E. Cassidy (SC-DC-67-1831 [1967] 1/4; C.A. **68** [1968] No. 3459). – [14] N. C. Fawcett, D. R. Anderson, P. E. Cassidy, M. F. Usaec (SC-DC 67-2060 [1967] 1/16; C.A. **69** [1968] No. 20109). – [15] I. Pavlík, R. Foniok, J. Grégr (Sb. Ved. Praci. Vys. Sk. Chemickotechnol. Pardubice **15** [1967] 25/78, 46).

[16] D. W. Hall, E. A. Hill, J. H. Richards (J. Am. Chem. Soc. **90** [1968] 4972/6). – [17] D. T. Roberts, W. F. Little, M. M. Bursey (J. Am. Chem. Soc. **90** [1968] 973/5). – [18] K. Sakai, M. Ishige, H. Kono, I. Motoyama, K. Watanabe, K. Hata (Bull. Chem. Soc. Japan **41** [1968] 1902/8). – [19] H. Patin, R. Dabard (Tetrahedron Letters **1969** 4971/4). – [20] F. M. Vigo (Diss. Univ. Illinois 1969, p. 143; Diss. Abstr. Intern. B **31** [1970] 598/9).

[21] I. J. Spilners, J. G. Larson (Org. Mass. Spectrom. **3** [1970] 915/24). – [22] M. P. Forost, S. N. Volyntseva, V. A. Emel'yanova (U.S.S.R. 317658 [1968/71]; C.A. **76** [1972] No. 85924). – [23] H. Patin, R. Dabard (Bull. Soc. Chim. France **1973** 2756/9). – [24] K. Yamakawa, M. Hisatome (J. Organometal. Chem. **52** [1973] 407/24). – [25] G. Cerichelli, B. Floris, G. Ortaggi (J. Organometal. Chem. **76** [1974] 73/9).

[26] A. N. Nesmeyanov, G. B. Shul'pin, M. I. Rybinskaya (Izv. Akad. Nauk SSSR Ser. Khim. **24** [1975] 2824/7; Bull. Acad. Sci. USSR Div. Chem. Sci. **24** [1975] 2714/7). – [27] A. N. Nesmeyanov, M. V. Tolstaya, G. B. Shul'pin, M. I. Rybinskaya (U.S.S.R. 539894 [1975/76]; C.A. **86** [1977] No. 171603). – [28] A. N. Nesmeyanov, G. B. Shul'pin, M. V. Tolstaya, M. I. Rybinskaya (U.S.S.R. 555114 [1975/77]; C.A. **87** [1977] No. 135954). – [29] A. Ratajcak, B. Misterkiewicz (Chem. Ind. [London] **1976** 902/3). – [30] A. E. Vodop'yanova, Yu. P. Losev, Ya. M. Paushkin, R. V. Chibrikova (Dokl. Akad. Nauk Belorussk.SSR **21** [1977] 611/3; C.A. **87** [1977] No. 68958).

[31] G. V. Shul'pin (Zh. Obshch. Khim. **51** [1981] 2152/3; J. Gen. Chem. [USSR] **51** [1981] 1854). – [32] P. J. Graham, R. V. Lindsey, G. W. Parshall, M. L. Peterson, G. M. Whitman (J. Am. Chem. Soc. **79** [1957] 3416/20). – [33] T. A. Mashburn, C. R. Hauser (J. Org. Chem. **26** [1961] 1671/2). – [34] M. Okawara, Y. Takemoto, H. Kitaoka, E. Haruki, E. Imoto (Kogyo Kagaku Zasshi **65** [1962] 685/8 from C.A. **58** [1963] 577). – [35] G. F. Grebenkina, M. A. Kruptsova, O. S. Popov (Deposited Doc. SPSTL-948-KhP-D81 [1981] 1/5 from C.A. **98** [1983] No. 143602).

[36] G. F. Grebenkina, M. A. Kruptsova, O. S. Popov (Deposited Doc. SPSTL-979-KhP-D81 [1981] 1/6 from C.A. **98** [1983] No. 107506). – [37] V. I. Sokolov, L. L. Troitskaya, N. S. Khrushcheva, O. A. Reutov (Dokl. Akad. Nauk SSSR **281** [1985] 861/3).

5.2.3.5.1.1.5.4 Substitution Reactions on the Cyclopentadienyl Rings

The CH_3CO groups considerably reduce the tendency of the cyclopentadienyl rings toward electrophilic substitution reactions. For the simplest electrophilic substitution reaction, the H/D exchange of the ring protons, the ratio of the average reaction rates of 1,1'-diacetylferrocene and ferrocene is 1/~2000 [5]. This is given by studies of the H/D exchange with CF_3CO_2D/C_6H_6 (mole ratio $Fe(C_5H_4{-}CO{-}CH_3)_2/CF_3CO_2D/C_6H_6$ 1/3/20) at 25°C after reprotonation of the CD_3CO groups with C_2H_5OH/OH^-; the rate constant for the isotope exchange of the ring protons is $k = 7.7 \times 10^{-8} s^{-1}$ [5]. In pyridine or $O(CH_2CH_2OCH_3)_2$ containing in each case 10 vol% D_2O, H/D exchange also occurs on the cyclopentadienyl rings. The distribution of the deuterium, however, does not occur statistically, for which a mechanism which rests on a $\pi \rightarrow \sigma \rightarrow \pi$ rearrangement of the ring has been held responsible [18, 19]. The introduction of a third CH_3CO group into the molecule cannot be forced under Friedel-Crafts conditions [2, 9]. The compound decomposes finally on long boiling in CS_2 with $CH_3COCl/AlCl_3$ [2]. The sulfonation attempted with SO_3 in $ClCH_2CH_2Cl$ miscarried, exchange of the CH_3CO groups by SO_3H groups took place [4]. No mercuration occurred with $Hg(O_2CCH_3)_2$ in C_6H_6/C_2H_5OH at the boiling point [2].

The arylation attempted under Gomberg-Bachmann-Hey conditions with diazonium salts failed, for the reaction proceeding with splitting of the compound see p. 25. Condensation with S_2Cl_2 in a 1/1 mole ratio in boiling benzene forms I [17].

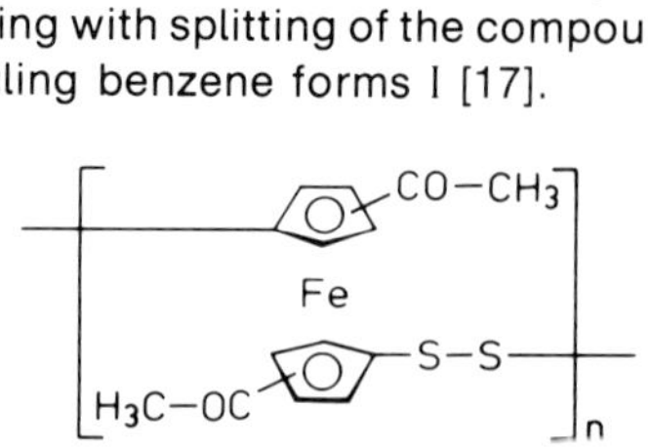

I

5.2.3.5.1.1.5.5 Basicity and Complex Formation

Basicity

Cf. p. 21 as well as "Ferrocene" 2, 1977, pp. 187/8 and 212.

Like all ferrocenyl ketones $Fe(C_5H_4\text{–}CO\text{–}CH_3)_2$ possesses in comparison to other aromatic ketones an uncommonly strong basicity, as a result of which it dissolves easily in strong mineral acids to form solutions with an intense violet color [1, 3]. With acids of the strength of CF_3CO_2H a smooth reaction forms the monoprotonated species $[Fe(C_5H_4\text{–}CO\text{–}CH_3)_2H]^+$ (see 5.2.3.5.1.2, p. 52), still stronger acids such as concentrated H_2SO_4 or CF_3SO_3H give the doubly protonated product $[Fe(C_5H_4\text{–}C(CH_3){=}OH)_2]^{2+}$ [24]; see also [15, 16] and p. 53.

In H_2SO_4/H_2O the pK_a is −3.86 for the conjugated acid $[Fe(C_5H_4\text{–}CO\text{–}CH_3)_2H]^+$ at 25°C [12], see also [11]. The equilibrium constant for reaction with CCl_3CO_2H from the electronic spectrum is at 25°C in benzene $K = [Fe(C_5H_4\text{–}CO\text{–}CH_3)_2][CCl_3CO_2H]/[Fe(C_5H_4\text{–}CO\text{–}CH_3)_2H^+ \cdot CCl_3CO_2^-] = (20.0 \pm 0.2) \times 10^{-3}$ [11, 13], in cyclohexane $(12.8 \pm 0.8) \times 10^{-3}$ M [13].

Complex Formation

As a compound with high basicity $Fe(C_5H_4\text{–}CO\text{–}CH_3)_2$ forms with Lewis acids strongly colored, difficultly soluble complexes, in which the carbonyl groups are coordinated as donor groups with the metal atoms as acceptors. In the IR spectrum appear instead of one ν(C=O) band the bathochromically shifted symmetric and antisymmetric ν(C$\cdots$O→) bands [8]. The quadrupole splitting in the ^{57}Fe Mössbauer spectrum shrinks [26].

BF_3 forms an indigo colored adduct [9].

The 1/1 donor-acceptor compounds obtained with AlX_3 (X = Cl, Br; mole ratio $Fe(C_5H_4\text{–}CO\text{–}CH_3)_2/AlX_3 = 1/(1\text{ to }2)$) [6], $SnCl_4$ (mole ratio 1/1) [8, 20], $SnBr_4$ [8], or $FeCl_3$ [25] represent chelate complexes in which the 1,1′-diacetylferrocene is bonded with the central metal atom as a bidentate ligand via its two CO groups [6 to 8, 20, 23, 25]. If $Fe(C_5H_4\text{–}CO\text{–}CH_3)_2$ in CCl_4 is mixed with $AlCl_3$ in ether in a 1/4 mole ratio, then in place of the precipitate there forms a dark, resinous mass, which sticks to the walls of the vessel [6].

The interaction enthalpy with liquid $SnCl_4$ is −13.0 kcal/mol (−54.4 kJ/mol) [23].

The only complexes $(Fe(C_5H_4\text{–}CO\text{–}CH_3)_2 \cdot TiX_4$, X = Cl, Br) formed with $TiCl_4$ in a 1/1 mole ratio in ether or from $Fe(C_5H_4\text{–}CO\text{–}CH_3)_2$ and $TiBr_4$ in a 1/1, 1/2, or 3/1 mole ratio in CS_2 with exclusion of air and moisture, are isostructural. They possess one free CO group and one coordinated with Ti [6, 7]. This structure, derived by Pavlik and co-workers [6, 7] from the IR spectrum is confirmed by the ^{57}Fe Mössbauer spectrum of $Fe(C_5H_4\text{–}CO\text{–}CH_3)_2 \cdot TiCl_4$; the quadrupole splitting in the complex is identical with that in $[C_5H_5FeC_5H_4\text{–}COH\text{–}CH_3]^+$ [26]. The designation of $Fe(C_5H_4\text{–}CO\text{–}CH_3)_2 \cdot TiCl_4$ in [24] as a chelate complex is thus obviously obsoleted. The 1/1 complex formed with $VOCl_3$ (1/1 mole ratio) also has one CO group coordinated with V and one free [20, 23].

In the 1/2 donor-acceptor complexes formed with $SbCl_5$ or $TiCl_4$ each CO group is coordinated in each case with one acceptor molecule [20, 23]. The enthalpy of interaction with liquid $TiCl_4$ amounts to −20.9 kcal/mol (−87.5 kJ/mol) [23].

Reaction of $Fe(C_5H_4\text{–}CO\text{–}CH_3)_2$ with $ZnCl_2$ (1/2), each in a saturated acetone solution, by heating of the mixture, evaporated to dryness in vacuum, for 1.5 h at 120°C/1 Torr yields with enolization **$Fe(C_5H_4\text{–}C(=CH_2)OH \cdot ZnCl_2)_2$** [22]. $Fe(C_5H_4\text{–}CO\text{–}CH_3)_2$ reacts in absolute C_2H_5OH or diethyl ether with HgX_2 (X = Cl, Br, I, or SCN) with a clear increase of the solubility of HgX_2, which possibly indicates an interaction of the two compounds. The UV spectrum is in all cases that of $Fe(C_5H_4\text{–}CO\text{–}CH_3)_2$ overlaid by that of the HgX_2 [14].

References on p. 32

$AgBF_4$ (mole ratio 1/1) in $C_2H_5OC_2H_5$ forms the red complex **[Fe(C$_5$H$_4$–CO–CH$_3$)$_2$·AgBF$_4$]$_x$**, whose IR spectrum indicates bonding to the Ag^+ through both O atoms. Due to its insolubility in all normal solvents a molecular weight determination has up til now been impossible; it appears to be oligomeric, if not even polymeric. Dissolution in liquid SO_2 forms the ferrocenium salt $[Fe(C_5H_4{-}CO{-}CH_3)_2]BF_4$ with deposition of Ag [10].

The charge transfer complex **Fe(C$_5$H$_4$–CO–CH$_3$)$_2$·CBr$_4$** forms in C_6H_6 in the presence of CBr_4; it has a UV maximum in benzene at 384 nm [21].

References:

[1] M. Rosenblum (Diss. Harvard Univ. 1953). – [2] A. N. Nesmeyanov, E. G. Perevalova, R. V. Golovnya, N. A. Simukova, O. V. Starovsky (Izv. Akad. Nauk SSSR Otd. Khim. Nauk **1957** 638/40; Bull. Acad. Sci. USSR Div. Chem. Sci. **1957** 647/9). – [3] M. Rosenblum, R. B. Woodward (J. Am. Chem. Soc. **80** [1958] 5443/9). – [4] N. A. Nesmeyanov, B. N. Strunin (Dokl. Akad. Nauk SSSR **137** [1961] 106/8; Proc. Acad. Sci. USSR Chem. Sect. **136/141** [1961] 275/6). – [5] A. N. Nesmeyanov, D. N. Kyrsanov, V. N. Setkina, N. V. Kislyakova, N. S. Kochetkova (Izv. Akad. Nauk SSSR Otd. Khim. Nauk **1962** 1932/6; Bull. Acad. Sci. USSR Div. Chem. Sci. **1962** 1845/8).

[6] I. Pavlík, K. Handlíř (Collection Czech. Chem. Commun. **31** [1966] 1958/74). – [7] I. Pavlík, J. Klikorka, K. Handlíř, P. Sedlmayer (Z. Chem. [Leipzig] **6** [1966] 427/8). – [8] I. Pavlík, J. Klikorka, K. Handlíř (Z. Chem. [Leipzig] **8** [1968] 390/1). – [9] P. Carty, M. F. A. Dove (J. Organometal. Chem. **21** [1970] 195/201). – [10] P. Carty, M. F. A. Dove (J. Organometal. Chem. **28** [1971] 125/35).

[11] B. Floris, G. Illuminati, P. E. Jones, G. Ortaggi (Coord. Chem. Rev. **8** [1972] 39/43). – [12] G. Cerichelli, B. Floris, G. Illuminati, G. Ortaggi (Gazz. Chim. Ital. **103** [1973] 911/22). – [13] G. Cerichelli, B. Floris, G. Illuminati, G. Ortaggi (Ann. Chim. [Rome] **64** [1974] 119/23). – [14] K. R. Mann, W. H. Morrison, D. N. Hendrickson (Inorg. Chem. **13** [1974] 1180/5). – [15] A. N. Nesmeyanov, G. B. Shul'pin, M. I. Rybinskaya, P. V. Petrovskii (Dokl. Akad. Nauk SSSR **215** [1974] 599/602; Dokl. Chem. Proc. Acad. Sci. USSR **214/219** [1974] 188/91).

[16] A. N. Nesmeyanov, G. B. Shul'pin, M. I. Rybinskaya (Dokl. Akad. Nauk SSSR **218** [1974] 1107/10; Dokl. Chem. Proc. Acad. Sci. USSR **214/219** [1974] 705/7). – [17] Yu. P. Losev, A. E. Vodop'yanova, Ya. M. Paushkin, A. L. Isakhanyan (Vestsi Akad. Navuk Belarusk. SSR Ser. Khim. Navuk **1975** No. 6, pp. 82/4). – [18] M. B. Moronski (Diss. Southern Illinois Univ. 1977 from Diss. Abstr. Intern. B **39** [1978] 754/5). – [19] D. W. Slocum, M. Moronski, J. Bencini, B. Conway (9th Intern. Conf. Organometal. Chem., Dijon 1979, Abstr. Papers P33T). – [20] L. M. Dyagileva, O. S. Morozov, N. N. Vyshinskii, V. G. Tsvetkov, Yu. A. Aleksandrov (Zh. Obshch. Khim. **50** [1980] 1859/62; J. Gen. Chem. [USSR] **50** [1980] 1516/8).

[21] A. N. Nesmeyanov, A. V. Vannikov, V. A. Zver'kov, L. V. Balabanova, M.-G. A. Shvekhgeimer, N. S. Kochetkova (Dokl. Akad. Nauk SSSR **252** [1980] 143/6; Dokl. Phys. Chem. Proc. Acad. Sci. USSR **250/255** [1980] 347/50). – [22] V. V. Korshak, S. L. Sosin, V. P. Alekseyeva, M. D. Litvinova (Vysokomol. Soedin. A **23** [1981] 780/8; Polym. Sci. [USSR] **23** [1981] 873/84). – [23] L. M. Dyagileva, O. S. Morosov, Yu. A. Alexandrov (22nd Intern. Conf. Coord. Chem., Budapest 1982, Abstr. Papers p. 757). – [24] G. Neshvad, R. M. G. Roberts, J. Silver (J. Organometal. Chem. **236** [1982] 349/58). – [25] R. M. G. Roberts, J. Silver, M. Clemance (unpublished results from [24, p. 357]).

[26] M. C. Clemance, R. M. G. Roberts, J. Silver (J. Organometal. Chem. **247** [1983] 219/22).

5.2.3.5.1.1.5.6 Reactions on the Substituents

See also the preceding Section 5.2.3.5.1.1.5.5, pp. 31/2, as well as "Oxidation" in Section 5.2.3.5.1.1.5.2, pp. 26/8, and "Reduction" in 5.2.3.5.1.1.5.3, pp. 28/30.

5.2.3.5.1.1.5.6.1 Nucleophilic Addition to the CO Group

Of N-Containing Nucleophiles

Addition and subsequent elimination reactions with $NH_2OH \cdot HCl$ in CH_3OH in the presence of KO_2CCH_3 give crystalline $Fe(C_5H_4\text{-}C(CH_3){=}NOH)_2$ as an orange powder which decomposes above 200°C [1, 91, 92]. The monooxime is obtained by reaction in a 1/1 mole ratio in boiling C_2H_5OH in the presence of NaO_2CCH_3 [35].

No reaction occurs with R_2NNH_2 ($R = CH_3$, C_2H_5) in C_2H_5OH under mild conditions. Under more severe conditions (increased time of heating, higher boiling solvents, greater excess of the hydrazine, presence of acids) oxidative degradation occurs. Only the monophenylhydrazone **$CH_3\text{-}CO\text{-}C_5H_4FeC_5H_4\text{-}C(CH_3){=}NNHC_6H_5$** (dark brown powder, m.p. 108 to 110°C) was recovered with $C_6H_5NHNH_2$ (even at a 1/2 mole ratio) in C_2H_5OH in 5 h at 50°C, then 10 d in the dark at room temperature [67]. In contrast $C_6H_5NHNH_2 \cdot HCl$ in 50/50 C_2H_5OH/H_2O saturated with NaO_2CCH_3 gives **$Fe(C_5H_4\text{-}C(CH_3){=}NNHC_6H_5)_2$**, which is reddish brown, m.p. 96 to 101°C (C_2H_5OH), with UV ($CHCl_3$) λ_{max} (ε) = 450 (1580) nm [46]. Condensation with $CH_3\text{-}CO\text{-}NHNH_2$ in the presence of catalytic amounts of CH_3CO_2H yields the bis(acetylhydrazone) [89]. The reaction with 2,4-dinitrophenylhydrazine (1/2 mole ratio) in C_2H_5OH with concentrated H_2SO_4 for 10 h at 50°C forms **$CH_3\text{-}CO\text{-}C_5H_4FeC_5H_4\text{-}C(CH_3){=}N\text{-}NH\text{-}C_6H_3(NO_2)_2$-2,4** with 198°C m.p., $\nu(C{=}O) = 1672\ cm^{-1}$, and **$Fe(C_5H_4\text{-}C(CH_3){=}N\text{-}NH\text{-}C_6H_3(NO_2)_2$-2,4$)_2$**, a black powder that does not melt under 300°C, shows no $\nu(C{=}O)$ bands, dissolves in hot $C_6H_5NO_2$ or $(CH_3)_2NCHO$, and can be recrystallized from the latter. The two compounds are obtained in 10 and 76% yields, respectively. They are separable by extraction with $CH_3CO_2CH_3$, in which only the first dissolves [72]. Also boiling with 4-$CH_3\text{-}C_6H_4\text{-}SO_2NHNH_2$ in the mole ratio 1/1.5 in absolute ethanol forms the mono- and the bishydrazone, which are separated by dissolution of the monohydrazone in hot 7/3 $CHCl_3/C_2H_5OH$. At a 1/2 mole ratio, in contrast, only the bishydrazone is obtained [53]. The reaction with thiosemicarbazide in C_2H_5OH/H_2O in the presence of glacial acetic acid yields $Fe(C_5H_4\text{-}C(CH_3){=}N\text{-}NH\text{-}CS\text{-}NH_2)_2$ [68, 77].

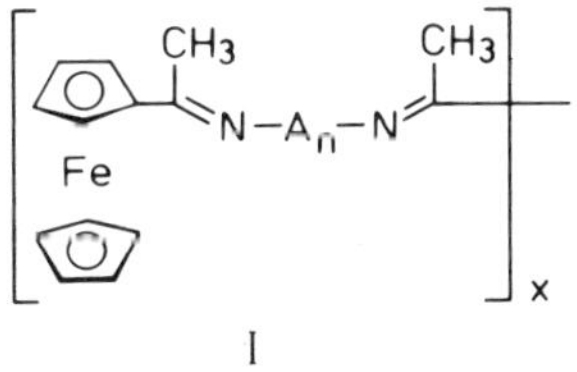

I

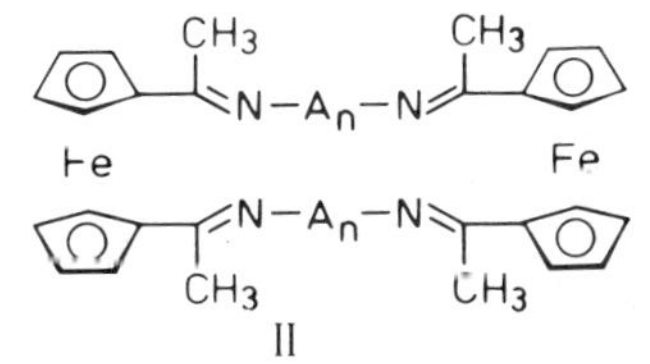

II

$H_2N\text{-}A_n\text{-}NH_2$ ($n = 0$, 1; $A = CH_2CH_2$) in boiling benzene with 4-$CH_3\text{-}C_6H_4\text{-}SO_3H \cdot H_2O$ as a catalyst produces polycondensation to I and II [56, 57, 63]. With H_2NNH_2 compound II with $n = 0$ constitutes a not inconsiderable part of the reaction product [54]. With $N_2H_4 \cdot H_2O$ (mole ratio 1/1) in C_2H_5OH $CH_3\text{-}CO\text{-}C_5H_4FeC_5H_4\text{-}C(CH_3){=}N\text{-}N{=}C(CH_3)\text{-}C_5H_4FeC_5H_4\text{-}C(CH_3){=}NNH_2$ forms on 5 h warming at 50°C and subsequent standing for 10 d in the dark at room temperature [67]. II with $n = 0$ forms in a yield of 38% on condensation with $N_2H_4 \cdot 2HCl$ in C_2H_5OH [74]. According to Göller [5] $O[{=}C(CH_3)\text{-}C_5H_4FeC_5H_4\text{-}C(CH_3){=}N\text{-}N{=}]_x{=}C(CH_3)\text{-}C_5H_4FeC_5H_4\text{-}CO\text{-}CH_3$ ($x = 4$) is formed by 1/1 $Fe(C_5H_4\text{-}CO\text{-}CH_3)_2$ and $N_2H_4 \cdot HCl$ in ethanol. Red-brown crystals of I with $n = 0$, molecular weight 1600, soluble in C_6H_6 or $CHCl_3$, form on 24 h boiling of diacetylferrocene and $N_2H_4 \cdot H_2SO_4$ in $(C_2H_5)_3N$ [24], while condensation with $N_2H_4 \cdot H_2O$ in butyric acid yields the black, X-ray amorphous I ($n = 0$) [20, 30]. In both cases I is paramagnetic and possesses semiconductor properties [20, 24, 30]. Polycondensation to $H_2N–CO–NH–CO–N[=C(CH_3)–C_5H_4FeC_5H_4–C(CH_3)=N–CO–NH–CO–N=]_x=C(CH_3)–C_5H_4FeC_5H_4–CO–CH_3$ takes place with biuret in a 1/1 mole ratio in dioxane or dimethyl sulfoxide in the presence of $ZnCl_2$ as a dehydrating agent or in glacial acetic acid (simultaneously a solvent and condensation agent). Raising the temperature favors the formation of insoluble polymers with a three dimensional structure (see III a, b, c with $R = CH_3$, A = CO–NH–CO) [83, 87].

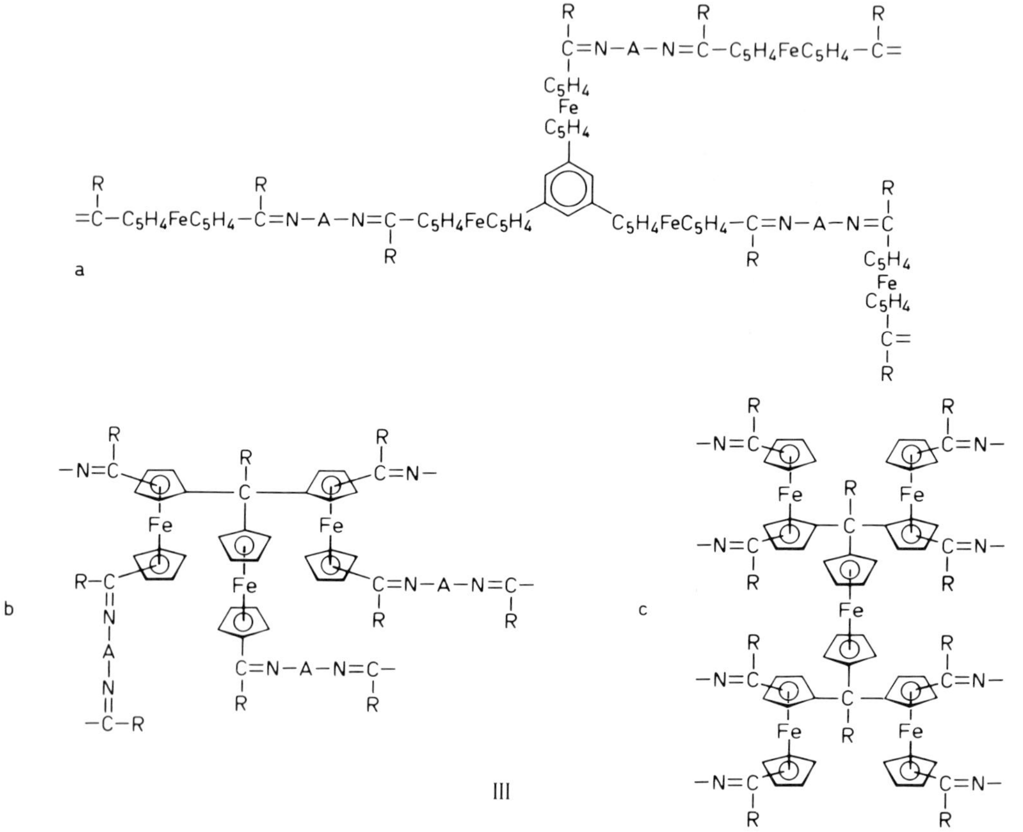

III

The polycondensation in C_6H_6/4-$CH_3–C_6H_4–SO_3H \cdot H_2O$ failed with aromatic diamines like 1,4-$(H_2N)_2C_6H_4$. It is in contrast successful in boiling toluene with Al_2O_3 as catalyst, forming I (n = 1; A = 1,4-C_6H_4) but no II [54, 56, 57, 63]. Polycondensation to the paramagnetic polyazomethines I (n = 1; A = 4,4′-$C_6H_4–C_6H_4$, 4,4′-C_6H_4–O–C_6H_4, 4,4′-C_6H_4–S–C_6H_4, 4,4′-C_6H_3–(NO_2-2)–S–S–(2-O_2N)C_6H_3) also takes place with $H_2N–A–NH_2$ in dioxane or in the melt in an inert atmosphere in the presence of $ZnCl_2$ or 4-$CH_3–C_6H_4–SO_3H$ as catalyst. In contrast to the solution polycondensation, which yields both soluble and insoluble polymeric products, the reaction in the melt leads only to insoluble three-dimensional structures [84, 87, 94]. These probably include in addition to units of I also those of III a, b, c ($R = CH_3$), since the diacetylferrocene also reacts in addition with itself to form polyferrocenylene vinylenes (cf. p. 42) or triferrocenyl benzene derivatives. If the dehydration takes place in the strongly acidic $H_2O–ZnCl_2–HCl$ system, the protonated ferrocene splits off the cyclopentadienyl anion, which alkylates the ferrocene groups inter- or intramolecularly [87, 94]. Due to the higher temperature the conversion in the melt is higher than in solution [84]. If equimolar quantities of $Fe(C_5H_4–CO–CH_3)_2$ and 4-$H_2NNH–C_6H_4–SO_2–C_6H_4–NHNH_2$-4′ in $(CH_3)_2NCHO$ with glacial acetic acid as catalyst are held

48 h in the dark in a sealed vessel, then subsequent addition of excess propan-2-ol precipitates the compound I ($n=1$; $A=4,4'$-NH–C_6H_4–SO_2–C_6H_4–NH; molecular weight in $(CH_3)_2NCHO$ 1300) in a 65% yield; the compound is soluble in $(CH_3)_2NCHO$ or hexamethylphosphoramide [46].

I with $n=1$, $A=4,4'$-$C_6H_3(NO_2$-2)–S–S–(2-$O_2N)C_6H_3$ is obtained with an exclusively linear structure on refluxing equimolar quantities of $Fe(C_5H_4$–CO–$CH_3)_2$ and IV in C_6H_6 with pyridine, which effects the activation of the N=S=O groups by formation of the nucleophile; for the mechanism of the reaction see scheme in V [84, 87, 94].

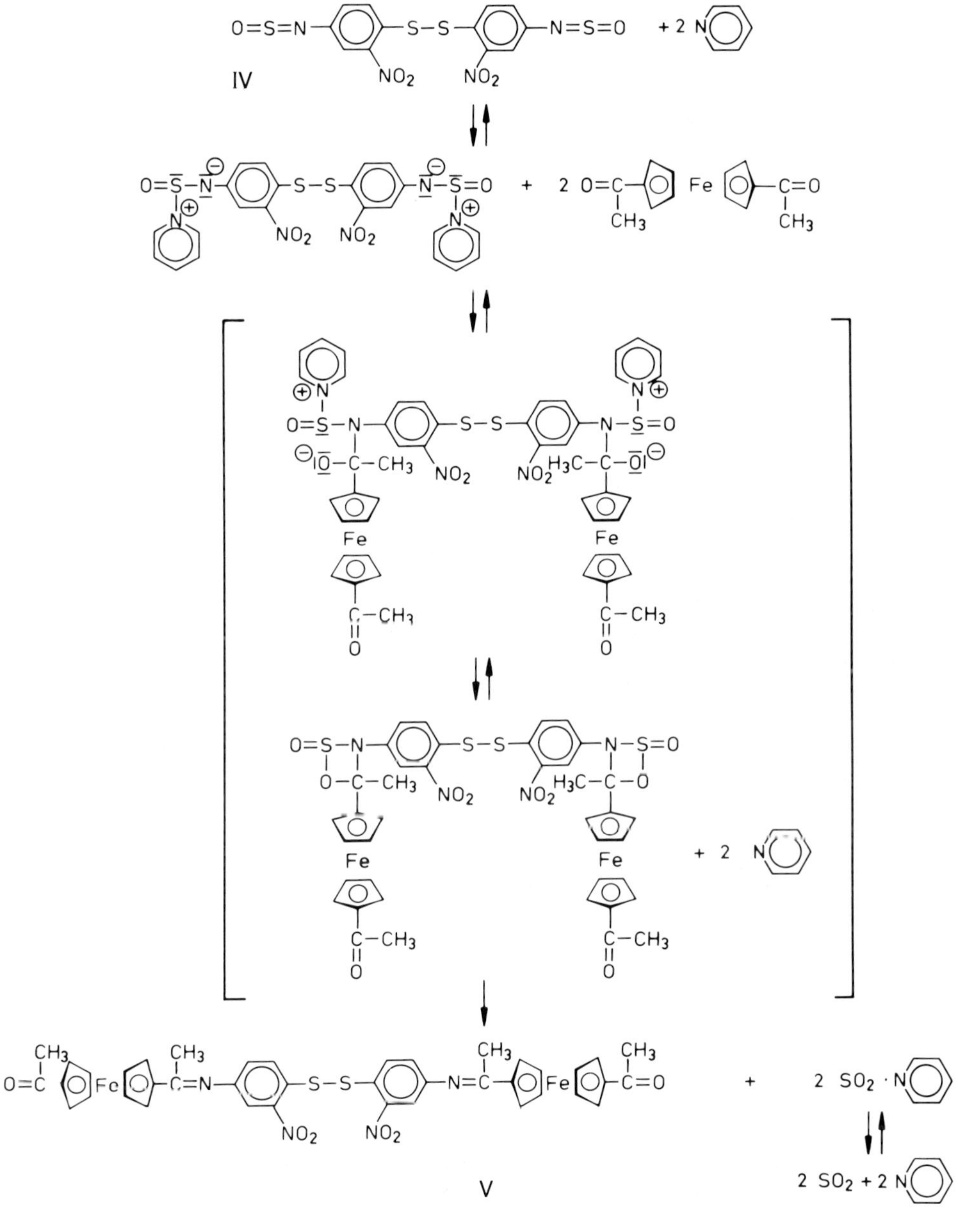

References on pp. 43/5

Of Organometallic Compounds

The addition of RLi (R = CH_3 [79], C_6H_5 [32, 33], m-carboranyl [97]) yields after subsequent hydrolysis $Fe(C_5H_4{-}CR(CH_3){-}OH)_2$ [32, 33, 79, 97]. In contrast to the reaction with R = m-carboranyl the reaction with R = o-carboranyl failed. This difference may be the result of the smaller resistance of the tertiary Li alcoholate compound with R = o-carboranyl versus that with R = m-carboranyl, in alkaline medium, which cleaves the C–C bond only in the former. Reaction with $Li[Si(C_6H_5)_2]_4Li$ gives $\{(Si(C_6H_5)_2)_4{-}C(CH_3)OH{-}C_5H_4FeC_5H_4{-}C(CH_3)OH\}_x$ in tetrahydrofuran [97].

Polymerization to VI (idealized unit) takes place with NaC≡C–A_n–C≡CNa (n = 0, 1; A = 1,4-C_6H_4) in liquid NH_3/tetrahydrofuran. The product is a light brown powder soluble in CH_2Cl_2 and $(CH_3)_2NCHO$ with a molecular weight of 1600 to 3000 [60]. For the reaction with NaC≡CH see p. 37; cf. also Section 5.2.3.5.1.1.5.6.3, p. 42.

The Reformatsky reaction with $BrCH_2CO_2C_2H_5$ in benzene under the action of metallic zinc dust and subsequent dehydration with $POCl_3$ in CH_2Cl_2 gives $Fe(C_5H_4{-}C(CH_3){=}CH{-}CO_2C_2H_5)_2$ [70].

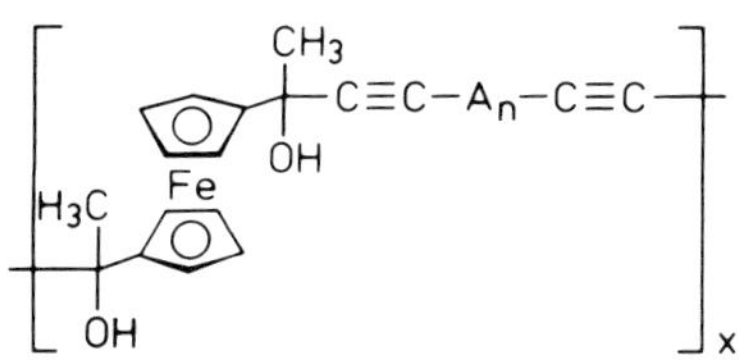

VI

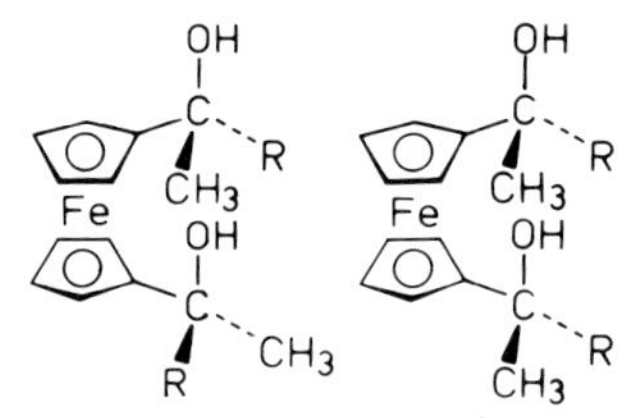

a b

VII

The reaction of $Fe(C_5H_4{-}CO{-}CH_3)_2$ with RMgX (X = Cl, Br) in benzene/ether proceeds with distinct stereoselectivity, as shown by the reaction with R ≠ CH_3. After hydrolysis the meso isomer VII a is recovered preferentially over the racemic isomer VII b. This is attributed to the fact that the nucleophilic attack of the first RMgX leads to formation of the oxyketone VIII with a chelate structure (see reaction scheme in IX) as an intermediate compound, to which the second RMgX approaches preferentially from the less hindered side, perpendicular to the C=O plane. The ratio of VII a and b in the reaction product amounts for the reaction with C_2H_5MgBr to 68/32, with C_6H_5MgBr to 82/18, with $C_6H_5CH_2MgCl$ to 80/20, and with cyclo-$C_6H_{11}MgBr$ to 85/15. C_6H_5MgBr and $Fe(C_5H_4{-}CO{-}CH_3)_2$ in a 1/1 ratio give $CH_3{-}CO{-}C_5H_4FeC_5H_4{-}C(CH_3)(C_6H_5)OH$ [78]. Regarding the reactions with $CH_2{=}CHMgCl$ [62], C_2H_5MgBr [37], or RMgX with R = n-C_4H_9, $C_6H_5CH_2$, $(CH_3)_3SiCH_2CH_2$ and X = Cl or Br [86] to form VII, see the papers cited.

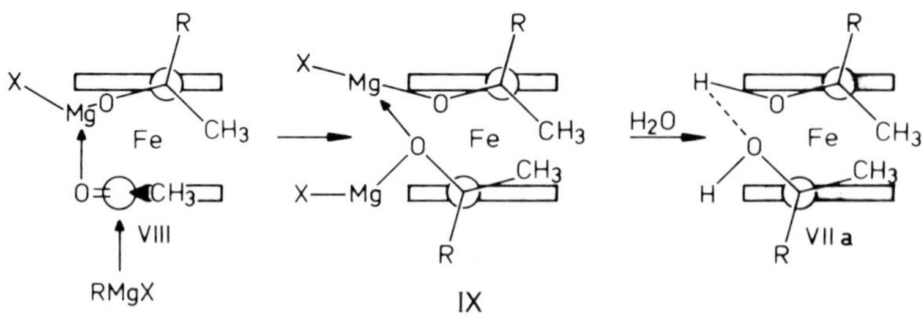

IX

VII with R = CH_3 is also obtained with CH_3MgI in C_6H_6/ether in 56% yield [2], see also [32, 33], or in absolute tetrahydrofuran in 83% yield [28]. While the reaction with RMgBr gives only the tertiary alcohol (VII) in high yields without a trace of the olefin, the reaction with RMgI nevertheless gives along with VII the diolefin X. If $Fe(C_5H_4{-}CO{-}CH_3)_2$ in benzene is added to an excess of RMgI in ether and the mixture stands overnight at room temperature, then thin layer chromatography shows that X has already formed in this stage of the reaction, that is, before the hydrolysis [37]. Possibly the alcoholate $Fe(C_5H_4{-}C(CH_3)(CH_2R')OMgI)_2$ splits off $IMgO^-$ ions with formation of the carbenium ion $[Fe(C_5H_4{-}C(CH_3)CH_2R')_2]^{2+}$, which with loss of a proton converts into X [50]. CH_3MgI after hydrolysis gave 56% VII ($R=CH_3$) and 32% X ($R'=H$), C_2H_5MgI gave 56% VII ($R=C_2H_5$) and 41% X ($R'=CH_3$) [37, 50]. If the reaction with CH_3MgI is conducted in the presence of free CH_3I (20 to 25%) with slow addition of the ferrocene derivative, then after 30 min at 60 to 65°C X with $R'=H$ is formed in 50 to 60% yield [42, 48] along with only slight traces of VII ($R=CH_3$) [42]. Apparently the reaction here proceeds via $Fe(C_5H_4{-}C(CH_3)_2I)_2$, which forms intermediately from the carbenium ion and the CH_3I and converts into X with $R'=H$ via splitting off of 2HI [48].

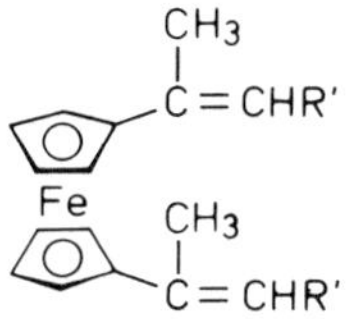

X

Of Other Nucleophiles

Reformatsky reactions are discussed under organometallic compounds in the preceding section.

In the presence of morpholine and $AlCl_3$ in anhydrous benzene at 0°C a second molecule of $Fe(C_5H_4{-}CO{-}CH_3)_2$ is added. Dimerization to XI in 17% yield takes place with splitting off of H_2O [51].

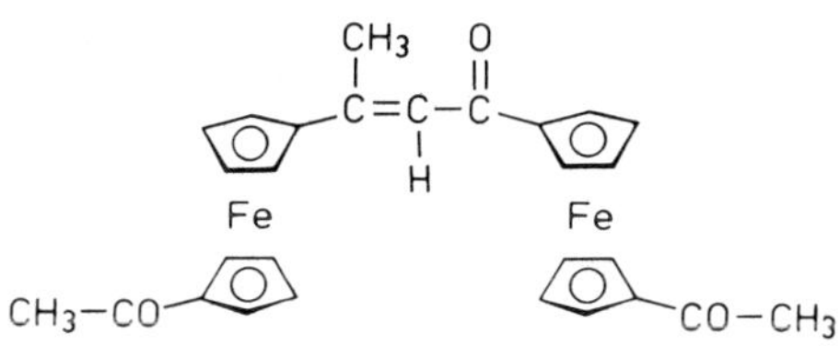

XI

The Knoevenagel condensation with cyclopentadiene goes in C_2H_5OH/C_2H_5ONa at the boiling point [49] and in $C_2H_5OH/H_2O/NaOH$ at room temperature [11]. A large excess of the reagent forms $Fe(C_5H_4{-}C(CH_3){=}C_5H_4\text{-cyclo})_2$ [11, 49], otherwise $CH_3{-}CO{-}C_5H_4FeC_5H_4{-}C(CH_3){=}C_5H_4$-cyclo [49].

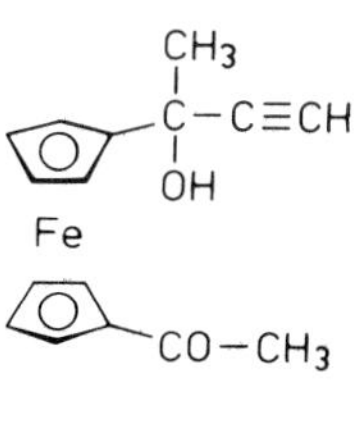

XII

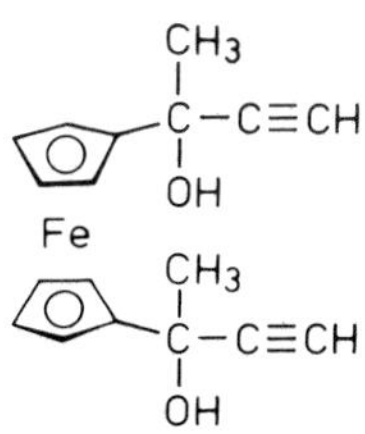

XIII

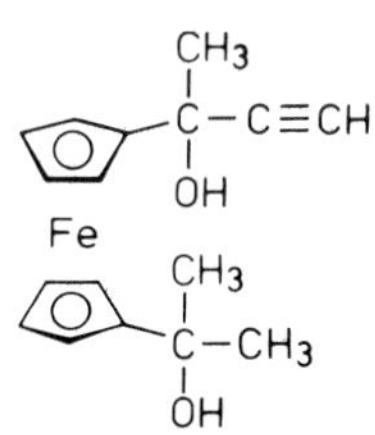

XIV

Addition without subsequent dehydration takes place with HC≡CH in liquid NH_3 in the presence of $NaNH_2$, to form XII, XIII, and XIV, which are separable from each other and the starting compound by preparative thin layer chromatography on Al_2O_3 with 50/1 C_6H_6/absolute C_2H_5OH [39]. All three compounds have also been obtained by the reaction with NaC≡CH in

References on pp. 43/5

liquid ammonia at $-37°C$. A 1/2 molar ratio for diacetylferrocene/NaC≡CH gives the yields 53, 12.7, and 3.5%, respectively; a 1/4 ratio gives 36, 11.6, and 14.5% [41]. For the reaction with $NaC{\equiv}C{-}A_n{-}C{\equiv}CNa$ ($n=0, 1$; $A=1,4\text{-}C_6H_4$), see p. 36.

No aldol condensation takes place with $CHCl_3$ in $CH_3OCH_2OCH_3$ in the presence of KOH at 0°C or room temperature [26].

$Fe(C_5H_4{-}C(CH_3)(CH_2CN)OH)_2$ forms in 62% yield with CH_3CN in liquid NH_3/ether/tetrahydrofuran in the presence of $NaNH_2$ after neutralization with NH_4Cl in liquid ammonia [55].

The Wittig reaction with $(C_6H_5)_3P{=}CH_2$ in dimethyl sulfoxide yields $Fe(C_5H_4{-}C(CH_3){=}CH_2)_2$ and $CH_3{-}CO{-}C_5H_4FeC_5H_4{-}C(CH_3){=}CH_2$ [37, 45, 79] plus XV as a successor product of the latter in the presence of excess $Na[CH_2SOCH_3]$ [38].

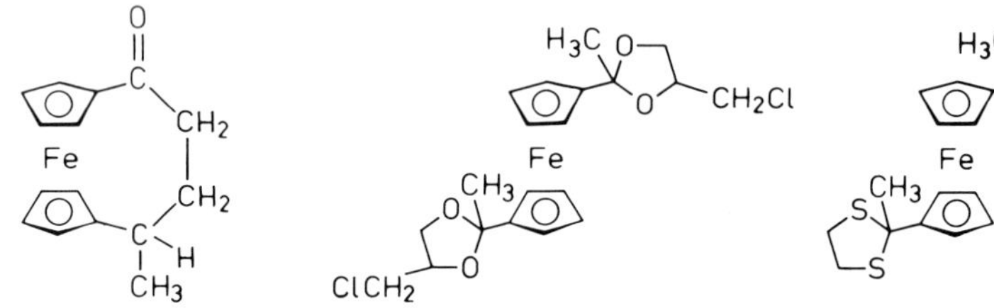

XV XVI XVII

$Fe(C_5H_4{-}CO{-}CH_3)_2$ reacts with epichlorhydrin in CCl_4 in the presence of $SnCl_4$ to give XVI [69]. With $HSCH_2CH_2SH$ in C_6H_6 at 0°C with introduction of dry HCl is formed the unstable dithiotolane XVII [76].

5.2.3.5.1.1.5.6.2 Behavior of the CH_3CO Groups as CH Acid Components

Reaction of the keto groups with themselves is discussed under "Of Other Nucleophiles" in Section 5.2.3.5.1.1.5.6.1, p. 37, and in Section 5.2.3.5.1.1.5.6.3, p. 42.

With Aldehydes

Aliphatic aldehydes RCHO (R = H, CH_3, C_2H_5, n-C_3H_7, i-C_3H_7, i-C_4H_9) form in boiling CH_3OH/H_2O/NaOH exclusively the ferrocenophanedione derivative I in yields of 9 to 18% of theoretical. The primarily occurring base-catalyzed aldol condensation produces the intermediate product II, that is cyclized to I by a spontaneous intramolecular Michael addition. The low yields are attributed to the relatively slight reactivity of the CO group in aliphatic aldehydes [17]. CH_2O in C_2H_5OH/H_2O/NaOH yields in addition to I (R = H, yield 13%) also III (yield 31%) [44]. Polycondensation with CH_2O in the presence of Na_2CO_3 in C_2H_5OH at 50°C yields a binuclear compound with $(HOCH_2)_2CH{-}CO{-}C_5H_4FeC_5H_4{-}CO{-}(CH_2)_3{-}CO{-}C_5H_4FeC_5H_4{-}CO{-}CH_2{-}CH_2OH$ as the idealized composition, which can be used for the impregnation of fibers of polyvinylalcohol [61, 64]; for the reaction with CH_2O in an alkaline medium, see also [5].

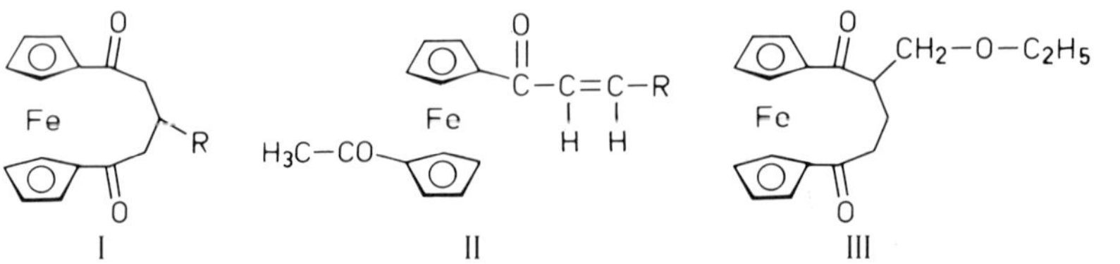

I II III

References on pp. 43/5

For the Mannich reaction with paraformaldehyde, see below. The condensation experiments conducted at the boiling point with CCl_3CHO in $(CH_3CO)_2O$ in the presence of NaO_2CCH_3 or in tetrahydrofuran in the presence of $(C_2H_5)_3N$ yielded exclusively polymeric resins [26].

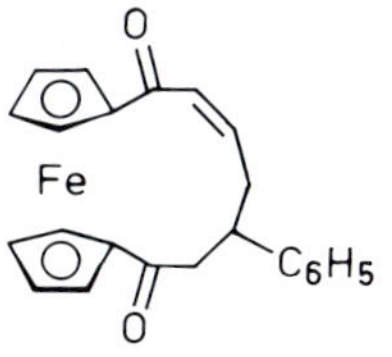

IV

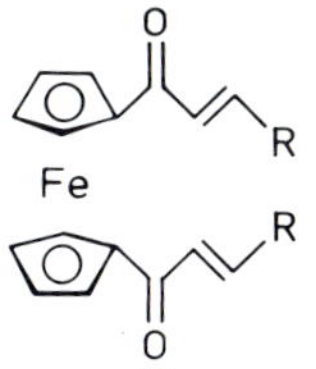

V

With aromatic aldehydes RCHO (R = aryl) in the presence of bases the compounds I, II, and/or V can be obtained in good to very good yields; the reaction No. 6 listed in Table 2 employs KOH as the catalyst, all the others NaOH:

Table 2

Reactions of $Fe(C_5H_4{-}CO{-}CH_3)_2$ with Aromatic Aldehydes.

No.	aldehyde RCHO	mole ratio $Fe(C_5H_4{-}CO{-}CH_3)_2$/aldehyde	solvent*)	t in °C**)	products	Ref.
1	$C_6H_5CH{=}CHCHO$	1/1	C_2H_5OH	boiling point	I and IV (mole ratio 15/1)	[96]
2	$C_6H_5CH{=}CHCHO$	1/2	C_2H_5OH	RT	V	[23, 29]
3	C_6H_5CHO	1/1	C_2H_5OH/H_2O	RT	I	[7]
4	C_6H_5CHO	1/1	DMF/H_2O	0 to RT	I	[36, 43]
5	C_6H_5CHO	1/2	CH_3OH/H_2O	40	I	[12]
6	C_6H_5CHO	1/2	C_2H_5OH/H_2O	RT	V	[13]
7	C_6H_5CHO	1/6	C_2H_5OH/H_2O	RT	V	[7, 13, 65]
8	$4\text{-}CH_3{-}C_6H_4{-}CHO$	1/6	C_2H_5OH/H_2O	RT	V	[82]
9	$4\text{-}Cl{-}C_6H_4{-}CHO$	1/1	DMF/H_2O	0 to RT	I	[36]
10	$2\text{-}Cl{-}C_6H_4{-}CHO$	1/1	DMF/H_2O	0 to RT	I	[36]
11	$2\text{-}Cl{-}C_6H_4{-}CHO$	1/2	CH_3OH/H_2O	40	I	[12]
12	$4\text{-}HO(CH_2)_2O{-}C_6H_4{-}CHO$	1/2	C_2H_5OH	RT	V	[23, 29]
13	$4\text{-}C_6H_5CH_2O{-}C_6H_4{-}CHO$	1/1	DMF/H_2O	0 to RT	I	[36]
14	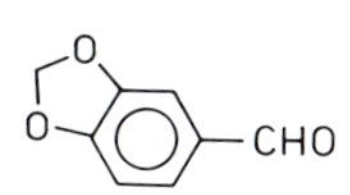	1/2	CH_3OH/H_2O	40	I	[12]

References on pp. 43/5

Table 2 (continued)

No.	aldehyde RCHO	mole ratio $Fe(C_5H_4{-}CO{-}CH_3)_2$/aldehyde	solvent*)	t in °C**)	products	Ref.
15	$4\text{-}(CH_3)_2N{-}C_6H_4{-}CHO$	1/2	C_2H_5OH	RT	V	[23, 29]
16	$4\text{-}(C_2H_5)_2N(CH_2)_2O{-}C_6H_4{-}CHO$	1/1	DMF/H_2O	0 to RT	I	[36]
17	$4\text{-}O_2N{-}C_6H_4{-}CHO$	1/2	CH_3OH/H_2O	40 to 50	II	[12]
18	$3\text{-}O_2N{-}C_6H_4{-}CHO$	1/2	CH_3OH/H_2O	40	II (78.6%) I (10%)	[12]
19	$2\text{-}O_2N{-}C_6H_4{-}CHO$	1/2	CH_3OH/H_2O	40 to 50	II	[12]
20	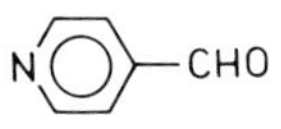	1/1	DMF/H_2O	0 to RT	I	[36]

*) DMF = $(CH_3)_2NCHO$. – **) RT = room temperature.

It has been established by [12, 95] that the compound obtained by [7] by reaction of $Fe(C_5H_4{-}CO{-}CH_3)_2$ with C_6H_5CHO in a 1/1 mole ratio, but not identified by them, is in fact I.

The first reaction step is the base-catalyzed aldol condensation with one RCHO molecule taking place with water elimination to form II. This can react further at a mole ratio of aldehyde/ketone ≧2 with a second aldehyde molecule to form V. It can, however, also cyclize to I by an intramolecular Michael addition. Prerequisite for this is the polarization of the C=C double bond by the neighboring CO group; it is favored, for example, by C_6H_5 or VI. Even the −I effect of Cl in the ortho position in $2\text{-}ClC_6H_4$ can not influence the +M effect so much that the intramolecular Michael addition does not take place. Only the NO_2 group in the ortho or para position hinders with its strong −I effect the formation of I; in meta position the −I effect in contrast is so far weakened that in addition to II (yield about 80%) I can also form in a yield of about 10% [12]. The compound I ($R = C_6H_5$) is also formed by the condensation with C_6H_5CHO catalyzed by acids (1/1 mole ratio; glacial acetic acid/conc. H_2SO_4) [71], and likewise with $R = 2\text{-}ClC_6H_4$ in the reaction with $2\text{-}ClC_6H_4{-}CHO$ (1/2 mole ratio; CH_3OH/conc. HCl at 55°C) [19].

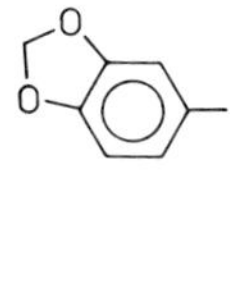

VI

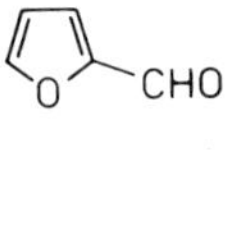

VII

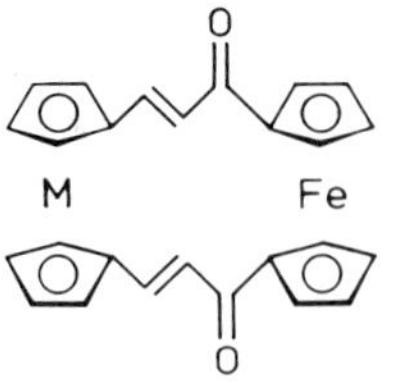

VIII

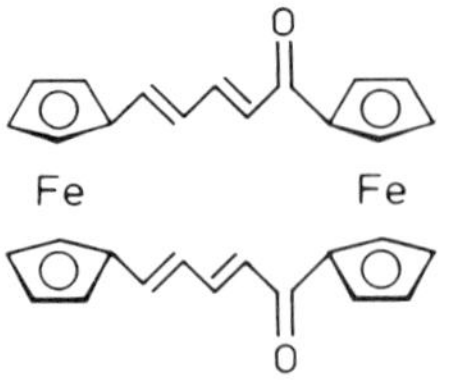

IX

With VII in a 1/2 mole ratio and catalysis by NaOH the reaction forms in CH_2Cl_2 at the boiling point [15] or in C_2H_5OH at 0°C [83] or room temperature [87] the compound V ($R = C_4H_3O$), in CH_3OH at 40 to 50°C the compound I ($R = C_4H_3O$, yield 67%) [12]. Reaction in the mole ratio 1/1 in C_2H_5OH at 100 to 130°C produces a resin which softens at 175°C [15], see also [66].

The reaction with $C_5H_5FeC_5H_4{-}CHO$ (1/2) in $CH_3OH/H_2O/NaOH$ at 45 to 50°C gives I ($R = C_5H_5FeC_5H_4$) [25].

With the dialdehyde $Fe(C_5H_4\text{–}CHO)_2$ (1/1) the condensation catalyzed by KOH produces in C_2H_5OH at 60°C compound VIII (M = Fe). With $Fe(C_5H_4\text{–}CH{=}CH\text{–}CHO)_2$ it forms under the same conditions IX [80]. VIII (M = Ru) is formed by the reaction with $Ru(C_5H_4\text{–}CHO)_2$ in C_2H_5OH/C_2H_5ONa at 60°C [81].

The Mannich reaction with paraformaldehyde and $(CH_3)_2NH \cdot HCl$ (1/4/4 mole ratio) in absolute C_2H_5OH/conc. HCl at the boiling point yields $Fe(C_5H_4\text{–}CO\text{–}CH_2\text{–}CH_2N(CH_3)_2 \cdot HCl)_2$ [14].

With Esters

The Claisen condensation takes place with esters such as $CH_3CO_2C_2H_5$, $C_2H_5CO_2C_2H_5$, or $C_6H_5CO_2CH_3$ with KNH_2 as the base on reaction first in liquid NH_3 and subsequently in ether. Acidifying with aq. HCl or NH_4Cl in liquid ammonia then forms $Fe(C_5H_4\text{–}CO\text{–}CH_2\text{–}CO\text{–}R)_2$ (R = CH_3, C_2H_5, or C_6H_5), which is obtained in a maximum yield (70% for R = CH_3 or C_6H_5; 50% for R = C_2H_5) on use of a 1/4/4 mole ratio of ketone/ester/KNH_2 [4, 10]. Since all attempts to isolate the monosubstituted diketone $CH_3\text{–}CO\text{–}C_5H_4FeC_5H_4\text{–}CO\text{–}CH_2\text{–}CO\text{–}R$ failed and the reaction required in its course at least two moles of KNH_2 per mole of ketone, it is assumed that the first acylation takes place on the intermediately formed carbanion X, the second on the anion XI thus formed, and that XI reacts more rapidly than X [4, 10]. The reaction can also be conducted with $NaNH_2$ instead of KNH_2, see for example [4, 88, 93]. With $HCO_2C_2H_5$ the salt $Na_2[Fe(C_5H_4\text{–}CO\text{–}CH{=}CH\text{–}O)_2]$ is obtained in C_6H_6 in the presence of Na metal; the ketoform $Fe(C_5H_4\text{–}CO\text{–}CH_2\text{–}CHO)_2$ cannot be prepared from this even under the mildest conditions, since it rapidly resinifies [22].

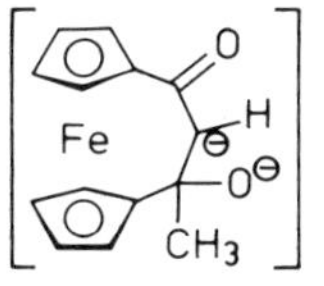

X

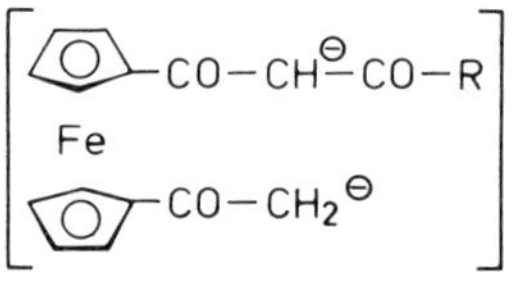

XI

The reaction with $C_2H_5O_2CCO_2C_2H_5$ in $C_2H_5OH/C_2H_5O^-$ produces after acidification with dilute acetic acid $Fe(C_5H_4\text{–}CO\text{–}CH_2\text{–}CO\text{–}CO_2C_2H_5)_2$ [58]. For the polycondensation with $CH_3O_2C\text{–}A\text{–}CO_2CH_3$ (A = $(CH_2)_4$, 1,4-C_6H_4, $C_5H_4FeC_5H_4$), see Section 5.2.3.5.1.1.5.6.3, p. 42.

The Claisen condensation with $(C_2H_5O)_2CO$ [9] in liquid NH_3/ether and with $NaNH_2$ as catalyst yields after decomposition of the Na salt with hydrochloric acid under ether in 34% yield the $Fe(C_5H_4\text{–}CO\text{–}CH_2\text{–}CO_2C_2H_5)_2$ (compound XII) [8] and considerable quantities of $C_2H_5O_2CNH_2$ [18]. The formation of the urethane can be suppressed by replacing the liquid NH_3 after formation of X from $Fe(C_5H_4\text{–}CO\text{–}CH_3)_2$ and $NaNH_2$ with ether for the subsequent reaction with $(C_2H_5O)_2CO$ (yield of XII 90%) [18]. XII is also obtained in C_6H_6 with NaH, at the boiling point, in 70 to 74% yield [90], 64% [73].

With Other Compounds

$Fe(C_5H_4\text{–}CO\text{–}CH(CH_2\text{–}C_6H_5)_2)_2$ is obtained with $C_6H_5CH_2Cl$ and KNH_2 (1/4/4 mole ratio) first in liquid NH_3, then in ether; even on addition of small quantities of chloride and amide no intermediate products can be isolated. Since the reaction with one equivalent KNH_2 does not go, here also X is indicated as the primary carbanion [13].

5.2.3.5.1.1.5.6.3 Polycondensation Reactions

For the reaction with CH_2O see pp. 38 and 41.

If $Fe(C_5H_4{-}CO{-}CH_3)_2$ is heated in the presence of $ZnCl_2$ for 3 h at 180°C, a polycondensation takes place to I, an insoluble polymer with semiconductor properties [21, 27, 40], see also [85].

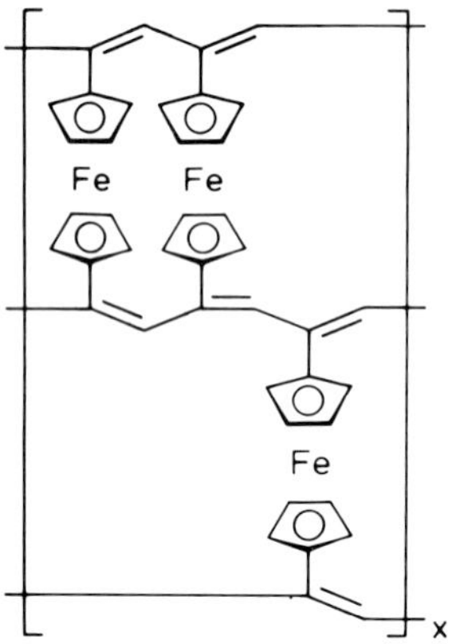

I

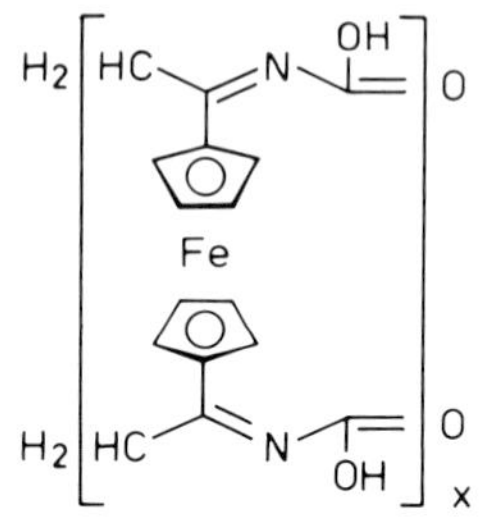

II

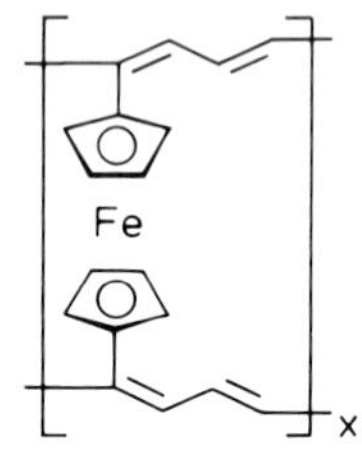

III

The action of $Al(OCH(CH_3)_2)_3$ forms $(C_{14}H_{14}O_{0.5}Fe)_x$ [2]. For dimerization see p. 37.

A ferrocene-containing poly(vinylenehydroxynitrile) of the approximate composition II (x = 2 to 3) forms with $NH_4[HCO_3]$ on heating to 170 to 190°C in the presence of $ZnCl_2$ [40]. The primary reaction step probably consists of the conversion of $NH_4[HCO_3]$ into carbamic acid NH_2CO_2H, which forms with $Fe(C_5H_4{-}CO{-}CH_3)_2$ predominantly oligomers of a linear and network structure with semiconductor properties [52].

Heating the ferrocene compound with CaC_2 in a 1/2 mole ratio at 150 to 200°C produces III. The CaC_2 acts here both as a condensation agent and as a reactant, in that it splits H_2O off of the ferrocene derivative and yields with it $HC{\equiv}CH$ [40].

Heteropolycondensation occurs with $CH_3O_2C{-}A{-}CO_2CH_3$ (A = $(CH_2)_4$, 1,4-C_6H_4, $C_5H_4FeC_5H_4$) in the presence of Na at 60°C [59].

Regarding simple and polycondensation with $H_2N{-}A_n{-}NH_2$ (n = 0, 1; A = CH_2CH_2, CONHCO, 4,4′-$C_6H_4{-}C_6H_4$, 4,4′-$C_6H_4{-}O{-}C_6H_4$, 4,4′-$C_6H_4{-}S{-}C_6H_4$, 4,4′-$C_6H_3(NO_2\text{-}2){-}S{-}S{-}(2\text{-}O_2N)C_6H_3$, 4,4′-$HN{-}C_6H_4{-}SO_2{-}C_6H_4{-}NH$) see 5.2.3.5.1.1.5.6.1, pp. 33/4.

5.2.3.5.1.1.5.6.4 Other Reactions on the Substituents

Refluxing with $CH_3CO_2C(CH_3){=}CH_2$ in conc. H_2SO_4 forms $Fe(C_5H_4{-}C({=}CH_2)O_2C{-}CH_3)_2$, the enol ester [6]. If a solution of $Fe(C_5H_4{-}CO{-}CH_3)_2$ in $HC(OC_2H_5)_3/CHCl_3$ is saturated with HCl gas at 20°C, then after 20 h standing a deposit forms of an tangled, asbestos-like, insoluble and infusible dark polymeric product, that probably is formed by polycyclotrimerization [47].

The Willgerodt-Kindler reaction with S and morpholine in boiling dioxane yields IV [3].

The Vilsmeier reaction with $(CH_3)_2NCHO$ and $POCl_3$ (mole ratio 1/5/1.5) at 20°C gives after 12 h reaction $Fe(C_5H_4{-}CCl{=}CH_2)_2$ (A, 28% yield) and $Fe(C_5H_4{-}CCl{=}CH{-}CHO)_2$ (B, 37% yield) [31]. The yield of B rises to 85% if the ketone (20 mmol) and $POCl_3$ (128 mmol) are reacted 2 h at 0°C and 1 h at room temperature in $(CH_3)_2NCHO$ (933 mmol, 70 mL) [34]. At a 1/2 mole ratio of ketone/$POCl_3$ it is possible to isolate also OHC–CH=CCl–$C_5H_4FeC_5H_4$–CCl=CH_2 (C) in addition to B [34]. According to [11] to obtain C the mole ratio must be 1/4, since otherwise only traces of C are formed along with the main product A.

On acetylation with $(CH_3CO)_2O$ in $BF_3 \cdot O(C_2H_5)_2$ at room temperature V and VI form; at a 1/27 mole ratio of ketone/anhydride the compound V represents the main product after a 1 h reaction. On lengthening the reaction time and increasing the anhydride excess, VI predominates [75].

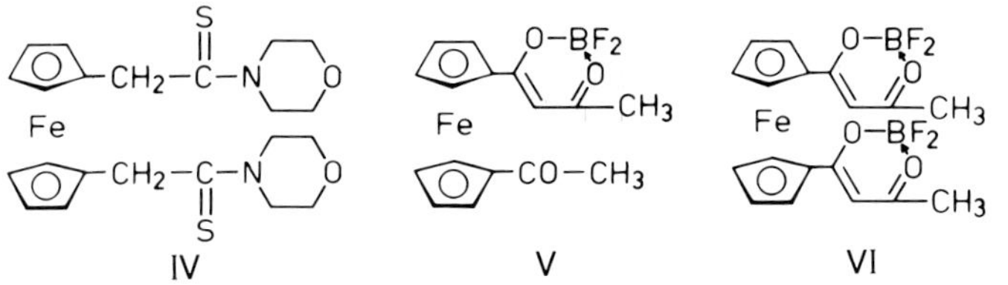

IV V VI

With SO_3 in $ClCH_2CH_2Cl$, first at 0°C then at 20°C, exchange occurs of the CH_3CO versus the SO_3H groups with formation of $Fe(C_5H_4–SO_3H)_2$; no other sulfonation takes place on the rings [16].

References:

[1] R. B. Woodward, M. Rosenblum, M. C. Whiting (J. Am. Chem. Soc. **74** [1952] 3458/9). – [2] R. Riemschneider, D. Helm (Chem. Ber. **89** [1956] 155/61). – [3] K. L. Rinehart, R. J. Curby, P. E. Sokol (J. Am. Chem. Soc. **79** [1957] 3420/4). – [4] C. R. Hauser, C. E. Cain (J. Org. Chem. **23** [1958] 1142/6). – [5] W. Göller (Diss. Stuttgart T.H. 1958).

[6] Union Carbide Corp., R. L. Pruett (U.S. 2947769 [1958/60]; C.A. **1961** 565). – [7] T. A. Mashburn, C. E. Cain, C. R. Hauser (J. Org. Chem. **25** [1960] 1982/6). – [8] K. Schlögl, H. Seiler (Monatsh. Chem. **91** [1960] 79/91). – [9] K. Schlögl, H. Seiler (Angew. Chem. **72** [1960] 38; 17th Intern. Congr. Pure Appl. Chem., Munich 1959, Vol. 1, Abstr. A 148). – [10] C. E. Cain, T. A. Mashburn, C. R. Hauser (J. Org. Chem. **26** [1961] 1030/4).

[11] M. Furdík, Š. Toma, J. Suchý (Chem. Zvesti **15** [1961] 547/53). – [12] M. Furdík, Š. Toma, J. Suchý, P. Elečko (Chem. Zvesti **15** [1961] 45/62). – [13] C. R. Hauser, T. A. Mashburn (J. Org. Chem. **26** [1961] 1795/800). – [14] C. R. Hauser, R. L. Pruett, T. A. Mashburn (J. Org. Chem. **26** [1961] 1800/1). – [15] V. N. Kotrelev, S. P. Kalinina, G. I. Kuznetsova (Plast. Massy **1961** No. 3, pp. 24/6; Soviet Plast. **1961** No. 3, pp. 22/3).

[16] A. N. Nesmeyanov, B. N. Strunin (Dokl. Akad. Nauk SSSR **137** [1961] 106/8; Proc. Acad. Sci. USSR Chem. Sect. **136/141** [1961] 275/6). – [17] M. Furdík, Š. Toma, J. Suchý (Chem. Zvesti **16** [1962] 449/57). – [18] K. Schlögl, M. Peterlik, H. Seiler (Monatsh. Chem. **93** [1962] 1309/27). [19] M. Furdík, Š. Toma (Acta Fac. Rerum. Nat. Comenianae Chim. No. 7 [1963] 545/7 from C.A. **61** [1964] 10706). – [20] Yu. V. Korshak, T. A. Pronyuk, B. E. Davydov (Neftekhimiya **3** [1963] 677/82; Petrol. Chem. [USSR] **3** [1964] 282/8).

[21] I. M. Paushkin, L. S. Polak, T. P. Vishnyakova, I. I. Patalakh, F. F. Machus, T. A. Sokolinskaya (J. Polym. Sci. C **4** [1963] 1481/94). – [22] K. Schlögl, H. Egger (Monatsh. Chem. **94** [1963] 1054/63). – [23] Kodak-Pathé, J.-P. C. G. Dubosc (Fr. 1361470 [1963/64]; C.A. **63** [1965] 13461). – [24] Hitachi Ltd., K. Hata, I. Motoyama, T. Masuko (Japan. 9679 [1963/67] from C.A. **68** [1968] No. 69582). – [25] M. Furdík, M. Dzurilla, Š. Toma, J. Suchý (Acta Fac. Rerum Nat. Univ. Comenianae Chim. No. 8 [1964] 569/79).

[26] M. Furdík, Š. Toma, M. Dzurilla, J. Suchý (Chem. Zvesti **18** [1964] 607/13). – [27] Ya. M. Paushkin, L. S. Polak, T. P. Vishnyakova, I. I. Patalakh, F. F. Machus, T. A. Sokolinskaya (Vyso-

komol. Soedin. **6** [1964] 545/50; Polym. Sci. [USSR] **6** [1964] 604/11). – [28] K. Schlögl, M. Fried (Monatsh. Chem. **95** [1964] 558/75). – [29] Eastman Kodak Co., J.-P. C. G. Dubosc (U.S. 3335008 [1964/67]; C.A. **67** [1967] No. 77884). – [30] B. E. Davydov, R. Z. Zakharyan, G. P. Karpacheva, B. A. Krentsel', G. A. Lapitskii, G. V. Khutareva (Dokl. Akad. Nauk SSSR **160** [1965] 650/3; Dokl. Phys. Chem. Proc. Acad. Sci. USSR **160/165** [1965] 90/3).

[31] K. Schlögl, W. Steyrer (Monatsh. Chem. **96** [1965] 1520/35). – [32] Colgate-Palmolive Co., J. E. Robertson (U.S. 3377248 [1965/68]; C.A. **69** [1968] No. 96878). – [33] Colgate-Palmolive Co., J. T. Suh (U.S. 3382267 [1965/68]; C.A. **69** [1968] No. 59378). – [34] M. Rosenblum, N. Brawn, J. Papenmeier, N. Applebaum (J. Organometal. Chem. **6** [1966] 173/80). – [35] K. Schlögl, H. Mechtler (Monatsh. Chem. **97** [1966] 150/67).

[36] Colgate-Palmolive Co., J. T. Suh, C. I. Judd (U.S. 3420865 [1966/69]; C.A. **70** [1969] No. 96961). – [37] W. M. Horspool, R. G. Sutherland (Tetrahedron Letters **1967** 4165/7). – [38] W. M. Horspool, R. G. Sutherland (Chem. Commun. **1967** 240). – [39] N. Ya. Mar'yashkin, T. P. Vishnyakova, A. A. Koridze (Tr. Mosk. Inst. Neftekhim. Gaz. Prom. No. 72 [1967] 67/70). – [40] Ya. M. Paushkin, T. P. Vishnyakova, S. A. Nisova, A. F. Lunin, O. Yu. Omarov, et al. (J. Polym. Sci. Polym. Chem. Ed. **5** [1967] 1203/21).

[41] T. P. Vyshnyakova, N. Ya. Mar'yashkin, M. E. Elyashberg (Zh. Org. Khim. **3** [1967] 947/9; J. Org. Chem. [USSR] **3** [1967] 910/1). – [42] V. V. Korshak, S. L. Sosin, B. A. Antipova (U.S.S.R. 234404 [1967/69]; C.A. **70** [1969] No. 106664). – [43] Colgate-Palmolive Co., J. T. Suh, C. I. Judd (U.S. 3557143 [1967/71]; C.A. **75** [1971] No. 6111). – [44] T. H. Barr, W. E. Watts (Tetrahedron **24** [1968] 3219/35). – [45] W. M. Horspool, R. G. Sutherland (Can. J. Chem. **46** [1968] 3453/6).

[46] E. W. Neuse (Macromolecules **1** [1968] 171/8). – [47] K. Schlögl, H. Soukup (Monatsh. Chem. **99** [1968] 927/46). – [48] S. L. Sosin, V. V. Korshak, T. M. Frunze (Dokl. Akad. Nauk SSSR **179** [1968] 1124/6; Dokl. Chem. Proc. Acad. Sci. USSR **178/183** [1968] 345/7). – [49] T. H. Barr, H. L. Lentzner, W. E. Watts (Tetrahedron **25** [1969] 6001/13). – [50] W. H. Horspool, R. G. Sutherland, J. R. Sutton (Can. J. Chem. **47** [1969] 3085/8).

[51] H. Kono, M. Shiga, I. Motoyama, K. Hata (Bull. Chem. Soc. Japan **42** [1969] 3267/9). – [52] Ya. M. Paushkin, T. P. Vishnyakova, F. F. Machus, F. A. Sokolinskaya, I. A. Golubeva (J. Polym. Sci. C **16** [1969] 4297/310). – [53] H. Alper, S. J. Brown, K. P. Jaffe (Org. Prep. Proced. **2** [1970] 167/9). – [54] R. G. Gamper, P. T. Funke, A. A. Volpe (AD-717648 [1970] 1/16; C.A. **75** [1971] No. 64382). – [55] A. A. Koridze, S. P. Gubin (J. Organometal. Chem. **22** [1970] 157/69).

[56] R. G. Gamper, P. T. Funke, A. A. Volpe (J. Polym. Sci. Polym. Chem. Ed. **9** [1971] 2137/46). – [57] R. G. Gamper, P. T. Funke, A. A. Volpe (Am. Chem. Soc. Div. Org. Coatings Plastics Chem. Papers **31** No. 2 [1971] 270/3). – [58] M. Laćan, R. Šarac-Arneri (Croat. Chem. Acta **43** [1971] 215/9). – [59] Ya. M. Paushkin, T. P. Vyshnyakova, I. D. Vlasova, I. A. Golubyeva, I. A. Yeryomina (5th Intern. Conf. Organometal. Chem., Moscow 1971, Vol. 1, Abstr., pp. 311/2). – [60] V. G. Shershun, T. P. Vishnyakova, Ya. M. Paushkin (Vysokomol. Soedin. B **13** [1971] 181/2).

[61] S. D. Slatina, Yu. K. Kirilenko, L. A. Vol'f, A. I. Meos, et al. (Radiokhimiya **13** [1971] 786/7; Soviet Radiochem. **13** [1971] 812/3). – [62] United States Dept. of the Army, C. W. Huskins, A. J. Thomas (U.S. 3770786 [1971/73]; C.A. **80** [1974] No. 49969). – [63] R. G. Gamper (Diss. Stevens Inst. Technol., Hoboken, N.J., 1972 from Diss. Abstr. Intern. B **33** [1972] 2528). – [64] S. D. Slatina, Yu. K. Kirilenko, L. A. Vol'f, A. I. Meos, I. B. Klimenko, V. I. Grachev, T. P. Vishnyakova, I. D. Vlasova (Zh. Prikl. Khim. **45** [1972] 446/7; J. Appl. Chem. [USSR] **45** [1972] 449/50). – [65] J. A. Winstead (J. Org. Chem. **37** [1972] 1271/2).

[66] Metals Research Inst., Tohoku Univ., S. Yajima, M. Oomori (Japan. Kokai 73-88200 [1972/73] from C.A. **81** [1974] No. 50493). – [67] G. S. Goldon, T. A. Balabina, T. V. Trynkina, A. N. Ushakova (Zh. Obshch. Khim. **43** [1973] 1271/4; J. Gen. Chem. [USSR] **43** [1973] 1262/4). –

[68] Y. Omote, R. Kobayashi, N. Sugiyama (Bull. Chem. Soc. Japan **46** [1973] 2896/7). – [69] Ya. M. Paushkin, A. M. Shevchik (Vestsi Akad. Navuk Belarusk.SSR Ser. Khim. Navuk **1973** No. 3, pp. 95/8). – [70] A. N. Nesmeyanov, G. B. Shul'pin, M. I. Rybinskaya, P. V. Petrovskii (Dokl. Akad. Nauk SSSR **215** [1974] 599/602; Dokl. Chem. Proc. Acad. Sci. USSR **214/219** [1974] 188/91).

[71] P. Elečko, E. Solčániová, Š. Toma (Chem. Zvesti **29** [1975] 411/7). – [72] M. G. Voronkov, A. K. Filippova, É. I. Dubinskaya, G. A. Prokop'eva (Zh. Obshch. Khim. **45** [1975] 706; J. Gen. Chem. [USSR] **45** [1975] 692). – [73] M. Hisatome, T. Sakamoto, K. Yamakawa (J. Organometal. Chem. **107** [1976] 87/101). – [74] N. S. Nametkin, V. D. Tyurin, I. K. Dobrov, I. V. Petrosyan, V. V. Trusov, A. Popal (7th Intern. Conf. Organometal. Chem., Kyoto, Japan, 1977, p. 30). – [75] M. N. Nefedova, V. N. Setkina, V. A. Kataev, T. V. Fisun, D. N. Kursanov (Dokl. Akad. Nauk SSSR **237** [1977] 352/5; Dokl. Chem. Proc. Acad. Sci. USSR **232/237** [1977] 662/5).

[76] A. Ratajćzak, B. Czech (Roczniki Chem. **51** [1977] 1735/40). – [77] H.-H. Wei, D.-M. Horng (J. Chinese Chem. Soc. [Taipei] [2] **25** [1978] 17/20 from C.A. **89** [1978] No. 43704). – [78] P. Canonne, G. Foscolos, R. Harder (J. Organometal. Chem. **178** [1979] 331/42). – [79] C. R. Jablonski (J. Organometal. Chem. **174** [1979] C3/C10). – [80] A. Kasahara, T. Izumi, I. Shimizu (Chem. Letters **1979** 1317/20).

[81] S.-I. Kamiyama, A. Kasahara, T. Izumi, I. Shimizu, H. Watanabe (Bull. Chem. Soc. Japan **54** [1981] 2079/82). – [82] A. Kasahara, T. Izumi, E. Yahagi, Y. Kamei (Yamagata Daigaku Kiyo Kogaku **16** [1981] 365/70). – [83] L. Tătaru, I. Mazilu, M. Vâtă, T. Lixandru, C. Simionescu (J. Organometal. Chem. **214** [1981] 107/14). – [84] L. Tătaru, M. Vâţă, I. Mazilu, T. Lixandru, S. Simionescu (Angew. Makromol. Chem. **101** [1981] 19/27). – [85] C. E. Carraher (J. Macromol. Sci. Chem. A **17** [1982] 1293/356, 1337).

[86] G. Wu, S. Liu, Z. Liao (Huaxue Xuebao **40** [1982] 67/72). – [87] C. Simionescu, T. Lixandru, L. Tătaru, I. Mazilu, M. Vâţă (J. Organometal. Chem. **238** [1982] 363/81). – [88] R. Yin, X. Zhen (Lanzhou Daxue Xuebao Ziran Kexueban **19** [1983] 41/5 from C.A. **100** [1984] No. 103578). – [89] M. Sugimoto, M. Nonoyama (Inorg. Nucl. Chem. Letters **15** [1979] 405/8). – [90] K. L. Rinehart, D. E. Bublitz, D. H. Gustafson (J. Am. Chem. Soc. **85** [1963] 970/82).

[91] M. Rosenblum (Diss. Harvard Univ. 1953). – [92] M. Rosenblum, R. B. Woodward (J. Am. Chem. Soc. **80** [1958] 5443/9). – [93] H. Imai (Bull. Chem. Soc. Japan **45** [1972] 1264/5). – [94] L. Tătaru, M. Dărânga, M. Vâţă, T. Lixandru, C. Simionescu (J. Therm. Anal. **22** [1981] 259/68). – [95] M. Furdík, Š. Toma, J. Suchý (Chem. Zvesti **15** [1961] 789/806).

[96] Š. Toma, M. Sališová (Acta Fac. Rerum Nat. Univ. Comenianae Chim. No. 21 [1975] 59/70). – [97] S. L. Sosin, V. P. Alekseeva, M. D. Litvinova, T. M. Ezhova, Yu. V. Shevchenko (Dokl. Akad. Nauk SSSR **271** [1983] 871/4; Dokl. Chem. Proc. Acad. Sci. USSR **268/273** [1983] 249/52).

5.2.3.5.1.1.6 Catalysis. Applications

$Fe(C_5H_4\text{–}CO\text{–}CH_3)_2$ catalyzes the decomposition of organic hydroperoxides to form the corresponding alcohols and ketones. It is, however, less effective than $C_5H_5FeC_5H_4\text{–}CO\text{–}CH_3$ or even $Fe(C_5H_5)_2$, and fails practically completely to work in the case of α-cumylhydroperoxide [25]. It catalyzes the hydroxylation of phenol by 59.4% H_2O_2 in acid solution to form a mixture of 1,2- and 1,4-dihydroxybenzene in a 2.3/1 mole ratio and ≦95% total yield referred to the phenol [11]. It catalyzes the formation of cyclo-$C_6H_{11}NHCO_2C_4H_9$-t from cyclo-$C_6H_{11}NCO$ and t-C_4H_9OH in CCl_4 under Hg lamp irradiation at 50°C. Conjectures concerning the reaction mechanism

References on pp. 47/8

are presented in the paper. The rate constant is 0.22 times that of the reaction catalyzed with $Fe(C_5H_5)_2$. With n- instead of the t-C_4H_9OH the reaction proceeds somewhat more slowly [21].

For temperature transitions in polyethylene terephthalate modified with $Fe(C_5H_4{-}CO{-}CH_3)_2$ see [34].

If the transesterification of 1,4-$C_6H_4(CO_2CH_3)_2$ with $HOCH_2CH_2OH$ is conducted in the presence of 0.05 to 1.0 wt% $Fe(C_5H_4{-}CO{-}CH_3)_2$, then not only is an additional catalyst unnecessary but also in the subsequent polycondensation a product of the composition I is obtained, which in contrast to the unmodified polyester possesses a high light resistance and thermal oxidation stability and is of an equally pure color (sandy to brownish red) [12, 18, 24, 26, 27, 32]. I modified with 0.05 to 0.1 wt% diacetylferrocene is suitable as the film base for a thermoplastic recording carrier [16]; modification with 0.1 to 0.5% increases the corrosion resistance of Al sheets coated with the polyester [31, 32].

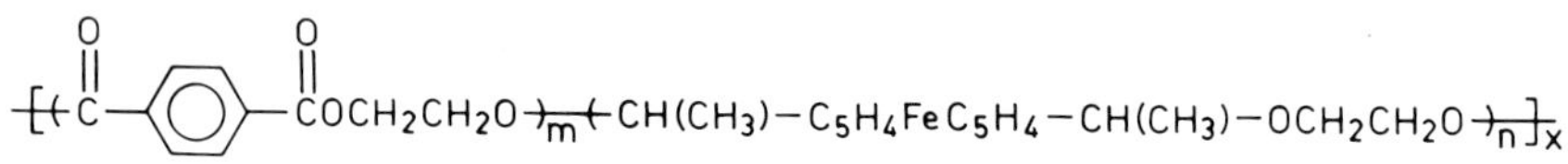

I

$Fe(C_5H_4{-}CO{-}CH_3)_2$ accelerates the polycondensation of maleic acid anhydride with $HOCH_2CH_2OH$ in the melt [19] as well as the cross linking of mixtures of II ($y \approx 130$; for example 100 parts) and III ($y \approx 250$, $z \approx 13$; for example 54 parts) by atmospheric oxygen at room temperature or 150°C to form elastomers [29]. With CH_2O it yields polycondensation products which serve for the impregnation of polyvinylalcohol fibers [9], see also p. 38. For the preparation of organic semiconductors by polycondensation with N_2H_4 salts see p. 33.

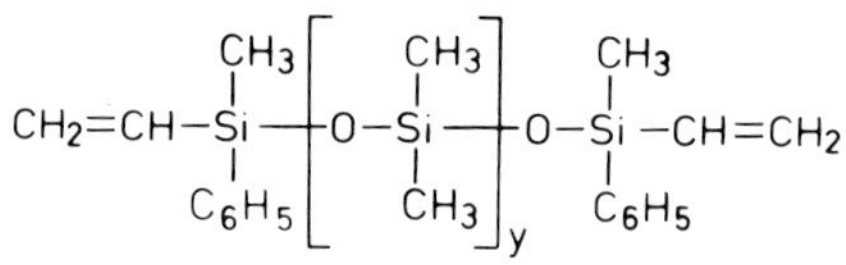

II

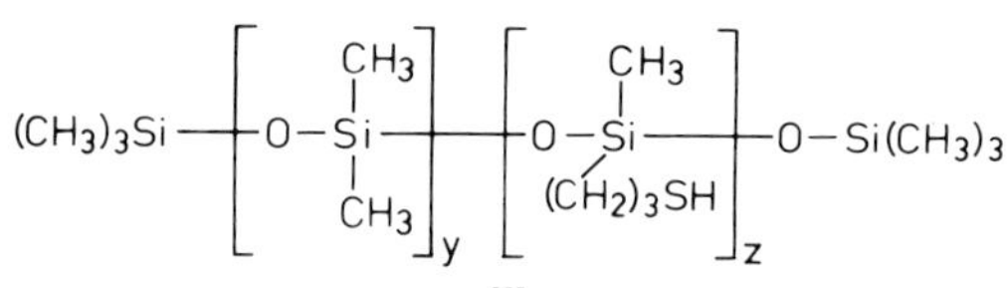

III

High electrode potential, low volatility, and good compatibility with rubber make $Fe(C_5H_4{-}CO{-}CH_3)_2$ an effective stabilizer of isoprene rubber [20]. It increases the thermal stability of polyorganosiloxanes [1]. Addition of ~2% amounts together with 0.86% benzoquinone to epoxy resins improves their stability and vulcanization properties [4]. As an additive to low pressure polyethylene films it retards aging due to heat or UV and γ irradiation. While its inhibiting effect on the thermal aging is superior to that of IV, it shows against aging by UV or γ irradiation a lower effectiveness than V or VI [14]. In contrast $Fe(C_5H_4{-}CO{-}CH_3)_2$ has been recommended by [5, 6, 15] as a prodegradant for the acceleration by sunlight of the decomposition of thermoplastic polymers made from $CH_2{=}CHCl$ [15], $CH_2{=}CH_2$, or $C_6H_5CH{=}CH_2$ [5, 6] in the presence of α-phenylindole [15] or benzophenone or its derivatives [6]. In this the benzophenone compound serves for the initiation of the degradation process, the diacetylferrocene for its propagation [6].

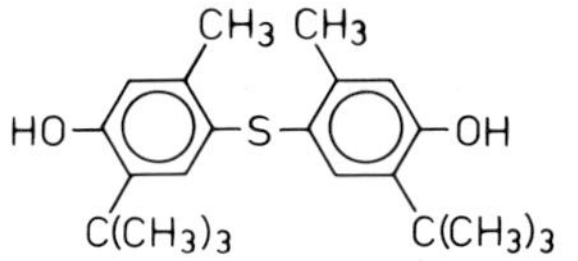

IV

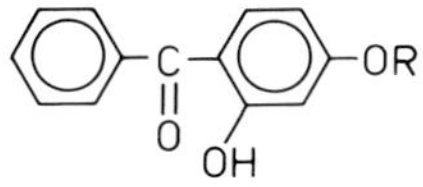

V

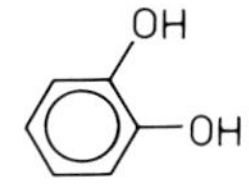

VI

Addition of 0.05 to 0.15 parts $Fe(C_5H_4{-}CO{-}CH_3)_2$ to mixtures of block polystyrene (30 to 70 parts) and polymethylmethacrylate (70 to 30 parts) generates a mother-of-pearl appearance [28]. The flammability of unsaturated, castable polyester resins (650 parts polyester made from maleic acid anhydride, tetrachlorophthalic acid, $HOCH_2CH_2OH$, and $CH_3CH(OH)CH_2OH$, 350 parts styrene, and 1 part hydroquinone) is reduced by addition of 1 wt% $Fe(C_5H_4{-}CO{-}CH_3)_2$ and 0.5 wt% $NCC(CH_3)_2N{=}NC(CH_3)_2CN$ [2]. Photosensitive layers for copying paper are produced using $Fe(C_5H_4{-}CO{-}CH_3)_2$ together with a diazonium salt [13], CBr_4 [3], or a chelating reagent for Fe such as o-phenanthroline [7, 8]. A photosensitive photoimaging material which provides images with increased contrast and stability comprises $Fe(C_5H_4{-}CO{-}CH_3)_2$ besides a halogenated hydrocarbon and a tanning compound such as indole, β-naphthol, or resorcinol [33]. As an antiknock agent it forms in comparison to $Fe(C_5H_5)_2$ fewer solid residues and possesses a lower solubility in the fuel [10].

$^{59}Fe(C_5H_4{-}CO{-}CH_3)_2$ has been employed for diagnostic purposes in nuclear medicine [22]. Doses of 0.15 wt% $Fe(C_5H_4{-}CO{-}CH_3)_2$ in the food of hepatectomized rats accelerates liver regeneration [30]. The insecticidal and acaricidal effects of carbamates, carboxylic acid esters, phosphoric acid esters, cycloalkanes, or tin compounds are synergistically strengthened by it [23]. It prevents or cures iron deficiency sicknesses of plants [17].

References:

[1] Dow Corning Corp., W. A. Piccoli (U.S. 2979482 [1956/61]; C.A. **1961** 18176). – [2] Badische Anilin- und Soda-Fabrik A.-G., I. Ziegler, R. D. Rauschenbach, R. Allwinn, K. Wintersberger (Belg. 643139 [1963/64]; C.A. **63** [1965] 3131). – [3] Kalle A.-G. (Neth. Appl. 65-04126 [1964/65]; C.A. **64** [1966] 10642). – [4] Hitachi Ltd., M. Sato, K. Hongo, T. Muroi (Japan. 70-38911 [1967/70] from C.A. **75** [1971] No. 118937). – [5] Imperial Chemical Industries Ltd., D. S. Brackman, D. J. Harper (Brit. 1382061 [1971/75]).

[6] A. Boberg, B. Aakermark, Aakerlund & Rausing (Swed. 11624/71 [1971]; Ger. Offen. 2244800 [1972/73]; C.A. **79** [1973] No. 19728). – [7] Ricoh Co., Ltd., T. Goto, K. Kojima, S. Kusakata (Japan. Kokai 74-01224 [1972/74] from C.A. **81** [1974] No. 8386). – [8] Ricoh Co., Ltd., T. Goto, K. Kojima, S. Kusakada (Japan. Kokai 74-05640 [1972/74] from C.A. **80** [1974] No. 151151). – [9] S. D. Slatina, T. N. Kalinina, L. A. Vol'f, E. A. Kalennikov (Sbornik Trudov Leningradskogo Instituta Tekstil'noi i Legkoi Promyshlennosti, Leningrad 1973, No. 14, Pt. 1, p. 103 from N. S. Kochetkova, Yu. K. Krynkina, Russ. Chem. Rev. **47** [1978] 486/94). – [10] V. Veselý, Š. Toma, J. Gurský, E. Patzelt (Ropa Uhlie **15** [1973] 194/7).

[11] Brichima S.p.A., P. Maggioni (Ger. Offen. 2407398 [1973/74]; C.A. **81** [1974] No. 151787). – [12] A. I. Volozhin, L. Yu. Anishchenko, N. R. Prokopchuk, Ya. M. Paushkin, et al. (U.S.S.R. 484752 [1973/77]; C.A. **88** [1978] No. 74727). – [13] Ricoh Co., Ltd., T. Goto (Japan. Kokai 76-53818 [1974/76] from C.A. **86** [1977] No. 131099). – [14] E. A. Kalennikov, V. S. Yuran, Ya. M. Paushkin, A. E. Pereverzev (Vestsi Akad. Navuk Belarusk.SSR Ser. Khim. Navuk **1975** No. 3, pp. 57/60). – [15] Solvay & Cie., P. Burgaud (Ger. Offen. 2513200 [1975]; C.A. **84** [1976] No. 45389).

[16] S. P. Gukalov, A. I. Volozhin, I. I. Globa, V. P. Chernyshev, N. F. Mironova, et al. (U.S.S.R. 532892 [1975/76]; C.A. **86** [1977] No. 36322). – [17] Bayer A.-G., V. Mues, J. Niggemann (Ger. Offen. 2553786 [1975/77]; C.A. **87** [1977] No. 67316). – [18] A. I. Volozhin, V. P. Chernyshev,

P. N. Zernov, V. A. Levdanskii, et al. (U.S.S.R. 608821 [1975/78]; C.A. **89** [1978] No. 61040). – [19] V. P. Red'ko, D. F. Kutepov, V. K. Skubin, M. D. Reshetova, S. A. Yankovskii, V. A. Mironov (Plasticheskie Massy **1976** No. 5, pp. 11/2). – [20] E. A. Kalennikov, T. E. Ognevskaya, T. N. Mitrokhina, R. I. Dashevskaya, Ya. M. Paushkin (Vestsi Akad. Navuk Belarusk. SSR Ser. Khim. Navuk **1977** No. 1, pp. 58/61).

[21] S. P. McManus, H. S. Bruner, H. D. Coble, M. Ortiz (J. Org. Chem. **42** [1977] 1428/33). – [22] E. A. Stadlbauer, E. Nipper, M. Wenzel (J. Labelled Compounds Radiopharm. **13** [1977] 491/508). – [23] Bayer A.-G., V. Mues, W. Behrenz (Ger. Offen. 2711546 [1977/78]; C.A. **90** [1979] No. 1705). – [24] A. I. Volozhin, L. Yu. Anishenko, N. R. Prokopchuk, Ya. M. Paushkin, E. M. Eizenshtein, et al. (Ger. Offen. 2702696 [1977/78]; C.A. **89** [1978] No. 164857). – [25] Yu. A. Aleksandrov, A. O. Kolmakov (Zh. Obshch. Khim. **48** [1978] 1653; J. Gen. Chem. [USSR] **48** [1978] 1510).

[26] Institut Fiziko-Organ. Khim. AN Belorussk. SSR Mogilevskii Komb. Sint. Volokna im. V.I. Lenina (U.S.S.R. 484752 [1973/78] from Organometal. Compounds **33** [1978] No. 4321; C.A. **88** [1978] No. 74727). – [27] A. I. Volozhin, L. Yu. Verkhovodka, Ya. M. Paushkin (Vysokomol. Soedin. B **20** [1978] 713/4). – [28] Sverdlovskii Zavod. Plastmass., I. B. Yuferova, I. S. Bykhovskaya, M. A. Knyaz'kina, et al. (U.S.S.R. 761514 [1978/80]; C.A. **94** [1981] No. 16812). – [29] Dow Corning Corp., G. R. Homan, C.-L. Lee (U.S. 4268655 [1979/81]; C.A. **95** [1982] No. 221132). – [30] L. L. Gershbein (Res. Commun. Chem. Pathol. Pharm. **27** [1980] 139/45).

[31] A. A. Trofimova, A. I. Volozhin, K. F. Knel'ts, Ya. M. Paushkin, I. I. Globa, E. L. Kantorovich, et al. (U.S.S.R. 730610 [1980] from C.A. **93** [1980] No. 73312). – [32] V. V. Abramov, A. A. Trofimova, A. I. Volozhin, N. V. Sibiryakova (Plasticheskie Massy **1981** No. 5, pp. 60/1). – [33] A. N. Nesmeyanov, L. V. Balabanova, N. S. Kochetkova, R. V. Lukyanova (U.S.S.R. 526205 [1974/82] from C.A. **98** [1983] No. 152867). – [34] L. Yu. Osmolovskaya et al. (Vestsi Akad. Navuk Belarusk. SSR **1985** No. 1, pp. 71/4).

5.2.3.5.1.2 Complexes, Ions, and Salts Derived from $Fe(C_5H_4{-}CO{-}CH_3)_2$

See also Sections 5.2.3.5.1.1.5.5 as well as 5.2.3.5.1.1.4 and 5.2.3.5.1.1.5.2 on pp. 31/2, 21/2, and 26/7, respectively.

Complexes

$[Fe(C_5H_4{-}COCH_3)_2 \cdot AgBF_4]_x$, $Fe(C_5H_4{-}C({=}CH_2)OH \cdot ZnCl_2)_2$, and $Fe(C_5H_4{-}CO{-}CH_3)_2 \cdot CBr_4$ are treated on pp. 31/2.

$Fe(C_5H_4{-}CO{-}CH_3)_2 \cdot 2SbCl_5$ forms as a dark orange complex when the donor and acceptor compounds react in a 1/2 mole ratio in benzene or toluene in an evacuated sealed tube. The IR spectrum (mineral oil) shows bands (in cm^{-1}) at 470 ($\nu O \rightarrow Sb$), 1130 ($\nu C{-}CH_3$), 1300 ($\nu C_{ring}{-}C_{acetyl}$), and 1620 ($\nu C \overset{...}{=} O \rightarrow Sb$), according to which the complex possesses structure I ($MX_n = SbCl_5$) with two fully equivalent, coordinated CO groups. The molecule has C_{2v} symmetry. Only one stretching vibration of the A_1 class of the coordinated CO groups is IR active [5, 6].

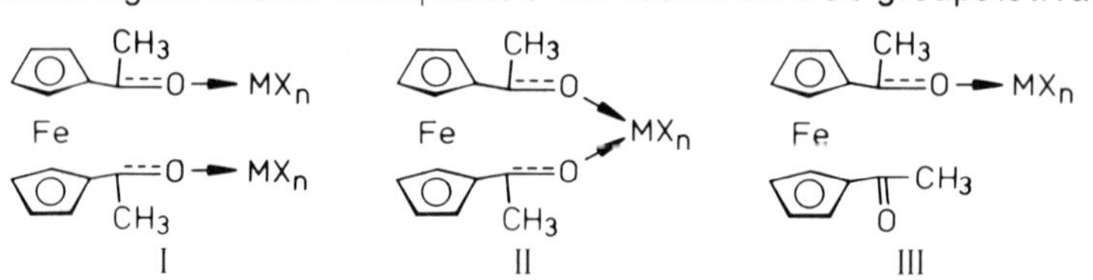

I II III

$Fe(C_5H_4–CO–CH_3)_2 \cdot AlX_3$ (X = Cl, Br) is obtained by addition of AlX_3 in ether to $Fe(C_5H_4–CO–CH_3)_2$ in CCl_4 as bright red crystals which decompose at 88 to 90°C for X = Cl or at 78 to 81°C for X = Br. The complexes are monomeric according to cryoscopic molecular weight determinations in $C_6H_5NO_2$ and display in Nujol the following IR bands in cm^{-1} [1]:

X = Cl	—	487	534	579	602	624
X = Br	405	488	534	578	600	628, 676, 710
assignment				ν(O → Al)		

X = Cl	765	792	840	900, 930, 965	985
X = Br	765	790	840, 807	902	982

X = Cl	1020	1044	1119, 1181	1218	1290 and 1305
X = Br	1017	1046, 1060	1120	1218	1295 and 1307
assignment			$\nu(C–CH_3)$		$\nu(C_{ring}–C_{acetyl})$

X = Cl	1345	1360	1406	1485	1610	1640
X = Br	1343	1365	1410	1485	1580	1610
assignment					ν_s(C$\cdots$O →)	ν_a(C$\cdots$O →)

Above all, the frequencies at 1610 and 1640 cm^{-1} for X = Cl and at 1580 and 1610 cm^{-1} for X = Br, which are assigned to the symmetrical and asymmetrical ν(C$\cdots$O→) vibrations, allow selection for the complex of the chelate structure II (M = Al, n = 3) with $Fe(C_5H_4–CO–CH_3)_2$ as a bidentate ligand [1].

The ^{57}Fe Mössbauer spectrum for $Fe(C_5H_4–CO–CH_3)_2 \cdot AlCl_3$ at 80 K yields (in mm/s) δ = 0.51(1) referred to natural Fe, Δ = 2.07(2), Γ = 0.16(1) (measured at half height). The smaller value for the quadrupole splitting versus that of the free compound is probably due to electron withdrawal via e_1 orbitals due to the Al [9]. Reaction of $Fe(C_5H_4–CO–CH_3)_2 \cdot AlCl_3$ with 1,2-diamino-benzene, 4-methyl-1,2-diamino-benzene, or 1,8-diamino-naphthalene in the melt forms IV with R = R′ = H; IV with R = CH_3, R′ = H or R = H, R′ = CH_3; and V, respectively [4].

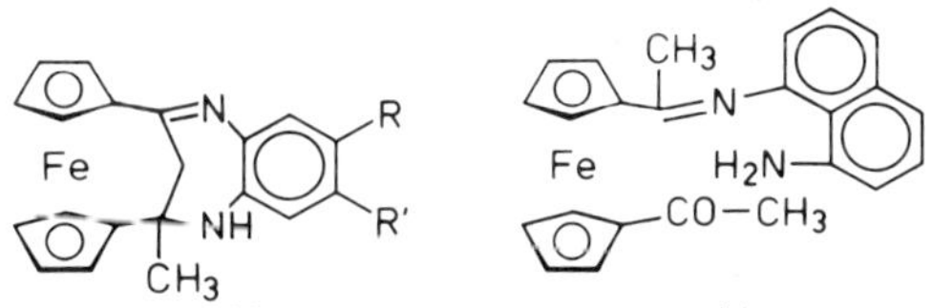

IV V

$Fe(C_5H_4–CO–CH_3)_2 \cdot TiCl_4$ precipitates as a dark violet deposit on reaction of the components in a 1/1 mole ratio in boiling ether. It decomposes at 115 to 118°C. A molecular weight determination was not possible due to the absence of a suitable solvent. In Nujol the compound exhibits the following IR bands (in cm^{-1}): 403, 427(sh), 463(sh), 484, 525, 545, 561 (νO → Ti?), 599, 622, 675, 717, 760, 850, 905, 989, 1023(sh), 1043, 1060(sh), 1120 ($\nu C–CH_3$), 1176, 1280(sh) ($\nu C_{ring}–C_{acetyl}$), 1310 ($\nu C_{ring}–C_{acetyl}$), 1595 (νC$\cdots$O →), 1678 (νC=O) [1]. The ^{57}Fe Mössbauer spectrum at 80 K gives (in mm/s) δ = 0.52(1) (referred to natural Fe), Δ = 2.10(1), and Γ = 0.14(1) (measured at half height) [9]. From the IR spectrum Pavlík, Handlíř derive for the complex the structure III ($MX_n = TiCl_4$) (see p. 48) with one coordinated and one free CO group [1]. According to Clemance et al. this is also supported by the equality of the quadrupole splitting for the complex and $[C_5H_5FeC_5H_4–COH–CH_3]^+$ (cf. "Ferrocene" 2, 1977, p. 212) [9]. Thus, the report by Neshvad et al. [7] of a chelate structure for the compound is apparently superseded.

$Fe(C_5H_4\text{–}CO\text{–}CH_3)_2 \cdot TiBr_4$ deposits independent of the mole ratio of the reactants as a dark green complex, which is insoluble in CS_2 and decomposes at 115 to 117°C, when $Fe(C_5H_4\text{–}CO\text{–}CH_3)_2$ and $TiBr_4$ react in CS_2 in the absence of air and moisture. The IR spectrum in Nujol shows bands (in cm^{-1}) at 467 (sh), 483, 492, 539 (sh), 559, 594, 622, 661 (sh), 672, 713, 748 ($\nu_a O \rightarrow Ti$?), 780 (sh), 830 (sh), 850, 890 (sh), 902, 1032 (sh), 1046, 1060, 1080, 1120 ($\nu C\text{–}CH_3$), 1175, 1222, 1260 (sh) ($\nu C_{ring}\text{–}C_{acetyl}$), 1300 and 1313 (sh) ($\nu C_{ring}\text{–}C_{acetyl}$), 1336, 1588 ($\nu C \overset{...}{=} O \rightarrow$), 1680 ($\nu C{=}O$), from which the complex has the structure III ($MX_n = TiBr_4$) (see p. 48). The complex decomposes in air or with H_2O with re-formation of $Fe(C_5H_4\text{–}CO\text{–}CH_3)_2$ [2].

$Fe(C_5H_4\text{–}CO\text{–}CH_3)_2 \cdot 2TiCl_4$ is obtained by reaction of the donor and acceptor compounds in a 1/2 mole ratio: [1] reports that reaction in CCl_4 gives a dark blue precipitate which decomposes at 107 to 109°C; [5] reports that reaction in an evacuated sealed tube in C_6H_6 or toluene gives a dark orange product. The ^{57}Fe Mössbauer spectrum at 80 K has (in mm/s) $\delta = 0.51(1)$ referred to natural Fe, $\Delta = 2.04(1)$, and $\Gamma = 0.15(1)$ measured at half height [9]. The IR bands in Nujol [1] (in mineral oil [5]) are located at 405 (sh), 465 (sh), 485, 540 (sh), 560 (565 [5]) ($\nu O \rightarrow Ti$?), 590, 620, 672, 760, 787, 848, 888, 903, 984, 1017 (sh), 1046, 1060 (sh), 1120 (1130 [5]) ($\nu C\text{–}CH_3$), 1182, 1219, 1290 (sh) and 1302 (1300 [5]) ($\nu C_{ring}\text{–}C_{acetyl}$), 1340, 1362, 1408, and 1586 (1610 [5]) ($\nu C \overset{...}{=} O \rightarrow$), all in cm^{-1}. The presence of only one absorption band for the coordinated CO groups indicates their equivalence as is given in structure I ($MX_n = TiCl_4$) (see p. 48) [1, 5, 6]. Also, the decrease in the quadrupole splitting in the complex versus the free compound indicates an electron withdrawal via both Ti atoms; it is nevertheless distinctly smaller than for $[Fe(C_5H_4\text{–}C(CH_3){=}OH)_2]^{2+}$ (cf. p. 53) [9]. As previously conjectured by [1], the two Ti atoms are connected with each other via two Cl bridges [9].

$Fe(C_5H_4\text{–}CO\text{–}CH_3)_2 \cdot SnCl_4$. The reaction of the components in equimolar quantities in CCl_4 leads to a dark violet, insoluble complex decomposing above 150°C, which according to cryoscopic molecular weight determinations is monomeric [3]. If the reaction is undertaken in a sealed, evacuated tube in C_6H_6 or toluene, a dark orange complex forms [5, 6]. At 80 K the ^{57}Fe Mössbauer spectrum has the parameters (in mm/s) $\delta = 0.55(1)$ referred to natural Fe, $\Delta = 2.08(1)$, and $\Gamma = 0.15(1)$ measured at half height. The ^{119}Sn Mössbauer spectrum has (in mm/s) $\delta = 0.37(1)$ referred to SnO_2 and $\Delta = 0.0$ [9]. The IR spectrum in mineral oil (Nujol) shows bands at 565 ($\nu O \rightarrow Sn$), 1130 ($\nu C\text{–}CH_3$), 1300 ($\nu C_{ring}\text{–}C_{acetyl}$) as well as the partially strongly overlapping symmetrical and antisymmetrical $\nu(C \overset{...}{=} O \rightarrow)$ bands at 1565 and 1600 (1544 and 1574 [3]) cm^{-1}, respectively, from which the complex has the chelate structure II ($MX_n = SnCl_4$) (see p. 48) [3, 5, 6]. The reduction of the Δ value in the ^{57}Fe Mössbauer spectrum versus that of the free compound is attributed to an electron withdrawal by Sn via e_1 orbitals and not to an inclination of the rings. The absence of the quadrupole splitting in the ^{119}Sn Mössbauer spectrum indicates a cis bonding for the O atoms of the ligands octahedrally arranged around the Sn [9].

$Fe(C_5H_4\text{–}CO\text{–}CH_3)_2 \cdot SnBr_4$ is a monomeric, dark red complex decomposing above 130°C, which precipitates on drop-wise addition of $SnBr_4$ in CCl_4 to the ferrocene derivative dissolved in CCl_4. It has structure II ($MX_n = SnBr_4$) (see p. 48). The symmetric and antisymmetric $\nu(C \overset{...}{=} O \rightarrow)$ bands in Nujol are located at 1552 and 1574 cm^{-1}, respectively [3].

$Fe(C_5H_4\text{–}CO\text{–}CH_3)_2 \cdot VOCl_3$ forms as a dark orange product on reaction of the components in equimolar quantities in C_6H_6 or toluene in a sealed, evacuated tube. It exhibits in mineral oil IR bands (in cm^{-1}) at 580 ($\nu O \rightarrow V$), 1000 ($\nu V{=}O$), 1130 ($\nu C\text{–}CH_3$), 1300 ($\nu C_{ring}\text{–}C_{acetyl}$), and 1630 ($\nu C \overset{...}{=} O \rightarrow$). The last is an unsymmetrical band with an extremely distinct shoulder at 1660 cm^{-1}. It belongs to a complex of the structure III ($MX_n = VOCl_3$) (see p. 48) with one coordinated (1630 cm^{-1}) and one free (1660 cm^{-1}) CO group [5, 6].

$Fe(C_5H_4\text{–}CO\text{–}CH_3)_2 \cdot FeCl_3 \cdot H_2O$ precipitates immediately in a practically quantitative yield as a black, extremely hygroscopic deposit on drop-wise addition of a solution of anhydrous $FeCl_3$

in ether to $Fe(C_5H_4–CO–CH_3)_2$ in CCl_4. The powder obtained after filtration and drying shows in the ^{57}Fe Mössbauer spectrum (all values in mm/s) at 80 K an isomer shift (referred to natural Fe) of $\delta = 0.39(1)$ for Fe^{III} and 0.53(1) for the ferrocene component, a quadrupole splitting of $\Delta \approx 0$ and 2.10(1), respectively, as well as a line width (measured at half height) of $\Gamma = 0.28(1)$ and 0.14(1) [9]. These data correspond to a chelate complex with structure II ($MX_n = FeCl_3$) (see p. 48) [8], in which the Fe^{III} is bonded in the same fashion as $SnCl_4$. The sixth coordination site in the octahedron is occupied by an H_2O molecule [9].

References:

[1] I. Pavlík, K. Handlíř (Collection Czech. Chem. Commun. **31** [1966] 1958/74). – [2] I. Pavlík, K. J. Klikorka, K. Handlíř, P. Sedlmayer (Z. Chem. [Leipzig] **6** [1966] 427/8). – [3] I. Pavlík, J. Klikorka, K. Handlíř (Z. Chem. [Leipzig] **8** [1968] 390/1). – [4] Y. Omote, R. Kobayashi, Y. Nakada, N. Sugiyama (Bull. Chem. Soc. Japan **46** [1973] 3315/6). – [5] L. M. Dyagileva, O. S. Morozov, N. N. Vyshinskii, V. G. Tsvetkov, Yu. A. Aleksandrov (Zh. Obshch. Khim. **50** [1980] 1859/62; J. Gen. Chem. [USSR] **50** [1980] 1516/8).

[6] L. M. Dyagileva, O. S. Morosov, Yu. A. Alexandrov (22nd Intern. Conf. Coord. Chem., Budapest 1982, p. 757). – [7] G. Neshvad, R. M. G. Roberts, J. Silver (J. Organometal. Chem. **236** [1982] 349/58). – [8] R. M. G. Roberts, J. Silver, M. Clemance (unpublished results from [7, p. 357]). – [9] M. C. Clemance, R. M. G. Roberts, J. Silver (J. Organometal. Chem. **247** [1983] 219/22).

Diacetylferrocenium Ion and Its Salts

For the electrochemical and chemical formation of the $\mathbf{[Fe(C_5H_4–CO–CH_3)_2]^+}$ ion see Sections 5.2.3.5.1.1.4 and 5.2.3.5.1.1.5.2, pp. 21/2 and 26/7, respectively.

Due to the electron withdrawing effect of the CH_3CO substituents this cation is unstable even in strongly acid solutions [10]. According to [5] this is why it is impossible to obtain a reproducible ESR spectrum. An ESR signal with $g_{\parallel} = 3.4$ and $g_{\perp} = 1.8$ has been reported [7]. The signals with $g_{\parallel} = 3.90$ and $g_{\perp} = 1.65$ obtained after the oxidation of $Fe(C_5H_4–CO–CH_3)_2$ with KIO_3, $[NH_4]_2[Ce(NO_3)_6]$, or $AgClO_4$ in acetone in the presence of several drops of H_2SO_4 resemble those of α-substituted 1,1′-dialkylferrocenium ions and indicate an alteration of the CH_3CO groups during the reaction [4]. The initially observed isotropic contact shifts vanish due to decomposition [2, 3].

Unlike with $C_5H_5FeC_5H_4–CO–CH_3$, the decomposition proceeds according to the equation

$$3[Fe(C_5H_4–CO–CH_3)_2]^+ \rightarrow 2\,Fe(C_5H_4–CO–CH_3)_2 + Fe^{3+} + 2\,C_5H_4–CO–CH_3$$

[10]. The rate is dependent on the temperature and the acid strength, with first order kinetics [10, 12]. At 25°C the rate constant k possesses as a function of the acidity function H_0 for the solution the following values:

H_0	0	−1	−2	−3	Ref.
k in s^{-1}	0.088	0.043	0.022	0.00142	[10]
	0.26	0.13	0.066	0.0042	[12]

On addition of oxidants the rate constant of the decomposition reaction drops; the rate constant in the presence of Ce^{IV} is ⅓ k [12]; see also p. 24.

The reduction potential determined chronopotentiometrically on a Pt electrode in CH_3CN (0.2 M $LiClO_4$) is $E_{0.22} = 0.805$ V versus the saturated calomel electrode [1]. For the electron transfer occurring in CH_3CN according to

$$[Fe(C_5H_4–CO–CH_3)_2]^+ + (4\text{-}CH_3OC_6H_4)_3N \rightleftharpoons Fe(C_5H_4–CO–CH_3)_2 + [(4\text{-}CH_3OC_6H_4)_3N]^{\ddagger}$$

the rate constant $k = 1.5 \times 10^9\ M^{-1} \cdot s^{-1}$ has been determined by means of computer-controlled

References on p. 52

internal reflection spectroelectrochemical measurements. It is in good agreement with the value calculated using Marcus' theory [9, 11].

[Fe(C_5H_4–CO–CH_3)$_2$]BF_4 is obtained as dark blue crystals by condensation of liquid SO_2 at −80°C onto Fe(C_5H_4–CO–CH_3)$_2$·$AgBF_4$, raising the temperature to −20°C and volatilization of the solvent from the filtrate freed from the precipitated silver. A shift of the ν(C=O) band by around 30 cm^{-1} to higher wave numbers is connected with the transformation of the complex into the salt [6].

References:

[1] T. Kuwana, D. E. Bublitz, G. Hoh (J. Am. Chem. Soc. **82** [1960] 5811/7). – [2] H. P. Fritz, H. J. Keller, K. E. Schwarzhans (Z. Naturforsch. **21 b** [1966] 809/10). – [3] H. P. Fritz, H. J. Keller, K. E. Schwarzhans (J. Organometal. Chem. **6** [1966] 652/9). – [4] R. Prins, A. R. Korswagen (J. Organometal. Chem. **25** [1970] C74/C76). – [5] R. Prins (Mol. Phys. **19** [1970] 603/20).

[6] P. Carty, M. F. A. Dove (J. Organometal. Chem. **28** [1971] 125/32). – [7] A. G. T. G. Kortbeek (personal communication from [8]). – [8] R. Prins, A. G. T. G. Kortbeek (J. Organometal. Chem. **33** [1971] C33/C34). – [9] N. Winograd, T. Kuwana (J. Am. Chem. Soc. **93** [1971] 4343/50). – [10] J. Holeček, K. Handlíř, J. Klikorka (Proc. Conf. Coord. Chem. **7** [1978] 65/70).

[11] J. C. Holland (Diss. Purdue Univ. 1979 from Diss. Abstr. Intern. B **40** [1979] 2637). – [12] J. Holeček, K. Handlíř, J. Klikorka (Collection Czech. Chem. Commun. **44** [1979] 1060/9).

Other Ions

[Fe(C_5H_4–CO–CH_3)$_2$]$^-$ and [Fe(C_5H_4–CO–CH_3)$_2$]$^{2-}$ are treated in Section 5.2.3.5.1.1.4, p. 22.

[Fe(C_5H_4–CO–CH_3)$_2$H]$^+$ forms by monoprotonation on dissolution of the compound in CF_3CO_2H or dilute H_2SO_4 [3, 4]. The ^{1}H NMR spectrum exhibits the following shifts δ in ppm:

solvent	CH_3	C_5H_4 H-3,4	C_5H_4 H-2,5	Ref.
CF_3CO_2H	2.03 s	4.33 t	4.50 t	[4]
CF_3CO_2H	2.15	4.50 and 4.60		[3]
H_2SO_4	2.25 s	4.47 and 5.30		[3]

The ^{13}C NMR spectrum in CF_3CO_2H yields δ = 25.5 (CH_3), 72.1 (C-2,5), 75.9 (C-3,4), 78.8 (C-1), 211.3 (CO) ppm [2, 3]. The ^{57}Fe Mössbauer spectrum (values in mm/s) in CF_3CO_2H at 80 K gives an isomer shift of δ = 0.52(1) referred to natural Fe and a quadrupole splitting of Δ = 2.09(1) [4]. The cation probably possesses the structure I, in which the proton forms a hydrogen bond to the two oxygen atoms of the acetyl groups arranged in the syn configuration. This structural proposal is supported by the sharp singlet of the CH_3 protons in the ^{1}H NMR spectrum as well as by the quadrupole splitting, which agrees with that of the various chelate complexes Fe(C_5H_4–CO–CH_3)$_2$·MCl_n (MCl_n = $AlCl_3$, $SnCl_4$, $FeCl_3$) [4, 5].

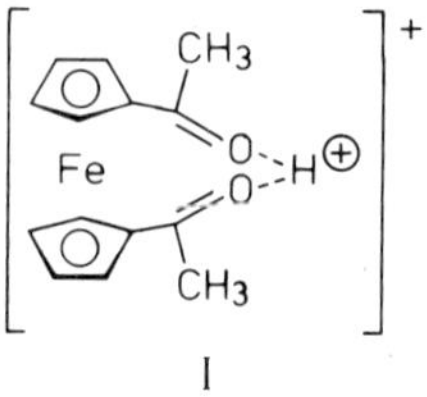

I

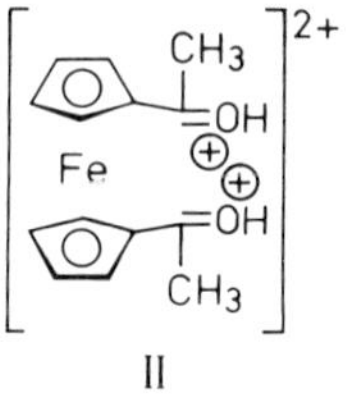

II

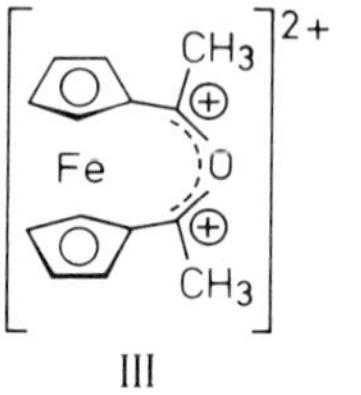

III

References on p. 53

[Fe(C_5H_4–C(CH_3)=OH_2]$^{2+}$ (see Formula II) represents the diprotonation product that forms on dissolution of the compound in CF_3SO_3H or concentrated sulfuric acid [4]. The ^{1}H NMR spectrum in 70% H_2SO_4: $\delta = 2.75$ (s, CH_3), 5.51 (br.s, H-3,4), 5.70 (br.s, H-2,5) ppm [4]. The ^{13}C NMR spectrum in conc. H_2SO_4: $\delta = 2.45$ (CH_3), 74.0 (C-1), 78.2 (C-2,5), 85.7 (C-3,4), 218.5 (CO) ppm [2]. The electronic spectrum in 90% H_2SO_4: $\lambda_{max}(\varepsilon) = 340$ sh, 524(1302) nm [1]. The ^{57}Fe Mössbauer spectrum (values in mm/s) at 80 K displays in 70% H_2SO_4 an isomer shift of $\delta = 0.55(1)$ versus natural Fe and a quadrupole splitting of $\Delta = 1.79(1)$. In CF_3SO_3H $\delta = 0.52(1)$ and $\Delta = 1.78(2)$. The Δ values are the lowest so far measured for ferrocene derivatives. Any electron withdrawal appears to occur via the ring-based orbitals (e_1) [4]. On the basis of the ^{13}C NMR spectrum Nesmeyanov et al. [2] did not completely exclude the possibility of formation of a bridged dication approximately of the Type III, see also [3].

References:

[1] G. Cerichelli, B. Floris, G. Illuminati, G. Ortaggi (Gazz. Chim. Ital. **103** [1973] 911/22). – [2] A. N. Nesmeyanov, G. B. Shul'pin, M. I. Rybinskaya, P. V. Petrovskii (Dokl. Akad. Nauk SSSR **215** [1974] 599/602; Dokl. Chem. Proc. Acad. Sci. USSR **214/219** [1974] 188/91). – [3] A. N. Nesmeyanov, G. B. Shul'pin, M. I. Rybinskaya (Dokl. Akad. Nauk SSSR **218** [1974] 1107/10; Dokl. Chem. Proc. Acad. Sci. USSR **214/219** [1974] 705/7). – [4] G. Neshvad, R. M. G. Roberts, J. Silver (J. Organometal. Chem. **236** [1982] 349/58). – [5] M. C. Clemance, R. M. G. Roberts, J. Silver (J. Organometal. Chem. **247** [1983] 219/22).

5.2.3.5.1.3 1,1′-Dibenzoylferrocene Fe(C_5H_4–CO–C_6H_5)$_2$

5.2.3.5.1.3.1 Preparation and Formation

The preparation of Fe(C_5H_4–CO–C_6H_5)$_2$ is effected predominantly by a Friedel-Crafts benzoylation of ferrocene with C_6H_5COCl in the presence of $AlCl_3$ (mole ratio 1/~2/~2 [3, 31, 32], 1/3/3 [8], or 1/2.8/2.45 [15]) in CH_2Cl_2 at room temperature [3, 15, 31] or at the boiling point [8, 32]. The ferrocene can either be added in small portions to the $C_6H_5COCl + AlCl_3$ [3, 15, 31] or inversely $C_6H_5COCl + AlCl_3$ can be added dropwise to the ferrocene [32, 35]. The yield is around 80 to 90% [3, 31, 32]. Boiling the mixture after adding the reactants to each other lowers the yield [3]. The product obtained with the reactants in a 1/~2/~2 molar ratio is free from $C_5H_5FeC_5H_4$–CO–C_6H_5 [3, 8] and tri- and polybenzoylated products [3], that obtained using a 1/3/3 ratio is free of $C_5H_5FeC_5H_4$–CO–C_6H_5 [3, 8]. The reaction can also be conducted in boiling CS_2 by addition of ferrocene + C_6H_5COCl to $AlCl_3$ (mole ratio 1/>2/<3 [9] or 1/>2/<4 [2]). The yields are 52 to 53% [2, 9], or 70% when the reaction is conducted at room temperature with a 1/1.7/1.9 mole ratio [1]. The benzoylation of ferrocene with $(C_6H_5CO)_2O$ in the presence of 85% H_3PO_4 as a catalyst is described by [29].

The compound is also formed by oxidation of Fe(C_5H_4–CHOH–C_6H_5)$_2$ with $[(C_6H_5)_3SiO]_2CrO_2$ in benzene at 50°C with stirring in the dark for 24 h in 85% yield [33]. The compound is produced from Fe(C_5H_4–CHOH–C_6H_5)$_2$ on an Al_2O_3 column in contact with air in 76.8% yield. In contrast on a SiO_2 column only traces were found along with a 55% yield of I [10].

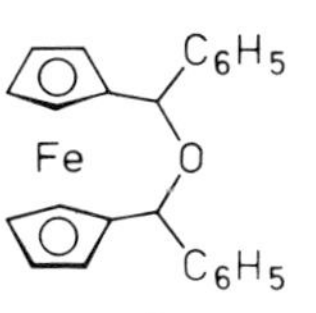

I

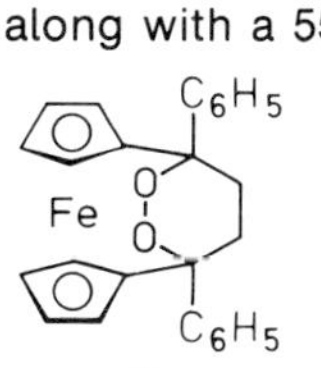

II

References on pp. 57/8

Trace amounts of $Fe(C_5H_4–CO–C_6H_5)_2$ also form during the photosensitized oxidation of $Fe(C_5H_4–C(C_6H_5)=CH_2)_2$ with molecular O_2 in CH_2Cl_2 or CH_3OH with Na-copperchlorophillin as a sensitizer, or in CH_2Cl_2 with Na vapor lamp irradiation in the presence of Na-copperchlorophillin and 1,4-diazobicyclo[2.2.2]octane, or in the dark by the autoxidation catalyzed by WCl_6 of $Fe(C_5H_4–C(C_6H_5)=CH_2)_2$ in CH_2Cl_2 [30]. It occurs as a by-product in thin layer chromatographic processing on alumina of the hydrolysis products of $C_6H_5–CH(Li)–C_5H_4FeC_5H_4–CH(OLi)–C_6H_5$ (obtained by cleavage of I with Li in tetrahydrofuran) [16].

The ion $[Fe(C_5H_4–CO–C_6H_5)_2]^+$ (see also p. 59) occurring in the mass spectrum of II does not form by thermal reaction from II but by splitting off of $CH_2=CH_2$ from $[II]^+$ by a reverse Diels-Alder reaction [41].

Purification is effected by column chromatography on Al_2O_3 with ether as the eluent [33, 35] or by recrystallization from CH_3OH [9], C_2H_5OH [15], C_6H_6 [2], C_6H_6/hexane [31], or hexane [19].

The **separation** from ferrocene and $C_5H_5FeC_5H_4–CO–C_6H_5$, the contaminants most often occurring in the Friedel-Crafts benzoylation of ferrocene, can be conducted by thin layer chromatography on Al_2O_3 in C_6H_6 or C_6H_6/absolute C_2H_5OH (30/1) [37] or by high pressure liquid chromatography on SiO_2 with ether/CH_3OH (30/1), cyclohexane/ether (5/4), or hexane/$CHCl_3$ (1/1) [39]. The elution sequence is ferrocene, monobenzoyl compound, and dibenzoyl compound. For thin-layer chromatography on SiO_2 gel in C_6H_6/C_2H_5OH (15/1) see [43, 44]. Gel permeation chromatography on cross-linked polyacryloylmorpholine gels such as Enzacryl K0 and K2, in $(CH_3)_2NCHO$ or $CHCl_3$, is described in [38]. The separation from ferrocene is possible on Enzacryl Gel K2 in $CHCl_3$, with the dibenzoyl compound eluting first. The R_f values are 0.0 in H_2O, 0.04 in pentane, 0.84 in 30/1 ether/CH_3OH for the thin layer chromatography on microcrystalline cellulose [40].

The identification of polycrystalline samples by means of X-ray diffraction photographs is covered in [6].

5.2.3.5.1.3.2 Physical Properties

Red [2, 9], dark red [1, 15], ruby red [42], purple [3], needle-like [3, 9, 42] crystals, that melt (in °C; solvent for recrystallization in brackets) at 100 to 103 (C_2H_5OH) [15], 101 to 102 [1], 102 to 103 [29], 104.0 to 105.0 [13, 14], 104 to 106 (CH_3OH) [9], 105 (hexane) [19], 105 to 106 (C_6H_6) [2, 21], 106.5 ± 0.5 [32, 35], 106 to 107 (C_6H_6/hexane) [28, 31], 106.5 to 107.5 [3].

Vapor Pressure. Measurements of the vapor pressure by a torsion-effusion technique between 358 and 381 K gave the vapor pressure equation, with P in kPa and T in K, $\log P = (9.29 \pm 0.24) - (4898 \pm 91)/T$ [35].

Molar heat capacity C_p in $cal \cdot mol^{-1} \cdot K^{-1}$. The following table presents for the temperatures t (±0.5°C) values $C_{p,\,meas}$ measured by differential calorimetry and values $C_{p,\,calc}$ calculated using the equation $C_{p,\,calc} = 94.17 + 0.693t$, with both $C_{p,\,meas}$ and $C_{p,\,calc}$ ±0.5 $cal \cdot mol^{-1} \cdot K^{-1}$ [32]:

t in °C	20.0	30.0	40.0	50.0	60.0	70.0	80.0
$C_{p,\,meas}$	108.2	114.5	121.8	128.9	136.3	142.9	149.1
$C_{p,\,calc}$	108.0	115.0	121.9	128.8	135.7	142.7	149.6

The **sublimation enthalpy** at 370 K calculated from the vapor pressure equation is $\Delta H^\circ_{subl} = 93.8 \pm 2.1$ kJ/mol. Calculated for 298 K from the 2nd law $\Delta H^\circ_{298,\,subl} = 109.3 \pm 6.0$ kJ/mol [35].

Thermodynamic Functions. The following table presents values for the gas phase calculated by the usual equations of statistical thermodynamics as a function of the temperature T in

K for the heat content ($H_T^\circ - H_{298}^\circ$) (±1 kJ/mol) and the free energy function $-(G_T^\circ - H_{298}^\circ)/T$ (±0.01 kJ·mol^{-1}·K^{-1}) [35]:

T in K	298.15	350	360	370	380	390
$H_T^\circ - H_{298}^\circ$	—	17.523	21.288	25.179	29.192	33.326
$-(G_T^\circ - H_{298}^\circ)/T$	0.5131	0.5170	0.5186	0.5203	0.5223	0.5243

Values for the solid compound, calculated from these data, the sublimation enthalpy at 298 K, and the C_p values are, ±0.015 kJ·mol^{-1}·K^{-1} [35]:

T in K	298.15	350	360	370	380	390
$-(G_T^\circ - H_{298}^\circ)/T$	0.3270	0.3328	0.3358	0.3388	0.3418	0.3447

The **dipole moment** μ from measurements of the dielectric constant determined in benzene [21]:

t in °C	20	30	45
μ in D	3.77	3.75	3.73

The implications regarding the molecular structure are discussed under "Crystal and Molecular Structure".

^{1}H NMR spectrum in $CDCl_3$: δ = 4.57 (H-3,4 of C_5H_4, apparently t; J = 1.93 Hz), 4.92 (H-2,5 of C_5H_4, apparently t; J = 1.93 Hz), both in ppm [12]. The line widths and the second moments derived from them for the solid state ^{1}H NMR spectrum of $Fe(C_5H_4\text{-}CO\text{-}C_6H_5)_2$ are reported by [7] for the range −196 to +25°C and of $Fe(C_5H_4\text{-}CO\text{-}C_6D_5)_2$ by [11] for the range −196 to +100°C. They make possible deductions concerning the reorientation process of the molecule [11].

^{57}Fe Mössbauer Spectrum. At 80 K referred to natural Fe the isomer shift δ = 0.53(1) mm/s and the quadrupole splitting Δ = 2.17(1) mm/s [34]. At 25°C δ = 0.702 ± 0.005 mm/s referred to $Na_2[Fe(CN)_5NO]\cdot 2H_2O$ and Δ = 2.185 ± 0.02 mm/s. Temperature dependence in mm·s^{-1}·K^{-1} is given by Δδ/ΔT = −0.043 ± 0.005 and ΔΔ/ΔT = −0.010 ± 0.003. The width of the left, narrower line of the doublet is Γ_{min} = 0.210 ± 0.01 mm/s. The right line is both some 10% broader and also somewhat more intense. Both effects can be attributed to an anisotropic Lamb-Mössbauer factor [27].

The **IR spectrum** in Nujol is shown in a figure in [13], in KBr for 666 to 5000 cm^{-1} in [6]. In KBr the ν(C=O) band is located at 1631 cm^{-1} [8]. In paraffin oil a shoulder appears at 1001 cm^{-1}, but no band around 1100 cm^{-1}, so that Rosenblum's rule is satisfied [20]. In CCl_4 bands occur at 1288 and 1652 cm^{-1}; addition of 5% CH_3OH to the solvent causes only a slight shift, to 1289 and 1648 cm^{-1}, respectively, but a distinct broadening of the bands, which indicates an exchange effect between the carbonyl groups and CH_3OH [19]. In dimethylsulfoxide bands occur at 718, 1155, 1264, 1700 [26], 1631 (νC=O) [22] cm^{-1}.

Electronic Absorption Spectrum and **Magnetic Circular Dichroism (MCD).** In isooctane $\lambda_{max}(\varepsilon_{max})$ = 247(22300), 276(12550), 354(1970), 470(700) nm [19]. For the spectrum in benzene see a figure in [9]. In cyclohexane $\lambda_{max}(\varepsilon_{max})$ = 471(747) nm. In the region of the d-d transition an S-shaped MCD band is observed with peaks at λ_1 = 520 and λ_2 = 443 nm ($\lambda_{\varepsilon=0}$ = 467 nm), which is partially overlapped by a Gaussian MCD curve with λ_{max} = 400 nm [25].

X-Ray photoelectron spectroscopy with Al Kα radiation gave 709.5 eV as the vertical ionization potential of the Fe-$2p_{3/2}$ band with a 1.8 eV half-width for the signal. A linear relationship exists between this half-width and the quadrupole splitting in the ^{57}Fe Mössbauer spectrum [23].

Crystal and Molecular Structure. $Fe(C_5H_4\text{-}CO\text{-}C_6H_5)_2$ forms monoclinic crystals, space group $P2_1/n\text{-}C_{2h}^2$ (No. 11), with the lattice parameters a = 11.69(2), b = 25.36(5), c = 6.27(1) Å, β = 90(1)°

References on pp. 57/8

[4, 5], or a = 11.60, b = 25.20, c = 6.28 Å, β = 90°05′ [14]. **Fig. 2** is a projection of the molecular structure on the a-b plane. In its upper part the atomic sites are given, while in the lower part the dense packing is illustrated [42]. The density calculated with Z = 4 [4, 5, 14] is 1.429 g/cm³, the measured density is 1.40 [42], 1.420 g/cm³ [14].

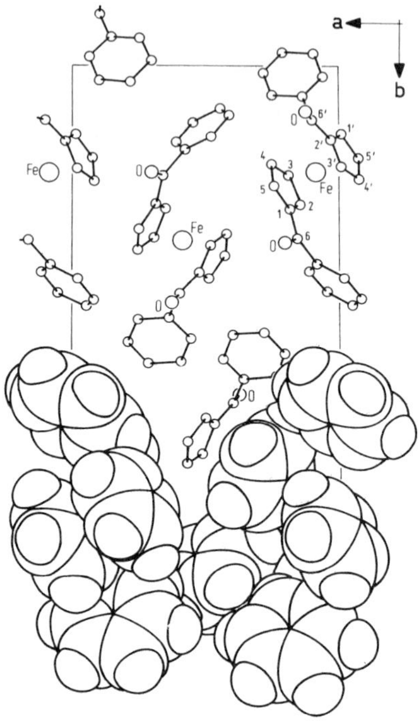

Fig. 2. Projection of the molecular structure of $Fe(C_5H_4{-}CO{-}C_6H_5)_2$ on the a-b plane [42].

The three-dimensional electron density distribution yields the following bond distances in Å: Fe–C = 2.05 ± 0.02; C–C = 1.41 ± 0.03 in the C_5H_4 ring, 1.39 ± 0.03 in the C_6H_5 ring, 1.52 ± 0.02 for C(1)–C(6); C–O = 1.21 ± 0.01. The C_6H_5CO groups are not found in the planes of the cyclopentadienyl rings but rotated from them by 60° around the C(1)–C(6) bond, in order to reduce the steric hindrance in the molecule. The molecule has a 1,2′-configuration in the crystal (see also I below) [4, 5, 42], which corresponds to a dihedral angle of φ = 108°; cf. [17, 18]. The activation energy has been calculated for the reorientation of the molecule in the crystalline state from the temperature dependence of the half-widths and second moments in the solid state ¹H NMR spectrum [11].

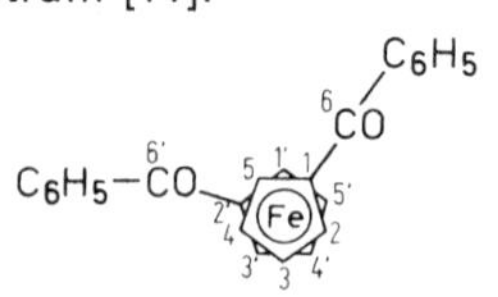

I

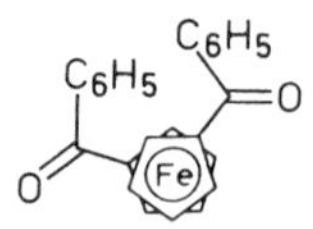

II

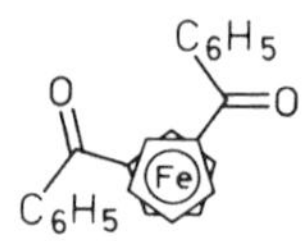

III

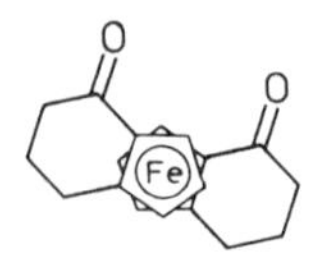

IV

Dissolved in benzene the free molecule can exist in the two rotationally isomeric conformations II and III. The correlation of the dipole moment with that of IV yields a preference for the racemic form II, in which the rotation of the cyclopentadienyl groups is nearly free, in contrast to the meso form III with slightly more restricted rotation [21], see also [24].

References on pp. 57/8

Dielectric relaxation measurements between 293 and 313 K at 7.5 to 38.2 GHz on the compound dissolved in p-xylene yield only one absorption in the GHz region, which can be assigned to the reorientation of the whole molecule and/or to the internal rotation around an axis perpendicular to the cyclopentadienyl rings. Debye-like behavior was observed. The macroscopic dielectric relaxation times are 63.2 ps at 293 K, 51.9 ps at 303 K, and 41.2 ps at 313 K. The Eyring diagram gives an activation enthalpy $\Delta H^{+}=2.3$ kcal/mol (9.6 kJ/mol). $\Delta S^{+}=-4.0\ cal\cdot mol^{-1}\cdot K^{-1}$ $(-16.7\ J\cdot mol^{-1}\cdot K^{-1})$. $\Delta G^{+}=3.47$, 3.51, and 3.55 kcal/mol (14.53, 14.70, 14.86 kJ/mol) at 293, 303, and 313 K, respectively [36].

References:

[1] A. N. Nesmeyanov, N. A. Vol'kenau (Dokl. Akad. Nauk SSSR **107** [1956] 262/4). – [2] R. Riemschneider, D. Helm (Chem. Ber. **89** [1956] 155/61). – [3] M. Rausch, M. Vogel, H. Rosenberg (J. Org. Chem. **22** [1957] 903/6). – [4] Yu. T. Struchkov [G. T. Stručkov], T. L. Khotsyanova [T. L. Khocjanova] (Kristallografiya **2** [1957] 382/3; Soviet Phys.-Cryst. **2** [1957] 376/7). – [5] G. T. Stručkov, T. L. Khocjanova (Acta Cryst. **10** [1957] 796/7).

[6] W. L. Baun (Anal. Chem. **31** [1959] 1308/11). – [7] N. Ockman (AD-229397 [1959] 1/62, 29; PB-147703 [1959] 1/59; U.S. Govt. Res. Rept. **34** [1960] 262). – [8] M. R. Rausch, E. O. Fischer, H. Grubert (J. Am. Chem. Soc. **82** [1960] 76/82). – [9] R. Riemschneider, D. Helm (Liebigs Ann. Chem. **646** [1961] 10/7). – [10] A. N. Nesmeyanov, I. I. Kritskaya, T. V. Antipina (Izv. Akad. Nauk SSSR Otd. Khim. Nauk **1962** 1777/83; Bull. Acad. Sci. USSR Div. Chem. Sci. **1962** 1685/90).

[11] A. Attalla (Diss. Univ. Cincinnati 1962 from Diss. Abstr. **24** [1963] 1843/4). – [12] M. D. Rausch, V. Mark (J. Org. Chem. **28** [1963] 3225/8). – [13] R. L. Schaaf, K. L. Rinehart (AD-403929 [1963] 1/80, 50). – [14] D. W. Fischer (Acta Cryst. **17** [1964] 619). – [15] J. E. Robertson, Colgate-Palmolive Co. (U.S. 3377248 [1965/68]; C.A. **69** [1968] No. 96878).

[16] A. N. Nesmeyanov, E. G. Perevalova, Yu. A. Ustynyuk, N. S. Prozorova, T. I. Tatashina, S. S. Churanov (Izv. Akad. Nauk SSSR Ser. Khim. **1966** 1646/8; Bull. Acad. Sci. Div. Chem. Sci. **1966** 1588/90). – [17] R. Riemschneider, W. Wucherpfennig (Z. Naturforsch. **21 b** [1966] 929/34). – [18] P. J. Wheatley (Perspect. Struct. Chem. **1** [1967] 1/40, 5, 13). – [19] A. M. Tarr, D. M. Wiles (Can. J. Chem. **46** [1968] 2725/31). – [20] A. Perjéssy, Š. Toma (Chem. Zvesti **23** [1969] 533/9).

[21] S. Sorriso, H. Lumbroso (Tetrahedron **26** [1970] 2695/701, [Erratum] 4674). – [22] L. H. Ali, A. Cox, T. J. Kemp (J. Chem. Soc. Chem. Commun. **5** [1972] 265/6). – [23] R. Gleiter, R. Seeger, H. Binder, E. Fluck, M. Cais (Angew. Chem. **84** [1972] 1107/8; Angew. Chem. Intern. Ed. Engl. **11** [1972] 1028/30). – [24] H. Lumbroso, C. Pigenet, H. L. Lentzner, W. E. Watts (Tetrahedron **28** [1972] 111/9). – [25] D. Nielson, D. Boone, H. Eyring (J. Phys. Chem. **76** [1972] 511/5).

[26] L. H. Ali, A. Cox, T. J. Kemp (J. Chem. Soc. Dalton Trans. **1973** 1468/75). – [27] E. Fluck, F. Hausser (Z. Anorg. Allgem. Chem. **396** [1973] 257/60). – [28] K. Yamakawa, M. Hisatome (J. Organometal. Chem. **52** [1973] 407/24). – [29] Bayer A.-G., V. Mues, W. Behrenz (Ger. Offen. 2711546 [1977/78]; C.A. **90** [1979] No. 1705). – [30] M. Hisatome, T. Hashiyama, K. Yamakawa (Tetrahedron Letters **1978** 3759/62).

[31] P. Canonne, G. Foscol, R. Harder (J. Organometal. Chem. **178** [1979] 331/42). – [32] M. Tomassetti, R. Curini, G. d'Ascenzo, G. Ortaggi (Thermochim. Acta **48** [1981] 333/41). – [33] J. Holeček, K. Handlíř, J. Klikorka, M. Nadvorník (J. Prakt. Chem. [4] **324** [1982] 345/8). – [34] G. Neshvad, R. M. G. Roberts, J. Silver (J. Organometal. Chem. **236** [1982] 349/58). – [35] M. Pelino, R. Gigli, M. Tomassetti (Thermochim. Acta **61** [1983] 301/5).

[36] E. Jakusek, P. Freundlich, H. A. Kołodziej, S. Sorriso (Advan. Mol. Relaxation Interact. Processes **24** [1982] 283/95 from C.A. **98** [1983] No. 82212). – [37] N. Ya. Mar'yashkin, T. P.

Vishnyakova, A. A. Koridze (Tr. Mosk. Inst. Neftekhim. Gazov. Prom. No. 72 [1967] 67/70). – [38] R. Epton, S. R. Holding, G. Marr, J. V. McLaren, G. J. Morgan (J. Organometal. Chem. **112** [1976] 89/94). – [39] D. T. Haworth, T. Liu (J. Chromatog. Sci. **14** [1976] 519/20). – [40] D. T. Haworth, T.-C. Liu (Separ. Sci. **11** [1976] 329/34).

[41] M. Hisatome, K. Yamakawa (Org. Mass. Spectrom. **13** [1978] 1/3). – [42] Yu. T. Struchkov [G. T. Stručkov] (Dokl. Akad. Nauk SSSR **110** [1956] 67/70). – [43] X.-J. Yang, Z.-Z. Xie, F.-S. Ni, S.-J. Shi (Gaodeng Xuexiao Huaxue Xuebao **1** [1980] 48/60 from C.A. **94** [1981] No. 219071). – [44] X. Yang, Z. Xie, F. Ni, S. Shi (Fundam. Res. Organometal. Chem. Proc. China-Japan-US Trilateral Semin. Organometal. Chem., Beijing 1980 [1982], pp. 765/9).

5.2.3.5.1.3.3 Solubility. Chemical Behavior

Solubility. $Fe(C_5H_4{-}CO{-}C_6H_5)_2$ is readily soluble in benzene, difficultly soluble in cold CH_3OH, practically insoluble in water [2] or concentrated hydrochloric acid [1, 2].

Electrochemical Behavior. Due to its height the formal redox potential cannot be determined by potentiometric titration with $K_2Cr_2O_7$ in 98% CH_3CO_2H/dilute $HClO_4$ (3/1) [9]. Electrochemical reduction in $(CH_3)_2NCHO$ gave the radical anion, $\mathbf{[C_6H_5{-}CO{-}C_5H_4FeC_5H_4{-}\dot{C}(C_6H_5){-}O^{(-)}]^{\bar{\cdot}}}$ [15], whose one line ESR spectrum yields a g value of 2.0086. There is no hyperfine splitting, and interannular electron transfer via the Fe is easy [16]. For the formation of this ketyl see also reduction with Na/K alloys.

The 30 eV **mass spectrum** has the $[M]^+$ peak at m/e = 394 (100% relative intensity) as the basic peak, with additional peaks at m/e = 366 (3.4%) and 289 (12%) [35].

Photolysis occurs on absorption of 354 and 470 nm light; its mechanism is still unclear. The photolysis is dependent on the solvent; apparently the interaction of the substance and solvent generates the photochemically unstable species. Solvents such as hydrocarbons, CCl_4, CS_2, or dimethyl phthalate are least favorable for the photodecomposition. While the photolysis in CH_3OH occurs very easily with the formation of $C_6H_5CO_2CH_3$ and a nonextractable, Fe-containing precipitate, possibly $(C_6H_5CO_2)_2Fe$, it proceeds only slowly in propan-2-ol. In general photodecomposition occurring in O- or N-containing solvents finally leads to a clear or weakly yellow solution and a finely pulverized or gelatinous precipitate of various shades of brown. In pyridine, $(CH_3)_2NCHO$, and $(CH_3)_2SO$ dark purple-brown solutions form, in contrast, but no precipitate [13]. On the other hand, according to [25] photoaquation for wavelengths > 280 nm proceeds independently of the solvent following a single reaction path in which the carbonyl convert into carboxylate groups; in $(CH_3)_2SO$ the ν(C=O) at 1641 cm^{-1} vanishes, $\nu(CO_2^-)$ at 1538 cm^{-1} appears. The intense purple color observed in moist $(CH_3)_2SO$, pyridine, or $(CH_3)_2NCHO$ is attributed to a secondary reaction which is caused by the presence of oxygen (from UV irradiation of the H_2O); see also [24]. In $(CH_3)_2CHOH/HCl$ a smooth photochemical reaction occurs with heterolytic splitting of the Fe–ring bond and formation of Fe^{II} and $C_5H_5{-}CO{-}C_6H_5$ [27].

With $C_6H_{6-n}(CH_3)_n$ ($\triangleq {}^6L$; $1 \leqq n \leqq 6$) in the ^6L as solvent when it is liquid or otherwise in nonane or decalin, **ligand exchange** takes place at 100 to 200°C in the presence of $AlCl_3$ (at least 2, better 4 mol per 1 mol $Fe(C_5H_4{-}CO{-}C_6H_5)_2$) to form after 5 h reaction $[Fe(C_6H_{6-n}(CH_3)_n)_2]^{2+}$ in a 60 to 80% yield [26, 32]. The optimum temperature range according to [40] lies at 115 to 130°C, since at lower temperatures the exchange does not occur [26, 32], at higher temperatures loss of the CH_3 groups and isomerization can occur [40]. See $Fe(C_5H_4{-}CO{-}CH_3)_2$, pp. 24/5, regarding the reaction mechanism.

$Fe(C_5H_4{-}CO{-}C_6H_5)_2$ like $Fe(C_5H_4{-}CO{-}CH_3)_2$ behaves differently than does ferrocene with respect to **diazonium salts**. Thus, not arylation but cleavage of the Fe–ring bond occurs with $[4{-}O_2N{-}C_6H_4{-}N_2]Cl$ in glacial acetic acid, during 100 min at room temperature and 150 min at 45 to 50°C [6]. According to [6] there is a 3% yield of a 1,2,3-oxadiazine derivative, see, however, p. 25.

The ferrocenium ion $\mathbf{[Fe(C_5H_4{-}CO{-}C_6H_5)_2]^+}$ forming on **oxidation** is even in very strongly acidic solution unstable and decomposes according to the equation

$$3\,[Fe(C_5H_4{-}CO{-}C_6H_5)_2]^+ \rightarrow 2\,Fe(C_5H_4{-}CO{-}C_6H_5)_2 + Fe^{3+} + 2\,C_5H_4{-}CO{-}C_6H_5.$$

The rate constant k of the reaction is dependent on the temperature and acid strength H_0 of the solution. At 25°C k (in s^{-1}) $= 4.84 \times 10^{-2}$ for $H_0 = 0$, 2.37×10^{-2} for $H_0 = -1$, 1.21×10^{-2} for $H_0 = -2$, and 2.68×10^{-3} for $H_0 = -3$ [36]. $AgBF_4$ does not oxidize the compound to the ferrocenium ion [21]. For its formation see also p. 54.

Reduction. $Fe(C_5H_4{-}CH_2{-}C_6H_5)_2$ forms smoothly on hydrogenolysis over Pt/C in $n{-}C_4H_9OH$ in a Parr bomb with 65 psi (448 kPa) H_2. Still better yields are obtained with Na in C_2H_5OH at 60 to 70°C [4], in the Clemmensen reduction with Zn amalgam and concentrated HCl in glacial acetic acid [1], with $NaBH_4$ in 1/8 mole ratio with addition of $AlCl_3$ in $CH_3O(CH_2)_2O(CH_2)_2OCH_3$ at 20°C (94% yield) [31], or with $LiAlH_4/AlCl_3$ in tetrahydrofuran [8]. On reduction with $(C_6H_5)_3SnH$ in the presence of CH_3COCl (mole ratio 1/2/4) in C_6H_6 the compound $C_6H_5{-}CO{-}C_5H_4FeC_5H_4{-}CH_2{-}C_6H_5$ (B, yield 21%) is obtained in addition to $Fe(C_5H_4{-}CH_2{-}C_6H_5)_2$ (A, yield 16%). These products are separable by layer chromatography on SiO_2. Even with excess reagent both A and B form. Only a really large excess after a sufficiently long reaction time finally gives essentially only A. Other reductants behaving like $(C_6H_5)_3SnH$ are $(C_4H_9)_3SnH$ and $(C_6H_5)_2SnH_2$, the latter of limited utility due to its instability [17, 29].

$Fe(C_5H_4{-}CHOH{-}C_6H_5)_2$ forms with $NaBH_4$ in $(CH_3)_2CHOH$ [10] or $LiAlH_4$ in ether/C_6H_6 [4, 30] as a mixture of the diastereoisomeric meso and racemic forms; cf. "Ferrocene" 7, 1980, p. 184. The smooth reduction observed by Nesmeyanov et al. [3] of $Fe(C_5H_4{-}CO{-}C_6H_5)_2$ with $LiAlH_4$ alone to $Fe(C_5H_4{-}CH_2{-}C_6H_5)_2$ may have been due to contamination of the $LiAlH_4$ with $AlCl_3$ [8].

Na/K alloys reduce the compound to the ketyl $[C_6H_5{-}CO{-}C_5H_4FeC_5H_4{-}\dot{C}(C_6H_5)O]^{\dot{-}}$ (see under "Electrochemical Behavior", p. 50) [1, 6]. The reduction with Mg/MgI_2 fails in ether/C_6H_6 even at the boiling point; preliminary experiments show that it is possible in higher boiling solvents [4].

Basicity. Like all ferrocenyl ketones $Fe(C_5H_4{-}CO{-}C_6H_5)_2$ also possesses an uncommonly high basicity. In CF_3CO_2H it dissolves smoothly with formation of the monoprotonated product $\mathbf{[Fe(C_5H_4{-}CO{-}C_6H_5)_2H]^+}$. This has an ^1H NMR spectrum in CF_3CO_2H with $\delta = 4.47$ (H-3,4 of C_5H_4), 4.55 (H-2,5 of C_5H_4), and 7.2 to 7.6 (C_6H_5) ppm. The ^{57}Fe Mössbauer spectrum in CF_3CO_2H at 80 K has $\delta = 0.53(1)$ mm/s referred to natural iron, $\Delta = 2.08(2)$ mm/s. The structure is probably that of I, in which the proton forms a hydrogen bond to the two O atoms in the benzoyl groups arranged in the syn configuration. This is supported by the low energy barrier for the partial rotation ("tick-tock" motion) around the C(1)–C(6) bond, which produces the magnetic equivalence of the H-2,5 protons at room temperature; it is lower than those of carbenium ions [39].

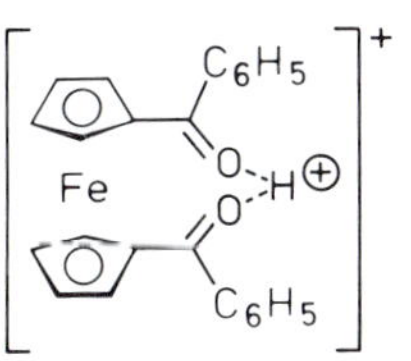

I

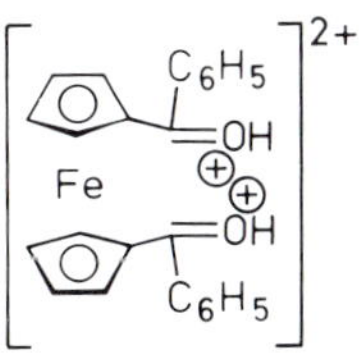

II

Diprotonation occurs in concentrated H_2SO_4 or CF_3SO_3H to form **$[Fe(C_5H_4{-}C(C_6H_5){=}OH)_2]^{2+}$** (see Formula II). Its ^1H NMR spectrum in 70% H_2SO_4 yields $\delta = 5.35$ (H-3,4 of C_5H_4), 5.52 (H-2,5 of C_5H_4), 7.2 to 8.0 (C_6H_5) ppm. In CF_3SO_3H $\delta = 5.53$ (H-3,4 of C_5H_4), 5.63 (H-2,5 of C_5H_4), and 7.2 to 8.0 (C_6H_5) ppm. The ^{57}Fe Mössbauer spectrum in 98% H_2SO_4 at 80 K gives $\delta = 0.55(1)$ mm/s referred to natural Fe, $\Delta = 1.85(2)$ mm/s [39].

No reaction appears to occur with $AgBF_4$, neither formation of a complex nor of the ferrocenium salt [21].

Additional Reactions on the Substituents. Heating with $NH_2OH \cdot HCl$ and KOH in C_2H_5OH causes nucleophilic addition on the CO groups and subsequent H_2O elimination to form $Fe(C_5H_4{-}C(C_6H_5){=}NOH)_2$ [4]. No solution polycondensation occurs with H_2NNH_2 or $H_2NCH_2CH_2NH_2$ in boiling benzene with $4\text{-}CH_3{-}C_6H_4{-}SO_3H \cdot H_2O$ as a catalyst [19, 22]; cf. in contrast $Fe(C_5H_4{-}CO{-}CH_3)_2$, p. 33.

Addition without subsequent dehydration occurs on passing HC≡CH through a mixture of a solution of the compound with Na in liquid ammonia. The reaction product formed consists of III and IV or only of III. If the reaction is conducted at −40°C with a 2/1 mole ratio of Na/ferrocene-derivative in tetrahydrofuran (THF) as the solvent, the yield of III is 30%. In THF/ether (1/3) at −60°C there is a 9.4% yield of III and 55% of IV. By using more sodium IV can be obtained in THF even at −40°C [12]. Preparative thin layer chromatography using Al_2O_3 with C_6H_6/absolute C_2H_5OH (50/1) separates $Fe(C_5H_4{-}CO{-}C_6H_5)_2$, III, and IV [11].

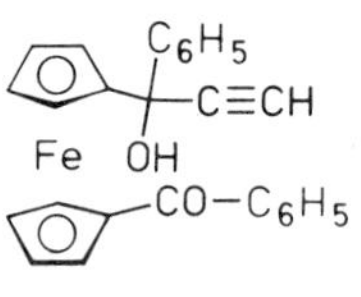

III

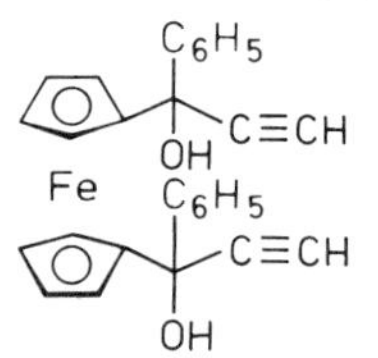

IV

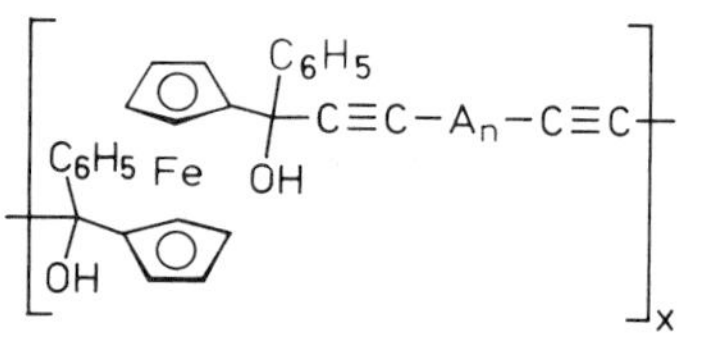

V

Na−C≡C−A_n−C≡C−Na, n = 0 or 1, A = 1,4-C_6H_4, in the presence of $NaNH_2$ in liquid NH_3/THF effects polymerization to V (idealized unit). The light brown product is soluble in CH_2Cl_2 and $(CH_3)_2NCHO$ and has a molecular weight between 1600 and 3000. The yield is 55% for n = 0 and 25% for n = 1 [23]. A 1/1/2 mole ratio of ferrocene/butadiyne/Na is necessary for the formation of V with n = 0. At a 1/4/8 mole ratio, in contrast, $Fe(C_5H_4{-}C(OH)(C_6H_5){-}C{\equiv}C{-}C{\equiv}CH)_2$ forms in 50% yield [18].

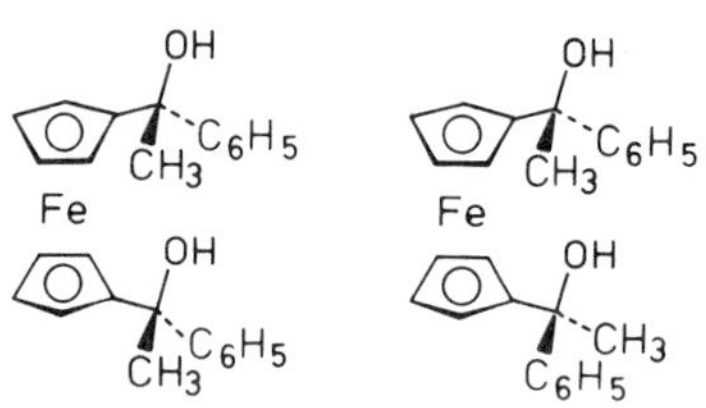

a b

VI

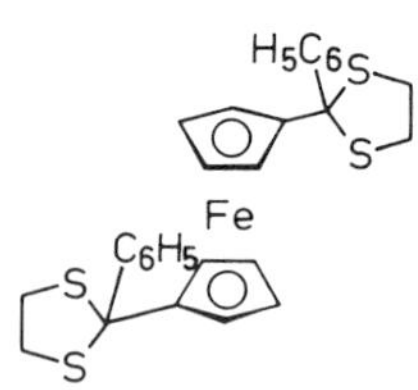

X

The reaction with CH_3MgBr in benzene/ether proceeds with distinct stereoselectivity. Hydrolysis of the reaction product gives the diastereoisomeric mixture VI, which contains the racemic and meso forms a and b in a 4/1 ratio. The reaction follows scheme in VII, wherein the primary step is the addition of the first CH_3MgBr molecule to the intermediate product VIII. The attack of the second CH_3MgBr on the interannular ring of VIII occurs preferably from the least hindered side with formation of the intermediate product IX, which yields VIa on hydrolysis [37]. $C_6H_5CH_2MgCl$ gives $Fe(C_5H_4{-}C(CH_2C_6H_5)OH{-}C_6H_5)_2$, m.p. 163 to 164°C [7]. The reaction with

C_6H_5MgI gives after hydrolysis $Fe(C_5H_4–C(C_6H_5)_2OH)_2$. The expected carbinol does not form with $(i-C_3H_7–O)_3Al$ but a diastereomeric mixture of $Fe(C_5H_4–CH(C_6H_5)–O–CH(CH_3)_2)_2$ [2].

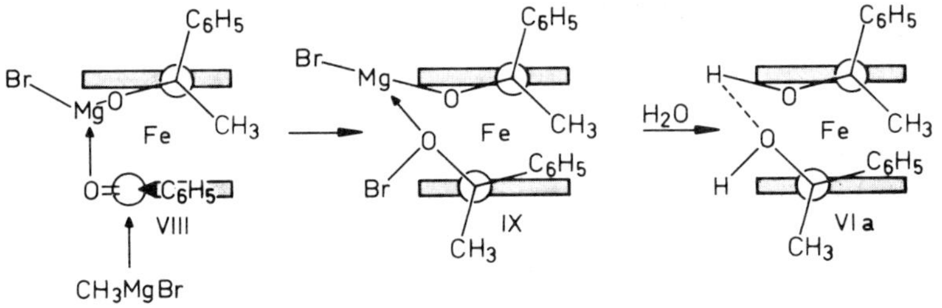

VII

The Knoevenagel condensation with cyclopentadiene in C_2H_5OH/C_2H_5ONa is found to yield $Fe(C_5H_4–C(C_6H_5)=C_5H_4\text{-cyclo})_2$ on use of a large excess of C_5H_6 (mole ratio 1/40), otherwise $C_6H_5–CO–C_5H_4FeC_5H_4–C(C_6H_5)=C_5H_4$-cyclo [14].

CH_3CN in liquid $NH_3/NaNH_2$ and subsequent neutralization with NH_4Cl give as product $Fe(C_5H_4–C(C_6H_5)OH–CH_2CN)_2$ in 75% yield [20].

The Wittig reaction with $(C_6H_5)_3P=CH_2$ (from $[CH_3P(C_6H_5)_3]I$ and $n-C_4H_9Li$) in tetrahydrofuran/hexane yields $Fe(C_5H_4–C(C_6H_5)=CH_2)_2$ [28]. The unstable dithiotolane X forms in benzene at 0°C with $HSCH_2CH_2SH$ on addition of dry HCl gas [34].

Substitution on the Cyclopentadienyl Rings. The C_6H_5CO groups considerably reduce the tendency of the cyclopentadienyl rings toward electrophilic substitution reactions. Thus, with SO_3 in $ClCH_2CH_2Cl$ no sulfonation occurs [5]. H/D exchange of the ring protons is observed in pyridine or $O(CH_2CH_2OCH_3)_2$ containing in each case 10 vol% D_2O. The distribution of the deuterium is not statistical, for which a speculative mechanism with a $\pi \rightarrow \sigma \rightarrow \pi$ rearrangement is held responsible [33, 38].

References:

[1] A. N. Nesmeyanov, N. A. Vol'kenau (Dokl. Akad. Nauk SSSR **107** [1956] 262/4). – [2] R. Riemschneider, D. Helm (Chem. Ber. **9** [1956] 155/61). – [3] A. N. Nesmeyanov, E. G. Perevalova, Z. A. Beinoravichute (Dokl. Akad. Nauk SSSR **112** [1957] 439/40). – [4] M. Rausch, M. Vogel, H. Rosenberg (J. Org. Chem. **22** [1957] 903/6). – [5] N. A. Nesmeyanov, B. N. Strunin (Dokl. Akad. Nauk SSSR **137** [1961] 106/8; Proc. Acad. Sci. USSR Chem. Sect. **136/141** [1961] 275/6).

[6] E. G. Perevalova, N. A. Simukova, T. V. Nikitina, P. D. Reshetov, A. N. Nesmeyanov (Izv. Akad. Nauk SSSR Otd. Khim. Nauk **1961** 77/83; Bull. Acad. Sci. USSR Div. Chem. Sci. **1961** 67/71). – [7] R. Riemschneider, D. Helm (Liebigs Ann. Chem. **646** [1961] 10/7). – [8] K. Schlögl, A. Mohar, M. Peterlik (Monatsh. Chem. **92** [1961] 921/6). – [9] E. G. Perevalova, S. P. Gubin, S. A. Smirnova, A. N. Nesmeyanov (Dokl. Akad. Nauk SSSR **155** [1964] 857/60; Dokl. Chem. Proc. Acad. Sci. USSR **154/159** [1964] 328/31). – [10] J. E. Robertson, Colgate-Palmolive Co. (U.S. 3377248 [1965/68]; C.A. **69** [1968] No. 96878).

[11] N. Ya. Mar'yashkin, T. P. Vishnyakova, A. A. Koridze (Tr. Mosk. Inst. Neftekhim. Gazov. Prom. No. 72 [1967] 67/70). – [12] T. P. Vishnyakova, N. Ya. Mar'yashkin, M. E. Elyashberg, I. A. Mel'nik, I. I. Zadko (Tr. Mosk. Inst. Neftekhim. Gazov. Prom. No. 72 [1967] 58/61; C.A. **68** [1968] No. 95919). – [13] A. M. Tarr, D. M. Wiles (Can. J. Chem. **46** [1968] 2725/31). – [14] T. H. Barr,

H. L. Lentzner, W. E. Watts (Tetrahedron **25** [1969] 6001/13). – [15] C. Elschenbroich, M. Cais (J. Organometal. Chem. **18** [1969] 135/43).

[16] J. J. McDonnell, G. Capen, R. Michaelson (Tetrahedron Letters **1969** 4251/4). – [17] H. Patin, R. Dabard (Tetrahedron Letters **1969** 4971/4). – [18] V. G. Shershun, T. P. Vishnyakova (Tr. Mosk. Inst. Neftekhim. Gazov. Prom. **1969** No. 3, pp. 55/8; C.A. **75** [1971] No. 49282). – [19] R. E. Gamper, P. T. Funke, A. A. Volpe (AD-717648 [1970] Govt. Rept. Announce U.S. **71** No. 6 [1971] 53; C.A. **75** [1971] No. 64382). – [20] A. A. Koridze, S. P. Gubin (J. Organometal. Chem. **22** [1970] 157/69).

[21] P. Carty, M. F. A. Dove (J. Organometal. Chem. **28** [1971] 125/32). – [22] R. G. Gamper, P. T. Funke, A. A. Volpe (J. Polym. Sci. Polym. Chem. Ed. **9** [1971] 2137/46). – [23] V. G. Shershun, T. P. Vishnyakova, Ya. M. Paushkin (Vysokomol. Soedin B **13** [1971] 181/2). – [24] L. H. Ali, A. Cox, T. J. Kemp (J. Chem. Soc. Chem. Commun. **5** [1972] 265/6). – [25] L. H. Ali, A. Cox, T. J. Kemp (J. Chem. Soc. Dalton Trans. **1973** 1468/75).

[26] D. Astruc (Tetrahedron Letters **1973** 3437/9). – [27] R. E. Bozak, H. Javaheripour (Chem. Ind. [London] **1973** 696/7). – [28] G. W. Gokel, J. P. Shepherd, W. P. Weber, H. G. Boettger, J. L. Holwick, D. J. McAdoo (J. Org. Chem. **38** [1973] 1913/8). – [29] H. Patin, R. Dabard (Bull. Soc. Chim. France **1973** 2756/9). – [30] K. Yamakawa, M. Hisatome (J. Organometal. Chem. **52** [1973] 407/24).

[31] A. N. Nesmeyanov, G. B. Shul'pin, M. V. Tolstaya, M. I. Rybinskaya (U.S.S.R. 555114 [1975/77]; C.A. **87** [1977] No. 135954). – [32] D. Astruc, R. Dabard (Tetrahedron **32** [1976] 245/9). – [33] M. B. Moronski (Diss. Southern Illinois Univ. 1977 from Diss. Abstr. Intern. B **39** [1978] 754/5). – [34] A. Ratajczak, B. Czech (Roczniki Chem. **51** [1977] 1735/40). – [35] M. Hisatome, K. Yamakawa (Org. Mass Spectrom. **13** [1978] 1/3).

[36] J. Holeček, K. Handlíř, J. Klikorka (Proc. Conf. Coord. Chem. **7** [1978] 65/70). – [37] P. Canonne, G. Foscolos, R. Harder (J. Organometal. Chem. **178** [1979] 331/42). – [38] D. W. Slocum, M. Moronski, J. Bencini, B. Conway (9th Intern. Conf. Organometal. Chem., Dijon 1979, Abstr. Papers No. P 33 T). – [39] G. Neshvad, R. M. G. Roberts, J. Silver (J. Organometal. Chem. **236** [1982] 349/58). – [40] D. Astruc (Diss. Univ. Rennes 1975 from [41]).

[41] D. Astruc (Tetrahedron **39** [1983] 4027/95, 4036).

5.2.3.5.1.3.4 Applications

$Fe(C_5H_4–CO–C_6H_5)_2$ acts as a UV radiation absorber in coatings made of synthetic resins or polymers [2]. Surfaces provided with these coatings possess a substantially enhanced resistance against degradation caused by the effects of UV radiation in space [1, 3]. Additions of 0.1 to 0.6 wt% to high pressure polyethylene increases its stability against UV radiation and oxidation [5].

$Fe(C_5H_4–CO–C_6H_5)_2$ has been used as the main component in hybrid propellant grains which are self-igniting with highly concentrated nitric acid as the oxidizer [4]. It represents an effective anti-knock agent, that in comparison with ferrocene forms less solid residue and possesses a smaller solubility in the fuel [6]. It strengthens synergistically the insecticidal and acaricidal actions of carbamates, carboxylic acid esters, phosphoric acid esters, cycloalkanes, or tin compounds [7].

References on p. 63

References:

[1] J. J. Mattice (Off. Dig. Fed. Soc. Paint Technol. **34** [1962] 602/21, 617). – [2] R. M. van Vliet (U.S. 3287314 [1962/66]; C.A. **66** [1967] No. 29600). – [3] R. G. Schmitt, R. C. Hirt (J. Appl. Polym. Sci. **7** [1963] 1565/80). – [4] Nitrochemie GmbH (Brit. 1282271 [1964/72]; C.A. **78** [1973] No. 6089). – [5] Z. Beinoravichute (Mater. Vses. Simp. Vopr. Proizvod. Primen. Trub Detalei Truboprovodov Polietilena, Vilnyus 1966, pp. 125/9; C.A. **68** [1968] No. 30546).

[6] V. Veselý, Š. Toma, J. Gurský, E. Patzelt (Ropa Uhlie **15** [1973] 194/7; C.A. **79** [1973] No. 81199). – [7] Bayer A.-G., V. Mues, W. Behrenz (Ger. Offen. 2711546 [1977/78]; C.A. **90** [1979] No. 1705).

5.2.3.5.1.4 Other Ketones of the Type $Fe(C_5H_4{-}CO{-}R)_2$

Protonation products and the salts $Na_2[Fe(C_5H_4COCHCHO)_2]$ (see "Further information" for No. 44, p. 77) and $Na_2[Fe(C_5H_4COCHCOC_6H_5)_2]$ (see "Further information" for No. 48, p. 80) are covered here in connection with the parent ferrocene ketone. All other ions and complexes derived from these ketones are treated in Section 5.2.3.5.1.8, pp. 153/5.

The preparation of the compounds listed in Table 3 takes place predominantly via the following methods:

Method I. Friedel-Crafts acylation of ferrocene:

a) Ferrocene is added as the solid in small portions [99] or dropwise in solution in CH_2Cl_2 [10, 110] or $ClCH_2CH_2Cl$ [86] with ice bath cooling [34] to the complex formed from RCOCl and $AlCl_3$, mole ratios 1/1/1 [110], <0.5/1/1 [10, 99] or 0.5/1.4/1.3 [48, 49], in CH_2Cl_2 or $ClCH_2CH_2Cl$. The reaction is continued with stirring at room temperature [10, 49, 99]. The low reaction temperature yields only a little tarry contamination in the product, which by use of an RCOCl/ferrocene ratio >2 is also free of $C_5H_5FeC_5H_4{-}CO{-}R$ [10].

b) Same as for Method Ia, except inverted with RCOCl + $AlCl_3$ in CH_2Cl_2 being added slowly to ferrocene in CH_2Cl_2 [75].

c) Same as for Method Ia, except with use of boiling temperatures instead of 0°C to room temperature [3, 20].

d) Same as Method Ic, but with RCOCl in CH_2Cl_2 being added to ferrocene + $AlCl_3$ in CH_2Cl_2 [71, 97].

e) $AlCl_3$ is added to a mixture cooled to 0 to 10°C of ferrocene + RCOCl in $ClCH_2CH_2Cl$. The reaction is continued at between 0 and 20°C [11, 17].

f) Ferrocene, RCOCl, and $AlCl_3$ in mole ratios of 1/>2/>2 [1, 4], 1/2/3 [116], 1/>2/3.5 [25], or 1/3/3 [61] in CS_2 are reacted for 20 h (Nos. 17, 34, 49, 50) [116] or 30 min with mild heating (Nos. 1 and 2) [1, 4] or at boiling temperatures for 4 d (No. 36) [61].

g) Same as Method Ic, but with RCOCl replaced by $(RCO)_2O$ [3].

h) Same as Method Ie, but with RCOCl replaced by $(RCO)_2O$ [11].

Method II. Catalytic hydrogenation of $Fe(C_5H_4{-}CO{-}CH{=}CHR')_2$, with R' = H for No. 1, CH_3 for No. 2, C_6H_5 for No. 18,

a) using Pd/C in benzene [55, 59] or absolute ethanol [16],

b) using Pt in ethanol [59].

Method III. From $Fe(C_5H_4{-}CO{-}CH_3)_2$:

a) By a base-catalyzed aldol condensation (cf. Section 5.2.3.5.1.1.5.6.2, pp. 39/40) at room temperature [16] in a $C_2H_5OH/H_2O/MOH$ solution, M = Na [10] or K [22], with R'CHO, where R' = C_6H_5 for No. 20, $C_6H_4CH_3$-4 for No. 21, $CH{=}CHC_6H_5$ for No. 22, $C_6H_4O(CH_2)_2OH$-4 for No. 41, with exclusion of light and radiation for Nos. 22 and 41 [43].

b) By oxidation with $C_6H_5I(O_2CCH_3)_2$ in CH_3OH/KOH at 0°C to room temperature and hydrolysis of the reaction product with 4-$CH_3C_6H_4SO_3H$ in acetone [105].

c) By oxidation with SeO_2 in boiling dioxane/H_2O [102].

d) By a Claisen condensation with acid esters $CH_3CO_2C_2H_5$ for No. 46, $C_2H_5CO_2C_2H_5$ for No. 47, and $C_6H_5CO_2CH_3$ for No. 48, in the presence of Na [118], KNH_2 [12, 19], or $NaNH_2$ [12, 74, 119], initially in liquid NH_3 and subsequently in ether [12, 19, 74] (cf. Section 5.2.3.5.1.1.5.6.2, p. 41) or in C_6H_6 [119] or without an additional solvent [116].

e) By acetylation with $(CH_3CO)_2O$ in the presence of BF_3 and subsequent treatment with NaF and HCl [68].

Purification. The compounds are purified by recrystallization [10] from hydrocarbons (Nos. 1 to 4, 10, 13, 17, 26, 33 to 35, 46, 50) [3, 4, 11, 17, 25, 30, 34, 35, 37, 74, 87, 99, 116], chlorinated-hydrocarbon/hydrocarbon mixtures (Nos. 19, 28, 30, 31, 36, 46, 47) [19, 22, 55, 61, 86, 118], alcohols (Nos. 12, 18, 20, 21, 26, 41) [15, 16, 23, 40, 41, 43, 45, 75, 81], acetone (No. 48) [19], ether or ether/hydrocarbon mixtures (Nos. 1 and 2) [45, 55, 59, 99]. The separation of the compounds Nos. 1, 2, 10, 15 to 17, 25, 42, and 50 from the reaction product is effected by column chromatography on Al_2O_3 [33, 44, 55, 59, 71, 97, 98, 116], that of Nos. 28, 29, and 45 by column chromatography on SiO_2 [59, 101], and that of No. 36 on Florisil [61]. Due to splitting off of HCl Al_2O_3 is unsuitable for Nos. 28 and 29 [1, 55, 59]. Gas-liquid chromatography between 100 and 200°C is unsuited for separation of the positional isomeric mixtures of $FeC_{10}H_8(CO{-}R)_2$ [109].

Identification. Polycrystalline samples of Nos. 5 to 9 can be identified by their X-ray powder diffraction patterns [13].

Electrochemical Behavior. The redox reaction – determined for Nos. 17, 34, 49, and 50 – is irreversible because the electron-withdrawing groups on cyclopentadienyl rings lead to destabilization of the ferrocene molecule. Under this condition some nucleophilic species, like H_2O (a contaminant in the CH_3CN), attack positively charged Fe atoms after oxidation during cyclic voltammetry. As a result the positive charge on Fe is decreased, thereby rendering its reduction difficult. The greater hyperconjugative donation of electron density to the cyclopentadienyl rings is responsible for the facile oxidation of Nos. 17 and 50 as compared to Nos. 34 and 49, respectively [116]. For the electrochemical behavior of No. 46 see p. 79.

Chemical Behavior. Clemmensen reduction with zinc amalgam in boiling hydrochloric acid can convert the CO groups into CH_2 groups. The reaction has been carried out with Nos. 1 [4], 2 [4, 67], 3 [30], 4 [20], 5 to 9 [10], 10 [11], 11 [18], 12 [11, 41], 13 [18], 14 [18], 17 [44, 110], 26 [11], 30, 31 [86], and 36 [61]. The compounds $Fe(C_5H_4{-}CH_2{-}R)_2$ with R = C_2H_5, n-C_3H_7, and n-C_4H_9 can be obtained smoothly and without problems from the compounds No. 1 to 3 with $LiAlH_4/AlCl_3$ in tetrahydrofuran [26, 30]. In contrast the catalytic hydrogenation with Pt/C succeeds only with No. 1 but not with the higher homologs No. 5 to 9, not even under more drastic conditions [10].

Applications. The compounds No. 20, 32, or 40 together with a benzophenone derivative [76] or α-phenylindole [88] are used as prodegradants for the accelerated decomposition by UV or sunlight of thermoplastics, primarily polyethylene and polystyrene [76] or polyvinylchloride [88]. The benzophenone compound serves for initiation of the degradation process, Nos. 20,

32, or 40 for its continuation [76]. The thermal stability of polyvinylchloride is not impaired by the additives [88]. Compounds No. 4, 11, 12, and 26 can also be admixed in 0.01 to 1 wt% quantities as prodegradants of polyolefins [73].

Exposure to the sun or a 500 W lamp will produce a white negative image on a blue to red background on baryta coated paper on which 0.5% solutions in ethanol of Nos. 22 and 41 have been absorbed and then dried. The light forms a new compound with a violet-blue to red color which is less soluble in ethanol than the parent compound, so that treatment with ethanol will fix the picture [40, 43].

Compounds No. 12 [11], 26 [91, 92], 11, 13, and 14 [17] are suitable for treatment of iron deficiency anemia in humans or animals, the last three particularly for oral use: they do not exhibit the drawbacks of preparations used up to now [17]. Nos. 14 and 26, however, are inferior to $C_5H_5FeC_5H_4$–CO–$CH_2CH(CH_3)CH_2C(CH_3)_3$ in their ability to increase the Fe content of the liver usable for blood formation [91, 92]. For applications of Nos. 1, 3, 7, 17, 28 see "Further information", pp. 71/81.

Table 3

Ketones of the Type $Fe(C_5H_4$–CO–$R)_2$.
Further information for numbers preceded by an asterisk is given at the end of the table, pp. 71/81.
For abbreviations and dimensions see p. VIII.
$E_{1/2}$ = half wave redox potential measured by cyclic voltammetry in CH_3CN (1 M $NaClO_4$) on a Pt anode vs. Ag/$AgNO_3$ cathode.

No.	R in $Fe(C_5H_4$–CO–$R)_2$	method of preparation (yield in %), properties, and remarks	Ref.
compounds with R containing only C and H:			
*1	C_2H_5	Ia (98), If (40), IIa (49), IIb see also No. 15, p. 74, and "Further information", p. 71 orange crystals, m.p. 52 to 53°, 52 to 54°, 53 to 54°, 62 to 63° ^{1}H NMR ($CDCl_3$): 1.17 (d, CH_3), 2.64 (q, CH_2; J = 6.9), 4.46 (t) and 4.77 (t) (C_5H_4) IR (Nujol): 1027, 1103, 1770, see "Further information", p. 71 UV (isooctane): $\lambda_{max}(\varepsilon)$ = 196 (24000), 224 (26000), 260 (10400), 318 (1600), 350 sh (690), 455 (350) UV (ethanol): 331 (1680), 464 (495), see "Further information", p. 71	[4, 38, 45, 49, 55, 59, 64, 65, 99]
*2	C_3H_7-n	Ia (77), Ic (59), If (40), Ig, IIb (82) see also No. 16, p. 74 lustrous orange to yellow crystals, m.p. 72 to 74°, 73.0 to 74.5°, 74 to 75° ^{1}H NMR (CCl_4): 4.34 (H-3,4 of C_5H_4), 4.63 (H-2,5 of C_5H_4), see "Further information", p. 72 IR ($CHCl_3$): 3099 to 3115 (νC–H of C_5H_4), see "Further information", p. 72 UV (isooctane): $\lambda_{max}(\varepsilon)$ = 196 (24000), 224 (26000), 260 (10400), 318 (1600), 350 sh (690), 455 (360) UV (isooctane): 318 (1580), 355 sh, 455 (360), see "Further information", p. 72	[3, 4, 31, 34, 38, 45, 50, 59, 64, 67]

Table 3 (continued)

No.	R in $Fe(C_5H_4{-}CO{-}R)_2$	method of preparation (yield in %), properties, and remarks	Ref.
*3	C_4H_9-n	Ib (86), see "Further information", p. 73 m.p. 33 to 36°	[30]
*4	C_5H_{11}-n	Ia (67), Ic (53.5) rust-red needles, m.p. 43 to 45°; orange solid, m.p. 38 to 39°	[3, 20, 48]
*5	C_7H_{15}-n	Ia (65) red-orange crystals, m.p. 54.8 to 56.0° ^{57}Fe Mössbauer data (78 K): $\delta = 0.31 \pm 0.02$ (^{57}Co/Cu), $\Delta = 2.22 \pm 0.05$, see "Further information", p. 73	[10, 42]
6	C_9H_{19}-n	Ia (41) red-orange crystals, m.p. 68.6 to 69.8° see also "Chemical Behavior", p. 64	[10]
*7	$C_{11}H_{23}$-n	Ia (44) red-orange crystals, m.p. 76.6 to 77.1° IR (KBr): see figure in [13]	[10, 13]
8	$C_{12}H_{25}$-n	Ia (15) red-orange crystals, m.p. 80.0 to 80.4° ^{57}Fe Mössbauer data (78 K): $\delta = 0.31 \pm 0.02$ (^{57}Co/Cu), $\Delta = 2.23 \pm 0.05$ IR (KBr): see figure in [13] see also "Chemical Behavior", p. 64	[10, 42]
9	$C_{15}H_{31}$-n	Ia (12) salmon-colored solid, m.p. 82.4 to 83.4° see also "Chemical Behavior", p. 64	[10]
*10	$CH(CH_3)_2$	Ie, see "Further information", p. 73 yellow-orange platelets, m.p. 49 to 51°, 50 to 51°, 53°	[11, 33, 80]
11	$CH_2CH(CH_3)_2$	Ie (54) m.p. 61° see also "Chemical Behavior" and "Applications", pp. 64 and 65	[17]
*12	$C(CH_3)_3$	Ie (31), Ih, see "Further information", p. 74 m.p. 126°, 127° ^{57}Fe NMR ($CDCl_3$): 526.1 ± 2.5, see "Further information", p. 74	[11, 15, 41, 103, 108]
*13	$CH_2C(CH_3)_3$	Ie (54), see "Further information", p. 74 m.p. 94°	[17]
14	$CH_2CH(CH_3)CH_2C(CH_3)_3$	Ie (25) m.p. 55° see also "Chemical Behavior" and "Applications", pp. 64 and 65	[17]

Table 3 (continued)

No.	R in $Fe(C_5H_4-CO-R)_2$	method of preparation (yield in %), properties, and remarks	Ref.
*15	$CH=CH_2$	see No. 28, p. 76 m.p. 71 to 71.5°, see "Further information", p. 74 ^{1}H NMR ($CDCl_3$): 4.52(t) and 4.80(t) (C_5H_4), 5.3 to 7.2 (ABX pattern, $CH=CH_2$) UV (C_2H_5OH): $\lambda_{max}(\log \varepsilon) = 230(4.30)$, 280(4.07), 3.64(3.33)	[1, 55]
*16	$\overset{7}{C}H=\overset{8}{C}H-CH_3$-(E)	see No. 29, p. 76 red solid, m.p. 157 to 158° ^{1}H NMR ($CDCl_3$): 1.98(dd, CH_3; J = 6.3, 1.0); 4.50(t) and 4.81(t) (C_5H_4); 6.43(dq, H-7; J = 15.0, 1.0), 7.12(dq, H-8; J = 15.0, 6.3) IR (Nujol): 1615 (νC=C), 1664 (νC=O) UV (ethanol): $\lambda_{max}(\varepsilon) = 358(2400)$, 478(1030)	[59]
*17	$CH_2C_6H_5$	Ia (97, 60, 50), If (69), see "Further information", p. 75 red to red-orange platelets, m.p. 143 to 144°, 146°, 146 to 147°, 148°, 150° ^{1}H NMR ($CDCl_3$): 3.96(s, CH_2); 4.43(m) and 4.76(m) (C_5H_4); 7.36(s, C_6H_5), see "Further information", p. 75 IR (KBr): 810 (δC–H of C_5H_4), 1590 (νC=C of C_6H_5), 1640 (νC=O) IR (CCl_4): 1673 (νC=O), see "Further information", p. 75 $E_{1/2}$ = 608 mV, see also p. 64 mass spectrum: m/e = 326 $[M]^+$, see "Further information", p. 75	[37, 44, 98, 99, 110, 111, 116, 117]
18	$CH_2CH_2C_6H_5$	IIa (51), see also No. 20, p. 75 red-orange needles, m.p. 130 to 130.5° IR (CCl_4): 887, 991, 1307, 1399, 1477, 1704 (νC=O) for chemical behavior see No. 19, p. 75	[16, 22]
*19	$CH(CH_2C_6H_5)_2$	see "Further information", p. 75 yellow needles, m.p. 185 to 186° IR: 704, 741, 754, 1397, 1689 (νC=O)	[22]
*20	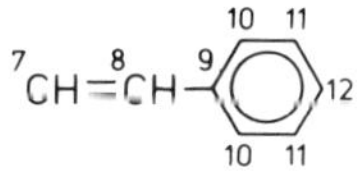	Ib (88), IIIa (75) red needles, m.p. 180.5 to 182°, 208 to 210° ^{1}H NMR ($CDCl_3$): 4.61(t) and 4.94(t) (C_5H_4), 6.03 to 7.94 (m, $CH=CHC_6H_5$) ^{13}C NMR (CH_2Cl_2): 71.4(C-2,5), 74.2(C-3,4), 82.2(C-1), 122.8 and 141.5(C-7, C-8), 128.5 and 129.2(C-10, C-11), 130.4(C-12), 135.0(C-9), 191.6(C=O) ^{13}C NMR (conc. H_2SO_4): see "Further information", p. 75 IR (KBr): 685, 699, 980, 1290, 1605, 1658, 1667 (νC=O) IR (Nujol): 1664 (νC=O) mol.wt: 462 (meas.), 446.3 (calc.)	[16, 22, 75, 83, 84]

Table 3 (continued)

No.	R in $Fe(C_5H_4–CO–R)_2$	method of preparation (yield in %), properties, and remarks	Ref.
*21	$CH{=}CH–C_6H_4–CH_3$-4-(E)	IIIa (38) red crystals, m.p. 212 to 214° ^{1}H NMR ($CDCl_3$): 2.31 (s, CH_3), 4.29 (m, H-3,4 of C_5H_4), 4.64 (m, H-2,5 of C_5H_4), 7.00 (d, CH=; J=16), 7.15 (d, 2H of C_6H_4; J=8), 7.49 (d, 2H of C_6H_4; J=8), 7.76 (d, =CH; J=16) IR (KBr): 965 (trans CH=CH), 1650 (νC=O) mass spectrum: m/e=474 $[M]^+$	[104]
22	$CH{=}CHCH{=}CHC_6H_5$	IIIa (73) sensitive to radiation, visible and UV light; see also "Applications", p. 65	[40, 43]
23	$C_6H_4–CH_3$-4	m.p. 175 to 177° dipole moment: $\mu(C_6H_6)$ = 3.96 (20°), 4.03 (30°), 3.98 (45°) D	[66]
*24	$C_6H_4–C_6H_5$-4	Ia (54)	[100]
25	C_3H_5-cyclo	Id (39) red granules, m.p. 100 to 102° ^{1}H NMR ($CDCl_3$): 0.8 to 1.3 (m, 8H), 2.0 to 2.4 (m, 2H), 4.53 (t, 4H), 4.85 (t, 4H) IR (Nujol): 1650 (νC=O)	[71]
*26	C_6H_{11}-cyclo	Ia (92), Ie m.p. 135°, 135 to 135.5°, 135 to 136° ^{1}H NMR ($CDCl_3$): 1.1 to 2.1 (m, 20H) and 2.45 to 3.1 (m, 2H) (C_6H_{11}); 4.5 (m) and 4.8 (m) (C_5H_4) IR (KBr): 810 (δC–H of C_5H_4), 1645 (νC=O)	[11, 23, 99]
compounds with R containing C, H, and halogen:			
27	$CHCl–CH_3$	small amounts enhance the thermal stability of organopolysiloxanes	[6]
*28	CH_2CH_2Cl	Ia (4), If, see No. 15, p. 74 orange-yellow solid, m.p. 117 to 121°, 119 to 121°	[1, 7, 55, 59]
*29	$CH_2CHCl–CH_3$	Ia (75) orange solid, m.p. 98 to 99° ^{1}H NMR ($CDCl_3$): 1.63 (d, CH_3; J=6.0), 3.09 (m, CH_2), 4.58 (m) and 4.82 (m) (C_5H_4 + CHCl) IR (Nujol): 1669	[59]

Table 3 (continued)

No.	R in $Fe(C_5H_4-CO-R)_2$	method of preparation (yield in %), properties, and remarks	Ref.
30	$(CH_2)_4Cl$	Ia (52) yellow crystals, m.p. 80° ^{1}H NMR ($CHCl_3$): 1.67 to 2.05 (m, 8 H, CH_2), 2.51 to 2.93 (m, 4 H, CH_2), 3.47 to 3.67 (m, CH_2Cl); 4.53 (t) and 4.80 (t) (C_5H_4) IR (KBr): 482, 500, 520, 535, 550, 645, 710, 730, 753, 765, 790, 835, 889, 973, 1025, 1050, 1070, 1088, 1230, 1255, 1280, 1300, 1340, 1380, 1405, 1452, 1670 (νC=O), 2870, 2890, 2950, 3120 for "Chemical Behavior" see p. 64	[86]
31	$(CH_2)_6Cl$	Ia (62) red-brown crystals, m.p. 77° ^{1}H NMR ($CHCl_3$): 1.23 to 2.05 (m, 16 H, CH_2), 2.54 to 2.92 (m, 4 H, CH_2), 3.88 (t, CH_2Cl); 4.72 (t) and 5.02 (t) (C_5H_4) IR (KBr): 500, 535, 550, 645, 715, 720, 740, 755, 815, 828, 840, 890, 925, 975, 1030, 1062, 1095, 1211, 1230, 1262, 1292, 1349, 1380, 1395, 1450, 1466, 1660 and 1674 (νC=O), 2860, 2890, 2930, 3090, 3120 IR (CCl_4): 1679 (νC=O) for "Chemical Behavior" see p. 64	[86]
32	$CH{=}CH-C_6H_4-Cl-4$	for "Applications" see p. 64	[76, 88]
*33	C_6H_4-F-4	Ia (76) m.p. 129 to 130.5° IR (Nujol): see figure in [35] UV: $\lambda_{max}(\log \varepsilon)$ = 250 (4.28), 350 (3.29), 472 (2.83)	[34, 35, 112]
*34	C_6H_4-Cl-4	If (72, 52) red needles, m.p. 185°, 186° ^{1}H NMR: 4.3 (d, H-3,4 of C_5H_4), 4.9 (d, H-2,5 of C_5H_4), 7.2 (d) and 7.4 (d) (C_6H_4) IR: (νC=O) between 1670 and 1680 UV (C_6H_6): see figure in [25] $E_{1/2}$ = 623 mV, see also p. 64	[25, 116]
*35	C_6H_4-Cl-2	Ia (96 referred to unrecovered ferrocene) crystals, m.p. 166 to 167° ^{1}H NMR ($CDCl_3$): 4.68 and 4.82 (AA′BB′, C_5H_4), 7.42 (m, C_6H_4) mol. wt: 463.1 (meas.)	[87]

References on pp. 81/4

Table 3 (continued)

No.	R in $Fe(C_5H_4{-}CO{-}R)_2$	method of preparation (yield in %), properties, and remarks	Ref.
*36	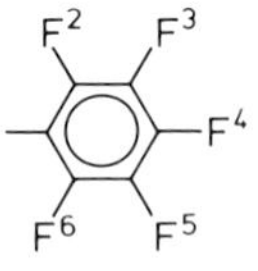	If (8.4), see "Further information", p. 77 purple-red needles, m.p. 188 to 189°; not volatile below 140°C/0.1 Torr ^{1}H NMR (CD_3COCD_3): 4.81(d, H-2,5), 5.10(s, H-3,4) ^{19}F NMR (THF): 141.1(F-2,6; |J(F-2,3)| = 22.63; |J(F-2,4)| = 2.75; |J(F-2,5)| = 8.54; |J(F-2,6)| = 0), 153.2(F-4; |J(F-3,4)| = 19.85), 161.4(F-3,5; |J(F-3,6)| = 8.54; |J(F-3,5)| = 4.8) IR ($CHCl_3$): 814, 837, 877, 896, 981, 995, 1026, 1052, 1123, 1256, 1315, 1342, 1370, 1403, 1447, 1494, 1517, 1666 (νC=O), see "Further information", p. 77	[61]
*37	$C_6H_4{-}CF_3$-4	for chemical behavior see "Further information", p. 77	[63]

compounds with R containing C, H, and O:

No.	R in $Fe(C_5H_4{-}CO{-}R)_2$	method of preparation (yield in %), properties, and remarks	Ref.
38	CH_2OH	IIIb (34) m.p. 170°(dec.) ^{1}H NMR ($CDCl_3$): 4.2(s, OH), 4.65(m, 8H), 4.85(m, 4H) IR (KBr): 1625, 1675, 3100, 3440	[105]
39	$C_6H_4{-}OH$-2	has unique stabilizing effects on coatings of silicone alkyd varnishes against degradation in space environment	[29]
40	$CH{=}CH{-}C_6H_4{-}OCH_3$-4	for "Applications" see p. 64	[76, 88]
41	$CH{=}CH{-}C_6H_4{-}$ OC_2H_4OH-4	IIIa (31) sensitive to radiation, visible and UV light, see also p. 65	[40, 43]
42	$C_6H_4{-}OCH_3$-4	Id reddish brown needles, m.p. 141.5 to 142° ^{1}H NMR ($CDCl_3$): 3.85(s, CH_3), 4.54(t) and 4.89(t) (A_2X_2, C_5H_4), 6.88(d) and 7.83(d) (C_6H_4) IR (KBr): 1613 and 1628 (νC=O) mass spectrum: m/e = 454 $[M]^+$	[97]
*43	$CHO \cdot H_2O$	IIIc would not crystallize IR ($CHCl_3$): see figure in [2]	[2, 102]
*44	CH_2CHO	see "Further information", p. 77	[36]
*45	COC_6H_5	see No. 17, p. 75 dark red crystals, m.p. 173° R_f: 0.53 (SiO_2 with $CHCl_3$)	[81, 101, 102]

Table 3 (continued)

No.	R in $Fe(C_5H_4{-}CO{-}R)_2$	method of preparation (yield in %), properties, and remarks	Ref.
*46	CH_2COCH_3	IIId (95, 72, 68), IIIe (35) small red crystals, m.p. 142.5 to 144°, 143 to 144° IR: 1290, 1351, 1471, 1538, 1562, 1613, see "Further information", p. 79 UV (CH_3CN): $\lambda_{max}(\varepsilon) = 315(27000), 478(810)$	[19, 57, 68, 74, 118, 119]
47	$CH_2COC_2H_5$	IIId (50) fine red needles, m.p. 114 to 115.5° IR: 1351, 1471, 1515, 1587 for chemical behavior with H_2NNH_2 see "Further information" for No. 46, p. 80	[19]
*48	$CH_2COC_6H_5$	IIId (72, 62) purple needles, m.p. 212 to 214°, 213.5 to 214° ^{57}Fe Mössbauer data (78, 298 K): $\delta = 0.77$, 0.75 ($Na_2[Fe(CN)_5NO] \cdot 2H_2O$), $\Delta = 2.11$, 2.10 IR: 933, 1020, 1042, 1074, 1119, 1188, 1233, 1305, 1351, 1377, 1488, 1567, 1602 IR (KBr or Nujol): 1530 mol. wt: 452 ± 45	[12, 19, 77, 79, 107]
supplements:			
49	$C_{10}H_{21}$-n	If (42) m.p. 74° IR: ($\nu C{=}O$) between 1670 and 1680 $E_{1/2} = 559$ mV, see also p. 64	[116]
50	$C_6H_4{-}CH_2{-}CH(CH_3){-}C_2H_5$-4	If (8) blood red crystals, m.p. 104° IR: ($\nu C{=}O$) between 1670 and 1680 $E_{1/2} = 528$ mV, see also p. 64	[116]

*Further information:

$Fe(C_5H_4{-}CO{-}C_2H_5)_2$ (Table 3, No. 1) forms in a 2% yield on 20 min boiling of $[FeC_{10}H_{10}]BF_4$ in $(C_2H_5CO)_2O$, along with 27% of $C_5H_5FeC_5H_4{-}CO{-}C_2H_5$ [54]. The occurrence of an IR band at 1103 cm^{-1} indicates an exception to the 9,10 μm Rosenblum's rule (see "Ferrocene" 1, 1974, p. 9) [65]. The assignment of the absorption maxima in the electronic spectrum corresponds to the transitions given in Table 1, p. 16, for $Fe(C_5H_4{-}CO{-}CH_3)_2$ [64].

The compound crystallizes in the monoclinic space group $P2_1/a$-C_{2h}^5 (No. 14); Z = 4. Lattice parameters a = 13.50 ± 0.04, b = 5.80 ± 0.04, c = 16.30 ± 0.12 Å, β = 89°20′ ± 40′ [9]. The same authors reported practically simultaneously the orthorhombic space group Pbca-D_{2h}^{15} (No. 61) with a = 13.40 ± 0.04, b = 5.8 ± 0.04, c = 37.92 ± 0.12 Å; Z = 8 [5, 8]. The molecule, like $Fe(C_5H_4{-}CO{-}C_6H_5)_2$, probably possesses a staggered conformation with the substituents in the 1,2′-positions [8, 9].

In concentrated hydrochloric acid the compound dissolves with a dark violet color [4]. It behaves with $[4\text{-}O_2N{-}C_6H_4N_2]Cl$ in glacial acetic acid like $Fe(C_5H_4{-}CO{-}CH_3)_2$ (see p. 25), i.e., not arylation but a cleavage of the Fe–cyclopentadienyl ring bond occurs. The reaction, however, proceeds less smoothly. There is considerable resin formation, and the yield of the iron-free compound drops to 2% [24]. Reduction with $NaBH_4$ in boiling isopropanol yields $Fe(C_5H_4–CHOH–C_2H_5)_2$ [49]. The reaction with CH_3MgBr in C_6H_6/ether takes place stereoselectively. The diastereoisomeric mixture I ($R = C_2H_5$) formed by hydrolysis of the reaction product contains the racemic and meso forms a and b in a 65/35 ratio. The reaction scheme corresponds to that of $Fe(C_5H_4–CO–C_6H_5)_2$ (see scheme in VII, p. 61) [99]. The reaction with R'MgI proceeds differently than that with R'MgBr: in addition to the tertiary alcohol the diolefin II is also obtained. With CH_3MgI the yields of I and II with $R' = CH_3$ are 47 and 49%, respectively. With C_2H_5MgI the yields are 62% $Fe(C_5H_4–C(C_2H_5)_2OH)_2$ and 26% II with $R' = C_2H_5$ [53, 62]. Reaction with 2-Cl–C_6H_4–CHO in a 1/2.5 molar ratio in $CH_3OH/H_2O/NaOH$ gives III in a 30% yield. The intermediate product primarily formed by the base-catalyzed aldol condensation is cyclized by a subsequent intramolecular Michael addition [21]. Compare this behavior with that of $Fe(C_5H_4–CO–CH_3)_2$, p. 40. For "Chemical Behavior" see also p. 64.

When used as an antiknock agent, this compound forms less residue than ferrocene [78]. Additions of <1% quantities to organopolysiloxanes such as dimethylpolysiloxane increases their thermal stability [6].

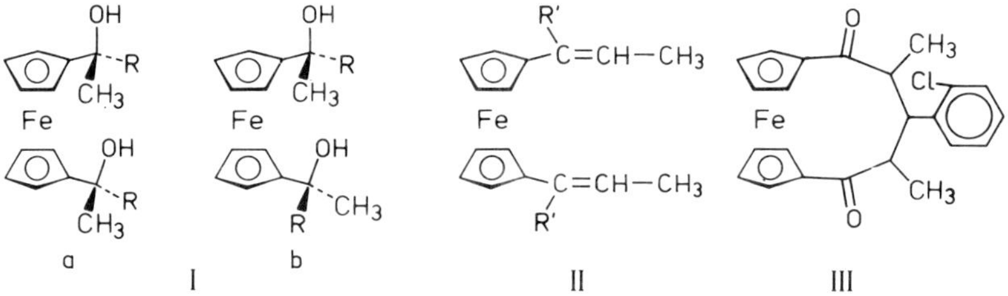

I II III

$Fe(C_5H_4–CO–C_3H_7$-n$)_2$ (Table **3**, No. **2**). The chemical ^1H NMR shifts for the cyclopentadienyl protons calculated by the additivity rule from the signals for $C_5H_5FeC_5H_4–CO–C_3H_7$-n agree well with the measured values given in Table 3, p. 65 [50]. The substituents on C-1 significantly affect the intensities of the (νC–H) bands of C_5H_4 but not the locations. These intensities are related to the Hammett constants σ_p of the substituents [31]. The assignments of the absorption maxima in the electronic spectrum correspond to the transitions given in Table 1, p. 16, for $Fe(C_5H_4–CO–CH_3)_2$ [64].

The compound crystallizes in the orthorhombic space group Acm-C_{2v}^{15} (No. 39) with Z = 4, lattice parameters $a = 11.91 \pm 0.11$, $b = 14.20 \pm 0.12$, and $c = 9.74 \pm 0.05$ Å [9]. One molecule of the four in the unit cell occupies a special site with symmetry 2 [8, 9]. The crystals are not isomorphous for the three members of the homologous series consisting of $Fe(C_5H_4–CO–CH_3)_2$, No. 1, and No. 2 [5].

No. 2, unlike No. 1, is insoluble in concentrated hydrochloric acid [4]. No. 2 (like No. 5 below) reacts with $[4\text{-}O_2N–C_6H_4–N_2]Cl$ in glacial acetic acid at 45 to 50°C with cleavage of the Fe–cyclopentadienyl ring bond and not with arylation. Moreover, >10% of the initial quantity of No. 2 is recovered unchanged. The yield of the Fe-free compound is low (cf., for $Fe(C_5H_4–CO–CH_3)_2$, p. 25) [24]. $Fe(C_5H_4–C_4H_9$-n$)_2$ forms in 80% yield on catalytic hydrogenation in absolute ethanol over $Cu(CrO_2)_2$ at 1500 psi (10342 kPa) H_2 and 185 to 190°C, and also on reduction with $NaBH_4/BF_3$ in dioxane in 75% yield [106]. Attempts to condense No. 2 with 2-Cl–C_6H_4–CHO in CH_3OH failed both with basic catalysis by 10% NaOH over a 10 h period and with acidic catalysis by concentrated HCl. In no case could a reaction be observed [21]; cf., in contrast No. 1. The Vilsmeier reaction with $(CH_3)_2NCHO$ and $POCl_3$ (molar ratio 1/4/2) at 40 to 50°C produces $Fe(C_5H_4–C(Cl)=CHC_2H_5)_2$ in 40% yield plus 17% of $Fe(C_5H_4–C(Cl)=C(CHO)C_2H_5)_2$ [47]. For "Chemical Behavior" see also p. 64.

$Fe(C_5H_4$–CO–C_4H_9-n$)_2$ (Table **3**, No. **3**) can also be prepared from IV by desulfurization in C_2H_5OH over Raney nickel. After column chromatography purification on Al_2O_3 the yield is 45% [30]. Rapid separation and identification of the compound are possible by thin layer chromatography on silicagel G in benzene along with ferrocene, in benzene or 30/1 benzene/ethanol with $C_5H_5FeC_5H_4$–CO–C_4H_9-n, and in 10/1 benzene/ethanol along with IV [27]. For "Chemical Behavior" see p. 64.

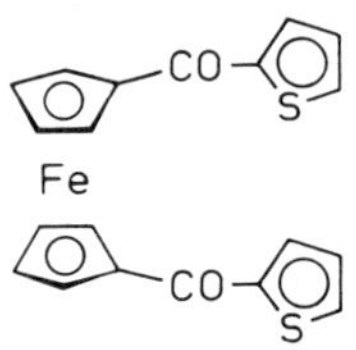

IV

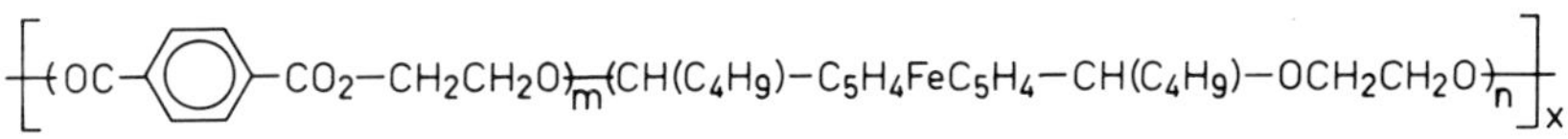

V

Transesterification of 1,4-$(CH_3O_2C)_2C_6H_4$ with $HOCH_2CH_2OH$ in the presence of 0.05 to 1.0 wt% $Fe(C_5H_4$–CO–C_4H_9-n$)_2$ not only eliminates the cost of the catalyst but also yields in the subsequent polycondensation a product of composition V, which in contrast to the unmodified polyester possesses high light resistance and thermal stability and is of an equally pure color (sandy to brown) [96].

$Fe(C_5H_4$–CO–C_5H_{11}-n$)_2$ (Table **3**, No. **4**) is hydrogenated in absolute ethanol over $Cu(CrO_2)_2$ to $Fe(C_5H_4$–C_6H_{13}-n$)_2$ at 1800 psi (12411 kPa) H_2 and 200 to 225°C, in 87% yield [3]. Reduction with $NaBH_4$ in boiling isopropanol forms $Fe(C_5H_4$–CHOH–C_5H_{11}-n$)_2$ [48]. For "Chemical Behavior" and "Applications" see pp. 64 and 65, respectively.

$Fe(C_5H_4$–CO–C_7H_{15}-n$)_2$ (Table **3**, No. **5**). In the ^{57}Fe Mössbauer spectrum the ratio of the recoil-free fraction at 298 K to that at 77 K is 0.36 [94]. No. 5 shows in the smectic H phase of the liquid crystals of PBPA (VI a) or BBEA (VI b) a Mössbauer effect only when it is suspended in the form of small crystallites, not when it is present in solution [85].

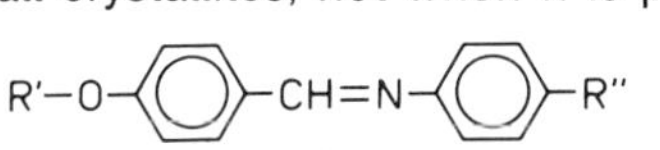

VI

a: R′ = n-C_5H_{11}, R″ = C_3H_7
b: R′ = C_4H_9, R″ = C_2H_5
c: R′ = n-C_4H_9, R″ = n-C_8H_{17}

A positive electrical field gradient perpendicular to the cyclopentadienyl rings is confirmed by the variation with the angle Θ between the orienting 9 kOe magnetic field and the incident γ radiation, of the ratio of the areas under the two peaks of the quadrupole doublet found in studies at 77 K of supercooled glasses having the smectic B liquid crystal structure of BBOA (VI c) and containing 1.5 to 2 wt% $Fe(C_5H_4$–CO–C_7H_{15}-n$)_2$. The intramolecular vibrational anisotropy $\varepsilon_M = 0.00 \pm 0.10$. From the change with Θ of the total intensity of the Mössbauer spectrum, the lattice contribution ε_L to the nuclear vibration ellipsoid is 0.07 ± 0.02. Obviously No. 5 is only weakly oriented in BBOA [90, 94]. For "Chemical Behavior" see p. 64.

$Fe(C_5H_4$–CO–$C_{11}H_{23}$-n$)_2$ (Table **3**, No. **7**) in 1 wt % quantities added together with 0.5 wt% $NCC(CH_3)_2N$=$NC(CH_3)_2CN$ reduces the flammability of unsaturated, castable polyester resins consisting of 650 parts of a polyester formed from maleic acid anhydride, tetrachlorophthalic acid, $HOCH_2CH_2OH$, and $CH_3CH(OH)CH_2OH$ plus 350 parts styrene and 1 part hydroquinone [39]. For "Chemical Behavior" see p. 64.

$Fe(C_5H_4$ CO–$CH(CH_3)_2)_2$ (Table **3**, No. **10**) can also be prepared by addition of ferrocene in CH_2Cl_2 to $(CH_3)_2C(COCl)_2$ in CH_2Cl_2 at 0°C with subsequent standing overnight at 0°C. $C_5H_5FeC_5H_4$–CO–$CH(CH_3)_2$, No. 10, and $C_5H_5FeC_5H_4$–CO–$C(CH_3)_2$–CO–$C_5H_4FeC_5H_5$ are isolated in that elution sequence on column chromatographic processing of the raw product on Al_2O_3 in

Skellysolve B, with benzene/Skellysolve B as eluent. The same products also form on conducting the reaction at high dilution in boiling CH_2Cl_2 [33].

$LiAlH_4$ reduces No. 10 to $Fe(C_5H_4\text{–}CHOH\text{–}CH(CH_3)_2)_2$, which shows no change of physical properties despite repeated recrystallizations [80]. For "Chemical Behavior" see also p. 64.

$Fe(C_5H_4\text{–}CO\text{–}C(CH_3)_3)_2$ (Table **3**, No. **12**) can according to [15] be prepared from ferrocene and $(CH_3)_3CCOCl$ in boiling $ClCH_2CH_2Cl$ more satisfactorily and economically if instead of $AlCl_3$ a non-Friedel-Crafts type catalyst, such as $FeCl_3$ + pivalic acid, is used in only small quantities. The preparation in the presence of Mg powder or by boiling of ferrocene with $(CH_3)_3CCOCl$ in $(n\text{-}C_4H_9)_2O$ is described in [15], preparation by Method Ie with $ZnCl_2$ instead of $AlCl_3$ in [11]. Fractional recrystallization from petroleum ether allows separation of $C_5H_5FeC_5H_4\text{–}CO\text{–}C(CH_3)_3$ and No. 12 [11].

In the ^{57}Fe NMR spectrum substituents with an sp-hybridized α C atom cause a distinct shift to higher frequencies; No. 12 has the highest shift so far found [103].

Hydrogenation in CH_3CO_2H over PtO_2 gives $Fe(C_5H_4\text{–}CH_2\text{–}C(CH_3)_3)_2$ [11]. For "Chemical Behavior" and "Applications" see also pp. 64 and 65, respectively.

$Fe(C_5H_4\text{–}CO\text{–}CH_2C(CH_3)_3)_2$ (Table **3**, No. **13**) also forms from $C_5H_5FeC_5H_4\text{–}C(CH_3)_3$ on acetylation together with $C_5H_5FeC_5H_3(CO\text{–}CH_3\text{-}1)C(CH_3)_3\text{-}2$ and -3 [115]. For "Chemical Behavior" and "Applications" see pp. 64 and 65, respectively.

$Fe(C_5H_4\text{–}CO\text{–}CH{=}CH_2)_2$ (Table **3**, No. **15**). In contrast to [1], Barr, Watts obtained from No. 28 (see p. 76) a deep red gum, unstable towards heat and light, which could not be crystallized [55].

Hydrogenation in C_6H_6 over Pd/C [55, 59] or in C_2H_5OH over Pt [59] yielded even at atmospheric pressure along with the expected No. 1 [55, 59] the compound VII (with R' = H). The creation of C–C bond under such mild conditions indicates the formation of a coordination between the double bond in No. 15 and the catalyst surface, which allows with the interannular ring closure a synchronously occurring H transfer from the metal to the end C atom of one double bond [59]. On passing dry HCl gas through a solution in CH_2Cl_2 No. 28 is formed along with VII (R' = Cl). The two compounds can be separated on SiO_2 with 2/1 ligroin/ether, which first elutes No. 28 (46% yield) and then VII (R' = Cl) (41% yield) [59]. The reaction with C_2H_5OH/C_2H_5ONa yields VII with $R' = OC_2H_5$ [55].

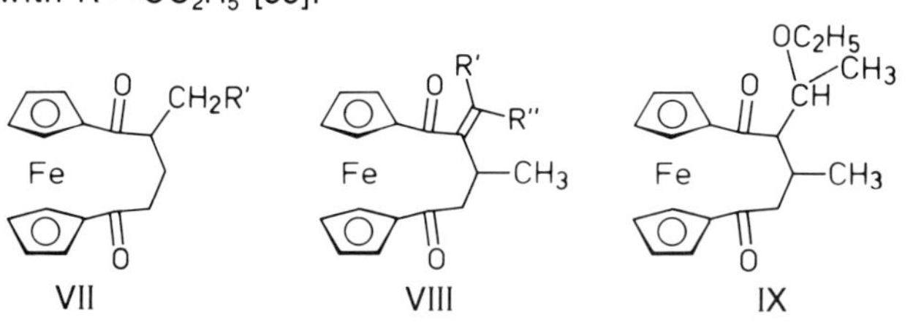

VII VIII IX

(E)-$Fe(C_5H_4\text{–}CO\text{–}CH{=}CHCH_3)_2$ (Table **3**, No. **16**) in contrast to No. 15 yields on catalytic hydrogenation by Method IIb no ferrocenophane derivative, but only No. 2. This can be isolated by column chromatography on Al_2O_3 with 10/1 ligroin/ether in an 82% yield [59]. Introduction of dry HCl into the CH_2Cl_2 solution gives VIII with $R' = CH_3$, R'' = H and vice versa. The yield of the compound melting at 153 to 155°C is 19% and is 10% for the one melting at 111 to 113°C. The two compounds and IX in addition form when a suspension of No. 16 in C_2H_5OH is added to $NaOC_2H_5/C_2H_5OH$. While the former cyclization reaction probably is enabled by protonation of one CO group, which follows a series of electron displacements, the latter begins with nucleophilic additon of the $C_2H_5O^-$ to one C(8) atom [59].

$Fe(C_5H_4{-}CO{-}CH_2C_6H_5)_2$ (Table **3**, No. **17**) is obtained in a 52% yield by hydration of $Fe(C_5H_4{-}C{\equiv}C{-}C_6H_5)_2$ on refluxing the latter with $HgCl_2$ and 10% H_2SO_4 in acetone for 4 d. After addition of H_2O, the reaction mixture is extracted with ether, and the concentrated extract is subjected to chromatography on Al_2O_3, with elution of the compound with 1/1 hexane/benzene [98].

Line widths and second moments derived from them for the solid state ^1H NMR spectra at various temperatures indicate that above −196°C a molecular motion sets in which is other than the free rotation of the cyclopentadienyl rings [14]. A linear correlation corresponding to a modified and extended Seth-Paul-Van Duyse equation has been observed between the ($\nu C{=}O$) vibration in CCl_4 and the $X^+(R^{A,B})$ constants of the structural fragments for 154 compounds with ferrocene structures of the general formula $R^A{-}CO{-}R^B$, among them No. 17 with $R^A = C_5H_4FeC_5H_4{-}COCH_2C_6H_5$ and $R^B = CH_2C_6H_5$ [117].

The oscillopolarogram in $C_2H_5OH/HClO_4$ on a flowing Hg electrode differs clearly from that of $C_5H_5FeC_5H_4{-}CO{-}CH_2C_6H_5$, so that contamination of one compound by the other is easily detectable [37]. Low voltage mass spectrometry with a 350°C inlet temperature and a nominal 8 eV ionizing voltage produces little fragmentation with a very intense molecular ion peak, which can be used for identification and semiquantitative determination of No. 17 [111].

Oxidation with $4{-}ON{-}C_6H_4{-}N(CH_3)_2$ in C_2H_5OH leads to No. 45 [81, 101] in 17% yield, when after 7 h refluxing in the dark the solvent is removed and the residue is hydrolyzed with boiling semiconcentrated hydrochloric acid. Isolation is effected chromatographically on SiO_2 columns with $CHCl_3$ [101]. Oxidation with SeO_2 in boiling dioxane/H_2O also leads to No. 45, according to [102]. In contrast, No. 45 can be obtained by [81] neither with SeO_2 nor with MnO_2. $HO_2C{-}C_5H_4FeC_5H_4{-}CO{-}CH_2C_6H_5$ forms with $OC(OC_2H_5)_2$ in boiling C_6H_6 under the action of $NaNH_2$ [32, 44] or NaH [44]. Reduction with KBH_4 in CH_3OH [110] or CH_3OH/H_2O [44] leads to $Fe(C_5H_4{-}CHOH{-}CH_2C_6H_5)_2$ [44, 110]. Regarding reduction see also p. 64. Heating with $NH_2OH \cdot HCl$ and K_2CO_3 in absolute ethanol forms the dioxime $Fe(C_5H_4{-}C(CH_2C_6H_5){=}NOH)_2$ [44, 110]. The reaction with CH_3MgBr in C_6H_6/ether is clearly stereoselective. Hydrolysis of the reaction product yields the diastereoisomer mixture I (with $R = CH_2C_6H_5$), which contains the racemic and meso forms a and b in a 71/29 ratio. The reaction scheme corresponds to that given for $Fe(C_5H_4{-}CO{-}C_6H_5)_2$ in scheme in VII on p. 61 [99]. Condensation with $Cr(CO)_6$ gives a red mixture separable by column chromatography on Al_2O_3, which consists of $(OC)_3CrC_6H_5CH_2{-}CO{-}C_5H_4FeC_5H_4{-}CO{-}CH_2C_6H_5$ and $Fe(C_5H_4{-}CO{-}CH_2C_6H_5Cr(CO)_3)_2$ [56], see "Chrom-Organische Verbindungen" 1971, Table 43, No. 15, and Table 73, No. 18, respectively.

$Fe(C_5H_4{-}CO{-}CH(CH_2C_6H_5)_2)_2$ (Table **3**, No. **19**) can be prepared by reaction of $Fe(C_5H_4{-}CO{-}CH_3)_2$ in liquid ammonia with $C_6H_5CH_2Cl$ in ether in a 1/4 mole ratio, in the presence of three equivalents of KNH_2. After stirring for 1 h the liquid ammonia is stripped off and replaced by ether, and the reaction product is subsequently hydrolyzed with water (yield 53%). The compound can also be prepared in the same fashion from No. 18 and $C_6H_5CH_2Cl$ in the presence of KNH_2 (molar ratio 1/2/2) [22].

$Fe(C_5H_4{-}CO{-}CH{=}CHC_6H_5)_2$ (Table **3**, No. **20**) gives in conc. H_2SO_4 the diprotonation product **$[Fe(C_5H_4{-}C(CH{=}CHC_6H_5){=}OH)_2]^{2+}$** with the following ^{13}C NMR spectrum: δ = 78.1 (C-2,5 of C_5H_4), 78.7 (C-1 of C_5H_4), 85.1 (C-3,4 of C_5H_4), 116.0 and 159.3 (CH=CH); 131.2, 133.5, and 138.3 (C_6H_5); 197.4 (CO) ppm [84]. No. 20 yields No. 18 on hydrogenation by Method IIa [16, 22]; it separates as a red-orange powder by concentrating the filtrate of the reaction mixture [16]. Refluxing with $Fe_3(CO)_{12}$ in benzene gives $Fe(C_5H_4{-}CO{-}CH{=}CH{-}C_6H_5Fe(CO)_3)_2$ (see Section 5.2.3.5.1.8) [83]. Reaction in ethanol/tetrahydrofuran with sodium hydroxide forms X. The mechanism seems to consist of an OH^--catalyzed reverse aldol condensation which follows an

References on pp. 81/4

intramolecular Michael addition [75]. Michael addition takes place in $(CH_3)_3COH/(CH_3)_3CO^-$ at 50°C with compounds which contain active methylene groups to form [7]-ferrocenophane derivatives in high yields. Thus XI with R″ = $CO_2C_2H_5$, $COCH_3$, or CN, respectively, is obtained with $CH_2(CO_2C_2H_5)_2$, $CH_3COCH_2CO_2C_2H_5$, or $NCCH_2CO_2C_2H_5$ [104].

With a cable-grade, diisooctylnaphthalate-plasticized polyvinylchloride, No. 20 at concentrations of 1 to 5% is not only poorly compatible but also induces exudation of other solid and liquid ingredients [51]. For "Applications" see p. 64.

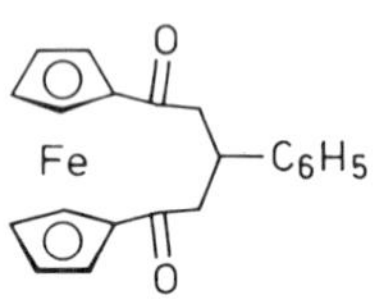

X

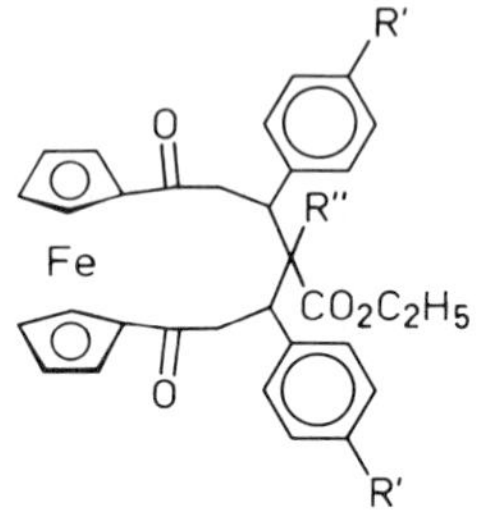

XI : R' = H

XII : R' = CH_3

(E)-$Fe(C_5H_4$–CO–CH=CH–C_6H_4–CH_3-4$)_2$ (Table **3**, No. **21**) behaves quite analogously to No. 20 with respect to compounds containing active methylene groups, to form XII with R″ = $CO_2C_2H_5$, $COCH_3$, or CN, respectively [104].

$Fe(C_5H_4$–CO–C_6H_4–C_6H_5-4$)_2$ (Table **3**, No. **24**). For thin-layer chromatography on SiO_2 see [113, 114]. The relative R_f value (related with ferrocene ≙ 1) in 15/1 C_6H_6/C_2H_5OH is 0.96, and in 15/1/3 toluene/petroleum ether/acetone is 0.71 (0.82 for $C_5H_5FeC_5H_4$–CO–C_6H_4–C_6H_5-4) [114].

No. 24 is reduced by KBH_4 in C_2H_5OH to $Fe(C_5H_4$–CHOH–C_6H_4–C_6H_5-4$)_2$ [100].

$Fe(C_5H_4$–CO–C_6H_{11}-cyclo$)_2$ (Table **3**, No. **26**) cannot be sulfonated on the cyclopentadienyl ring with SO_3 in $ClCH_2CH_2Cl$; 45% of the inserted quantity of No. 26 is instead decomposed [23]. The reaction with CH_3MgBr in C_6H_6/ether proceeds with strong stereoselectivity. The diastereoisomeric mixture I with R = cyclo-C_6H_{11} formed by hydrolysis of the reaction product contained the racemic and meso forms a and b in a >98/<2 ratio. The reaction scheme corresponds to that for $Fe(C_5H_4$–CO–$C_6H_5)_2$ (see scheme in VII, p. 61) [99]. For reduction and "Applications" see pp. 64 and 65, respectively.

$Fe(C_5H_4$–CO–$CH_2CH_2Cl)_2$ (Table **3**, No. **28**) does not remain unchanged on neutral Al_2O_3 in contrast to the behavior on SiO_2 (see "Further information" to No. 15, p. 74). HCl is split off with formation of No. 15 in 64% yield by column chromatography on Al_2O_3 in C_6H_6 with ether as eluent [1, 55].

Reduction with $NaBH_4/AlCl_3$ in $CH_3OCH_2CH_2OCH_3$ at room temperature gives a high yield of $Fe(C_5H_4$–CH_2–$CH_2CH_2Cl)_2$ [89]. Refluxing with KCN in C_2H_5OH yields $Fe(C_5H_4$–CO–$CH_2CH_2CN)_2$ [7].

No. 28 is active against bacteria, yeasts, and fungi. In lower concentrations it is bacteriostatic, in higher bacteriocidal [93, 95].

$Fe(C_5H_4$–CO–$CH_2CHClCH_3)_2$ (Table **3**, No. **29**) yields No. 16 (91%) during column chromatography on Al_2O_3 in ether [59].

$Fe(C_5H_4$–CO–C_6H_4–F-4$)_2$ (Table **3**, No. **33**) crystallizes in the monoclinic space group C2/m-C_{2h}^3 (No. 12). Lattice parameters a = 12.69, b = 21.11, c = 13.65 Å, β = 91°45′. The density calculated with Z = 8 is 1.564 g/cm³; the density measured is 1.560 g/cm³ [112].

$Fe(C_5H_4$–CO–C_6H_4–Cl-4$)_2$ (Table **3**, No. **34**) yields $Fe(C_5H_4$–$C(CH_2$–C_6H_4–Cl-4)OH–C_6H_4–Cl-4$)_2$ with 4-Cl–C_6H_4–CH_2MgCl in ether/benzene after subsequent hydrolysis [25].

Fe(C_5H_4–CO–C_6H_4–Cl-2)$_2$ (Table **3**, No. **35**). Reaction in $CH_3OCH_2CH_2OCH_3$ in a 1/2 mole ratio with $KOC(CH_3)_3$ plus small amounts of water gives a practically quantitative yield of $Fe(C_5H_4–CO_2H)_2$ [87].

Fe(C_5H_4–CO–C_6F_5)$_2$ (Table **3**, No. **36**) forms in a 21% yield when ferrocene lithiated with n-$C_4H_9Li\cdot(CH_3)_2NCH_2CH_2N(CH_3)_2$ in hexane is reacted at the boiling point with C_6F_5COCl. After the usual processing the ferrocene (32%) is separated from the reaction product by column chromatography on Florisil by elution with petroleum ether; ether elutes No. 36 [61].

The IR bands occurring in the 9,10 μm region are attributed to F–C absorptions [61], compare "Ferrocene" 1, 1974, p. 9.

In the mass spectrum the molecular ion $[M]^+$ possesses the highest relative abundance (set at 100%), followed by $[C_5H_4CO]^+$ (72.5%), $[C_5H_4COFeF]^+$ (55%), $[FeF]^+$ (20%), $[C_3H_3]^+$ (19%), $[Fe]^+$ (13%), $[C_5H_3]^+$ (6.5%), $[C_3F_3]^+$ (6.3%), $[C_5H_4C_5H_4]^+$ (4%), $[C_5H_3C_6F_4]^+$ (3.7%), $[C_6F_2C_5H_3]^+$ (3.6%), and $[C_{12}H_4F_2]^+$ (2.8%). Metastable peaks at m/e = 44.6, 115.7, 519.5, and ~288 indicate a cleavage of CO (transition from m/e 92 → 64, 167 → 139, 574 → 546) or C_6F_5 (transition from m/e 574 → 407). $[M]^{2+}$ appears as a doubly charged ion (2.5%) [60]. Regarding reduction see p. 64.

Fe(C_5H_4–CO–$C_6H_4CF_3$-4)$_2$ (Table **3**, No. **37**) is reduced electrochemically in $(CH_3)_2NCHO$ or by means of Na/K alloys to the ketyl $[4\text{-}CF_3–C_6H_4–CO–C_5H_4FeC_5H_4–\dot{C}(C_6H_4CF_3\text{-}4)O^{(-)}]^{\dot{-}}$ (see Section 5.2.3.5.1.8) [63].

Fe(C_5H_4–CO–CHO)$_2\cdot$2H_2O (Table **3**, No. **43**) condenses in boiling $(CH_3)_2NCHO$ or m-cresol with XIII (n = 0) or XIV to the ferromagnetic XV (R′ = H, n = 0) or XVI (R′ = H), respectively [102]. XVII probably forms with o-phenylenediamine [2].

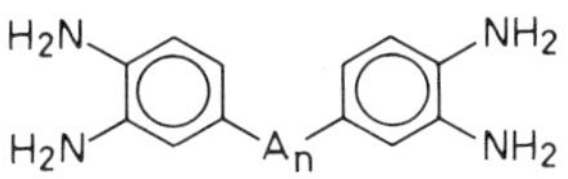

XIII

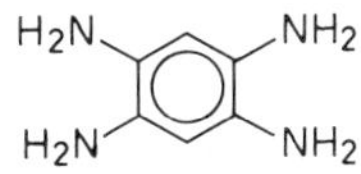

XIV

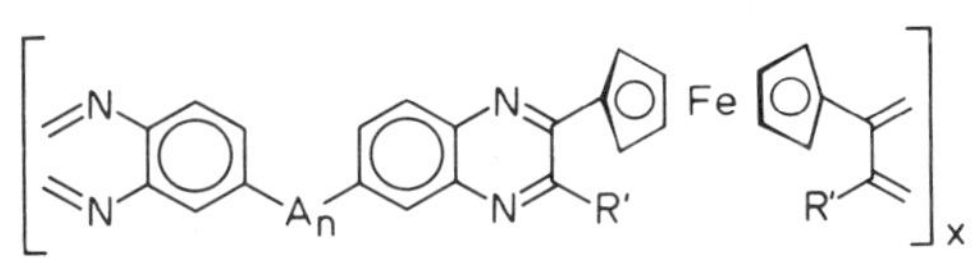

XV

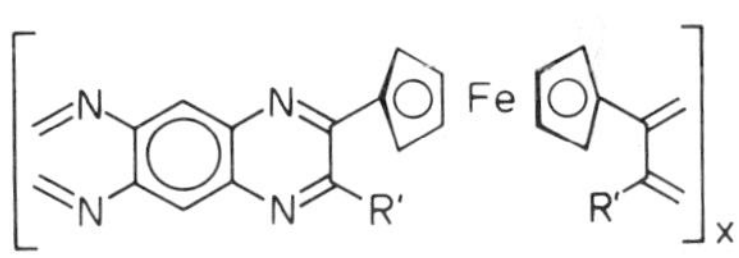

XVI

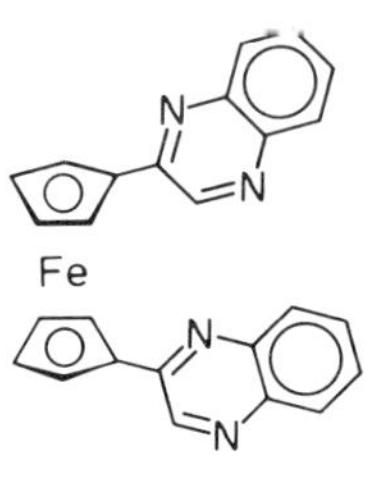

XVII

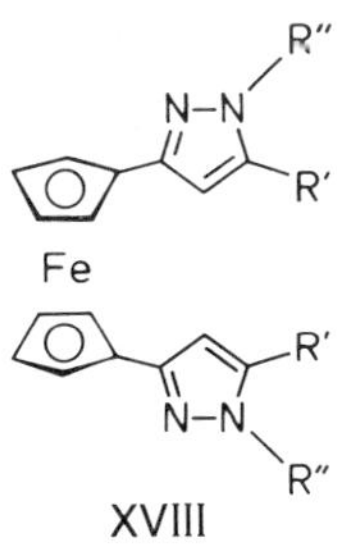

XVIII

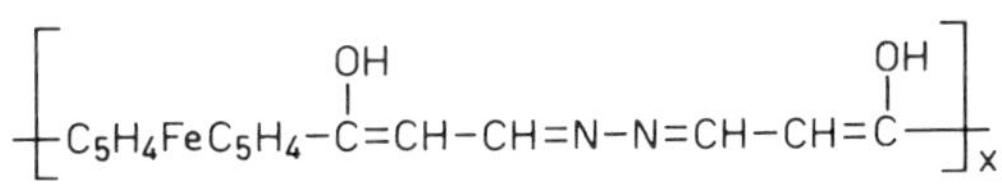

XIX

Fe(C_5H_4–CO–CH_2CHO)$_2$ (Table **3**, No. **44**). Reaction of $Fe(C_5H_4–CO–CH_3)_2$ in benzene first with Na and subsequently with $HCO_2C_2H_5$ forms with self-heating the Na salt of the enol form of No. 44, **Na$_2$[Fe(C_5H_4–CO–CH=CHO$^-$)$_2$]**, which precipitates out practically quantitatively as an orange crystal slurry. This slurry is hydrolyzed by adding small portions with ice cooling to 10%

 References on pp. 81/4

H_2SO_4 covered by an ether layer. The ether layer is immediately decanted, neutralized, dried, and evaporated. Nevertheless, even under the most careful conditions, $Fe(C_5H_4\text{–}CO\text{–}CH_2CHO)_2$ cannot be obtained pure, since it rapidly becomes resinous [36].

This compound reacts with $N_2H_4 \cdot H_2O$ in absolute C_2H_5OH to form XVIII (R' = R'' = H) [36]. Polyconjugated, soluble polyazines with the idealized structure XIX form on condensation with H_2NNH_2 in sulfolane solution at 70°C, while substantially insoluble products form at 90 to 100°C with involvement of the enolic OH groups in the reaction [46]. Polyhydrazones are obtained with aromatic or pseudoaromatic dihydrazines [58].

$Fe(C_5H_4\text{–}CO\text{–}CO\text{–}C_6H_5)_2$ (Table **3**, No. **45**) crystallizes in the monoclinic space group $P2_1/c\text{-}C^5_{2h}$ (No. 14) with $a = 7.338 \pm 0.002$, $b = 16.872 \pm 0.004$, $c = 8.046 \pm 0.002$ Å, $\beta = 87.0° \pm 0.01°$; $Z = 2$. The calculated density is 1.503, measured density 1.48 g/cm³. **Fig. 3** shows the molecular structure in an x–z projection with bond distances (in Å) and angles. The cyclopentadienyl rings in the molecule lie in parallel planes as do the phenyl groups, with a dihedral angle of 62.5° between the two sets of planes. The C(6)–C(7) bond distance of 1.528 Å is unusually long for a C–C sp^2 single bond. This is possibly due to the parallel excess charge on the two C atoms, which lengthens the bond due to the additional Coulomb repulsion. The shortest intermolecular H···H distances are larger than the shortest intermolecular H···O distances, which indicates for the latter an interaction exceeding the Van der Waals forces. The intermolecular interactions are portrayed in **Fig. 4** [101].

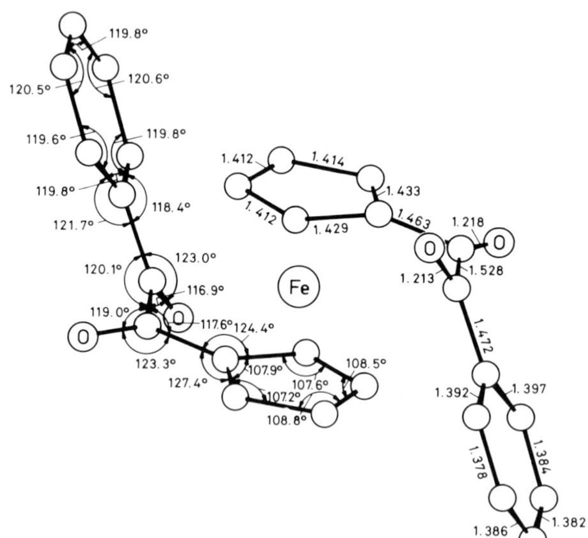

Fig. 3. Molecular structure of $Fe(C_5H_4\text{–}CO\text{–}CO\text{–}C_6H_5)_2$, projected on the x–z plane [101].

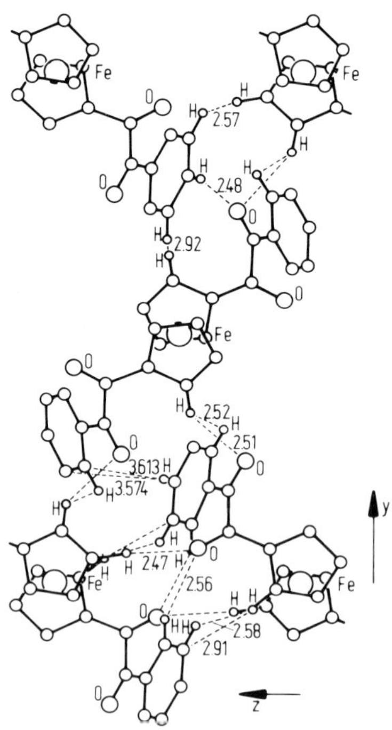

Fig. 4. Intermolecular interactions in the $Fe(C_5H_4\text{–}CO\text{–}CO\text{–}C_6H_5)_2$ crystal [101].

References on pp. 81/4

The compound is well soluble in $CHCl_3$ and less so in cresol, benzene, ethanol, or CCl_4 [81]. It reacts with organic tetramines like XIII (n = 0) or XIV (see p. 77) in $(CH_3)_2NCHO$ at 100 to 200°C with formation of the high temperature stable [81], ferromagnetic polyquinoxalines XV (R′ = C_6H_5, n = 0) or XVI (R′ = C_6H_5) (see p. 77) in practically quantitative yield [102]. The compounds XV (R′ = C_6H_5, n = 0 or 1) were obtained with XIII (n = 0 or 1 and A = CH_2, O, S, SO_2, CO) in m-cresol starting at room temperature and then at 180°C [82]. According to [101] there form with XIII (n = 1, A = O) in solution only oligomers with a molecular weight of 1800 to 2900 instead of the high molecular weight polyphenylquinoxalines; there is no basis for this evident from the molecular structure.

$Fe(C_5H_4$–CO–CH_2CO–$CH_3)_2$ (Table **3**, No. **46**). The IR spectrum shows no bands belonging to free CO groups. The bands occurring at 1562 and 1613 cm^{-1} indicate a conjugated chelate structure [19]. The half-wave potential measured polarographically at 20°C in CH_3CN/0.1 M $NaClO_4$ on a rotating Pt electrode is $E_{1/2}$ = 790 mV vs. the saturated calomel electrode. The redox reaction is connected with a transition of more than two electrons [57].

In the mass spectrum the molecular ion $[M]^+$ with m/e = 354 possesses the greatest relative abundance. The fragmentation takes place following the scheme in XX (with abundances in % relative to that of the molecular ion).

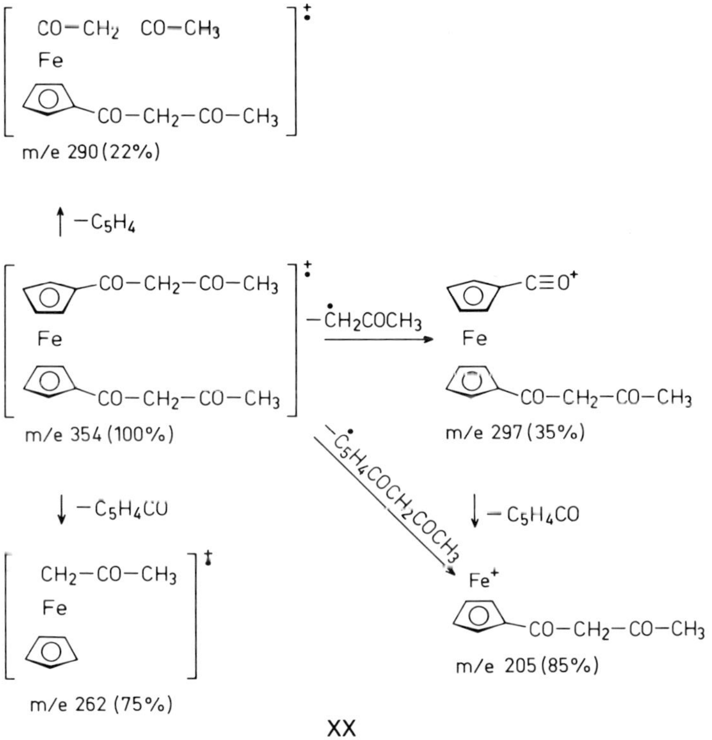

XX

The sequence m/e = 354 → 297 → 205 is confirmed by metastable peaks at m/e = 249.2 and 141.5, the direct transition m/e 354 → 205 by one at m/e = 118.7, and the transitions m/e 354 → 290 and 354 → 262 by such at m/e = 235.9 and 193.9, respectively. The ions at m/e = 290 and 262 possibly likewise fragment to m/e 205. In addition the peaks at m/e 163 (46%), 178 (20%), and 270 (12%) are assigned to the ions $[CH_3COC_5H_4Fe]^+$, $[C_5H_5FeCH_2COCH_3]^{+\cdot}$, and $[C_5H_5FeC_5H_4COCH_2COCH_3]^{+\cdot}$ or $[Fe(C_5H_4COCH_3)_2]^{+\cdot}$, respectively [74].

Condensation with $R'NH_2$ (mole ratio 1/3) in n-$C_{16}H_{34}$ (for $R' = C_6H_5$) or isooctane (for $R' = C_4H_9$, C_5H_{11}, β-naphthyl) at 100°C yields $Fe(C_5H_4{-}CO{-}CH_2C(CH_3){=}NR')_2$ [70]. The reaction with the diamines $H_2N{-}A{-}NH_2$ ($A = (CH_2)_2$, $(CH_2)_6$, $CH(CH_3)CH_2$, 1,3-C_6H_4, 4,4'-$C_6H_4{-}C_6H_4$) in CH_3OH at the boiling point leads for a 1/1 mole ratio to XXI, at a 2/1 mole ratio to XXII [68, 69, 72].

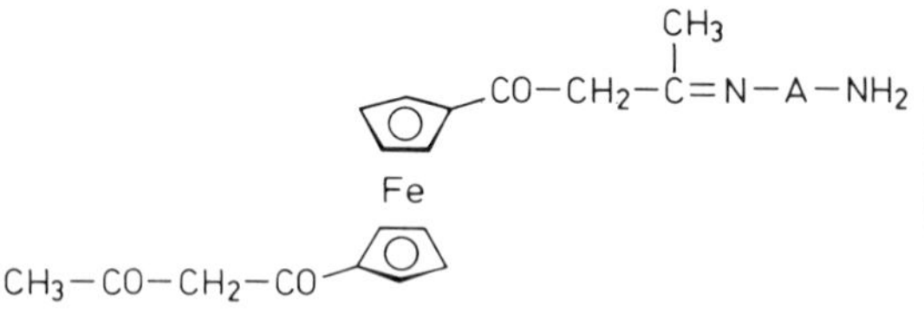

XXI

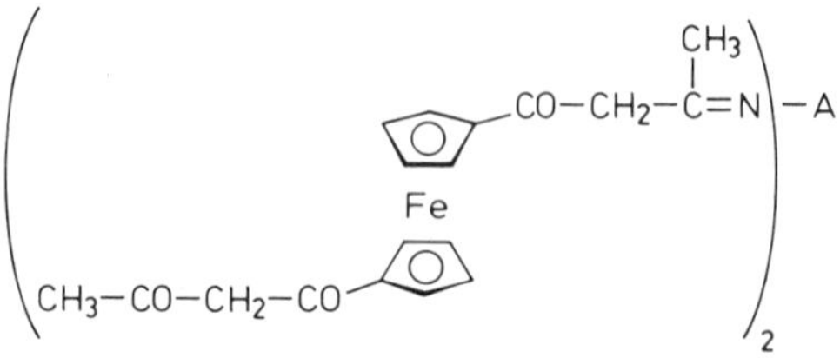

XXII

By reaction with $NH_2OH \cdot HCl$ in 1/1 pyridine/absolute ethanol at 70°C XXIII is obtained [118]. Compound XVIII ($R' = CH_3$, $R'' = H$) (see p. 77) forms with H_2NNH_2 in absolute ethanol [19, 118], from No. 47 the corresponding XVIII with $R' = C_2H_5$ and $R'' = H$ [19]. Reaction with H_2NNHR'' ($R'' = C_6H_5$, $CONH_2$) at 60°C in C_2H_5OH leads to XVIII with $R' = CH_3$, $R'' = C_6H_5$ and $CONH_2$, respectively [118]. Solution condensation with aromatic or pseudoaromatic dihydrazines yields soluble polyhydrazones [58]. On heating with $[Ni_2(H_2NCH_2CH_2NH_2)_4Cl_2]Cl_2$ (cf. "Nickel", Part C, Chapter 1, 1968, p. 153) and pyridine in CH_3OH one obtains XXIV, a cyclic trimeric complex of high purity [52]. No. 46 is useful as complexing agent for heavy metals, e.g., Pb, Ce [119].

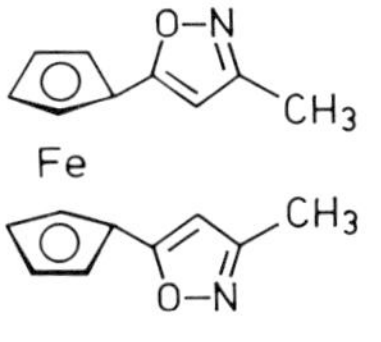

XXIII

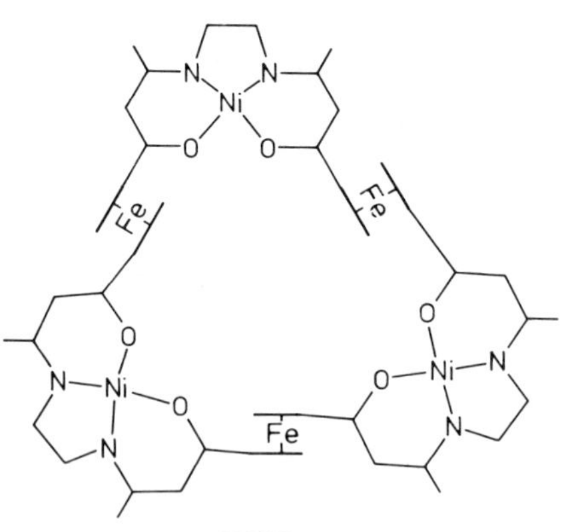

XXIV

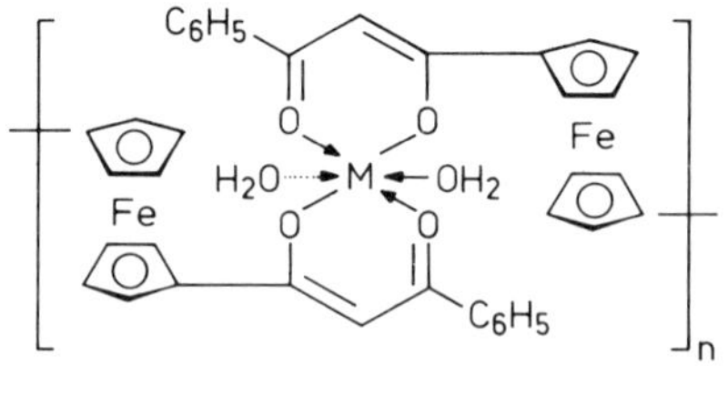

XXV

$Fe(C_5H_4{-}CO{-}CH_2CO{-}C_6H_5)_2$ (Table **3**, No. **48**). CO bands at 1567 and 1602 (1575 and 1613 [19]) cm^{-1} instead of those characteristic for free C=O groups indicate that the compound is present in an enol form [12].

The compound is moderately soluble in acetone, insoluble in ether or H_2O [12]. If a methanol solution is added to an aqueous solution of $M^{II}(O_2CCH_3)_2 \cdot xH_2O$, the polymeric chelate complex XXV precipitates, with M^{II} = Fe, Co, or Ni [107]. With M^{II} = Cu, in contrast, there forms the monomeric complex $Fe(C_5H_4–CO–CH=C(O)C_6H_5)_2Cu$ (see Section 5.2.3.5.1.8, p. 154) [107], which also precipitates on conducting the reaction in acetone [12]. Fe^{3+} can be determined in the chelates of No. 48 by titration with ferrocene (only Cu^{2+} interferes) [28].

A large excess of hydrazine in C_2H_5OH gives XVIII ($R'=C_6H_5$, $R''=H$) (see p. 77) [12, 19]. Solution condensation with aromatic or pseudoaromatic dihydrazines leads to formation of soluble polyhydrazones [58].

The sodium salt of No. 48, **$Na_2[Fe(C_5H_4–CO–CH=C(O)C_6H_5)_2]$** exhibits a broad C=O band at 1540 cm^{-1} [77, 79]. If an acetone solution of the salt is added to a solution of $UO_2(NO_3)_2 \cdot 6H_2O$ in acetone/H_2O, a brown-red precipitate forms, which has according to molecular weight determinations by the Rast method the polymeric structure XXVI. The coordinatively bound H_2O can be replaced in benzene easily by other monodentate ligands, such as for instance CH_3COCH_3, pyridine, $(CH_3)_2SO$, $(C_4H_9)_3PO$, $(C_6H_5)_3PO$, $(C_6H_5)_3AsO$, or pyridine-N-oxide [77, 79].

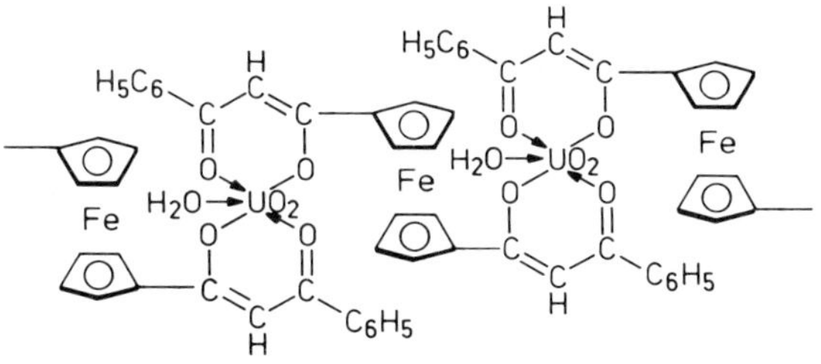

XXVI

References:

[1] R. B. Woodward, M. Rosenblum, M. C. Whiting (J. Am. Chem. Soc. **74** [1952] 3458/9). – [2] M. Rosenblum (Diss. Harvard Univ. 1953). – [3] A. C. Nixon, T. P. Rudy, N.V. de Bataafsche Petroleum Maatschappij (Brit. 763550 [1954/56]; C.A. **1957** 10580; Ger. 1003727 [1954/57]). – [4] A. N. Nesmeyanov, N. A. Vol'kenau (Dokl. Akad. Nauk SSSR **107** [1956] 262/4). – [5] Yu. T. Struchkov [G. T. Stručkov] (Dokl. Akad. Nauk SSSR **110** [1956] 67/70).

[6] W. A. Piccoli, Dow Corning Corp. (U.S. 2979482 [1956/61]; C.A. **1961** 18176). – [7] P. J. Graham, P. V. Lindsey, G.W. Parshall, M. L. Peterson, G. M. Whitman (J. Am. Chem. Soc. **79** [1957] 3416/20). – [8] Yu. T. Struchkov [G. T. Stručkov], T. L. Khotsyanova [T. L. Khocjanova] (Kristallografiya **2** [1957] 382/3; Soviet. Phys. Cryst. **2** [1957] 376/7). – [9] G.T. Stručkov, T. L. Khocjanova (Acta Cryst. **10** [1957] 796/7). – [10] M. Vogel, M. Rausch, H. Rosenberg (J. Org. Chem. **22** [1957] 1016/8).

[11] T. Leigh, Imperial Chemical Industries, Ltd. (Brit. 819108 [1957/59]; C.A. **1960** 7732; Ger. 1150077 [1958/63]; Belg. 570298 [1958/62]; Austrian 212331 [1958/60]; Austrian 210435 [1958/60]). – [12] C. R. Hauser, C. E. Cain (J. Org. Chem. **23** [1958] 1142/6). – [13] W. L. Baun (Anal. Chem. **31** [1959] 1308/11). – [14] N. Ockman (AD-229397 [1959] 1/62, 29). – [15] R. J. Stephenson, Imperial Chemical Industries, Ltd. (Brit. 861833 [1959/61]; C.A. **1961** 25981).

[16] T. A. Mashburn, C. E. Cain, C. R. Hauser (J. Org. Chem. **25** [1960] 1982/6). – [17] T. Leigh, Imperial Chemical Industries, Ltd. (Brit. 869504 [1959/61]; C.A. **1961** 24790; Ger. 1161560 [1960/64]). – [18] T. Leigh, Imperial Chemical Industries Ltd. (Ger. 1163821 [1960/64]; C.A. **1961** 24790). – [19] C. E. Cain, T. A. Mashburn, C. R. Hauser (J. Org. Chem. **26** [1961] 1030/4). – [20] E. L. De Young (J. Org. Chem. **26** [1961] 1312).

[21] M. Furdík, Š. Toma, J. Suchý (Chem. Zvesti **15** [1961] 789/806). – [22] C. R. Hauser, T. A. Mashburn (J. Org. Chem. **26** [1961] 1795/800). – [23] N. A. Nesmeyanov, B. N. Strunin (Dokl. Akad. Nauk SSSR **137** [1961] 106/8; Proc. Acad. Sci. USSR Chem. Sect. **136/141** [1961] 275/6). – [24] E. G. Perevalova, N. A. Simukova, T. V. Nikitina, P. D. Reshetov, A. N. Nesmeyanov (Izv. Akad. Nauk SSSR Otd. Khim. Nauk **1961** 77/83; Bull. Acad. Sci. USSR Div. Chem. Sci. **1961** 67/71). – [25] R. Riemschneider, D. Helm (Liebigs Ann. Chem. **646** [1961] 10/7).

[26] K. Schlögl, A. Mohar, M. Peterlik (Monatsh. Chem. **92** [1961] 921/6). – [27] K. Schlögl, H. Pelousek, A. Mohar (Monatsh. Chem. **92** [1961] 533/41). – [28] L. Wolf, H. Franz, H. Hennig (Z. Chem. [Leipzig] **1** [1961] 220/1). – [29] J. J. Mattice (Off. Dig. Fed. Soc. Paint Technol. **34** [1962] 602/21). – [30] K. Schlögl, H. Pelousek (Liebigs Ann. Chem. **651** [1962] 1/10).

[31] G. G. Dvoryantseva, M. I. Struchkova, Yu. N. Sheinker (Dokl. Akad. Nauk SSSR **152** [1963] 617/20; Dokl. Chem. Proc. Acad. Sci. USSR **148/153** [1963] 740/3). – [32] B. Gautheron, R. Dabard (Bull. Soc. Chim. France **1963** 2009). – [33] M. Rosenblum, A. K. Banerjee, N. Danieli, R. W. Fish, V. Schlatter (J. Am. Chem. Soc. **85** [1963] 316/24). – [34] R. L. Schaaf, C. T. Lenk (J. Org. Chem. **28** [1963] 3238/40). – [35] R. L. Schaaf, K. L. Rinehart (AD-403929 [1963] 1/80, 50; PB-147703 [1959] 1/59; U.S. Govt. Res. Rept. **34** [1960] 262).

[36] K. Schlögl, H. Egger (Monatsh. Chem. **94** [1963] 1054/63). – [37] J. Tirouflet, R. Dabard, E. Laviron (Bull. Soc. Chim. France **1963** 1655/60). – [38] B. M. Yavorskii, N. S. Kochetkova, G. B. Zaslavskaya, A. N. Nesmeyanov (Dokl. Akad. Nauk SSSR **149** [1963] 111/3; Dokl. Chem. Proc. Acad. Sci. USSR **148/153** [1963] 202/4). – [39] I. Ziegler, R. D. Rauschenbach, R. Allwinn, K. Wintersberger, Badische Anilin- und Sodafabrik A.-G. (Belg. 643139 [1963/64]; C.A. **63** [1965] 3131). – [40] Kodak-Pathé, J. P. Dubosc (Fr. 1361470 [1963/64]; C.A. **63** [1965] 13461/2).

[41] T. Leigh (J. Chem. Soc. **1964** 3294/302). – [42] A. V. Lesikar (J. Chem. Phys. **40** [1964] 2746/7). – [43] J. P. C. Dubosc, Eastman Kodak Co. (U.S. 3335008 [1964/67]; C.A. **67** [1967] No. 77884). – [44] B. Gautheron (Diss. Dijon 1965, pp. 1/100, 57/8). – [45] A. N. Nesmeyanov, B. M. Yavorskii, G. B. Zaslavskaya, N. S. Kochetkova (Dokl. Akad. Nauk SSSR **160** [1965] 837/40; Dokl. Chem. Proc. Acad. Sci. USSR **160/165** [1965] 131/4).

[46] R. F. Orloski, E. W. Neuse (unpublished results from E. W. Neuse, H. Rosenberg, J. Macromol. Sci. Rev. Macromol. Chem. C **4** [1970] 1/145, 91). – [47] K. Schlögl, W. Steyrer (Monatsh. Chem. **96** [1965] 1520/35). – [48] J. E. Robertson, Colgate-Palmolive Co. (U.S. 3377248 [1965/68]; C.A. **69** [1968] No. 96878). – [49] J. T. Suh, Colgate-Palmolive Co. (U.S. 3382267 [1965/68]; C.A. **69** [1968] No. 59378). – [50] G. G. Dvoryantseva, S. L. Portnova, Yu. N. Sheinker, L. P. Yur'eva, A. N. Nesmeyanov (Dokl. Akad. Nauk SSSR **169** [1966] 1083/6; Dokl. Chem. Proc. Acad. Sci. USSR **166/171** [1966] 773/5).

[51] E. W. Neuse, N. Sonnenberg, H. D. Chandler (Eur. Polym. J. **20** [1984] 1107/11). – [52] E. J. Olszewski, D. F. Martin (J. Organometal. Chem. **5** [1966] 203/4). – [53] W. M. Horspool, R. G. Sutherland (Tetrahedron Letters **1967** 4165/7). – [54] A. N. Nesmeyanov, E. G. Perevalova, L. N. Yur'eva (Izv. Akad. Nauk SSSR Ser. Khim. **1967** 694/6; Bull. Acad. Sci. USSR Div. Chem. Sci. **1967** 675/6). – [55] T. H. Barr, W. E. Watts (Tetrahedron **24** [1968] 3219/35).

[56] J. Besançon, J. Tirouflet (Rev. Chim. Minerale **5** [1968] 363/86). – [57] H. Hennig, O. Gürtler (J. Organometal. Chem. **11** [1968] 307/16). – [58] E. W. Neuse (Macromolecules **1** [1968] 171/8). – [59] T. H. Barr, W. E. Watts (Tetrahedron **25** [1969] 861/9). – [60] M. I. Bruce (Org. Mass Spectrom. **2** [1969] 997/1003).

[61] M. I. Bruce, M. J. Melvin (J. Chem. Soc. C **1969** 2107/12). – [62] W. H. Horspool, R. G. Sutherland, J. R. Sutton (Can. J. Chem. **47** [1969] 3085/8). – [63] J. J. McDonnell, G. Capen, R. Michealson (Tetrahedron Letters **1969** 4251/4). – [64] A. N. Nesmeyanov, G. B. Zaslavskaya, B. M. Yavorskii, N. S. Kochetkova, N. P. Gambaryan (Opt. Spektrosk. **26** [1969] 945/9; Opt.

Spectrosc. [USSR] **26** [1969] 512/5). – [65] A. Perjéssy, Š. Toma (Chem. Zvesti **23** [1969] 533/9).

[66] S. Sorriso, H. Lumbroso (Tetrahedron **26** [1970] 2695/701, [Erratum] 4674). – [67] I. J. Spilners, J. G. Larson (Org. Mass Spectrom. **3** [1970] 915/24). – [68] T. P. Vishnyakova, I. D. Vlasova, I. A. Eremina (Zh. Org. Khim. **6** [1970] 1837/40; J. Org. Chem. [USSR] **6** [1970] 1847/9). – [69] I. A. Eremina (Neft i Gaz i Ikh Produkty, Moscow 1971, pp. 144/5). – [70] I. A. Eremina, A. A. Zaveleva (Neft i Gaz i Ikh Produkty, Moscow 1971, pp. 143/4).

[71] W. M. Horspool, R. G. Sutherland, B. J. Thomson (J. Chem. Soc. C **1971** 1558/63). – [72] Ya. M. Paushkin, T. P. Vyshnyakova, I. D. Vlasova, I. A. Golubyeva, I. A. Yeryomina (5th Intern. Conf. Organometal. Chem., Moscow 1971, Vol. 1, Abstr., pp. 311/2). – [73] D. S. Brackman, D. J. Harper, Imperial Chemical Industries, Ltd. (Brit. 1382061 [1971/75]). – [74] H. Imai (Bull. Soc. Chem. Japan **45** [1972] 1264/5). – [75] J. A. Winstead (J. Org. Chem. **37** [1972] 1271/2).

[76] A. Boberg, B. Aakermark, AB. Aakerlund & Rausing (Swed. 11624/71 [1971]; Ger. Offen. 2244800 [1972/73]; C.A. **79** [1973] No. 19728). – [77] U. Casellato, M. Vidali, P. A. Vigato, B. Bandoli, D. A. Clemente (Inorg. Nucl. Chem. Letters **9** [1973] 299/302). – [78] V. Veselý, Š. Toma, J. Gurský, E. Patzelt (Ropa Uhlie **15** [1973] 194/7). – [79] P. A. Vigato, U. Casellato, D. A. Clemente, G. Bandoli, M. Vidali (J. Inorg. Nucl. Chem. **35** [1973] 4131/9). – [80] K. Yamakawa, M. Hisatome (J. Organometal. Chem. **52** [1973] 407/24).

[81] H. Raubach, B. Falk, U. Wilhelm (Ger. [East] 109878 [1973/74]; C.A. **83** [1975] No. 179310). – [82] H. Raubach, B. Falk, U. Stoffer, H. Oehlert (Ger. [East] 110890 [1973/75]; C.A. **84** [1976] No. 5654). – [83] A. N. Nesmeyanov, G. B. Shul'pin, L. V. Rybin, N. T. Gubenko, M. I. Rybinskaya, P. V. Petrovskii, V. I. Robas (Zh. Obshch. Khim. **44** [1974] 2032/41; J. Gen. Chem. [USSR] **44** [1974] 1994/2001). – [84] A. N. Nesmeyanov, G. B. Shul'pin, M. I. Rybinskaya, P. V. Petrovskii (Dokl. Akad. Nauk SSSR **215** [1974] 599/602; Dokl. Chem. Proc. Acad. Sci. USSR **214/219** [1974] 188/91). – [85] J. M. Wilson, D. L. Uhrich (Mol. Cryst. Liquid Cryst. **25** [1974] 113/21).

[86] A. N. Nesmeyanov, M. I. Rybinskaya, G. B. Shul'pin, A. A. Pogrebnyak (J. Organometal. Chem. **92** [1975] 341/67). – [87] V. Rapić, K. Schlögl, B. Steinitz (J. Organometal. Chem. **94** [1975] 87/98). – [88] P. Burgaud, Solvay & Cie. (Ger. Offen. 2513200 [1975] 1/12, 5; C.A. **84** [1976] No. 45389). – [89] A. N. Nesmeyanov, G, B. Shul'pin, M. V. Tolstaya, M. I. Rybinskaya (U.S.S.R. 555114 [1975/77]; C.A. **87** [1977] No. 135954). – [90] V. O. Aimiuwu (Diss. Kent State Univ. 1976 from Diss. Abstr. Intern. B **37** [1976] 2351).

[91] R. Lattrell, H. Kief, H. Baehr, Hoechst A.-G. (Ger. Offen. 2623486 [1976/77]; C.A. **88** [1978] No. 89852). – [92] R. Lattrell, H. Kief, H. Baehr, Hoechst A.-G. (Ger. Offen. 2623487 [1976/77]; C.A. **88** [1978] No. 89851). – [93] Š. Toma, P. Foltínová (Czech. 183387 [1976/80] from C.A. **95** [1981] No. 109476). – [94] V. O. Aimiuwu, D. L. Uhrich (Mol. Cryst. Liquid Cryst. **43** [1977] 295/312). – [95] Š. Toma, P. Foltínová, E. Zíkova (Acta Fac. Rerum Nat. Univ. Comenianae Chim. No. 25 [1977] 9/19).

[96] A. I. Voloshin, L. Yu. Anishenko, N. R. Prokopchuk, Ya. M. Paushkin, E. M. Eizenshtein, et al. (Ger. Offen. 2702696 [1977/78] 1/10; C.A. **89** [1978] No. 164857). – [97] M. Hisatome, S. Koshikawa, K. Chimura, H. Hashimoto, K. Yamakawa (J. Organometal. Chem. **145** [1978] 225/39). – [98] A. Siegel, M. D. Rausch (Syn. Reactiv. Inorg. Metal-Org. Chem. **8** [1978] 209/26). – [99] P. Canonne, G. Foscolos, R. Harder (J. Organometal. Chem. **178** [1979] 331/42). – [100] L.-C. Song (Hua Hsueh Tung Pao **1979** 313/5 from C.A. **92** [1980] No. 6652).

[101] K. Szulzewsky, I. Seidel, S. Kulpe, E. Höhne, H. Raubach (Krist. Tech. **14** [1979] 37/45). – [102] M. Vernois, H. L. Williams (J. Appl. Polym. Sci. **23** [1979] 1601/6). – [103] E. Haslinger, W. Robien, K. Schlögl, W. Weissensteiner (J. Organometal. Chem. **218** [1981] C11/C14). –

[104] A. Kashahara, T. Izumi, E. Yahagi, Y. Kamei (Yamagata Daigaku Kiyo Kogaku **16** [1981] 365/70; C.A. **95** [1981] No. 133100). – [105] R. M. Moriarty, H. Hu, S. C. Gupta (Tetrahedron Letters **22** [1981] 1283/6).

[106] G.V. Shul'pin (Zh. Obshch. Khim. **51** [1981] 2152/3; J. Gen. Chem. [USSR] **51** [1981] 1854). – [107] H.-H. Wei, D.-M. Horng (Transition Metal Chem. [Weinheim] **6** [1981] 319/20). – [108] E. Haslinger, K. Koci, W. Robien, K. Schlögl (Monatsh. Chem. **114** [1983] 495/9). – [109] A. N. Nesmeyanov [Nessmejanow], L. P. Yur'eva [Jurjewa], N. S. Kochetkova [Kotschetkowa], S. V. Vitt [Witt] (Izv. Akad. Nauk SSSR Ser. Khim. **1966** 560/2; Bull. Acad. Sci. USSR Div. Chem. Sci. **1966** 830/2; Abhandl. Deut. Akad. Wiss. Berlin Kl. Chem. Geol. Biol. **1966** No. 2, pp. E265/E268). – [110] R. Dabard, B. Gautheron (Compt. Rend. **254** [1962] 2014/6).

[111] D. J. Clancy, I. J. Spilners (Anal. Chem. **34** [1962] 1839). – [112] D. W. Fischer (Acta Cryst. **17** [1964] 619). – [113] X.-J. Yang, Z.-Z. Xie, F.-S. Ni, S.-J. Shi (Gaodeng Xuexiao Huaxue Xuebao **1** [1980] 48/60 from C.A. **94** [1981] No. 219071). – [114] X. Yang, Z. Xie, F. Ni, S. Shi (Fundam. Res. Organometal. Chem. Proc. China-Japan-US Trilateral Semin. Organometal. Chem., Beijing 1980 [1982], pp. 765/9). – [115] L. I. Tolstykh, L. Errais (Tr. Mosk. Inst. Neftekhim. Gazov. Prom. No. 158 [1981] 144/9 from C.A. **99** [1983] No. 5775).

[116] M. A. El-Hashash, S. El-Nagdy, R. Saleh (Indian J. Chem. A **22** [1983] 605/6). – [117] A. Perjéssy, Š. Toma (Collection Czech. Chem. Commun. **48** [1983] 1635/46). – [118] B.V. Polyakov, V. P. Tverdokhlebov, I.V. Tselinskii, N. M. Bakstova, G. M. Frolova (Zh. Obshch. Khim. **53** [1983] 2046/50; J. Gen. Chem. [USSR] **53** [1983] 1847/50). – [119] R. Yin, X. Zhen (Lanzhou Daxue Xuebao Ziran Kexueban **19** [1983] 41/5 from C.A. **100** [1984] No. 103578).

5.2.3.5.1.5 Ketones of the Type $R'–CO–C_5H_4FeC_5H_4–CO–R''$

For the compounds listed in Table 5, pp. 92/103, the substituents R appear in the sequence R = alkyl, alkenyl, alkinyl, arylalkyl, arylalkenyl, arylalkinyl, aryl, carbocyclyl, C–H–halogen radical, and C–H–O groups. The usual Gmelin "principle of the last position" leads here to placing R' as the substituent in the compound formula located in the last position, with R'' located as the first substituent. Thus, for example the compound $C_6H_5–CO–C_5H_4FeC_5H_4–CO–CH{=}CHC_6H_5$ (No. 37) appears after the compound $C_6H_5CH{=}CH–CO–C_5H_4FeC_5H_4–CO–CH_2C(CH_3)_3$ (No. 24), but before the compound 4-Cl–$C_6H_4CH{=}CH–CO–C_5H_4FeC_5H_4–CO–C_6H_5$ (No. 57).

Preparation. For preparation of the compounds see the following methods and chemical behavior of the listed parent compound:

Method I. Friedel-Crafts acylation of $C_5H_5FeC_5H_4–CO–R$ using $AlCl_3$. The individual processes differ primarily in the sequence of addition of the specific reaction constituents, their molar ratio (given in parentheses below), their composition, and the subsequent column chromatographic isolation of the compound.

a) In a weak N_2 flow the $C_5H_5FeC_5H_4–CO–R'$ compound dissolved in a minimum quantity of CH_2Cl_2 is slowly added at room temperature to a solution of $AlCl_3$ + R''COCl (1/4/2). After the usual processing the reaction product is filtered through an Al_2O_3 column and eluted with ether. $C_5H_5FeC_5H_4–CO–R'$ is separated from the eluate by column chromatography on Al_2O_3 with petroleum ether, subsequently the diketone is eluted with ether. For the lower melting compounds No. 33 and 34 the total reaction product must be chromatographed on SiO_2 or Al_2O_3. The products so produced are very pure [28].

b) The same as Method Ia, but with $C_5H_5FeC_5H_4–CO–R''$ plus R'COCl [28].

c) $C_5H_5FeC_5H_4$–CO–R′ in CH_2Cl_2 is slowly added to $AlCl_3$ + R″COCl (1/6/3) in CH_2Cl_2 and subsequently reacted 2h at room temperature and then 1 to 2h at 35°C. Separation on SiO_2 by the gradient elution with benzene/$CH_3CO_2C_2H_5$ at 19/1 to 3/1. For Nos. 67 and 69 small quantities of I with n = 5 or 7 are isolated in addition to the main products [48].

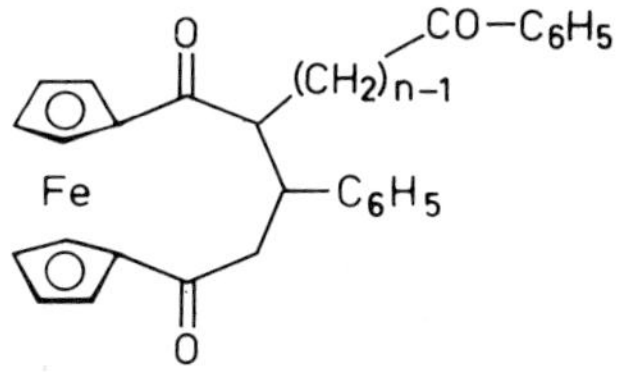

I

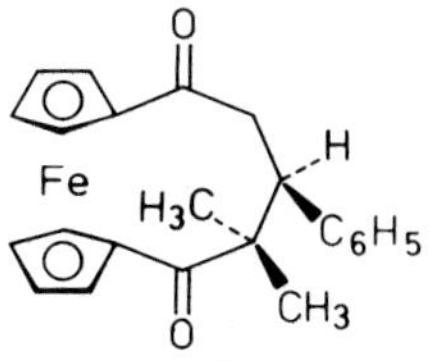

II

d) $C_5H_5FeC_5H_4$–CO–R″ in CH_2Cl_2 is added to $AlCl_3$ + R′COCl (1/(>2 to 2.5)/1) at room temperature. After the usual processing No. 28 is isolated by column chromatography on SiO_2 with 97/3 benzene/$CH_3CO_2C_2H_5$ [39]. No. 32 is separated on activated Al_2O_3 with 1/1 hexane/ether after previous elution with hexane of $C_5H_5FeC_5H_4$–CO–CH_3 [4].

e) $C_5H_5FeC_5H_4$–CO–CH_3, $AlCl_3$, and C_6H_5COCl (1/2/1) react in CS_2 first at room, then at boiling temperatures. $C_5H_5FeC_5H_4$–CO–CH_3 is eluted first from Al_2O_3 by benzene, subsequently No. 32 by benzene containing 0.03% ethanol [1].

f) $C_5H_5FeC_5H_4$–CO–R″ + R′COCl in CH_2Cl_2 are added dropwise to $AlCl_3$ (1/2.3/2.7) in CH_2Cl_2. No. 5 is separated on SiO_2 with 90/10 hexane/ether from the 1,2-compound and $C_5H_5FeC_5H_4$–CO–R″ [27].

g) As for Method If but in the mole ratio 1/1.6/2 in CS_2, with gentle boiling and with column chromatographic processing on Al_2O_3 with 2/1 benzene/benzine as eluent [2].

h) $C_5H_5FeC_5H_4$–CO–R′ + $(R''CO)_2O$ in CH_2Cl_2 is added to $AlCl_3$ (1/(2.2 to 5.6)/(4.5 to 8)) in CH_2Cl_2 at room temperature. Isolation takes place on SiO_2 with 90/10 Skelly B/ether as eluent, in the case of No. 40 on Al_2O_3 with 75/25 Skelly B/ether. In this latter case $C_5H_5FeC_5H_4$–CO–$C_{10}H_{15}$ passes through into the eluate and the No. 40 is retained on the column, from which it is eluted with ether [26].

i) Addition of $AlCl_3$ in small portions at room temperature (for No. 43 with cooling to 0°C) to $C_5H_5FeC_5H_4$–CO–R′ + R″COCl (2/1/>1) in CH_2Cl_2. Separation on SiO_2 with 19/1 to 0/1 benzene/$CH_3CO_2C_2H_5$ [50] or C_6H_6 [7].

j) Like Method Ii, but with $C_5H_5FeC_5H_4$–CO–R″ + R′COCl [50].

k) Like Method Ii, but with a mole ratio of 1.6/1/<1.5 and separation on Al_2O_3 with 19/1 benzene/$CH_3CO_2C_2H_5$ as eluent [51].

l) $AlCl_3$ dissolved in CH_2Cl_2 is added to $C_5H_5FeC_5H_4$–CO–R″ + $(R'CO)_2O$ (2.7/1/1.8) in CH_2Cl_2 at room temperature. Isolation on SiO_2 with 75/25 hexane/ether as eluent [27].

m) $AlCl_3$ + R″COCl in CH_2Cl_2 is added to $C_5H_5FeC_5H_4$–CO–R′ (3/1.6/1) in CH_2Cl_2. Column chromatographic processing on Al_2O_3 with 97/3 benzene/$CH_3CO_2C_2H_5$ [19, 20] or on SiO_2; elution by benzene follows that with the preceding mixture [42].

n) Like Method Im, but with R′COCl and $C_5H_5FeC_5H_4$–CO–R″ plus separation on SiO_2 in benzene [34] or with benzene/ether [8].

References on pp. 109/10

o) $R'COCl + AlCl_3$ in CH_2Cl_2 is added with ice-cooling to $C_5H_5FeC_5H_4$–CO–R″ + $AlCl_3$ (1.5/1.5/1/1.5) [46].

p) $R'COCl$ in CH_2Cl_2 is added to $AlCl_3 + C_5H_5FeC_5H_4$–CO–R″ (~1/~2.5/1) in boiling CH_2Cl_2 and reacted for 35 min. Separation on activated Al_2O_3 with ether eluent [4].

q) As for Method Ip but in a ~2/~2/1 mole ratio and at room temperature [8].

r) Like Method Ip, but with water-cooling and processing on SiO_2 in 4/1 benzene/ $CH_3CO_2C_2H_5$ [13].

s) Same as Method Ir, but with R″COCl and $C_5H_5FeC_5H_4$–CO–R′ [13].

t) As for Method Iq but with bubbling N_2 through the solution. The product recovered on SiO_2 with 97/3 benzene/$CH_3CO_2C_2H_5$ is rechromatographed in a 95/5 mixture of the same solvents [35].

u) Same as Method It but with R″COCl and $C_5H_5FeC_5H_4$–CO–R′ [35].

v) To $C_5H_5FeC_5H_4$–CO–R″ in CH_2Cl_2 at temperatures below 5°C are added $AlCl_3$ in small portions, R′COCl, and finally once more $AlCl_3$ (mole ratio 1/1/1/1). After completion of the reaction for 1 h at 0°C and 1 h at room temperature, processing with H_2O and K_2CO_3 follows. The product is purified by column chromatography on SiO_2 with benzene [56].

Method II. Boiling the ferrocenophanedione II (p. 85) with $C_2H_5OH/H_2O/NaOH$; No. 22 is isolated from the reaction mixture by column chromatography on SiO_2 with 98/2 benzene/$CH_3CO_2C_2H_5$ as eluent [39].

Re-formation from Dioximes. The dioximes HON=C(C_6H_5)–$C_5H_4FeC_5H_4$–CR″=NOH with R″ = CH_3, C_2H_5, C_3H_7, or $CH_2C_6H_5$ in acetone after addition of $C_6H_5SO_2Cl$ in the presence of OH^- and subsequent hydrolyses give in 10 to 15% yields both the compounds No. 32 to 35 and C_6H_5NH–CO–$C_5H_4FeC_5H_4$–CO–R″. If R′ ≠ C_6H_5, the Beckmann rearrangement fails and the hydrolysis of the intermediately formed imminium salts leads only to the diacyl compounds [37].

Purification of the compounds is accomplished by recrystallization from the solvents given in parentheses following the melting points in Table 5.

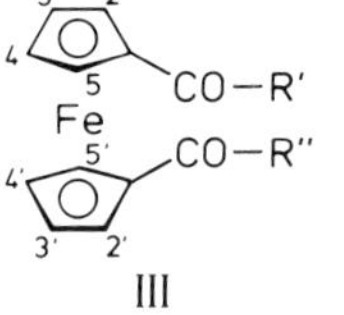

III

Properties

For description of the ^{1}**H NMR spectra** protons are numbered as in Formula III. The H-2,5 and H-2′,5′ protons show lower screening than the H-3,4 and H-3′,4′ protons. All four signals appear as triplets. If R′ and R″ are very similar, then only two signals appear for the cyclopentadienyl protons [28]. The compounds No. 19, 29, 54, 56, 58 to 60, 64, and 66 exhibit beside the signals of $C_5H_5FeC_5H_4$–CO–R′ the signal of the CH_3CO protons at $\delta = 2.83$ ppm [51]. In the compounds No. 67 to 70 the two activated CH_2 groups in R′ (7 and 11, 12, 13, or 14, respectively) are distinguished clearly by separate signals. The down field signal around $\delta = 2.95$ ppm for the CH_2 group neighboring the C_6H_5CO group (11 to 14) shows its stronger acidic character in comparison to CH_2-7 with $\delta \approx 2.65$ ppm [48].

The **IR spectra** contain two ν(C=O) bands, corresponding to those of the two individual ferrocenyl monoketones [28]. For the compounds No. 37, 38, 55, 57, and 65, which exist in CCl_4 in only a single conformation, the ν(C=O) bands (in cm^{-1}) of the cinnamoyl groups vary with the Hammett constants σ of the substituents as $\nu(C{=}O) = 7.28(\pm 0.62)\sigma + 1663.4(\pm 0.1)$ [17]. In the compounds No. 19, 29, 54, 56, 58 to 60, 64, and 66, also, the location of ν(C=O) for R′ varies with the substituents on the C_6H_4, while ν(C=O) for CH_3CO is always found at 1680 cm^{-1} [51]. The compounds $R'{-}CO{-}C_5H_4FeC_5H_4{-}CO{-}CH_3$ show in paraffin oil a weak band in the region 1113 to 1122 cm^{-1} [18].

A linear correlation corresponding to a modified and extended Seth-Paul-Van Duyse equation has been observed between the ν(C=O) vibration in CCl_4 and the $X^+(R^{A,B})$ constants of the structural fragments for 154 compounds with ferrocene structures of the general formula $R^A{-}CO{-}R^B$, among them the compound No. 32 and the s-cis compounds No. 8, 10, 11, 19 to 22, 24, 29, 54 to 57, 59, 60, and 64 to 66 with $R^A = C_5H_4FeC_5H_4{-}CO{-}R''$ and $R^B = R'$, and the compounds No. 2 and 32 and the s-cis compounds No. 8, 10, 11, 19 to 22, 24, and 37 with $R^A = C_5H_4FeC_5H_4{-}CO{-}R'$ and $R^B = R''$ (s-cis denotes the conformation of the C=C bond with respect to the C=O bond studied or present in the given structural fragment) [55].

Electrochemical and Chemical Behavior

The following half-wave potentials (in V vs. the saturated calomel electrode) have been measured for the **polarographic reduction** in ethanol/H_2O at pH 2.48: $E_{1/2} = -0.71$ for No. 37, −0.70 for No. 55, −0.69 for No. 57, and −0.71 for No. 65. The values show only a very slight dependence on the substituents of the C_6H_4 groups [3].

Reduction. The compounds are reduced smoothly and practically quantitatively by $LiAlH_4$/$AlCl_3$ in ether or tetrahydrofuran to the $R'{-}CH_2{-}C_5H_4FeC_5H_4{-}CH_2{-}R''$, as shown by studies with the compounds No. 1 to 3, 14 to 16, and 32 to 35 [6, 22, 29]. Clemmensen reduction with zinc amalgam and concentrated hydrochloric acid in CH_3CO_2H led in contrast as a function of R′ and R″ to differing products. When R′ and R″ are alkyl groups (Nos. 1 to 3) or arylalkyl groups (Nos. 14 to 16), only the hydrocarbon derivatives $R'{-}CH_2{-}C_5H_4FeC_5H_4{-}CH_2{-}R''$ (IV) occur. If $R' = C_6H_5$ and R″ = alkyl or arylalkyl (Nos. 32 to 35), then V (see formula) forms besides IV as a second product, in 8% yield for No. 32, 10% for Nos. 33 and 34, and 16% for No. 35. The two compounds are separable by thin layer chromatography on SiO_2 with petroleum ether [6, 22, 39].

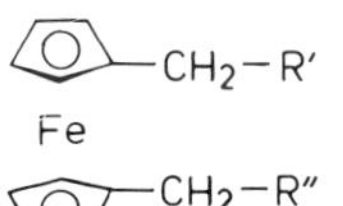

IV

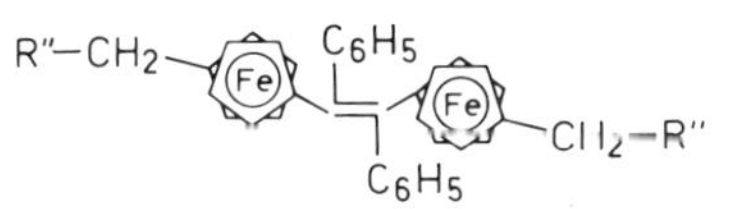

V

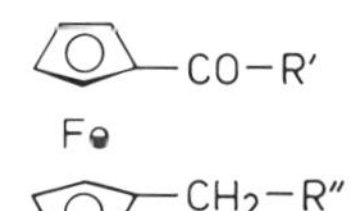

VI

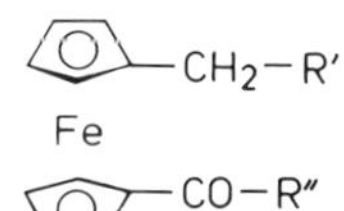

VII

$(C_6H_5)_3SnH$ in the presence of CH_3COCl reduces the diketones (molar ratio 2/4/1) in benzene selectively to $R'{-}CO{-}C_5H_4FeC_5H_4{-}CH_2{-}R''$ (VI), $R'{-}CH_2{-}C_5H_4FeC_5H_4{-}CO{-}R''$ (VII), and the hydrocarbon derivative IV. The more different R′ and R″ are, the greater is the selectivity. Thus, for No. 14 the yield of VI is 48%, with no VII and 12% IV. For No. 32 it is 48% VI, no VII, and 13% IV. For No. 33, 20% VI, 16% VII, and 19% IV. For No. 35 besides traces of VI, 38% VII and 17% IV. A slight excess of $(C_6H_5)_3SnH$ increases the yield of IV only a little; only a large excess leads after a sufficiently long reaction time to complete reduction. The mechanism is still unclear, but does not appear to involve radicals. The presence of CH_3COCl is indispensable for the process [16, 28, 30]. For behavior of No. 37 see p. 107. $(C_4H_9)_3SnH$ acts much less selectively: Thus, No. 32

gives in addition to 30% VI and 15% IV also 20% VII. No. 35 yields in addition small quantities (7%) of $C_6H_5CH_2$–$C_5H_4FeC_5H_4$–$C(O_2CCH_3)$=CHC_6H_5, the enol acetate derived from VII [30]. $(C_6H_5)_2SnH_2$ possesses to be sure the same selectivity as $(C_6H_5)_3SnH$, but due to its instability a limited usefulness [30].

Reduction with KBH_4 in CH_3OH for Nos. 1 to 3, 14 to 16, 32 to 35 [6, 22], with $NaBH_4$ in $(CH_3)_2CHOH$ for Nos. 1 and 32 [4, 5], or with $LiAlH_4$ in ether for Nos. 32 [33, 38] and 36 [23] leads to the dialcohols VIII, which are obtained as mixtures of the two diastereoisomers a (erythro forms) and b (threo forms) in 1/1 ratios [33].

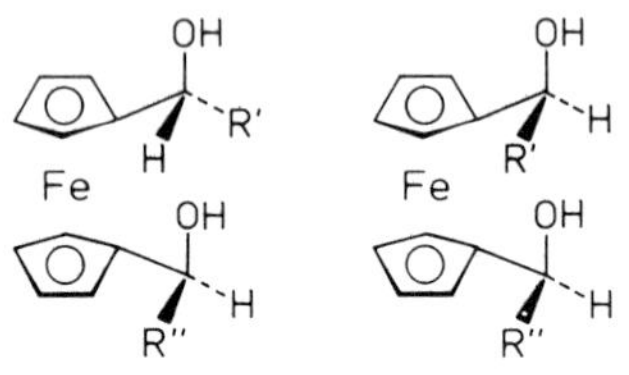

a b

VIII

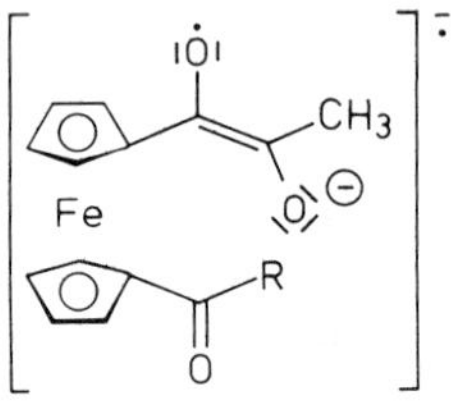

IX

Oxidation. Compounds of the R–CO–$C_5H_4FeC_5H_4$–CO–CH_2CH_3 type, such as Nos. 1, 3, 6, 7, 39, and 40, which possess a CH_2 group in the α position to the C=O, are oxidized by O_2 in $(CH_3)_2SO/KOC(CH_3)_3$ to the stable radical anion IX (see Section 5.2.3.5.1.8, p. 155). In the presence of a larger excess of $KOC(CH_3)_3$ or on its replacement by $CsOC(CH_3)_3$, No. 3 gives a mixture of $[n\text{-}C_3H_7$–CO–$C_5H_4FeC_5H_4$–CO=$COCH_3]^{\dot{-}}$ and $[C_2H_5$–CO–$C_5H_4FeC_5H_4$–CO=$COC_2H_5]^{\dot{-}}$ in an 8.6/1 ratio. Apparently the oxidation takes place via the primarily formed dianion $[CH_3CH$–CO–$C_5H_4FeC_5H_4$–CO–$CHC_2H_5]^{2-}$. Both $[(CH_3)_2CHCH_2$–CO–C_5H_4Fe–C_5H_4–CO=$COCH_3]^{\dot{-}}$ and a second radical anion $[C_2H_5$–CO–$C_5H_4FeC_5H_4$–CO=$COCH(CH_3)_2]^{\dot{-}}$ are formed by No. 6 [26, 27].

No. 71 is oxidized to HO_2C–$C_5H_4FeC_5H_4$–CO–C_6H_5 by boiling with $KOC(CH_3)_3$ in $CH_3OCH_2OCH_3/H_2O$ and subsequent hydrolysis [56].

Monoximes. Dioximes. Reaction of the compounds with $NH_2OH \cdot HCl$ and KOH in a 1/5/10 molar ratio in boiling ethanol gives for compounds with R', R'' = alkyl or arylalkyl only dioximes, for compounds with R' or R'' = aryl in contrast predominantly monoximes. Only at a molar ratio of 1/20/50 are these latter also completely converted to the dioximes [6, 43]. For re-formation from the dioximes see p. 86.

Michael Addition. The compounds C_6H_5–CO–$C_5H_4FeC_5H_4$–CO–CH=CHC_6H_4–R-4 with R = H (No. 37), Cl (No. 57), and OCH_3 (No. 65) react with $NCCH_2CO_2C_2H_5$ in $CH_3OH/NaOCH_3$ with nucleophilic addition to form C_6H_5–CO–$C_5H_4FeC_5H_4$–CO–$CH_2CH(C_6H_4$–R-4)CH(CN)$CO_2C_2H_5$. Under first order conditions with respect to the ferrocene derivative the following rate constants have been measured at 60.0 ± 0.1°C: $k \cdot 10^3 = 26.7 \pm 1.8$ for No. 37, 30.7 ± 1.6 for No. 57, and 12.4 ± 1.3 for No. 65, all in min^{-1}. These show in contrast to the reduction potentials of Nos. 37, 57, and 65 (cf. p. 87) a distinct dependence on R [10].

Compounds with the grouping X (see formula) allow during an intramolecular Michael addition cyclization to derivatives of the [5]-ferrocenophane-6,10-dione XI. In the base-catalyzed reaction, part D of X acts as a donor, part A as an acceptor. Under first order conditions with respect to X, the rate-determining step (for which the rate constants k listed in

Table 4, pp. 90/1, are valid) consists in the addition of the carbanion formed by proton cleavage due to the action of bases on D, to the β C atom (C-8) of the double bond [14, 50, 51]. For kinetic measurements in HCl/CH_3OH for No. 56 see p. 109.

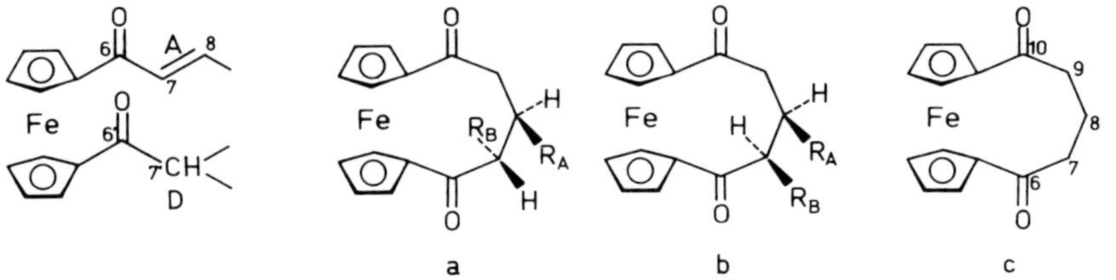

X XI

If X contains the substituents R_A and R_B on two different atoms from among C-7, C-7′, and C-8, then XI can be obtained as the two diastereoisomers, the trans form a and the cis form b, see for example [36, 39]. While reaction at lower room temperatures and higher dilutions favors the cis isomers, higher concentrations and boiling temperatures give nearly exclusively the trans isomers [39].

With variation of the donor part for $C_6H_5CH{=}CH{-}CO{-}C_5H_4FeC_5H_4{-}CO{-}R''$ with $R''=CH_3$ (No. 19), C_2H_5 (No. 20), C_3H_7 (No. 21), $CH(CH_3)_2$ (No. 22), $CH_2CH(CH_3)_2$ (No. 23), or $CH_2C_6H_5$ (No. 25) in $CH_3OH/NaOCH_3$, kinetic studies conducted spectrophotometrically between 35 and 55°C show a distinct increase of the activation parameter with the size of R″, such that the activation entropy $\Delta S^{\neq}$ increases even more strongly than the activation enthalpy $\Delta H^{\neq}$ [50]. Investigations conducted in $CH_3OH/NaOH$ at 25 to 40°C on $C_6H_5CH{=}CH{-}CO{-}C_5H_4FeC_5H_4{-}CO{-}CH_2C_6H_4{-}R$ with R = H (No. 25), CH_3-4 (No. 26), CH_3-3 (No. 27), F-4 (No. 47), Cl-4 (No. 49), Cl-3 (No. 50), Br-4 (No. 51), Br-3 (No. 52), I-3 (No. 53), OCH_3-4 (No. 62), and OCH_3-3 (No. 63) show the rate constants to increase with the increasing acceptor action of R. The rate constants k and Hammett constants σ of R are connected via $\log k/k_o = \varrho\sigma$, where k_o = the rate constant for R = H, ϱ = the specific reaction constant. The activation energy remains relatively unaffected by R [14].

Kinetic studies with variation of the acceptor portion of $CH_3{-}CO{-}C_5H_4FeC_5H_4{-}CO{-}R'$ (with $R'=CH{=}CH_2$ (No. 8), $C(CH_3){=}CH_2$ (No. 9), $CH{=}CHCH_3$ (No. 10), $CH{=}C(CH_3)_2$ (No. 11), and $CH{=}CHC(CH_3)_3$ (No. 12)) show an impeding of the reaction by these substituents, but as a whole a smaller effect on the reaction than that produced by changes in the donor portion [50]. This holds also for studies on $CH_3{-}CO{-}C_5H_4FeC_5H_4{-}CO{-}CH{=}CHC_6H_4{-}R$ with R = H (No. 19), CH_3-4 (No. 29), F-4 (No. 54), Cl-4 (No. 56), Cl-3 (No. 58), Br-4 (No. 59), Br-3 (No. 60), OCH_3-4 (No. 64), and OCH_3-3 (No. 66), in $CH_3OH/NaOCH_3$ between 35 and 55°C. R still affects $\Delta S^{\neq}$ more than $\Delta H^{\neq}$. Here, also, $\log k = \varrho\sigma + c$ is valid, where ϱ is temperature-independent in the given range [51].

It is also possible to catalyze the intramolecular Michael addition with acids instead of bases. With substituents on two different C atoms from among C-7, 7′, 8, a mixture of XIa and b is obtained in which in general XIb predominates [36].

Aldol Condensation. The compounds $R'{-}CO{-}C_5H_4FeC_5H_4{-}CO{-}CH_3$ react with the aromatic aldehydes RCHO in a 1/1 molar ratio to form $R'{-}CO{-}C_5H_4FeC_5H_4{-}CO{-}CH{=}CHR$ with release of H_2O in the presence of bases. If $R'=C_6H_5$, the chalcone-analogous compound constitutes the end product of the reaction [28]. If R′ is an alkyl or arylalkyl radical, then the internal Michael addition follows spontaneously, forming XI [28, 39]. The same reaction conditions are valid for the preferential formation of cis or trans isomers as with the Michael addition [39]. For evidence that the two cyclization products can in fact be treated as [5]-ferrocenophane-6,10-dione derivatives, see [2]; other structures proposed by Hauser et al. [52] are thus excluded.

References on pp. 109/10

Table 4

Rate Constants k for Intramolecular Michael Addition Cyclization in $NaOCH_3/CH_3OH$ (under first order conditions with respect to the compounds listed) and Related Activation Parameters.

compound No.	$k \cdot 10^4$ in s^{-1} 15°C	20°C	25°C	30°C	35°C	40°C	45°C	50°C	55°C	$\Delta H^{\neq}$ in kJ/mol	$\Delta S^{\neq}$ in $J \cdot mol^{-1} \cdot K^{-1}$	Ref.
8	1023.60	1275.20	1683.90	1992.20	—	—	—	—	—	33.05±3.77	−157.65±36.43	[50]
9	141.71	198.39	291.47	397.45	—	—	—	—	—	50.17±1.17	−114.77±11.30	[50]
10	20.15	33.21	49.39	68.00	—	—	—	—	—	59.16±3.77	−108.46±2.08	[50]
11	8.32	13.55	22.42	36.01	—	—	—	—	—	72.01±2.43	−65.23±5.44	[50]
12	1.39	2.57	4.75	8.38	—	—	—	—	—	86.82±1.63	−25.77±0.38	[50]
17	5.79	9.77	16.61	26.55	—	—	—	—	—	73.97±0.80	−58.66±1.72	[50]
18	0.97	0.97	2.22	4.62	—	—	—	—	—	110.75±2.34	+45.19±2.50	[50]
19	—	—	—	—	0.7886	1.511	2.826	4.222	6.312	87.03±5.08	−48.79±3.81	[51]
19	—	—	—	—	1.071	—	2.866	4.197	6.363	33.05±3.77	−157.65±36.43	[50]
19	—	—	—	—	—	—	—	—	—	77.08±2.09*)	−81.35±10.89*)	[41]
20	—	—	—	—	1.873	—	5.282	8.657	13.093	79.71±2.10	−58.74± 4.20	[50]
21	—	—	—	—	1.600	—	4.803	8.264	13.608	88.20±2.10	−39.66±2.30	[50]
22	—	—	—	—	0.132	—	0.423	0.925	1.458	103.51±5.40	−11.05±1.50	[50]
23	—	—	—	—	0.0158	—	0.075	0.165	0.267	116.02±7.20	+13.22±2.43	[50]
25	—	—	8.05±0.12	10.35±1.15	13.78±0.68	19.87±0.28	—	—	—	46.05±0.42	—	[14]
25	2817.0	4078.6	6197.4	8106.6	—	—	—	—	—	52.09±2.51	−102.17±2.92	[50]
26	—	—	5.72±1.63	7.23±0.3	10.92±1.15	14.45±0.42	—	—	—	47.73±0.84	—	[14]
27	—	—	6.28±0.12	8.58±0.28	13.05±0.28	17.28±0.47	—	—	—	51.50±1.26	—	[14]

References on pp. 109/10

29	—	—	—	—	0.6368	1.220	1.864	3.149	4.755	83.29±3.00	−62.94±3.12	[51]
31	—	—	—	—	—	—	—	—	—	79.42±2.05*)	−71.09±14.24*)	[41]
47	—	—	16.92±0.[illegible]1	23.25±0.12	34.25±0.78	46.53±0.78	—	—	—	52.75±0.42	—	[14]
49	—	—	34.38±1.[illegible]2	47.23±4.97	66.88±1.15	91.17±2.72	—	—	—	49.82±0.42	—	[14]
50	—	—	43.37±0.6	59.95±1.58	82.2 ±1.85	97.6 ±2.33	—	—	—	42.29±0.42	—	[14]
51	—	—	41.27±0.[illegible]5	50.6 ±1	74.1 ±1.85	91.97±1.63	—	—	—	46.89±6.28	—	[14]
52	—	—	47.17±1.2	60.05±0.0	97.65±2.3	102.85±6.8	—	—	—	41.03±0.0	—	[14]
53	—	—	46 ±0.38	62.73±2.15	80.07±2.05	100.28±2.86	—	—	—	43.96±2.51	—	[14]
54	—	—	—	—	0.9396	1.626	2.590	5.062	5.956	80.94±6.58	−67.47±7.51	[51]
56	—	—	—	—	1.1535	2.204	3.880	6.179	11.677	94.77±2.48	−21.05±0.71	[51]
56	—	—	—	—	—	—	—	—	—	78.54±2.09*)	−75.15±22.19*)	[41]
58	—	—	—	—	1.8875	3.047	6.107	9.738	13.455	85.59±5.32	−46.80±3.79	[51]
59	—	—	—	—	1.2408	2.439	4.450	7.339	9.547	87.01±6.73	−44.95±4.56	[51]
60	—	—	—	—	1.9075	3.355	4.619	10.020	14.034	85.37±6.18	−47.65±4.50	[51]
62	—	—	6.73±0.03	8.72±0.23	13.53±0.38	15.48±0.12	—	—	—	42.71±1.67	—	[14]
63	—	—	9.27±0.07	13.15±0.88	19.48±0.25	24.05±0.65	—	—	—	46.48±2.51	—	[14]
64	—	—	—	—	0.7521	1.275	2.221	3.755	5.645	85.62±1.89	−54.36±1.62	[51]
66	—	—	—	—	0.9706	1.454	2.935	4.229	6.743	82.81±4.48	−61.63±4.53	[51]

*) Values for 314 K from measurements between 20 and 60°C.

References on pp. 109/10

Condensation with C_6H_5CHO in a 1/1 molar ratio is also effected by concentrated H_2SO_4 in glacial acetic acid. With $R'=C_2H_5$ (No. 1), C_3H_7 (No. 2), or CH_2Cl (No. 41) in addition to the cis compound XIb (p. 89) ($R_A=C_6H_5$, $R_B=CH_3$, C_2H_5, or Cl), the compounds XIIa and b are obtained with $R_B=CH_3$ or C_2H_5 for Nos. 1 and 2, respectively, and XIIa with $R_B=Cl$ for No. 41. Since neither on cyclization experiments with Nos. 20 or 28 in conc. CH_3CO_2H/H_2SO_4 nor in the attempted isomerization of XIb ($R_A=C_6H_5$, $R_B=CH_3$) do derivatives of XII appear under these conditions, R′–CO–$C_5H_4FeC_5H_4$–CO–$CH(COCH_3)_2$ and $C_6H_5CH{=}CR_B$–CO–$C_5H_4FeC_5H_4$–CO–$CH(COCH_3)_2$ are assumed as the primary and secondary intermediate products of the reaction leading to XII [40].

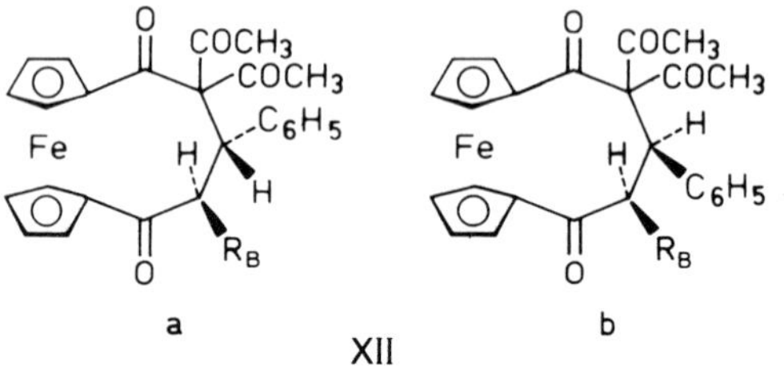

XII

Basicity of C_6H_5–CO–$C_5H_4FeC_5H_4$–COCH=CHC$_6H_4$–R-4. These compounds with R = H (No. 37), CH_3 (No. 38), F (No. 55), Cl (No. 57), or OCH_3 (No. 65) form a hydrogen bond with phenol. A measure of the basicity of the compounds is the valence bonding shift of the bound phenolic OH groups. A linear relationship exists between this and the σ^+ and σ constants of R, in each case. The intermolecular hydrogen bonding enthalpies at room temperature are $-\Delta H=4.6$ for No. 37, 4.7 for No. 38, 4.6 for Nos. 55 and 57, and 4.8 for No. 65, in kcal/mol (19.3, 19.7, 19.3, and 20.1, respectively, in kJ/mol) [21].

Applications

No. 37 is an effective antiknock agent. It forms less residue and possesses a lower solubility in the fuel than does ferrocene [32]. Compounds such as No. 14, 32, and 39 can be used together with a diazonium salt for preparation of photosensitive copying material coated on a paper support [53, 54]. No. 71 serves as parent compound for the synthesis of potentially catalytic bimetallic compounds containing the $(CO)_3Cr$ group [56]. For possible applications of Nos. 41, 42, 44, and 45 see "Further information", p. 107.

Table 5

Ketones of the Type R′–CO–$C_5H_4FeC_5H_4$–CO–R″.
Further information for numbers preceded by an asterisk is given at the end of the table, pp. 103/9.
For abbreviations and dimensions see p. VIII.

No.	R′–CO–$C_5H_4FeC_5H_4$–CO–R″ R′	 R″	method of preparation (yield in %), properties, and remarks	Ref.
compounds with R′ and R″ containing only C and H:				
*1	C_2H_5	CH_3	Ia (85), Ig (62.8, 57), Ip orange solid, yellow-red needles, m.p. 54 to 55° (petroleum ether), 58.5 to 59° (ether), 64° (hexane) ^{1}H NMR ($CDCl_3$): 1.19 (t, CH_3 of C_2H_5), 2.35 (s, $COCH_3$), 2.70 (m, CH_2); 4.49 (t, 4H) and 4.77 (m, 4H) (C_5H_4) IR (KBr): see figure in [2]	[2, 4 to 6, 26, 28]

Table 5 (continued)

No.	$R'-CO-C_5H_4FeC_5H_4-CO-R''$ R'	 R''	method of preparation (yield in %), properties, and remarks	Ref.
*2	n-C_3H_7	CH_3	Ia (77), Ig (45), Il (70) yellow-red crystals, m.p. 49 to 50° (petroleum ether), 54°, 58° (hexane) ^{1}H NMR ($CDCl_3$): 1.01 (t, CH_3 of C_3H_7), 1.72 (m, CH_2 of CH_2CH_3), 2.34 (s, $COCH_3$), 2.65 (t, $COCH_2$); 4.48 (t, 4H) and 4.74 (m, 4H) (C_5H_4) IR (KBr): see figure in [2] IR (CCl_4): 1686 (νC=O of COR'')	[2, 6, 27, 28, 55]
3	n-C_3H_7	C_2H_5	Ia (82), Ih (73.2) m.p. 39.5°, 46°, b.p. 180 to 182°/0.3 Torr ^{1}H NMR ($CDCl_3$): 1.00 (t, CH_3 of C_3H_7), 1.17 (t, CH_3 of C_2H_5), 1.73 (m, middle CH_2 of C_3H_7), 2.63 (t, $COCH_2$ of C_3H_7), 2.69 (m, $COCH_2$ of C_2H_5); 4.45 (t, 4H) and 4.76 (t, 4H) (C_5H_4)	[6, 26, 28]
*4	$(CH_3)_2CH$	CH_3	Ig (50) yellow-red needles, m.p. 45 to 46° (petroleum ether) IR (KBr): see figure in [2]	[2]
5	$(CH_3)_2CHCH_2$	CH_3	If (72) m.p. 76 to 77° ^{1}H NMR ($CDCl_3$): 1.01 (d, CH_3 of i-C_4H_9), 2.28 (m, CH of i-C_4H_9), 2.35 (s, $COCH_3$), 2.55 (d, $COCH_2$); 4.51 (t, 4H) and 4.78 (t, 4H) (C_5H_4)	[27]
6	$(CH_3)_2CHCH_2$	C_2H_5	Ih (79.5) m.p. 42°, b.p. 193 to 194°/0.3 Torr ^{1}H NMR ($CDCl_3$): 1.00 (d, CH_3 of i-C_4H_9), 1.19 (t, CH_3 of C_2H_5), 2.25 (m, CH of i-C_4H_9), 2.54 (d, CH_2 of i-C_4H_9), 2.70 (m, CH_2 of C_2H_5); 4.47 (m, 4H) and 4.77 (m, 4H) (C_5H_4)	[26]
7	$(CH_3)_3C$	C_2H_5	Ih (81.4) m.p. 32° ^{1}H NMR ($CDCl_3$): 1.20 (t, CH_3 of C_2H_5), 1.32 (s, C_4H_9), 2.75 (m, CH_2); 4.48 (t, 4H), 4.78 (t, 2H), and 4.84 (t, 2H) (C_5H_4)	[26]
*8	$\overset{8}{C}H_2{=}\overset{7}{C}H$	CH_3	Ii, see No. 43, p. 107 red solid, m.p. 60 to 61° (benzene/ petroleum ether or ligroin) ^{1}H NMR ($CDCl_3$): 2.29 (s, CH_3), 4.39 to 4.94 (m, 8H, C_5H_4), 5.57 to 7.35 (ABX pattern, $CH{=}CH_2$)	[7 to 9, 50, 55]

References on pp. 109/10

Table 5 (continued)

No.	$R'-CO-C_5H_4FeC_5H_4-CO-R''$ R'	R''	method of preparation (yield in %), properties, and remarks	Ref.
			^1H NMR ($CDCl_3$): 2.34 (s, CH_3); 4.50 (m), 4.56 (t), and 4.83 (t) (C_5H_4); 5.69 (dd, H-8_a), 6.47 (dd, H-8_b; J(H-8_a, 8_b) = 2.6), 6.83 (dd, H-7; J (H-7, 8_a) = 17, J (H-7, 8_b) = 9.8) IR (Nujol): 1661 ($\nu C{=}O$) IR (CCl_4): 1663 ($\nu C{=}O$ of COR') for s-cis, 1682 ($\nu C{=}O$ of COR'') UV (C_2H_5OH): $\lambda_{max}(\varepsilon)$ = 348sh (1645), 362 (1680), 476 (855) UV (absolute C_2H_5OH): $\lambda_{max}(\log \varepsilon)$ = 221 (4.35), 270 (4.08), 366 (3.23) UV ($CH_3OH/NaOCH_3$): $\lambda_{max}(\log \varepsilon)$ = 277 (3.662)	
*9	$\overset{8}{C}H_2{=}\overset{7}{C}(CH_3)$	CH_3	Ij (12) m.p. 80 to 82° (benzene/petroleum ether) ^1H NMR ($CDCl_3$): 2.03 (m, CH_3-7), 2.35 (s, $COCH_3$); 4.50 (m), 4.74 (t), and 4.84 (t) (relative intensity 2/1/1, C_5H_4); 5.66 (m, H-8_a; J (H-8_a, CH_3-7) = J (H-8_a, 8_b) = 1), 5.75 (m, H-8_b) UV ($CH_3OH/NaOCH_3$): $\lambda_{max}(\log \varepsilon)$ = 265 (3.903)	[50]
*10	$CH_3\overset{8}{C}H{=}\overset{7}{C}H$	CH_3	Ij (47) m.p. 108 to 109° (benzene/petroleum ether) ^1H NMR ($CDCl_3$): 1.95 (dd, CH_3-8), 2.34 (s, $COCH_3$); 4.51 (m), 4.74 (t), and 4.81 (t) (relative intensity 2/1/1, C_5H_4); 6.47 (dq, H-7 or 8; J (H-7, 8) = 15.2, J (H-7, CH_3-8) = 1), 7.09 (m, H-8 or 7; J (H-8, CH_3-8) = 6.2) IR (CCl_4): 1672 ($\nu C{=}O$ of COR') for s-cis, 1681 ($\nu C{=}O$ of COR'') UV ($CH_3OH/NaOCH_3$): $\lambda_{max}(\log \varepsilon)$ = 265 (3.645)	[50, 55]
*11	$(CH_3)_2\overset{8}{C}{=}\overset{7}{C}H$	CH_3	Ij (58) m.p. 57 to 58° (benzene/petroleum ether) ^1H NMR ($CDCl_3$): 1.97 (d, CH_3-8_a), 2.23 (d, CH_3-8_b; J (CH_3-8_b, H-7) = 1), 2.32 (s, $COCH_3$); 4.46 (m) and 4.75 (m) (same intensity, C_5H_4); 6.30 (m, H-7) IR (CCl_4): 1659.5 ($\nu C{=}O$ of COR') for s-cis, 1681 ($\nu C{=}O$ of COR'') UV ($CH_3OH/NaOCH_3$): $\lambda_{max}(\log \varepsilon)$ = 267 (3.984)	[50, 55]
*12	$(CH_3)_3C\overset{8}{C}H{=}\overset{7}{C}H$	CH_3	Ii (79) m.p. 80 to 82° (benzene/petroleum ether) ^1H NMR ($CDCl_3$): 1.10 (s, C_4H_9), 2.28 (s, $COCH_3$); 4.44, 4.50, 4.70, and 4.79 (all t of same intensity, C_5H_4); 6.31 (d, H-7 or 8; J (H-7, 8) = 15.9), 7.08 (d, H-8 or 7) UV ($CH_3OH/NaOCH_3$): $\lambda_{max}(\log \varepsilon)$ = 266 (4.283)	[50]

References on pp. 109/10

Table 5 (continued)

No.	$R'–CO–C_5H_4FeC_5H_4–CO–R''$ R'	 R''	method of preparation (yield in %), properties, and remarks	Ref.
*13	n-$C_8H_{17}CH=$ $CH(CH_2)_7$	CH_3	Ig (53) red needles, m.p. 115 to 117° IR (KBr): see figure in [2]	[2]
*14	$C_6H_5CH_2$	CH_3	Ia (80), Ig (55) yellow crystals, m.p. 121 to 123° (benzene), 126° (hexane) ^{1}H NMR ($CDCl_3$): 4.38 (t, H-3′,4′), 4.50 (t, H-3,4), 4.70 (t, H-2′,5′), 4.82 (t, H-2,5) IR (KBr): see figure in [2]	[2, 6, 28]
15	$C_6H_5CH_2$	C_2H_5	Ia (76) m.p. 71° (hexane) ^{1}H NMR ($CDCl_3$): 4.37 (t, H-3′,4′), 4.48 (t, H-3,4), 4.72 (t, H-2′,5′), 4.81 (t, H-2,5)	[6, 28]
16	$C_6H_5CH_2$	C_3H_7-n	Ia (77) m.p. 90°, 91° (hexane) ^{1}H NMR ($CDCl_3$): 4.39 (t, H-3′,4′), 4.50 (t, H-3,4), 4.70 (t, H-2′,5′), 4.82 (t, H-2,5)	[6, 28]
*17	$C_6H_5CH_2$	$\overset{7'}{C}H=C(CH_3)_2$	Ij (47) m.p. not reported ^{1}H NMR ($CDCl_3$): 1.92 (d, CH_3), 2.18 (d, CH_3), 3.90 (s, CH_2); 4.32, 4.44, 4.68, and 4.75 (all same intensity t, C_5H_4); 6.26 (m, H-7′; J(H-7′, CH_3) = 1), 7.26 (m, C_6H_5) UV ($CH_3OH/NaOCH_3$): $\lambda_{max}(\log \varepsilon) = 277 (4.004)$	[50]
*18	$C_6H_5CH_2$	$\overset{7'}{C}H–\overset{8'}{C}HC(CH_3)_3$	Ii (41) m.p. 128 to 129° (benzene/petroleum ether) ^{1}H NMR ($CDCl_3$): 1.14 (s, C_4H_9), 3.93 (s, CH_2); 4.40, 4.46, 4.73, and 4.78 (all same intensity t, C_5H_4); 6.32 (d, H-7′ or 8′), 7.08 (d, H-8′ or 7′; J(H-7′,8′) = 15.8), 7.32 (m, C_6H_5) UV ($CH_3OH/NaOCH_3$): $\lambda_{max}(\log \varepsilon) = 268 (4.328)$	[50]
*19	$C_6H_5\overset{8}{C}H=\overset{7}{C}H$	CH_3	Ii (54), Ik (54), Im (34), Iq (80), Is (35) red solid, m.p. 125 to 127° (3/1 acetone/petroleum ether), 131 to 132° (benzene/petroleum ether, or ligroin/ether) ^{1}H NMR ($CDCl_3$): 2.30 (s, CH_3); 4.48, 4.55, 4.75, and 4.88 (all same intensity t, C_5H_4); 7.08 (d, H-7 or 8; J(H-7,8) = 15.5), 7.3 to 7.7 (m, C_6H_5), 7.80 (d, H-8 or 7) ^{1}H NMR ($CDCl_3$): 2.34 (s, CH_3); 4.51, 4.58, 4.79, and 4.81 (all t, C_5H_4); 7.01 (d) and 7.87 (d), (CH=CH; J = 15.5), 7.2 to 7.9 (m, C_6H_5)	[8, 13, 18, 19, 50, 51, 55]

References on pp. 109/10

Table 5 (continued)

No.	$R'{-}CO{-}C_5H_4FeC_5H_4{-}CO{-}R''$ R'	R''	method of preparation (yield in %), properties, and remarks	Ref.
			IR (paraffin oil): 691 (γC–H of C_6H_5), 1010 (δC–H of C_5H_4), 1116 ($\nu_sC_{ring}{-}C_{acetyl}$), 1670 ($\nu$C=O) IR (Nujol): 1653 ($\nu$C=O), 1669 ($\nu$C=O) IR ($CCl_4$): 1661 ($\nu$C=O of COR') for s-cis, 1679.5 (νC=O of COR'') UV (C_2H_5OH): λ_{max} (log ε) = 223 (4.37), 264 sh (4.13), 309 (4.34), 365 sh (3.33) UV (C_2H_5OH): λ_{max} (log ε) = 226 (3.19), 320 (4.05) UV ($CH_3OH/NaOCH_3$): λ_{max} (log ε) = 307 (4.154) R_f: 0.26 on Al_2O_3 in $C_6H_6/CH_3CO_2C_2H_5$ 95/5 and 0.33 on SiO_2 in $C_6H_6/CH_3CO_2C_2H_5$ 4/1	
*20	$C_6H_5\overset{8}{C}H{=}\overset{7}{C}H$	C_2H_5	Ii (49), Is (27), see No. 1, p. 103 m.p. 132 to 134° (benzene/petroleum ether), 133 to 135° (acetone/petroleum ether, 3/1) ^{1}H NMR ($CDCl_3$): 1.11 (t, CH_3), 2.65 (q, CH_2); 4.52, 4.55, 4.75, and 4.85 (all same intensity t, C_5H_4); 7.05 (d, H-7 or 8; J(H-7,8) = 16.1), 7.3 to 7.7 (m, C_6H_5), 7.78 (d, H-8 or 7) IR (CCl_4): 1663 (νC=O of COR') for s-cis, 1682 (νC=O of COR'') UV ($CH_3OH/NaOCH_3$): λ_{max} (log ε) = 310(4.475)	[13, 36, 39, 50, 55]
*21	$C_6H_5\overset{8}{C}H{=}\overset{7}{C}H$	C_3H_7-n	Ii (22) m.p. 110 to 112° (benzene/petroleum ether) ^{1}H NMR ($CDCl_3$): 0.94 (t, CH_3), 1.72 (quint, CH_2), 2.62 (t, CH_2); 4.47, 4.55, 4.77, and 4.88 (all same intensity t, C_5H_4); 7.08 (d, H-7 or 8; J(H-7,8) = 16.1), 7.3 to 7.7 (m, C_6H_5), 7.79 (d, H-8 or 7) IR (CCl_4): 1663 (νC=O of COR') for s-cis, 1676 (νC=O of COR'') UV ($CH_3OH/NaOCH_3$): λ_{max}(log ε) = 308 (4.051)	[50, 55]
*22	$C_6H_5\overset{8}{C}H{=}\overset{7}{C}H$	$CH(CH_3)_2$	Ic (63), Is (36), II (67), see No. 4, p. 104 m.p. 94 to 95° (benzene/petroleum ether), 95 to 96° (acetone/petroleum ether 3/1) ^{1}H NMR ($CDCl_3$): 1.21 (d, CH_3), 3.05 (s, CH of i-C_3H_7); 4.52, 4.58, 4.81, and 4.91 (all same intensity t, C_5H_4); 7.09 (d, H-7 or 8; J(H-7,8) = 15.6), 7.3 to 7.7 (m, C_6H_5), 7.80 (d, H-8 or 7) IR (CCl_4): 1663 (νC=O of COR') for s-cis, 1675.5 (νCO of COR'') UV ($CH_3OH/NaOCH_3$): λ_{max}(log ε) = 307 (4.148)	[13, 39, 50, 55]

Table 5 (continued)

No.	$R'-CO-C_5H_4FeC_5H_4-CO-R''$ R'	 R''	method of preparation (yield in %), properties, and remarks	Ref.
*23	$C_6H_5\overset{8}{C}H{=}\overset{7}{C}H$	$CH_2CH(CH_3)_2$	Ii (29) m.p. 110 to 113° (benzene/petroleum ether) ^1H NMR ($CDCl_3$): 2.15 (d, CH_3), 2.23 (m, CH of i-C_4H_9), 2.50 (d, CH_2); 4.47, 4.55, 4.76, and 4.88 (all same intensity t, C_5H_4); 7.07 (d, H-7 or 8; J(H-7,8) = 15.7), 7.3 to 7.8 (m, C_6H_5), 7.80 (d, H-8 or 7) UV ($CH_3OH/NaOCH_3$): $\lambda_{max}(\log \varepsilon) = 310(4.464)$	[50]
*24	$C_6H_5\overset{8}{C}H{=}\overset{7}{C}H$	$CH_2C(CH_3)_3$	Ii (41) m.p. 124° (benzene/petroleum ether) ^1H NMR ($CDCl_3$): 1.02 (s, C_4H_9), 2.52 (s, CH_2); 4.49, 4.57, 4.75, and 4.88 (all same intensity t, C_5H_4); 7.09 (d, H-7 or 8; J(H-7,8) = 15.7), 7.3 to 7.7 (m, C_6H_5), 7.80 (d, H-8 or 7) IR (CCl_4): 1663 (νC=O of COR') for s-cis, 1682 (νC=O of COR'')	[50, 55]
*25	$C_6H_5\overset{8}{C}H{=}\overset{7}{C}H$	$CH_2C_6H_5$	Ii (47), Is (47) m.p. 150 to 152° (acetone/petroleum ether 3/1 or benzene/petroleum ether) ^1H NMR ($CDCl_3$): 3.95 (s, CH_2); 4.47 and 4.81 (same intensity m, C_5H_4); 7.07 to 7.7 (m, C_6H_5), 7.05 (d, H-7 or 8; J(H-7,8) = 15.8), 7.81 (d, H-8 or 7) IR (paraffin oil): 1011sh (δC–H of C_5H_4) UV ($CH_3OH/NaOCH_3$): $\lambda_{max}(\log \varepsilon) = 305(4.468)$	[13, 18, 50]
*26	$C_6H_5CH{=}CH$	$CH_2C_6H_4CH_3$-4	Is (35.7) m.p. 169 to 171° (acetone/petroleum ether 3/1)	[13]
*27	$C_6H_5CH{=}CH$	$CH_2C_6H_4CH_3$-3	Is (35.7) m.p. 144 to 144.5° (acetone/petroleum ether 3/1)	[13]
*28	$C_6H_5CH{=}C(CH_3)$	CH_3	Id (30.6), Is m.p. 100 to 101° (acetone/petroleum ether) ^1H NMR ($CDCl_3$): 2.25 (d, $=CCH_3$), 2.4 (s, $COCH_3$); 4.52 (m, 4H), 4.77 (m, 2H), and 4.9 (m, 2H) (C_5H_4); 7.45 (m, s, CHC_6H_5)	[36, 39]
*29	4-CH_3– C_6H_4CH–CH	CH_3	Ik (39) m.p. 132 to 134° (benzene/petroleum ether) IR (CCl_4): 1658 (νC=O of COR') for s-cis UV ($CH_3OH/NaOCH_3$): $\lambda_{max}(\log \varepsilon) = 317(4.424)$	[51, 55]

References on pp. 109/10

Table 5 (continued)

No.	$R'-CO-C_5H_4FeC_5H_4-CO-R''$ R'	R''	method of preparation (yield in %), properties, and remarks	Ref.
*30	$C_6H_5CH{=}CH-CH{=}CH$	CH_3	Im (51) red-violet crystals, m.p. 132 to 135° (benzene/petrol) ^{1}H NMR: 2.34 (s, CH_3), 4.50 (t, H-3′,4′), 4.55 (t, H-3,4), 4.76 (t, H-2′,5′), 4.86 (t, H-2,5), 6.5 to 7.7 (m, 9H, $CH{=}CHCH{=}CHC_6H_5$)	[42]
*31	$C_6H_5C{\equiv}C$	CH_3	It (20), Iu (58.8) crystals, m.p. 83 to 84° (benzene/petroleum ether) ^{1}H NMR ($CDCl_3$): 2.39 (s, CH_3), 4.6 (m, H-3,4, H-3′,4′), 4.85 (t, H-2′5′ or H-2,5; J≈1.2), 5.6 (t, H-2,5 or H-2′,5′), 7.3 to 7.8 (m, C_6H_5) IR (Nujol): 1638 (νC=O), 1671 (νC=O), 2210 (νC≡C)	[35]
*32	C_6H_5	CH_3	Ia (75), Id, Ie (48), Im (57) bright red powder, red-brown to deep red needles with trapezoidal base, m.p. 69 to 69.5°, 69 to 70° (ether), 71 to 72° (benzene/petroleum ether 1/2), 72° (hexane), 72 to 73° (acetone/petroleum ether) ^{1}H NMR ($CDCl_3$): 4.47 (t, H-3′,4′), 4.55 (t, H-3,4), 4.73 (t, H-2′,5′), 4.89 (t, H-2,5) IR (paraffin oil): 1007 sh (δC–H of C_5H_4), 1117 ($\nu_s C_{ring}-C_{acetyl}$) IR ($CCl_4$): 1671.9 (νC=O) IR ($CCl_4$): 1649 (νC=O of COR′), 1686 (νC=O of COR″) UV (C_2H_5OH): $\lambda_{max}(\log \varepsilon) = 220 (4.50)$, 266 (4.35), 345 sh (3.50) R_f: 0.25 on SiO_2 with 4/1 $C_6H_6/CH_3CO_2C_2H_5$ dipole moment: $\mu(C_6H_6) = 3.98$ D (25°)	[1, 4, 12, 15, 18, 20, 24, 25, 28, 33, 55]
33	C_6H_5	C_2H_5	Ia (77) m.p. 38°, 40° (hexane) ^{1}H NMR ($CDCl_3$): 4.44 (t, H-3′,4′), 4.52 (t, H-3,4), 4.74 (t, H-2′,5′), 4.87 (t, H-2,5)	[6, 28]
34	C_6H_5	C_3H_7-n	Ia (71) m.p. 38°, 46° (hexane) ^{1}H NMR ($CDCl_3$): 4.47 (t, H-3′,4′), 4.57 (t, H-3,4), 4.75 (t, H-2′,5′), 4.90 (t, H-2,5)	[6, 28]
35	C_6H_5	$CH_2C_6H_5$	Ib (75) m.p. 86° (hexane), 87° ^{1}H NMR ($CDCl_3$): 4.43 (t, H-3′,4′), 4.47 (t, H-3,4), 4.78 (t, H-2′,5′), 4.85 (t, H-2,5)	[6, 28]

References on pp. 109/10

Table 5 (continued)

No.	$R'-CO-C_5H_4FeC_5H_4-CO-R''$ R'	R''	method of preparation (yield in %), properties, and remarks	Ref.
36	C_6H_5	$CH_2CH_2C_6H_5$	see No. 37, p. 107 liquid at room temperature ^{1}H NMR ($CDCl_3$): 2.97 (s, 4H), 4.42 (m, 4H), 4.78 (m, 4H), 7.3 (s, 5H), 7.3 to 8 (m, 5H) IR (Nujol): 1625 (νC=O), 1661 (νC=O)	[23, 31]
*37	C_6H_5	$\overset{7'}{C}H=\overset{8'}{C}HC_6H_5$-(E)	see No. 32, p. 107 m.p. 127° (ethanol), 129° (ethanol/water) ^{1}H NMR ($CDCl_3$): 7.03 (d, H-8'; J(H-8', 7') = 15), 7.75 (d, H-7') IR (paraffin oil): 1015 (δC–H of C_5H_4) IR (Nujol): 1629 (νC=O), 1641 (νC=O) IR (CCl_4): 1597 ν(C=C), 1665 ν(C=O of COR'') for s-cis UV ($CH_3OH/NaOCH_3$): $\lambda_{max}(\log \varepsilon) = 310$ (4.38)	[3, 10, 17, 18, 28, 55]
38	C_6H_5	$CH=CHC_6H_4-CH_3$-4	see No. 32, p. 107 m.p. 143 to 145° (ethanol) IR (CCl_4): 1662 (νC=O of COR'') for s-cis	[17, 21, 55]
39	$4-CH_3C_6H_4$	C_2H_5	Ih (90) m.p. 103° ^{1}H NMR ($CDCl_3$): 1.10 (t, CH_3 of C_2H_5), 2.43 (s, CH_3 of $C_6H_4CH_3$), 2.60 (m, CH_2); 4.46, 4.53, 4.75, and 4.89 (all t, 2H) (C_5H_4); 7.51 (m, C_6H_4)	[26]
40	$C_{10}H_{15}$ (adamantyl)	C_2H_5	Ih (71) m.p. 89° ^{1}H NMR ($CDCl_3$): 1.17 (t, CH_3), 1.75 (m, 5H of $C_{10}H_{15}$), 2.00 (s, 10H of $C_{10}H_{15}$), 2.60 (m, CH_2); 4.43 (m, 4H), 4.77 (t, 2H), and 4.87 (t, 2H) (C_5H_4)	[26]
compounds with R' containing C, H, and halogen:				
*41	$ClCH_2$	CH_3	Io (82.4) red crystals, m.p. 107 to 109° (benzene) ^{1}H NMR ($CDCl_3$): 2.35 (s, CH_3), 4.40 (s, CH_2Cl); 4.51 and 4.85 (both m, 4H, C_5H_4)	[46]
*42	$ClCH_2$	$CH=CHC_6H_5$-(E)	In (31) crystals, m.p. 134 to 135° (benzene/acetone) ^{1}H NMR ($CDCl_3$): 4.35 (s, CH_2Cl), 4.59 (m, H-2' to 5'), 4.80 (t, H-3, 4), 4.92 (t, H-2, 5), 7.04 (complex m, C_6H_5), 7.82 (q, CH=CH; J = 16.2) IR (Nujol): 1656 (νC=O of COR''), 1692 (νC=O of COR')	[34]

Table 5 (continued)

No.	R′–CO–$C_5H_4FeC_5H_4$–CO–R″ R′	R″	method of preparation (yield in %), properties, and remarks	Ref.
*43	$ClCH_2CH_2$	CH_3	Ii, In (73) red solid, m.p. 97 to 97.5° (ligroin/ether) ^{1}H NMR ($CDCl_3$): 2.32 (s, CH_3), 3.11 (m, CH_2Cl), 3.88 (m, $COCH_2$); 4.54 and 4.80 (both t, 4H) (C_5H_4) IR (Nujol): 1664 (νC=O), 1675 (νC=O)	[7, 8]
*44	$ClCH_2CH_2$	$CH{=}CHC_6H_5$-(E)	In (63.4) crystals, m.p. 117 to 119° (benzene/petroleum ether) ^{1}H NMR ($CDCl_3$): 3.11 (t, CH_2Cl), 3.28 (t, $COCH_2$), 4.53 (t, H-3′,4′), 4.61 (t, H-2′,5′), 4.77 (t, H-3,4), 4.94 (t, H-2,5), 7.09 (m, C_6H_5), 7.83 (q, CH=CH; J = 15.6) IR (Nujol): 1668 (νC=O of COR″), 1684 (νC=O of COR′)	[34]
*45	CH_3CHCl	$CH{=}CHC_6H_5$-(E)	In (29.3) crystals, m.p. 118 to 120° (benzene/petroleum ether) ^{1}H NMR ($CDCl_3$): 1.70 (d, CH_3), 4.60 (m, H-2′ to 5′), 4.88 (m, H-2 to 5), 4.88 (m, COCH), 7.05 (complex m, C_6H_5), 7.81 (q, CH=CH; J = 16) IR (Nujol): 1660 (νC=O of COR″), 1680 (νC=O of COR′)	[34]
46	$ClCH_2CH_2CH_2$	$CH{=}CHC_6H_5$-(E)	In (43.1) crystals, m.p. 108 to 111° (benzene/petroleum ether) ^{1}H NMR ($CDCl_3$): 2.14 (quint, CH_2), 3.64 (t, $COCH_2$), 4.51 (t, H-3′,4′), 4.59 (t, H-2′,5′), 4.68 (t, H-3,4), 4.95 (t, H-2,5), 7.07 (m, C_6H_5), 7.82 (q, CH=CH; J = 15.3), 8.86 (t, CH_2Cl) IR (Nujol): 1655 (νC=O of COR″), 1678 (νC=O of COR′)	[34]
*47	$4\text{-}FC_6H_4CH_2$	$CH{=}CHC_6H_5$	Ir (33.2) m.p. 154 to 155° (acetone/petroleum ether 3/1)	[13, 36]
*48	$3\text{-}FC_6H_4CH_2$	$CH{=}CHC_6H_5$	Ir	[36]
*49	$4\text{-}ClC_6H_4CH_2$	$CH{=}CHC_6H_5$	Ir (32) m.p. 143 to 144.5° (acetone/petroleum ether 3/1)	[13]
*50	$3\text{-}ClC_6H_4CH_2$	$CH{=}CHC_6H_5$	Ir (30) m.p. 129 to 139.5° (acetone/petroleum ether 3/1)	[13]

References on pp. 109/10

Table 5 (continued)

No.	R′–CO–$C_5H_4FeC_5H_4$–CO–R″ R′	R″	method of preparation (yield in %), properties, and remarks	Ref.
*51	4-$BrC_6H_4CH_2$	$CH=CHC_6H_5$	Ir (31.2) m.p. 147 to 149° (acetone/petroleum ether 3/1)	[13]
*52	3-$BrC_6H_4CH_2$	$CH=CHC_6H_5$	Ir (39.1) m.p. 136.5 to 138° (acetone/petroleum ether 3/1)	[13]
*53	3-$IC_6H_4CH_2$	$CH=CHC_6H_5$	Ir (35.7) m.p. 131 to 132° (acetone/petroleum ether 3/1)	[13]
*54	4-$FC_6H_4CH=CH$	CH_3	Ik (57) crystals, m.p. 135 to 139° (benzene/petroleum ether) IR (CCl_4): 1661(νC=O of COR′) for s-cis UV ($CH_3OH/NaOCH_3$): $\lambda_{max}(\log \varepsilon) = 307 (4.183)$	[51, 55]
55	4-$FC_6H_4CH=CH$	C_6H_5	see No. 32, p. 107 m.p. 157° (ethanol) IR (CCl_4): 1665 (νC=O of COR′) for s-cis	[3, 17, 55]
*56	4-$ClC_6H_4CH=CH$	CH_3	Ik (34), Im (79) red crystals, m.p. 165 to 167° (benzene/petroleum ether), 167° (benzene) IR (CCl_4): 1661 (νC=O of COR′) for s-cis UV ($CH_3OH/NaOCH_3$): $\lambda_{max}(\log \varepsilon) = 311(4.173)$	[41, 51, 55]
57	4-$ClC_6H_4CH=CH$	C_6H_5	see No. 32, p. 107 m.p. 131° (ethanol) IR (CCl_4): 1666 (νC=O of COR′) for s-cis UV ($CH_3OH/NaOCH_3$): $\lambda_{max}(\log \varepsilon) = 316 (4.44)$	[3, 10, 17, 55]
*58	3-$ClC_6H_4CH=CH$	CH_3	Ik (37) crystals, m.p. 144 to 146° (benzene/petroleum ether) UV ($CH_3OH/NaOCH_3$): $\lambda_{max}(\log \varepsilon) = 298\ (4.427)$ see "Further information" for No. 54, p. 108	[51]
*59	4-$BrC_6H_4CH=CH$	CH_3	Ik (54) crystals, m.p. 169 to 171° (benzene/petroleum ether) IR (CCl_4): 1661 (νC=O of COR′) for s-cis UV ($CH_3OH/NaOCH_3$): $\lambda_{max}(\log \varepsilon) = 312 (4.239)$ see "Further information" for No. 54, p. 108	[51, 55]

References on pp. 109/10

Table 5 (continued)

No.	$R'-CO-C_5H_4FeC_5H_4-CO-R''$ R'	R''	method of preparation (yield in %), properties, and remarks	Ref.
*60	$3-BrC_6H_4CH{=}CH$	CH_3	Ik (41) crystals, m.p. 134 to 136° (benzene/petroleum ether) IR (CCl_4): 1662 ($\nu C{=}O$ of COR') for s-cis UV ($CH_3OH/NaOCH_3$): $\lambda_{max}(\log \varepsilon) = 300(4.095)$ see "Further information" for No. 54, p. 108	[51, 55]
*61	$2-ClC_6H_4$	CH_3	not reported	[45]
compounds with R' containing C, H, and O:				
*62	$4-CH_3OC_6H_4CH_2$	$CH{=}CHC_6H_5$	Ir (32.3) m.p. 141 to 143° (3/1 acetone/petroleum ether)	[13]
*63	$3-CH_3OC_6H_4CH_2$	$CH{=}CHC_6H_5$	Ir (34.5) m.p. 109 to 110° (3/1 acetone/petroleum ether)	[13]
*64	$4-CH_3O-C_6H_4CH{=}CH$	CH_3	Ik (35) crystals, m.p. 157 to 159° (benzene/petroleum ether) IR (CCl_4): 1657 ($\nu C{=}O$ of COR') for s-cis UV ($CH_3OH/NaOCH_3$): $\lambda_{max}(\log \varepsilon) = 342(4.081)$	[51, 55]
65	$4-CH_3O-C_6H_4CH{=}CH$	C_6H_5	see No. 32, p. 107 m.p. 98 to 100° (ethanol) IR (CCl_4): 1661 ($\nu C{=}O$ of COR') for s-cis UV ($CH_3OH/NaOCH_3$): $\lambda_{max}(\log \varepsilon) = 346(4.39)$	[3, 10, 17, 55]
*66	$3-CH_3O-C_6H_4CH{=}CH$	CH_3	Ik (29) crystals, m.p. 137 to 139° (benzene/petroleum ether) IR (CCl_4): 1661 ($\nu C{=}O$ of COR') for s-cis UV ($CH_3OH/NaOCH_3$): $\lambda_{max}(\log \varepsilon) = 301(4.346)$	[51, 55]
*67	$C_6H_5CO\overset{11}{C}H_2-(CH_2)_3\overset{7}{C}H_2$	$CH{=}CHC_6H_5$	Ic (25) m.p. 98 to 101° (benzene/petroleum ether) ^1H NMR ($CDCl_3$): 1.0 to 2.0 (m, $(CH_2)_3$), 2.65 (t, CH_2-7), 2.95 (t, CH_2-11), 4.50 (t, H-3, 4), 4.57 (t, H-3', 4'), 4.78 (t, H-2, 5), 4.90 (t, H-2', 5'), 6.8 to 8.1 (m, $C_6H_5 + CH{=}CHC_6H_5$)	[48]

Table 5 (continued)

No.	$R'-CO-C_5H_4FeC_5H_4-CO-R''$ R'	R''	method of preparation (yield in %), properties, and remarks	Ref.
*68	$C_6H_5CO\overset{12}{C}H_2-(CH_2)_4\overset{7}{C}H_2$	$CH=CHC_6H_5$	Ic (66) m.p. 127 to 130° (benzene/petroleum ether) ^1H NMR ($CDCl_3$): 1.0 to 2.0 (m, $(CH_2)_4$), 2.65 (t, CH_2-7), 2.96 (t, CH_2-12), 4.50 (t, H-3, 4), 4.58 (t, H-3′, 4′), 4.80 (t, H-2, 5), 4.92 (t, H-2′, 5′), 6.9 to 9.1 (m, $C_6H_5+CH=CHC_6H_5$)	[48]
*69	$C_6H_5CO\overset{13}{C}H_2-(CH_2)_5\overset{7}{C}H_2$	$CH=CHC_6H_5$	Ic (78) m.p. 95 to 97° (benzene/petroleum ether) ^1H NMR ($CDCl_3$): 1.0 to 2.0 (m, $(CH_2)_5$), 2.63 (t, CH_2-7), 2.95 (t, CH_2-13), 4.50 (t, H-3, 4), 4.57 (t, H-3′, 4′), 4.78 (t, H-2, 5), 4.90 (t, H-2′, 5′), 6.9 to 8.1 (m, $C_6H_5+CH=CHC_6H_5$)	[48]
*70	$C_6H_5CO\overset{14}{C}H_2-(CH_2)_6\overset{7}{C}H_2$	$CH=CHC_6H_5$	Ic (64) m.p. 112 to 114° (benzene/petroleum ether) ^1H NMR ($CDCl_3$): 1.0 to 1.9 (m, $(CH_2)_6$), 2.63 (t, CH_2-7), 2.93 (t, CH_2-14), 4.48 (t, H-3, 4), 4.56 (t, H-3′, 4′), 4.78 (t, H-2, 5), 4.88 (t, H-2′, 5′), 6.9 to 8.0 (m, $C_6H_5+CH=CHC_6H_5$)	[48]
supplement:				
71	2-ClC_6H_4	C_6H_5	Iv (94.4) cherry red oil ^1H NMR: 4.23 (s), 4.32 (s), 4.60 (q; J = 2), 4.75 (t; J = 2), 5.0 (t; J = 1.5), 7.41 (s), 7.56 (q; J = 1.8), 7.82 (m) IR: 1110 (w), 1293 (s), 1350 (m), 1378 (m), 1450 (s), 1610 (m), 1645 (s), 1735 (s), 2900 (m), 2975 (m), 3050 (w), 3080 (w), 3100 (w) see "Oxidation", p. 88, and "Applications", p. 92	[56]

* Further information:

C_2H_5–CO–C_5H_4FeC$_5H_4$–CO–CH$_3$ (Table **5**, No. **1**) gives in 17 h at room temperature on condensation in a 1/1 molar ratio with C_6H_5CHO in 70/12 C_2H_5OH/10% aqueous NaOH a 2.7% yield of compound No. 20 plus 45.3% of XIb and traces of XIa (see p. 89), with $R_A = C_6H_5$ and $R_B = CH_3$; separation of No. 20 from XIa, b by column chromatography on SiO_2 with 98/2 benzene/$CH_3CO_2C_2H_5$. If the reaction is conducted at the boiling point, XIa is obtained in 69.3% yield after reaction for 1 h [39]. C_6H_5CHO in ethanol with a slight excess of 15% aqueous sodium hydroxide forms at 50 to 60°C XIb as above [28]. 2-ClC_6H_4CHO in CH_3OH/H_2O/NaOH at 65°C gives in 47% yield XI with $R_A = C_6H_4Cl$-2 and $R_B = CH_3$ (m.p. 209 to 210°C) [2]. Condensation with 1/1 C_6H_5CHO in 4/1 glacial acetic acid/conc. H_2SO_4 gives along with a 10% yield of XIb a 10.4% yield also of XIIb and 6.9% of XIIa (see p. 92) ($R_B = CH_3$) with a 52.5% recovery of No. 1 [40].

References on pp. 109/10

n-C_3H_7–CO–$C_5H_4FeC_5H_4$–CO–CH_3 (Table **5**, No. **2**) gives with 1/1 C_6H_5CHO in C_2H_5OH/H_2O/NaOH at room temperature [39] or at 50 to 60°C [28] XIb (see p. 89) with $R_A = C_6H_5$, $R_B = C_2H_5$ [28, 39]. 2-ClC_6H_4CHO in 14/1 CH_3OH/10% aqueous NaOH forms at 65°C XI with R_A = 2-ClC_6H_4, $R_B = C_2H_5$, m.p. 262 to 264°C [2]. Condensation with C_6H_5CHO in 4/1 CH_3CO_2H/conc. H_2SO_4 leads to XIb in 5.7% yield, 4.6% XIIb (see p. 92), and XIIa with $R_B = C_2H_5$ along with 69% recovery of No. 2 [40].

$(CH_3)_2CH$–CO–$C_5H_4FeC_5H_4$–CO–CH_3 (Table **5**, No. **4**) condenses with 1/1 C_6H_5CHO in C_2H_5OH/H_2O/NaOH in 17 h at room temperature to form No. 22 and XIII with $R_A = C_6H_5$, each in a 33.5% yield. Both compounds are isolated in the sequence stated by column chromatography on SiO_2 in 98/2 benzene/$CH_3CO_2C_2H_5$ [39]. Reaction with 2-ClC_6H_4CHO in CH_3OH/H_2O/NaOH at 65°C forms XIII with R_A = 2-ClC_6H_4 [2].

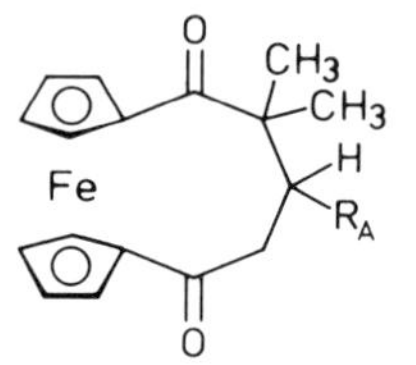

XIII

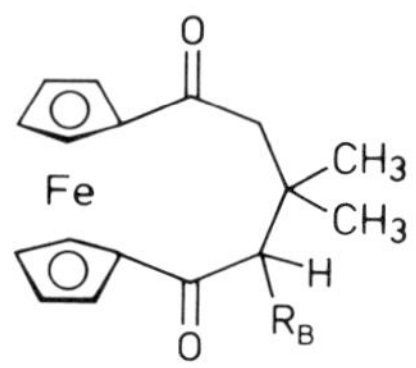

XIV

R′–CO–$C_5H_4FeC_5H_4$–CO–CH_3 (Table **5**, Nos. **8** to **12** with R′ = CH_2=CH, CH_2=C(CH_3), CH_3CH=CH, $(CH_3)_2$C=CH, $(CH_3)_3$CCH=CH). The base-catalyzed internal Michael addition of No. 8 yields XIc (see p. 89) [7, 8, 50]. No. 9 gives XI with R_A = H, $R_B = CH_3$, No. 10 XI with $R_A = CH_3$, R_B = H, No. 11 XIV with R_B = H, and No. 12 XI with $R_A = C(CH_3)_3$ and R_B = H. If during the preparation of No. 9 or No. 10 the reaction mixture is allowed to stand for a long time after evaporation of the solvent and before the column chromatographic processing, spontaneous cyclization takes place to XI with R_A = H, $R_B = CH_3$ or vice-versa, respectively [50]. For kinetic studies and related activation parameters see Table 4, p. 90.

n-C_8H_{17}CH=CH$(CH_2)_7$–CO–$C_5H_4FeC_5H_4$–CO–CH_3 (Table **5**, No. **13**). The alkali-catalyzed condensation with 2-ClC_6H_4CHO failed, probably due to steric hindrance [2].

$C_6H_5CH_2$–CO–$C_5H_4FeC_5H_4$–CO–CH_3 (Table **5**, No. **14**) gives in 17 h at room temperature on 1/1 condensation with C_6H_5CHO in C_2H_5OH/H_2O/NaOH both XIa and b (see p. 89), with $R_A = R_B = C_6H_5$ in 17 and 30% yields, respectively [39]. A mixture of the two also forms at 50 to 60°C [28]. 2-ClC_6H_4CHO (1/1) in CH_3OH/10% aqueous NaOH (14/1) at 65°C produces XIa and b with R_A = 2-ClC_6H_4, $R_B = C_6H_5$ in yields of 72 and 27%, respectively [39]. That only XIa was recovered by [2] under these conditions is attributed by [39] to the isolation method selected.

$C_6H_5CH_2$–CO–$C_5H_4FeC_5H_4$–CO–R″ (Table **5**, Nos. **17** and **18** with R″ = CH=C$(CH_3)_2$, CH=CHC$(CH_3)_3$) and **C_6H_5CH=CH–CO–$C_5H_4FeC_5H_4$–CO–$CH_2C(CH_3)_3$** (Table **5**, No. **24**). Nos. 17 and 18 cyclize in $CH_3OH/NaOCH_3$ during an intramolecular Michael addition to form XIV (see above) with $R_B = C_6H_5$ or XI (p. 89) with $R_A = C(CH_3)_3$, $R_B = C_6H_5$, respectively. The fact that No. 24, which has the same substituents as No. 18 but interchanged with respect to the donor and acceptor groups, cannot be cyclized under these conditions, may be attributed neither to steric grounds nor to a deficient acidity of the CH_2 group, but has its origins only in the structure of the carbanion formed as an intermediate. Only the carbanion formed from No. 18 possesses a planar structure with the possibility of charge delocalization in the C_6H_5 ring, supported by the rigid bonding of the t-C_4H_9 to the sp^2 C atom, and with it a small rotation barrier [50]. For kinetic studies of the cyclization and related activation parameters see Table 4, p. 90.

$C_6H_5CH{=}CH{-}CO{-}C_5H_4FeC_5H_4{-}CO{-}CH_3$ (Table **5**, No. **19**) cyclizes on standing for several hours in $C_6H_6/CH_3CO_2C_2H_5$ (49/1) on Al_2O_3 (pH 6.5 to 7.5) or on heating with KOH [13] or NaOH [8] in C_2H_5OH to form XI (see p. 89) with $R_A=C_6H_5$, $R_B=H$. For kinetic studies with $NaOCH_3/CH_3OH$ see Table 4, p. 90.

$C_6H_5CH{=}CH{-}CO{-}C_5H_4FeC_5H_4{-}CO{-}C_2H_5$ (Table **5**, No. **20**) cyclizes on standing for several days in $C_6H_6/CH_3CO_2C_2H_5$ (49/1) on Al_2O_3 (pH 6.5 to 7.5) forming XIb (see p. 89) with $R_A=C_6H_5$, $R_B=CH_3$. In contrast heating with KOH in C_2H_5OH gives XIa. If the latter reaction is undertaken at greater dilution (threefold quantity of C_2H_5OH), a mixture of XIa and b is produced [13]. For kinetic measurements in $NaOCH_3/CH_3OH$ see Table 4, p. 90.

Cyclization in 4/1 glacial acetic acid/conc. H_2SO_4 gives 15 and 83% yields, respectively, of XIa and b [36].

$C_6H_5CH{=}CH{-}CO{-}C_5H_4FeC_5H_4{-}CO{-}R''$ (Table **5**, Nos. **21** to **23** with $R''=C_3H_7$-n, $CH(CH_3)_2$, $CH_2CH(CH_3)_2$). The compounds cyclize with $NaOCH_3/CH_3OH$ during the intramolecular Michael addition to XI (see p. 89) with $R_A=C_6H_5$ and $R_B=C_2H_5$ for No. 21 or i-C_3H_7 for No. 23, or to XIII (p. 104) with $R_A=C_6H_5$ for No. 22 [50]. For kinetic data for the reactions see Table 4, p. 90. No. 22 also cyclizes on standing for several hours in 49/1 $C_6H_6/CH_3CO_2C_2H_5$ on Al_2O_3 (pH 6.5 to 7.5) forming XIII with $R_A=C_6H_5$. Heating No. 22 with KOH in C_2H_5OH in contrast yields only the initial compound plus a polymeric substance [13].

$C_6H_5CH{=}CH{-}CO{-}C_5H_4FeC_5H_4{-}CO{-}CH_2C(CH_3)_3$ (Table **5**, No. **24**). See "Further information" for Nos. 17 and 18.

$C_6H_5CH{=}CH{-}CO{-}C_5H_4FeC_5H_4{-}CO{-}CH_2C_6H_4R$ (Table **5**, Nos. **25** to **27** with R=H, CH_3-4, CH_3-3). On standing in 49/1 $C_6H_6/CH_3CO_2C_2H_5$ on Al_2O_3 (pH 6.5 to 7.5) all three compounds cyclize to the cis isomers XIb (see p. 89) with $R_A=C_6H_5$, $R_B=C_6H_4R$. On Al_2O_3 previously washed with H_2SO_4 there is on the contrary no reaction. On heating with KOH in ethanol the trans isomers XIa are formed, with $R_A=C_6H_5$, $R_B=C_6H_4R$ [13]. For kinetic data for the reaction in $NaOCH_3/CH_3OH$ see Table 4, p. 90.

Cyclization in glacial acetic acid/conc. H_2SO_4 (4/1) yields XIa and b with $R_A=C_6H_5$ and $R_B=C_6H_4R$, in 17 and 41% yields, respectively, for No. 25 and 25 and 40% for No. 27 [36].

$C_6H_5CH{=}C(CH_3){-}CO{-}C_5H_4FeC_5H_4{-}CO{-}CH_3$ (Table **5**, No. **28**) is soluble in benzene. With 1/5 NaOH in C_2H_5OH it cyclizes at room temperature to XIb (see p. 89) with $R_A=C_6H_5$, $R_B=CH_3$, in a 64.8% yield after 24 h reaction. At boiling temperatures it forms XIa with $R_A=C_6H_5$, $R_B=CH_3$, in 64.8% yield after 1 h. On standing in benzene on Al_2O_3 (pH 6.5 to 7.5) after 14 d XIa and b are obtained in 8.1 and 67.8% yields, respectively [39]. The acid-catalyzed Michael addition in 4/1 glacial acetic acid/conc. H_2SO_4 likewise yields XIa and b, in yields of 14 and 82%, respectively [36].

4-$CH_3C_6H_4CH{=}CH{-}CO{-}C_5H_4FeC_5H_4{-}CO{-}CH_3$ (Table **5**, No. **29**) cyclizes to XI with $R_A=$ 4-$CH_3C_6H_4$, $R_B=H$ [51]. For rate constants and related activation parameter see Table 4, p. 91.

$C_6H_5CH{=}CHCH{=}CH{-}CO{-}C_5H_4FeC_5H_4{-}CO{-}CH_3$ (Table **5**, No. **30**) yields as products of the intramolecular Michael addition XI (see p. 89) with $R_A=CH{=}CH{-}C_6H_5$ and $R_B=H$, and XV in a 15/1 ratio with NaOH in boiling C_2H_5OH and in a ~20/1 ratio with conc. H_2SO_4 in CH_3CO_2H at room temperature [42].

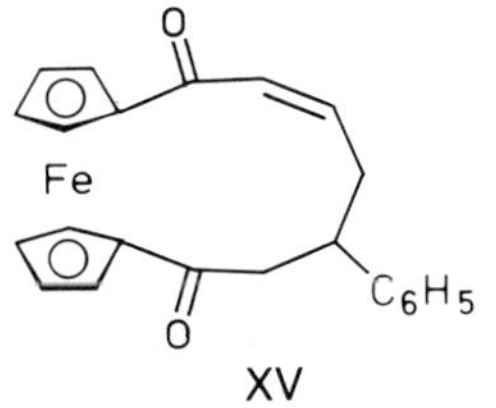

XV

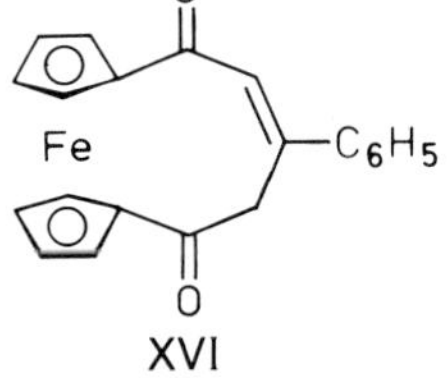

XVI

References on pp. 109/10

$C_6H_5C{\equiv}C$–CO–$C_5H_4FeC_5H_4$–CO–CH_3 (Table **5**, No. **31**) cyclizes to XVI with KOH in boiling C_2H_5OH, but not on standing in 49/1 $C_6H_6/CH_3CO_2C_2H_5$ on Al_2O_3 [35]. For kinetic studies in $CH_3OH/NaOCH_3$ see Table 4, p. 91.

C_6H_5–CO–$C_5H_4FeC_5H_4$–CO–CH_3 (Table **5**, No. **32**) is a trace product formed from $CH_2{=}C(C_6H_5)$–$C_5H_4FeC_5H_4$–$C(CH_3){=}CH_2$ during its photosensitized oxygenation by molecular O_2 in CH_2Cl_2 in the presence of sodium copper chlorophyllin as a sensitizer with Na vapor lamp irradiation, and also in the dark in benzene during autoxidation catalyzed by 7N HCl [47].

The air-stable compound crystallizes in the monoclinic space group $P2_1/c$-C^5_{2h} (No. 14) with the lattice parameters a=17.581, b=7.042, c=11.858 Å, β=90°02′; Z=4. The calculated density is 1.50, the measured density 1.47 g/cm³. **Fig. 5** gives the projection of the molecule on the xy0 plane, **Fig. 6** that on the x0z plane; from these the 1,2′-configuration is evident. The two planar parallel cyclopentadienyl rings are rotated with respect to each other by the angle φ=14°, so that the molecule assumes an intermediate position between the prismatic (φ=0°) and antiprismatic (φ=36°) configurations. Projection of the molecule on a cyclopentadienyl ring plane shows that the centers of gravity of the two rings do not coincide, but have in each case a distance of 0.16 Å from the Fe atom.

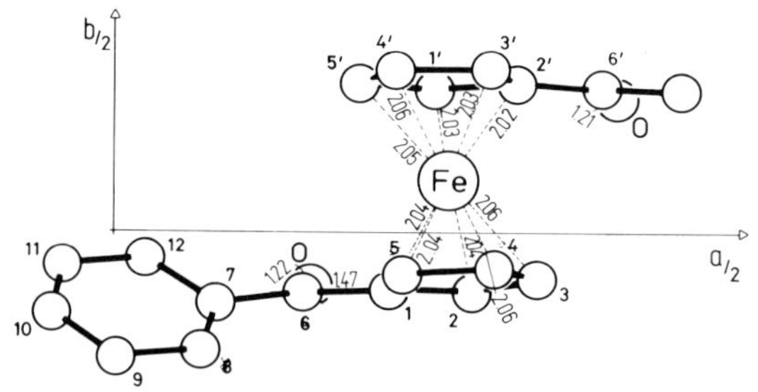

Fig. 5. Molecular structure of C_6H_5–CO–$C_5H_4FeC_5H_4$–CO–CH_3 in the crystal, projected on the xy0 plane [25].

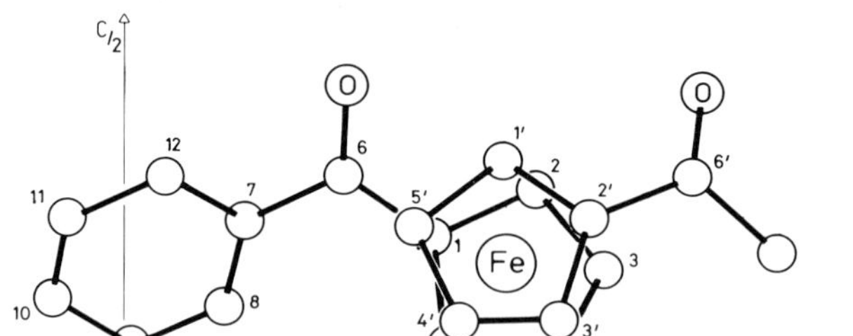

Fig. 6. Molecular structure of C_6H_5–CO–$C_5H_4FeC_5H_4$–CO–CH_3 in the crystal, projected on the x0z plane [12].

The steric hindrance between H-5 and H-8 (see Fig. 5) produces a rotation around the C(6)-C(7) axis, so that the planes of the C_6H_5 and C_5H_4 rings include an angle of 43°37′. The CO groups form an angle of 11°30′ with the C_5H_4 ring and one of 32°05′ with the C_6H_5 ring. Each molecule is surrounded in the crystal by 14 nearest neighbors [12, 25]. The size of the dipole moment in benzene shows that the free molecule occurs preferentially in a racemic form [24].

No. 32 undergoes in ethanol an aldol condensation to form the chalcone-analogue compounds C_6H_5–CO–$C_5H_4FeC_5H_4$–CO–CH=CHR^{III} [3, 21] on reaction in a 1/1/1 molar ratio with 10% NaOH and the aldehydes R^{III}CHO where R^{III} = RC_6H_4 with R = H [3, 28], 4-NC, 4-H_2N, 4-$(CH_3)_2N$ [3], 3-O_2N [21], 4-O_2N [3] or where R^{III} = 2-furyl, 2-thienyl [3], or $C_5H_5FeC_5H_4$ [3, 28]. The compounds Nos. 37, 55 (yield 82%), 57 (yield 92.5%), and 65 (yield 44.4%) are obtained from 4-R–C_6H_4CHO with R = H, F, Cl, and CH_3O, respectively, in 2.5 h at 10 to 15°C and subsequently in a refrigerator [3]. Warming No. 32 with C_6H_5CHO for a short time up to 50 to 60°C until the precipitate appears yields No. 37 practically quantitatively [28]. With 4-$CH_3C_6H_4$CHO first at room temperature, then in a refrigerator, No. 38 is produced in a 67% yield [21]. No. 32 is more reactive than $C_5H_5FeC_5H_4$–CO–CH_3 but less than $Fe(C_5H_4–CO–CH_3)_2$; thus, in contrast to the latter it is unable to undergo an autocondensation [3]. With 1/7 cyclopentadiene in C_2H_5OH/$NaOC_2H_5$ it forms C_6H_5–CO–$C_5H_4FeC_5H_4$–C(CH_3)=C_5H_4-cyclo [11].

(E)-C_6H_5–CO–$C_5H_4FeC_5H_4$–CO–CH=CHC_6H_5 (Table **5**, No. **37**) plus $(C_6H_5)_3SnH$ in a 1/3 mole ratio for 5 h at 120°C gives a selective reduction of the double bond to form No. 36 in a 57% yield. Addition of 10% $NCC(CH_3)_2N$=$NC(CH_3)_2CN$ as a catalyst accelerates the reaction at temperatures as low as 60°C, yielding 61% No. 36 in 2.3 h. In the column chromatography on basic, deactivated Al_2O_3, $(C_6H_5)_3SnSn(C_6H_5)_3$ is eluted first with petroleum ether, subsequently No. 36 with petroleum ether/ether [23, 31].

$ClCH_2$–CO–$C_5H_4FeC_5H_4$–CO–CH_3 (Table **5**, No. **41**). Neither a cyclization product nor the starting compound could be isolated after attempts to condense the compound with C_6H_5CHO in the presence of bases at either room or elevated temperatures. In contrast, the acid-catalyzed condensation in glacial acetic acid/conc. H_2SO_4 yielded XIb (see p. 89) with R_A = C_6H_5, R_B = Cl and XIIa (see p. 92) with R_B = Cl along with an 80% recovery of the starting No. 41 [40].

Due to its content of active Cl it possesses in low concentrations a bacteriostatic capability and is in higher concentrations bacteriocidal [46].

(E)-$ClCH_2$–CO–$C_5H_4FeC_5H_4$–CO–CH=CHC_6H_5 (Table **5**, No. **42**) cyclizes with KOH in ethanol to form XVIIb. Standing in 49/1 C_6H_6/$CH_3CO_2C_2H_5$ on Al_2O_3 forms XVIIa and b in a 1/1 ratio [35]. Cyclization in glacial acetic acid/conc. H_2SO_4 proceeds smoothly to form XIb (see p. 89) with R_A = C_6H_5, R_B = Cl in 85% yield plus a 5.6% yield of XVII [40].

The compound easily splits off Cl and possesses a strong bacteriostatic power [34].

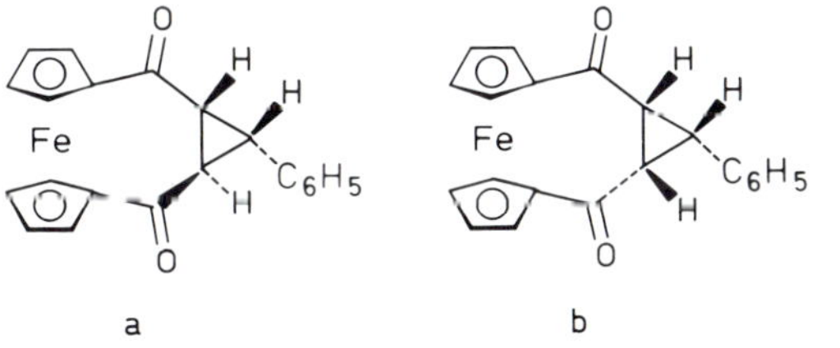

XVII

$ClCH_2CH_2$–CO–$C_5H_4FeC_5H_4$–CO–CH_3 (Table **5**, No. **43**) splits HCl off on chromatography in CH_2Cl_2 on neutral Al_2O_3 with a benzene/ether eluent, forming No. 8 [8, 50] in a 52% yield [8]. No. 43 cyclizes to XIc (p. 89) with $NaOCH_3$ in CH_3OH [7] or with NaOH in C_2H_5OH [8].

(E)-$ClCH_2CH_2$–CO–$C_5H_4FeC_5H_4$–CO–CH=CHC_6H_5 and **(E)-CH_3CHCl–CO–$C_5H_4FeC_5H_4$–CO–CH=CHC_6H_5** (Table **5**, Nos. **44** and **45**) possess antibiotic effectiveness against bacteria, yeasts, and fungi [44], but only at a level which is 20% of that of $C_5H_5FeC_5H_4$–CO–CH_2CH_2Cl or $C_5H_5FeC_5H_4$–CO–CHClCH_3 [34].

References on pp. 109/10

$RC_6H_4CH_2$–CO–$C_5H_4FeC_5H_4$–CO–CH=CHC_6H_5 (Table **5**, Nos. **47** to **53** with R = 4-F, 3-F, 4-Cl, 3-Cl, 4-Br, 3-Br, 3-I). Compounds No. 47 and 49 to 53 cyclize on standing in 49/1 C_6H_6/$CH_3CO_2C_2H_5$ on Al_2O_3 (pH 6.5 to 7.5) to XIb (see p. 89) with $R_A = C_6H_5$, $R_B = RC_6H_4$. These reactions do not occur if the Al_2O_3 is previously washed with acid. The trans isomers XIa with $R_A = C_6H_5$, $R_B = RC_6H_4$ form with KOH in boiling ethanol [13]. For kinetics measurements with $NaOCH_3$ in CH_3OH see Table 4, p. 91.

The cyclization in 4/1 glacial acetic acid/conc. H_2SO_4 yields a mixture of the two isomers XIa and b with $R_A = C_6H_5$, $R_B = RC_6H_4$. The yields for No. 47 are 13 and 63%, respectively, and 49 and 48% for No. 48 [36].

RC_6H_4CH=CH–CO–$C_5H_4FeC_5H_4$–CO–CH_3 (Table **5**, Nos. **54**, **58** to **60** with R = 4-F, 3-Cl, 4-Br, 3-Br) cyclize with NaOH in CH_3OH to XI (see p. 89) with $R_A = RC_6H_4$, $R_B = H$ [51]. For kinetic parameters of the reaction see Table 4, p. 91.

4-ClC_6H_4CH=CH–CO–$C_5H_4FeC_5H_4$–CO–CH_3 (Table **5**, No. **56**) cyclizes during the intramolecular Michael addition to form XIX (see scheme in XVIII) [41, 49, 51]. For rate constants measured in $NaOCH_3$/CH_3OH and related activation parameters see Table 4, p. 91.

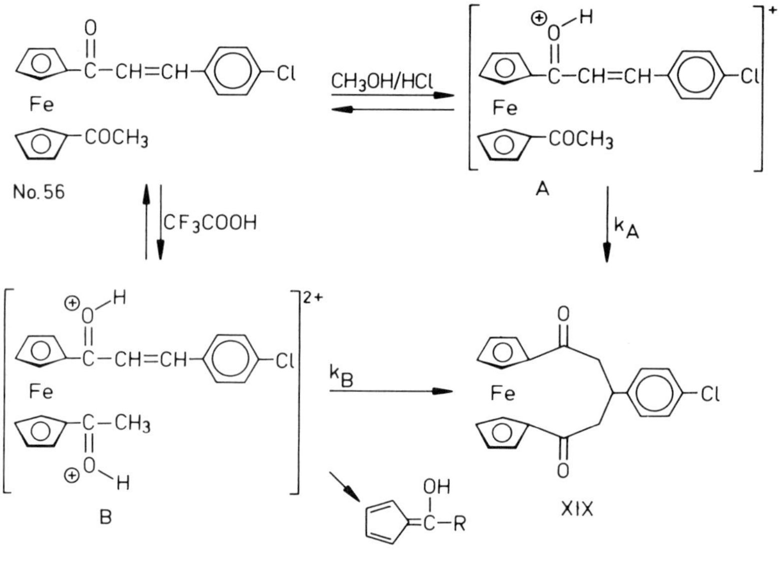

XVIII

Acid-catalyzed cyclization after 20 h at room temperature gave for XIX yields of 95% with HCl in CH_3OH, 85% with $BF_3 \cdot O(C_2H_5)_2$ in dioxane, 68% with conc. H_2SO_4 in glacial acetic acid, and 21% in pure CF_3CO_2H. In the latter case decomposition products of the ferrocene derivatives occur along with a 58% recovery of the starting No. 56. Addition of CCl_4 accelerates the decomposition in CF_3CO_2H so strongly that at above 40°C it is a reaction competitive with cyclization. The cyclization in CF_3CO_2H proceeds via the diprotonation product **[H–$\overset{(+)}{O}$=C(CH_3)–$C_5H_4FeC_5H_4$–C(CH=CH–C_6H_4–Cl-4)=$\overset{(+)}{O}$–H]$^{2+}$** (see B in XVIII; electronic spectrum in CF_3CO_2H shows a flat maximum at $\lambda = 600$ nm, which is ascribed to the d-d* transition of B). This ion either reacts further to form XIX (at 40°C $k_B \approx 6.5 \times 10^{-5} s^{-1}$) or decomposes. In HCl/$CH_3OH$ only the CO group with the higher basicity is protonated with formation of

$[CH_3\text{–}CO\text{–}C_5H_4FeC_5H_4\text{–}C(CH\text{=}CH\text{–}C_6H_4\text{–}Cl\text{-}4)\text{=}\overset{(+)}{O}\text{–}H]^+$ (A). For this cyclization the rate constants are, in s^{-1} (temperatures in °C), $k_A \cdot 10^4 = 1.564$ (28.0°), 3.086 (36.6°), 8.748 (46.5°), and 21.660 (54.4°). $\Delta H^{\ddagger} = 81.76 \pm 2.00$ kJ/mol, $\Delta S^{\ddagger} = 55.73 \pm 10.04\, J \cdot mol^{-1} \cdot K^{-1}$ [49].

$2\text{-}ClC_6H_4\text{–}CO\text{–}C_5H_4FeC_5H_4\text{–}CO\text{–}CH_3$ (Table **5**, No. **61**). Cyclization failed with $KOC(C_6H_5)_3$ in boiling xylene [45].

4- and **$3\text{-}CH_3OC_6H_4CH_2\text{–}CO\text{–}C_5H_4FeC_5H_4\text{–}CO\text{–}CH\text{=}CHC_6H_5$** (Table **5**, Nos. **62**, **63**) cyclize on standing in $C_6H_6/CH_3CO_2C_2H_5$ (49/1) on Al_2O_3 (pH 6.5 to 7.5) to form the cis compound XIb (see p. 89) with $R_A = C_6H_5$, $R_B = 4\text{-}$ or $3\text{-}CH_3OC_6H_4$, respectively. Previous washing of the Al_2O_3 with acid prevents the reaction. In contrast, with KOH in boiling C_2H_5OH cyclization occurs to the trans isomers XIa (see p. 89) (R_A and R_B as above). If the latter reaction is conducted with No. 62 in high dilution, XIa and b form in a 1/3 ratio [13]. For kinetics measurements with $NaOCH_3$ in CH_3OH see Table 4, p. 91.

No. 63 gives in glacial acetic acid/conc. H_2SO_4 XIa and b in 14 and 45% yields, respectively [36].

4- and **$3\text{-}CH_3OC_6H_4CH\text{=}CH\text{–}CO\text{–}C_5H_4FeC_5H_4\text{–}CO\text{–}CH_3$** (Table **5**, Nos. **64** and **66**) react with $NaOCH_3$ in CH_3OH to form XI (see p. 89) with $R_A = 4\text{-}$ or $3\text{-}CH_3OC_6H_4$, respectively, $R_B = H$. For kinetics data for the reaction see Table 4, p. 91.

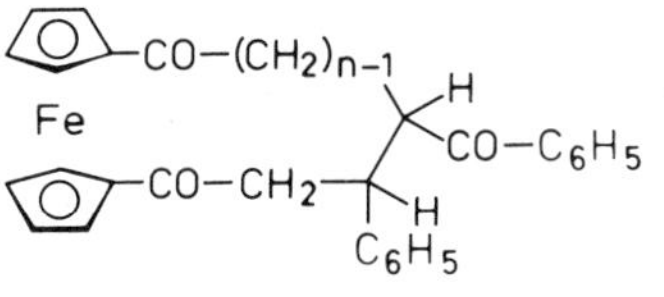

XIX

$C_6H_5CO(CH_2)_n\text{–}CO\text{–}C_5H_4FeC_5H_4\text{–}CO\text{–}CH\text{=}CHC_6H_5$ (Table **5**, Nos. **67** to **70** with n = 5 to 8). The thermodynamically controlled Michael addition with NaOH in C_2H_5OH takes place at room temperature with formation of the cis isomers XIb (see p. 89) with $R_A = C_6H_5$, $R_B = C_6H_5CO(CH_2)_{n-1}$. The product of the kinetically controlled reaction would be chiefly XIX, which would form by reaction at the acid CH_2 group. XIb with $R_A = C_6H_5$, $R_B = C_6H_5CO(CH_2)_7$ is also obtained from No. 70 with an ether solution of $NaOCH_3$ (mole ratio 1/1), while the reaction with NaOH in boiling C_2H_5OH yields the trans isomer XIa with R_A and R_B as above [48].

References:

[1] A. N. Nesmeyanov, N. A. Volkenau (Dokl. Akad. Nauk SSSR **111** [1956] 605/8; Proc. Acad. Sci. USSR Chem. Sect. **111** [1956] 689/91). – [2] M. Furdík, Š. Toma, J. Suchý (Chem. Zvesti **15** [1961] 789/806). – [3] Š. Toma (Chem. Zvesti **19** [1965] 703/10). – [4] Colgate-Palmolive Co., J. E. Robertson (U.S. 3377248 [1965/68]; C.A. **69** [1968] No. 96878). – [5] Colgate-Palmolive Co., J. T. Suh (U.S. 3415859 [1965/68]; C.A. **70** [1969] No. 78158).

[6] R. Dabard, H. Patin (Compt. Rend. C **263** [1966] 1153/5). – [7] M. Furdík, P. Elečko, Š. Kováč (Chem. Zvesti **21** [1967] 501/6). – [8] T. H. Barr, W. E. Watts (Tetrahedron **24** [1968] 3219/35). – [9] T. H. Barr, W. E. Watts (J. Organometal. Chem. **15** [1968] 177/85). – [10] Š. Toma (Acta Fac. Rerum Nat. Univ. Comenianae Chim. No. 13 [1968] 31/5).

[11] T. H. Barr, H. L. Lentzner, W. E. Watts (Tetrahedron **25** [1969] 6001/13). – [12] G. Calvarin, J. Bouvaist, D. Weigel (Compt. Rend. C **268** [1969] 2288/90). – [13] P. Elečko (Chem. Zvesti **23** [1969] 198/211). – [14] P. Elečko (Chem. Zvesti **23** [1969] 212/8). – [15] W. F. Little, H. T. Busby (4th Intern. Conf. Organometal. Chem., Bristol 1969, p. G 2).

[16] H. Patin, R. Dabard (Tetrahedron Letters **1969** 4971/4). – [17] A. Perjéssy (Chem. Zvesti **23** [1969] 905/15). – [18] A. Perjéssy, Š. Toma (Chem. Zvesti **23** [1969] 533/9). – [19] Š. Toma (Collection Czech. Chem. Commun. **34** [1969] 2235/48). – [20] Š. Toma, E. Kalužayová (Chem. Zvesti **23** [1969] 540/52).

[21] Š. Toma, A. Perjéssy (Chem. Zvesti **23** [1969] 343/51). – [22] H. Patin (Compt. Rend. C **270** [1970] 243/5). – [23] H. Patin, L. Roullier, R. Dabard (Compt. Rend. C **271** [1970] 1103/6). – [24] S. Sorriso, H. Lumbroso (Tetrahedron **26** [1970] 2695/701; misprint correction p. 4674). – [25] G. Calvarin, D. Weigel (Acta Cryst. B **27** [1971] 1253/63).

[26] J. J. McDonnell, D. J. Pochopien (J. Org. Chem. **36** [1971] 2092/8). – [27] J. J. McDonnell, D. J. Pochopien (J. Org. Chem. **37** [1972] 4064/7). – [28] R. Dabard, H. Patin (Bull. Soc. Chim. France II **1973** 2158/64). – [29] H. Patin, R. Dabard (Bull. Soc. Chim. France II **1973** 2413/7). – [30] H. Patin, R. Dabard (Bull. Soc. Chim. France II **1973** 2756/9).

[31] H. Patin, R. Dabard (Bull. Soc. Chim. France II **1973** 2764/8). – [32] V. Veselý, Š. Toma, J. Gurský, E. Patzelt (Ropa Uhlie **13** No. 4 [1973] 194/7; C.A. **79** [1973] No. 81199). – [33] K. Yamakawa, M. Hisatome (J. Organometal. Chem. **52** [1973] 407/24). – [34] P. Elečko, P. Foltínova, M. Sališová, E. Solčániová, Š. Toma (Chem. Zvesti **28** [1974] 94/9). – [35] P. Elečko, E. Solčániová, Š. Toma (Chem. Zvesti **28** [1974] 396/401).

[36] P. Elečko, E. Solčániová, M. Vida (Collection Czech. Chem. Commun. **39** [1974] 3684/8). – [37] H. Patin (Tetrahedron Letters **1974** 2893/6). – [38] S. S. Churanov, N. A. Tikhonia (Vestn. Mosk. Univ. Khim. **30** [1975] 597/9; Moscow Univ. Chem. Bull. **30** [1975] No. 5, pp. 63/4). – [39] P. Elečko (Acta Fac. Rerum Nat. Univ. Comenianae Chim. No. 21 [1975] 49/58). – [40] P. Elečko, E. Solčániová, Š. Toma (Chem. Zvesti **29** [1975] 411/7).

[41] M. Livař, M. Sališova, Š. Toma (Collection Czech. Chem. Commun. **40** [1975] 415/9). – [42] Š. Toma, M. Sališová (Acta Fac. Rerum Nat. Univ. Comenianae Chim. No. 21 [1975] 59/70). – [43] D. Mourot, H. Patin (J. Organometal. Chem. **114** [1976] 89/97). – [44] Š. Toma, P. Foltínova (Czech. 183387 [1976/80] 1/2 from C.A. **95** [1981] No. 109476). – [45] M. Sališová, Š. Toma, E. Solčániová (J. Organometal. Chem. **132** [1977] 419/27).

[46] Š. Toma, P. Foltínová, E. Zíkova (Acta Fac. Rerum Nat. Univ. Comenianae Chim. No. 25 [1977] 9/19). – [47] M. Hisatome, T. Hashiyama, K. Yamakawa (Tetrahedron Letters **1978** 3759/62). – [48] M. Sališová, Š. Toma, E. Solčániová (Chem. Zvesti **34** [1980] 507/13). – [49] Š. Toma, A. Gáplovský (Collection Czech. Chem. Commun. **47** [1982] 1988/96). – [50] Š. Toma, A. Gáplovský (Collection Czech. Chem. Commun. **47** [1982] 2246/57).

[51] Š. Toma, A. Gáplovský, J. Federič (Collection Czech. Chem. Commun. **47** [1982] 1997/2003). – [52] T. A. Mashburn, C. E. Cain, C. R. Hauser (J. Org. Chem. **25** [1960] 1982/6). – [53] Ricoh Co., Ltd., T. Goto, K. Kojima, S. Kusakada (Japan. Kokai 73-81522 [1972/73] from C.A. **80** [1974] No. 151183). – [54] Ricoh Co., Ltd., T. Goto (Japan. Kokai 76-53818 [1974/76] from C.A. **86** [1977] No. 131099). – [55] A. Perjessy, Š. Toma (Collection Czech. Chem. Commun. **48** [1983] 1635/46).

[56] M. J. Golightly (AD-A124405 [1982] 1/60, 34/5; C.A. **99** [1983] No. 88334).

5.2.3.5.1.6 Ketones of the Type R′–CO–$C_5H_4FeC_5H_4$–R^2 (R′–CO = R^1)

Compounds of the type R′–CO–$C_5H_4FeC_5H_4$–A–CO–R″ are treated in Section 5.2.3.1.7, pp. 139/52.

The compounds listed in Table 6 are prepared predominantly by the following methods:

Method I. Selective reduction of R′–CO–$C_5H_4FeC_5H_4$–CO–R″ (CH_2–R″ = R^2):

a) With $(C_6H_5)_3SnH$ plus CH_3COCl (molar ratio 1/2/4) in benzene at room temperature. In the subsequent thin layer chromatography on SiO_2 elution occurs first for R′–CH_2–$C_5H_4FeC_5H_4$–CH_2–R″ and the main part of the $(C_6H_5)_3SnCl$ with petroleum ether, then for the desired compounds No. 2, 52, 71, or 75 with petroleum ether/ether, and finally for the starting compound with ether. No. 40 is not isolated alone but together with No. 72, No. 54 with traces of No. 76. Both mixtures are separable by preparative thin layer chromatography [48, 68].

b) With $(C_4H_9)_3SnH$ as for Method Ia. The No. 71 appearing in the same fraction with No. 7 is very difficult to separate from it by preparative thin layer chromatography; the yield is determined by ^{1}H NMR spectroscopy. In addition to No. 54 C_6H_5–CO–$C_5H_4FeC_5H_4$–CO–CH_2–C_6H_5 gives its enol acetate C_6H_5–CH_2–$C_5H_4FeC_5H_4$–C(=CH–C_6H_5)O_2CCH_3 in a 7% yield [68].

Method II. From Fe(C_5H_4–CO–R′)$_2$:

a) By partial hydrogenation over PtO_2 in glacial acetic acid at room temperature [4, 16, 52].

b) By reduction in dioxane with $NaBH_4$ plus concentrated H_2SO_4 [97].

c) By reduction with powdered $NaBH_4$ in CH_3OH at room temperature [40, 72]. The reaction product is chromatographed on basic [72] Al_2O_3 deactivated with 4.6% H_2O. No. 25 is eluted with benzene [72] or 9/1 benzene/ether [40] following Fe(C_5H_4–CO–CH_3)$_2$ and before Fe(C_5H_4–CHOH–CH_3)$_2$.

d) By ionic hydrogenation with $(C_2H_5)_3SiH$ after dissolution in excess CF_3CO_2H (molar ratio 1/2/240) for 20 h at 20°C. The extract in CH_3Cl recovered after the decomposition of the reaction mixture with H_2O yields after the normal processing a mixture of No. 2 and siloxane, from which the latter is removed by column chromatography on Al_2O_3 or by washing with petroleum ether [81].

e) By a 3 h reaction with NaC≡CH in a 2/1 molar ratio, at −37°C in ether/liquid NH_3 in a C_2H_2 stream [38]. The by-products Fe(C_5H_4–C(CH_3)(C≡CH)OH)$_2$ (A) and HO(HC≡C)C(CH_3)–$C_5H_4FeC_5H_4$–CHOH–CH_3 (B) are removed by thin layer chromatography on Al_2O_3 with 50/1 benzene/ethanol as eluent. First Fe(C_5H_4–CO R′)$_2$, then No. 26, and finally (A) (12.7%) and (B) (3.5%) are eluted [35].

f) As with Method IIe, but with tetrahydrofuran instead of ether and at −60°C. Under these conditions the formation of Fe(C_5H_4–C(C_6H_5)(C≡CH)OH)$_2$ is avoided. At −40°C the yield of No. 80 drops to 33%; no Fe(C_5H_4–C(C_6H_5)(C≡CH)OH)$_2$ still forms [39].

g) By reaction with RMgBr (R = C_6H_5 for No. 27, CH_3 for No. 79) in benzene ether with cooling. The compounds are separated from the viscous product formed after hydrolysis with H_2O and the usual processing by chromatography on SiO_2 columns with 4/1 benzene/ether as eluent [92].

Method III. By the Knoevenagel reaction of R′–CO–$C_5H_4FeC_5H_4$–CO–CH_3 with cyclopentadiene in the presence of $NaOC_2H_5$:

a) In a 5/12/25 mole ratio in ethanol first for 30 min at room temperature and then for 2 h at boiling. During the column chromatography on partially deactivated

References on pp. 137/9

Al_2O_3 2/1 ligroin/ether elutes first $Fe(C_5H_4{-}C(CH_3){=}C_5H_4\text{-cyclo})_2$ (9.5% yield) and then 1/1 ligroin/benzene No. 14 [45].

b) In the molar ratio 1/7/14 as for Method IIIa. Ligroin elutes from Al_2O_3 a red band, which yields a dark red gum, probably I, and 5/1 ligroin/ether subsequently No. 77 [45].

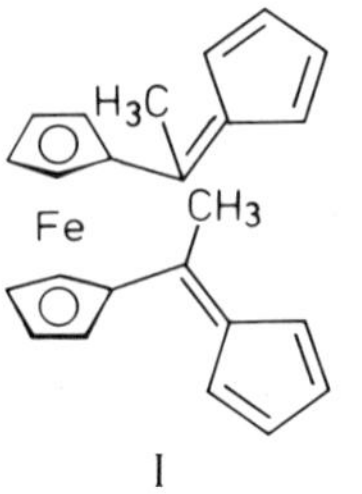

I

Method IV. Condensation of $R^2{-}C_5H_4FeC_5H_4{-}CO{-}CH_3$ with C_6H_5CHO and NaOH (molar ratio 2/3/20) in ethanol/water for 12 h at room temperature. When the starting materials are not the pure compounds No. 9 to 13, 17 to 23, 31, or 35 but their isomer mixtures, the compounds No. 56 to 69 obtained are also mixed with their homoannular isomers, which can be separated by chromatography on SiO_2 columns with benzene as the eluent [87].

Method V. Thermal condensation of $C_5H_5FeC_5H_4{-}CO{-}R'$ with CF_3COCF_3 and $AlCl_3$ (molar ratio 2/3/4) in CH_2Cl_2 in a Carius tube at room temperature (reaction time 3 h for No. 81, 4 h for No. 28) with subsequent column chromatography on Al_2O_3. During the preparation of No. 28 toluene eluted first traces of a red substance (m/e = 394 $[M]^+$ in the mass spectrum), probably a homoannular disubstituted isomer, then CH_2Cl_2 eluted the $C_5H_5FeC_5H_4{-}CO{-}CH_3$ (36% recovery), and finally 1/1 CH_2Cl_2/acetone No. 28. During the separation of No. 81 it is eluted with CH_2Cl_2 following the elution of $C_5H_5FeC_5H_4{-}CO{-}C_6H_5$ (61.5% recovery) with 2/1 toluene/CH_2Cl_2 [70].

Method VI. Friedel-Crafts acylation of $C_5H_5FeC_5H_4{-}R^2$:

a) With $(R'CO)_2O$ plus 85% H_3PO_4 (molar ratio ca. 1/20/4) at 90 to 95°C [51, 87, 88]. The separation following the usual processing is effected by column chromatography on Al_2O_3 [51] or SiO_2 [87, 88] with benzene as eluent. In several cases pure 1,1'-isomers could be isolated by a combination of column chromatography and fractional crystallization [87, 88].

b) With $(R'CO)_2O$ (1/0.6) and polyphosphoric acid at 45°C for 2 h [36].

c) As for Method VIb, but in 10 min on a boiling water bath. During the column chromatography on Al_2O_3 in petroleum ether No. 15 is eluted with ether [27].

d) As for Method VIc, but at a molar ratio of 1/>10 [10] or 1/>60 [24]. No. 16 can be eluted with ether [10] or benzene [10, 24] following the elution of $C_5H_5FeC_5H_4{-}Br$ with hexane [24].

e) By addition of $BF_3\cdot O(C_2H_5)_2$ at 0°C to $C_5H_5FeC_5H_4{-}R^2 + (R'CO)_2O$ in a molar ratio of (12 to 15)/5/(8 to 12), in CH_2Cl_2, and continuation of the reaction for 30 min at 0°C and then 30 min to several hours at room temperature [23, 26, 50]. The fractions of the isomers in the gross product were determined by vapor phase chromatography at 215°C [23]. Separations of the isomer mixtures are described under "Further information" for the individual compounds.

f) Addition of $(R'CO)_2O$ and BF_3 in ether to $C_5H_5FeC_5H_4{-}R^2$ (molar ratio 2/2/1) in CH_2Cl_2 at 0°C and warming the mixture within 6 h up to room temperature.

During the column chromatography on Al_2O_3 are eluted in succession $C_5H_5FeC_5H_4$–R^2, the 1,2-isomer, and No. 8 with petroleum ether, 9/1 petroleum ether/ether, and 4/1 petroleum ether/ether, respectively [76].

g) By saturation of the solution of $C_5H_5FeC_5H_4$–R^2 and $(R'CO)_2O$ (1/1) in CH_2Cl_2 at 0°C with BF_3 [9].

h) By reaction with $(C_2H_5CO)_2O$ and $AlCl_3$ (molar ratio 1/1/1) in CH_2Cl_2 for 3 h. Four bands develop on the Al_2O_3 column with 9/1 Skelly B/ether, the second yielding No. 38 [59].

i) As for Method VIh but at 0°C. $C_5H_5FeC_5H_4$–Br and No. 41 are separated by column chromatography on SiO_2 with Skelly B [59].

j) $AlCl_3$ in ether is mixed in succession with $C_5H_5FeC_5H_4$–R^2 and Al powder in R'COCl in the molar ratio 12/10/(5 to 7)/(13 to 15), and the reaction is continued with stirring at room temperature for 16 to 24 h (2 h for No. 39) [6 to 8]. The compounds No. 39, 46, 47, and 50 are separated from their 1,3-isomers by column chromatography on Al_2O_3 in petroleum ether [6, 7]. No. 55 crystallizes from 9/1 ethanol/water [8].

k) A solution of $C_5H_5FeC_5H_4$–R^2 and $AlCl_3$ in ether is mixed with R'COCl (molar ratio 10/12/13) and the reaction is continued for 3 h at room temperature for Nos. 51 or 84 and for 20 h at boiling for Nos. 86 and 88. Column chromatography on Al_2O_3 in petroleum ether isolates the compounds No. 51 and 86. Nos. 84 and 88 are in addition separated from their 1,3-isomers [8].

l) The 1/1 complex obtained from $AlCl_3$ and R'COCl in CH_2Cl_2 is immediately added dropwise at 0°C to $C_5H_5FeC_5H_4$–R^2 in CH_2Cl_2 and the reaction is continued at 0°C (Nos. 32, 36 [36], 78 [84]), at 5 to 10°C (No. 45 [7]), at room temperature (Nos. 37, 70 [60]), or first at 0°C and then at room temperature (Nos. 15, 16 [20]).

m) As for Method VI l, but from the beginning at room temperature, for 2.5 to 20 h [4, 8, 16, 37, 58, 63, 78, 83, 95].

n) As for Method VI l, but with 3 h heating (No. 7) [17], to 40 to 50°C (No. 53) [8]. No. 53 and small quantities of a not clearly identified isomer are separated by column chromatography on Al_2O_3 in petroleum ether [8].

o) As for Method VI l, but in $CHCl_3$. The reaction temperature extends from 0°C to room temperature, where the reaction mixture is allowed to stand overnight. Column chromatography on Al_2O_3 in 1/1 ligroin/benzene yields No. 15 from the second band [34].

p) $AlCl_3$ in CH_3COCl is added dropwise to $C_5H_5FeC_5H_4$–R^2 in 1/1 CH_3COCl/CH_2Cl_2, then reacted 1 h, all at 0°C [15].

q) $C_5H_5FeC_5H_4$–R^2 is refluxed for 2 h with $AlCl_3$ and CH_3COCl (molar ratio 1/1/~7) in CH_2Cl_2. $C_5H_5FeC_5H_4$–$C(CF_3)_2OH$ is eluted from Al_2O_3 columns with CH_2Cl_2, then No. 28 with 50/1 CH_2Cl_2/acetone [70].

r) R'COCl in CH_2Cl_2 is added immediately following finely divided $AlCl_3$ at 0°C to a solution of $C_5H_5FeC_5H_4$–R^2 (molar ratio 1/0.56/1) in CH_2Cl_2, the usual processing after 15 min. No. 48 is separated from $C_5H_5FeC_5H_4$–$C(CH_3)_3$ and its 1,3-isomer by preparative thin layer chromatography on SiO_2 [78].

s) The solution of $AlCl_3$ in CH_2Cl_2/R_2O (R = $CH(CH_3)_2$ or n-C_4H_9) is stirred with $C_5H_5FeC_5H_4$–$CH_2C(CH_3)_3$ and $(CH_3)_3CCOCl$ (molar ratio 12/5/5) for 2 h at 20°C and is evaporated to dryness in vacuum before the usual processing. No. 49 and its 1,3-isomer are separated by chromatography on Al_2O_3 columns with benzene eluent [6].

References on pp. 137/9

t) $AlCl_3$ is added at −10 to −20°C in portions to $C_5H_5FeC_5H_4$–R^2 + R′COCl (molar ratio 1/1/1) in CH_2Cl_2 and the reaction mixture is warmed over 64 h to room temperature. No. 74 is purified by column chromatography on Al_2O_3 in benzene [8].

u) As for Method VIt, but first at 0 to 5°C and after 40 min at 20°C [83].

Method VII. Reaction of Ag–$C_5H_4FeC_5H_4$–X with C_6H_5COCl in benzene in the presence of CuX (X = Cl or Br). Separation from the main product X–$C_5H_4FeC_5H_4$–$C_5H_4FeC_5H_4$–X (50% yields for X = Cl, 36% for Br) takes place by column chromatography on Al_2O_3 with petroleum ether; compounds No. 92 or 78 are subsequently eluted with 1/1 benzene/petroleum ether [102].

Method VI yields in general mixtures of the three positional isomers, whose total yields are designated by "m" in the table when no separations were conducted. The numbers following the solidus after the "m" give the percentage fractions of the 1,1′-isomers, whose determination without previous separation can, for example, be conducted by vapor phase chromatography [27] or by ^1H NMR spectroscopy [19, 26, 87], if necessary with the addition of a shift reagent such as $Eu(dpm)_3$ (see Formula II) [62]. The reaction temperature and the sequence and rate of reagent addition affect both the total yield and also the isomer ratio [80].

Acylation always occurs in a substantial extent in the 1′-position, normally more than in the 3-position; the fraction for the 1,2-isomer is in general the least, see, for example [80].

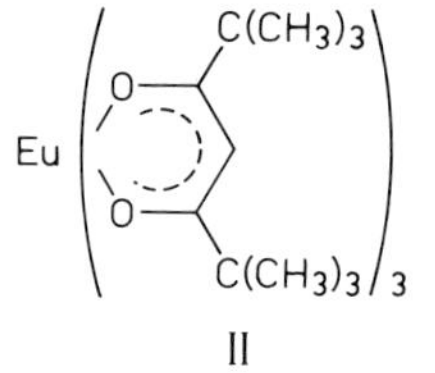

II

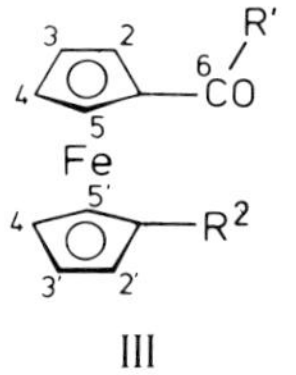

III

Separation of the isomers is conducted mainly by chromatography, see [79]. The sequence of increasing absorbability is 1,2≪1,1′<1,3. Due to the similarity of the absorption properties, especially, those of the 1,1′- and 1,3-isomers, the separation is often incomplete or requires a considerable effort. In some cases the possibility exists of converting the inseparable isomer mixture chemically into a separable mixture, see, for example, Nos. 2 and 32 [36]. Special data are given in each case under the "Further information" for the individual compounds.

Properties

The ^1H NMR spectra (general numbering scheme shown in III) of Nos. 1 to 3 clearly display a transfer of the donor effect of the R^2 group to the shift of the cyclopentadienyl protons 2 to 5. In No. 4 in contrast it appears to be missing completely for the $(CH_3)_3C$ group; the CH_3CO group effects a shift of the protons 2′ to 5′ toward lower field by around 10^{-2} ppm [42].

The compounds CH_3–CO–$C_5H_4FeC_5H_4$–R^2 with R^2 = alkyl show in contrast to their 1,2- and 1,3-isomers two medium intensity bands in the IR spectra around 895 and 1115 cm^{-1} [3]. Correlation of the ν(C=O) data for CH_3–CO–$C_5H_4FeC_5H_4$–R^2 with $R^2 = C_2H_5$ (No. 2), Br (No. 16), OCH_3 (No. 32), CO–CH_3 (see p. 13), CO–C_6H_5 (see p. 98), and $CO_2C_2H_5$ (see p. 300) in CCl_4 with Hammett σ_p-constants of R^2 suggests that there is no significant transannular conjugation of substituents on the two rings in ferrocene [47]. A linear correlation corresponding to a modified and extended Seth-Paul-Van Duyse equation has been observed between the ν(C=O) vibrations in CCl_4 and the $X^+(R^{A,B})$ constants of the structural fragments for 154 compounds with ferrocene structures of the general formula $R^A–CO–R^B$, among them the compounds No. 7, 15, 16, 56 to 59, 61 to 67, 69, 90, and 91 with $R^A = C_5H_4FeC_5H_4–R^2$ and $R^B = R'$. The $X^+(R^{A,B})$ used have been confirmed by the recently derived empirical relationship for calculation of the substituent constants and transmission factors [103].

For the compounds $CH_3–CO–C_5H_4FeC_5H_4–R^2$ with R^2 = alkyl the UV band around 230 nm shifts to higher wavelengths on going from the 1,2- to the 1,1'- to the 1,3-isomer. The same is true for $R^2 = C_6H_5$ with still greater differences [5].

Chemical Behavior

The R'–CO group is converted in the Clemmensen reaction by amalgamated Zn and concentrated HCl in CH_3CO_2H [50, 66] or C_2H_5OH [36] into the $R'–CH_2$ group. Thus, Nos. 15, 16, and 32 form $C_2H_5–C_5H_4FeC_5H_4–R$ with R = Cl [50, 66], Br [10], and OCH_3 [36]; Nos. 39, 45, 49, 53, 84, and 88 give $R'–CH_2–C_5H_4FeC_5H_4–CH_2–C(CH_3)_3$ with $R' = C_2H_5$, i-C_3H_7, t-C_4H_9, $CH_2C_6H_5$, cyclo-C_6H_{11}, and 2-ClC_6H_4 [7]; and Nos. 46, 47, and 50 yield $(CH_3)_3C–CH_2–C_5H_4FeC_5H_4–R$ with R = n-C_3H_7, $CH_2CH(CH_3)_2$, and $CH_2CH_2C_6H_5$, respectively [7].

$NaBH_4$ [20, 40, 76] or KBH_4 [60] in CH_3OH and $LiAlH_4$ in boiling ether [27, 67, 78] reduce the keto groups to R'–CHOH groups. Thus, the compound $R'–CHOH–C_5H_4FeC_5H_4–R^2$ with $R' = R^2 = CH_3$ is won from No. 1 [40, 77], with $R' = CH_3$ and R^2 = Cl from No. 15 [22, 27, 40], with $R' = CH_3$ and R^2 = Br from No. 16 [27, 40], with $R' = CH_3$ and $R^2 = CH(CH–CHC_6H_5)OCH_3$ or $CH=CHCH(C_6H_5)OCH_3$ from Nos. 33 and 34, respectively [76], with $R' = R^2 = C(CH_3)_3$ from No. 48 [78], and with $R' = C_6H_5$ and $R^2 = CH_3$ or C_2H_5 from Nos. 70 [60] and 71 [67], respectively.

Compounds with the grouping $R–CH_2–CO–C_5H_4FeC_5H_4–R^2$, such as Nos. 37, 38, 41, and 42, are oxidized in dry $(CH_3)_2SO$ in the presence of a three- to fourfold excess of $KOC(CH_3)_3$ by small quantities of oxygen (1 s in air) to the stable paramagnetic radical anions IV [59] (see 5.2.3.5.1.8, p. 155).

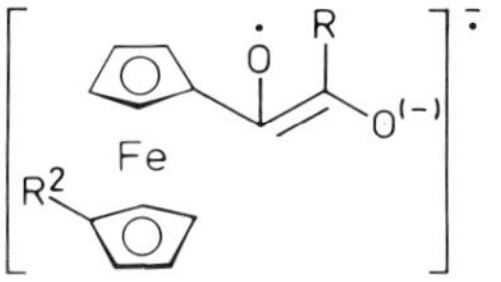

IV

$R^2–C_5H_4FeC_5H_4–CO–CH_3$ condenses with C_6H_5CHO in $NaOH/H_2O/C_2H_5OH$ (cf. preparation Method IV) to form $R^2–C_5H_4FeC_5H_4–CO–CH=CHC_6H_5$ [87].

The substituent in the 1'-position (R^2) exerts a considerable influence on the basicity of $C_5H_5FeC_5H_4–CO–CH_3$. A linear relationship exists between the Hammett constants σ_M for R^2 and log K, where $K = [R^2–C_5H_4FeC_5H_4–CO–CH_3][H_3O^+]/[R^2–C_5H_4FeC_5H_4–C(CH_3)=OH^+][H_2O]$, and $R^2 = COCH_3$, C_2H_5, Br, or CN. The transmission of the polar effect of R^2 appears in essence to occur by an inductive mechanism through space and less via the Fe [64].

Applications

The compounds No. 5, 39, 45, 51, 53, 55, 74, 84, 86, and 88 are intermediates in syntheses of blood-building substances for the relief of iron deficiency anemia [8].

The mixture containing No. 70 as the main component, obtained by benzoylation of $C_5H_5FeC_5H_4–CH_3$, is an effective photostabilizer for polyolefins, such as polyethylene [100]. Added in quantities of 0.2 to 1.3 wt% to high pressure polyethylene films, it increases their tensile strength and breaking elongation by 20 to 35 and 25 to 30%, respectively, after 100 h UV irradiation in air at 25°C. Apparently the addition inhibits cross-linking produced in untreated films by the irradiation [101].

Table 6

Ketones of the Type $R'-CO-C_5H_4FeC_5H_4-R^2$.
Further information for numbers preceded by an asterisk is given at the end of the table, pp. 127/37. For abbreviations and dimensions see p. VIII.

No.	$R^1-C_5H_4FeC_5H_4-R^2$ $R^1 = CO-R'$	R^2	method of preparation (yield in %, "m" indicates isomeric mixture, number following solidus is fraction present of 1,1'-compound), properties, and remarks	Ref.
*1	CO–CH_3	CH_3	VIe (73 m/49%; 22.6; m/47 to 49%), VIg, see "Further information", p. 127 m.p. 42 to 43°, 42.7 to 43.2°, 59 to 60° ^{1}H NMR (CCl_4): 1.89(1.91) (s, R^2), 3.99 (H-2' to 5'), 4.30 (H-3,4), 4.58 (H-2,5) ^{1}H NMR (C_6H_6): 1.80 (s, R^2) IR(Nujol): 890.5, 921.7, 1116, 1278.8	[9, 19, 23, 26, 37, 44]
*2	CO–CH_3	C_2H_5	Ia (33), IIa (30), IIb (32), IId (55), VIe (12.4; m/37 to 48%), VIg, VIm (m/31%), see "Further information", p. 128 red oil, b.p. 128°/0.2, 130°/0.3 Torr, long small orange needles, m.p. 12.4 to 13.2° (petroleum ether) ^{1}H NMR (CCl_4): 2.27 (CH_2 of C_2H_5), 4.01 (H-2' to 5'), 4.31 (H-3,4), 4.59 (H-2,5) ^{1}H NMR ($CDCl_3$): 2.37 ($COCH_3$), 4.08 (s, H-2' to 5'), 4.43 (t, H-3,4), 4.70 (t, H-2,5; J(α-H) = J(β-H) = 2) ^{1}H NMR ($CDCl_3$): 1.10 (CH_3 of C_2H_5), 2.33 (CH_2 of C_2H_5), 2.40 ($COCH_3$), 4.05 (H-2' to 5'), 4.40 (H-3,4), 4.65 (H-2,5) ^{1}H NMR (C_6H_6): 3.87 (s, H-2' to 5'), 4.14 (t, H-3,4), 4.60 (t, H-2,5; J(α-H) = J(β-H) = 2) IR (CCl_4): 893, 906, 1116, 1662.5 (νC=O), see figure in [16] IR (film): 892.9, 907.4, 1114.8, 1278.8 UV (CCl_4): λ_{max} (log ε) = 228 (4.18), 270 (3.76), 340 (3.11), 460 (2.73) UV (C_6H_6): λ_{max} (ε) = 365 (1465), 455 (398) UV (19/1 C_2H_5OH/H_2O): λ_{max}(ε) = 228 (15100), 270 (5800) UV (H_2O): λ_{max} (log ε) = 278 (3.85), 346 (3.29), 4.72 (2.99)	[1, 3 to 5, 16, 19, 23, 26, 47, 48, 52, 56, 62, 64, 68, 81, 93, 97, 104]
*3	CO–CH_3	$CH(CH_3)_2$	VIe (15; m/45.6%) b.p. 146°/1 Torr ^{1}H NMR (CCl_4): 2.57 (CH of $CH(CH_3)_2$), 4.02 (H-2' to 5'), 4.34 (H-3,4), 4.61 (H-2,5) IR (film): 892.9, 920.8, 1114.8, 1278.8	[19, 23, 26]

References on pp. 137/9

Table 6 (continued)

No.	R^1–$C_5H_4FeC_5H_4$–R^2 R^1 = CO–R′	R^2	method of preparation (yield in %, "m" indicates isomeric mixture, number following solidus is fraction present of 1,1′-compound), properties, and remarks	Ref.
*4	CO–CH_3	$C(CH_3)_3$	VIe (33 m/26%; m/41%) b.p. 144°/3 Torr ^{1}H NMR (CCl_4): 4.02 (H-2′ to 5′), 4.39 (H-3,4), 4.68 (H-2,5) IR (film): 892.9, 909.1, 1114.8, 1273.9	[19,23,26]
*5	CO–CH_3	$CH_2C(CH_3)_3$	VIm m.p. 73 to 74° (petroleum ether)	[8]
*6	CO–CH_3	$C(CH_3)$=CH_2	see "Further information", p. 129 dark red, unstable crystals, m.p. 35 to 38.5° ^{1}H NMR ($CDCl_3$): 1.99 (s, CH_3), 2.29 (s, CH_3); 4.22 (m), 4.35 (m), and 4.64 (m) (8 H, C_5H_4); 4.9 (s) and 5.09 (s) (=CH_2)	[41]
*7	CO–CH_3	$CH_2C_6H_5$	Ib (20), VId (43), VIn (11), see "Further information", p. 129 m.p. 79 to 81° (petroleum ether or acetone/light petroleum ether) IR (paraffin oil): 1008 (δC–H of C_5H_4), 1119 (νC_{ring}–C_{acetyl}) IR (CCl_4): 1675 (νC=O) UV (C_2H_5OH): λ_{max} (log ε) = 226 (4.53), 264 (4.09), 337 (3.41) R_f: 0.52 (on SiO_2 with $C_6H_6/CH_3CO_2C_2H_5$, 8/2)	[17, 49, 51, 68, 103]
*8	CO–CH_3	CH_2CH=CHC_6H_5-(E)	VIf (35 related to unrecovered $C_5H_5FeC_5H_4$–R^2) orange solid, m.p. 58 to 59° ^{1}H NMR ($CDCl_3$): 2.34 (s, CH_3), 3.13 (m, CH_2); 4.1 (m), 4.47 (m), and 4.72 (m) (8 H, C_5H_4); 6.2 to 6.5 (m, CH=CH), 7.3 (m, C_6H_5) UV (C_2H_5OH): λ_{max} (ε) = 253 (22500)	[76]
*9	CO–CH_3	C_6H_5	VIa (18.6 m/67%), VIp, see "Further information", p. 130 m.p. 75 to 83° (m), 88.5 to 89.5° (petroleum ether/ether, 2/1) orange rods, m.p. 92.5 to 93.5° (ether/petroleum ether), m.p. 113 to 114° (pure isomer) ^{1}H NMR ($CDCl_3$): 2.09 (s, CH_3), 4.32 (m, H-3,3′,4,4′), 4.62 (m, H-2,2′,5,5′), 7.0 to 7.6 (m, C_6H_5) IR ($CHCl_3$): 887.3, 893.7, 1113.6 UV (19/1 C_2H_5OH/H_2O): λ_{max} (ε) = 236 (16500), 267 (10500)	[3, 5, 11, 15, 87, 88]

References on pp. 137/9

Table 6 (continued)

No.	R^1–$C_5H_4FeC_5H_4$–R^2 R^1 = CO–R'	R^2	method of preparation (yield in %, "m" indicates isomeric mixture, number following solidus is fraction present of 1,1'-compound), properties, and remarks	Ref.
10	CO–CH_3	$C_6H_4CH_3$-4	VIa (32.25 m/50%) m.p. 70 to 78° (m) ^{1}H NMR ($CDCl_3$): 2.13 (s, $COCH_3$), 2.33 (s, CH_3), 4.35 (m, H-3,3',4,4'), 4.65 (m, H-2,2',5,5'), 7.0 to 7.5 (m, C_6H_4)	[87, 88]
11	CO–CH_3	$C_6H_4CH_3$-3	VIa (60.7 m/65%) m.p. 52 to 59° (m), m.p. 63° (pure isomer) ^{1}H NMR $(CDCl)_3$: 2.13 (s, $COCH_3$), 2.40 (s, CH_3), 4.49 (m, H-3,3',4,4'), 4.65 (m, H-2,2',5,5'), 6.9 to 7.4 (m, C_6H_4)	[87, 88]
12	CO–CH_3	$C_6H_3(CH_3)_2$-3,4	VIa (42.3 m/52%) oil (m) ^{1}H NMR ($CDCl_3$): 2.05 (s, $COCH_3$), 2.22 (s, CH_3), 4.23 (m, H-3,3',4,4'), 4.54 (m, H-2,2',5,5'), 6.9 to 7.4 (m, C_6H_3)	[87, 88]
13	CO–CH_3	$C_6H_3(CH_3)_2$-3,5	VIa (41.8 m/52%) oil (m) ^{1}H NMR ($CDCl_3$): 2.03 (s, $COCH_3$), 2.28 (s, CH_3), 4.23 (m, H-3,3',4,4'), 4.55 (m, H-2,2',5,5'), 6.8 to 7.2 (m, C_6H_3)	[87, 88]
14	CO–CH_3	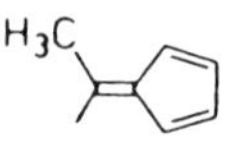	IIIa (46) unstable red oil, uncrystallizable ^{1}H NMR ($CDCl_3$): 2.28 (s, $COCH_3$), 2.51 (s, CH_3), 4.4 to 4.8 (m, 8H, C_5H_4), 6.56 (s, vinyl protons)	[45]
*15	CO–CH_3	Cl	VIc (61), VIe (48.6), VI l (81), VIo (57), see "Further information", p. 130 m.p. 53 to 55°, 53.5 to 54.5° (hexane), 55.5 to 56.7°, 62 to 63° (hexane) ^{1}H NMR (CCl_4): 4.10 (t, H-3'4'), 4.41 (t, H-2',5'), 4.51 (t, H-3,4), 4.77 (t, H-2,5) ^{1}H NMR ($CDCl_3$): 2.40 (CH_3), see "Further information", p. 130 IR (CCl_4): 1118, 1281, 1690 (spectrum appears to be shifted slightly due to improper tracking [20]), no bands at 1000 and 1100 IR (CCl_4): 1680 (νC=O) UV (CH_3OH): λ_{max} = 224 to 225, 270	[20, 27, 28, 34, 50, 56, 91, 103]

References on pp. 137/9

Table 6 (continued)

No.	R^1–$C_5H_4FeC_5H_4$–R^2 R^1 = CO–R′	R^2	method of preparation (yield in %, "m" indicates isomeric mixture, number following solidus is fraction present of 1,1′-compound), properties, and remarks	Ref.
*16	CO–CH_3	Br	VId (54 to 56), VI l (75), see "Further information", p. 131 m.p. 56 to 58°, 57 to 58°, 61.5 to 63° (hexane) ^{1}H NMR ($CHCl_3$): 2.41 (s, CH_3), 4.14 (t, H-3′,4′; J = 2), 4.47 (t, H-2′,5′; J = 2), 4.58 (t, H-3,4; J = 2), 4.78 (t, H-2,5; J = 2) ^{1}H NMR (CCl_4): 4.10 (t, H-3′,4′), 4.35 (t, H-2′,5′), 4.41 (t, H-3,4), 4.64 (t, H-2,5) IR (CCl_4): 1114, 1276, 1682 (νC=O) IR (CCl_4): 1681 (νC=O) IR (CCl_4): 1668.6 (νC=O) UV (n-heptane): $\lambda_{max}(\varepsilon)$ = 220 (2300), 255 sh (5500), 300 to 330 (ca. 1000), 348 sh (400), 448 (280) UV (CH_3OH): λ_{max} = 223 to 224, 250 to 260 (inflection)	[10, 20, 24 28, 29, 47, 103]
17	CO–CH_3	C_6H_4Cl-4	VIa (31.15 m/49%) m.p. 85 to 90° (m), m.p. 101° (pure isomer) ^{1}H NMR ($CDCl_3$): 2.13 (s, CH_3), 4.34 (m, H-3,3′,4,4′), 4.61 (m, H-2,2′,5,5′), 7.2 to 7.6 (m, C_6H_4)	[87, 88]
18	CO–CH_3	C_6H_4Cl-3	VIa (54.01 m/71%) oil (m) ^{1}H NMR ($CDCl_3$): 2.21 (s, CH_3); 4.32 (t) and 4.40 (t) (H-3,3′,4,4′); 4.66 (t) and 4.75 (t) (H-2,2′,5,5′); 7.0 to 7.7 (m, C_6H_4) ^{1}H NMR ($CDCl_3$): 2.21 (s, CH_3), 4.32 (t, H-3,3′,4,4′), 4.66 (t, H-5,5′), 4.75 (t, H-2,2′), 7.0 to 7.7 (m, C_6H_4)	[87, 88]
19	CO–CH_3	C_6H_4Br-4	VIa (44.15 m/72%) m.p. 87 to 93° (m), m.p. 105° (pure isomer) ^{1}H NMR ($CDCl_3$): 2.13 (s, CH_3), 4.35 (m, H-3,3′,4,4′), 4.62 (m, H-2,2′5,5′), 7.2 to 7.6 (m, C_6H_4)	[87, 88]
20	CO–CH_3	C_6H_4Br-3	VIa (49.4 m/73%) oil (m) ^{1}H NMR ($CDCl_3$): 2.11 (s, CH_3), 4.32 (m, H-3,3′,4,4′), 4.60 (m, H-2,2′,5,5′), 7.0 to 7.6 (m, C_6H_4)	[87, 88]

References on pp. 137/9

Table 6 (continued)

No.	R^1–$C_5H_4FeC_5H_4$–R^2 R^1 = CO–R′	R^2	method of preparation (yield in %, "m" indicates isomeric mixture, number following solidus is fraction present of 1,1′-compound), properties, and remarks	Ref.
21	CO–CH_3	$C_6H_3Cl_2$-3,4	VIa (44.9 m/75%) oil (m) ^{1}H NMR ($CDCl_3$): 2.13 (s, CH_3), 4.31 (m, H-3,3′,4,4′), 4.56 (m, H-2,2′,5,5′), 7.0 to 7.6 (m, C_6H_3)	[87, 88]
22	CO–CH_3	$C_6H_4CF_3$-3	VIa (47.5 m/78%) oil (m) ^{1}H NMR ($CDCl_3$): 2.09 (s, CH_3), 4.34 (m, H-3,3′,4,4′), 4.63 (m, H-2,2′,5,5′), 7.3 to 7.7 (m, C_6H_4)	[87, 88]
23	CO–CH_3	$C_6H_3(CF_3$-3)Cl-4	VIa (60.3 m/75%) oil (m) ^{1}H NMR ($CDCl_3$): 2.14 (s, CH_3), 4.35 (m, H-3,3′,4,4′), 4.59 (m, H-2,2′,5,5′), 7.2 to 7.7 (m, C_6H_3)	[87, 88]
*24	CO–CH_3	OH	see No. 16, p. 132 m.p. 115.5 to 118.5° (in N_2)	[10]
*25	CO–CH_3	$CHOHCH_3$	IIc (30), for formation see No. 2, p. 128 ^{1}H NMR ($CDCl_3$): 1.40 (d) and 1.46 (d) (CH_3; $\Delta\nu$ = 4.5 Hz; J = 6), 2.35 (s, $COCH_3$; $\Delta\nu$ = 1.3 Hz), 4.15 (CH), 4.4 to 4.8 (H-2 to 5) IR (CCl_4): 1282, 1660 (1667, 1677), 3500 UV (H_2O): $\lambda_{max}(\varepsilon)$ = 230 (13800), 268 (6800), 342 (1440), 455 (660)	[40, 72]
26	CO–CH_3	$C(CH_3)$–(C≡CH)OH	IIe (53) m.p. 80 to 81° (heptane) IR: 2130 (νC≡C), between 3330 and 3360 (νOH) R_f: 0.254 (on Al_2O_3 with 50/1 C_6H_6/absolute C_2H_5OH)	[35, 38]
*27	CO–CH_3	$\overset{6'}{C}(CH_3)$–(C_6H_5)OH	IIg (49) m.p. 53 to 54° (ether/hexane) ^{1}H NMR ($CDCl_3$): 1.84 (s, CH_3-6′), 2.42 (s, $COCH_3$), 2.87 (s, OH); 4.08 to 4.14 (m, 1H) and 4.2 to 4.32 (m, 3H), 4.54 to 4.62 (m, 2H), 4.77 to 4.93 (dm, 2H) (C_5H_4); 7.16 to 7.58 (m, C_6H_5) IR (CCl_4): 835 (δC–H of C_5H_4), 1610 (νC=C of C_6H_5), 1685 (νC=O), 3510 (νOH) mass spectrum: m/e 348 $[M]^+$, 330 $[M-H_2O]^+$	[92]

Table 6 (continued)

No.	R^1–$C_5H_4FeC_5H_4$–R^2 R^1 = CO–R′	R^2	method of preparation (yield in %, "m" indicates isomeric mixture, number following solidus is fraction present of 1,1′-compound), properties, and remarks	Ref.
28	CO–CH_3	$C(CF_3)_2OH$	V (46), VIq (19) stable red crystals, m.p. 104 to 106°, subl. at 100°/0.01 Torr ^{1}H NMR ($CDCl_3$): 2.37 (s, CH_3); 4.33 (m, 2H), 4.43 (t, 2H), 4.63 (t, 2H), and 4.90 (t, 2H) (C_5H_4, A_2B_2 patterns); 4.76 (s, OH) ^{19}F NMR ($CDCl_3$): 75.21 (s, CF_3) IR: 830, 850, 1030, 1050, and 1060 (ferrocenyl); 1200 to 1300 (C–F); 1640 (C=O), 3100 (OH)	[70]
*29	CO–CH_3	$O_2CC_6H_5$	see No. 24, p. 132 m.p. 84 to 84.5° (2/1 cyclohexane/hexane) ^{1}H NMR (CCl_4): 3.95 (t, H-3′,4′), 4.43 (t, H-3,4), 4.58 (t, H-2′,5′), 4.74 (t, H-2,5) (A_2B_2 patterns)	[10, 28]
*30	CO–CH_3	$CH(CH_3)$–O_2CCH_3	see No. 25, p. 132 clear red-orange oil, n_D^{25}: 1.5850 UV (CH_3OH): $\lambda_{max}(\varepsilon)$ = 226 (17480), 269 (6220), 337 (1222) UV (cyclohexane): $\lambda_{max}(\varepsilon)$ = 223 (23200), 269 (5280), 325 sh (884)	[40]
31	CO–CH_3	$C_6H_4O_2CCH_3$-4	VIa (43.75 m/53%) oil (m) ^{1}H NMR ($CDCl_3$): 2.08 (s, $COCH_3$), 2.25 (s, O_2CCH_3), 4.29 (m, H-3,3′,4,4′), 4.58 (m, H-2,2′,5,5′), 6.8 to 7.6 (m, C_6H_4)	[87, 88]
*32	CO–CH_3	OCH_3	VIb (45 m/–), VI l (51 m/–), see "Further information", p. 133 deep red oil (m) IR (CCl_4): 1660.8 (νC=O)	[36]
*33	CO–CH_3	$\overset{6'}{C}H(CH{=}CHC_6H_5)$–$OCH_3$	see "Further information", p. 133 ^{1}H NMR ($CDCl_3$): 2.39 (s, $COCH_3$), 3.34 (s, OCH_3), 4.15 to 4.75 (m, 9H, C_5H_4 + H-6′), 5.9 to 6.8 (m, CH=CH), 7.40 (br. s, C_6H_5)	[76]
*34	CO–CH_3	CH=CH–$\overset{8'}{C}H$–$(C_6H_5)OCH_3$	see "Further information", p. 133 ^{1}H NMR ($CDCl_3$): 2.22 (s, $COCH_3$), 3.32 (s, OCH_3), 4.15 to 4.75 (m, 9H, C_5H_4 + H-8′), 5.9 to 6.6 (m, CH=CH), 7.39 (br. s, C_6H_5)	[76]

References on pp. 137/9

Table 6 (continued)

No.	R^1–$C_5H_4FeC_5H_4$–R^2 R^1 = CO–R′	R^2	method of preparation (yield in %, "m" indicates isomeric mixture, number following solidus is fraction present of 1,1′-compound), properties, and remarks	Ref.
*35	CO–CH_3	$C_6H_4OCH_3$-4	VIa (23.01 m/69%), VIp m.p. 88 to 93° (m), orange-red plates, m.p. 97 to 98° (petroleum ether) (1,1′- + 1,3-isomers) ^{1}H NMR ($CDCl_3$): 2.15 (s, $COCH_3$), 3.83 (s, OCH_3), 4.35 (m, H-3,3′,4,4′), 4.60 (m, H-2,2′,5,5′), 6.8 to 7.5 (m, C_6H_4)	[15,87,88]
*36	CO–CH_3	CHO	VI l (41.1), see "Further information", p. 133 m.p. 90 to 93° (C_6H_6/petroleum ether), 94 to 95° (n-hexane/ether) IR (CCl_4): 1675.2 (νC=O of R^1)	[18,47,56]
*37	CO–C_2H_5	CH_3	VI l (45), see "Further information", p. 134 oil (oxime, m. p. 100°) ^{1}H NMR ($CDCl_3$): 1.17 (t, CH_3 of C_2H_5), 1.88 (s CH_3-1′), 2.70 (q, CH_2); 4.03 (s, 4H), 4.36 (t, 2H), and 4.66 (t, 2H) (C_5H_4)	[53, 60]
*38	CO–C_2H_5	$C(CH_3)_3$	VIh (22.0) ^{1}H NMR ($CDCl_3$): 1.17 (t, CH_3 of C_2H_5), 1.22 (s, $C(CH_3)_3$), 2.70 (m, CH_2); 4.05 (m, 2H), 4.15 (d, 2H), 4.47 (m, 2H), and 4.75 (m, 2H) (C_5H_4)	[59]
39	CO–C_2H_5	$CH_2C(CH_3)_3$	VIj m.p. 40 to 42°	[7, 8]
40	CO–C_2H_5	$CH_2C_6H_5$	Ia (16) b.p. 190°/1 Torr	[48, 68]
41	CO–C_2H_5	Br	VIi (20.9) m.p. 31° ^{1}H NMR ($CDCl_3$): 1.19 (t, CH_3), 2.78 (m, CH_2); 4.12 (t, 2H), 4.41 (t, 2H), 4.52 (t, 2H), and 4.82 (t, 2H) (C_5H_4)	[59]
*42	CO–C_2H_5	C_6H_4Cl-4	see "Further information", p. 134 m.p. 102 to 103° ^{1}H NMR ($CDCl_3$): 1.07 (t, CH_3), 2.47 (m, CH_2); 4.33 (t, 4H) and 4.62 (m, 4H) (C_5H_4); 7.28 (s, C_6H_4)	[59]
*43	CO–C_3H_7-n	CH_3	VIm (67 m/52%), see "Further information", p. 135 ^{1}H NMR ($CDCl_3$ or CCl_4): 1.00 (t, CH_3 of C_3H_7), 1.75 (m, –CH_2–), 1.82 (s, CH_3-1′), 2.59 (t, $COCH_2$); 3.91 (s, 4H), 4.24 (m, 2H), and 4.60 (m, 2H) (A_2B_2 pattern) (C_5H_4)	[58]

Table 6 (continued)

No.	R^1–$C_5H_4FeC_5H_4$–R^2 R^1 = CO–R′	R^2	method of preparation (yield in %, "m" indicates isomeric mixture, number following solidus is fraction present of 1,1′-compound), properties, and remarks	Ref.
*44	CO–CH$(CH_3)_2$	C$(CH_3)_3$	VIm (7), see "Further information", p. 135 deep orange liquid ^{1}H NMR: 1.16 (d, CH_3 of i-C_3H_7), 1.17 (s, C$(CH_3)_3$), 2.8 to 3.3 (m, CH of i-C_3H_7), 3.85 to 4.15 (m, H-2′ to 5′); 4.44 (t, 2H) and 4.73 (t, 2H) (H-2 to 5)	[78]
*45	CO–CH$(CH_3)_2$	$CH_2C(CH_3)_3$	VI l b.p. 150 to 152°/1.0 Torr	[7, 8]
46	CO–C$(CH_3)_3$	C_3H_7-n	VIj b.p. 166°/2 Torr	[6, 7]
47	CO–C$(CH_3)_3$	$CH_2CH(CH_3)_2$	VIj b.p. 150°/1 Torr	[6, 7]
48	CO–C$(CH_3)_3$	C$(CH_3)_3$	VIr (24) deep orange plates (pentane), m.p. 61 to 63° ^{1}H NMR: 1.19 (s, C$(CH_3)_3$), 1.31 (s, C$(CH_3)_3$), 3.95 to 4.15 (m, H-2′ to 5′); 4.44 (t, 2H) and 4.84 (t, 2H) (H-2 to 5)	[78]
*49	CO–C$(CH_3)_3$	$CH_2C(CH_3)_3$	VIs m.p. 54, 56 to 57°	[2, 6]
50	CO–C$(CH_3)_3$	$CH_2CH_2C_6H_5$	VIj m.p. 93 to 94°	[6, 7]
51	CO–CH_2CH–$(CH_3)CH_2$–C$(CH_3)_3$	$CH_2C(CH_3)_3$	VIk b.p. 154°/2 Torr	[8]
*52	CO–$CH_2C_6H_5$	C_2H_5	Ia (48) b.p. 197°/0.4 Torr, m.p. 45°	[48, 65, 68]
53	CO–$CH_2C_6H_5$	$CH_2C(CH_3)_3$	VIn m.p. 100 to 101°	[7, 8]
*54	CO–$CH_2C_6H_5$	$CH_2C_6H_5$	Ia (38), Ib m.p. 125°	[48, 65, 68]
55	CO–CH=CHC_6H_5	$CH_2C(CH_3)_3$	VIj m.p. 125 to 126° (9/1 C_2H_5OH/H_2O)	[8]
56	CO–CH=CHC_6H_5	C_6H_5	IV (41.2) m.p. 153 to 155° ^{1}H NMR ($CDCl_3$): 4.329 (t, H-3′,4′), 4.440 (t, H-3,4), 4.630 (t, H-2′,5′), 4.688 (t, H-2,5) IR (CCl_4): 1663 (νC=O) for s-cis	[87, 103]

References on pp. 137/9

Table 6 (continued)

No.	R^1–$C_5H_4FeC_5H_4$–R^2 R^1 = CO–R′	R^2	method of preparation (yield in %, "m" indicates isomeric mixture, number following solidus is fraction present of 1,1′-compound), properties, and remarks	Ref.
57	CO–CH= CHC_6H_5	$C_6H_4CH_3$-4	IV (26.1) m.p. 155 to 157° IR (CCl_4): 1662.3 (νC=O) for s-cis	[87, 103]
58	CO–CH= CHC_6H_5	$C_6H_4CH_3$-3	IV (48) m.p. 158 to 160° IR (CCl_4): 1664 (νC=O) for s-cis	[87, 103]
59	CO–CH= CHC_6H_5	$C_6H_3(CH_3)_2$-3, 4	IV (57.1) m.p. 122 to 123° IR (CCl_4): 1663 (νC=O) for s-cis	[87, 103]
60	CO–CH= CHC_6H_5	$C_6H_3(CH_3)_2$-3, 5	IV (63.2) m.p. 115 to 116°	[87]
61	CO–CH= CHC_6H_5	C_6H_4Cl-4	IV (11) m.p. 149° IR (CCl_4): 1663.6 (νC=O) for s-cis	[87, 103]
62	CO–CH= CHC_6H_5	C_6H_4Cl-3	IV (66.1) m.p. 58° IR (CCl_4): 1664 (νC=O) for s-cis	[87, 103]
63	CO–CH= CHC_6H_5	C_6H_4Br-4	IV (24) m.p. 154 to 155° IR (CCl_4): 1663.4 (νC=O) for s-cis	[87, 103]
64	CO–CH= CHC_6H_5	C_6H_4Br-3	IV (59.6) m.p. 122° IR (CCl_4): 1664.1 (νC=O) for s-cis	[87, 103]
65	CO–CH= CHC_6H_5	$C_6H_3Cl_2$-3, 4	IV (55.2) m.p. 136 to 137° IR (CCl_4): 1663 (νC=O) for s-cis	[87, 103]
66	CO–CH= CHC_6H_5	$C_6H_4CF_3$-3	IV (58.2) m.p. 118 to 119° IR (CCl_4): 1664 (νC=O) for s-cis	[87, 103]
67	CO–CH= CHC_6H_5	$C_6H_3(CF_3\text{-}3)Cl$-4	IV (64.9) m.p. 131 to 131.5° IR (CCl_4): 1663 (νC=O) for s-cis	[87, 103]
68	CO–CH= CHC_6H_5	$C_6H_4O_2CCH_3$-4	IV (39.7) m.p. 66 to 69°	[87]
69	CO–CH= CHC_6H_5	$C_6H_4OCH_3$-4	IV (56.5) m.p. 134 to 135° IR (CCl_4): 1662.7 (νC=O) for s-cis	[87, 103]

References on pp. 137/9

Table 6 (continued)

No.	$R^1–C_5H_4FeC_5H_4–R^2$ $R^1 = CO–R'$	R^2	method of preparation (yield in %, "m" indicates isomeric mixture, number following solidus is fraction present of 1,1′-compound), properties, and remarks	Ref.
*70	$CO–C_6H_5$	CH_3	VI l (36), VIm (51 m/46%), see "Further information", p. 135 m.p. 40° (pentane) ^1H NMR ($CDCl_3$): 1.83 (s, CH_3); 4.01 (s, 4H), 4.46 (t, 2H), and 4.78 (t, 2H) (C_5H_4); 7.3 to 8 (m, C_6H_5) ^1H NMR (CCl_4): 1.83 (s, CH_3)	[37,53,60, 95]
*71	$CO–C_6H_5$	C_2H_5	Ia (48), Ib (30), see "Further information", p. 135 b.p. 192°/0.6 Torr	[48,65,68]
72	$CO–C_6H_5$	C_3H_7-n	Ia (20) b.p. 185°/0.6 Torr	[48, 68]
*73	$CO–C_6H_5$	$C(CH_3)_3$	VIm (84 m/28%), see "Further information", p. 136 mass spectrum: m/e = 137 $[M-(C_6H_5CO+C_3H_4+C_5H_4)]^+$	[63]
74	$CO–C_6H_5$	$CH_2C(CH_3)_3$	VIt m.p. 85 to 86°	[8]
75	$CO–C_6H_5$	$CH_2C_6H_5$	Ia (21) m.p. 78° ^1H NMR ($CDCl_3$): 3.51 (s, CH_2), 4.12 (s, H-2′ to 5′), 4.45 (t) and 4.81 (t) (H-2 to 5)	[48,68,94]
76	$CO–C_6H_5$	$CH_2CH_2C_6H_5$	Ia (traces only)	[48, 68]
77	$CO–C_6H_5$	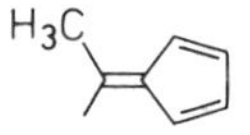	IIIb (44) red solid, m.p. 81 to 82° (ligroin/ether) ^1H NMR ($CDCl_3$): 2.48 (s, CH_3); 4.4 to 4.7 (m) and 4.9 (t) (C_5H_4); 6.5 (s, vinyl protons), 7.2 to 8.0 (m, C_6H_5)	[45]
*78	$CO–C_6H_5$	Br	VI l (55), VII (21) red oil, m.p. 36 to 37° ^1H NMR ($CDCl_3$): 4.10 (t) and 4.38 (t) (H-2′ to 5′), 4.57 (t) and 4.92 (t) (H-2 to 5), 7.35 to 8.0 (m, C_6H_5) IR (liquid film): 494 (νC–Br), 1597 (νC=C of C_6H_5), 1639 (νC=O)	[84, 102]

References on pp. 137/9

Table 6 (continued)

No.	$R^1–C_5H_4FeC_5H_4–R^2$ $R^1 = CO–R'$	R^2	method of preparation (yield in %, "m" indicates isomeric mixture, number following solidus is fraction present of 1,1'-compound), properties, and remarks	Ref.
*79	$CO–C_6H_5$	$C(CH_3)–(C_6H_5)OH$	IIg (36) viscous oil ^{1}H NMR ($CDCl_3$): 1.76 (s, CH_3), 3.71 (s, OH); 4.04 (m, 3H), 4.14 to 4.17 (m, 1H), 4.44 to 4.55 (m, 2H), 4.78 (m, 1H), and 5.00 (m, 1H) (C_5H_4); 7.02 to 7.47 (m, 8H), and 7.73 to 7.84 (m, 2H) (C_6H_5) IR (film): 815 (δC–H of C_5H_4), 1585 and 1603 (νC=C of C_6H_5), 1635 (νC=O), 3450 (νOH)	[92]
80	$CO–C_6H_5$	$C(C{\equiv}CH)–(C_6H_5)OH$	IIf (36) m.p. 128 to 130° R_f: 0.44 (on Al_2O_3 with 50/1 C_6H_6/abs. C_2H_5OH)	[35, 39]
81	$CO–C_6H_5$	$C(CF_3)_2OH$	V (20) initially isolated as red oil (b.p. 180°/0.25 Torr), which solidified after distillation, stable red crystals, m.p. 82 to 83° (CS_2) ^{1}H NMR (CS_2): 4.12 (m, 2H), 4.22 (t, 2H), 4.58 (t, 2H), and 4.88 (m, 2H) (C_5H_4, A_2B_2-pattern), 6.28 (s, OH); 7.3 to 7.65 (m, 3H) and 7.70 to 7.95 (m, 2H) (C_6H_5) ^{19}F NMR (CS_2): 74.81 (s, CF_3) IR (film): 710 and 740 (C_6H_5); 840, 850, 860, 1030, 1040, and 1060 (ferrocenyl); 1200 to 1300 (νC–F), 1630 (νC=O), 3230 (νOH, intramolecularly associated)	[70]
*82	$CO–C_6H_5$	CH_2OCH_3	see "Further information", p. 136 red oil	[86]
*83	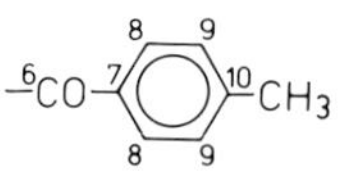	CH_3	VIm (63 m/44.8%), see "Further information", p. 136 ^{1}H NMR ($CDCl_3$): 1.83 (s, CH_3-1'), 2.42 (s, CH_3-10), 4.05 (s, H-2' to 5'), 4.48 (m, H-3, 4), 4.81 (m, H-2, 5), 7.1 to 7.9 (m, C_6H_4) ^{13}C NMR ($CDCl_3$): 13.7 (0.5 C, CH_3-1'), 69.3 (C-3', 4'), 71.1 (C-2', 5'), 72.0 (C-2, 5), 73.1 (C-3, 4), 78.7 (C-1), 85.5 (C-1'); 128.1 (0.6 C), 128.2 (1.0 C), 128.4 (0.4 C), and 128.8 (2.0 C) (C-8, 8, 9, 9); 137.2 (C-7), 141.7 (C-10), 198.3 or 198.9 (C=O)	[95]

Table 6 (continued)

No.	R^1–$C_5H_4FeC_5H_4$–R^2 R^1 = CO–R′	R^2	method of preparation (yield in %, "m" indicates isomeric mixture, number following solidus is fraction present of 1,1′-compound), properties, and remarks	Ref.
			mass spectrum: m/e = 318 $[M]^+$, 224, 202, 199, 121, 119, 91, 56 R_f: 0.58 (on SiO_2 with 2/1 hexane/$CH_3CO_2C_2H_5$)	
84	CO–C_6H_{11}-cyclo	$CH_2C(CH_3)_3$	VIk m.p. 71 to 73°	[7, 8]
*85	CO–CH_2–CH_2CF_3	C(CH_2CH_2–$CF_3)_2OH$	see "Further information", p. 136 m.p. 82° IR: 1666 (νC=O), 3500 (νOH)	[46]
86	CO–C_6H_4Cl-4	$CH_2C(CH_3)_3$	VIk m.p. 118 to 120°	[8]
*87	CO–C_6H_4Cl-2	CH_3	VIu (35 related to unrecovered $C_5H_5FeC_5H_4$–R^2), see "Further information", p. 137 ^1H NMR ($CDCl_3$): 1.88 (s, CH_3) IR (CH_2Cl_2): 1650 (νC=O)	[83]
88	CO–C_6H_4Cl-2	$CH_2C(CH_3)_3$	VIk b.p. 190 to 195°/0.4 Torr	[7, 8]
*89	CO–C_6H_4Cl-2	C_6H_5	VIm (76 related to unrecovered $C_5H_5FeC_5H_4$–R^2m/58%)	[83]
supplements:				
90	CO–CH=CH–C_6H_5	$C_6H_3Cl_2$-3,5	IR (CCl_4): 1662.4 (νC=O) for s-cis	[103]
91	CO–CH=CH–C_6H_5	C_6H_4OH-4	IR (CCl_4): 1662.6 (νC=O) for s-cis	[103]
92	CO–C_6H_5	Cl	VII (40) m.p. 33.5 to 35° (pentane in dry ice/acetone)	[102]

* Further information:

CH_3–CO–$C_5H_4FeC_5H_4$–CH_3 (Table **6**, No. **1**) is prepared from ClCO–$C_5H_4FeC_5H_4$–CH_3 in a 44% yield, when it is added dropwise as a benzene solution to an ether solution of $Cd(CH_3)_2$. After decomposition with ice and acidification with dilute HCl the compound is extracted with ether, with purification by thin layer chromatography with benzene [44].

The positional isomers from Method VIe (p. 112) can be separated by column chromatography on neutral [26] Al_2O_3 [9, 26] in benzene/hexane [9] with 1/1 pentane/ether [26] or a benzene/ether mixture [9] as eluent. The 42 to 43.2°C melting point of the compound thus obtained, however, deviates so considerably from the 59 to 60°C found for the No. 1 prepared by the preceding method, that some isomer contamination of this product is probable.

Oscillopolarography on a rotating Pt electrode in 1/1 H_2O/C_2H_5OH, 1 M in $HClO_4$, shows reversible oxidation per $CH_3{-}CO{-}C_5H_4FeC_5H_4{-}CH_3 - e^- \rightleftharpoons [CH_3{-}CO{-}C_5H_4FeC_5H_4{-}CH_3]^+$. The increments to the potential contributed by the substituents correlate with the σ_p Hammett constants [12, 22, 25]. Since the potential for all three positional isomers is the same, they cannot be distinguished by polarography [22].

$CH_3{-}CROH{-}C_5H_4FeC_5H_4{-}CH_3$ with R = H, CH_3, or $C(CH_3)_3$ is obtained by reduction with $LiAlH_4$ [40, 77], $LiCH_3$ [76, 77], or $LiC(CH_3)_3$ [75], respectively. The Vilsmeier complex of $(CH_3)_2NCHO$ and $POCl_3$ forms $OHC{-}CH{=}CCl{-}C_5H_4FeC_5H_4{-}CH_3$ [44].

$CH_3{-}CO{-}C_5H_4FeC_5H_4{-}C_2H_5$ (Table **6**, No. **2**) is also obtained from $Fe(C_5H_4{-}C_2H_5)_2$ by oxidation for 12 h with a five- to tenfold excess of $BaMnO_4$ in boiling $CHCl_3$, in 38% yield [99], or with MnO_2 in methylcyclohexane for 30 min, in 19% yield [52]. Separation from $Fe(C_5H_4{-}C_2H_5)_2$ and $Fe(C_5H_4{-}CO{-}CH_3)_2$ is treated in the following section.

Acetylation of $C_5H_5FeC_5H_4{-}C_2H_5$ with the silicoanhydride of acetic acid in the presence of $SnCl_4$ (molar ratio 1/1/2) is effected by boiling $SiCl_4$ with glacial acetic acid in benzene for 3 h and then adding successively benzene solutions of $C_5H_5FeC_5H_4{-}C_2H_5$ and $SnCl_4$ and reacting for 4 h. After the usual processing $FeC_{10}H_8(CO{-}CH_3)C_2H_5$ and $FeC_{10}H_7(CO{-}CH_3)_2C_2H_5$ are separated chromatographically in 2/1 petroleum ether/benzene on Al_2O_3 columns by elution of the former with benzene [1]. Boiling $Fe(C_5H_4{-}CO{-}CH_3)_2$ in 2/1 H_2O/dioxane for 12 h with Ni deposited on Zn dust forms according to gas chromatographic analysis 40% of No. 2, 10% of $Fe(C_5H_4{-}C_2H_5)_2$, and only small traces of $CH_3{-}CO{-}C_5H_4FeC_5H_4{-}CHOH{-}CH_3$ (No. 25), with a recovery of $Fe(C_5H_4{-}CO{-}CH_3)_2$ amounting to more than 40% [43]. Catalytic hydrogenolysis of $Fe(C_5H_4{-}CO{-}CH_3)_2$ in 1/9 $HClO_4/CH_3CO_2H$ at 1 atm H_2 and $\leqq 55°C$ gives No. 2, with 10% Pd/C as catalyst; the ratio of catalyst to diketone must not exceed 1/1 [104].

Friedel-Crafts acetylation of $C_5H_5FeC_5H_4{-}C_2H_5$ with $CH_3COCl/AlCl_3$ always yields a mixture of all three isomers [73, 80], which can be subdivided by column chromatography in benzene on Al_2O_3 into 21 fractions of which the fractions 6 to 9 contain No. 2 [4]. Elution from Al_2O_3 also works with 1/1 pentane/ether [26] or initially with benzene (No. 2 after the 1,2-isomer) and then with 9/1 $C_6H_6/CHCl_3$ (1,3-isomer) [16]. Preparative thin layer chromatography on Al_2O_3 plates with 6/1 benzene/ether as eluent separates the 1,2-isomer with certainty, but No. 2 and the 1,3-isomer are eluted as a mixture; they can be separated by thin layer chromatography only after oxidation with MnO_2 to the diacetylferrocenes [80]. Separation can be effected by gas-liquid chromatography on polyethylene glycol M-20 or neopentylglycol adipate; for the latter relative to $C_5H_5FeC_5H_4{-}CN$ as 1.00 the retained volumes at 150°C for the 1,2-, 1,1′-, and 1,3-isomers are 1.33, 1.76, and 2.02, respectively [30]. Separation as the semicarbazones via their successive crystallization from 2/1 C_2H_5OH/H_2O (No. 2) and C_2H_5OH (1,2- + 1,3-compounds) is recommended [1]. Without previous separation of the isomers their ratio can be determined by ^1H NMR spectroscopy in the presence of $Eu(dpm)_3$ (see Formula II, p. 114), which causes a strong paramagnetic shift for the substituent protons [62].

From $Fe(C_5H_4{-}C_2H_5)_2$ it is separated chromatographically in Skelly B on SiO_2 columns; No. 2 is eluted with benzene after $Fe(C_5H_4{-}C_2H_5)_2$ with cyclohexane. If $Fe(C_5H_4{-}C_2H_5)_2$, No. 2, and $Fe(C_5H_4{-}CO{-}CH_3)_2$ are present simultaneously, then the three phases developed with C_6H_6 on SiO_2 are eluted in succession, respectively, with C_6H_6 and then 19/1 and 4/1 $C_6H_6/CH_3CO_2C_2H_5$ [52].

The relative shift of the CH_3CO protons in $CDCl_3$ (vs. $C_5H_5FeC_5H_4{-}CO{-}CH_3$ as 0) is to higher field, $\Delta\delta_{CH_3} = -1.1$ Hz [56]. Regarding the dependence of $\Delta\delta_{CH_3}$ on the Hammett constants σ_p of the heteroannular substituents R^2 see $Fe(C_5H_4{-}CO{-}CH_3)_2$, p. 9.

No. 2 is oxidized in a reversible one electron reaction to the ferrocenium cation **$[CH_3{-}CO{-}C_5H_4FeC_5H_4{-}C_2H_5]^+$**. In 9/1 C_2H_5OH/H_2O, 0.01 M in $NaClO_4$ and 0.001 M in $HClO_4$, the half-wave potential at 20 ± 1°C is $E_{1/2}$ = 0.66 V vs. the saturated calomel electrode (SCE) and thus lies around 0.32 V above that of ferrocene [57]. In CH_3CN (0.1 M[$(C_4H_9)_4N$]BF_4) on a dropping-mercury electrode reduction to Fe and the two anions $[C_2H_5\text{–}C_5H_4]^-$ and $[CH_3CO\text{–}C_5H_4]^-$ takes place at −2.32 V (vs. SCE) with the overall consumption of $2e^-$ and breaking of the two Fe-cyclopentadienyl ring bond systems [96].

No. 2 is strongly basic and is protonated in H_2SO_4/H_2O to $\mathbf{[C_2H_5\text{–}C_5H_4FeC_5H_4\text{–}C(CH_3){=}OH]^+}$, which in 90% H_2SO_4/H_2O solution has $\lambda_{max}(\varepsilon)$ = 360 (2422) and 518 (2168) nm; half protonation occurs in 40.0% H_2SO_4. At 25°C pK_a = −2.53 measured spectrometrically for the conjugated acid [64, 93]. The electronic spectrum gives for the reaction with CCl_3CO_2H the equilibrium constant $K = [C_2H_5\text{–}C_5H_4FeC_5H_4\text{–}CO\text{–}CH_3]\cdot[CCl_3CO_2H]/[C_2H_5\text{–}C_5H_4FeC_5H_4\text{–}C(CH_3){=}OH\cdot O_2CCCl_3]$ at 25°C in benzene as $(10.7 \pm 0.4) \times 10^{-3}$ mol/L [71].

$\mathbf{2,4\text{-}(O_2N)_2C_6H_3\text{–}NH\text{–}N{=}C(CH_3)\text{–}C_5H_4FeC_5H_4\text{–}C_2H_5}$ is formed by 2,4-dinitrophenylhydrazine as blue-black rodlets with tetragonal basal faces which after multiple recrystallization from $CH_3CO_2C_2H_5/C_2H_5OH$ melt at 179 to 180°C [4, 16]. The reaction with $NH_2OH\cdot HCl$ in C_2H_5OH/H_2O in the presence of NaO_2CCH_3 yields the oxime [31]. On passing BF_3 through a solution with $(CH_3CO)_2O$ in CH_2Cl_2 No. 2 is acetylated in the positions α and β to the C_2H_5 group in the ratio 1/2.87; the total yield of $CH_3\text{–}CO\text{–}C_5H_4FeC_5H_3(C_2H_5)CO\text{–}CH_3$ is 95% [52].

$\mathbf{CH_3\text{–}CO\text{–}C_5H_4FeC_5H_4\text{–}CH(CH_3)_2}$ (Table **6**, No. **3**). The resolution of the isomer mixture recovered following Method VIe (p. 112) is more difficult than for Nos. 1 and 2. The 1,2-isomer is eluted with 9/1 pentane/ether as a faster band before the main band from a neutral Al_2O_3 column. On subsequent elution with a mixture more and more enriched with ether the main band is spread out into ten fractions without discrete bands appearing. From the first two No. 3 is won, from the last the 1,3-isomer [26].

$\mathbf{CH_3\text{–}CO\text{–}C_5H_4FeC_5H_4\text{–}C(CH_3)_3}$ (Table **6**, No. **4**). The isomer mixture obtained by Method VIe (p. 112) is resolved by the procedure reported for No. 3. In spite of trying a multiplicity of columns No. 4 and the 1,3-isomer could not be separated [26].

$HO(CH_3)_2C\text{–}C_5H_4FeC_5H_4\text{–}C(CH_3)_3$ is obtained with $LiCH_3$ in ether [77].

$\mathbf{CH_3\text{–}CO\text{–}C_5H_4FeC_5H_4\text{–}CH_2C(CH_3)_3}$ (Table **6**, No. **5**) crystallizes out on treatment of the reaction product prepared by Method VIm (p. 113) with petroleum ether, while the mother liquor retains the $C_5H_5FeC_5H_3(CH_2C(CH_3)_3)CO\text{–}CH_3$ [8].

$\mathbf{CH_3\text{–}CO\text{–}C_5H_4FeC_5H_4\text{–}C(CH_3){=}CH_2}$ (Table **6**, No. **6**) is prepared by the Wittig reaction from $Fe(C_5H_4\text{–}CO\text{–}CH_3)_2$ with $(C_6H_5)_3P{=}CH_2$ (prepared from NaH, $(CH_3)_2SO$, and $[CH_3P(C_6H_5)_3]Br$) in $(CH_3)_2SO$ and is separated from $Fe(C_5H_4\text{–}C(CH_3){=}CH_2)_2$ by chromatography on Al_2O_3 columns with elution of the latter with 1/1 petroleum ether/ether. No. 6 is subsequently eluted with 1/3 petroleum ether/ether; yield 53.5% [41]. Under the action of bases [41], such as NaH in $(CH_3)_2SO$ [33], it cyclizes to V [33, 41].

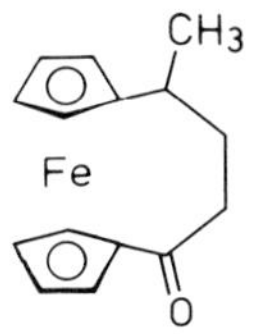

V

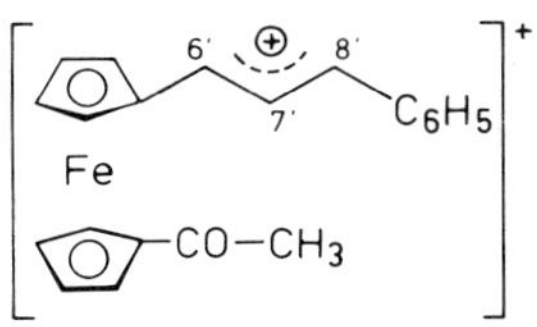

VI

$\mathbf{CH_3\text{–}CO\text{–}C_5H_4FeC_5H_4\text{–}CH_2C_6H_5}$ (Table **6**, No. **7**) is separable in a pure form from the positional isomer mixture obtained by Methods VIa or n (pp. 112 or 113), by column chromatography on Al_2O_3 with benzene as eluent [17, 51]. From No. 58, which appears by preparation of

No. 7 by Method Ib (p. 111) in the same fraction on chromatographic processing, it is separated only with great difficulty, by subsequent preparative thin layer chromatography [68].

(E)-CH_3–CO–$C_5H_4FeC_5H_4$–CH_2CH=CHC_6H_5 (Table **6**, No. **8**). The preparation by Method VIf (p. 112) but with a doubled excess of $(CH_3CO)_2O$ and BF_3 failed just as did that with CH_3COCl/ $AlCl_3$ [76].

The carbenium ion **[CH_3–CO–$C_5H_4FeC_5H_4$–$CHCHCHC_6H_5$]$^+$** (see Formula VI) is formed with a dark brown coloration by reaction of No. 8 in CH_2Cl_2 with $[(C_6H_5)_3C]BF_4$. This reacts with $CH_3OH/NaOCH_3$ to give a mixture of Nos. 33 and 34 (total yield 83%). For isolation, the orange-yellow solution given by addition of the reagents is poured into H_2O, and the residue remaining after the usual processing of the organic phase is separated chromatographically on Al_2O_3 previously exposed 24 h in the atmosphere. After elution of $(C_6H_5)_3CH$ with petroleum ether, first unreacted VI (ca. 9%) and subsequently the mixture of Nos. 33 and 34 (ratio 35/65) are eluted with 4/1 petroleum ether/ether. The preferential formation of No. 34 is caused by the CH_3CO substituent, which reduces the ability of the ferrocenyl group to delocalize positive charge, so that the CH_3O^- addition is easier on the C-8′ than on C-6′ [76].

CH_3–CO–$C_5H_4FeC_5H_4$–C_6H_5 (Table **6**, No. **9**). On addition in a molar ratio 1.3/1/1 of CH_3COCl in CH_2Cl_2 to $C_5H_5FeC_5H_4$–C_6H_5 + $AlCl_3$ in CH_2Cl_2 at 0°C and reaction for 3 h at 0°C and 1 h at room temperature, predominantly the mixture of the three isomers is obtained. No. 9 can be isolated by column chromatography on Al_2O_3 in Skelly B, with rechromatographing on Al_2O_3 with ether as eluent [5], and finally with multiple recrystallization from 2/1 petroleum ether/ ether [5, 11], in >5% yield [5]. According to [15] No. 9 and its 1,3-isomer are not separable chromatographically on acid-washed Al_2O_3 columns (activity degree 1) with ether. If a mixture of $C_5H_5FeC_5H_4$–C_6H_5 and ferrocene is acetylated with CH_3COCl + $AlCl_3$ (molar ratio 1/1/2/2) in ether by the Friedel-Crafts method, then $FeC_{10}H_8(C_6H_5)COCH_3$ is recovered in a 5% yield; 64% of the $C_5H_5FeC_5H_4$–C_6H_5 but only 30% of the ferrocene remain unchanged [11].

CH_3–CO–$C_5H_4FeC_5H_4$–Cl (Table **6**, No. **15**). While Hall, Richards [20] obtained exclusively No. 15 by Method VIl (p. 113) (without recovery of $C_5H_5FeC_5H_4$–Cl), Morrison, Pauson observed formation of ferrocene during Friedel-Crafts acetylation of $C_5H_5FeC_5H_4$–Cl [13]. If following the reaction by Method VI l after 1.5 h stirring at 0°C and 1 h at room temperature, there is another 1 h of boiling, the yield of No. 15 still amounts to only 5%; in addition chromatography on SiO_2 columns with 19/1 $C_6H_6/CH_3CO_2C_2H_5$ eluent gives CH_3–CO–$C_5H_4FeC_5H_3$(Cl-1)CO–CH_3-3 (29%), $C_5H_5FeC_5H_4$–Cl (12%), and $Fe(C_5H_4$–CO–$CH_3)_2$ (6.1%) [91].

From the ^1H NMR spectrum it is evident that the influence of a substituent on the proton screening in its own C_5H_4 ring is greater than for the other ring. The shifts calculated by the additivity rule from the signals for $C_5H_5FeC_5H_4$–Cl and $C_5H_5FeC_5H_4$–CO–CH_3 deviate more than usual from those measured in CCl_4 (see Table 6, p. 118) [28]. The shift in $CDCl_3$ for the CH_3 protons relative to that for $C_5H_5FeC_5H_4$–CO–CH_3 is $\Delta\delta_{CH_3}$ = 0.9 Hz to lower field [56]. Regarding the dependence of $\Delta\delta_{CH_3}$ on the Hammett constants σ_p of the heteroannular substituents R^2, see $Fe(C_5H_4$–CO–$CH_3)_2$, p. 9.

The quarter wave potential has been measured at 25°C chronopotentiometrically as $E_{\tau/4}$ = 0.49 V (vs. Ag/(0.01 M $AgClO_4$ + 0.2 M $[(C_2H_5)_4N]ClO_4$ in CH_3CN)) for the reversible oxidation to the ferrocenium ion **[CH_3–CO–$C_5H_4FeC_5H_4$–Cl]$^+$** in CH_3CN (0.2 M $NaClO_4$). It is located at $\Delta E_{\tau/4}$ = 0.427 V above that of ferrocene. The dependence of $\Delta E_{\tau/4}$ on the sums of the two substituent constants $\Sigma\sigma_P$, $\Sigma\sigma_M$, and $(\Sigma\sigma_M + 2\Sigma\sigma_P)/3$ of a number of 1,1′-disubstituted ferrocene derivates, among them No. 15, is treated in [32].

The ionic hydrogenation of No. 15 with $(C_2H_5)_3SiH$ and CF_3CO_2H (molar ratio 2/5/10) leads to C_2H_5–$C_5H_4FeC_5H_4$–Cl, in a 45% yield after 2 h at 20°C. The rate-determining step appears to be the reaction of the hydride anion H^- with the protonated carbonyl group, whose positive charge is localized on the C atom due to the electron acceptor effect of the Cl [66].

$2,4\text{-}(O_2N)_2C_6H_3\text{–}NH\text{–}N{=}C(CH_3)\text{–}C_5H_4FeC_5H_4\text{–}Cl$ is obtained with 2,4-dinitrophenylhydrazine in the usual way, melting point 159 to 161°C [27]. The Vilsmeier reaction with $(CH_3)_2NCHO$ and $POCl_3$ first at 0°C and then at room temperature gives $OHC\text{–}CH{=}C(Cl)\text{–}C_5H_4FeC_5H_4\text{–}Cl$ in 68% yield [61]. CuCN in γ-picoline forms $NC\text{–}C_5H_4FeC_5H_4\text{–}CO\text{–}CH_3$; the electron withdrawing effect of the CH_3CO group retards the reaction [50].

$CH_3\text{–}CO\text{–}C_5H_4FeC_5H_4\text{–}Br$ (Table **6**, No. **16**) is obtained by Methods VId [10] and VI l [20] (pp. 112, 113) free of the homoannular disubstituted isomers and by the latter method also without recovery of unreacted $C_5H_5FeC_5H_4\text{–}Br$ [20]. The reaction of $C_5H_5FeC_5H_4\text{–}Br$ with $(CH_3CO)_2O$ in the presence of $MgBr_2 \cdot O(C_2H_5)_2$ (molar ratio 2/3/6) in benzene/ether after 2 h heating and the subsequent usual processing gives only a 10% yield of No. 16, after thin layer chromatography on Al_2O_3. In contrast, the presence of $MgBr_2$ increases the yield achieved in the Friedel-Crafts acetylation of $C_5H_5FeC_5H_4\text{–}Br$ with $CH_3COCl/AlCl_3$ via formation of the reactive CH_3COBr [90].

Also, the ^1H NMR spectrum indicates, by the absence of a signal at $\delta = 4$ ppm for the protons of an unsubstituted cyclopentadienyl ring, that the product is free of the 1,2- and 1,3-isomers [24]. The shifts calculated by the additivity rule from the signals for $C_5H_5FeC_5H_4\text{–}Br$ and $C_5H_5FeC_5H_4\text{–}CO\text{–}CH_3$ agree very well with the values measured in CCl_4 (see Table 6, p. 119) [28]. Further, the values calculated for λ_{max} and ε of the "ferrocene band" under the assumption of the additivity of the effects of the substituents correspond to those measured. On the contrary, the effect of the substituents of the location of the remaining bands is not additive [29].

No. 16 is well soluble in benzene, ether, or ethanol, in contrast difficultly soluble in petroleum ether [10]. The quarter wave potential $E_{\tau/4} = 0.469$ V (vs. Ag/(0.01 M $AgClO_4$ + 0.2 M $[(C_2H_5)_4N]ClO_4$ in CH_3CN)) was measured chronopotentiometrically for the reversible oxidation to the ferrocenium ion **$[CH_3\text{–}CO\text{–}C_5H_4FeC_5H_4\text{–}Br]^+$** at 25°C. It is located at $\Delta E_{\tau/4} = 0.406$ V higher than that for ferrocene. Regarding the dependence of $\Delta E_{\tau/4}$ on the sum of the two substituent constants, $\Sigma\sigma_P$, $\Sigma\sigma_M$, and $(\Sigma\sigma_M + 2\Sigma\sigma_P)/3$ for a number of 1,1′-disubstituted ferrocene derivatives, among them No. 16, see [32].

Oxidation with I_2 in pyridine on a hot water bath forms $HO_2C\text{–}C_5H_4FeC_5H_4\text{–}Br$ [14]. No. 16 has considerable basicity and is protonated in H_2SO_4/H_2O to **$[Br\text{–}C_5H_4FeC_5H_4\text{–}C(CH_3){=}OH]^+$**, whose pK_a value in H_2SO_4/H_2O at 25°C has been determined spectrometrically as −3.97. In 55.1% H_2SO_4 half protonation occurs [64]. The equilibrium constant for the reaction with CCl_3CO_2H $K = [Br\text{–}C_5H_4FeC_5H_4\text{–}CO\text{–}CH_3] \cdot [HO_2CCCl_3]/[Br\text{–}C_5H_4FeC_5H_4\text{–}C(CH_3){=}OH \cdot O_2CCCl_3]$ has been determined spectrometrically at 25°C in benzene to be $(23.2 \pm 0.2) \times 10^{-3}$ mol/L, in cyclohexane as $(5.8 \pm 0.5) \times 10^{-3}$ mol/L [71]. With $NH_2OH \cdot HCl$ is formed the oxime, which melts at 109.5 to 111.5°C (C_2H_5OH/H_2O) [10]. $OHC\text{–}CH{=}C(Cl)\text{–}C_5H_4FeC_5H_4\text{–}Br$ in 82% yield is formed by the Vilsmeier reaction with $(CH_3)_2NCHO/POCl_3$ [55].

The Ullmann reaction is made difficult by the electronegative CH_3CO group. Nevertheless, at 125°C and a reaction time of 16 to 24 h the yield of VII (see „Ferrocene" 6, 1977, Table 3, No. 13, p. 20) is satisfactory. On increasing the temperature the reduction becomes the dominant reaction, so that on heating with Cu to 150°C $C_5H_5FeC_5H_4\text{–}CO\text{–}CH_3$ is formed in 80% yield within 30 min [21]. Heating for 18 h with activated Cu bronze gives VII and $C_5H_5FeC_5H_4\text{–}CO\text{–}CH_3$ with yields of 31 and 39%, respectively. If an ~1/2 mixture of No. 16 and $C_5H_5FeC_5H_4\text{–}Br$ is heated for 28 h with activated Cu bronze to 90 and 92°C, then one obtains VII (11.4% yield), VIII (16.3% yield, cf. "Ferrocene" 6, 1977, p. 15), IX (cf. "Ferrocene" 6, 1977, p. 5), and $C_5H_5FeC_5H_4\text{–}CO\text{–}CH_3$, the latter two in yields of 29.6 and 61.6%, respectively (calculated assuming No. 16 to be the only source), and finally ferrocene in 42.4% yield calculated with the assumption of $C_5H_5FeC_5H_4\text{–}Br$ as the only source. The compounds are separable by chromatography on Al_2O_3 columns with hexane, hexane/benzene, and benzene/ether as eluents [24]. Heating of an intimate mixture of No. 16, cymantrenyl silver, and CuBr to 140 to 150°C forms VII and X (yield 29% referred to $AgC_5H_4Mn(CO)_3$) [74].

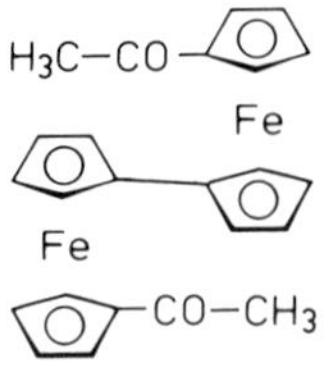

VII

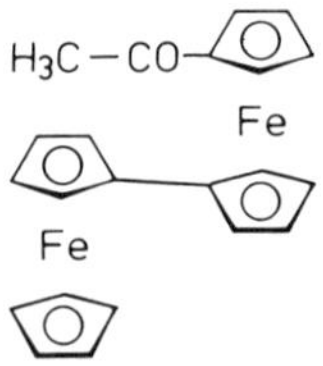

VIII

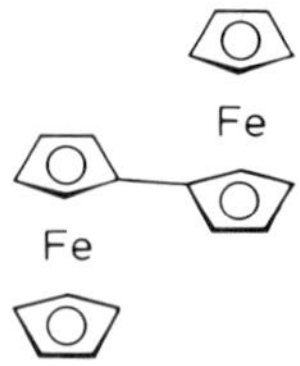

IX

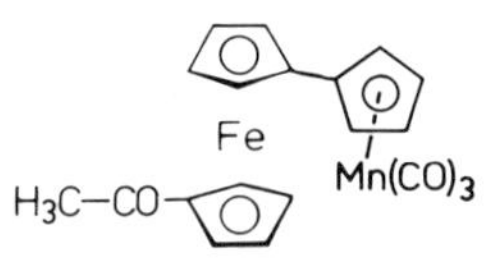

X

$Cu(O_2CCH_3)_2$ in boiling C_2H_5OH/H_2O reacts incompletely to CH_3–CO–$C_5H_4FeC_5H_4$–O_2CCH_3, which is not isolated, but for the separation of No. 16 is hydrolyzed to No. 24 by heating with KOH in C_2H_5OH/H_2O. No. 16 is immediately extracted in ether from the residue remaining after the evaporation of the ethanol, No. 24 subsequently after introduction of CO_2. The ether is distilled off under N_2 and No. 24 is recrystallized from H_2O in 26% yield [10].

CH_3–CO–$C_5H_4FeC_5H_4$–OH (Table **6**, No. **24**) is in the solid state somewhat more stable in air than $C_5H_5FeC_5H_4$–OH. In solution it is easily oxidized. It reacts with C_6H_5COCl in 10% potassium hydroxide following the Schotten-Baumann process to form No. 29 in 77% yield [10].

CH_3–CO–$C_5H_4FeC_5H_4$–CHOH–CH_3 (Table **6**, No. **25**). The ^1H NMR spectrum does not change with temperature up to 120°C in $Cl_2HCCHCl_2$. In $CDCl_3$ the CH_3CO protons give two main signals with $\Delta\nu = 1.3$ Hz and the cyclopentadienyl protons of the CH_3CO–C_5H_4 ring four different signal groups with their respective centers at $\delta = 4.39$, 4.46, 4.65, and 4.70 ppm. The area under the two methyl proton doublets of the CH_3CHOH group are nearly equally large [72]. In the IR spectrum no peak is observed for secondary hydroxyl groups. The ^1H NMR and IR spectra indicate the presence of an interannular hydrogen bond (see XI). The resulting hindered rotation leads to the conformers A and B, which exist as the respective racemates [72].

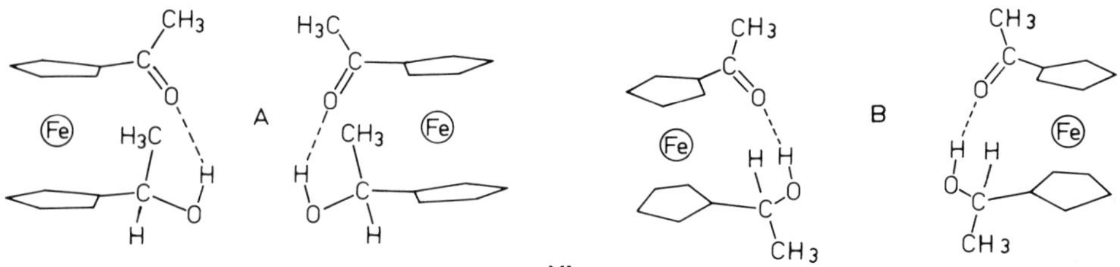

XI

The carbenium ion **[CH_3–CO–$C_5H_4FeC_5H_4$–$^{\oplus}$CH–CH_3]$^+$** (XII) occurs in strong acids (>40% H_2SO_4). In it the hydrogen bonding is disrupted and a free rotation of the cyclopentadienyl rings around the orthogonal axis is possible. ^1H NMR in 2/1 $CF_3CO_2H/CDCl_3$, notation shown in XII: $\delta = 1.63$ (d, CH_3-6′; J = 6 Hz), 2.43 (s, $COCH_3$), 4.23 (H-2′ to 5′), 4.6 to 4.78 (H-2 to 5) ppm. UV spectrum in >40% H_2SO_4: $\lambda_{max}(\varepsilon) = 260$ (15200) nm [72].

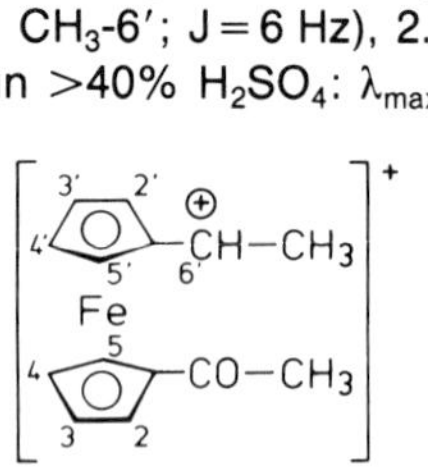

XII

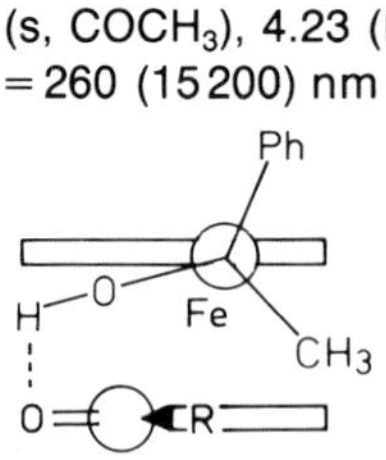

XIII

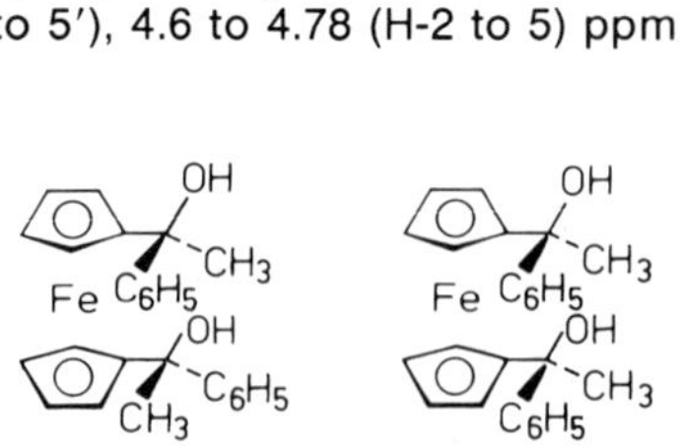

a (meso) b (rac)

XIV

The acetate No. 30 is obtained in 90% yield when No. 25 is allowed to stand for 12 h with $(CH_3CO)_2O$ and pyridine at room temperature [40]. Oxidation with $[(C_6H_5)_3SiO]_2CrO_2$ in benzene at 50°C occurs selectively at the CHOH with formation of $Fe(C_5H_4$–CO–$CH_3)_2$ in 78% yield [98].

CH_3–CO–$C_5H_4FeC_5H_4$–C(CH_3)(C_6H_5)OH (Table **6**, No. **27**). The single OH absorption band which No. 27 exhibits in CCl_4 indicates the conformation XIII (R = CH_3) with intramolecular hydrogen bonding. The reaction with C_6H_5MgBr (molar ratio 1/3) in benzene/ether takes place stereoselectively. The diastereoisomer mixture XIV obtained by hydrolysis of the reaction product contains the meso and racemic compounds a and b in an 82/18 ratio [92]. For the reaction scheme see IX, p. 36.

CH_3–CO–$C_5H_4FeC_5H_4$–$O_2CC_6H_5$ (Table **6**, No. **29**). The ^{1}H NMR spectrum clearly shows that the influence of the substituents on the proton screening in their own rings is by far greater than for the other ring. The chemical shifts calculated according to the additivity rule from the spectra of $C_5H_5FeC_5H_4$–CO–CH_3 and $C_5H_5FeC_5H_4$–$O_2CC_6H_5$ differ by less than 0.1 ppm from the measured values [28].

CH_3–CO–$C_5H_4FeC_5H_4$–CH(CH_3)–O_2CCH_3 (Table **6**, No. **30**) is purified by distillation at 60°C/ 0.005 Torr. The rate constant at 45°C for hydrolysis proceeding in 4/1 acetone/H_2O by the S_N1 mechanism is $k = (1.242 \pm 0.012) \times 10^{-5} s^{-1}$; $k_{rel} = 0.0157$ referred to the rate constant of 1.00 for $C_5H_5FeC_5H_4$–CH(CH_3)–O_2CCH_3 at 45°C. The k_{rel} values for X–$C_5H_4FeC_5H_4$–CH(CH_3)–O_2CCH_3 (X = CH_3, Cl, Br, $COCH_3$, CO_2CH_3, or CN) indicate that the hydrolysis is affected primarily by the inductive and field effects of X, while resonance effects, if any at all, are of subordinate importance [40].

CH_3–CO–$C_5H_4FeC_5H_4$–OCH_3 (Table **6**, No. **32**). While only the mixture of the three positional isomers is obtained by Method VIb (p. 112), Method VI l (p. 113) yields in addition to these also the 1,2- and 1,3-dimethoxyferrocenes. On chromatographic processing of the reaction products in benzene on Al_2O_3 columns benzene elutes these and $C_5H_5FeC_5H_4$–OCH_3 before $FeC_{10}H_8(OCH_3)$CO–CH_3, which is eluted with 1/1 benzene/ether. No. 32 is not separable from the latter even by multiple rechromatographing. Even after conversion of the mixture into $FeC_{10}H_8(C_2H_5)OCH_3$ by a Clemmensen reduction or with $LiAlH_4/AlCl_3$, the 1,1′-isomer, while separable from the 1,2-isomer, can not be gotten completely free from the 1,3-isomer [36].

CH_3–CO–$C_5H_4FeC_5H_4$–CH(CH=CHC_6H_5)OCH_3, CH_3–CO–$C_5H_4FeC_5H_4$–CH=CHCH(C_6H_5)OCH_3 (Table **6**, Nos. **33, 34**) can each be concentrated to 80%, if the mixture of Nos. 33 and 34 (ratio 35/65) obtained from No. 8 (see there for additional data, p. 130) is subjected to preparative thin layer chromatography on SiO_2 with 9/1 petroleum ether/ether as eluent, No. 33 appearing in the first and No. 34 in the last fraction. The two compounds decompose on standing, do not isomerize in the presence of $NaOCH_3$ in CH_3OH, and are reduced by $NaBH_4$ on the CH_3CO groups to CH_3CHOH groups [76].

CH_3–CO–$C_5H_4FeC_5H_4$–$C_6H_4OCH_3$-4 (Table **6**, No. **35**) is separable by column chromatography with ether on acid-washed Al_2O_3 (activity degree 1) from the 1,2- but not from the 1,3-isomer [15].

CH_3–CO–$C_5H_4FeC_5H_4$–CHO (Table **6**, No. **36**) forms from CH_3–CHOH–$C_5H_4FeC_5H_4$–CH_2OH on oxidation with activated MnO_2 in $CHCl_3$ for 3 d at room temperature with occasional shaking, followed by recrystallization from benzene/petroleum ether of the evaporation residue of the filtrate, in 64% yield [18]. Preparation by Method VI l (p. 113) must be conducted in the dark with ice cooling in order to form only No. 36. It is separated chromatographically from the reaction product on deactivated Al_2O_3 columns with ether following elution of $C_5H_5FeC_5H_4$–CHO with 1/1 ether/hexane. On the other hand, reaction first for 1 h with ice-cooling and then 3 h at room temperature, gives on chromatographic processing four fractions, two each with ether/hexane and pure ether. The second fraction yields traces of the 1,3-isomer and the fourth fraction XV in 1.2% yield [56].

References on pp. 137/9

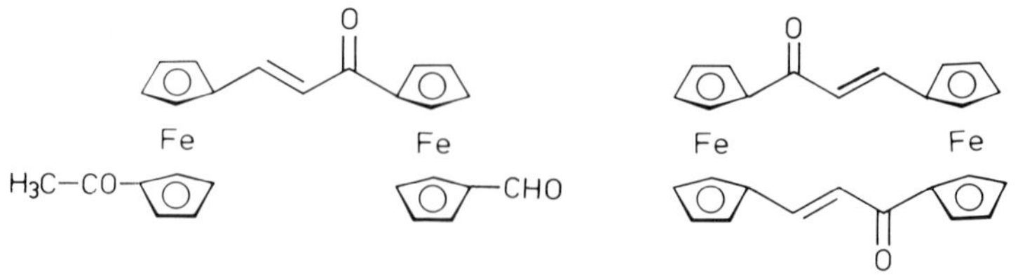

XV XVI

The occurrence of small quantities of XVI on cyclization with KOH in boiling ethanol of C_6H_5–CO–CH=CH–$C_5H_4FeC_5H_4$–CO–CH_3 leads one to suspect that No. 36 could have formed from the starting compound by a retro-Claisen-Schmidt condensation and then been converted by autocondensation into XVI; No. 36 was, however, not identified during the course of the reaction [69].

The radical anion **[CH_3–CO–$C_5H_4FeC_5H_4$–CHO]$^-$** and the **[CH_3–CO–$C_5H_4FeC_5H_4$–CHO]$^{2-}$** dianion form on electrochemical reduction with in each case a one electron transition. The half-wave potentials measured polarographically in $(CH_3)_2NCHO$/0.1 M $[(C_4H_9)_4N]PF_6$ are $E_{1/2} = -1.87$ and -2.40 V, respectively, against the saturated calomel electrode. When the compound is electrolyzed in this solution at a controlled potential, it forms with the consumption of 2F metallic iron, $[CH_3CO–C_5H_4]^-$, and $[OHC–C_5H_4]^-$ [89].

The bis-semicarbazone **$H_2NCONHN$=C(CH_3)–$C_5H_4FeC_5H_4$–CH=NNHCONH$_2$** melts at 243°C [18]. Reduction of No. 36 with $LiAlH_4/AlCl_3$ in ether gives C_2H_5–$C_5H_4FeC_5H_4$–CH_3 in a quantitative yield [56]. The Knoevenagel condensation gives HO_2C–CH=CH–$C_5H_4FeC_5H_4$–CO–CH_3 with malonic acid [18].

C_2H_5–CO–$C_5H_4FeC_5H_4$–CH_3 (Table **6**, No. **37**). The mixture prepared by Method VI l (p. 113) contains No. 37, the 1,2-, and the 1,3-isomer in a 14/5/6 ratio. It can be separated chromatographically on Al_2O_3 columns with 10/1 benzene/ether as eluent. Reaction with $NH_2OH \cdot HCl$ and KOH in C_2H_5OH yields the oxime [60].

C_2H_5–CO–$C_5H_4FeC_5H_4$–C(CH_3)$_3$ (Table **6**, No. **38**). By addition of CH_2=CHCOCl in CH_2Cl_2 to $AlCl_3$ and $C_5H_5FeC_5H_4$–C$(CH_3)_3$ (molar ratio 1/1/1) in CH_2Cl_2 at −78°C with subsequent stirring at this temperature overnight followed by the usual processing is obtained an isomer mixture consisting predominantly of No. 38 and its 1,3-isomer in 16% yield (referred to unrecovered $C_5H_5FeC_5H_4$–C$(CH_3)_3$) along with a mixture of the homoannular ketones XVII (12%) and the heteroannular ketones XVIII and XIX (61%). During the chromatography on Al_2O_3 columns with 3/1 light petroleum ether/ether the isomer mixture with No. 38 is eluted as a viscous oil after XVII and before XVIII and XIX. This oil can be separated neither by thin layer nor by further column chromatography [^{1}H NMR ($CDCl_3$): δ = 1.05 to 1.35 (m, $C_4H_9 + CH_3$ of C_2H_5), 2.75 (br. q, $COCH_2$), 4.0 to 4.2 (m), 4.20 (s), 4.4 to 4.6 (m), and 4.7 to 4.9 (m) (cyclopentadienylprotons) ppm] [85].

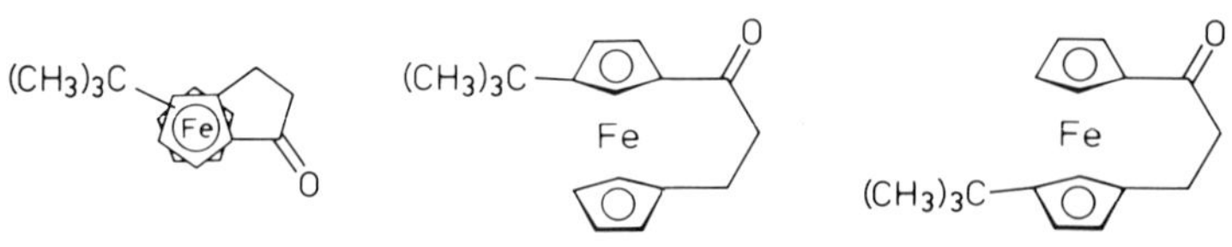

XVII XVIII XIX

C_2H_5–CO–$C_5H_4FeC_5H_4$–C_6H_4Cl-4 (Table **6**, No. **42**) is recovered from the mixture of the three positional isomers produced by the Friedel-Crafts reaction of $C_5H_5FeC_5H_4$–C_6H_4Cl-4 with $(C_2H_5CO)_2O$ and $AlCl_3$ (molar ratio 1/1.3/1) in CH_2Cl_2 for 4.5 h, by selective recrystallization after column chromatography on SiO_2 with Skelly B to remove unreacted $C_5H_5FeC_5H_4$–C_6H_4Cl-4; yield 30.2% [59].

n-C_3H_7–CO–$C_5H_4FeC_5H_4$–CH_3 (Table **6**, No. **43**). The product from Method VIm (p. 113) is chromatographed in hexane on an acid-washed Al_2O_3 column. A mixture strongly enriched in No. 43 and the 1,3-isomer is eluted with hexane/benzene after elution of $C_5H_5FeC_5H_4$–CH_3 with hexane and most of the 1,2-isomer with 2/1 hexane/benzene. Rechromatographing in benzene on neutral Al_2O_3 with 2/1 CH_2Cl_2/ether as eluent yields No. 43 and $C_5H_5FeC_5H_3$(CO–C_3H_7-n-1)CH_3-3 as a 94/6 mixture. Reduction with $LiAlH_4/AlCl_3$ in ether produces n-C_4H_9–$C_5H_4FeC_5H_4$–CH_3 [58].

$(CH_3)_2CH$–CO–$C_5H_4FeC_5H_4$–$C(CH_3)_3$ (Table **6**, No. **44**). The mixture of No. 44 and its 1,3-isomer obtained by Method VIm (p. 113) is resolved into its components by repeated column and thin layer chromatography on Al_2O_3 and SiO_2, respectively [78]. No. 44 yields with CH_3Li $(CH_3)_2CH$–$C(CH_3)OH$–$C_5H_4FeC_5H_4$–$C(CH_3)_3$ [78].

$(CH_3)_2CH$–CO–$C_5H_4FeC_5H_4$–$CH_2C(CH_3)_3$ (Table **6**, No. **45**) is separated from its 1,3-isomer by chromatography on Al_2O_3 columns in petroleum ether [8].

$(CH_3)_3C$–CO–$C_5H_4FeC_5H_4$–$CH_2C(CH_3)_3$ (Table **6**, No. **49**) is used in tablets as an oral remedy for iron-deficiency anemia in humans and animals [2].

$C_6H_5CH_2$–CO–$C_5H_4FeC_5H_4$–CH_2–R″ (Table **6**; Nos. **52** and **54** with R″ = CH_3 or C_6H_5). These compounds are acetylated on the R″–CH_2–C_5H_4 ring in the 2- and 3-positions by CH_3COCl and $AlCl_3$ (molar ratio 1/2/4). The ratio of the yields of the 2- and 3-isomers is 1/3.2 for No. 52 and 1/10.7 for No. 54 [65].

C_6H_5–CO–$C_5H_4FeC_5H_4$–CH_3 (Table **6**, No. **70**). The three positional isomers produced by Method VI l (p. 113) have been successfully separated by column chromatography on Al_2O_3 using 10/1 benzene/ether as the eluent [60]. On the other hand, only the 1,2-isomer could be separated from the mixture prepared by Method VIm (p. 113), using preparative thin layer chromatography on Kieselgel-G (SiO_2) with benzene/petroleum ether [37]; the residual mixture of No. 70 and $C_5H_5FeC_5H_3$(CO–C_6H_5-1)CH_3-3 (molar ratio 5/4) boils between 75 and 80°C (air bath temperature) at 0.005 Torr [37]. No. 70 yields with $NH_2OH \cdot HCl$ and KOH in boiling C_2H_5OH the oxime [60], which melts at 118°C [53].

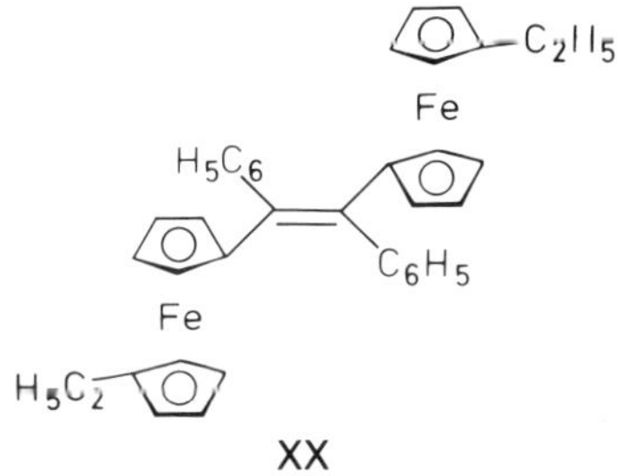

XX

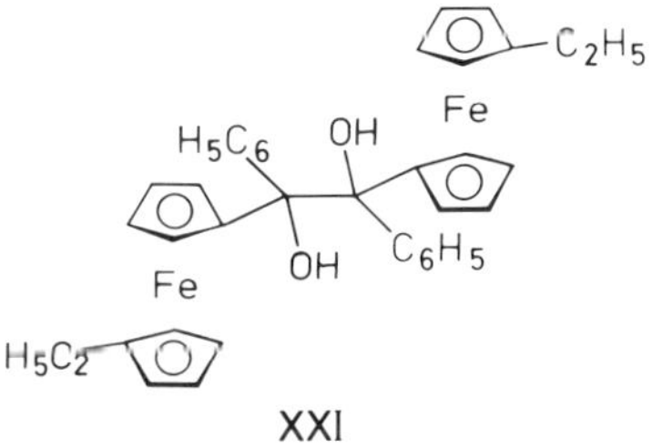

XXI

C_6H_5–CO–$C_5H_4FeC_5H_4$–C_2H_5 (Table **6**, No. **71**) is formed in a quantitative yield by the action of bases on [C_2H_5–$C_5H_4FeC_5H_4$–$C(C_6H_5)$=NH_2]PF_6 in C_2H_5OH/H_2O [82].

Reduction with CH_3MgBr in the presence of $CoCl_2$ [67] or $FeCl_3$ and subsequent treatment with a few drops of concentrated HCl in benzene produces XX [54]. The reaction appears to proceed via the very unstable XXI, which either is rapidly oxidized again to No. 71 or, in acidic media, converts into XX [54, 67]. XX arises also in ca. 15% yield from the action of $NaOP(C_6H_5)_2$ on No. 71. The two products are separable by preparative thin layer chromatography [54, 67]. Acetylation with CH_3COCl and $AlCl_3$ (molar ratio 1/2/4) in CH_2Cl_2 occurs on the 2- and 3-positions on the C_2H_5–C_5H_4 ring. The yields of C_6H_5–CO–$C_5H_4FeC_5H_3$(CO–CH_3-1)C_2H_5-2 and -3 amount to 14 and 36%, respectively, with 30% recovery of No. 71 [65].

C_6H_5–CO–C_5H_4FeC$_5H_4$–C(CH$_3$)$_3$ (Table **6**, No. **73**). Only the 1,2-isomer can be separated from the $FeC_{10}H_8(COC_6H_5)C(CH_3)_3$ mixture obtained by Method VIm (p. 113) by chromatography on SiO_2 gel with benzene as eluent ($R_f = 0.8$). The product appearing at $R_f = 0.55$ yields a viscous red liquid, which consists of a 1/2 mixture of No. 73 and the 1,3-isomer and which is not further separable by chromatography [63]. The properties and chemical reactions of the mixture are discussed with the 1,3-isomer in Section 5.2.3.5.3.2, Table 11, No. 49, p. 198.

C_6H_5–CO–C_5H_4FeC$_5H_4$–Br (Table **6**, No. **78**). Two bands are obtained by column chromatography on Al_2O_3. No. 78 is eluted from the second by 10/1 $C_6H_6/CH_3CO_2C_2H_5$, following elution of the mixture of ferrocene and $C_5H_5FeC_5H_4$–Br with hexane, and is purified by rechromatographing [84].

The mass spectrum exhibits the following peaks (with relative intensities in parentheses): m/e = 370 (49; $[M]^+$ with ^{81}Br), 368 (52; $[M]^+$ with ^{79}Br), 105 (100; $[C_6H_5CO]^+$). The Ullmann reaction with activated Cu at 120 to 130°C gives XXII (53% yield) and $C_5H_5FeC_5H_4$–CO–C_6H_5 (29%). $NH_2OH \cdot HCl$ in boiling $KOH/H_2O/C_2H_5OH$ forms the oxime (m.p. 146 to 147°C) [84].

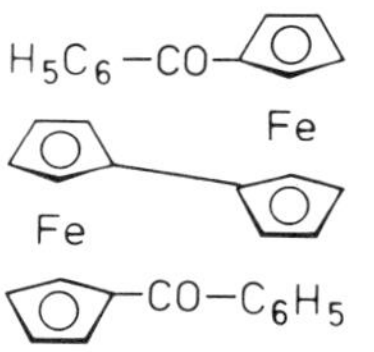

XXII

C_6H_5–CO–C_5H_4FeC$_5H_4$–C(CH$_3$)(C$_6H_5$)OH (Table **6**, No. **79**) cannot be crystallized. The single OH absorption band indicates the conformation XIII (R = C_6H_5) (see p. 132) with intramolecular hydrogen bonding. The reaction with CH_3MgBr (1/3) takes place with stereoselectivity. The diastereomeric mixture XIV (p. 132) formed after hydrolysis of the reaction product contains a and b in a 1/4 ratio [92]. For the reaction scheme, see $Fe(C_5H_4–CO–C_6H_5)_2$, scheme in VII p. 61.

C_6H_5–CO–C_5H_4FeC$_5H_4$–CH$_2$OCH$_3$ (Table **6**, No. **82**). Li–$C_5H_4FeC_5H_4$–CH_2OCH_3 is condensed with C_6H_5CN in 2 h at room temperature, followed by hydrolysis with H_2O. On chromatographing the reaction product on an Al_2O_3 column, No. 82 is eluted with 7/3 petroleum ether/CH_2Cl_2 in 30% yield after elution of $C_5H_5FeC_5H_4$–CH_2OCH_3 (65% yield) with 9/1 petroleum ether/CH_2Cl_2 [86].

4-CH$_3$C$_6$H$_4$–CO–C$_5$H$_4$FeC$_5$H$_4$–CH$_3$ (Table **6**, No. **83**). No. 83 and its 1,3-, and 1,2-isomers are present in a 13/11/5 ratio in the red syrup obtained from the main band with 6/1 hexane/ $CH_3CO_2C_2H_5$ on a dry SiO_2 column on chromatographing the product from Method VIm (p. 113). Recrystallization from 1/10 $CHCl_3$/hexane results in a brass-colored powder (m.p. 80 to 85°C), that consists predominantly of No. 83 and its 1,3-isomer (ratio 1.4/1). The red syrup is separable into the isomers by gas-liquid chromatography on a column filled with 3% OV-255 on Chromosorb W HP. The retention times at a 225°C column temperature are for the 1,2-, 1,1-′, and 1,3-isomers 6.7, 8.6, and 9.3 min; at 215°C, 9.1, 11.8, and 12.9 min, respectively [95].

CF$_3$CH$_2$CH$_2$–CO–C$_5$H$_4$FeC$_5$H$_4$–C(CH$_2$CH$_2$CF$_3$)$_2$OH (Table **6**, No. **85**) forms by the action of $CF_3(CH_2)_2MgCl$ on $Fe(C_5H_4–CO_2CH_3)_2$ in boiling ether, in 31.3% yield, along with 27.5% yield of $Fe(C_5H_4–C(CH_2CH_2CF_3)_2OH)_2$, 9.8% of $CH_3O_2C–C_5H_4FeC_5H_4–C(CH_2CH_2CF_3)_2OH$, and 12.8% of $CH_3O_2C–C_5H_4FeC_5H_4–CO–CH_2CH_2CF_3$. The raw product obtained after decomposition with an ice/NH_4Cl solution is separated chromatographically with petroleum ether, benzene, and ethanol as eluents on an Al_2O_3 column shielded from light [46].

2-ClC_6H_4–CO–$C_5H_4FeC_5H_4$–CH_3 (Table **6**, No. **87**). From the reaction mixture obtained by Method VIu (p. 114) the 1,2-isomer is separated by preparative thin layer or column chromatography. No. 87 can be satisfactorily won from the remaining mixture of the 1,1'- and 1,3-isomers by slow crystallization from a concentrated solution in 4/1 n-hexane/$CHCl_3$.

$KOC(CH_3)_3$ plus a little water form in $CH_3OCH_2CH_2OCH_3$ the compound HO_2C–$C_5H_4FeC_5H_4$–CH_3 in more than 80% yield. The formation of possible traces of $C_5H_5FeC_5H_4$–CH_3 has not been established [83].

2-ClC_6H_4–CO–$C_5H_4FeC_5H_4$–C_6H_5 (Table **6**, No. **89**). Only the 1,2-isomer can be separated chromatographically from the isomer mixture obtained by Method VIm (p. 113). The isomer ratio in the raw product was determined by converting it with $KOC(CH_3)_3$ in H_2O/CH_3OCH_2–CH_2OCH_3 into the isomer mixture $FeC_{10}H_8(CO_2H)C_6H_5$, which after esterification is converted into and is analyzed as $FeC_{10}H_8(C_6H_5)CH_3$ by NMR spectroscopy [83].

References:

[1] A. N. Nesmeyanov, N. A. Volkenau (Dokl. Akad. Nauk SSSR **111** [1956] 605/8; Proc. Acad. Sci. USSR Chem. Sect. **106/111** [1956] 689/91). – [2] W. G. M. Jones, T. Leigh, J. L. Madinaveitia, Imperial Chemical Industries Ltd. (Brit. 841710 [1957/60]; C.A. **1961** 25173). – [3] M. Rosenblum (Chem. Ind. [London] **1958** 953). – [4] M. Rosenblum, R. B. Woodward (J. Am. Chem. Soc. **80** [1958] 5443/9). – [5] M. Rosenblum (J. Am. Chem. Soc. **81** [1959] 4530/6).

[6] R. J. Stephenson, Imperial Chemical Industries Ltd. (Brit. 864197 [1959/61]; C.A. **1961** 17647). – [7] L. A. Day, Imperial Chemical Industries Ltd. (Brit. 864198 [1959/61]; C.A. **1961** 17647). – [8] S. Birtwell, Imperial Chemical Industries Ltd. (Brit. 861834 [1959/61]; Ger. 1116663 [1960/62]; C.A. **1961** 16565). – [9] E. A. Hill, J. H. Richards (J. Am. Chem. Soc. **83** [1961] 4216/21). – [10] A. N. Nesmeyanov, V. A. Sazonova, V. N. Drozd (Dokl. Akad. Nauk SSSR **137** [1961] 102/5; Proc. Acad. Sci. USSR Chem. Sect. **136/141** [1961] 272/4).

[11] A. N. Nesmeyanov, E. G. Perevalova, S. P. Gubin, T. V. Nikitina, A. A. Ponomarenko, L. S. Shilovtseva (Dokl. Akad. Nauk SSSR **139** [1961] 888/91; Proc. Acad. Sci. USSR Chem. Sect. **136/141** [1961] 767/70). – [12] J. Komenda, J. Tirouflet (Compt. Rend. **254** [1962] 3093/5). – [13] I. G. Morrison, P. L. Pauson (Proc. Chem. Soc. **1962** 177). – [14] A. N. Nesmeyanov, V. A. Sazonova, V. N. Drozd (Izv. Akad. Nauk SSSR Otd. Khim. Nauk **1962** 45/7; Bull. Acad. Sci. USSR Div. Chem. Sci. **1962** 38/9). – [15] M. Rosenblum, W. G. Howells (J. Am. Chem. Soc. **84** [1962] 1167/72).

[16] M. Rosenblum (Diss. Harvard Univ. 1953, pp. 140, 146). – [17] K. Schlögl, H. Pelousek (Liebigs Ann. Chem. **651** [1962] 1/10). – [18] K. Schlögl, M. Peterlik, H. Seiler (Monatsh. Chem. **93** [1962] 1309/27). – [19] R. A. Benkeser, Y. Nagai, J. Hooz (Bull. Chem. Soc. Japan **36** [1963] 482/3). – [20] D. W. Hall, J. H. Richards (J. Org. Chem. **28** [1963] 1549/54).

[21] A. N. Nesmeyanov, V. N. Drozd, V. A. Sazonova, V. I. Romanenko, A. K. Prokof'ev, L. A. Nikonova (Izv. Akad. Nauk SSSR Otd. Khim. Nauk **1963** 667/74; Bull. Acad. Sci. USSR Div. Chem. Sci. **1963** 597/603). – [22] J. Tirouflet, E. Laviron, R. Dabard, J. Komenda (Bull. Soc. Chim. France **1963** 857/62). – [23] R. A. Benkeser, Y. Nagai, J. Hooz (J. Am. Chem. Soc. **86** [1964] 3742/6). – [24] S. I. Goldberg, R. L. Matteson (J. Org. Chem. **29** [1964] 323/6). – [25] J. Komenda (Chem. Zvesti **18** [1964] 378/84).

[26] Y. Nagai, J. Hooz, R. A. Benkeser (Bull. Chem. Soc. Japan **37** [1964] 53/60). – [27] M. D. Reshetova, L. M. Yarysheva, E. G. Perevalova, A. N. Nesmeyanov (Izv. Akad. Nauk SSSR Ser. Khim. **1965** 2196/8; Bull. Acad. Sci. USSR Div. Chem. Sci. **1965** 2160/1). – [28] G. G. Dvoryantseva, S. L. Portnova, Yu. N. Sheinker, L. P. Yur'eva, A. N. Nesmeyanov (Dokl. Akad. Nauk SSSR **169** [1966] 1083/6; Dokl. Chem. Proc. Acad. Sci. USSR **166/171** [1966] 773/5). – [29] K. I. Grandberg, S. P. Gubin (Izv. Akad. Nauk SSSR Ser. Khim. **1966** 551/4; Bull. Acad. Sci. USSR Div.

Chem. Sci. **1966** 519/21). – [30] A. N. Nesmeyanov [Nessmejanow], L. P. Yur'eva [Jurjewa], N. S. Kochetkova, S. V. Vitt [Witt] (Izv. Akad. Nauk SSSR Ser. Khim. **1966** 560/2; Bull. Acad. Sci. USSR Div. Chem. Sci. **1966** 530/2; Abhandl. Deut. Akad. Wiss. Berlin Kl. Chem. Geol. Biol. **1966** No. 2, pp. E265/E268).

[31] K. Schlögl, H. Mechtler (Monatsh. Chem. **97** [1966] 150/67). – [32] D. W. Hall, C. D. Russell (J. Am. Chem. Soc. **89** [1967] 2316/22). – [33] W. M. Horspool, R. G. Sutherland (Chem. Commun. **1967** 240). – [34] G. R. Knox, I. G. Morrison, P. L. Pauson, M. A. Sandhu, W. E. Watts (J. Chem. Soc. C **1967** 1853/6). – [35] N. Ya. Mar'yashkin, T. P. Vishnyakova, A. A. Koridze (Tr. Mosk. Inst. Neftekhim. Gaz. Prom. No. 72 [1967] 67/70).

[36] S. McVey, I. G. Morrison, P. L. Pauson (J. Chem. Soc. C **1967** 1847/50). – [37] K. Schlögl, H. Falk, G. Haller (Monatsh. Chem. **98** [1967] 82/94). – [38] T. P. Vishnyakova, N. Ya. Mar'yashkin, M. E. Elyashberg (Zh. Org. Khim. **3** [1967] 947/9; J. Org. Chem. [USSR] **3** [1967] 910/1). – [39] T. P. Vishnyakova, N. Ya. Mar'yashkin, M. E. Elyashberg, I. A. Mel'nik, I. I. Zadko (Tr. Mosk. Inst. Neftekhim. Gaz. Prom. No. 72 [1967] 58/61). – [40] D. W. Hall, E. A. Hill, J. H. Richards (J. Am. Chem. Soc. **90** [1968] 4972/6).

[41] W. M. Horspool, R. G. Sutherland (Can. J. Chem. **46** [1968] 3453/6). – [42] A. N. Nesmeyanov, N. S. Kochetkova, E. I. Fedin, E. V. Leonova, P. V. Petrovskii (Dokl. Akad. Nauk SSSR **178** [1968] 362/4; Dokl. Chem. Proc. Acad. Sci. USSR **178/183** [1968] 41/3). – [43] K. Sakai, M. Ishige, H. Kono, I. Motoyama, K. Watanabe, K. Hata (Bull. Chem. Soc. Japan **41** [1968] 1902/8). – [44] K. Schlögl, H. Soukup (Monatsh. Chem. **99** [1968] 927/46). – [45] T. H. Barr, H. L. Lentzner, W. E. Watts (Tetrahedron **25** [1969] 6001/13).

[46] G. P. Chalykh, E. B. Sokolova, O. D. Shalygina, N. V. Noskova (Tr. Mosk. Khim. Tekhnol. Inst. **1969** No. 61, p. 164). – [47] W. F. Little, H. T. Busby (Proc. 4th Intern. Conf. Organometal. Chem., Bristol 1969, Abstr. G 2). – [48] H. Patin, R. Dabard (Tetrahedron Letters **1969** 4971/4). – [49] A. Perjéssy, Š. Toma (Chem. Zvesti **23** [1969] 533/9). – [50] M. Sato, T. Ito, I. Motoyama, K. Watanabe, K. Hata (Bull. Chem. Soc. Japan **42** [1969] 1976/81).

[51] Š. Toma, E. Kalužayová (Chem. Zvesti **23** [1969] 540/52). – [52] F. M. Vigo (Diss. Univ. Illinois 1969; Diss. Abstr. Intern. B **31** [1970] 598/9). – [53] C. Moïse, J. Tirouflet, D. Sautrey (Compt. Rend. C **271** [1970] 951/4). – [54] H. Patin (Compt. Rend. C **270** [1970] 243/5). – [55] M. Rosenblum, N. M. Brawn, D. Ciappenelli, J. Tancrede (J. Organometal. Chem. **24** [1970] 469/77).

[56] M. Sato, M. Koga, I. Motoyama, K. Hata (Bull. Chem. Soc. Japan **43** [1970] 1142/7). – [57] J. E. Gorton, H. L. Lentzner, W. E. Watts (Tetrahedron **27** [1971] 4353/60). – [58] J. W. Huffman, J. F. Cope (J. Org. Chem. **36** [1971] 4068/72). – [59] J. J. McDonnell, D. J. Pochopien (J. Org. Chem. **36** [1971] 2092/8). – [60] C. Moïse, D. Sautrey, J. Tirouflet (Bull. Soc. Chim. France **1971** 4562/76).

[61] T. S. Lixandru (Zh. Obshch. Khim. **42** [1972] 1991/2; J. Gen. Chem. [USSR] **42** [1972] 1985/6). – [62] J. Paul, K. Schlögl, W. Silhan (Monatsh. Chem. **103** [1972] 243/54). – [63] L. H. Ali, A. Cox, T. J. Kemp (J. Chem. Soc. Dalton Trans. **1973** 1468/75). – [64] G. Cerichelli, B. Floris, G. Illuminati, G. Ortaggi (Gazz. Chim. Ital. **103** [1973] 911/22). – [65] R. Dabard, H. Patin (Bull. Soc. Chim. France **1973** 2158/64).

[66] L. I. Kazakova, N. M. Loim, Z. N. Parnes (Zh. Obshch. Khim. **43** [1973] 1577/80; J. Gen. Chem. [USSR] **43** [1973] 1561/3). – [67] H. Patin, R. Dabard (Bull. Soc. Chim. France **1973** 2413/7). – [68] H. Patin, R. Dabard (Bull. Soc. Chim. France **1973** 2756/9). – [69] Š. Toma, M. Sališová (J. Organometal. Chem. **57** [1973] 191/8). – [70] M. I. Bruce, F. G. A. Stone, B. J. Thomson (J. Organometal. Chem. **77** [1974] 77/81).

[71] G. Cerichelli, B. Floris, G. Illuminati, G. Ortaggi (Ann. Chim. [Rome] **64** [1974] 119/23). – [72] G. Cerichelli, B. Floris G. Ortaggi (J. Organometal. Chem. **76** [1974] 73/9). – [73] E. V.

Leonova, N. S. Kochetkova, E. I. Fedin, P. V. Petrovskii (Khim. Metal. Met. Khim. Tekhnol. Silikatov 11th Mendeleevsk. S'ezd. Obshch. Prikl. Khim., Moscow 1974 [1975], pp. 98/9; C.A. **84** [1976] No. 164987). – [74] A. N. Nesmeyanov, N. N. Sedova, V. A. Sazonova, I. F. Leshcheva, I. S. Rogozhin (Dokl. Akad. Nauk SSSR **218** [1974] 356/8; Dokl. Chem. Proc. Acad. Sci. USSR **218/223** [1974] 647/9). – [75] T. D. Turbitt, W. E. Watts (J. Chem. Soc. Perkin Trans. II **1974** 177/84).

[76] T. D. Turbitt, W. E. Watts (J. Chem. Soc. Perkin Trans. II **1974** 185/9). – [77] T. D. Turbitt, W. E. Watts (J. Chem. Soc. Perkin Trans. II **1974** 189/95). – [78] T. S. Abram, W. E. Watts (J. Chem. Soc. Perkin Trans. I **1975** 116/21). – [79] F. Jursík (J. Chromatog. Libr. **3** [1975] 1087/114, 1108). – [80] A. N. Nesmeyanov, E. V. Leonova, N. S. Kochetkova, A. I. Malkova, A. G. Makarovskaya (J. Organometal. Chem. **96** [1975] 275/8).

[81] A. N. Nesmeyanov, G. B. Shul'pin, M. I. Rybinskaya (Izv. Akad. Nauk SSSR Ser. Khim. **1975** 2824/7; Bull. Acad. Sci. USSR Div. Chem. Sci. **1975** 2714/7). – [82] H. Patin, D. Mourot (Compt. Rend. C **281** [1975] 737/40). – [83] V. Rapić, K. Schlögl, B. Steinitz (J. Organometal. Chem. **94** [1975] 87/98). – [84] K. Yamakawa, M. Hisatome, Y. Sako, S. Ichida (J. Organometal. Chem. **93** [1975] 219/30). – [85] W. Crawford, T. D. Turbitt, W. E. Watts (J. Organometal. Chem. **105** [1976] 341/8).

[86] D. W. Slocum, B. P. Koonsvitsky (J. Org. Chem. **41** [1976] 3664/8). – [87] Š. Toma, A. Maholányiová, E. Solčaniová (Acta Fac. Rerum Nat. Univ. Comenianae Chim. No. 24 [1976] 53/80). – [88] Š. Toma, A. Maholányiová, E. Solčaniová (Collection Czech. Chem. Commun. **42** [1977] 1013/26). – [89] N. El Murr, A. Chaloyard, E. Laviron (Nouv. J. Chim. **2** [1978] 15/7). – [90] A. N. Nesmeyanov, A. N. Pushin, V. A. Sazonova (Izv. Akad. Nauk SSSR Ser. Khim. **1978** 1685; Bull. Acad. Sci. USSR Div. Chem. Sci. **1978** 1474/5).

[91] Š. Toma, M. Paričiová, E. Solčaniová (Acta Fac. Rerum Nat. Univ. Comenianae Chim. No. 26 [1978] 61/71). – [92] P. Canonne, G. Foscolos, R. Harder (J. Organometal. Chem. **178** [1979] 331/42). – [93] G. Ortaggi, P. Riccio, I. Tritto (J. Org. Chem. **44** [1979] 2920/3). – [94] G. B. Shul'pin, A. N. Kitaigorodskii (Izv. Akad. Nauk SSSR Ser. Khim. **1980** 168/70; C.A. **92** [1980] No. 214425). – [95] B. E. Maryanoff (J. Org. Chem. **46** [1981] 70/4).

[96] M. G. Peterleitner, L. I. Denisovich, A. Z. Kreindlin, M. I. Rybinskaya (Izv. Akad. Nauk SSSR Ser. Khim. **1981** 2700/4; Bull. Acad. Sci. USSR Div. Chem. Sci. **1981** 2316/9). – [97] G. V. Shul'pin (Zh. Obshch. Khim. **51** [1981] 2152/3; J. Gen. Chem. [USSR] **51** [1981] 1854). – [98] J. Holeček, K. Handlíř, J. Klikorka, M. Nádvorník (J. Prakt. Chem. **324** [1982] 345/8). – [99] A. Z. Kreindlin, M. I. Rybinskaya (Izv. Akad. Nauk SSSR Ser. Khim. **1982** 174/7; Bull. Acad. Sci. USSR Div. Chem. Sci. **1982** 164/7). – [100] Z. A. Beinoravichute, I. I. Kryauchyunas (U.S.S.R. 228938 [1966/68]; C.A. **70** [1968] No. 78821).

[101] Z. Kaikarene (Polim. Mater. Ikh Issled. No. 12 [1971] 57/60; C.A. **78** [1973] No. 73020). – [102] A. N. Nesmeyanov, V. A. Sazonova, N. N. Sedova, L. S. Borodina, Yu. I. Volgin (Dokl. Akad. Nauk SSSR **213** [1973] 1099/102; Dokl. Chem. Proc. Acad. Sci. USSR **208/213** [1973] 929/31). – [103] A. Perjéssy, Š. Toma (Collection Czech. Chem. Commun. **48** [1983] 1635/46). – [104] V. I. Sokolov, L. L. Troitskaya, N. S. Khrushcheva, O. A. Reutov (Dokl. Akad. Nauk SSSR **281** [1985] 861/3).

5.2.3.5.1.7 Ketones of the Type R'–CO–A–$C_5H_4FeC_5H_4$–R^2 (R'–CO–A = R^1)

The preparation of the compounds listed in Table 7 takes place by the following methods as well as by the procedures described in connection with the chemical reactions of the starting compounds (see "Further information" listed for these):

Method I. The mixture of $Fe(C_5H_4{-}I)_2$, $(C_2H_5)_3N$, $(C_6H_5)_3P$, $Pd(O_2CCH_3)_2$, and $CH_2{=}CH{-}R$ (molar ratio 1/2.8/0.04/0.02/2.8) with $R = CHOHCH_3$ for No. 1 and $COCH_3$ for No. 2 is heated in CH_3CN in a sealed tube for 8 h at 100°C. After addition of H_2O and ether the ethereal phase is separated, concentrated, and subjected to column chromatography on Al_2O_3 [19].

Method II. Li_2PdCl_4 is reacted in C_2H_5OH with $Fe(C_5H_4{-}HgCl)_2$ (5 mmol) and $CH_2{=}CH{-}CO{-}R$ (20 mmol), where $R = CH_3$ for No. 2, C_6H_5 for No. 4, for 12 h at room temperature with stirring. After filtering off the precipitated Pd and distilling away the C_2H_5OH the remaining raw product is chromatographed in benzene on an Al_2O_3 column. I appears first (yield 12% along with No. 2, 9% with No. 4) and finally the desired compound, which also can be eluted with $CHCl_3$ [11].

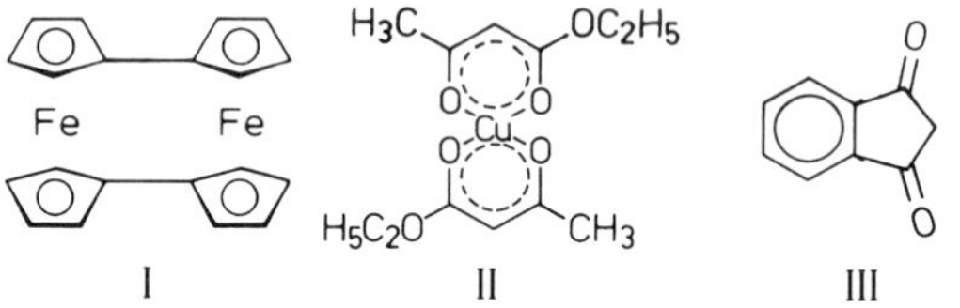

I II III

Method III. The mixture of $Fe(C_5H_4{-}Li)_2 + C_5H_5FeC_5H_4{-}Li$ (obtained from ferrocene and $n\text{-}C_4H_9Li$ in ether) is reacted with $R{-}CO{-}CH{=}CHN(CH_3)_2$, where $R = n\text{-}C_3H_7$ for No. 3 and C_6H_5 for No. 4, in tetrahydrofuran for 2.5 h at room temperature and then for 1 h at 50 to 60°C. Chromatography on a short Al_2O_3 column in benzene yields in succession ferrocene (petroleum ether), $C_5H_5FeC_5H_4{-}CH{=}CH{-}CO{-}R$ (benzene), and No. 3 or No. 4 (ether eluent) [7].

Method IV. Condensation at room temperature:

a) Of $Fe(C_5H_4{-}CHO)_2$ with $CH_3COC_6H_5$ using 20% NaOH in C_2H_5OH [8].

b) Of $R^2{-}C_5H_4FeC_5H_4{-}CHO$ with $CH_3COC_6H_5$ and NaOH (molar ratio 1/1/10) in 2/1 ethanol/water for 12 h. When starting not from the pure compound but its positional isomer mixture, an isomer mixture is obtained which is separable by column chromatography on SiO_2 with benzene as the eluent, in the elution sequence 1,2-, 1,1′-, and 1,3-isomers [18].

c) Of $R^2{-}C_5H_4FeC_5H_4{-}CHO$ with indan-1,3-dione (see Formula III) in benzene in the presence of several drops of $(C_2H_5)_3N$ (reaction time 12 h). Starting from the positional isomer mixture yields a mixture of isomers, which can be separated as given under Method IV b. The elution sequence is here, however, 1,2-, 1,3-, and the 1,1′-isomers [18].

Method V. The Gomberg-Bachmann-Hey reaction of ferrocene with $[4\text{-}CH_3COC_6H_4N_2]_2SO_4$ in CH_3CO_2H/H_2O at 10 to 15°C and subsequent reduction with $NaHSO_3$ [1, 2]. Separation from the 1,2- and 1,3-isomers [2], $C_5H_5FeC_5H_4{-}C_6H_4COCH_3\text{-}4$ [1, 2], and an unidentified colorless compound (m.p. 121 to 123°C) [2] is conducted chromatographically on Al_2O_3 columns in $CHCl_3/C_6H_6$ with 4/1 C_6H_6/Skelly B as eluent [1, 2]. According to [22] No. 6 is prepared in a corresponding fashion.

Method VI. Cleavage of II in the presence of $Fe(C_5H_4{-}CH_2OH)_2$ [4].

Method VII. Friedel-Crafts acylation of $C_5H_5FeC_5H_4{-}R^1$:

a) $AlCl_3 + R^2Cl$ in CH_2Cl_2 are added drop-wise to $C_5H_5FeC_5H_4{-}R^1$ in CH_2Cl_2 and reacted for 4 h at room temperature [10, 12, 13, 17]. Column chromatography on SiO_2 [12] or Al_2O_3 with $C_6H_6/CH_3CO_2C_2H_5$ (ratio from 97/3 to 94/6) as eluent concludes the processing [10, 13, 17]. No. 9 is separated in the same way from its 1,3-isomer (yield 13.5%) [12], No. 50 from its 1,2-isomer (10%) [13].

In the preparation of No. 30 it elutes from Al_2O_3 in the third position, $C_5H_5FeC_5H_4$–R^1 in the first [10, 17], and No. 32 last (yield 1.7%) [17]. According to [17] $C_5H_5FeC_5H_4$–CH=C($COCH_3$)COC_6H_5 (9.4%) is isolated in the second position and not the $C_5H_5FeC_5H_3$($COCH_3$-2)CH=CHCOC_6H_5-1 assumed by [10]. The yield of No. 32 rises at the expense of $C_5H_5FeC_5H_4$–CH=C($COCH_3$)COC_6H_5 when the acylation is conducted at a higher $AlCl_3$/$C_5H_5FeC_5H_4$–R^1 ratio [17].

In the preparation of No. 31 the elution sequence from SiO_2 with 3 to 10% $CH_3CO_2C_2H_5$ in C_6H_6 is the starting compound $C_5H_5FeC_5H_4$–R^1 (37.5% recovery), $C_5H_5FeC_5H_4$–CH=C($COCH_3$)COCH=CHC_6H_5 (7%), and No. 31 [16, 17]. Correspondingly, in the preparation of No. 33 were isolated the compounds $C_5H_5FeC_5H_4$–R^1 (41%), $C_5H_5FeC_5H_4$–CH=C($COCH_3$)CH=CHCOC_6H_5 (13.5%), No. 33, and finally No. 34 (5%) [16, 17]. The yield of No. 34 is inversely proportional to that of $C_5H_5FeC_5H_4$–CH=C($COCH_3$)CH=CH–CO–C_6H_5, since No. 34 forms by acetylation of the latter (and not from No. 33) [16].

b) As with Method VIIa, from $C_5H_5FeC_5H_4$–R^1, CH_3COCl, and $AlCl_3$ in the molar ratio 1/>2/>4 [17].

c) $AlCl_3$ is added in portions at 0 to 5°C to $C_5H_5FeC_5H_4$–R^1 (n = 2 to 9) and R^2Cl (molar ratio 2/1/>1) and reacted for 4 h at room temperature. Column chromatography on SiO_2 with C_6H_6/$CH_3CO_2C_2H_5$ (ratio 19/1 to 10/1) can separate the 1,2-isomer (7 to 10% yield), while the 1,1′- and 1,3-isomers are only isolable as the mixture "m" (yield 20 to 25%), which cannot be separated either chromatographically or by recrystallization [20].

d) $C_5H_5FeC_5H_4$–R^1 in CH_2Cl_2 is added at 0 to 5°C to a solution of $R^2Cl + AlCl_3$ (molar ratio 1/5/10) in CH_2Cl_2 and the reaction is continued for n = 3 to 6 for 1 h at 35°C, for n = 7 to 9 for 2 to 3 h at room temperature. The separation corresponding to that in Method VIIc yields besides the 1,2-isomer (5 to 7%) the mixture "m" of the 1,1′- and 1,3-isomers (10 to 15% collective yield) together with C_6H_5CH=CHCO–$C_5H_4FeC_5H_3$(COCH=CHC_6H_5-1)R^1-2 and -3 in yields of 7 to 10 and 10 to 15%, respectively [20].

Method VIII. Acylation of Fe(C_5H_4–C_6H_5)$_2$:

a) With $AlCl_3$ and CH_3COCl (molar ratio 1/1/1.3) in CH_2Cl_2 initially for ca. 5 h at 0°C and then 1.5 h at room temperature. No. 48 is separated by column chromatography from C_6H_5–$C_5H_4FeC_5H_3$(C_6H_5-1)$COCH_3$-2 and -3 on neutral Al_2O_3 with 1/1 benzene/Skellysolve B [1].

b) With $AlCl_3$ and 2-ClC_6H_4COCl (molar ratio 2/3/2.2) in boiling CH_2Cl_2. At room temperature No. 54 does not form. It is separated by thin layer chromatography from C_6H_5–$C_5H_4FeC_5H_3$(C_6H_5-1)(COC_6H_4Cl-2)-2 and -3 (collective yield 67%) [15].

Chemical Behavior. Compounds containing CH_3CO– undergo in the presence of bases the Claisen-Schmidt reaction with aldehydes, as for example 4-ClC_6H_4CHO, with formation of the –CO–CH=CH– group. Compounds endowed with this group can reform the aldehyde and CH_3CO compound under the same conditions via the retro-Claisen-Schmidt reaction; however, they can also cyclize via an intramolecular Michael addition to 1,1′-ferrocenophane derivatives. Compounds which contain in one substituent the –CO–CH=CH– group and in the other the CH_3CO– group can exchange the two groups in the substituents by a trans-Claisen-Schmidt reaction. For additional details see "Further information" for the individual compounds.

References on pp. 152/3

Table 7

Ketones of the Type $R'-CO-A-C_5H_4FeC_5H_4-R^2$.
Further information for numbers preceded by an asterisk is given at the end of the table, pp. 150/2.
For abbreviations and dimensions see p. VIII.

No.	$R^1-C_5H_4FeC_5H_4-R^2$ $R^1 = A-CO-R'$ R^2	method of preparation (yield in %), properties, and remarks	Ref.
compounds with $R^1 = R^2$:			
1	$CH_2CH_2-CO-CH_3$	I (53) m.p. 153 to 154° ^{1}H NMR: 2.09 (s, CH_3), 2.55 (m, CH_2–CO), 2.64 (m, C(1)–CH_2), 4.08 (m) and 4.11 (m) (C_5H_4) IR: 1718 (νC=O) mass spectrum: m/e 310 $[M]^+$	[19]
*2	$CH{=}CH-CO-CH_3$	I (76), II (27) m.p. 112 to 113° ^{1}H NMR ($CDCl_3$): 2.29 (s, CH_3), 4.47 (s, C_5H_4), 6.34 (d, =CH–CO), 7.46 (d, –CH=) IR: 1617, 1685	[11, 19]
3	$\overset{6}{C}H{=}\overset{7}{C}H-CO-\overset{9}{C}H_2\overset{10}{C}H_2CH_3$-(E)	III (4) red crystals, m.p. 65 to 66° (pentane) ^{1}H NMR (CCl_4): 0.91 (CH_3), 1.57 (H-10), 2.34 (H-9), 4.30 (H-3, 4), 4.40 (H-2, 5), 6.12 (H-7; J(H-6, 7) = 17.1), 7.10 (H-6) IR (KBr): 820 and 837 (γC–H of C_5H_4); 978 (γ =C–H); 1385, 1402, and 1418 (γC–H of C_5H_4); 1632 (νC=C); 1678 and 1697 (νC=O)	[7]
*4	$\overset{6}{C}H{=}\overset{7}{C}H-CO-C_6H_5$	II (25), III (6), IVa, see "Further information", p. 150 red crystals (darker than No. 3, E,E-compound), m.p. 145 to 146°, 178 to 180° (CH_3OH, E,E-compound), 192° (C_2H_5OH) ^{1}H NMR ($CDCl_3$): 4.43 (H-3, 4), 4.50 (H-2, 5), 7.00 (H-7; J(H-6, 7) = 15.4), 7.57 (H-6), 7.60 (C_6H_5) (E,E-compound) ^{1}H NMR ($CDCl_3$): 4.49 (s, C_5H_4), 5.57 (d, H-7), 7.71 (d, H-6), 7.28 (s, C_6H_5) IR: 1619, 1680	[5, 7, 8, 11]

References on pp. 152/3

No.	R^1	R^2	method of preparation (yield in %), properties and remarks	Ref.
			IR (KBr): 828 and 860 (γC–H of C_5H_4); 978 (γ =C–H); 1368 and 1400 (γC–H of C_5H_4); 1598 and 1610 (νC=C); 1662 (νC=O) (E, E-compound)	
*5	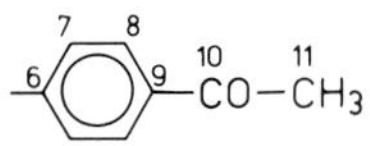		V (ca. 4 to 5) m.p. 243 to 245° (dec.), 246 to 248° IR (paraffin oil): 1017 (C_5H_4), 1120 (ν_sC(9)–C_{acetyl}) IR ($CHCl_3$): 889 UV (C_2H_5OH/H_2O, 95/5): $\lambda_{max}(\varepsilon)$ = 286 (30000), ca. 308 (19000), 390 (4220), ca. 465 (2000)	[1 to 3, 9]
*6	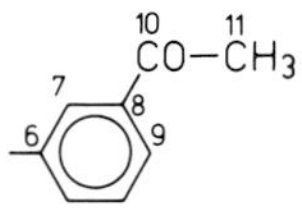		(V) m.p. 145 to 147° IR (paraffin oil): 1025 (C_5H_4), 1117 (ν_sC(8)–C_{acetyl})	[9, 22]
7	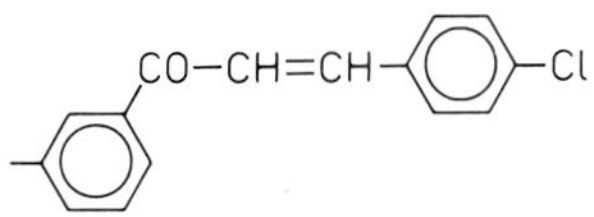		see Nos. 6, p. 150, and 59, p. 152 orange-red crystals, m.p. 215 to 217° (C_6H_6/petroleum ether) IR (Nujol): 1013 (C_5H_4); 1572, 1600, and 1615 (γC=C of C_6H_5); 1664 (νC=O)	[13]
8	$CH_2O_2CCH_2$–CO–CH_3		VI	[4]

compounds with $R^1 \neq R^2$:

No.	R^1	R^2	method of preparation (yield in %), properties and remarks	Ref.
*9	CH_2CH_2–CO–CH_3	$COCH_3$	VIIa (30), see Nos. 10 and 15, p. 151 red oil IR (Nujol): 1006 and 1109 (C_5H_4); 1114 (ν_sC(1′)–C_{acetyl}); 1682 and 1728 (νC=O)	[12]
*10	CH_2CH_2–CO–CH_3	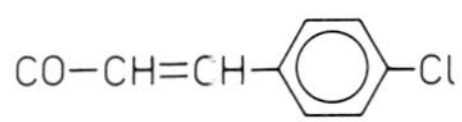	VIIa (29.4), see No. 15, p. 151 m.p. 105 to 106° (acetone/petroleum ether) IR (Nujol): 1012 and 1108 (C_5H_4); 1573 and 1593 (γC=C of C_6H_5); 1657 and 1718 (νC=O)	[12]
11	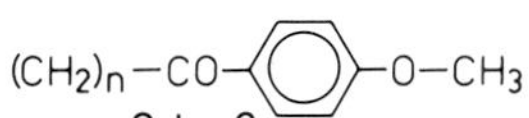 n = 2 to 6	$COCH=CHC_6H_5$	VIIc, VIId	[20]

References on pp. 152/3

Table 7 (continued)

No.	$R^1–C_5H_4FeC_5H_4–R^2$ R^1 = A–CO–R′	R^2	method of preparation (yield in %), properties, and remarks	Ref.
12	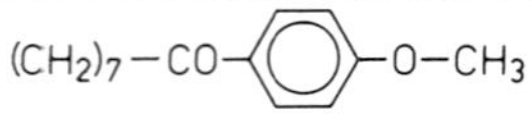	$COCH_3$	see No. 13, p. 151 red-orange oil (m) ^{1}H NMR: plot in [20]	[20]
*13	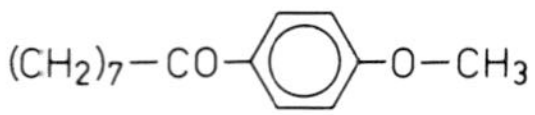	$COCH{=}CHC_6H_5$	VIIc (19.2 m), VIId m.p. 92 to 102° (C_6H_6/petroleum ether) (m) ^{1}H NMR: plot in [20]	[20]
14	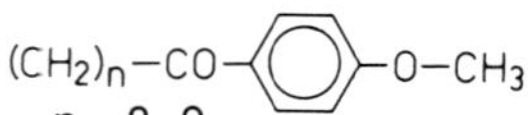 n = 8, 9	$COCH{=}CHC_6H_5$	VIIc, VIId	[20]
*15	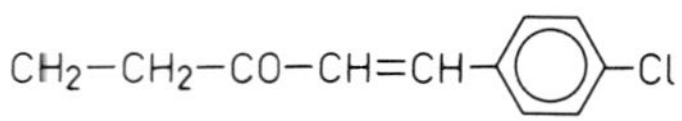	$COCH_3$	VIIa (43), see No. 9, p. 150 yellow crystals, m.p. 89 to 95° (C_2H_5OH/petroleum ether) IR (Nujol): 1014 and 1106 (C_5H_4); 1115 ($\nu_sC(1')–C_{acetyl}$); 1595 and 1620 (γC=C of C_6H_4); 1658 and 1695 (νC=O)	[12]
16	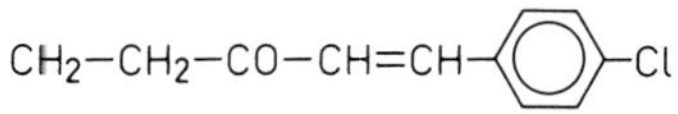	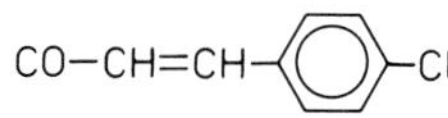	see No. 9, p. 150 deep red crystals, m.p. 155 to 157° (C_2H_5OH/petroleum ether) IR (Nujol): 1015 (C_5H_4); 1595 and 1615 (γC=C of C_6H_4); 1658 and 1692 (νC=O)	[12]
17	$CH{=}CH–CO–C_6H_5$	C_6H_5	IVb (55.5) m.p. 105 to 107° ^{1}H NMR ($CDCl_3$): 4.308 (t, H-3′,4′), 4.366 (s, H-2 to 5), 4.638 (t, H-2′,5′)	[18]
18	$CH{=}CH–CO–C_6H_5$	$C_6H_4CH_3$-4	IVb (52.4) m.p. 149 to 150°	[18]
19	$CH{=}CH–CO–C_6H_5$	$C_6H_4CH_3$-3	IVb (42.4) m.p. 83°	[18]
20	$CH{=}CH–CO–C_6H_5$	$C_6H_3(CH_3)_2$-3,4	IVb (54) m.p. 104 to 105°	[18]

References on pp. 152/3

21	CH=CH–CO–C_6H_5	$C_6H_3(CH_3)_2$-3,5	IVb (40.5) m.p. 63 to 64°	[18]
22	CH=CH–CO–C_6H_5	C_6H_4Cl-4	IVb (53) m.p. 131 to 132°	[18]
23	CH=CH–CO–C_6H_5	C_6H_4Cl-3	IVb (45.6) oil	[18]
24	CH=CH–CO–C_6H_5	C_6H_4Br-4	IVb (30.4) m.p. 137 to 138°	[18]
25	CH=CH–CO–C_6H_5	C_6H_4Br-3	IVb (51) m.p. 94 to 95°	[18]
26	CH=CH–CO–C_6H_5	$C_6H_3Cl_2$-3,4	IVb (43.6) m.p. 120 to 121°	[18]
27	CH=CH–CO–C_6H_5	$C_6H_4CF_3$-3	IVb (53.5) m.p. 105 to 106°	[18]
28	CH=CH–CO–C_6H_5	$C_6H_3(CF_3$-3)Cl-4	IVb (59) m.p. 109 to 110°	[18]
29	CH=CH–CO–C_6H_5	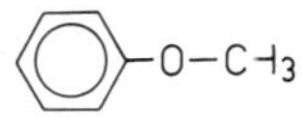	IVb (45.5) m.p. 157 to 160°	[18]
*30	CH=CH–CO–C_6H_5	$COCH_3$	VIIa (15.6, 18.7) m.p. 128 to 130° (C_2H_5OH), 129 to 130° IR (paraffin oil): 694 (γC–H of C_6H_5); 1018 (C_5H_4); 1117 (ν_sC(1')–C_{acetyl}); 1650 and 1663 (νC=O) UV (C_2H_5OH): λ_{max}(log ε) = 270 (4.05), 334 (4.19) R_f: 0.41 (on SiO_2 with 8/2 $C_6H_6/CH_3CO_2C_2H_5$), 0.29 (on SiO_2 with 19/1 $C_6H_6/CH_3CO_2C_2H_5$)	[9, 10, 17]
*31	CH=CH–CO–CH=CHC$_6H_5$	$COCH_3$	VIIa (26) reddish violet crystals, m.p. 128 to 132° (C_6H_6/petroleum ether) ^{1}H NMR ($CDCl_3$): 2.29 (s, $COCH_3$), 4.47 [16] (4.49 [17]) (m, H-3,3',4,4'), 4.57 (t, H-2,5 [17], 2',5' [16]), 4.77 (t, H-2',5' [17], 2,5 [16]), 6.5 to 7.9 (CH=CHCOCH=CHC$_6H_5$)	[16, 17]

References on pp. 152/3

Table 7 (continued)

No.	$R^1-C_5H_4FeC_5H_4-R^2$ $R^1 = A-CO-R'$	R^2	method of preparation (yield in %), properties, and remarks	Ref.
32	$CH{=}C(COCH_3)CO{-}C_6H_5$	$COCH_3$	VIIb (5), see preparation and "Further information" for No. 30, pp. 141 and 151 m.p. 115 to 117° (C_2H_5OH) ^1H NMR ($CDCl_3$): 2.30 (s, CH_3), 4.23 (s, H-2 to 5), 4.35 (m, H-3′,4′), 4.70 (t, H-2′,5′), 7.1 to 7.9 (C_6H_5 + CH=)	[17]
*33	$CH{=}CHCH{=}CH{-}CO{-}C_6H_5$	$COCH_3$	VIIa (7.8) violet crystals, m.p. 128 to 130° (petroleum ether) ^1H NMR ($CDCl_3$): 2.30 (s, $COCH_3$), 4.49 (m, H-3,3′,4,4′), 4.59 (t, H-2,5 [17], 2′,5′ [16]), 4.77 (t, H-2′,5′ [17], 2,5 [16]), 7.0 to 8.1 (m, CH=CH–CH=CH, C_6H_5)	[16, 17]
34	$CH{=}C(COCH_3)CH{=}CH{-}CO{-}C_6H_5$	$COCH_3$	see preparation of No. 33, p. 141 violet crystals, m.p. 72 to 77° (petroleum ether) ^1H NMR ($CDCl_3$): 2.25 (s, $COCH_3$), 2.35 (s, $COCH_3$); 4.35 (m, 6H) and 4.66 (m, 2H) (C_5H_4); 6.4 to 8.0 (CH=C–CH=CH, C_6H_5)	[16, 17]
35	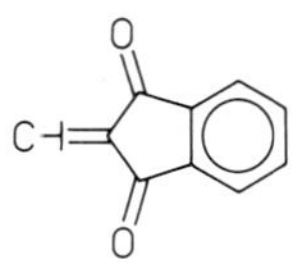	C_6H_5	IVc (63.5) m.p. 152 to 153° ^1H NMR ($CDCl_3$): 4.308 (t, H-3′,4′), 4.638 (t, H-2′,5′), 4.775 (t, H-3,4), 5.336 (t, H-2,5)	[18]
36	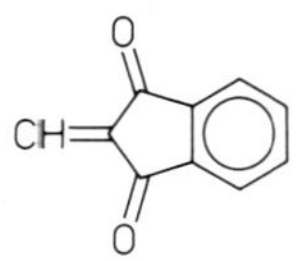	$C_6H_4CH_3$-4	IVc (54.7) m.p. 152 to 153°	[18]
37	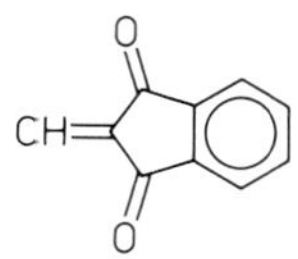	$C_6H_4CH_3$-3	IVc (49.5) m.p. 136 to 138°	[18]
38	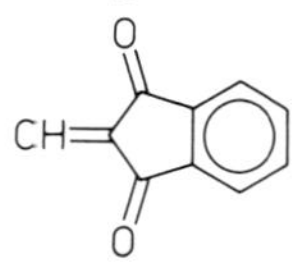	$C_6H_3(CH_3)_2$-3,4	IVc (57.4) m.p. 186 to 188°	[18]

References on pp. 152/3

39	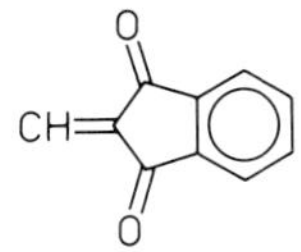	$C_6H_3(CH_3)_2$-3,5	IVc (51.4) m.p. 176 to 176.5°	[18]
40	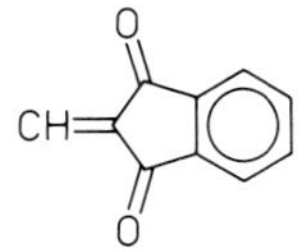	C_6H_4Cl-4	IVc (55) m.p. 184 to 185°	[18]
41	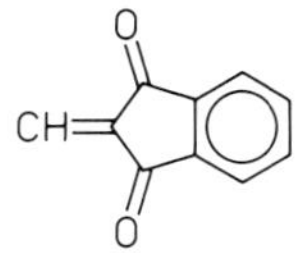	C_6H_4Cl-3	IVc (72.4) m.p. 177 to 179°	[18]
42	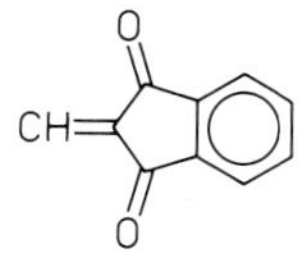	C_6H_4Br-4	IVc (60.4) m.p. 193 to 195°	[18]
43	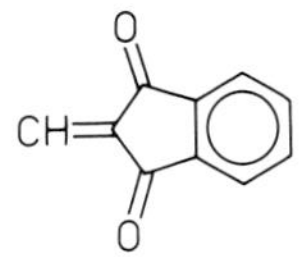	C_6H_4Br-3	IVc (44.5) m.p. 156 to 157°	[18]
44	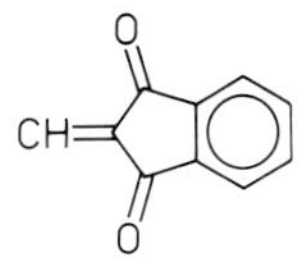	$C_6H_3Cl_2$-3,4	IVc (60) m.p. 191 to 193°	[18]
45	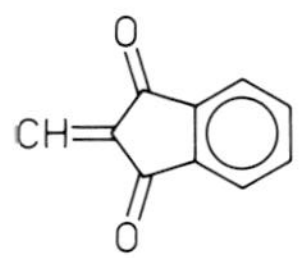	$C_6H_4CF_3$-3	IVc (57) m.p. 179 to 180°	[18]
46	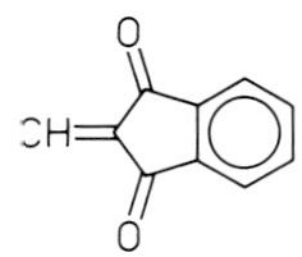	$C_6H_3(CF_3$-3)Cl-4	IVc (24) m.p. 168 to 170°	[18]

References on pp. 152/3

Table 7 (continued)

No.	$R^1-C_5H_4FeC_5H_4-R^2$ $R^1 = A-CO-R'$	R^2	method of preparation (yield in %), properties, and remarks	Ref.
47	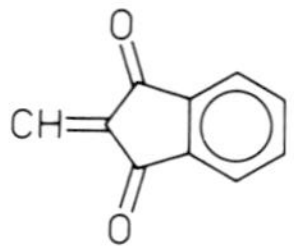	$C_6H_4OCH_3$-4	IVc (55.6) m.p. 147 to 149°	[18]
48	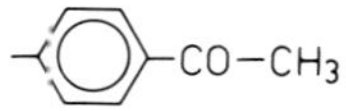	C_6H_5	VIIIa (2) orange crystals, m.p. 151 to 152° (ether/Skellysolve B) IR: as for that of $Fe(C_5H_4-C_6H_5)_2 + C_5H_5FeC_5H_4-C_6H_4COCH_3$-4	[1]
*49	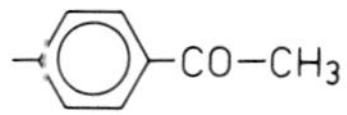	$COCH_3$	VIIa (35), see No. 50, p. 151 orange-red crystals, m.p. 140 to 143° (C_6H_6/petroleum ether) IR (Nujol): 1023 (C_5H_4); 1113 ($\nu_sC(1')-C_{acetyl}$); 1570 (γC=C of C_6H_4); 1608 and 1680 (νC=O)	[13]
*50	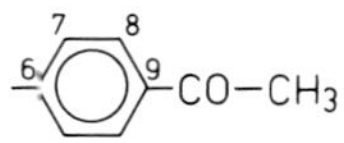	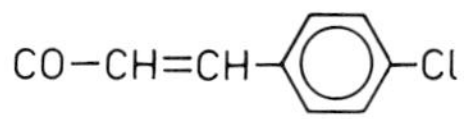	VIIa (45), see No. 49, p. 151 red crystals, m.p. 158 to 160° (benzene) IR (Nujol): 1012 (C_5H_4); 1115 ($\nu_sC(9)-C_{acetyl}$); 1569 and 1605 (γC=C of C_6H_4); 1659 and 1681 (νC=O)	[13]
51	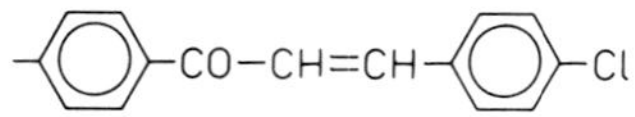	$COCH_3$	VIIa (45), for formation see No. 49, p. 151 m.p. 187 to 188° (C_6H_6) IR (Nujol): 1015 (C_5H_4); 1118 ($\nu_sC(1')-C_{acetyl}$); 1604 and 1618 (γC=C of C_6H_4); 1670 (νC=O)	[13]
52	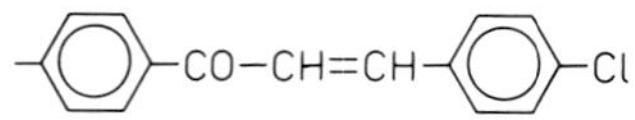	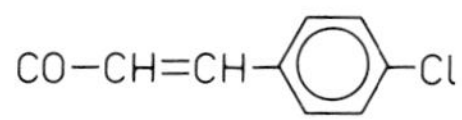	see No. 49, p. 151 crystals, m.p. 241 to 243° ($OCHN(CH_3)_2$) IR (Nujol): 1012 (C_5H_4); 1568 and 1595 (γC=C of C_6H_4); 1665 and 1676 (νC=O)	[13]
*53	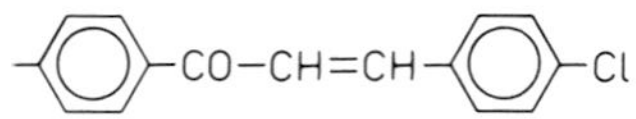	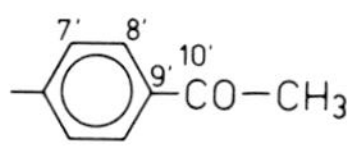	see No. 5, p. 150 deep red crystals, m.p. 209 to 210° (C_6H_6) IR (Nujol): 1017 (C_5H_4); 1118 ($\nu_sC(9')-C_{acetyl}$); 1573 and 1618 (γC=C of C_6H_4); 1664 and 1680 (νC=O) IR (CCl_4): 1665.5 (νC=O of R^1) for s-cis	[13, 23]

References on pp. 152/3

54	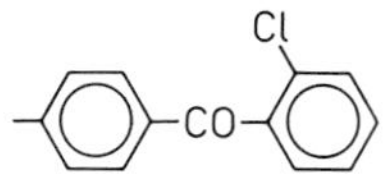	C_6H_5	VIIIb (8) ^{1}H NMR ($CDCl_3$): 4.25 (pseudotriplett, 2H), 4.35 (pseudotriplett, 2H), and 4.55 (m, 4H) (C_5H_4), 7.1 to 7.8 (13H, phenyl) mass spectrum: m/e = 476/478 $[M]^+$	[15]
*55	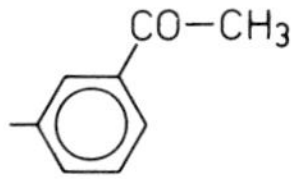	$COCH_3$	VIIa (90), see No. 57, p. 152 m.p. 76 to 77° (petroleum ether) IR (Nujol): 1018 (C_5H_4); 1115 ($\nu_sC(1')-C_{acetyl}$); 1605 (γC=C of C_6H_4); 1659 and 1685 (νC=O)	[13]
*56	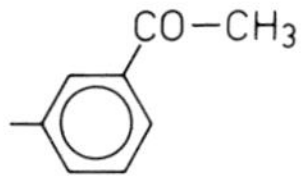	$COCH{=}CHC_6H_5$	for chemical behavior see "Further information", p. 152	[14]
*57	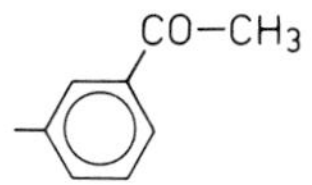	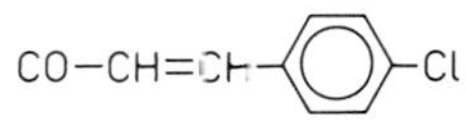	VIIa (54.3) m.p. 99 to 101° (C_6H_6/petroleum ether) IR (Nujol): 1012 (C_5H_4); 1568 and 1595 (γC=C of C_6H_4); 1655 and 1683 (νC=O)	[13]
58	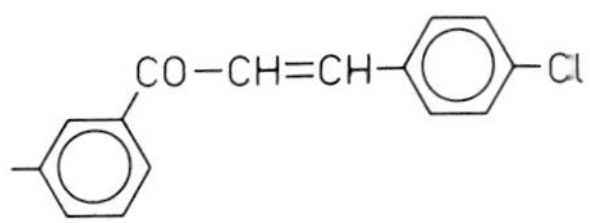	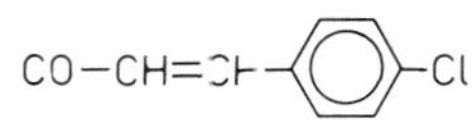	see No. 55, p. 152 violet crystals, m.p. 169 to 172° (acetone/petroleum ether) IR (Nujol): 1010 and 1111 (C_5H_4); 1570, 1595, and 1615 (γC=C of C_6H_4); 1656 and 1666 (νC=O)	[13]
*59	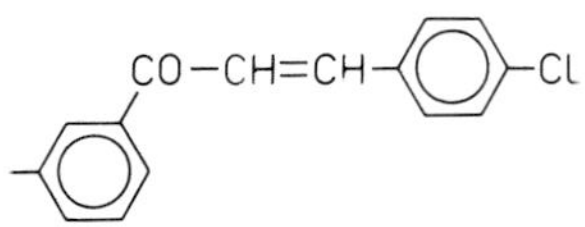	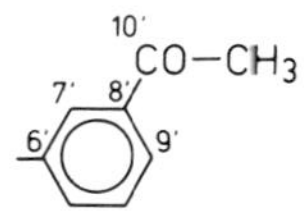	see No. 6, p. 150 m.p. 154 to 157° (C_6H_6/petroleum ether) IR (Nujol): 1012 (C_5H_4); 1113 ($\nu_sC(8')-C_{acetyl}$); 1596 and 1613 (γC=C of C_6H_4); 1670 and 1687 (νC=O)	[13]

* Further information:

$Fe(C_5H_4\text{–}CH{=}CH\text{–}CO\text{–}CH_3)_2$ (Table **7**, No. **2**) yields the compound IV via the base-catalyzed condensation with $Ru(C_5H_4\text{–}CHO)_2$ [21].

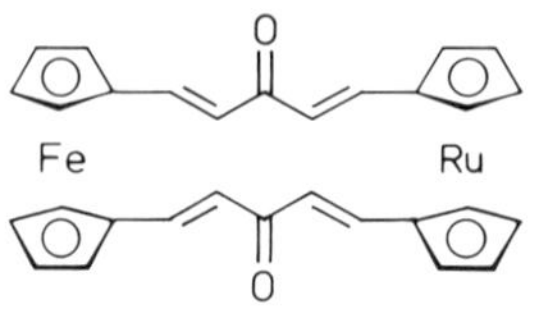

IV

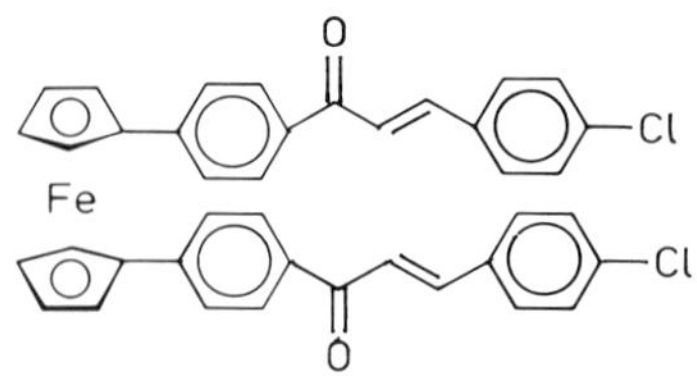

V

$Fe(C_5H_4\text{–}CH{=}CH\text{–}CO\text{–}C_6H_5)_2$ (Table **7**, No. **4**) forms in a 14% yield (referred to C_5H_6) along with ferrocene (11%) and $C_5H_5FeC_5H_4\text{–}CH{=}CHCOC_6H_5$ (9%) by addition in portions of $FeCl_2$ with ice-cooling to the cyclopentadienide ion $[C_6H_5COCH{=}CHC_5H_4]^-$ (obtained from C_5H_5Na and $C_6H_5COCH{=}CHN(CH_3)_2$ in tetrahydrofuran). The reaction is conducted 1 h at room temperature and 2 h at 40 to 50°C. In contrast, addition of $FeCl_2$ suspended in $(C_2H_5)_2NH$ gives only traces of No. 4 [6].

No. 4 dissolves at −70°C in liquid $(CH_3)_2NH$ with a dark red coloration. At −10°C a color change to yellow-orange takes place after 15 min, and after distillation off of the amine **$Fe(C_5H_4\text{–}CH{=}CH\text{–}CO\text{–}C_6H_5)_2\cdot 1.65(CH_3)_2NH$** is left. It is yellow, difficultly soluble in acetone or ether, and splits off $(CH_3)_2NH$ with re-formation of No. 4, slowly on standing, rapidly on recrystallization from a hot solution. It also decomposes considerably during chromatography, where it is stubbornly retained on Al_2O_3. The adduct melts over a wide range (150 to 178°C) and shows no melting point depression with No. 4. It dissolves in acid solutions, from which it can be recovered by rendering them alkaline. Possibly the amine is added at the double bond [6].

$Fe(C_5H_4\text{–}C_6H_4\text{–}CO\text{–}CH_3\text{-}x)_2$ (Table **7**, Nos. **5** and **6** with x = 4 and 3). No. 5 is condensed with $4\text{-}ClC_6H_4CHO$ (molar ratio 1/1) using 10% NaOH in boiling ethanol. Column chromatography on SiO_2 with 97/3 $C_6H_6/CH_3CO_2C_2H_5$ as eluent gives No. 53 (10% yield) and small quantities of an unidentified compound, probably **$Fe(C_5H_4\text{–}C_6H_4\text{–}CO\text{–}CH{=}CHC_6H_4Cl\text{-}4)_2$** (see Formula V), with a 20% recovery of No. 5. In contrast, conducting the reaction in $(CH_3)_2NCHO$ at a bath temperature of 80 to 90°C leads to cyclization of the No. 53 primarily formed via an intramolecular Michael addition to form VI in 80% yield. The reaction does not proceed further in ethanol due to the insolubility of No. 53 in that reaction medium [13].

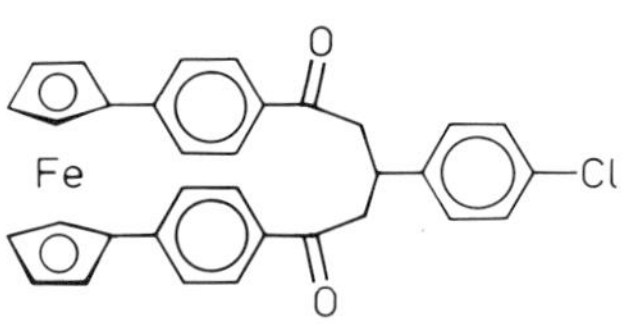

VI

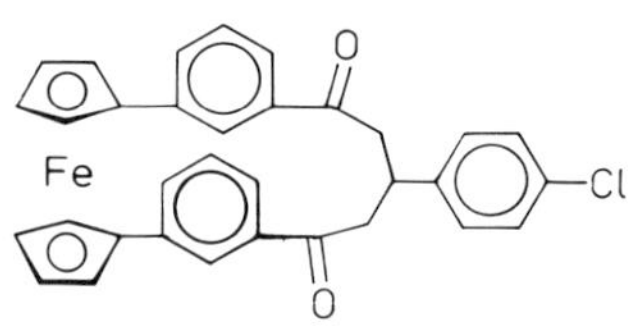

VII

No. 6 yields under the corresponding conditions in ethanol No. 59 (14.7% yield), No. 7 (4.3%), and VII (6.2%) with 6.6% recovery. In $(CH_3)_2NCHO$ the yield of VII rises to 25% [13].

$CH_3\ CO\ CH_2CH_2\text{–}C_5H_4FeC_5H_4\text{–}CO\text{–}CH_3$ (Table **7**, No. **9**) condenses with $4\text{-}ClC_6H_4CHO$ and NaOH (molar ratio 1/1/50) in C_2H_5OH at room temperature to form No. 15 in 29.2% yield and 3.7% of No. 16. These products are separable on SiO_2 chromatographic columns using 19/1 $C_6H_6/CH_3CO_2C_2H_5$ as eluent [12].

References on pp. 152/3

CH_3–CO–CH_2CH_2–$C_5H_4FeC_5H_4$–CO–CH=CHC_6H_4Cl-4 (Table **7**, No. **10**). With NaOH (molar ratio 1/7) in 1 h in boiling ethanol occur both intramolecular Michael addition to form VIII (38.7% yield) and retro-Claisen-Schmidt reaction to form No. 9 (43.4% yield). The mixture is separated on SiO_2 chromatographic columns with 97/3 $C_6H_6/CH_3CO_2C_2H_5$ with 16.6% recovery of No. 10 [12].

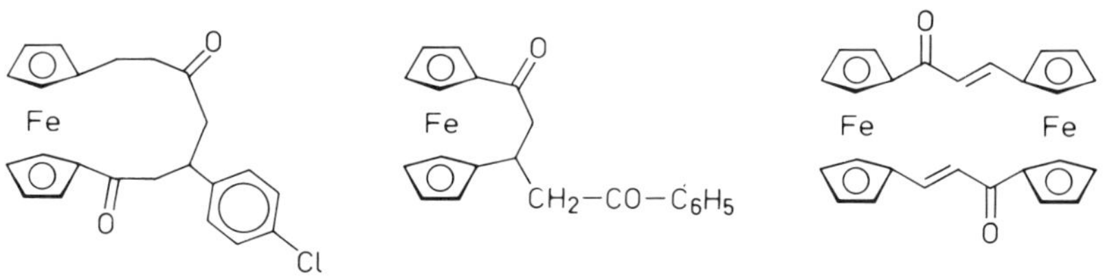

VIII IX X

4-$CH_3OC_6H_4$–CO–$(CH_2)_7$–$C_5H_4FeC_5H_4$–CO–CH=CHC_6H_5 (Table **7**, No. **13**). Refluxing No. 13 (mixed with its 1,3-isomer) with NaOH in ethanol for 2 h produces no intramolecular cyclization. Column chromatography on SiO_2 with $C_6H_6/CH_3CO_2C_2H_5$ (20/1 to 10/1) as eluent yields instead along with 30% of the starting mixture the compounds No. 12 and its 1,3-isomer (32% joint yield) formed by a retro-Claisen-Schmidt reaction [20].

4-ClC_6H_4CH=CH–CO–CH_2CH_2–$C_5H_4FeC_5H_4$–CO–CH_3 (Table **7**, No. **15**) yields with NaOH in boiling ethanol a number of products, among which are No. 9 (27.5% yield) formed by a retro-Claisen-Schmidt reaction, traces of No. 10 from a trans-Claisen-Schmidt reaction, and traces of VIII, the product of an intramolecular Michael addition [12].

C_6H_5–CO–CH=CH–$C_5H_4FeC_5H_4$–CO–CH_3 (Table **7**, No. **30**). KOH in boiling ethanol gives slight traces of IX from an intramolecular Michael addition plus X. The latter can form by autocondensation of CH_3–CO–$C_5H_4FeC_5H_4$–CHO, which has been neither isolated or identified, but could have formed from No. 30 by a retro-Claisen-Schmidt reaction. Changing the reaction medium to $(CH_3)_2NCHO$ raised the yield of IX somewhat [12]. Acetylation experiments with CH_3COCl and $AlCl_3$ (molar ratio 1/2/4) lead neither to No. 32 nor to any other acetylation product. Only two unidentified compounds in small yields and a 60% recovery of No. 30 were found [17].

C_6H_5CH=CH–CO–CH=CH–$C_5H_4FeC_5H_4$–CO–CH_3 (Table **7**, No. **31**) can not be cyclized by a base-catalyzed reaction. Instead an exclusively polymeric material forms as a result of a retro-Claisen-Schmidt reaction. Trial of a catalysis by acids led only to recovery of the starting compound [16].

C_6H_5–CO–CH=CHCH=CH–$C_5H_4FeC_5H_4$–CO–CH_3 (Table **7**, No. **33**). In order to avoid the retro-Claisen-Schmidt reaction which possibly would proceed in the presence of a basic catalyst instead of the desired intramolecular Michael addition, cyclization was attempted using an acidic catalyst. However, no ferrocenophane derivatives formed, and only small quantities of the starting material were recovered [16].

4-CH_3–CO–C_6H_4–$C_5H_4FeC_5H_4$–CO–CH_3 (Table **7**, No. **49**) on condensation with 4-ClC_6H_4CHO (molar ratio 1/1) for 4 h in boiling alcohol using 10% NaOH gives along with 26% recovery of No. 49 the compounds No. 52 (25% yield), No. 50 (13%), and small quantities of No. 51. The red precipitate formed was extracted with benzene, the residue (No. 52) recrystallized from $(CH_3)_2NCHO$, and the extract chromatographed on SiO_2. Small amounts of Nos. 52, 50, 51, and 49 were eluted in succession by 97/3 $C_6H_6/CH_3CO_2C_2H_5$ [13].

4-CH_3–CO–C_6H_4–$C_5H_4FeC_5H_4$–CO–CH=CHC_6H_4Cl-4 (Table **7**, No. **50**). No intramolecular cyclization takes place with NaOH in boiling ethanol, rather a retro-Claisen-Schmidt reaction with

References on pp. 152/3

formation of No. 49 in 23% yield. Autocondensation of No. 50 could, according to the elemental analysis, have formed the additional, unidentified products eluted with 97/3 $C_6H_6/CH_3CO_2C_2H_5$ after column chromatographic separation on Al_2O_3 (15.7% recovery of No. 50) [13].

4-(4-ClC$_6$H$_4$CH=CH–CO)–C$_6$H$_4$–C$_5$H$_4$FeC$_5$H$_4$–C$_6$H$_4$–CO–CH$_3$-4 (Table **7**, No. **53**) is difficultly soluble in C_2H_5OH. Bases cyclize it to VI (see p. 150) in a large volume of C_2H_5OH. Conducting the reaction in $(CH_3)_2NCHO$ gives excellent yields [13]. Kinetic studies in $CH_3OH/NaOH$ in the region from 20 to 60°C yield for the cyclization at 314 K the following activation parameters: $\Delta H^{\ddagger} = 20.59 \pm 2.40$ kcal/mol (86.21 ± 10.05 kJ/mol), $\Delta S^{\ddagger} = -19.89 \pm 12.7$ cal·mol^{-1}·K^{-1} (-83.27 ± 53.17 J·mol^{-1}·K^{-1}) [14].

3-CH$_3$–CO–C$_6$H$_4$–C$_5$H$_4$FeC$_5$H$_4$–CO–CH$_3$ (Table **7**, No. **55**). Condensation with 4-ClC_6H_4CHO in boiling $NaOH/H_2O/CH_3OH$ and chromatographic separation on Al_2O_3 columns with 97/3 $C_6H_6/CH_3CO_2C_2H_5$ as eluent yield compound XI with R = Cl (2.3% yield), small quantities of unidentified products, and No. 58 (23% yield), with only a slight recovery of No. 55 [13].

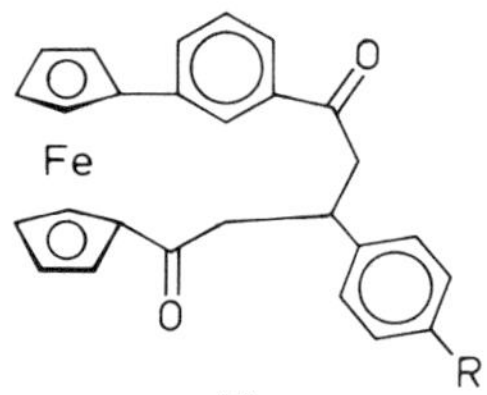

XI

3-CH$_3$–CO–C$_6$H$_4$–C$_5$H$_4$FeC$_5$H$_4$–CO–CH=CHC$_6$H$_5$ (Table **7**, No. **56**) cyclizes to XI (R = H) during the intramolecular Michael addition. Kinetic studies at 20 to 60°C in $CH_3OH/NaOCH_3$ yield the activation parameters $\Delta H^{\ddagger} = 16.37 \pm 1.19$ kcal/mol (68.54 ± 4.98 kJ/mol), $\Delta S^{\ddagger} = -28.79 \pm 8.5$ cal·mol^{-1}·K^{-1} (-120.54 ± 35.59 J·mol^{-1}·K^{-1}) for the cyclization at 314 K [14].

3-CH$_3$–CO–C$_6$H$_4$–C$_5$H$_4$FeC$_5$H$_4$–CO–CH=CHC$_6$H$_4$Cl-4 (Table **7**, No. **57**). Reaction with NaOH (molar ratio 2/15) for 30 min in boiling C_2H_5OH involves both an intramolecular Michael addition to form XI, with R = Cl, in 11% yield and also a retro-Claisen-Schmidt reaction to form No. 55 in 16.5% yield. The products are separated from each other and No. 57 (16.5% recovery) by column chromatography on Al_2O_3 with 97/3 $C_6H_6/CH_3CO_2C_2H_5$ as eluent. Conducting the reaction in $(CH_3)_2NCHO$ in the presence of 4-ClC_6H_4CHO at 80 to 90°C bath temperature increases the yield of XI (R = Cl) to 33% [13]. Studies of the kinetics of the cyclization in $CH_3OH/NaOCH_3$ at 20 to 60°C yield for the cyclization at 314 K the activation parameters $\Delta H^{\ddagger} = 18.72 \pm 0.80$ kcal/mol (78.38 ± 3.35 kJ/mol) and $\Delta S^{\ddagger} = -20.17 \pm 6.7$ cal·mol^{-1}·K^{-1} (-84.45 ± 28.05 J·mol^{-1}·K^{-1}) [14].

3-(4-ClC$_6$H$_4$CH=CH–CO)–C$_6$H$_4$–C$_5$H$_4$FeC$_5$H$_4$–C$_6$H$_4$–CO–CH$_3$-3 (Table **7**, No. **59**) is cyclized intramolecularly by NaOH to give VII (see p. 150). The yield is better in $(CH_3)_2NCHO$ than in C_2H_5OH, but even in the former only moderate. Condensation with 4-ClC_6H_4CHO gives No. 7 [13].

References:

[1] M. Rosenblum (J. Am. Chem. Soc. **81** [1959] 4530/6). – [2] M. Rosenblum, W. G. Howells, A. K. Banerjee, C. Bennett (J. Am. Chem. Soc. **84** [1962] 2726/32). – [3] M. Rosenblum, J. O. Santer, W. G. Howells (J. Am. Chem. Soc. **85** [1963] 1450/8). – [4] C. U. Pittman, J. A. Eikenberry (J. Alabama Acad. Sci. **36** [1965] 338/40). – [5] J. Tirouflet, C. Moïse (Compt. Rend. C **262** [1966] 1889/90).

[6] A. N. Nesmeyanov, M. I. Rybinskaya, L. M. Korneva (Izv. Akad. Nauk SSSR Ser. Khim. **1967** 2649/52; Bull. Acad. Sci. USSR Div. Chem. Sci. **1967** 2523/6). – [7] A. N. Nesmeyanov, M. I. Rybinskaya, L. M. Korneva, M. P. Kumpolova (Izv. Akad. Nauk SSSR Ser. Khim. **1967** 2642/8; Bull. Acad. Sci. USSR Div. Chem. Sci. **1967** 2516/22). – [8] C. Moïse, J. Tirouflet, H. Singer (Bull. Soc. Chim. France **1969** 1182/7). – [9] A. Perjéssy, Š. Toma (Chem. Zvesti **23** [1969] 533/9). – [10] Š. Toma (Collection Czech. Chem. Commun. **34** [1969] 2235/48).

[11] A. Kasahara, T. Izumi, G. Saito, M. Yodono, R. Saito, Y. Goto (Bull. Chem. Soc. Japan **45** [1972] 895/900). – [12] Š. Toma, M. Sališová (J. Organometal. Chem. **57** [1973] 191/8). – [13] Š. Toma, M. Sališová (J. Organometal. Chem. **57** [1973] 199/205). – [14] M. Livař, M. Sališová, Š. Toma (Collection Czech. Chem. Commun. **40** [1975] 415/9). – [15] V. Rapić, K. Schlögl, B. Steinitz (J. Organometal. Chem. **94** [1975] 87/98).

[16] Š. Toma, M. Sališová (Acta. Fac. Rerum Nat. Univ. Comenianae Chim. No. 21 [1975] 59/70). – [17] Š. Toma, M. Sališová, E. Solčiánová (J. Organometal. Chem. **90** [1975] 335/41). – [18] Š. Toma, A. Maholányiová, E. Solčaniová (Acta Fac. Rerum Nat. Univ. Comenianae Chim. No. 24 [1976] 53/80). – [19] A. Kasahara, T. Izumi, M. Maemura (Bull. Chem. Soc. Japan **50** [1977] 1021/2). – [20] M. Sališová, E. Solčaniová, Š. Toma (Acta Fac. Rerum Nat. Univ. Comenianae Chim. No. 28 [1980] 103/26).

[21] S.-I. Kamiyama, A. Kasahara, T. Izumi, I. Shimizu, H. Watanabe (Bull. Chem. Soc. Japan **54** [1981] 2079/82). – [22] Š. Toma, M. Sališová (unpublished results from [13]). – [23] A. Perjéssy, Š. Toma (Collection Czech. Chem. Commun. **48** [1983] 1635/40).

5.2.3.5.1.8 Complexes and Ions Derived from 1,1′-Ketones

Protonation products are discussed in connection with the individual ketones. The salts $Na_2[Fe(C_5H_4{-}COCHCHO)_2]$ and $Na_2[Fe(C_5H_4{-}COCHCOC_6H_5)_2]$ are treated along with the ketones $Fe(C_5H_4{-}CO{-}CH_2CHO)_2$ (Table 3, No. 44, p. 70) and $Fe(C_5H_4{-}CO{-}CH_2COC_6H_5)_2$ (Table 3, No. 48, p. 71). Complexes, ions, and salts derived from $Fe(C_5H_4{-}CO{-}CH_3)_2$ are covered in Section 5.2.3.5.1.1.5.2, pp. 48/53. The adduct $Fe(C_5H_4{-}CH{=}CH{-}CO{-}C_6H_5)_2 \cdot x(CH_3)_2NH$ is found on p. 150.

Complexes

$Fe(C_5H_4{-}CO{-}CH{=}CHC_6H_5Fe(CO)_3)_2$ (see Formula I) forms from $Fe(C_5H_4{-}CO{-}CH{=}CHC_6H_5)_2$ on heating with $Fe_3(CO)_{12}$ in C_6H_6 in an Ar stream in 36% yield. The solution is concentrated and the complex is recovered by thin layer chromatography on Al_2O_3 with 10/1 petroleum ether/acetone as eluent. Recrystallization from hexane gives orange-brown needles that melt at 130 to 135°C with decomposition. The IR spectrum in KBr shows ν(C≡O) bands at 1990, 2010, and 2070 cm^{-1} [5].

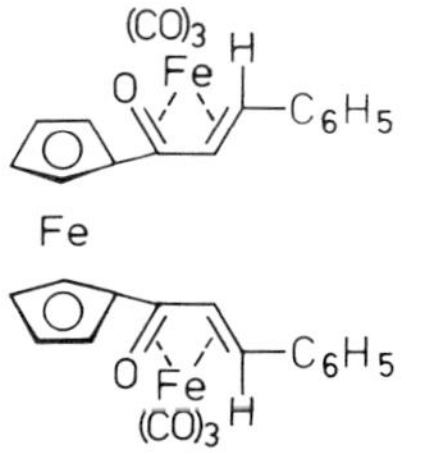

I

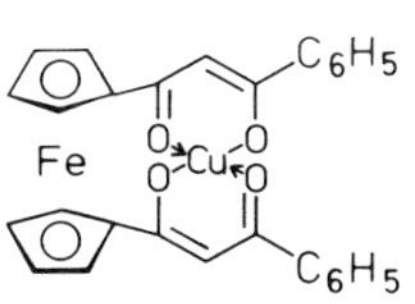

II

References on p. 156

$Fe(C_5H_4\text{–}CO\text{–}CH{=}C(O)C_6H_5)_2Cu$ (see Formula II). This chelate complex precipitates when $Fe(C_5H_4\text{–}CO\text{–}CH_2COC_6H_5)_2$ reacts in boiling acetone with a warm, saturated, aqueous solution of $Cu(O_2CCH_3)_2$ in a large excess and is refluxed for 15 min (88% yield) [1] or when the aqueous $Cu(O_2CCH_3)_2$ solution is added to the ketone in CH_3OH in a 1/1 molar ratio [6]. The yellow powder recrystallized from petroleum ether melts above 300°C and has a molecular weight of 545 (determined by the Rast method in camphor) [6], 518.25 (average of four measurements) [1], 538.3 (calculated value) [6].

The ^{57}Fe Mössbauer spectrum shows the following isomer shifts δ (referred to $Na_2[Fe(CN)_5NO]\cdot 2H_2O$) and quadrupole splittings Δ: at 78 K $\delta = 0.78$, $\Delta = 2.32$ mm/s; at 298 K $\delta = 0.75$, $\Delta = 2.21$ mm/s [6]. IR bands are found at 829.9, 948.8, 1023.5, 1037.3, 1067.2, 1084.6, 1183.4, 1233, 1303.8, 1349.5, 1379.3, 1394.7, 1410.4, 1453.5, 1490.3, 1538.5, 1594.5, and 1718.2 cm^{-1} [1]. In KBr $\nu(C{=}O) = 1525\ cm^{-1}$. The electronic spectrum in $CHCl_3$: λ_{max} (log ε) = 354 (4.81) (ligand), 487 (3.74), 625 (1.70) ($B_{2g}(D) \leftarrow B_{3g}$ or $A_g \leftarrow$) nm. The dipole moment measured at 298 K is $\mu_{eff} = 1.90\ \mu_B$ (Gouy balance) vs. a theoretical value of 1.78 μ_B. A distorted quadratic plane symmetry (C_{2h}) has been assumed for Cu^{II} [6].

The starting ketone is recovered by stirring 12 h with 20% H_3PO_4 [1].

Ions

Ferrocenium Ions

For $[Fe(C_5H_4\text{–}CO\text{–}C_6H_5)_2]^+$ see p. 59, for $[R'\text{–}CO\text{–}C_5H_4FeC_5H_4\text{–}R^2]^+$ see the parent ketones in 5.2.3.5.1.6, pp. 111/37.

$[Fe(C_5H_4\text{–}CO\text{–}C_3H_7\text{-n})_2]^+$ displays isotropic contact shifts in acetone and 2 N D_2SO_4 at room temperature of 0.5, 8.0, and −18.1 (line width 1 kHz) ppm, which were assigned to the CH_3, $COCH_2$, and ring protons, respectively. The ion exists in equilibrium with $Fe(C_5H_4\text{–}CO\text{–}C_3H_7\text{-n})_2$ in solution and is substantially more unstable than $[Fe(C_5H_4\text{–}C_4H_9)_2]^+$ [2].

Ketyls $[R\text{–}CO\text{–}C_5H_4FeC_5H_4\text{–}C(R)O]^{\dot{-}}$

See p. 58 for $[C_6H_5\text{–}CO\text{–}C_5H_4FeC_5H_4\text{—}C(C_6H_5)O]^{\dot{-}}$.

$[4\text{-}CF_3C_6H_4\text{–}CO\text{–}C_5H_4FeC_5H_4\text{–}C(C_6H_4CF_3\text{-}4)O]^{\dot{-}}$ is formed from $Fe(C_5H_4\text{–}CO\text{–}C_6H_4CF_3\text{-}4)_2$ by reduction with Na/K alloy or electrochemically in $(CH_3)_2NCHO$. The hfs splitting in the ESR spectrum indicates six equivalent F atoms; $A_F = 1.0$ G, $g = 2.0086$. The equivalence of the six F atoms presupposes a rapid interannular conduction of the electrons via the Fe atom (see III) [3].

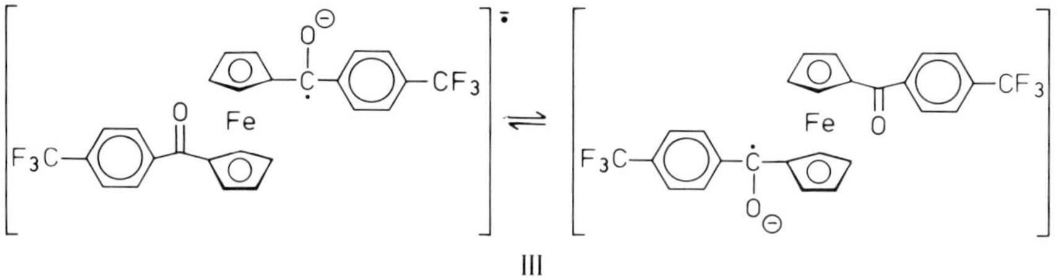

III

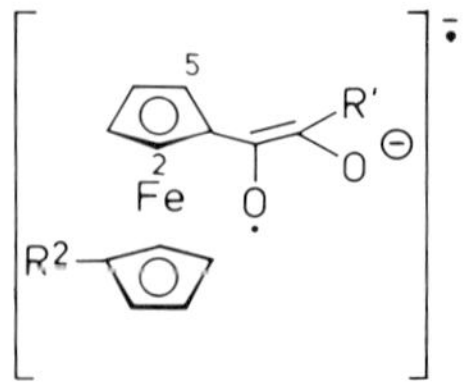

IV

Semidiones of the Type $[R^2–C_5H_4FeC_5H_4–CO{=}C(R')O]^{\dot{-}}$

The stable radical anions IV listed in the following table were obtained in an H-cell by the 1s action of atmospheric oxygen on the enolate anion of $R^2–C_5H_4FeC_5H_4–CO–CH_2–R'$ in dry $(CH_3)_2SO$ in the presence of a three- to fourfold excess of $KOC(CH_3)_3$. Oxidation of $n\text{-}C_3H_7–CO–C_5H_4FeC_5H_4–CO–C_2H_5$ with a larger excess of $KOC(CH_3)_3$ or with its replacement by $CsOC(CH_3)_3$ forms Nos. 5 and 10 in an 8.6/1 ratio. $(CH_3)_2CHCH_2–CO–C_5H_4FeC_5H_4–CO–C_2H_5$ gives Nos. 6 and 11. ESR studies on IV yield the following g values and hfs coupling constants A_H for the α protons of the R′ and the cyclopentadienyl protons H-2 and H-5 [4]:

ion	$[R^2–C_5H_4FeC_5H_4–CO{=}C(R')O]^{\dot{-}}$		g value	A_H in Gauss	
No.	R′	R^2		α–H of R′	H-2,5 of C_5H_4
1	CH_3	$C(CH_3)_3$	2.00706	4.20	0.50
2	CH_3	Br	2.00700	3.95	0.55
3	CH_3	$C_6H_4Cl\text{-}4$	2.00749	4.00	0.52
4	CH_3	$CO–CH_3$	2.00728	3.80	0.55
5	CH_3	$CO–C_3H_7\text{-}n$	2.00728	3.80	0.55
6	CH_3	$CO–CH_2CH(CH_3)_2$	2.00728	3.82	0.55
7	CH_3	$CO–C(CH_3)_3$	2.00728	3.83	0.52
8	CH_3	$CO–C_6H_4CH_3\text{-}4$	2.00739	3.72	0.55
9	CH_3	$CO–C_{10}H_{15}$ (adamantyl)	2.00728	3.85	0.52
10	CH_2CH_3	$CO–C_2H_5$	2.00729	3.55	0.55
11	$CH(CH_3)_2$	$CO–C_2H_5$	2.00729	1.55	0.55

Protons of the $R^2–C_5H_4$ ring do not interact with the unpaired spin; nevertheless, polar effects of R^2 are transmitted via the cyclopentadienyl ring [4].

The following times have been measured for the complete H/D exchange of the α protons of R′ in $(CD_3)_2SO/KOC(CH_3)_3$ at room temperature [4]:

ion No.	R′	R^2	time
3	$CH_3 \rightarrow CD_3$	$C_6H_4Cl\text{-}4$	40 ± 5 min
5	$CH_3 \rightarrow CD_3$	$CO–C_3H_7\text{-}n$	15 ± 5 min
7	$CH_3 \rightarrow CD_3$	$CO–C(CH_3)_3$	15 ± 5 min
9	$CH_3 \rightarrow CD_3$	$CO–C_{10}H_{15}$	15 ± 5 min
10	$CH_2CH_3 \rightarrow CD_2CH_3$	$CO–C_2H_5$	10 h
11	$CH(CH_3)_2 \rightarrow CD(CH_3)_2$	$CO–C_2H_5$	72 h

The exchange generally occurs slowly via the intermediate product V. It undergoes a distinct acceleration for Nos. 5, 7, and 9 with $R^2 = CO–R''$, since these can be stabilized interannularly by the formation of VI. The small exchange rates of Nos. 10 and 11 are due to the formation necessary of $–\overset{(-)}{C}HCH_3$ and $–\overset{(-)}{C}(CH_3)_2$, respectively [4].

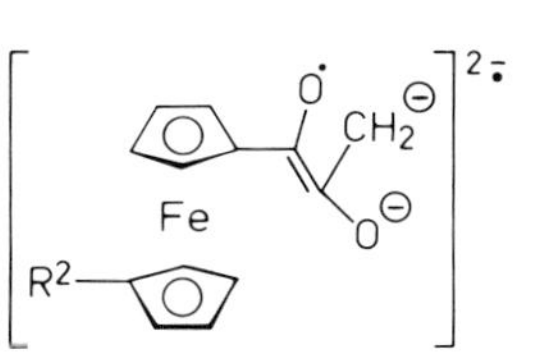

V

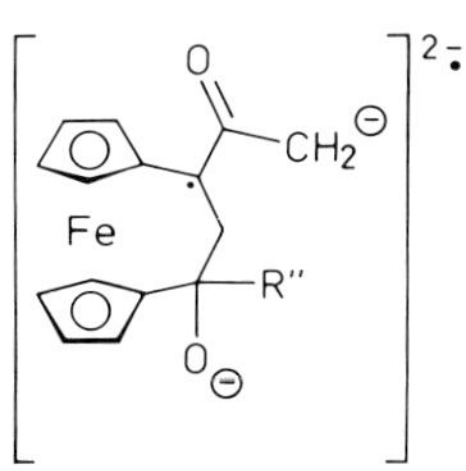

VI

References on p. 156

References:

[1] C. R. Hauser, C. E. Cain (J. Org. Chem. **23** [1958] 1142/6). – [2] H. P. Fritz, H. J. Keller, K. E. Schwarzhans (J. Naturforsch. **21 b** [1966] 809/10). – [3] J. J. McDonnell, G. Capen, R. Micheal-son (Tetrahedron Letters **1969** 4251/4). – [4] J. J. McDonnell, D. J. Pochopien (J. Org. Chem. **36** [1971] 2092/8). – [5] A. N. Nesmeyanov, G. B. Shul'pin, L. V. Rybin, N. T. Gubenko, M. I. Rybinskaya, P. V. Petrovskii, V. I. Robas (Zh. Obshch. Khim. **44** [1974] 2032/41; J. Gen. Chem. [USSR] **44** [1974] 1994/2001).

[6] H.-H. Wei, D.-M. Horng (Transition Metal Chem. [Weinheim] **6** [1981] 319/20).

5.2.3.5.2 1,2-Compounds

Regarding ferrocene chirality and the centrochiral nomenclature used throughout, see "Ferrocene" 7, 1980, pp. 4/7 [German text], 1/4 [English translation].

5.2.3.5.2.1 1,2-Diacetylferrocene $C_5H_5FeC_5H_3(CO–CH_3)_2$-1,2

Preparation

Oxidation of $C_5H_5FeC_5H_3(CHOH–CH_3)_2$-1,2 as a diastereomeric mixture of I, II, and III (cf. "Ferrocene" 7, 1980, p. 205) with excess activated MnO_2 for 4 h in CH_2Cl_2 gives a quantitative yield [12]. If not the complete mixture but only the racemate III is reacted with MnO_2 in a 1/5 molar ratio, the 1,2-diacetylferrocene forms in small amounts along with IV [11]. Starting from the meso form II gives traces along with V [12]. Both IV and V can be further oxidized to $C_5H_5FeC_5H_3(COCH_3)_2$-1,2, the compound V reacting more easily than IV [11]. The oxidation of $C_5H_5FeC_5H_3(COCH_3$-1)C_2H_5-2 is conducted with a fivefold excess of MnO_2 in boiling methylcy-clohexane [5].

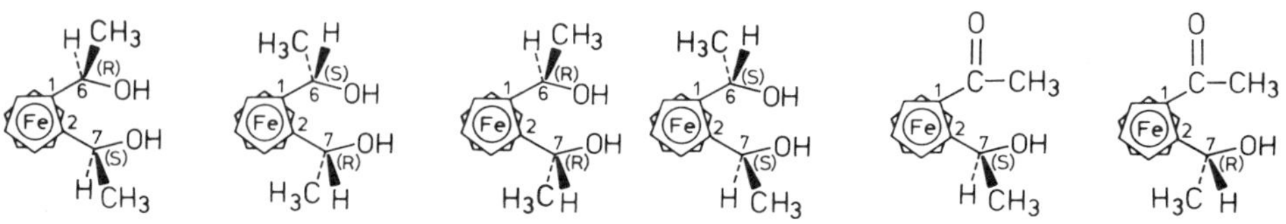

m.p. 152°C — I; m.p. 134°C — II; a m.p. 110°C b — III; m.p. 72°C — IV; m.p. 108°C — V

Friedel-Crafts acetylation of ferrocene by adding ferrocene in CH_2Cl_2 to a mixture of CH_3COCl and $AlCl_3$ (molar ratio 1/2.8/2.5) in CH_2Cl_2 at room temperature forms in addition to the main product, $Fe(C_5H_4–COCH_3)_2$, as the only homoannular disubstituted compound $C_5H_5FeC_5H_3(COCH_3)_2$-1,2 [13], see also [3], and not, as conjectured by [1], the 1,3-isomer. The yield amounts to 1 to 2% [2, 4, 10]. It rises to above 8% when a 4.72 M solution of ferrocene in CH_2Cl_2 is added at a rate of 6 mL/min to a CH_2Cl_2 solution of $AlCl_3 + CH_3COCl$ (molar ratio 1/~11/~12) and the reaction is subsequently continued for 4 h, all at room temperature [15]. Acetylation of ferrocene with $(CH_3CO)_2O$ (molar ratio 1/2) at 0°C in CH_2Cl_2 by saturating the solution with BF_3 is recommended due to the easier processing of the raw reaction product. However, even small concentrations of HF hinder the formation of the diacetylation products of the ferrocene [8].

References on p. 158

Ferrocene, $C_5H_5FeC_5H_4$–$COCH_3$, $C_5H_5FeC_5H_3(COCH_3)_2$-1,2, and its 1,1′-isomer are separated by chromatography on Al_2O_3 columns using as eluent an ether/benzene mixture [4] or in succession petroleum ether, 1/1, and 1/3 ether/petroleum ether [8]. In thin layer chromatography on Al_2O_3 with 60/1 ether/H_2O the R_f values are 0.68 for $C_5H_5FeC_5H_4$–$COCH_3$, 0.32 for the 1,2-compound, and 0.19 for its 1,1′-isomer [15]. Due to the large excess of the 1,1′-isomer on preparation by the Friedel-Crafts method, it is recommended to crystallize out the major part of it from the CH_2Cl_2 solution by addition of cyclohexane and concentrating the solution [4]. Alternatively, the 1,2-isomer can be extracted from the raw product with petroleum ether together with $C_5H_5FeC_5H_4$–$COCH_3$ and ferrocene, with only small amounts of the 1,1′-compound appearing in the extract [8]. The solvents are removed from both the mother liquor and the extract by vacuum distillation before the chromatographic processing [4, 8].

Properties

The bright yellow [15] crystalline compound melts at 92 to 94°C (hexane) [12], 93 to 94°C [15], 94 to 95°C (petroleum ether) [8], 96 to 97°C (ether) [4], 105°C [10].

The ^1HNMR spectrum in $CDCl_3$ exhibits the following shifts δ in ppm:

solvent	CH_3	C_5H_5	C_5H_3 H-4	C_5H_3 H-3,5	Ref.
$CDCl_3$	2.49(s)	4.28(s)	4.63(t)	4.91(d)	[14]
$CDCl_3$	2.50(s)	4.28(s)	4.62(t)	4.88(d)	[10]
$CDCl_3$	2.55(s)	4.30(s)	4.65(t)	4.95(d)	[12]
liquid SO_2	2.5(s)	4.41(s)	4.81(t)	5.205(d)	[7]

IR bands in $CHCl_3$ are located at 917, 1004, 1109, and 1114 cm^{-1} [4]. The electronic spectrum shows absorptions at $\lambda_{max}(\log\varepsilon) = 260(4.05), 330(3.16), 370(3.19), 450(2.80)$ nm [1]. In absolute C_2H_5OH $\lambda_{max}(\varepsilon) = 340$sh (1080), 370(1310), and 451(690) nm [6].

Solubility. Chemical Behavior

The compound is more easily soluble in aliphatic hydrocarbons than its 1,1′-isomer [8].

In contrast to $Fe(C_5H_4$–$COCH_3)_2$ it is decomposed by a 632.8 nm He/Ne laser beam [8]. No pyrolysis occurs in the vapor phase at 10 Torr N_2 within 1.5 h at a 100°C vessel temperature [1]. Sunlight photolyzes a solution in benzene to form VI with $R = CH_3$ [16]. Extended Hückel theory quantum mechanical calculations for $C_5H_5FeC_5H_3(COCH_3)_2$-1,2 in acidic media also yield this decomposition product along with $[C_5H_5Fe]^+$; they indicate that in contrast there should be no decomposition in neutral media [17].

Oxidation with NaOCl in H_2O/CH_3OH produces $C_5H_5FeC_5H_3(CO_2H)_2$-1,2; in the absence of water $C_5H_5FeC_5H_3(CO_2H$-1)$COCH_3$-2 forms as the main product [2]. From the ability of the dicarboxylic acid to form an anhydride Richards, Curphey [2], independent of [3] (see p. 158), have identified the 1,2-positions of the substituents in $C_5H_5FeC_5H_3(COCH_3)_2$, for which Rosenblum [1] initially conjectured a 1,3-structure. Exposure of a solution in $(CH_3)_2SO$ in the presence of $KOC(CH_3)_3$ for 1 s to atmospheric air forms the stable radical anion VII [10].

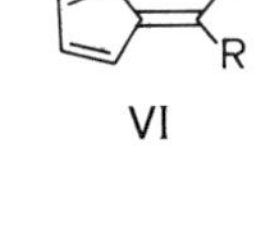

VI

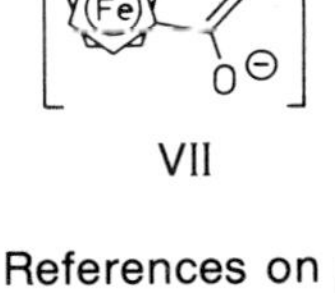

VII

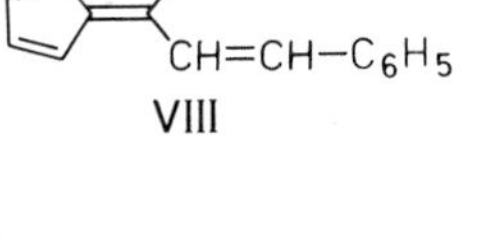

VIII

References on p. 158

Reduction with $LiAlH_4/AlCl_3$ in ether yields $C_5H_5FeC_5H_3(C_2H_5)_2$-1,2 [15]. Aqueous KBH_4 solution forms in CH_3OH at room temperature $C_5H_5FeC_5H_3(CHOH–CH_3)_2$-1,2, which consists of 5% I, 50% II, and 45% IIIa + b [11, 12]. Since the ν(OH) bands at 3460 and 3585 cm^{-1} of II and III presuppose the 1,2-positions, the same position must also be valid for the CH_3CO groups [3], see also p. 157.

Further acetylation fails with a gross excess of $CH_3COCl/AlCl_3$ in CH_2Cl_2 both at room temperature and at 40°C [1, 4, 8].

Condensation with C_6H_5CHO in a 1/1 molar ratio in $NaOH/H_2O/C_2H_5OH$ proceeds smoothly at room temperature to form $C_5H_5FeC_5H_3(COCH{=}CHC_6H_5$-1)$COCH_3$-2 (25% yield) and $C_5H_5FeC_5H_3$–$(COCH{=}CHC_6H_5)_2$-1,2 (22% yield) as well as their decomposition products VIII with R = CH_3 (7%) and $CH{=}CHC_6H_5$ (19%). The latter form even in the dark [16]. The corresponding reaction with $C_6H_5CH{=}CHCHO$ leads only to VI with R = $CH{=}CHCH{=}CHC_6H_5$. An analogous primary product, $C_5H_5FeC_5H_3(COCH{=}CHCH{=}CHC_6H_5$-1)$COCH_3$-2, apparently also forms here; it, however, is less stable than $C_5H_5FeC_5H_3(COCH{=}CHC_6H_5$-1)$COCH_3$-2 and decomposes spontaneously to the fulvene derivative [16].

Gaseous BF_3 forms a red-brown adduct [8].

$AgBF_4$ (molar ratio 1/1) in dry ether at room temperature immediately precipitates the red complex **[$C_5H_5FeC_5H_3(COCH_3)_2$-1,2·$AgBF_4$]$_n$**, which due to its insolubility in all common solvents and its nonvolatility appears to be oligomeric if not polymeric. Its formation displaces the ν(C=O) band of the $C_5H_5FeC_5H_3(COCH_3)_2$-1,2 by around 50 cm^{-1} toward lower wave numbers, which indicates a coordination of the Ag with both O atoms. With use of a slight excess of $AgBF_4$ the complex decomposes immediately with formation of a brown, Ag-free solid, which dissolves in liquid SO_2 with a brown coloration. This solution is also obtained with liquid SO_2 plus [$C_5H_5FeC_5H_3(COCH_3)_2$-1,2·$AgBF_4$]$_n$, with the Ag depositing out as a mirror. The brown, extremely moisture-sensitive solid is diamagnetic at room temperature, the ν(C=O) vibration is shifted vs. that of the complex by around a further 69 cm^{-1} toward lower wave numbers. The ^1H NMR spectrum in liquid SO_2 with signals at δ = 2.85 (s, 6H), 4.87 (s, 5H), and 6.17 (broad m, 3H) ppm resembles that of $C_5H_5FeC_5H_3(COCH_3)_2$-1,2. Apparently this tetrafluoroborate is possibly a carbenium salt and not a ferrocenium salt like $[Fe(C_5H_4–COCH_3)_2]BF_4$ (see p. 52) and $[C_5H_5FeC_5H_3(COCH_3)_2$-1,3$]BF_4$ [9].

References:

[1] M. Rosenblum (Diss. Harvard Univ. 1953, p. 108). – [2] J. H. Richards, T. J. Curphey (Chem. Ind. [London] **1956** 1456). – [3] D. S. Trifan, J. L. Weinmann, L. P. Kuhn (J. Am. Chem. Soc. **79** [1957] 6566/7). – [4] M. Rosenblum, R. B. Woodward (J. Am. Chem. Soc. **80** [1958] 5443/9). – [5] A. F. Ellis (Diss. Univ. Illinois 1963 from Diss. Abstr. **24** [1963] 510).

[6] T. H. Barr, W. E. Watts (J. Organometal. Chem. **15** [1968] 177/85). – [7] P. Carty (Diss. Nottingham Univ. 1968, p. 82 from [9]). – [8] P. Carty, M. F. A. Dove (J. Organometal. Chem. **21** [1970] 195/201). – [9] P. Carty, M. F. A. Dove (J. Organometal. Chem. **28** [1971] 125/32). – [10] J. J. McDonnell, D. J. Pochopien (J. Org. Chem. **37** [1972] 4064/7).

[11] C. Moïse, Y. Mugnier (Tetrahedron Letters **1972** 1845/8). – [12] C. Moïse, Y. Mugnier (Bull. Soc. Chim. France **1973** 1382/4). – [13] E. V. Leonova, N. S. Kochetkova, E. I. Fedin, P. V. Petrovskii (Khim. Metal. Met. Khim. Tekhnol. Silikatov 11th Mendeleevsk. S'ezd Obshch. Prikl. Khim., Moscow 1974 [1975], pp. 98/9; C.A. **84** [1976] No. 164987). – [14] A. N. Nesmeyanov, E. V. Leonova, E. I. Fedin, P. V. Petrovsky, N. S. Kochetkova (J. Organometal. Chem. **96** [1975] 279/82). – [15] A. N. Nesmeyanov, E. V. Leonova, N. S. Kochetkova, A. I. Malkova (J. Organometal. Chem. **96** [1975] 271/4).

[16] Š. Toma, M. Paričiová, E. Solčániová, K. Leško (Collection Czech. Chem. Commun. **43** [1978] 295/303). – [17] P. Zahradník, Š. Toma (Acta. Fac. Rerum Nat. Univ. Comenianae Chim. No. 29 [1981] 1/7).

5.2.3.5.2.2 Additional Ketones of the Type $C_5H_5FeC_5H_3(CO{-}R)_2$-1,2

$C_5H_5FeC_5H_3(CO{-}C_2H_5)_2$. Nesmeyanov et al. [1] obtained the homoannular disubstituted dipropionylferrocene in a 5% yield by reaction of ferrocene and $(C_2H_5CO)_2O$ in the presence of 85% H_3PO_4 at 110 to 120°C along with a 48% yield of $C_5H_5FeC_5H_4{-}CO{-}C_2H_5$ and separated it from the latter by column chromatography on Al_2O_3 with petroleum ether. After recrystallization from pentane the product melts at 43.5 to 44.5°C and has IR bands at 1008 and 1104 (C_5H_5) as well as 1662 (C=O) cm^{-1} [1].

$C_5H_5FeC_5H_3(CO{-}CH{=}CHC_6H_5)_2$-1,2 has been obtained in a 22% yield by condensation of $C_5H_5FeC_5H_3(CO{-}CH_3)_2$-1,2 with C_6H_5CHO in ethanol at room temperature with 10% NaOH. It is separated from VIII (see p. 157) (R = CH=CHC$_6$H$_5$) (19%), VI (see p. 157) (R = CH=CHC$_6$H$_5$) (7%), and $C_5H_5FeC_5H_3(CO{-}CH{=}C_6H_5$-1)CO-$CH_3$-2 (A; 25%) by column chromatography on SiO_2 with benzene as eluent. VIII and VI appear before the compound, A after it. Recrystallization from benzene/light petroleum ether gives red crystals melting at 159 to 160°C. The ^{1}H NMR spectrum in $CDCl_3$ shows chemical shifts at $\delta = 4.22$ (s, C_5H_5), 4.78 (t, H-4), 5.10 (d, H-3,5), 7.28 (dd, CH=CH; J = 15.8 Hz), 7.3 to 7.65 (m, C_6H_5) ppm. The IR bands in CCl_4 are located for ν(C=C) and ν(C=O) at 1610 and 1670 cm^{-1}, respectively. Photolysis forms VIII (R = CH=CHC$_6$H$_5$) [2].

References:

[1] A. N. Nesmeyanov, E. G. Perevalova, V. D. Tyurin, S. P. Gubin (Izv. Akad. Nauk SSSR Ser. Khim. **1966** 1938/43; Bull. Acad. Sci. USSR Div. Chem. Sci. **1966** 1874/8). – [2] Š. Toma, M. Paričiová, E. Solčániová, K. Leško (Collection Czech. Chem. Commun. **43** [1978] 295/303).

5.2.3.5.2.3 Ketones of the Type $C_5H_5FeC_5H_3(CO{-}R$-1)CO–R′-2

The compounds compiled in Table 8 possess on one cyclopentadienyl ring two different substituents and are therefore chiral. They are prepared in the racemic forms by the following methods:

Method I. Friedel-Crafts acylation of $C_5H_5FeC_5H_4{-}CO{-}CH_3$:

a) With C_2H_5COCl and $AlCl_3$ in a 1/1.6/2 mole ratio in boiling CS_2. No. 1 is eluted with 1/3 ether/Skelly B from an SiO_2 chromatographic column [1].

b) By dropwise addition of $AlCl_3$ in CH_2Cl_2 to the mixture of $C_5H_5FeC_5H_4{-}CO{-}CH_3$ and $(C_3H_7CO)_2O$ (2.8/1/1.8 mole ratio) in CH_2Cl_2 at room temperature followed after 4 to 5 h by hydrolysis at 5 to 6°C. Separation of No. 2 takes place on an SiO_2 chromatographic column with 1/3 ether/hexane [2].

c) As for Method Ib except that the $(CH_3)_2CHCH_2COCl/C_5H_5FeC_5H_4{-}CO{-}CH_3$ mixture in CH_2Cl_2 is added dropwise at room temperature to $AlCl_3$ (mole ratio 2.3/1/2.7) in CH_2Cl_2. In the column chromatography on SiO_2 No. 3 is eluted with 1/9 ether/hexane [2].

Method II. Condensation of $C_5H_5FeC_5H_3(CO{-}CH_3)_2$-1,2 with C_6H_5CHO using 10% NaOH in ethanol at room temperature and subsequent column chromatography on SiO_2 with benzene eluent [3].

Chemical Behavior. The compounds No. 1 to 3 have been oxidized to the stable, paramagnetic radical anions I with R = CH_3 [2] (see also [1]), C_2H_5, or $CH(CH_3)_2$ [2] by a 1 s exposure to atmospheric oxygen in dry $(CH_3)_2SO$ in the presence of a three- or fourfold excess of $KOC(CH_3)_3$. The fact that the oxidation of No. 1 in the presence of $(CD_3)_2SO$ leads to II and not to III indicates that oxidation takes place on the CH_3CO and not on the C_2H_5CO group [2].

 References on p. 160

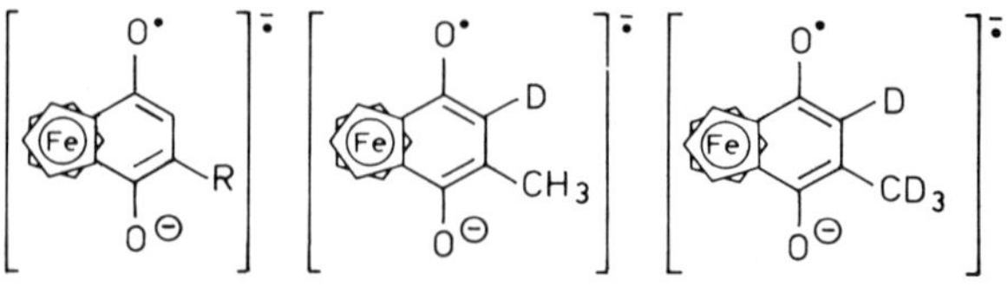

I II III

No. 4 is a very labile compound that decomposes photolytically to IV in sunlight [3].

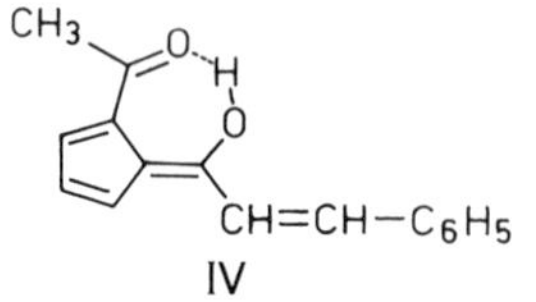

IV

Table 8

Ketones of the Type $C_5H_5FeC_5H_3$(CO–R-1)CO–R′-2.
For abbreviations and dimensions see p. VIII.

No.	$C_5H_5FeC_5H_3$(CO–R-1)CO–R′-2 R	R′	method of preparation (yield in %), and properties	Ref.
1	C_2H_5	CH_3	Ia (5.3) m.p. 46° ^{1}H NMR ($CDCl_3$): 1.17 (t, CH_3 of C_2H_5), 2.47 (s, $COCH_3$), 2.86 (m, CH_2 of C_2H_5), 4.25 (s, C_5H_5); 4.58 (t, 1H) and 4.88 (d, 2H) (C_5H_3)	[1, 2]
2	$\overset{7}{C}H_2$–$\overset{8}{C}H_2$–$\overset{9}{C}H_3$	CH_3	Ib (<2) m.p. 103 to 103.5° ^{1}H NMR ($CDCl_3$): 1.0 (t, CH_3-9), 1.74 (m, CH_2-8), 2.48 (s, $COCH_3$), 2.81 (t, CH_2-7), 4.27 (s, C_5H_5); 4.61 (t, 1H) and 4.89 (d, 2H) (C_5H_3)	[2]
3	CH_2–$CH(CH_3)_2$	CH_3	Ic (>6) m.p. 59 to 60° ^{1}H NMR ($CDCl_3$): 1.00 (d, $(CH_3)_2$), 2.28 (m, CH), 2.49 (s, $COCH_3$), 2.72 (d, CH_2), 4.28 (s, C_5H_5); 4.62 (t, 1H) and 4.90 (d, 2H) (C_5H_3)	[2]
4	CH=CH–C_6H_5	CH_3	II (25%) red oil ^{1}H NMR ($CDCl_3$): 2.54 (s, $COCH_3$), 4.30 (s, C_5H_5), 4.70 (t, H-4), 5.02 (m, H-3, 5), 7.30 to 7.70 (m, C_6H_5); 7.31 (d) and 7.69 (d) (CH=CH; J=15.8) IR (CCl_4): 1605 (νC=C)	[3]

References:

[1] J. J. McDonnell, D. J. Pochopien (J. Org. Chem. **36** [1971] 2092/8). – [2] J. J. McDonnell, D. J. Pochopien (J. Org. Chem. **37** [1972] 4064/7). – [3] Š. Toma, M. Paričiová, E. Solčániová, K. Leško (Collection Czech. Chem. Commun. **43** [1978] 295/303).

5.2.3.5.2.4 Ketones of the Type $C_5H_5FeC_5H_3(CO–R'\text{-}1)R^2\text{-}2$ (CO–R' = R^1)

The compounds $C_5H_5FeC_5H_3(CO–R'\text{-}1)A–CO–R''\text{-}2$ are covered in Section 5.2.3.5.2.5, pp. 182/6.

All of the compounds collected in Table 9 have a plane of chirality. See "Ferrocene" 7, 1980, pp. 1/4 (English version), pp. 4/7 (German text) regarding the "ferrocene chirality" thus produced and its manner of notation. The compounds No. 18, 21, and 22 possess in addition in the asymmetrically substituted C atom of R^2 a center of chirality, so that they can occur in the form of two diastereoisomers.

The preparation of the compounds is predominantly conducted as follows:

Method I. Reaction of $C_5H_5FeC_5H_3(CO_2Li\text{-}1)R^2\text{-}2$ with R'Li in a 1/3 mole ratio, R' = C_2H_5, at the boiling point. Separation takes place by column chromatography on Al_2O_3 with C_6H_6 as the eluent [32, 38]. With C_6H_5Li the alcohol IV and an unidentified, viscous red oil were obtained in addition to No. 30 [38].

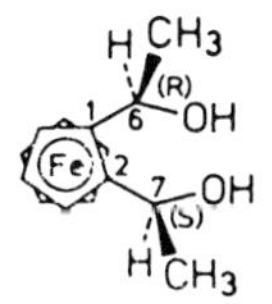

m.p. 152°C

I

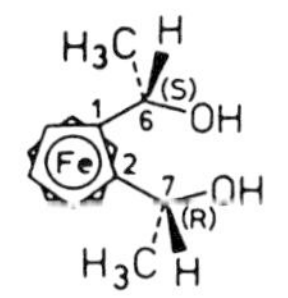

m.p. 134°C

II

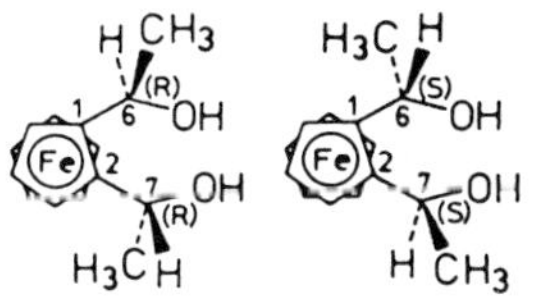

a b

m.p. 110°C

III

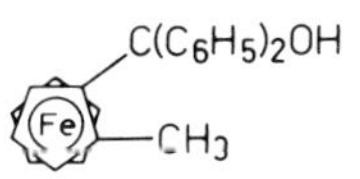

IV

Method II. Oxidation of $C_5H_5FeC_5H_3(CHOH–CH_3)_2\text{-}1,2$ (cf. "Ferrocene" 7, 1980, 5.2.3.1.2.2.1, pp. 205/6) with activated MnO_2 in CH_2Cl_2 at room temperature:

a) Of the meso form I [43, 48].

b) Of the meso form II [43, 48]. The reaction product always contains in addition traces of $C_5H_5FeC_5H_3(CO–CH_3)_2\text{-}1,2$ and is purified by preparative thin layer chromatography with ether as eluent [48].

c) Of the racemic form III a + b. Contrary to theoretical expectations only No. 18a (1SR, 7SR) is so formed [47, 48] along with $C_5H_5FeC_5H_3(CO–CH_3)_2\text{-}1,2$ in small quantities, but not No. 18b (1SR, 7RS) [43].

Method III. Condensation of $C_5H_5FeC_5H_3(CO–CH_3\text{-}1)R^2\text{-}2$ with C_6H_5CHO and NaOH (mole ratio 2/3/20) in C_2H_5OH/H_2O at room temperature. Since the starting materials are not the pure compounds No. 9, 14, or 16 but mixtures with their positional isomers, the compounds No. 27 to 29 are obtained mixed with their 1,1'-isomers and are separated from these by column chromatography on SiO_2 with benzene as eluent [61].

Method IV. Reaction of $[C_5H_5FeC_5H_3(CO–C_6H_5\text{-}1)CH_2CH_2N(CH_3)_3\text{-}2]I$ for 3 d with excess 10 to 25% KOH in boiling 1/1 $CH_3OCH_2CH_2OCH_3$/KOH, with subsequent column chromatographic purification of the raw product on Al_2O_3 [39], see also [40].

Method V. Friedel-Crafts acylation of $C_5H_5FeC_5H_4–R^2$:

a) With $(CH_3CO)_2O$ in the presence of H_3PO_4 [60, 65]. No. 2 is obtained mixed with its 1,3-isomer, which predominates [60, 65].

b) Like Method Va. $C_5H_5FeC_5H_4–R^2$, $(CH_3CO)_2O$, and 85% H_3PO_4 in a 1/20/4 mole ratio are reacted at 90 to 95°C, with the anhydride serving simultaneously as the reaction medium [29, 61, 64]. The mixture, which acts more mildly than the Perrier complex, extensively suppresses side reactions such as the acetylation of the phenyl group for $R^2=C_6H_5$ [61]. Reaction products are processed by column chromatography with benzene as the eluent on Al_2O_3 [29] or SiO_2 [61, 64]. The positional isomer mixtures obtained can in general not be separated by usual means. The amount of 1,2-isomer was determined by NMR spectroscopy. Only for Nos. 8, 10, and 16 isolation was successful by a combination of column chromatography and crystallization [61, 64].

c) With $(CH_3CO)_2O$ (mole ratio 1/0.6) and polyphosphoric acid at 45°C for 2 h [21].

d) $BF_3 \cdot O(C_2H_5)_2$ at 0°C is added to the mixture of $C_5H_5FeC_5H_4–R^2$ and $(CH_3CO)_2O$ in the mole ratio (2.4 to 3)/1/(1.6 to 2), and the reaction is continued at room temperature for 30 [11, 23] to 120 [13] min. The portions of the individual isomers in the total mixture were determined by gas chromatography at 215°C [11]. For separation of the mixtures see "Further information" for the individual compounds.

e) The CH_2Cl_2 solution of $C_5H_5FeC_5H_4–R^2$ and $(CH_3CO)_2O$ (mole ratio 1/1) is saturated at 0°C with BF_3, and the reaction mixture after 1.25 h at 0°C is warmed up to room temperature over a 3 h period [6].

f) $C_5H_5FeC_5H_4–R^2$ in CH_2Cl_2 is reacted at 0°C with $(CH_3CO)_2O$ and an ether/BF_3 solution (molar ratio 1/2/2) and the mixture is warmed to room temperature over a 6 h period. In chromatography on an Al_2O_3 column $C_5H_5FeC_5H_4–R^2$ and No. 7 are eluted in succession with petroleum ether and 9/1 petroleum ether/ether, respectively, before the 1,1'- and 1,3-isomers [52].

g) The complex obtained from R'COCl and $AlCl_3$ (1/1) in CH_2Cl_2 is added dropwise at 0°C to $C_5H_5FeC_5H_4–R^2$ in CH_2Cl_2, and the reaction is continued for 5 min at 0°C for No. 20 [21] or for 3 h at room temperature for Nos. 24 and 30 [38].

h) As for Method Vg, but at room temperature from the beginning. After completion of the addition the reaction is continued for 1 to 1.5 h for Nos. 2 and 30 [4, 23], 4 h for No. 26 [36], or 18 h for No. 35 [66]. No. 26, free of its 1,1'- and 1,3-isomers, is separated by chromatographing the reaction product in hexane on acid-washed Al_2O_3 with elution by 2/1 hexane/benzene after elution of $C_5H_5FeC_5H_4–CH_3$ with hexane. Later fractions eluted with hexane/benzene mixtures still contain No. 26, from which they are freed by repeated chromatography [36].

i) $C_5H_5FeC_5H_4–C_2H_5$ in CH_2Cl_2 is added at a constant rate over 7 h at −5°C to the mixture of CH_3COCl and $AlCl_3$ (molar ratio 1/1/1) in CH_2Cl_2, and the mixture is let stand an additional 30 min at this temperature. This procedure yields optimum conditions for the homoannular acetylation, which, however, due to the activation of the position 3 by the C_2H_5 group leads predominantly to the 1,3-isomer of No. 2 [57], see also [50].

j) $C_5H_5FeC_5H_4–R^2$ is reacted at 20°C with R'COCl and $AlCl_3$ (molar ratio 2/3/3), and No. 38 separated from the mixture of the three positional isomers by chromatography [58].

k) Reaction of $C_5H_5FeC_5H_4–R^2$ at room temperature over a 2 h period with $AlCl_3$ in CH_2Cl_2 and subsequently with R'COCl (molar ratio 1/<1/1.4), and continuation of the reaction for 1 h at room temperature. No. 31 is separated from its 1,1'- and 1,3-isomers by thin layer chromatography on silica gel with benzene [45].

l) Addition of CH_3COCl in CH_2Cl_2 at 0°C to the mixture of $C_5H_5FeC_5H_4–C_6H_5$ and $AlCl_3$ (mole ratio 1.3/1/1) in CH_2Cl_2 and continuation of the reaction first for 3 h at 0°C, then 1 h at room temperature [5].

m) Addition of $AlCl_3$ [29, 58] to the mixture of $C_5H_5FeC_5H_4–R^2$ and R′COCl (mole ratio 1/1/1) in CH_2Cl_2 at 0 to 5°C and continuation of the reaction 40 min for No. 37 [58] or 3 h for No. 6 [29, 58] at room temperature. The separation of the compounds from the reaction mixtures occurs by column chromatography on Al_2O_3 [29, 58] with benzene/petroleum ether as eluent for No. 6 [29] or by preparative thin layer chromatography on SiO_2 for No. 37 [58].

n) Addition of $AlCl_3$ in CH_3COCl to $C_5H_5FeC_5H_4–R^2$ in CH_3COCl/CH_2Cl_2 (1/1) at 0°C and continuation of the reaction at this temperature for 1 h. No. 23 is separable from the mixture of its positional isomers by column chromatography with ether on acid-washed Al_2O_3 (activity level 1) [22].

Method V yields in general a mixture of the three positional isomers, whose total yield is denoted with "m" in Table 9 if no separation followed. The number following the solidus gives the fractional amount in % of the 1,2-isomer, as determined, for example, by ^1H NMR spectroscopy. The amount of 1,2-isomer generally is considerably less than that of the 1,1′-isomer and with few exceptions, such as for Nos. 9 and 14, also below that of the 1,3-isomer.

Separation. The separation of the three positional isomers is difficult and is conducted primarily by chromatographic methods, among which gas chromatography [12, 23] yields better results than thin layer or column chromatography [23]. The 1,2-compound is, nevertheless, relatively the easiest to separate from the mixture, since it exhibits a distinctly smaller absorptivity than the other two isomers; it appears first in the eluate. For details see under the "Further information" for the compounds.

Properties. The ^1H NMR spectra (for numbering see Formula V) of the 1,2-compounds show in comparison with the corresponding signals of the 1,1′- and 1,3-isomers the lowest-field shifts for the protons on C-7 [7, 23] and the highest-field shifts for the ring proton in the α position to the CO group [7]. Calculation of the chemical shift of the ring protons by means of an additivity relationship (average deviation 0.04 ppm) is treated in [13]. Compounds with R^2 = alkyl show in addition to the singlet for the C_5H_5 protons three separated multiplets for the C_5H_3 protons [13].

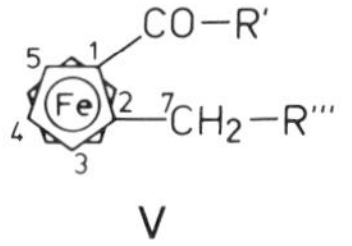

V

According to Rosenblum [3] the compounds $C_5H_5FeC_5H_3(CO–CH_3\text{-}1)R^2\text{-}2$ with R^2 = alkyl, in contrast to their 1,1′- and 1,3-isomers, exhibit in the IR spectrum only a medium to weak intensity band in the vicinity of 922 cm^{-1}, which is displaced for $R^2 = C_6H_5$ to 912 cm^{-1}. On the other hand according to [13], with R^2 = alkyl both the 1,2- and 1,3-isomers possess in the region between 889 and 935 cm^{-1} two bands, of which the higher frequency band is the more intense for the 1,2-isomer, in contrast the lower frequency band for the 1,3-isomer.

Chemical Behavior. The R′CO group is reduced by $LiAlH_4/AlCl_3$ in ether to the $R'CH_2$ group. Thus, No. 2 gives $C_5H_5FeC_5H_3(C_2H_5)_2\text{-}1,2$ [12], and No. 26 forms $C_5H_5FeC_5H_3(C_4H_9\text{-}n\text{-}1)CH_3\text{-}2$ [36]. $LiAlH_4$ [6, 53] or KBH_4 [32, 38] reduce the R′CO to the R′CHOH group. This forms in a 1/1 ratio the diastereoisomers VI and VII (with R′ = R** and R* = H), which in turn represent, respectively, the racemic mixtures of the enantiomers a and b [6, 32, 38, 53]. (The formulas are so presented that, according to the sequence rules of Cahn, Ingold, Prelog [69], arranged in decreasing priority OH is placed before R** before R*.)

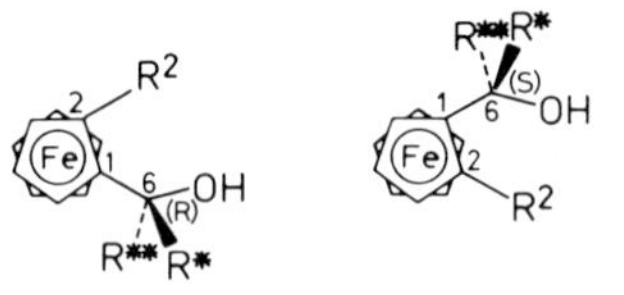

a (1R, 6R)

b (1S, 6S)

(1RS, 6RS)

VI

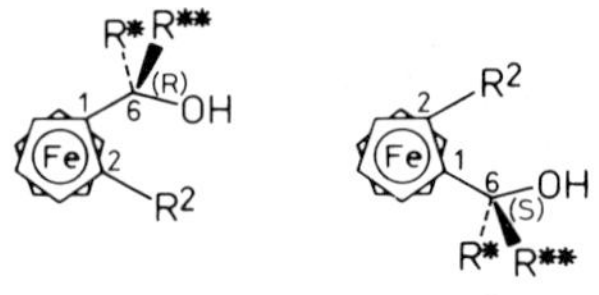

a (1S, 6R)

b (1R, 6S)

(1SR, 6RS)

VII

The reaction with the Grignard reagent R″MgX proceeds preferentially with addition on the C-6 and the oxygen atom following scheme in VIII and leads after hydrolysis likewise to the alcohols VI and VII. Since the achiral reagent, however, behaves differently with respect to the

VIII

diastereogenic sides of the ketone, the reaction is stereoselective and the more pronouncedly so the more voluminous R″ is. If $R^2 = CH_3$ and R′ is small, the intermediate product A is preferred, for voluminous R′ in contrast B is preferred. Oxygen arranged in the endo position with respect to the ferrocene molecule would give the corresponding A′ or B′ as the preferred intermediate products [38]. This means that VI and VII do not form in a 1/1 ratio; in general (1RS, 6RS)-VI predominates versus the diastereoisomeric (1SR, 6RS)-VII.

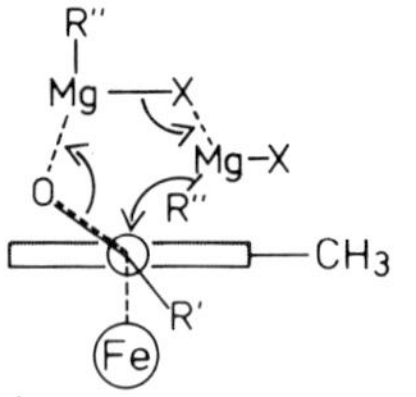

A

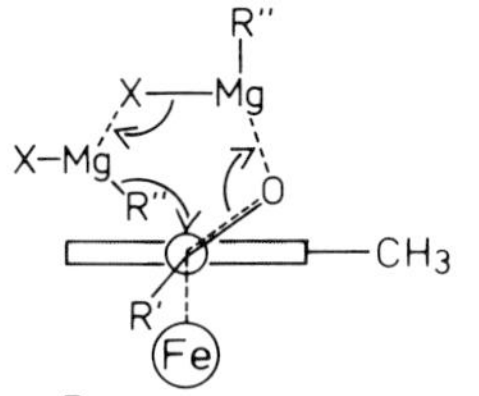

B

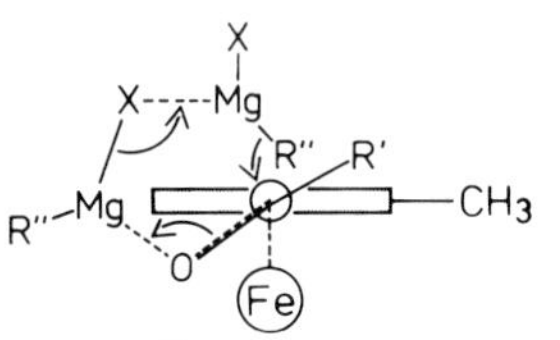

A′

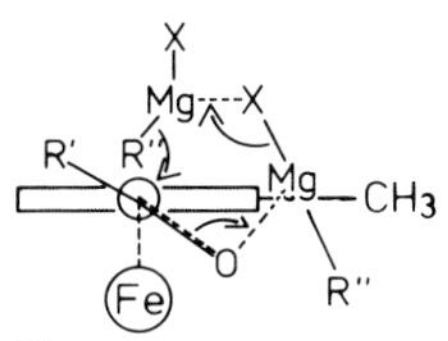

B′

If the carbonyl function is sterically shielded by a bulky alkyl group, as for example in No. 4 by $R^2 = C(CH_3)_3$, then instead of the addition the enolization according to scheme in IX represents the preferred reaction step, which is followed after hydrolysis by H_2O loss to form the CH≡C group [54].

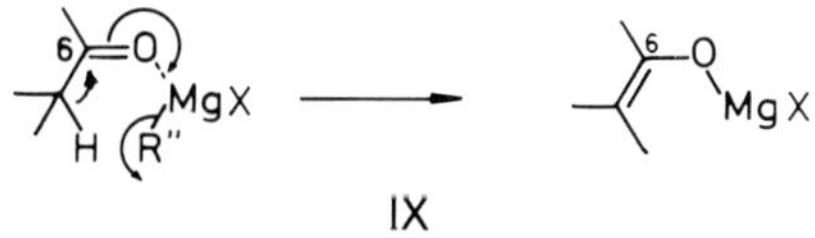

IX

The compounds $C_5H_5FeC_5H_3(CO{-}CH_3\text{-}1)R^2\text{-}2$ are able to condense with C_6H_5CHO following Method III to form $C_5H_5FeC_5H_3(CO{-}CH{=}CHC_6H_5\text{-}1)R^2\text{-}2$ [61].

References on pp. 181/2

Table 9

Ketones of the Type $C_5H_5FeC_5H_3(CO{-}R'\text{-}1)R^2\text{-}2$.
Further information for numbers preceded by an asterisk is given at the end of the table, pp. 173/81.
For abbreviations and dimensions see p. VIII.

No.	$C_5H_5FeC_5H_3(R^1\text{-}1)R^2\text{-}2$ $R^1 = CO{-}R'$	R^2	method of preparation (yield in %, "m" indicates isomeric mixture, number following solidus is fraction present of 1,2-compound), properties, and remarks	Ref.
*1	$CO{-}CH_3$	CH_3	I (ca. 50), Vd (11; 73m/20%), Ve (m/32%), see "Further information", p. 173 m.p. 61 to 61.5° (petroleum ether), 62.2 to 63.2° ^{1}H NMR (CCl_4): 2.24 or 2.30 or 2.33 (s, CH_3-2), 4.05 (s, C_5H_5), 4.18 (m, H-3), 4.28 (m, H-4), 4.44 (m, H-5) ^{1}H NMR (C_6H_6): 2.33 (s, CH_3-2) IR (Nujol): 895.2(w), 932.8(s), 1004(s), 1105(s), 1273.9(s)	[6, 7, 13, 15, 23, 33, 38]
*2	$CO{-}CH_3$	C_2H_5	Va, Vd (7; m/16%), Vh, Vi (m/20%), see "Further information", p. 175 orange crystals, m.p. 42.5 to 44.5° (petroleum ether), 44.5° ^{1}H NMR (CCl_4): 2.72 (q, CH_2 of C_2H_5), 4.07 (s, C_5H_5), 4.21 (m, H-3), 4.31 (m, H-4), 4.47 (m, H-5), see "Further information", p. 176 IR (Nujol): 892(w), 926(s), 1002(s), 1106.2(s), 1270.6(s) IR (CCl_4): 920, 1004, 1109.9, see figure in [1] and p. 163 UV (C_2H_5OH/H_2O, 19/1): $\lambda_{max}(\log \varepsilon) = 225$ (4.27), 268 (3.85), 335 (3.05), 460 (2.63) UV (H_2O): $\lambda_{max}(\varepsilon) = 335$ sh, 466 (677) UV (C_6H_6): $\lambda_{max}(\varepsilon) = 310$ sh, 470 (324)	[1, 3 to 5, 7, 11, 13, 25, 44, 46, 50, 57, 65]

References on pp. 181/2

Table 9 (continued)

No.	$C_5H_5FeC_5H_3(R^1\text{-}1)R^2\text{-}2$ $R^1 = CO\text{–}R'$	R^2	method of preparation (yield in %, "m" indicates isomeric mixture, number following solidus is fraction present of 1,2-compound), properties, and remarks	Ref.
*3	CO–CH_3	$CH(CH_3)_2$	Vd (7.6; m/10.6%) b.p. 145°/1 Torr ^{1}H NMR (CCl_4): 3.41 (CH of i-C_3H_7), 4.08 (s, C_5H_5), 4.24 (m, H-3), 4.36 (m, H-4), 4.49 (m, H-5) IR (film): 928.5 (s), 1004 (s), 1108.6 (s), 1265.8 (s)	[7, 11, 13]
*4	CO–CH_3	$C(CH_3)_3$	Vd (2.7; m/4%), see "Further information", p. 176 m.p. 99 to 100° ^{1}H NMR (CCl_4): 4.15 (s, C_5H_5), 4.21 (m, H-3), 4.29 (m, H-4), 4.54 (m, H-5) IR (Nujol): 888.9 (w), 923.4 (s), 1005 (s), 1109.9 (s), 1246.9 (s)	[7, 11, 13]
*5	CO–CH_3	$CH_2C_6H_5$	Vb (7) m.p. 76° (acetone/light petroleum ether) IR (paraffin oil): 1006 (δ C–H of cyclopentadienyl), 1108 (antisymectric C_5-ring breathing) UV (C_2H_5OH): λ_{max} (log ε) = 224sh (4.53), 264 (4.04), 335 (3.20) R_f: 0.60 (on SiO_2 with C_6H_6/$CH_3CO_2C_2H_5$, 4/1)	[28, 29]
6	CO–CH_3	$CH{=}CHC_6H_5$	Vb (26), Vm (9) m.p. 139 to 140° (acetone/light petroleum ether) IR (paraffin oil): 1008 (δC–H of cyclopentadienyl), 1106 (antisymetric C_5-ring breathing) UV (C_2H_5OH): λ_{max}(log ε) = 205 (4.84), 258 (4.21), 310 (4.44) R_f: 0.64 (on SiO_2 with C_6H_6/$CH_3CO_2C_2H_5$, 4/1)	[28, 29]
*7	CO–CH_3	$CH_2CH{=}$ CHC_6H_5-(E)	Vf (11 related to unrecovered $C_5H_5FeC_5H_4$–R^2) orange liquid	[52]

Table 9 (continued)

No.	$C_5H_5FeC_5H_3(R^1\text{-}1)R^2\text{-}2$ $R^1 = CO\text{–}R'$	R^2	method of preparation (yield in %, "m" indicates isomeric mixture, number following solidus is fraction present of 1,2-compound), properties, and remarks	Ref.
			^1H NMR ($CDCl_3$): 2.41 (s, $COCH_3$), 3.68 (m, CH_2), 4.19 (s, C_5H_5); 4.37 (m), 4.49 (m), and 4.65 (m) (C_5H_3); 6.45 to 6.55 (m, CH=CH), 7.3 (m, C_6H_5) UV (C_2H_5OH): $\lambda_{max}(\varepsilon)$ = 253 (22500)	
*8	$CO\text{–}CH_3$	C_6H_5	Vb (19m/9%), Vl (5) amber colored, octahedral crystals, m.p. 72 to 74°, 72.5 to 74.0° (petroleum ether) ^1H NMR ($CDCl_3$): 2.29 (s, $COCH_3$), 4.19 (s, C_5H_5), 4.57 (m, H-4), 4.62 (m, H-3), 4.82 (m H-5), 7.0 to 7.6 (m, C_6H_5) IR ($CHCl_3$): 912.4, 1001, 1105 UV (C_2H_5OH/H_2O, 19/1): $\lambda_{max}(\varepsilon)$ = 227 (17400), 265 (11500)	[3, 5, 61, 64]
9	$CO\text{–}CH_3$	$C_6H_4CH_3$-4	Vb (32m/34%) m.p. 70 to 78° (m) ^1H NMR ($CDCl_3$): 2.32 (s, $COCH_3$), 2.33 (s, CH_3), 4.23 (s, C_5H_5), 4.53 (m, H-4), 4.60 (m, H-3), 4.81 (m, H-5), 7.0 to 7.5 (m, C_6H_4) for chemical behavior see Method III, p. 161	[61, 64]
10	$CO\text{–}CH_3$	$C_6H_4CH_3$-3	Vb (61m/11%) m.p. 80 to 82° ^1H NMR ($CDCl_3$): 2.31 (s, $COCH_3$), 2.40 (s, CH_3), 4.25 (s, C_5H_5), 4.53 (m, H-4), 4.61 (m, H-3), 4.84 (m, H-5), 6.9 to 7.4 (m, C_6H_4)	[61, 64]
11	$CO\text{–}CH_3$	$C_6H_3(CH_3)_2$-3,4	Vb (42m/17%) oil (m) ^1H NMR ($CDCl_3$): 2.17 (s, $COCH_3$), 2.22 (s, CH_3), 4.12 (s, C_5H_5), 4.74 (m, H-5), 6.9 to 7.4 (m, C_6H_3)	[61, 64]

References on pp. 181/2

Table 9 (continued)

No.	$C_5H_5FeC_5H_3(R^1\text{-}1)R^2\text{-}2$ $R^1=CO\text{–}R'$	R^2	method of preparation (yield in %, "m" indicates isomeric mixture, number following solidus is fraction present of 1,2-compound), properties, and remarks	Ref.
12	$CO\text{–}CH_3$	$C_6H_3(CH_3)_2$-3,5	Vb (42m/20%) oil (m) ^{1}H NMR ($CDCl_3$): 2.28 (s, $COCH_3$, CH_3), 4.13 (s, C_5H_5), 4.74 (m, H-5), 6.8 to 7.2 (s, C_6H_3)	[61, 64]
*13	$CO\text{–}CH_3$	I	see "Further information", p. 178 orange crystals, m.p. 75 to 76° (CH_3OH/H_2O) ^{1}H NMR ($CDCl_3$): 2.50 (s, $COCH_3$), 4.26 (s, C_5H_5), 4.54 (t, H-4), 4.82 (d, H-3, 5), see "Further information", p. 178 IR (KBr): 825, 1000, 1110, 1265, 1320, 1355, 1375, 1430, 1675 (νC=O)	[42]
14	$CO\text{–}CH_3$	C_6H_4Cl-4	Vb (31m/32%) m.p. 85 to 90° (m) ^{1}H NMR ($CDCl_3$): 2.35 (s, $COCH_3$), 4.22 (s, C_5H_5), 4.55 (m, H-4), 4.63 (m, H-3), 4.83 (m, H-5), 7.2 to 7.6 (m, C_6H_4) for chemical behavior see Method III, p. 161	[61, 64]
15	$CO\text{–}CH_3$	C_6H_4Cl-3	Vb (55m/10%) oil (m) ^{1}H NMR ($CDCl_3$): 2.21 (s, $COCH_3$), 4.25 (s, C_5H_5), 4.91 (m, H-5), 7.0 to 7.7 (m, C_6H_4)	[61, 64]
16	$CO\text{–}CH_3$	C_6H_4Br-4	Vb (44m/12%) m.p. 97 to 98° ^{1}H NMR ($CDCl_3$): 2.37 (s, $COCH_3$), 4.21 (s, C_5H_5), 4.58 (m, H-4), 4.63 (m, H-3), 4.83 (m, H-5), 7.2 to 7.6 (m, C_6H_4) for chemical behavior see Method III, p. 161	[61, 64]

Table 9 (continued)

No.	$C_5H_5FeC_5H_3(R^1\text{-}1)R^2\text{-}2$ $R^1 = CO\text{-}R'$	R^2	method of preparation (yield in %, "m" indicates isomeric mixture, number following solidus is fraction present of 1,2-compound), properties, and remarks	Ref.
17	$CO\text{-}CH_3$	$C_6H_4Br\text{-}3$	Vb (49m/6%) oil (m) ^1H NMR ($CDCl_3$): 7.0 to 7.6 (m, C_6H_4)	[61, 64]
*18a	$CO\text{-}CH_3$ (1SR, 7SR)	$\overset{7}{C}HOH\text{-}CH_3$	IIa (80), IIc (80) m.p. 72° (hexane) ^1H NMR ($CDCl_3$): 1.45 (d, CH_3-7), 2.45 (s, $COCH_3$), 4.20 (s, C_5H_5); 4.50 (m, 1H) and 4.70 (m, 2H)(C_5H_3); 4.85 (q, H-7)	[43, 48]
*18b	$CO\text{-}CH_3$ (1SR, 7RS)	$\overset{7}{C}HOH\text{-}CH_3$	IIb m.p. 108° ^1H NMR ($CDCl_3$): 1.45 (d, CH_3-7), 2.40 (s, $COCH_3$), 4.30 (s, C_5H_5); 4.45 (m, 1H) and 4.60 (m, 2H)(C_5H_3); 4.70 (q, H-7)	[43, 48]
19	$CO\text{-}CH_3$	$C_6H_4O_2CCH_3\text{-}4$	Vb (43m/20%) oil (m) ^1H NMR ($CDCl_3$): 2.25 (s, $COCH_3$, CH_3), 4.16 (s, C_5H_5), 4.76 (m, H-5), 6.8 to 7.6 (s, C_6H_4)	[61, 64]
*20	$CO\text{-}CH_3$	OCH_3	Vc (45m), Vg (51m) m.p. 35 to 36° IR (neat): 1680 (νC=O)	[21]
*21a	$CO\text{-}CH_3$	$\overset{7}{C}H(CH{=}CHC_6H_5)\text{-}OCH_3\text{-}(E)$	see No. 7, p. 177, and "Further information", p. 178 orange oil ^1H NMR ($CDCl_3$): 2.40 (s, $COCH_3$), 3.58 (s, OCH_3), 4.26 (s, C_5H_5), 4.2 to 4.7 (m, C_5H_3), 5.2 to 5.4 (m, H-7), 6.0 to 6.8 (m, CH=CH), 7.1 to 7.5 (m, C_6H_5)	[52]
*21b	$CO\text{-}CH_3$	$\overset{7}{C}H(CH{=}CHC_6H_5)\text{-}OCH_3\text{-}(E)$	see No. 7, p. 177, and "Further information", p. 178 ^1H NMR ($CDCl_3$): 2.32 (s, $COCH_3$), 3.28 (s, OCH_3), 4.05 (s, C_5H_5), 4.2 to 4.6 (m, C_5H_3), 5.0 to 5.5 (m, H-7), 6.4 to 6.7 (m, CH=CH), 7.1 to 7.5 (m, C_6H_5)	[52]

References on pp. 181/2

Table 9 (continued)

No.	$C_5H_5FeC_5H_3(R^1\text{-}1)R^2\text{-}2$ R^1 = CO–R′	 R^2	method of preparation (yield in %, "m" indicates isomeric mixture, number following solidus is fraction present of 1,2-compound), properties, and remarks	Ref.
*22	CO–CH_3 1/1 mixture of (1SR, 9SR)- and (1SR, 9RS)-diastereomers	$\overset{7}{C}H{=}\overset{8}{C}H\overset{9}{C}H(C_6H_5)$– OCH_3-(E)	see No. 7, p. 177, and "Further information", p. 178 orange oil ^1H NMR ($CDCl_3$): 2.38 (s, $COCH_3$), 3.35 (s) and 3.42 (s) (OCH_3), 4.03 (s) and 4.13 (s) (C_5H_5), 4.3 to 4.9 (m, C_5H_3 + H-9), 5.7 to 6.2 (m, CH=CH), 7.35 (br. s, C_6H_5)	[52]
23	CO–CH_3	$C_6H_4OCH_3$-4	Vb (23 m/7%), Vn orange, microcrystalline powder, m.p. 137 to 139° (Skellysolve B) IR ($CHCl_3$): 913	[22, 61, 64]
*24	CO–C_2H_5	CH_3	I (ca. 50), Vg (17) m.p. 45° (pentane) ^1H NMR ($CDCl_3$): 1.18 (t, CH_3 of C_2H_5), 2.30 (s, CH_3-2), 2.74 (q) and 2.78 (q) (CH_2), 4.11 (s, C_5H_5); 4.29 (q, 1H), 4.38 (t, 1H), and 4.58 (q, 1H) (C_5H_3)	[32, 38]
*25	CO–C_2H_5	C_6H_4Cl-4	see "Further information", p. 179 m.p. 103° ^1H NMR ($CDCl_3$): 1.17 (t, CH_3), 2.74 (m, CH_2), 4.20 (s, C_5H_5), 4.60 (m, 2H) and 4.84 (m, 1H) (C_5H_3), 7.40 (m, C_6H_4)	[37]
26	CO–C_3H_7-n	CH_3	Vh (10) red oil ^1H NMR ($CDCl_3$ or CCl_4): 0.99 (t, CH_3 of C_3H_7-n), 1.70 (m, –CH_2–), 2.21 (s, CH_3-2), 2.55 (m, $COCH_2$), 3.98 (s, C_5H_5); 4.15 (m, 2H) and 4.39 (m, 1H) (C_5H_3)	[36]
27	CO–CH=CHC_6H_5	$C_6H_4CH_3$-4	III (66) m.p. 141 to 142°	[61]

References on pp. 181/2

Table 9 (continued)

No.	$C_5H_5FeC_5H_3(R^1\text{-}1)R^2\text{-}2$ $R^1 = CO\text{–}R'$	R^2	method of preparation (yield in %, "m" indicates isomeric mixture, number following solidus is fraction present of 1,2-compound), properties, and remarks	Ref.
28	$CO\text{–}CH{=}CHC_6H_5$	$C_6H_4Cl\text{-}4$	III (32) m.p. 162 to 163°	[61]
29	$CO\text{–}CH{=}CHC_6H_5$	$C_6H_4Br\text{-}4$	III (69.5) m.p. 150°	[61]
*30	$CO\text{–}C_6H_5$	CH_3	I (ca. 50), Vg (14), Vh (9), see "Further information", p. 179 m.p. 87 to 89° (CH_3OH/H_2O), 118 to 120° (hexane), 119° ^1H NMR ($CDCl_3$): 2.35 (s, CH_3), 4.11 (s, C_5H_5), 4.31 (q, 1H) and 4.46 (d, 2H) (C_5H_3), 7.2 to 8.0 (m, C_6H_5) ^1H NMR (CCl_4): 2.26 (s, CH_3), 3.99 (s, C_5H_5), 4.12 (m, 1H) and 4.29 (m, 2H) (C_5H_3) ^1H NMR (CCl_4): 2.32 (s, 3.0H), 4.02 (s, 5.4H), 4.15 (m, 1.0H), 4.32 (m, 2.0H), 7.16 to 7.87 (m, 4.8H)	[23, 32, 38, 51]
*31	$CO\text{–}C_6H_5$	$C(CH_3)_3$	Vk (4) viscous liquid ^1H NMR ($OS(CD_3)_2$): 1.05 (s, C_4H_9), 4.05 to 4.15 (m, C_5H_3), 4.08 (s, C_5H_5), 7.33 (m, 2H) and 7.80 (m, 3H) (C_6H_5) IR ($OS(CH_3)_2$): 1638 ($\nu C{=}O$) "ferrocene band" ($OS(CH_3)_2$): $\lambda_{max} = 460$ R_f: 0.8 (on SiO_2 with C_6H_6) mass spectrum: m/e (relative intensity in %; ion) = 346 (100; M^+), 331 (0.8; $[M-CH_3]^+$), 316 (0.7; $[M-2CH_3]^+$), 3.01 (0.6; $[M-3CH_3]^+$), 289 (0.9; $[M-C_4H_9]^+$), 226 (40; $[M-C_4H_6-C_5H_3]^+$), 121 (12; $C_5H_5Fe^+$), 105 (44; $C_6H_5CO^+$), 77 (20; $C_6H_5^+$), 65 (3; $C_5H_5^+$), 63 (2.4; $C_5H_3^+$), 56 (8; Fe^+), 28 (24; CO^+)	[41, 45]

References on pp. 181/2

Table 9 (continued)

No.	$C_5H_5FeC_5H_3(R^1\text{-}1)R^2\text{-}2$ $R^1 = CO–R'$	 R^2	method of preparation (yield in %, "m" indicates isomeric mixture, number following solidus is fraction present of 1,2-compound), properties, and remarks	Ref.
32	$CO–C_6H_5$	$CH{=}CH_2$	IV (80) oil ^1H NMR ($CDCl_3$): 4.06 (s, C_5H_5), 5.17 and 5.48 (ABX with the third proton apparently situated under the C_6H_5 protons, C_5H_3) IR (neat): 1005 (m), 1109.9 (m), 1639.3 (s, νC=O)	[39]
*33	$CO–C_6H_5$	OCH_3	see "Further information", p. 180 oil ^1H NMR (CCl_4): 3.63 (s, CH_3), 4.04 (m, 1H of C_5H_3), 4.18 (s and m, C_5H_5 and 1H of C_5H_3), 4.46 (m, 1H of C_5H_3), 7.49 and 7.82 (m, C_6H_5) IR: 1005 and 1110 (C_5H_5), 1635 (νC=O)	[67]
*34	$CO–C_6H_5$	CH_2OCH_3	see "Further information", p. 180 oil	[59]
*35	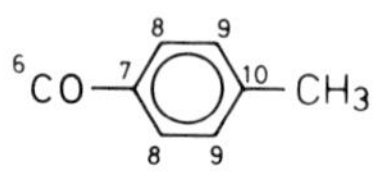	CH_3	Vh (63m/18%) ^{13}C NMR ($CDCl_3$): 15.1 (0.2C, CH_3-2), 70.8 (C_5H_5), 88.3 (0.2C, C-2); 128.1 (0.6C), 128.2 (1.0C), 128.4 (0.4C), and 128.8 (2.0C) (C-8,8,9,9); 137.2 (C-7), 141.7 (C-10), ca. 201 (C-6) mass spectrum: m/e = 318 $[M]^+$, 224, 202, 199, 121, 119, 91, 56	[66]
36	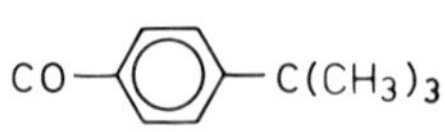	CH_3	see No. 30, p. 180 m.p. 124° (hexane) ^1H NMR ($CDCl_3$): 1.35 (s, C_4H_9), 2.33 (s, CH_3-2), 4.11 (s, C_5H_5); 4.29 (m, 1H) and 4.46 (m, 2H) (C_5H_3); 7.43 (d, 2H; J = 8) and 7.84 (d, 2H; J = 8) (C_6H_4)	[38]

References on pp. 181/2

Table 9 (continued)

No.	$C_5H_5FeC_5H_3(R^1\text{-}1)R^2\text{-}2$ $R^1 = CO\text{–}R'$	R^2	method of preparation (yield in %, "m" indicates isomeric mixture, number following solidus is fraction present of 1,2-compound), properties, and remarks	Ref.
*37	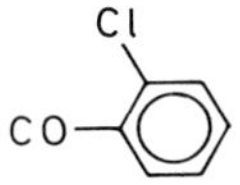	CH_3	V m (10.5 related to unrecovered $C_5H_5FeC_5H_4\text{–}CH_3$) m.p. 75 to 78° ^{1}H NMR ($CDCl_3$): 2.30 (s, CH_3) IR (CH_2Cl_2): 1650 (νC=O)	[58]
*38	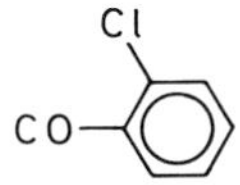	C_6H_5	V j (13 related to unrecovered $C_5H_5FeC_5H_4\text{–}C_6H_5$) oil ^{1}H NMR ($CDCl_3$): 4.25 (s, C_5H_5), 4.55 (m, 2H) and 4.72 (m, 1H) (C_5H_3), 7.20 to 7.75 ($C_6H_5 + C_6H_4$)	[58]

* Further information:

$C_5H_5FeC_5H_3(CO\text{–}CH_3\text{-}1)CH_3\text{-}2$ (Table **9**, No. **1**) results on 5 h boiling of X (p. 174) in absolute ethanol over Raney Ni in a 60% yield, on 6.5 h boiling in a 25% yield. The separation of the coproduct $C_5H_5FeC_5H_3(C_2H_5\text{-}1)CH_3\text{-}2$ (yield 28 or 51%, respectively) takes place by thin layer chromatography on Al_2O_3 with 5/3 benzene/petroleum ether [15].

No. 1 can be separated from the positional isomeric mixture obtained by Methods V c or d by column chromatography on Al_2O_3 with 1/1 pentane/ether [13] or in hexane/benzene with elution by a benzene/ether mixture [6], see also [56]. High pressure liquid-liquid chromatography with 1/9 CH_2Cl_2/isooctane on a column filled with Carbowax 400 on Corasil has been used to separate it from its 1,3-isomer [47].

Differentiation of the three positional isomers polarographically is not possible, since the potential for the reversible one-electron oxidation to the ferrocenium cation is the same for all three isomers [10].

The ^{1}H NMR signal for the CH_3 protons on C-2 is shifted toward lower field versus those for $C_5H_5FeC_5H_4\text{–}CH_3$, in CCl_4 by $\Delta\delta_{CH_3} = 0.285$ [33], 0.35 [23], in C_6H_6 by 0.49 ppm [23].

Clemmensen reduction with Zn and concentrated HCl in CH_3CO_2H gives the product $C_5H_5FeC_5H_3(C_2H_5\text{-}1)CH_3\text{-}2$ [15, 33]. $LiAlH_4$ in ether [6, 53] or KBH_4 in CH_3OH/H_2O [32, 38] form in a 1/1 ratio the diastereoisomers (1RS, 6RS)- and (1SR, 6RS)-$C_5H_5FeC_5H_3(CHOH\text{–}CH_3\text{-}1)CH_3\text{-}2$ (see VI and VII, respectively, (p. 164) with $R^{**} = R^2 = CH_3$, $R^* = H$). CH_3Li in ether yields $C_5H_5FeC_5H_3(C(CH_3)_2OH\text{-}1)CH_3\text{-}2$ [52, 53]. C_2H_5MgBr in ether gives only one alcohol, namely (1SR, 6RS)-$C_5H_5FeC_5H_3(C(CH_3)OH\text{–}C_2H_5\text{-}1)CH_3\text{-}2$ (VII with $R^* = R^2 = CH_3$, $R^{**} = C_2H_5$), mixed with the $C_5H_5FeC_5H_3(C(CH_3){=}CHCH_3\text{-}1)CH_3\text{-}2$ and $C_5H_5FeC_5H_3(C(C_2H_5){=}CH_2\text{-}1)CH_3\text{-}2$ hydrocarbons. In contrast C_6H_5MgBr forms the expected diastereoisomeric mixture of (1RS, 6RS)- and (1SR, 6RS)-$C_5H_5FeC_5H_3(C(CH_3)OH\text{–}C_6H_5\text{-}1)CH_3\text{-}2$ (VI and VII, respectively, with $R^* = R^2 = CH_3$, $R^{**} = C_6H_5$), in which the share of the first amounts to 90 to 96% [32, 38]. The reaction with the bulky $(CH_3)_3CMgCl$ yields XI [38].

References on pp. 181/2

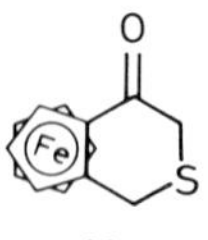

X

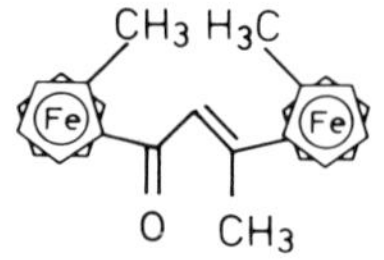

XI

Optically Active Forms. (+)(1R)-$C_5H_5FeC_5H_3(CO{-}CH_3\text{-}1)CH_3\text{-}2$ is prepared with CH_3Li from (−)-$C_5H_5FeC_5H_3(CO_2Li\text{-}1)CH_3\text{-}2$ by Method I (p. 161) [38], or from (−)(1R, 6S)-$C_5H_5FeC_5H_3$-($CHOH{-}CH_3\text{-}1)CH_3\text{-}2$ (VIIb, see p. 164, with $R^{**} = R^2 = CH_3$, $R^* = H$) by oxidation with MnO_2 at room temperature in petroleum ether. It is purified over deactivated silica gel with 10/1 cyclohexane/$CH_3CO_2C_2H_5$ as solvent; the yield is 81% [63]. Resolution of the racemate on a column filled with a slurry of cellulose acetate in benzene leads only to an optical purity p of 2.3% for the first fraction, while the last is enriched in the (−)enantiomer approximately equally strongly [16]. The compound, obtained by [32] only as an oil, melts at 37 to 38°C and boils at (60 to 65)°C/0.03 Torr [63]. The specific rotation $[\alpha]_D^{25} = +626° \pm 10°$ (C_2H_5OH, c = 1.5) [38]; $[\alpha]_D^{20} = +540°$ (C_6H_6, c = 0.72) [63], +544° (C_6H_6, c = 1; extrapolated to p = 100%) [16]; $[\alpha]_{578}^{20} = +626°$ (C_6H_6, c = 0.72) [63]. The circular dichroism curve is shown in a figure in [32]. The ^1H NMR spectrum in $CDCl_3$ shows the following shifts (in ppm): δ = 2.29(s) and 2.40(s) (CH_3-2, $COCH_3$), 4.12 (s, C_5H_5), ca. 4.3(m) and ca. 4.6(m) (C_5H_3) [63].

(−)(1S)-$C_5H_5FeC_5H_3(CO{-}CH_3\text{-}1)CH_3\text{-}2$ is prepared from (+)(1S)-$C_5H_5FeC_5H_3(CO_2Li\text{-}1)CH_3\text{-}2$ by Method I [38], or from the free acid (p = 80%) by converting it into the acid chloride with PCl_3/pyridine in benzene and reacting this with $Cd(CH_3)_2$ in ether. The yield is 71% after purification by preparative thin layer chromatography on SiO_2 with benzene [16, 20]. Regarding resolution of the racemate see the (+) enantiomer above. The (−) enantiomer is recovered as an oil [20, 38]. The specific rotation $[\alpha]_D^{20} = -544°$ (C_6H_6, c = 1.0; extrapolated to p = 100%), −562° (C_2H_5OH, c = 0.78; p = 100%) [16, 20]; $[\alpha]_D^{25} = -676° \pm 10°$ (C_2H_5OH, c = 1.7) [38]. The compound fulfills the semiempirical rule of Brewster [67, 68] concerning the relationship between the sign of the rotation and the absolute configuration [16]. In the IR(CCl_4) $\nu_{C=O} = 1670\ cm^{-1}$. The electronic spectrum in C_2H_5OH shows absorptions at $\lambda_{max}(\varepsilon)$ = 221 (14825), 265 (6425), 330 (924), 366 sh (480), 458 (404) nm. The optical rotatory dispersion in C_2H_5OH, for c ≈ 0.1 and p = 100%, is $[M]_{342} = -23400°$, $[M]_{390} = +4540°$, $[M]_{397} = +4320°$, $[M]_{450} = +10570°$, $[M]_{525} = -6600°$ [20].

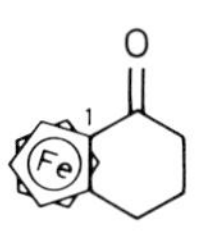

(+)(1S)-XII

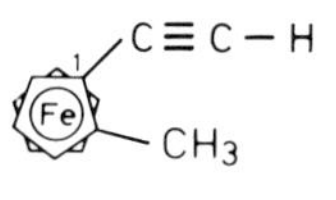

(−)(1S)-XIII

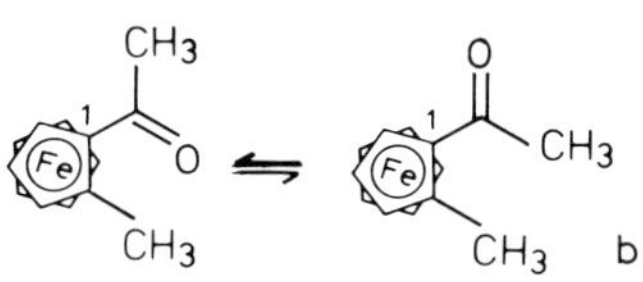

a b

s-cis-form s-trans-form

XIV

An enthalpy difference ΔG° = 0.67 ± 0.2 kcal/mol (2.81 kJ/mol) is found at 283 K between the conformations a and b of (−)(1S)- No. 1 (see Formula XIV). This is deduced from the temperature dependence between 283 and 343 K of the molar amplitude of the Cotton effect in the region of the ferrocene bands and comparison with the corresponding values for (+)(1S)-XII and (−)(1S)-XIII as reference substances. In equilibrium the compound occurs 76.5% as XIVa [19]. A band analysis using a computer curve fitting program of the circular dichroism curve measured in C_2H_5OH between 350 and 600 nm shows it to consist of a superposition of three components. These possess at the temperatures T and the wavelengths λ the following rotatory strengths R [26]:

	R in 10^{-40} erg·cm³		
T in K	λ 458 nm	433 nm	530 nm
203.2	−27.3	+22.1	+7.6
293.2	−18.6	+14.3	+6.0

The estimation of the thermodynamic quantities from the data for the temperature dependence of the first two bands yields for the conformational equilibrium at 20°C a contribution of XIVa of about 87%. The very high value may be traced back to the fact that the temperature dependence of the Lorentz factor and solvation equilibria were not taken into account [26]. A conformational equilibrium with 82% XIVa, in good agreement, can be derived from the dipole moment 2.89±0.05 D measured in cyclohexane; for free rotation of the CH_3CO group and an included angle of 57° between the acetyl partial moment and the rotational axis C(1)–C(6) μ should be 3.18 D. The potential change on rotation of the CH_3CO is determined dominantly by the Van der Waals and steric interactions and shows two minima with an energy difference 0.8 kcal/mol (3.3 kJ/mol), at the resonance-stabilized conformations XIVa and b [18]. For the influence of the solvents on the conformational equilibrium and the optical activity see [24].

Racemization is accelerated by acids, inhibited by bases [30]. The rate constant is $1.3\times10^{-4}\,s^{-1}$ in CH_3NO_2 at 368.2 K for the racemization catalyzed by an equimolar quantity of $HClO_4$ in the presence of ascorbic acid as an antioxidant [35].

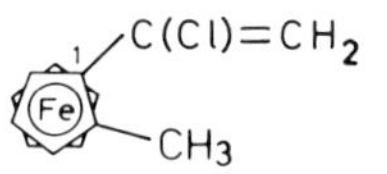

(−)(1S)-XV

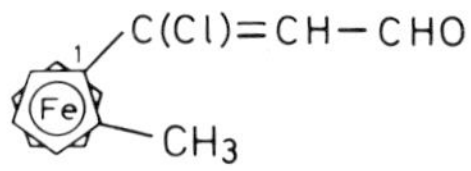

(1S)-XVI

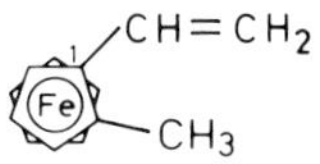

(+)(1S)-XVII

(−)(1S)-XV forms in 81% yield in 15 min at 0°C with the Vilsmeier complex of $(CH_3)_2NCHO$ and $POCl_3$. At 20°C with a reaction time of 1.5 h the yield drops to 12%, and in addition a 48% yield of (1S)-XVI is obtained [16].

Reduction with $LiAlH_4$ in ether leads to (+)-$C_5H_5FeC_5H_3(CHOH{-}CH_3\text{-}1)CH_3\text{-}2$, a 1/1 mixture of the diastereoisomers (1S, 6S) and (1S, 6R) (see VIb and VIIa, p. 164, respectively, with $R^{**}=R^2=CH_3$, $R^*=H$) [27]. This mixture reacts further with $KHSO_4$ [16] or acidic Al_2O_3 to (+)(1S)-XVII [16, 20]. The absolute configuration (1S) is found from the latter correlation and independently from the preparation path [16].

$C_5H_5FeC_5H_3(CO{-}CH_3\text{-}1)C_2H_5\text{-}2$ (Table **9**, No. **2**). Its preparation along with the two other positional isomers by acetylation of $C_5H_5FeC_5H_4{-}C_2H_5$ with the silicoanhydride of acetic acid in the presence of $SnCl_4$ (molar ratio 1/1/2) and separation from $FeC_{10}H_7(CO\,CH_3)_2C_2H_5$ of their mixture is treated in connection with $CH_3{-}CO{-}C_5H_4FeC_5H_4{-}C_2H_5$ (Table 6, No. 2, p. 128).

The mixture of the three positional isomers has been separated by column chromatography on Al_2O_3 with benzene as solvent and eluent [1, 4] in 21 fractions, of which the first three contained No. 2 [4], see also [56]. By repeated thin layer chromatography on Al_2O_3 plates with 6/1 benzene/ether No. 2 ($R_f=0.48$) can be separated from the 1,1′- and 1,3-isomer mixture ($R_f=0.32$). Repetition of the process yields the latter free of No. 2 [57]. Quantitative separation of all three isomers is possible by gas-liquid chromatography [57], for example on poly(ethylene glycol) M-20 or on neopentylglycoladipate at 150°C. For the latter the retained volumes referred to that of $C_5H_5FeC_5H_4{-}CN$ as unity amount to 1.33, 1.76, and 2.02 for the 1,2-, 1,1′-, and 1,3-isomers, respectively [17]. After conversion into the semicarbazones, fractional crystallization from 2/1 C_2H_5OH/H_2O is able to separate from $H_2NCONHN{=}C(CH_3){-}C_5H_4FeC_5H_4{-}C_2H_5$ (m.p. 141 to 142°C) only a mixture of the 1,2-isomer with the 1,3-isomer (m.p. 190 to 192°C) [2]. Without

References on pp. 181/2

previous separation of the isomers their ratio is determined by ^{1}H NMR spectrometry in CCl_4 in the presence of $Eu(dpm)_3$(dpm = dipivalomethanate), which causes a strong paramagnetic displacement of the substituent protons [44].

The ^{1}H NMR spectrum in $CDCl_3$ shows at room temperature, due to the magnetic nonequivalence of the CH_2 protons, a splitting of their signal into a double quartet with the coupling constant J = (2 to 3.5) Hz. The doubling disappears at −60°C, where only a normal quartet is present [25].

Activated MnO_2 oxidizes the compound to $C_5H_5FeC_5H_3(CO{-}CH_3)_2$-1,2 in boiling methylcyclohexane [8, 57]. Protonation on the oxygen to form **$[C_5H_5FeC_5H_3(C(CH_3){=}OH\text{-}1)C_2H_5\text{-}2]^+$** takes place in H_2SO_4/H_2O. The electronic spectrum of this cation in 90% H_2SO_4 solution is $\lambda_{max}(\varepsilon)$ = 370(2342), 516(813) nm. Half protonation occurs in 46.2% H_2SO_4. At 25°C pK_a = −3.05 for the conjugated acid in H_2SO_4/H_2O. No. 2 thus has a comparatively small basicity, which is distinctly lower than that of the 1,1′-isomer and also than that of $C_5H_5FeC_5H_4{-}CO{-}CH_3$ [46]. The (−)-menthydrazine (see Formula XVIII) forms in C_2H_5OH plus NaO_2CCH_3/CH_3CO_2H the (−)-$C_5H_5FeC_5H_3(C(CH_3){=}NNHCO_2C_{10}H_{19}$-1)$C_2H_5$-2 (m.p. 140 to 145°C, $[\alpha]_D$ = −15° for c = 1.0), for which resolution of the racemate – albeit not complete – is successful [9, 14], see the following paragraph.

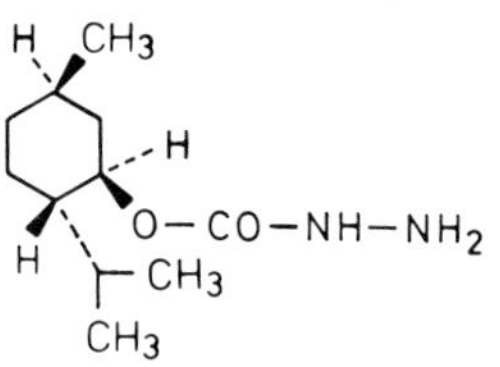

(−)-XVIII

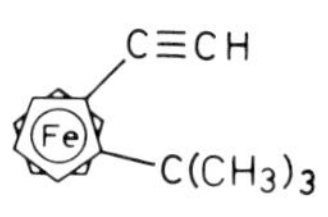

XIX

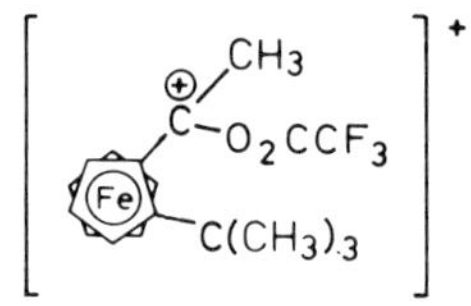

XX

Optically Active Forms. (−)(1S)-$C_5H_5FeC_5H_3(CO{-}CH_3$-1)C_2H_5-2 is obtained by oxidation of (1S, 6R)-$C_5H_5FeC_5H_3(CHOH{-}CH_3$-1)$C_2H_5$-2 (Formula VIIa, p. 164, with $R^2 = C_2H_5$, $R^{**} = CH_3$, $R^* = H$) with activated MnO_2 in petroleum ether at room temperature. After purification on silica gel which has been treated with concentrated NH_3/H_2O, with 10/3 petroleum ether/$CH_3CO_2C_2H_5$ as the solvent, the yield is 67% [63].

The boiling point is 80 to 90°C at 0.1 Torr. The specific rotation in benzene (c = 7 g/L) is $[\alpha]_D^{20} = -496°$, $[\alpha]_{578}^{20} = -572°$. The ^{1}H NMR spectrum in $CDCl_3$ shows shifts at δ = 1.17 (t, CH_3 of C_2H_5; J = 7.5 Hz), 2.40 (s, $COCH_3$), 2.78 (m, CH_2 of C_2H_5; J = 7.5 Hz; J_{gem} = 2.5 Hz), 4.12 (s, C_5H_5), and 4.4 and 4.6 (m, C_5H_3) ppm [63].

(−)(1S)-No. 2 is correlated with the (+)(1S)-$C_5H_5FeC_5H_3(CH{=}CH_2$-1)C_2H_5-2 via the (1S, 6SR)-$C_5H_5FeC_5H_3(CHOH{-}CH_3$-1)$C_2H_5$-2 (diastereomeric mixture of VIb and VIIa with $R^{**} = CH_3$, $R^* = H$, $R^2 = C_2H_5$) which forms with $LiAlH_4$ and gives the vinyl compound on dehydration [9, 14, 63].

$C_5H_5FeC_5H_3(CO{-}CH_3\text{-}1)CH(CH_3)_2\text{-}2$ (Table **9**, No. **3**) is separable from the mixture of the positional isomers by column chromatography on Al_2O_3 with 9/1 pentane/ether, as a faster band before the main band of the 1,1′- and 1,3-isomers [13].

$C_5H_5FeC_5H_3(CO{-}CH_3\text{-}1)C(CH_3)_3\text{-}2$ (Table **9**, No. **4**) is obtained in a quantitative yield on dissolution of XIX in CF_3CO_2H and subsequent hydrolysis of the resulting XX by pouring the solution into a great excess of saturated aqueous $NaHCO_3$ solution [49, 54, 62], cf. "Ferrocene" 7, 1980, p. 98. The compound is purified by preparative thin layer chromatography on SiO_2 with petroleum ether/ether as eluent [62]. Friedel-Crafts acetylation of $C_5H_5FeC_5H_4{-}C(CH_3)_3$ with CH_3COCl gives No. 4 together with its 1,3-isomer and $Fe(C_5H_4{-}CO{-}CH_2C(CH_3)_3)_2$ [70].

The separation from the mixture of its positional isomers proceeds as for No. 3, but with greater difficulty [13].

The reaction with CH_3Li in ether yields $C_5H_5FeC_5H_3(C(CH_3)_2OH\text{-}1)C(CH_3)_3\text{-}2$ [53]. Predominantly dehydration to XIX (68% yield) occurs with i-C_3H_7MgBr in 60 h in boiling (70°C) benzene/ether [54, 55]; $C_5H_5FeC_5H_3(C(CH_3)OH\text{–}C_3H_7\text{-i-}1)C(CH_3)_3\text{-}2$ forms in a 3% yield [55]. After only a 15 min reaction time the latter alone can be isolated, again in a 3% yield, along with unreacted No. 4 [54]. Reaction in boiling benzene also gives XIX as the main product, in 57% yield, along with $C_5H_5FeC_5H_3(C(CH_3)OH\text{–}C_3H_7\text{-i-}1)C(CH_3)_3\text{-}2$ (9%) and $C_5H_5FeC_5H_3(CHOH\text{–}CH_3\text{-}1)C(CH_3)_3\text{-}2$ (2%) [54]. See in this regard the general discussion, p. 164.

$C_5H_5FeC_5H_3(CO\text{–}CH_3\text{-}1)CH_2C_6H_5\text{-}2$ (Table **9**, No. **5**) is separated from the 1,1′-isomer, the main product of Method Vb (p. 162), by column chromatography on Al_2O_3 with benzene as eluent [29].

(E)-$C_5H_5FeC_5H_3(CO\text{–}CH_3\text{-}1)CH_2CH{=}CHC_6H_5\text{-}2$ (Table **9**, No. **7**). The preparation following Method Vf, but with a doubled excess of $(CH_3CO)_2O$ and BF_3, failed as did that with CH_3COCl/$AlCl_3$ [52]. Reaction of the solution of No. 7 in CH_2Cl_2 with $[C(C_6H_5)_3]BF_4$ in CH_2Cl_2 forms, with a dark brown coloration, the carbenium ion **$[C_5H_5FeC_5H_3(CO\text{–}CH_3\text{-}1)CHCHCHC_6H_5\text{-}2]^+$** (see Formula XXI), which reacts with $NaOCH_3/CH_3OH$ in an instantaneously quantitative yield to form the isomeric ethers No. 21 and 22 in a 39/61 ratio. The fact that the kinetically controlled addition of CH_3O^- occurs more often on C-9 to give No. 22 than on C-7 to give No. 21 is due to the CH_3CO group, which reduces the ability of the ferrocenyl group to delocalize positive charge. Isolation take place by column chromatography on Al_2O_3 (exposed for 24 h to the atmosphere). After elution of $(C_6H_5)_3CH$ with petroleum ether, first unreacted No. 7 and then the mixture of Nos. 21 and 22 are eluted with 4/1 petroleum ether/ether [52].

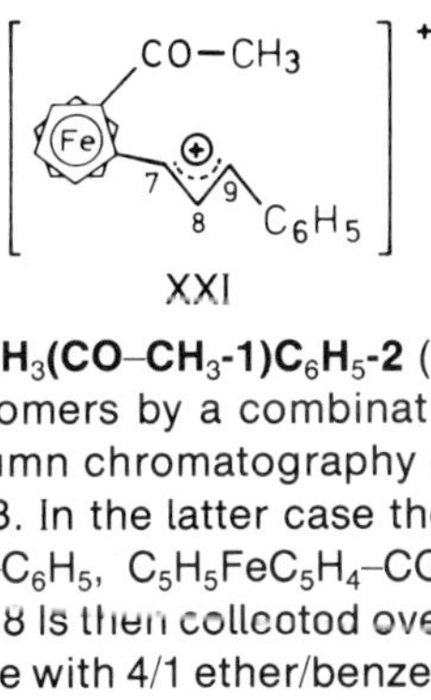

XXI

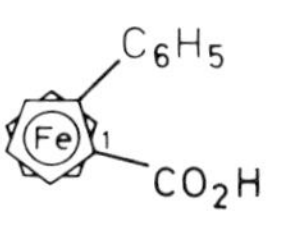

(+)(1R)-XXII

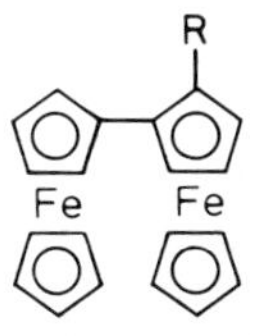

XXIII

$C_5H_5FeC_5H_3(CO\text{–}CH_3\text{-}1)C_6H_5\text{-}2$ (Table **9**, No. **8**) can be obtained pure from the mixture of the positional isomers by a combination of column chromatography and recrystallization [5, 61, 64]. The column chromatography can be conducted on SiO_2 with C_6H_6 [61, 64] or on Al_2O_3 with Skellysolve B. In the latter case the elution of No. 8 begins after eight weeks following that of $C_5H_5FeC_5H_4\text{–}C_6H_5$, $C_5H_5FeC_5H_4\text{–}CO\text{–}CH_3$, and sometimes also $C_5H_5FeC_5H_4\text{–}C_6H_4COCH_3\text{-}4$ by-product. No. 8 is then collected over two weeks in four fractions, which are rechromatographed on Al_2O_3 once with 4/1 ether/benzene and once with ether. Trituration with ether and recrystallization from low boiling petroleum ether then yields pure No. 8 [5]. No. 8 is also separable from its 1,3-isomer by high pressure liquid-liquid chromatography at room temperature on a column filled with Carbowax 400 on Corasil with isooctane containing 10 to 20% CH_2Cl_2 as eluent [47].

Optically Active Forms. (+)(1R)-$C_5H_5FeC_5H_3(CO\text{–}CH_3\text{-}1)C_6H_5\text{-}2$ is prepared in a 94% yield from (+)(1R)-$C_5H_5FeC_5H_3(CO_2H\text{-}1)C_6H_5\text{-}2$ (XXII) via reaction of its acid chloride with $Cd(CH_3)_2$ in ether. The product purified by preparative thin layer chromatography on SiO_2 with C_6H_6 is oily and possesses an optical purity p of 71.5%. The specific rotation $[\alpha]_D^{20} = +92°$ (C_2H_5OH, c = 0.2 to 0.3; p = 100%). The electronic spectrum (C_2H_5OH): $\lambda_{max}(\varepsilon) = 450$ (460) nm. Circular dichroism (C_2H_5OH, c = 0.02 to 0.06; p = 100%): $\Delta\varepsilon_{445} = -1.73$, $\Delta\varepsilon_{515} = +0.48$. Analysis of the experimental curve by a computer curve-fitting program shows it to be the result of a superposition of three bands with the rotatory strengths (in 10^{-40} erg·cm³) of −7.4 (450 nm), +1.8 (480 nm), and +1.3 (510 nm) [31]. A relationship between the sum of the absolute values of the rotatory strengths and the dipole strengths shows the transition to be electrically and magnetically dipole-forbidden; a dissymmetrically distorted, inherently symmetric chromophor is present [34]. Mass spectrum: m/e = 304 $[M]^+$ [31]. The absolute configuration is given from the correlation with (+)(1R)-XXII [31]. $LiAlH_4$ forms the diastereoisomeric mixture (1R, 6RS)-$C_5H_5FeC_5H_3$–($CHOH$–CH_3-1)C_6H_5-2 (see Formulas VIa and VIIb, p. 164, with $R^{**} = CH_3$, $R^* = H$, $R^2 = C_6H_5$) [31].

$C_5H_5FeC_5H_3(CO$–CH_3-1)I-2 (Table **9**, No. **13**) is prepared in 82% yield from the soluble dark complex formed from $C_5H_5FeC_5H_3$(CO–CH_3-1)HgCl-2 with I_2 in CH_2Cl_2 by treatment with aqueous $Na_2S_2O_3$ solution and purified by preparative thin layer chromatography on SiO_2 with CH_2Cl_2. The compound is unstable at room temperature. It shows an anomalous A_2B pattern for the protons of the C_5H_3 ring in the ^1H NMR spectrum, possibly due to steric effects which prevent a planar location for the CO groups. The Ullmann reaction with $C_5H_5FeC_5H_4$–I plus activated Cu bronze at a 130°C bath temperature yields after subsequent column chromatography three bands, of which the first gives XXIII with R = H, the second XXIII with R = $COCH_3$ (yield 37%), and the third finally a very complex mixture (red powder, m.p. 148 to 149°C), that possibly contains 2,7- and 2,10-diacetylbiferrocene. The same mixture also forms during the Ullmann coupling of No. 13 with activated Cu bronze [42].

(1SR, 7SR)- and (1SR, 7RS)-$C_5H_5FeC_5H_3$(CO–CH_3-1)$\overset{7}{C}HOH$–CH_3-2 (Table **9**, Nos. **18a** and **b**, see Formulas XXIV und XXV, respectively, which only show one enantiomer of the racemates) are both oxidized by MnO_2 to $C_5H_5FeC_5H_3(CO$–$CH_3)_2$-1,2, with No. 18b reacting more rapidly than No. 18a.

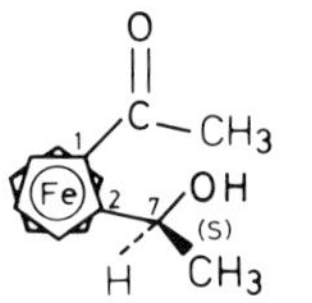

m.p. 72°C
XXIV

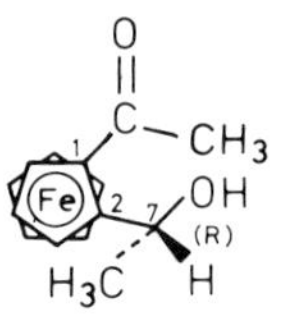

m.p. 108°C
XXV

KBH_4 in CH_3OH yields $C_5H_5FeC_5H_3(CHOH$–$CH_3)_2$-1,2; while No. 18a gives the racemic form IIIa and b (see p. 161) along with traces of the meso form I, No. 18b gives the meso form II along with traces of the racemic form IIIa and b [43, 48].

$C_5H_5FeC_5H_3$(CO–CH_3-1)OCH_3-2 (Table **9**, No. **20**). Method Vd (p. 162) yields only the mixture of the three positional isomers. Method Vg yields in addition $C_5H_5FeC_5H_3(OCH_3)_2$-1,2 and -1,3. On column chromatographic processing of the reaction products in benzene on alumina these latter two compounds and $C_5H_5FeC_5H_4$–OCH_3 elute out first with benzene and then with 1/1 benzene/ether the $FeC_{10}H_8(COCH_3)OCH_3$ mixture. After multiple rechromatographing No. 20 can be obtained from the first fraction as a solid and purified by subsequently subliming it twice [21].

(E)-$C_5H_5FeC_5H_3$(CO–CH_3-1)$\overset{7}{C}H$(CH=CHC_6H_5)OCH_3-2 and **(E)-$C_5H_5FeC_5H_3$(CO–CH_3-1)CH=CH$\overset{9}{C}H$(C_6H_5)OCH_3-2** (Table **9**, Nos. **21** and **22**). The mixture won in a 39/61 ratio from No. 7 (see p. 177) contains the two compounds as a mixture of the diastereomers, (1SR, 7SR)- and (1SR, 7RS)-No. 21 and (1SR, 9SR)- and (1SR, 9RS)-No. 22, respectively. It is resolved on Al_2O_3 chromatographic columns into three bands, which on elution with 4/1 light petroleum ether/ether yield in succession No. 21a, the 1/1 mixture of (1SR, 9SR)- and (1SR, 9RS)-No. 22, and finally No. 21b. The diastereomers a and b of No. 21 form in a 4/9 ratio.

The compounds No. 21 and 22 decompose on standing but do not isomerize in the presence of $NaOCH_3$ under the preparation conditions. $NaBH_4$ reduces the CH_3CO groups to CH_3CHOH [52].

$C_5H_5FeC_5H_3(CO{-}C_2H_5\text{-}1)CH_3\text{-}2$ (Table **9**, No. **24**) is separated from the mixture of the three positional isomers by Al_2O_3 column chromatography with 10/1 benzene/ether as eluent [28]. CH_3MgI in ether gives not the expected alcohols, but essentially a mixture of the unsaturated hydrocarbons XXVI and XXVII; the XXVI can be isolated by column chromatography on SiO_2 with 3/2 hexane/pentane. C_6H_5MgBr yields in a ca. 9/1 ratio the alcohols (1RS, 6RS)- and (1SR, 6RS)-$C_5H_5FeC_5H_3(C(C_2H_5)OH{-}C_6H_5\text{-}1)CH_3\text{-}2$ (see Formulas VI and VII, p. 164, with $R^{**} = C_6H_5$, $R^* = C_2H_5$, $R^2 = CH_3$) with the melting points 91 to 94 and 108 to 110°C, respectively [32, 38].

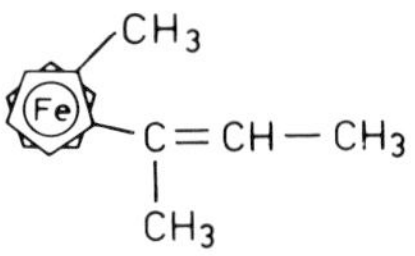

XXVI

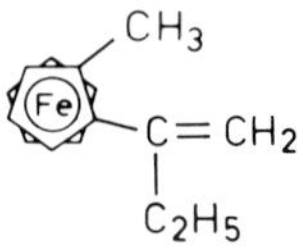

XXVII

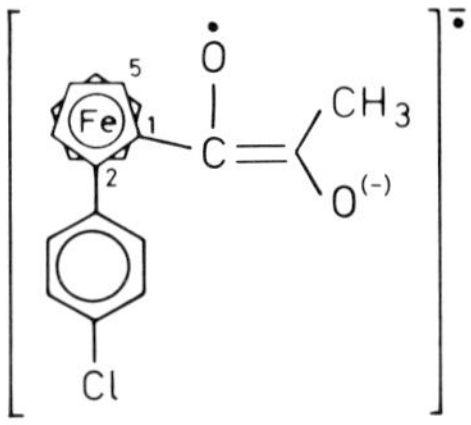

XXVIII

Optically Active Forms. (+)(1R)-$C_5H_5FeC_5H_3(CO{-}C_2H_5\text{-}1)CH_3\text{-}2$ is obtained using C_2H_5Li by Method I (p. 161) from (−)-$C_5H_5FeC_5H_3(CO_2Li\text{-}1)CH_3\text{-}2$ as an oil with the specific rotation $[\alpha]_D^{25} = +404° \pm 10°$ (C_2H_5OH, c = 0.4) [38].

(−)(1S)-$C_5H_5FeC_5H_3(CO{-}C_2H_5\text{-}1)CH_3\text{-}2$ is also formed in the same manner from (+)-$C_5H_5FeC_5H_3(CO_2Li\text{-}1)CH_3\text{-}2$ as an oil; $[\alpha]_D^{25} = -404° \pm 10°$ (C_2H_5OH, c = 0.4). The circular dichroism curve is plotted in the paper [38].

$C_5H_5FeC_5H_3(CO{-}C_2H_5\text{-}1)(C_6H_4Cl\text{-}4)\text{-}2$ (Table **9**, No. **25**) is obtained by a Friedel-Crafts reaction of $C_5H_5FeC_5H_4{-}C_6H_4Cl\text{-}4$ with $(C_2H_5CO)_2O$ and $AlCl_3$ in a 1/1.3/1 mole ratio in CH_2Cl_2 (reaction time 4.5 h) as a mixture with the 1,1'- and 1,3-isomers after column chromatographic separation on SiO_2 with Skelly B of the unreacted $C_5H_5FeC_5H_4{-}C_6H_4Cl\text{-}4$. After removal of the 1,1'-compound by selective recrystallization the remaining 1,2- and 1,3-mixture is subjected to column chromatography on SiO_2 with 93/7 cyclohexane/benzene, which gives first the 1,3-isomer and then No. 25 (yield 15.4%). A three- to fourfold excess of $KOC(CH_3)_3$ in $(CH_3)_2SO$ forms the enolate anion, which on 1 s exposure to air is oxidized to the stable radical anion **$[C_5H_5FeC_5H_3(CO{=}C(CH_3)O\text{-}1)(C_6H_4Cl\text{-}4)\text{-}2]^{\dot{-}}$** (see Formula XXVIII). This has a g value of 2.00694 and a hyperfine splitting coupling constant for the CH_3 protons of $A_H = 4.17$ G and for the H-5 of the C_5H_3 ring of $A_H = 0.50$ G [37].

$C_5H_5FeC_5H_3(CO{-}C_6H_5\text{-}1)CH_3\text{-}2$ (Table **9**, No. **30**) has been prepared in a 52% yield by reduction of $[C_5H_5FeC_5H_3(CO{-}C_6H_5\text{-}1)CH_2N(CH_3)_3\text{-}2]I$ with a 20-fold excess of Na in liquid ammonia and isolated after column chromatography on Al_2O_3 by gas chromatography and vacuum distillation [51]. The Methods Vg and Vh (p. 162) yield No. 30 mixed with the two other positional isomers, from which it is separated by preparative thin layer chromatography on Kieselgel G with benzene/petroleum ether [23] or by chromatography on an Al_2O_3 column with 10/1 benzene/ether [38].

In the ^1H NMR spectrum the signal of the CH_3 protons in CCl_4 is shifted by 0.31 ppm to lower field with respect to that in $C_5H_5FeC_5H_4{-}CH_3$ [23].

Reduction with 10% Na amalgam in a 3/2 H_2O/C_6H_6 mixture (vigorously stirred for 1 d) [51] or with KBH_4 in CH_3OH/H_2O [32, 38] form in a 1/1 ratio (1RS, 6RS)- and (1SR, 6RS)-$C_5H_5FeC_5H_3{-}(CHOH{-}C_6H_5\text{-}1)CH_3\text{-}2$ (see Formulas VI and VII (p. 164), respectively, with $R^{**} = C_6H_5$, $R^* = H$, $R^2 = CH_3$) [32, 38, 51]. In contrast CH_3MgI in ether leads to (1RS, 6RS)- and (1SR, 6RS)-$C_5H_5FeC_5H_3(C(CH_3)OH{-}C_6H_5\text{-}1)CH_3\text{-}2$ (Formulas VI and VII, respectively, with $R^{**} = C_6H_5$,

$R^* = R^2 = CH_3$) in a 6/4 to 7/3 ratio [32, 38] and C_2H_5MgBr gives (1RS, 6RS)- and (1SR, 6RS)-$C_5H_5FeC_5H_3(C(C_2H_5)OH–C_6H_5\text{-}1)CH_3\text{-}2$ (Formulas VI and VII with $R^{**} = C_6H_5$, $R^* = C_2H_5$, $R^2 = CH_3$) in a ratio between 85/15 and 87/13 [32, 38]. The reaction with $(CH_3)_3CMgCl$ proceeds differently; in ether first at room temperature and then at the boiling point it gives No. 36, which is purified chromatographically in benzene on an Al_2O_3 column [38].

Optically Active Forms. (+)(1S)-$C_5H_5FeC_5H_3(CO–C_6H_5\text{-}1)CH_3\text{-}2$ has been prepared from (+)-$C_5H_5FeC_5H_3(CO_2Li\text{-}1)CH_3\text{-}2$ by Method I (p. 161) [38] or in a 5% yield from the acid chloride of (+)(1S)-$C_5H_5FeC_5H_3(CO_2H\text{-}1)CH_3\text{-}2$ (p = 62%) by reaction for 10 h with $AlCl_3$ in boiling benzene and subsequent purification by thin layer chromatography on SiO_2 in benzene. From the latter reaction the absolute configuration is deduced [31].

The specific rotation $[\alpha]^{20}_D = +370°$ (C_2H_5OH, c = 0.1; extrapolated to p = 100%) [31], +448° ± 10° (C_2H_5OH, c = 0.4) [38]. IR (CCl_4): 1650 cm^{-1} (νC=O). Electronic spectrum in C_2H_5OH: $\lambda_{max}(\varepsilon) = 480(1033)$ nm. Circular dichroism (C_2H_5OH, c = 0.03 to 0.05; extrapolated to p = 100%): $\Delta\varepsilon_{495} = +1.22$ [31]. Analysis of the experimental circular dichroism curve by a computer curve fitting program indicates that it results from the superpositioning of three bands with the rotatory strengths $R \cdot 10^{40} = +8.6$ (480 nm), −4.7 (490 nm), and +1.1 (510 nm) erg·cm^3 [31]. A relationship between the sum of the absolute values of the rotatory strength and the dipole strength shows the transition to be electrically and magnetically dipole forbidden; the inherently symmetrical chromophore is dissymmetrically distorted [34].

Mass spectrum: m/e = 304 $[M]^+$ [31]. Reduction with KBH_4 leads to (+)(1S, 6S)- and (−)(1S, 6R)-$C_5H_5FeC_5H_3(CHOH–C_6H_5\text{-}1)CH_3\text{-}2$ (see Formulas VIb and VIIa (p. 164) with $R^{**} = C_6H_5$, $R^* = H$, $R^2 = CH_3$) [38].

(−)(1R)-$C_5H_5FeC_5H_3(CO–C_6H_5\text{-}1)CH_3\text{-}2$ is obtained from (−)-$C_5H_5FeC_5H_3(CO_2Li\text{-}1)CH_3\text{-}2$ by Method I (p. 161). It melts at 121°C, has a specific rotation $[\alpha]^{25}_D = -445° \pm 10°$ (C_2H_5OH, c = 0.35) (circular dichroism curve given in the paper), and yields with KBH_4 (−)(1R, 6R)- and (+)(1R, 6S)-$C_5H_5FeC_5H_3(CHOH–C_6H_5\text{-}1)CH_3\text{-}2$ (see Formulas VIa and VIIb (p. 164), respectively, with $R^{**} = C_6H_5$, $R^* = H$, $R^2 = CH_3$) [38].

$C_5H_5FeC_5H_3(CO–C_6H_5\text{-}1)C(CH_3)_3\text{-}2$ (Table **9**, No. **31**) reacts with H_2O in moist, O_2-free $(CH_3)_2SO$ to form $[\pi\text{-}C_5H_5Fe(OS(CH_3)_2)_n]^+$, $C_6H_5CO_2^-$, and tert-butylcyclopentadiene on irradiation with UV of λ > 280 nm at room temperature [41, 43].

$C_5H_5FeC_5H_3(CO–C_6H_5\text{-}1)OCH_3\text{-}2$ (Table **9**, No. **33**) is prepared by reaction with C_6H_5COCl of $C_5H_5FeC_5H_3(Cu\text{-}1)OCH_3\text{-}2$ in C_6H_6 at 25°C for 4 h and then subjecting the reaction product to thin layer chromatography on Al_2O_3 with 1/1 petroleum ether/ether. One obtains No. 33 in a 22% yield, $C_5H_5FeC_5H_4–OCH_3$ and its dimer, and traces of an isomer of No. 33 of unknown structure. The same products, with No. 33 in 19% yield, have also been won by the reaction of $C_5H_5FeC_5H_3(Ag\text{-}1)OCH_3\text{-}2$ with C_6H_5COCl in the presence of CuCl in benzene [67].

$C_5H_5FeC_5H_3(CO–C_6H_5\text{-}1)CH_2OCH_3\text{-}2$ (Table **9**, No. **34**) forms in a ca. 30% yield on reaction of $C_5H_5FeC_5H_3(Li\text{-}1)CH_2OCH_3\text{-}2$ with C_6H_5CN in ether/hexane and subsequent hydrolysis. The compound is free from the 1,1′-isomer if the Li compound has been prepared by transmetallation of $C_5H_5FeC_5H_3(HgCl\text{-}1)CH_2OCH_3\text{-}2$ with n-C_4H_9Li. No. 34 is isolated by chromatography on an Al_2O_3 column with CH_2Cl_2/petroleum ether as eluent [59].

$C_5H_5FeC_5H_3(CO–(C_6H_4CH_3\text{-}4)\text{-}1)CH_3\text{-}2$ (Table **9**, No. **35**) is isolated together with its 1,1′- and 1,3-isomers in a 1/2.6/2 ratio as a red syrup, when the product obtained by Method Vh (p. 162) is chromatographed on dry SiO_2 columns with 6/1 hexane/$CH_3CO_2C_2H_5$ and, thus, separated from the ditoluoylation products. The syrup is separable into the isomers by gas-liquid chromatography on a column filled with 3% OV-225 on Chromosorb W. The retention times for No. 35 are 6.7 min at a 225°C column temperature and 9.1 min at 215°C (for the 1,1′-isomer 8.6 and 11.6 min, respectively, and for the 1,3-isomer 9.3 and 12.9 min) [66].

$C_5H_5FeC_5H_3(CO–(C_6H_4Cl\text{-}2)\text{-}1)CH_3\text{-}2$ and **$C_5H_5FeC_5H_3(CO–(C_6H_4Cl\text{-}2)\text{-}1)C_6H_5\text{-}2$** (Table **9**, Nos. **37** and **38**) are cleaved with $KOC(CH_3)_3$ and a little H_2O in $CH_3OCH_2CH_2OCH_3$ with formation of $C_5H_5FeC_5H_3(CO_2H\text{-}1)R^2\text{-}2$ ($R^2 = CH_3$ for No. 37, C_6H_5 for No. 38). The reaction goes in over 80% yield for No. 37, traces of $C_5H_5FeC_5H_4–CH_3$ possibly formed cannot be detected. For No. 38 the yield of the carboxylic acid is 70%, that of $C_5H_5FeC_5H_4–C_6H_5$ 10% [58].

References:

[1] M. Rosenblum (Diss. Harvard Univ. 1953, p. 140). – [2] A. N. Nesmeyanov, N. A. Volkenau (Dokl. Akad. Nauk SSSR **111** [1956] 605/8; Proc. Acad. Sci. USSR Chem. Sect. **106/111** [1956] 689/91). – [3] M. Rosenblum (Chem. Ind. [London] **1958** 953). – [4] M. Rosenblum, R. B. Woodward (J. Am. Chem. Soc. **80** [1958] 5443/9). – [5] M. Rosenblum (J. Am. Chem. Soc. **81** [1959] 4530/6).

[6] E. A. Hill, J. H. Richards (J. Am. Chem. Soc. **83** [1961] 4216/21). – [7] R. A. Benkeser, Y. Nagai, J. Hooz (Bull. Chem. Soc. Japan **36** [1963] 482/3). – [8] A. F. Ellis (Diss. Univ. Illinois 1963, pp. 1/152 from Diss. Abstr. **24** [1963] 510). – [9] K. Schlögl, M. Fried (Tetrahedron Letters **1969** 1473/8). – [10] J. Tirouflet, E. Laviron, R. Dabard, J. Komenda (Bull. Soc. Chim. France **1963** 857/62).

[11] R. A. Benkeser, Y. Nagai, J. Hooz (J. Am. Chem. Soc. **86** [1964) 3742/6). – [12] D. E. Bublitz (Can. J. Chem. **42** [1964] 2381/4). – [13] Y. Nagai, J. Hooz, R. A. Benkeser (Bull. Chem. Soc. Japan **37** [1964] 53/60). – [14] K. Schlögl, M. Fried, H. Falk (Monatsh. Chem. **95** [1964] 576/97). – [15] A. N. Nesmeyanov, E. A. Perevalova, L. I. Leont'eva, Yu. A. Ustynyuk (Izv. Akad. Nauk SSSR Ser. Khim. **1965** 1882/4; Bull. Acad. Sci. USSR Div. Chem. Sci. **1965** 1847/9).

[16] H. Falk, K. Schlögl (Tetrahedron **22** [1966] 3047/53). – [17] A. N. Nesmeyanov [Nessmejanow], L. P. Yur'eva [Jurjewa], N. S. Kochetkova, S. V. Vitt [Witt] (Izv. Akad. Nauk SSSR Ser. Khim. **1966** 560/2; Bull. Acad. Sci. USSR Div. Chem. Sci. **1966** 530/2; Abhandl. Deut. Akad. Wiss. Berlin Kl. Chem. Geol. Biol. **1966** No. 2, pp. E265/E268). – [18] H. Falk, G. Haller (Monatsh. Chem. **98** [1967] 2290/301). – [19] H. Falk, G. Haller, K. Schlögl (Monatsh. Chem. **98** [1967] 2058/69). – [20] G. Haller, K. Schlögl (Monatsh. Chem. **98** [1967] 603/18).

[21] S. McVey, I. G. Morrison, P. L. Pauson (J. Chem. Soc. C **1967** 1847/50). – [22] M. Rosenblum, W. G. Howells (J. Am. Chem. Soc. **84** [1962] 1167/72). – [23] K. Schlögl, H. Falk, G. Haller (Monatsh. Chem. **98** [1967] 82/94). – [24] H. Falk, G. Haller (Monatsh. Chem. **99** [1968] 279/88). – [25] H. Egger, A. Nikiforov (Monatsh. Chem. **100** [1969] 242/53).

[26] H. Falk, O. Hofer (Monatsh. Chem. **100** [1969] 1507/14). – [27] H. Falk, C. Krasa, K. Schlögl (Monatsh. Chem. **100** [1969] 254/73). – [28] A. Pérjessy, Š. Toma (Chem. Zvesti **23** [1969] 533/9). – [29] Š. Toma, E. Kalužayová (Chem. Zvesti **23** [1969] 540/52). – [30] K. Bauer, H. Falk, H. Lehner, K. Schlögl, U. Wagner (Monatsh. Chem. **101** [1970] 941/3).

[31] H. Lehner, K. Schlögl (Monatsh. Chem. **101** [1970] 895/911). – [32] C. Moïse, J. Tirouflet, D. Sautrey (Compt. Rend. C **271** [1970] 951/4). – [33] M. Sato, M. Koga, I. Motoyama, K. Hata (Bull. Chem. Soc. Japan **43** [1970] 1142/7). – [34] H. Falk, H. Lehner (Tetrahedron **27** [1971] 2279/91). – [35] H. Falk, H. Lehner, J. Paul, U. Wagner (J. Organometal. Chem. **28** [1971] 115/24).

[36] J. W. Huffman, J. F. Cope (J. Org. Chem. **36** [1971] 4068/72). – [37] J. J. McDonnell, D. J. Pochopien (J. Org. Chem. **36** [1971] 2092/8). – [38] C. Moïse, D. Sautrey, J. Tirouflet (Bull. Soc. Chim. France **1971** 4562/76). – [39] D. W. Slocum, C. A. Jennings, T. R. Engelmann, B. W. Rockett, C. R. Hauser (J. Org. Chem. **36** [1971] 377/81). – [40] D. W. Slocum, R. Marchal, C. R. Ernst (Am. Chem. Soc. Div. Org. Coat. Plast. Chem. Papers **31** [1971] 285/96).

[41] L. H. Ali, A. Cox, T. J. Kemp (Chem. Commun. **5** [1972] 265/6). – [42] R. F. Kovar, M. D. Rausch (J. Organometal. Chem. **35** [1972] 351/66). – [43] C. Moïse, Y. Mugnier (Tetrahedron

Letters **1972** 1845/8). – [44] J. Paul, K. Schlögl, W. Silhan (Monatsh. Chem. **103** [1972] 243/54). – [45] L. H. Ali, A. Cox, T. J. Kemp (J. Chem. Soc. Dalton Trans. **1973** 1468/75).

[46] G. Cerichelli, B. Floris, G. Illuminati, G. Ortaggi (Gazz. Chim. Ital. **103** [1973] 911/22). – [47] R. Eberhardt, H. Lehner, K. Schlögl (Monatsh. Chem. **104** [1973] 1409/20). – [48] C. Moïse, Y. Mugnier (Bull. Soc. Chim. France **1973** 1382/4). – [49] T. S. Abram, W. E. Watts (Chem. Commun. **1974** 857/8). – [50] E. V. Leonova, N. S. Kochetkova, E. I. Fedin, P. V. Petrovskii (Ref. Dokl. Soobshch. 11th Mendeleevsk. S'ezd Obshch. Prikl. Khim., Alma-Ata 1974 [1975], Vol. 2, pp. 98/9; Ref. Zh. Khim. **1976** No. 4 V 112; C.A. **84** [1976] No. 164987).

[51] D. W. Slocum, R. L. Marchal, W. E. Jones (J. Organometal. Chem. **72** [1974] 227/37). – [52] T. D. Turbitt, W. E. Watts (J. Chem. Soc. Perkin Trans. II **1974** 185/9). – [53] T. D. Turbitt, W. E. Watts (J. Chem. Soc. Perkin Trans. II **1974** 189/95). – [54] T. S. Abram, W. E. Watts (J. Organometal. Chem. **86** [1975] 109/18). – [55] T. S. Abram, W. E. Watts (J. Chem. Soc. Perkin Trans. I **1975** 116/21).

[56] F. Jursík (J. Chromatog. Libr. **3** [1975] 1087/114, 1108). – [57] A. N. Nesmeyanov, E. V. Leonova, N. S. Kochetkova, A. I. Malkova, A. G. Makarovskaya (J. Organometal. Chem. **96** [1975] 275/8). – [58] V. Rapić, K. Schlögl, B. Steinitz (J. Organometal. Chem. **94** [1975] 87/98). – [59] D. W. Slocum, B. P. Koonsvitsky (J. Org. Chem. **41** [1976] 3664/8). – [60] S. L. Sosin, B. A. Antipova, T. M. Yezhova, L. V. Dzhashi, V. V. Korshak (Vysokomol. Soedin. A **18** [1976] 34/9; Polym. Sci. [USSR] **18** [1976] 40/6).

[61] Š. Toma, A. Maholányiová, E. Solčániová (Acta Fac. Rerum Nat. Univ. Comenianae Chim. No. 24 [1976] 53/80). – [62] T. S. Abram, W. E. Watts (J. Chem. Soc. Perkin Trans. I **1977** 1522/6). – [63] D. Marquarding, H. Burghard, I. Ugi, R. Urban, H. Klusacek (J. Chem. Res. S **1977** 82/3; J. Chem. Res. M **1977** 0915/0958). – [64] Š. Toma, A. Maholányiová, E. Solčaniová (Collection Czech. Chem. Commun. **42** [1977] 1013/26). – [65] V. V. Korshak, S. L. Sosin (Organometal. Polym. Symp., New Orleans 1977 [1978], pp. 25/38).

[66] B. E. Maryanoff (J. Org. Chem. **46** [1981] 70/4). – [67] A. N. Nesmeyanov, N. N. Sedova, S. K. Moiseev, V. A. Sasonova (Dokl. Akad. Nauk SSSR **256** [1981] 98/101; Dokl. Chem. Proc. Acad. Sci. USSR **256/261** [1981] 7/9). – [67] J. H. Brewster (J. Am. Chem. Soc. **81** [1959] 5475/83, 5483/93, 5493/500). – [68] J. H. Brewster (Tetrahedron **13** [1961] 106/22). – [69] R. S. Cahn, C. Ingold, V. Prelog (Angew. Chem. **78** [1966] 413/47, 424; Angew. Chem. Intern. Ed. Engl. **5** [1966] 385/415, 394). – [70] L. I. Tolstykh, L. Errais (Tr.-Mosk. Inst. Neftekhim. Gazov. Prom-Sti. Im. I. M. Gubkina **158** [1981] 144/9 from C.A. **99** [1983] No. 5775).

5.2.3.5.2.5 Ketones of the Type $C_5H_5FeC_5H_3(A{-}CO{-}R'\text{-}1)R^2\text{-}2$ ($A{-}CO{-}R' = R^1$)

The compounds collected in Table 10 have been prepared by the following methods:

Method I. Condensation of $C_5H_5FeC_5H_3(CHO)_2$-1,2 with $C_6H_5COCH_3$ (mole ratio 1/1) in C_2H_5OH at room temperature in the presence of a fivefold molar quantity of NaOH (3 h reaction time). In the chromatographic separation of the reaction products on SiO_2 with 9/1 $C_6H_6/CH_3CO_2C_2H_5$, Nos. 9, 1, and 7 elute in succession [7].

Method II. Gomberg-Bachmann-Hey reaction of ferrocene with $[4\text{-}CH_3COC_6H_4N_2]_2SO_4$ (molar ratio 1/5) in dilute aqueous acetic acid at 10 to 15°C for 12 h, with subsequent reduction with $NaHSO_3$ [2]. The separation of $C_5H_5FeC_5H_4{-}C_6H_4COCH_3$-4, an unidentified colorless compound (m.p. 121 to 123°C), the 1,1'-, and the 1,3-isomers takes place by chromatography on an Al_2O_3 column in $CHCl_3/C_6H_6$. The first two compounds are eluted with 4/1 benzene/Skellysolve B before No. 2, the latter two after it [2].

References on p. 186

Method III. Acylation of $C_5H_5FeC_5H_4-(CH_2)_nCOC_6H_4OCH_3$-4 with $C_6H_5CH{=}CHCOCl$ in CH_2Cl_2:

a) $AlCl_3$ is added in portions to a mixture of the ferrocenyl compound (n = 2 to 9) and acid chloride (molar ratio 2/1/1.2) in CH_2Cl_2 at 0 to 5°C, and the reaction is continued for 4 h at room temperature. The 1,2-compounds are separated in 7 to 10% yields from the mixture of 1,1'- and 1,3-isomers (joint yield 20 to 25%) by column chromatography on SiO_2 with $C_6H_6/CH_3CO_2C_2H_5$ in a 19/1 to 10/1 volume ratio [9].

b) As for Method IIIa but with dropwise addition at 0 to 5°C of the ferrocenyl compound (n = 3 to 9) in CH_2Cl_2 to the solution in CH_2Cl_2 of the mixture of acid chloride and $AlCl_3$ (molar ratio 1/5/10). The reaction is continued for compounds with n = 3 to 6 for 1 h at 35°C, for those with n = 7 to 9 for 2 to 3 h at room temperature. In addition to the desired compounds (yields 5 to 7%) and the 1,1'- and 1,3-isomers (10 to 15%), the compunds I (7 to 10%) and II (10 to 15%) also form [7].

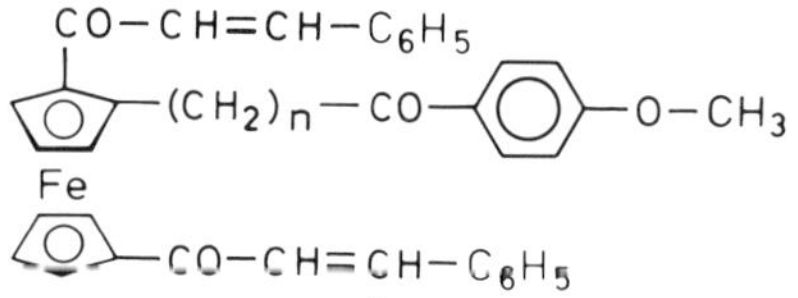

I

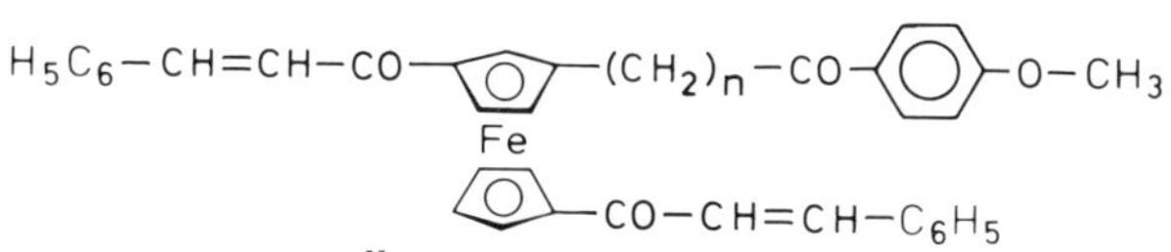

II

Method IV. $C_5H_5FeC_5H_3(CH{=}CHCOC_6H_5\text{-}1)CH(CH_3)N(CH_3)_2$-2 is refluxed for 20 min with CH_3I in acetone, followed by dilution with ether and evaporation of the washed and dried solution. The compound is purified by column chromatography on SiO_2 with C_6H_6 [8].

a) The reaction starts with (1R, 7R)-III [8].

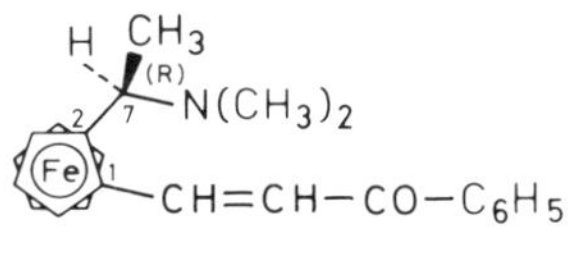

(1R, 7R)-III

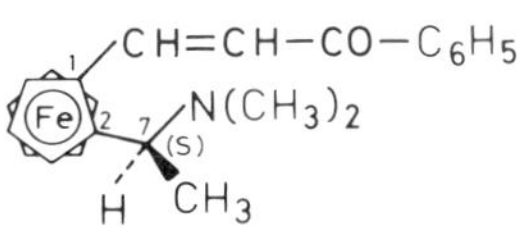

(1S, 7S)-IV

b) The reaction starts with (1S,7S)-IV.

The absolute configurations of Nos. 6a and 6b are given by these preparative correlations [8].

Method V. Acylation of $C_5H_5FeC_5H_4-R^1$ in CH_2Cl_2 with a solution of R^2Cl and $AlCl_3$ (molar ratio 2/3/6) in CH_2Cl_2 slowly added at room temperature and continuation of the reaction for 4 h at this temperature. No. 10 is separated from the 1,1'-isomer by column chromatography on Al_2O_3 with 97/3 $C_6H_6/CH_3CO_2C_2H_5$ [5].

While the compounds No. 1 and 2 with $R^1 = R^2$ are achiral, Nos. 3 to 10 with $R^1 \neq R^2$ have a chirality plane (for the nomenclature used throughout for the ferrocene chirality see "Ferrocene" 7, 1980, pp. 1/4 for the English version, pp. 4/7 for the German text). With the exception of No. 6 these ketones are prepared only in the racemic form.

References on p. 186

Table 10

Ketones of the Type $C_5H_5FeC_5H_3(A–CO–R'-1)R^2$-2.
Further information for numbers preceded by an asterisk is given at the end of the table, p. 186.
For abbreviations and dimensions see p. VIII.

No.	$C_5H_5FeC_5H_3(R^1-1)R^2$-2 $R^1 = A–CO–R'$	R^2	method of preparation (yield in %), properties, and remarks	Ref.
compounds with $R^1 = R^2$:				
*1	CH=CH–CO–C_6H_5		I (14) red crystals, m.p. 169 to 170° (C_6H_6/light petroleum ether) ^{1}H NMR ($CDCl_3$): 4.22 (s, C_5H_5); 4.78 (t, H-4), 4.97 (d, H-3,5) (C_5H_3); 7.15 to 8.10 (m, C_6H_5, CH=CH) IR (CCl_4): 1585 (νC=C), 1665 (νC=O)	[7]
2	C_6H_4–CO–CH_3-4		II (0.7) orange plates, m.p. 146 to 147° (ether/petroleum ether) IR ($CHCl_3$): 923, 1003, 1109 UV (C_2H_5OH/H_2O, 95/5): $\lambda_{max}(\varepsilon)$ = 264 (21600), 291 (25500), 363 (3440), 460 (1540)	[1,2]
compounds with $R^1 \neq R^2$:				
3	$(CH_2)_n$–CO–C_6H_4–OCH_3-4 n = 2 to 4, 6, 7, 9	CO–CH=CHC_6H_5	IIIa, IIIb	[9]
4	$(CH_2)_5$–CO–C_6H_4–OCH_3-4	CO–CH=CHC_6H_5	IIIa, IIIb m.p. 113 to 116° (C_6H_6/petroleum ether)	[9]
5	$(CH_2)_8$–CO–C_6H_4–OCH_3-4	CO–CH=CHC_6H_5	IIIa, IIIb ^{1}H NMR see figure in [9]	[9]
6	CH=CH–CO–C_6H_5-(E)	$\overset{7}{C}H=CH_2$	IV	[8]
a	CH=CH–CO–C_6H_5-(E) (−)(1R)	CH=CH_2	IVa (34) m.p. 91 to 92° $[\alpha]_D^{20} = -2229°$ ($CHCl_3$, c = 0.084) ^{1}H NMR: 4.08, 4.53, and 4.73 (m, $C_5H_5FeC_5H_3$), 5.18 and 5.43 (m, =CH_2), 6.85 (m, H-7), 7.16 (d, =CHCO), 7.45 (m, C_6H_5), 7.92 (d, CH=) IR: 900 and 990 (CH=CH_2); 910, 1000, and 1100 (C_5H_3); 980 and 1600 (E-CH=CH); 1660 (νC=O) mass spectrum: m/e = 342 $[M]^+$	[8]

References on p. 186

Table 10 (continued)

No.	$C_5H_5FeC_5H_3(R^1\text{-}1)R^2\text{-}2$ $R^1 = A\text{–}CO\text{–}R'$	R^2	method of preparation (yield in %), properties, and remarks	Ref.
b	$CH{=}CH\text{–}CO\text{–}C_6H_5$-(E) (+)(1S)	$CH{=}CH_2$	IVb (40) m.p. 91 to 92° $[\alpha]^{20}_D = +2238°$ ($CHCl_3$, c = 0.082) ^1H NMR: 4.08, 4.53, and 4.72 (m, $C_5H_5FeC_5H_3$), 5.18 and 5.43 (m, $=CH_2$), 6.85 (m, H-7), 7.16 (d, =CHCO), 7.45 (m, C_6H_5), 7.92 (d, CH=) IR: 900 and 990 ($CH{=}CH_2$); 910, 1000, and 1100 (C_5H_3); 980 and 1600 (E-CH=CH), 1660 (νC=O) mass spectrum: m/e = 342 $[M]^+$	[8]
*7	$CH{=}CH\text{–}CO\text{–}C_6H_5$	CHO	I red crystals, m.p. 121 to 123° (C_6H_6) ^1H NMR ($CDCl_3$): 4.31 (s, C_5H_5); 4.84 (t, H-4), 5.06 (d, H-3, 5) (C_5H_3); 7.5 to 8.1 (m, C_6H_5), 7.38 (d) and 8.10 (d) (CH=CH; J = 15.3), 10.19 (s, CHO) IR (CCl_4): 1504 (νC=C), 1660 (νC=O)	[7]
*8	$CH{=}CH\text{–}CO\text{–}C_6H_5$	$COCH_3$	IR (paraffin oil): 1006 (δC–H of cyclopentadienyl), 1106 (antisymmetric cyclopentadienyl ring breathing)	[3]
9	$\overset{6}{C}H(OC_2H_5)\overset{7}{C}H_2\text{–}CO\text{–}C_6H_5$	$CH{=}CHCOC_6H_5$	I (25), see No. 1, p. 186 red crystals, m.p. 142° (C_6H_6/light petroleum ether) ^1H NMR ($CDCl_3$): 1.15 (t, CH_3 of C_2H_5), 2.5 to 4.8 (m, OCH_2, H-6, 7, 7), 4.08 (br. s, C_5H_3), 4.35 (s, C_5H_5), 7.30 to 8.20 (m, C_6H_5, CH=CH) IR (CCl_4): 1030 (ν_sC–O–C), 1120 (ν_aC–O–C), 1610 (νC=C), 1660 and 1670 (νC=O)	[7]
10	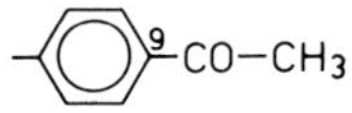	$COCH{=}CHC_6H_4Cl\text{-}4$	V (10) deep red crystals, m.p. 224 to 226° (C_6H_6/petroleum ether) IR (Nujol): 1013 and 1108 (cyclopentadienyl), 1120 (ν_sC(9)–C_{acetyl}); 1568, 1595, and 1608 (γC=C of C_6H_4); 1655 and 1678 (νC=O)	[5]

References on p. 186

* Further information:

$C_5H_5FeC_5H_3(CH{=}CH{-}CO{-}C_6H_5)_2$-1,2 and **$C_5H_5FeC_5H_3(CH{=}CH{-}CO{-}C_6H_5$-1)CHO-2** (Table **10**, Nos. **1** and **7**) are thermally and photochemically stable compounds [7]. In No. 1 one double bond experiences such a strong activation by the two CO groups that it is able to add the weak nucleophile C_2H_5OH in only 2 h with formation of No. 9 [7].

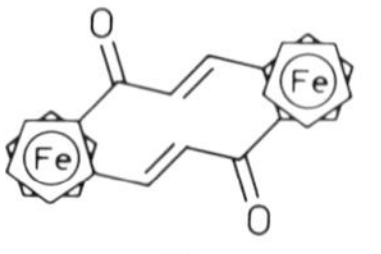

V

$C_5H_5FeC_5H_3(CH{=}CH{-}CO{-}C_6H_5$-1)$COCH_3$-2 (Table **10**, No. **8**) in contrast to its 1,1′-isomer exhibits no IR band around 1117 cm^{-1} [3]. Attempts to cyclize No. 8 with KOH in boiling ethanol or with $NaOCH_3$ in $(CH_3)_2NCHO$ have failed. $C_5H_5FeC_5H_4{-}CH{=}CHCOC_6H_5$ in 30% yield [4] and $C_5H_5FeC_5H_4{-}CH{=}CHCOCH{=}CH{-}C_5H_4FeC_5H_5$ [6] in 12% yield [4] are obtained. The result in [4], that the latter compound should be regarded as V, is superseded [6].

References:

[1] M. Rosenblum, W. G. Howells, A. K. Banerjee, C. Bennett (J. Am. Chem. Soc. **84** [1962] 2726/32). – [2] M. Rosenblum, J. O. Santer, W. G. Howells (J. Am. Chem. Soc. **85** [1963] 1450/8). – [3] A. Perjéssy, Š. Toma (Chem. Zvesti **23** [1969] 533/9). – [4] Š. Toma, M. Sališová (J. Organometal. Chem. **57** [1973] 191/8). – [5] Š. Toma, M. Sališová (J. Organometal. Chem. **57** [1973] 199/205).

[6] Š. Toma, M. Sališová, E. Solčániová (J. Organometal. Chem. **90** [1975] 335/41). – [7] Š. Toma, M. Paričiová, E. Solčániová, K. Leško (Collection Czech. Chem. Commun. **43** [1978] 295/303). – [8] A. Kasahara, T. Izumi, H. Watanabe (Bull. Chem. Soc. Japan **52** [1979] 957/8). – [9] M. Sališová, E. Solčániová, Š. Toma (Acta Fac. Rerum Nat. Univ. Comenianae Chim. No. 28 [1980] 103/26).

5.2.3.5.2.6 Ions of the Type $[C_5H_5FeC_5H_3(CO{=}CO{-}R\text{-}1)O\text{-}2]^{2\dot{-}}$

The new paramagnetic species I are formed by oxidation of the cyclopentadienyl ring at the C-2 position during reaction in excess O_2 of the radical anions $[C_5H_5FeC_5H_4{-}CO{=}CO{-}R]^{\dot{-}}$ obtained from the $C_5H_5FeC_5H_4{-}CO{-}R$ (R = CH_3, C_2H_5, $CH(CH_3)_2$) by a brief action of O_2 in the presence of a three- to fourfold molar excess of $KOC(CH_3)_3$ in $(CH_3)_2SO$ (cf. "Ferrocene" 2, 1977, p. 302).

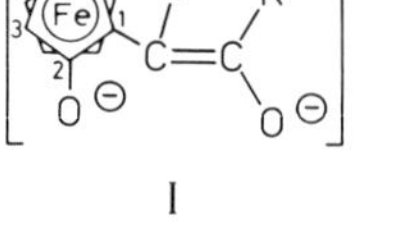

I

ESR studies yield g = 2.00708 and the hyperfine splitting coupling constant for the cyclopentadienyl protons H-3,4,5 A_H = 0.5 G for all R studied, while for the α protons of the R the values of A_H are 4.25, 3.70, and 1.71 G, respectively, for R = CH_3, C_2H_5, and $CH(CH_3)_2$.

Reference:

J. J. McDonnell, D. J. Pochopien (J. Org. Chem. **36** [1971] 2092/8).

5.2.3.5.3 1,3-Compounds

5.2.3.5.3.1 Ketones of the Type $C_5H_5FeC_5H_3(CO{-}R'\text{-}1)CO{-}R''\text{-}3$ (with R' = or ≠ R'')

5.2.3.5.3.1.1 1,3-Diacetylferrocene $C_5H_5FeC_5H_3(CO{-}CH_3)_2\text{-}1,3$

The compound is prepared by oxidizing $C_5H_5FeC_5H_3(CO{-}CH_3\text{-}1)C_2H_5\text{-}3$ with activated MnO_2 in the boiling solvents $CHCl_3$ (17 d) [5], CH_2Cl_2 [7], methylcylohexane [2], or benzene (48 h) [3]. The yields are 82% [7] or 38% (referred to the reacted starting compound) [3]. If a mixture of the 1,3-isomer with $CH_3CO{-}C_5H_4FeC_5H_4{-}C_2H_5$ (cf. Table 6, No. 2, p. 218) is used as the starting material, then the subsequent thin layer chromatography on Al_2O_3 with 60/1 ether/H_2O yields with $R_f = 0.52$ at the head of the band a $C_5H_5FeC_5H_3(CO{-}CH_3)_2\text{-}1,3$ contaminated with 10% $Fe(C_5H_4{-}CO{-}CH_3)_2$ [5]. The two compounds are, due to the similarity of their absorption properties, not completely separable by thin layer chromatography on Al_2O_3 or SiO_2 [3]. Acetylation of ferrocene with $CH_3COCl/AlCl_3$ (molar ratio 1/2.8/2.5) in CH_2Cl_2 at room temperature gives a mixture of the 1,3- and 1,1'-diacetylferrocene in a 1/60 ratio. The 1,1'-isomer is crystallized out, and the 1,3-isomer is isolated from the boiled down mother liquor by repeated column chromatography on Al_2O_3 in C_6H_6 with ether/C_6H_6 as eluent, in 1.2% yield [1]. The extremely low melting point of this product, 96 to 97°C, suggests that here, too, the separation was not completely successful.

The orange crystals [3] melt at 96 to 97°C (ether) [1], 180 to 181°C (petroleum ether) [3], 187 to 188°C [7]. The ^1H NMR spectra in $CDCl_3$ exhibit the following signals (δ in ppm):

CH_3	C_5H_5	H-4,5 of C_5H_3	H-2 of C_5H_3	Ref.
2.44 (s)	4.78 (s)	5.06 (d)	5.40 (t)	[3]
2.37	4.21	5.05	5.30	[4]

The IR spectrum in KBr shows strong bands at 1005 and 1105 (C_5H_5), 1656 (νC=O), and additional bands at 450, 485, 498, 512, 545, 565, 832, 850, 892, 905, 969, 1073, 1198, 1235, 1305, 1328, 1365, 1410, and 1460 cm^{-1} [3]. Electronic absorption spectrum in absolute C_2H_5OH: λ_{max} (ε) = 232 (16240), 259 (10450), 345 (6293), and 450 (4422) nm [3]; in $CHCl_3$: λ_{max} (log ε) = 260 (4.05), 330 (3.16), 370 (3.19), 450 (2.80) nm [1].

1,3-Diacetylferrocene is slightly soluble in petroleum ether, moderately soluble in ether or benzene, and very well soluble in $CHCl_3$, CH_2Cl_2, and concentrated acids [1]. It is a stable compound [6], as is also shown by quantum chemical calculations by the extended Hückel-theory method, according to which no decomposition in acid or neutral media is to be expected [8]. No pyrolysis occurs within 1.5 h in the vapor phase at 10 Torr N_2 pressure, vessel temperature 100°C [1]. Oxidation with NaOCl in $C_2H_5OH/CH_2Cl_2/H_2O$ gives $C_5H_5FeC_5H_3(CO_2H)_2\text{-}1,3$ in a quantitative yield [7]. With NaOBr in dioxane/H_2O below 10°C the yield is 50% [9].

$AgBF_4$ in a 1/1 mole ratio in dry ether forms the red $[C_5H_5FeC_5H_3(CO{-}CH_3)_2\text{-}1,3 \cdot AgBF_4]_n$ complex, in which the Ag is coordinated with the two O atoms. Due to its insolubility in all normal solvents no molecular weight determination is possible. It appears to be oligomeric, if not even polymeric. In its formation the ν(C=O) IR band is displaced around 46 cm^{-1} toward lower wave numbers versus that of $C_5H_5FeC_5H_3(CO{-}CH_3)_2\text{-}1,3$. If the complex is dissolved in liquid SO_2, on warming to −20°C $[C_5H_5FeC_5H_3(CO{-}CH_3)_2\text{-}1,3]BF_4$ forms, a dark blue, crystalline ferrocenium salt, with precipitation of Ag and displacement of the ν(C=O) band around 34 cm^{-1} toward higher wave numbers [3].

No further acetylation occurs with $CH_3COCl/AlCl_3$ in CH_2Cl_2 at room temperature or 40°C [1]. Base-catalyzed condensation with C_6H_5CHO leads to $C_5H_5FeC_5H_3(CO{-}CH{=}CHC_6H_5)_2\text{-}1,3$ and $C_5H_5FeC_5H_3(CO{-}CH{=}CHC_6H_5\text{-}1)COCH_3\text{-}3$ [6]. Reaction with $C_5H_5FeC_5H_3(CHO)_2\text{-}1,3$ in KOH/C_2H_5OH at 55°C under conditions of high dilution yields I as well as the polycondensation product II [9]. If $C_5H_5FeC_5H_3(CO{-}CH_3)_2\text{-}1,3$ in KOH/C_2H_5OH is mixed dropwise with ethanolic

References on p. 188

$C_5H_5FeC_5H_4$–CHO solution, $C_5H_5FeC_5H_3(CO–CH{=}CH–C_5H_4FeC_5H_5)_2$-1,3 is formed, in a 60% yield at 55°C for a reaction time of 2 h [9].

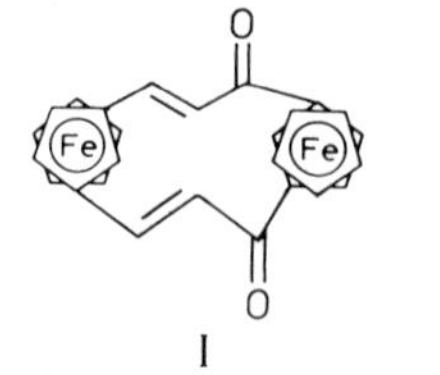

I

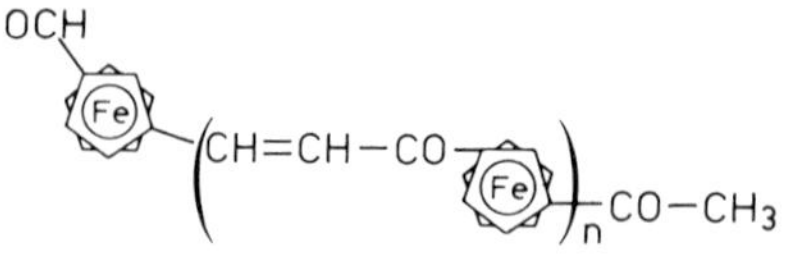

II

5.2.3.5.3.1.2 Additional Ketones of the Type $C_5H_5FeC_5H_3$(CO–R′-1)CO–R″-3

For $C_5H_5FeC_5H_3(CO–C_2H_5)_2$ see 5.2.3.5.2.2, p. 159.

$C_5H_5FeC_5H_3(CO–CH{=}CHC_6H_5)_2$-1,3 forms on addition at room temperature of 10% NaOH to a mixture of $C_5H_5FeC_5H_3(CO–CH_3)_2$-1,3 and C_6H_5CHO in C_2H_5OH. It is isolated on SiO_2 with benzene following $C_5H_5FeC_5H_3(CO–CH{=}CHC_6H_5$-1)$COCH_3$-3, in the form of red crystals which melt at 300°C (from benzene) with decomposition. The stable compound is too insoluble to take an ^1H NMR spectrum. The IR spectrum in CCl_4 has a ν(C=C) band at 1610 and a ν(C=O) band at 1670 cm^{-1} [6].

$C_5H_5FeC_5H_3(CO–CH{=}CHC_6H_5$-1)$COCH_3$-3. For preparation see the previous paragraph. The stable compound forms red crystals melting at 260°C (from benzene) with decomposition. The IR spectrum in CCl_4 has bands at 1605 (νC=C) and 1665 (νC=O) cm^{-1}. Due to its slight solubility no ^1H NMR spectrum could be observed [6].

References:

[1] M. Rosenblum (Diss. Harvard Univ. 1953, p. 108). – [2] A. F. Ellis (Diss. Univ. Illinois 1963, pp. 1/152 from Diss. Abstr. **24** [1963] 510). – [3] P. Carty, M. F. A. Dove (J. Organometal. Chem. **28** [1971] 125/32). – [4] A. N. Nesmeyanov, E. V. Leonova, E. I. Fedin, P. V. Petrovsky, N. S. Kochetkova (J. Organometal. Chem. **96** [1975] 279/82). – [5] A. N. Nesmeyanov, E. V. Leonova, N. S. Kochetkova, A. I. Malkova, A. G. Makarovskaya (J. Organometal. Chem. **96** [1975] 275/8).

[6] Š. Toma, M. Paričiová, E. Solčániová, K. Leško (Collection Czech. Chem. Commun. **43** [1978] 295/303). – [7] M. Hisatome, O. Tachikawa, M. Sasho, K. Yamakawa (J. Organometal. Chem. **217** [1981] C17/C20). – [8] P. Zahradník, Š. Toma (Acta Fac. Rerum Nat. Univ. Comenianae Chim. No. 29 [1981] 1/7). – [9] A. Kasahara, T. Izumi, Y. Yoshida, I. Shimizu (Bull. Chem. Soc. Japan **55** [1982] 1901/6).

5.2.3.5.3.2 Ketones of the Type $C_5H_5FeC_5H_3$(CO–R′-1)R^2-3 (CO–R′ = R^1)

Compounds of the compositions $C_5H_5FeC_5H_3$(CO–R′-1)A–CO–R″-3 are treated in Section 5.2.3.5.3.3, pp. 205/9.

Each of the compounds compiled in Table 11 possesses a plane of chirality. (For the centrochiral RS nomenclature used for the consequent ferrocene chirality see "Ferrocene" 7, 1980, pp. 1/4 for the English version, pp. 4/7 for the German text.) Apart from No. 1, however, these compounds have been prepared only in the racemic form. The compounds No. 21 and 22 in addition have in an asymmetrically substituted C atom of R^2 (C-7 and C-9, respectively) a chirality center, so that they can occur in two diastereoisomeric forms; these have not been investigated.

In general the compounds have been prepared by the following two methods, of which Method I always yields a mixture with the positional 1,1′-isomer and in the majority of cases also with the 1,2-isomer. The total yield of the positional isomer mixture is denoted with "m" and is reported when no separation was conducted. The numbers following the solidus give the percentages of the 1,3-isomers in the mixtures, determined for example by gas chromatography [14], ^{1}H NMR spectroscopy [11, 16, 44], if necessary with addition of a shift reagent [32]. The relative abundances of the isomers in general are $1,2<1,3<1,1'$. In exceptional cases – especially after conducting the reaction corresponding to the Methods Ia, Ij, or Ik – the 1,3-isomers are the most abundant.

Method I. Friedel-Crafts acylation of $C_5H_5FeC_5H_4{-}R^2$:

a) $BF_3 \cdot O(C_2H_5)_2$ is added at 0°C to $C_5H_5FeC_5H_4{-}R^2$ and $(CH_3CO)_2O$ in a mole ratio (2.4 to 3)/1/(1.6 to 2) in CH_2Cl_2, and the reaction is continued first for 30 min at 0°C, then for 30 to 120 min at room temperature [14, 16]. The fraction of the 1,3-isomer in the mixture of positional isomers is determined by gas chromatography at 215°C [14].

b) The solution of $C_5H_5FeC_5H_4{-}R^2$ and $(CH_3CO)_2O$ (molar ratio 1/1) in CH_2Cl_2 at 0°C is saturated with BF_3 and subsequently warmed up to room temperature over a 2.75 h period [9].

c) $C_5H_5FeC_5H_4{-}R^2$ in CH_2Cl_2 is mixed at 0°C with $(CH_3CO)_2O$ and a solution of BF_3 in ether (mole ratio 1/2/2), and the mixture is warmed up to room temperature with stirring over a 6 h period. Column chromatography on Al_2O_3 isolates in succession $C_5H_5FeC_5H_4{-}R^2$, the 1,2-isomer, the 1,1'-isomer, and finally No. 6 with the eluents petroleum ether and then petroleum ether/ether in the ratios 9/1 and 4/1, respectively [36].

d) $C_5H_5FeC_5H_4{-}R^2$ is reacted with $(CH_3CO)_2O$ in the presence of H_3PO_4. No. 2 predominates in the mixture of the homoannular disubstituted compounds [43].

e) $C_5H_5FeC_5H_4{-}R^2$ is heated with $(CH_3CO)_2O$ and 85% H_3PO_4 (molar ratio 1/20/4) for 10 min up to 90 to 95°C [44, 46]. The use of a reaction mixture which works more mildly than the Perrier complex substantially suppresses side reactions such as the acetylation of the phenyl group when $R^2 = C_6H_5$ [44]. The isomer mixture obtained cannot be separated by column chromatography on SiO_2 with benzene as eluent; therefore the fraction of the 1,3-isomer has been determined by NMR spectroscopy [44,46].

f) $C_5H_5FeC_5H_4{-}OCH_3$ and $(CH_3CO)_2O$ (mole ratio 5/3) are reacted with polyphosphoric acid for 2 h at 45°C [20].

g) $C_5H_5FeC_5H_4{-}R^2$ is reacted with R'COCl and $AlCl_3$ (molar ratio 2/3/3) in CH_2Cl_2 at 20°C [41].

h) The complex obtained from R'COCl and $AlCl_3$ in CH_2Cl_2 is immediately added dropwise with cooling [31] at 0 to 5°C to $C_5H_5FeC_5H_4{-}R^2$ (molar ratio ca. 1/1/1) in CH_2Cl_2, and the reaction is continued for 5 min at 0°C (No. 20) [20], 75 min at 5 to 10°C (No. 30) [8], or 3 h at room temperature (Nos. 24 and 48) [31].

i) The reaction as in Method Ih is begun at room temperature and continued at this temperature for an additional 1 to 1.5 h (Nos. 2, 5, 48) [1, 4, 8, 21], 4 h (No. 28) [29], 18 h (No. 50) [48], or 20 h (No. 29) [38]; see also [35].

j) $C_5H_5FeC_5H_4{-}C_2H_5$ in CH_2Cl_2 is dropped over a 7 h period into $CH_3COCl/AlCl_3$ (mole ratio 1/1/1) in CH_2Cl_2 at −5°C, and the reaction is continued 30 min more at this temperature. This procedure yields the optimum conditions for the preparation of the homoannular disubstituted compounds and especially for that of No. 2 due to the activation of the position 3 in the cyclopentadienyl ring by the C_2H_5 group [40].

References on pp. 203/4

k) $C_5H_5FeC_5H_4–C(CH_3)_3$ is mixed at room temperature over a 2 h period with $AlCl_3$ in CH_2Cl_2, subsequently with C_6H_5COCl (molar ratio ca. 1/1/1.4), and the reaction is continued 1 h at this temperature. Thin layer chromatography on silica gel with benzene separates unreacted $C_5H_5FeC_5H_4–C(CH_3)_3$ ($R_f = 0.9$) and the 1,2-isomer ($R_f = 0.8$) from the mixture of No. 49 and its 1,1′-isomer (2/1; $R_f = 0.55$). The latter mixture appears to be inseparable; with none of the eluents used could a resolution be observed on the silica gel column [33].

l) $AlCl_3$ in CH_2Cl_2 is reacted successively with $C_6H_5CH_2COCl$ and $C_5H_5FeC_5H_4–R^2$ (molar ratio 1.2/1.3/1), and the reaction is continued for 3 h at 40 to 50°C. The main product, $C_6H_5CH_2CO–C_5H_4FeC_5H_4–CH_2C(CH_3)_3$, is separated by chromatography on an Al_2O_3 column with petroleum ether [8].

m) CH_3COCl in CH_2Cl_2 at 0°C is added over a 30 min period to the mixture of $C_5H_5FeC_5H_4–C_6H_5$ and $AlCl_3$ (molar ratio 1.3/1/1), and the reaction is continued for 3 h at 0°C, then 1 h at room temperature [5].

n) $AlCl_3$ dissolved in CH_3COCl is added dropwise at 0°C to $C_5H_5FeC_5H_4–R^2$, dissolved in CH_3COCl/CH_2Cl_2, and the reaction is continued for 1 h at 0°C [10].

o) $AlCl_3$ is added in portions to $C_5H_5FeC_5H_4–R^2$ and R′COCl (mole ratio 1/1/1) in CH_2Cl_2 at 0 to 5°C, and the reaction is continued another 40 min at room temperature. The 1,2-isomer can be separated by preparative thin layer chromatography or by column chromatography [41]. For separation of the mixture of No. 52 and its 1,1′-isomer see "Further information", p. 203.

p) Finely divided $AlCl_3$ and R′COCl in CH_2Cl_2 are added in rapid succession at 0°C to a solution of $C_5H_5FeC_5H_4–R^2$ (mole ratio 1/0.56/1) in CH_2Cl_2, and the mixture is then worked up after 15 min. Separation of $C_5H_5FeC_5H_4–R^2$ and the 1,1′-isomer takes place by preparative thin layer chromatography on SiO_2 [38].

q) The solution of $AlCl_3$ in CH_2Cl_2 plus isopropyl or n-butyl ether is stirred at 20°C with $C_5H_5FeC_5H_4–CH_2C(CH_3)_3$ and $(CH_3)_3CCOCl$ (mole ratio 2.4/1/1) for 2 h and then evaporated to dryness in vacuum before the usual processing. The separation of No. 34 from its 1,1′-isomer takes place by column chromatography on Al_2O_3 with benzene as eluent [6].

r) $AlCl_3$ in ether is mixed in succession with $C_5H_5FeC_5H_4–R^2$ and R′COCl in the mole ratio 1.2/1/>1.2 and reacted for 3 h (No. 51) [7, 8] or 16 h (No. 36) [7] at room temperature or for 22 h at the boiling point (No. 53) [8]. Nos. 51 and 53 are separated from their respective 1,1′-isomers by column chromatography on Al_2O_3 in petroleum ether [7, 8].

s) $C_5H_5FeC_5H_4–R^2$, R′COCl, $AlCl_3$, and Al powder in the molar ratio 1/1.3/1.2/(0.5 to 0.75) are reacted in ether at room temperature for 16 to 17 [6], 20 h [7]. The 1,3-compound is separated from its 1,1′-isomer by chromatography on an Al_2O_3 column [6].

Method II. Condensation of $C_5H_5FeC_5H_4$ $(CO–CH_3\text{-}1)R^2\text{-}3$ with C_6H_5CHO and NaOH (mole ratio 2/3/20) in C_2H_5OH/H_2O at room temperature (reaction time 12 h). When starting not with the pure compounds No. 7, 9, 10, 11, 13, 15, 16, 17, or 18 but with the isomer mixtures obtained by Method Ie, the compounds here produced must be separated from $C_6H_5CH{=}CH–CO–C_5H_4FeC_5H_4–R^2$ and if necessary from small amounts of their 1,2-isomers, which happens by column chromatography on SiO_2 with benzene as eluent, in the isomer elution sequence 1,2, 1,1′, and 1,3 [44].

The **separation** of the isomer mixtures is effected dominantly by chromatography, in rare cases by fractional crystallization. Since the difference in adsorptivity between the 1,3- and

1,1′-isomers is even less than between the 1,3- and 1,2-isomers, the first separation in particular seldom succeeds completely and is often connected with considerable experimental effort. At times the possibility exists of chemically converting the inseparable mixture (see, for example, Nos. 2 and 30) into a better separable one. To the extent that the separation has not already been treated along with the preparation method, see the special results under the "Further information" for the individual compounds.

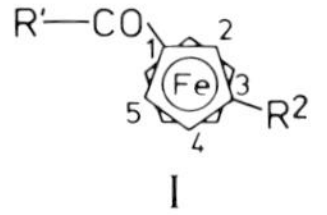

I

Properties. The ^{1}H NMR spectra (enumeration as for Formula I) for the compounds $C_5H_5FeC_5H_3(CO{-}CH_3\text{-}1)R^2\text{-}3$ (R^2 = alkyl) exhibit as a rule besides the singlet of the C_5H_5 protons two lower field peaks with a 2/1 intensity ratio, for which the latter (H-4) is more shielded than the former (H-2,5). The chemical shift of the ring protons can be calculated with an average deviation of 0.04 ppm by means of an additivity relationship [16].

In the IR spectrum the compounds $C_5H_5FeC_5H_3(CO{-}CH_3\text{-}1)R^2\text{-}3$ (R^2 = alkyl) show two average to weak intensity bands around 905 and 920 cm^{-1}, of which the former is the more intense [16]. With $R^2 = C_6H_5$ they shift to the lower frequences 897 and 905 cm^{-1} [3]. In this the 1,3-compounds differ from the corresponding 1,2-compounds, which according to [3] exhibit in this region only one band, or according to [16] also two bands, but with the higher frequency one as the more intense. A linear correlation corresponding to a modified and extended Seth-Paul-Van Duyse equation has been observed between the ν(C=O) vibration in CCl_4, and the $X^+(R^{A,B})$ constants of the structural fragments for 154 compounds with ferrocene structures of the general formula $R^A{-}CO{-}R^B$, among them the compounds No. 42, 43, 45, 47, and 55, with $R^A = 3\text{-}R^2{-}C_5H_3FeC_5H_5$ and $R^B = CH{=}CH{-}C_6H_5$ s-cis with respect to the C=O bond. The $X^+(R^{A,B})$ used have been confirmed by the recently derived empirical relationship for calculation of the substituent constants and transmission factors [26].

Chemical Reactions. Reduction of the R′–CO to an $R'{-}CH_2$ group can be effected by the Clemmensen method with Zn and concentrated hydrochloric acid in CH_3CO_2H, as reported for the compounds No. 2 [2], 30 [7], a mixture of No. 32 and $(CH_3)_3CCO{-}C_5H_4FeC_5H_4{-}CH_2CH(CH_3)_2$ [7], Nos. 34, 36, or 51 [7] or with $LiAlH_4/AlCl_3$ as reported for Nos. 2 [22] or 28 [29]. The R′–CHOH group forms with $LiAlH_4$ in a nonstereoselective reaction. Thus the diastereomers (1SR, 6SR)-II and (1SR, 6RS)-III in a 1/1 ratio are obtained with $R^{**} = R^2 = CH_3$, $R^* = H$ from No. 1 [37], with $R^{**} = CH_3$, $R^2 = C_2H_5$, $R^* = H$ from No. 2 [43], or with $R^{**} = R^2 = C(CH_3)_3$, $R^* = H$ from No. 33. The condensation of $C_5H_5FeC_5H_3(CO{-}CH_3\text{-}1)R^2\text{-}3$ with C_6H_5CHO and NaOH (see Method II) leads to $C_5H_5FeC_5H_4(CO{-}CH{=}CHC_6H_5\text{-}1)R^2\text{-}3$ (conducted with compounds No. 7, 9, 10, 11, 13, 15, 16, 17, and 18) [44].

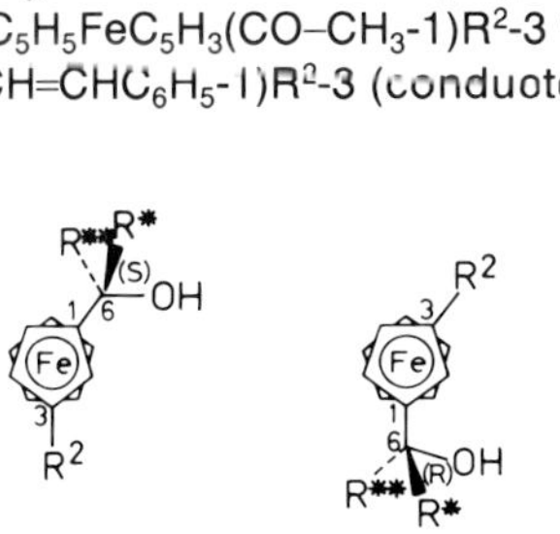

a (1S, 6S) b (1R, 6R)

II

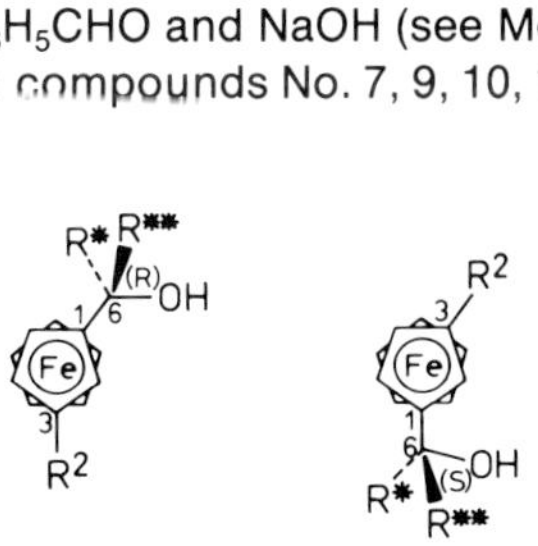

a (1S, 6R) b (1R, 6S)

III

Applications. The compounds No. 5, 37, 51, and 53 are intermediate products in the synthesis of blood-building medications for the elimination of iron deficiency anemia [8].

References on pp. 203/4

Table 11

Ketones of the Type $C_5H_5FeC_5H_3(CO{-}R'{-}1)R^2{-}3$.

Further information for numbers preceded by an asterisk is given at the end of the table, pp. 199/203.
For abbreviations and dimensions see p. VIII.

No.	$C_5H_5FeC_5H_3(CO{-}R'{-}1)R^2{-}3$ R′ of COR′	R^2	method of preparation (yield in %, "m" indicates isomeric mixture, number following solidus is fraction present of the 1,3-compound), properties, and remarks	Ref.
*1	CH_3	CH_3	Ia (10; 73 m/31%), Ib (m/23.5%) m.p. 57.5 to 61.5°, 58°, 61.5° ^{1}H NMR (CCl_4): 2.03 (s, CH_3-3), 4.08 (s, C_5H_5), 4.31 (H-4), 4.6 (H-2,5) ^{1}H NMR (CCl_4): 4.01 (C_5H_5), 4.32 (H-4; J(H-2,4) = 1.0), 4.59 (H-5; J(H-4,5) = 2.5), 4.6 (H-2; J(H-2,5) = 1.4) ^{1}H NMR ($CHCl_3$): 4.17 (C_5H_5), 4.48 (H-4; J(H-2,4) = 1.1), 4.72 (H-5; J(H-4,5) = 3.1), 4.74 (H-2; J(H-2,5) = 1.5) ^{1}H NMR (C_6H_6): 3.88 (C_5H_5), 4.09 (H-4; J(H-2,4) = 1.45), 4.55 (H-2; J(H-2,5) = 1.4), 4.56 (H-5; J(H-4,5) = 2.45) IR (Nujol): 905 (m), 935.5 (w), 1005 (s), 1299 (s)	[9, 11, 14 to 17, 21, 31]
*2	CH_3	C_2H_5	Ia (30; 4; m/36.2%), Id, Ii, Ij (m/56.3%), see "Further information", p. 199 fine orange needles, m.p. 11.6 to 12.6° (petroleum ether), b.p. 144°/0.8 Torr ^{1}H NMR (CCl_4): 2.38 (CH_2 of C_2H_5), 4.09 (C_5H_5), 4.31 (H-4), 4.63 (H-2,5), see "Further information", p. 200 IR (Nujol): 901 (s), 919 (m), 1003 (s), 1106 (s), 1290 (s) IR (CCl_4): 901, 919, 1004, 1110, see figure in [1] UV (CCl_4): $\lambda_{max}(\log \varepsilon)$ = 230 (4.18), 272 (3.81), 340 (3.05), 460 (2.69) UV (C_2H_5OH/H_2O, 95/5): $\lambda_{max}(\varepsilon)$ = 230 (15100), 272 (6500)	[1, 3 to 5, 11, 14, 16, 35, 40, 43, 47]
*3	CH_3	$CH(CH_3)_2$	Ia (7; m/43.8%) b.p. 143°/1 Torr ^{1}H NMR (CCl_4): 2.66 (CH of C_3H_7), 4.09 (C_5H_5), 4.29 (H-4), 4.59 (H-2,5) IR (film): 901 (s), 934.6 (m), 1004 (s), 1107.4 (s), 1282 (s)	[11, 14, 16]
*4	CH_3	$C(CH_3)_3$	Ia (33 m/66%; m/55%), see "Further information", p. 200 m.p. 83.5 to 84.5° ^{1}H NMR (CCl_4): 4.1 (C_5H_5), 4.28 (H-4), 4.57 (H-2,5) IR (Nujol): 900 (s), 923.4 (w), 1005 (s), 1105 (s), 1242.2 (s)	[11, 14, 16]

*5	CH_3	$CH_2C(CH_3)_3$	II m.p. 76 to 78°	[8]
*6	CH_3	$CH_2CH{=}CHC_6H_5$-(E)	Ic (21 related to unrecovered $C_5H_5FeC_5H_4{-}R^2$) yellow solid, m.p. 111 to 112° ^{1}H NMR ($CDCl_3$): 2.35 (s, $COCH_3$), 3.28 (m, CH_2), 4.19 (s, C_5H_5); 4.50 (m) and 4.76 (m) (C_5H_3); 6.3 to 6.5 (m, CH=CH), 7.33 (m, C_6H_5) UV (C_2H_5OH): $\lambda_{max}(\varepsilon)$ = 252 (22000)	[36]
*7	CH_3	C_6H_5	Ic (19 m/24%), Im (3), In orange rods, m.p. 113 to 114° (ether/petroleum ether) ^{1}H NMR ($CDCl_3$): 2.41 (s, $COCH_3$), 4.05 (s, C_5H_5), 4.90 (m, H-4,5), 5.23 (t, H-2), 7.0 to 7.6 (m, C_6H_5) IR ($CHCl_3$): 895.3, 905, 1000, 1103.8 UV (C_2H_5OH/H_2O, 95/5): $\lambda_{max}(\varepsilon)$ = 245 (23600), 270sh (11300)	[3, 5, 10, 44, 46]
8	CH_3	$C_6H_4CH_3$-4	Ie (32 m/16%) m.p. 70 to 78° (m) ^{1}H NMR ($CDCl_3$): 2.33 (s, CH_3), 2.43 (s, $COCH_3$), 4.08 (s, C_5H_5), 4.92 (m, H-4,5), 5.22 (m, H-2), 7.0 to 7.4 (m, C_6H_4)	[44, 46]
9	CH_3	$C_6H_4CH_3$-3	Ie (61 m/24%) m.p. 52 to 59° (m) ^{1}H NMR ($CDCl_3$): 2.40 (s, CH_3), 2.43 (s, $COCH_3$), 4.08 (s, C_5H_5), 4.91 (m, H-4,5), 5.23 (m, H-2), 6.9 to 7.4 (m, C_6H_4) for chemical behavior see Method II, p. 190	[44, 46]
10	CH_3	$C_6H_3(CH_3)_2$-3,4	Ie (42 m/31%) oil (m) ^{1}H NMR ($CDCl_3$): 2.22 (s, CH_3), 2.35 (s, $COCH_3$), 3.98 (s, C_5H_5), 4.83 (m, H-4,5), 5.16 (m, H-2), 6.9 to 7.4 (m, C_6H_3) for chemical behavior see Method II, p. 190	[44, 46]
11	CH_3	$C_6H_3(CH_3)_2$-3,5	Ie (42 m/28%) oil (m) ^{1}H NMR ($CDCl_3$): 2.28 (s, CH_3, $COCH_3$), 3.98 (s, C_5H_5), 4.83 (m, H-4,5), 5.17 (m, H-2), 6.8 to 7.2 (m, C_6H_3) for chemical behavior see Method II, p. 190	[44, 46]

References on pp. 203/4

Table 11 (continued)

No.	$C_5H_5FeC_5H_3(CO{-}R'{-}1)R^2{-}3$ R′ of COR′	R^2	method of preparation (yield in %, "m" indicates isomeric mixture, number following solidus is fraction present of the 1,3-compound), properties, and remarks	Ref.
12	CH_3	C_6H_4Cl-4	Ie (31m/19%) m.p. 85 to 90° (m) ^{1}H NMR ($CDCl_3$): 2.42 (s, $COCH_3$), 4.05 (s, C_5H_5), 4.90 (m, H-4,5), 5.20 (m, H-2), 7.2 to 7.6 (m, C_6H_4)	[44, 46]
13	CH_3	C_6H_4Cl-3	Ie (55m/19%) oil (m) ^{1}H NMR ($CDCl_3$): 2.40 (s, $COCH_3$), 4.13 (s, C_5H_5), 5.01 (m, H-4,5), 5.30 (m, H-2), 7.0 to 7.7 (m, C_6H_4) for chemical behavior see Method II, p. 190	[44, 46]
14	CH_3	C_6H_4Br-4	Ie (44m/16%) m.p. 87 to 93° (m) ^{1}H NMR ($CDCl_3$): 2.42 (s, $COCH_3$), 4.06 (s, C_5H_5), 4.91 (m, H-4,5), 5.21 (m, H-2), 7.2 to 7.6 (m, C_6H_4)	[44, 46]
15	CH_3	C_6H_4Br-3	Ie (49m/21%) oil (m) ^{1}H NMR ($CDCl_3$): 2.39 (s, $COCH_3$), 4.05 (s, C_5H_5), 4.89 (m, H-4,5), 5.19 (m, H-2), 7.0 to 7.6 (m, C_6H_4) for chemical behavior see Method II, p. 190	[44, 46]
16	CH_3	$C_6H_3Cl_2$-3,4	Ie (45m/25%) oil (m) ^{1}H NMR ($CDCl_3$): 2.40 (s, $COCH_3$), 4.03 (s, C_5H_5), 4.88 (m, H-4,5), 5.16 (m, H-2), 7.0 to 7.6 (m, C_6H_3) for chemical behavior see Method II, p. 190	[44, 46]
17	CH_3	$C_6H_4CF_3$-3	Ie (47m/22%) oil (m) ^{1}H NMR ($CDCl_3$): 2.40 (s, $COCH_3$), 4.04 (s, C_5H_5), 4.93 (m, H-4,5), 5.25 (m, H-2), 7.3 to 7.7 (m, C_6H_4) for chemical behavior see Method II, p. 190	[44, 46]

18	CH_3	$C_6H_3(CF_3\text{-}3)Cl\text{-}4$	Ie (50 m/25%) oil (m) ^{1}H NMR ($CDCl_3$): 2.43 (s, $COCH_3$), 4.05 (s, C_5H_5), 4.93 (m, H-4, 5), 5.23 (m, H-2), 7.2 to 7.7 (m, C_6H_3) for chemical behavior see Method II, p. 190	[44, 46]
19	CH_3	$C_6H_4O_2CCH_3\text{-}4$	Ie (43 m/27%) oil (m) ^{1}H NMR ($CDCl_3$): 2.39 (s, $COCH_3$), 4.03 (s, C_5H_5), 4.86 (m, H-4, 5), 5.18 (m, H-2), 6.8 to 7.6 (m, C_6H_4)	[44, 46]
*20	CH_3	OCH_3	If (45 m), Ih (51 m) deep red oil (m)	[20]
*21	CH_3 (impure)	$\overset{7}{C}H(CH{=}CHC_6H_5)OCH_3$-(E)	see No. 6, p. 200 ^{1}H NMR ($CDCl_3$): 2.22 (s, $COCH_3$), 3.32 (s, OCH_3), 4.14 (s, C_5H_5), 4.5 to 4.9 (m, C_5H_3, H-7), 6.0 to 6.8 (m, CH=CH), 7.29 (br.s, C_6H_5)	[36]
*22	CH_3 (impure)	$CH{=}CH\text{-}\overset{9}{C}H(C_6H_5)OCH_3$-(E)	see No. 6, p. 200 ^{1}H NMR ($CDCl_3$): 2.22 (s, $COCH_3$), 3.30 (s, OCH_3), 4.04 (s, C_5H_5), 4.4 to 4.9 (m, C_5H_3, H-9), 5.9 to 6.5 (m, CH=CH), 7.29 (br.s, C_6H_5)	[36]
*23	CH_3	CHO	see "Further information", p. 201 orange crystals, m.p. 106 to 107.5° ^{1}H NMR ($CDCl_3$): see figure in [28]	[28]
*24	C_2H_5	CH_3	Ih (19) oil (oxime: m.p. 145°) ^{1}H NMR ($CDCl_3$): 1.17 (t, CH_3 of C_2H_5), 2.03 (CH_3-3), 2.70 (q, CH_2 of C_2H_5), 4.12 (s, C_5H_5); 4.45 (m, 1H) and 4.70 (d, 2H) (C_5H_3)	[27, 31]
*25	C_2H_5	C_4H_9-n	for chemical behavior see "Further information", p. 201	[25]
*26	C_2H_5	$C(CH_3)_3$	see "Further information", p. 202	[42]
*27	C_2H_5	$C_6H_4Cl\text{-}4$	see "Further information", p. 202 m.p. 113° ^{1}H NMR ($CDCl_3$): 1.22 (t, CH_3 of C_2H_5), 2.77 (m, $COCH_2$), 4.04 (s, C_5H_5); 4.90 (d, 2H) and 5.23 (m, 1H)(C_5H_3); 7.34 (m, C_6H_4)	[30]

Table 11 (continued)

No.	$C_5H_5FeC_5H_3(CO{-}R'\text{-}1)R^2\text{-}3$ R′ of COR′	R^2	method of preparation (yield in %, "m" indicates isomeric mixture, number following solidus is fraction present of the 1,3-compound), properties, and remarks	Ref.
*28	C_3H_7-n	CH_3	Ii (67 m/30%) ^{1}H NMR ($CDCl_3$ or CCl_4?): 0.98 (t, CH_3 of C_3H_7), 1.68 (m, $-CH_2-$), 1.99 (s, CH_3-3), 2.55 (m, $COCH_2$), 3.96 (s, C_5H_5); 4.21 (m, 2H) and 4.59 (m, 2H) (cyclopentadienyl protons) (for a 84/16 mixture of No. 28 and its 1,1′-isomer)	[29]
*29	$CH(CH_3)_2$	$C(CH_3)_3$	Ii (8) deep orange colored plates, m.p. 68 to 70° (pentane) ^{1}H NMR: 1.20 (d, CH_3 of i-C_3H_7), 1.30 (s, C_4H_9), 2.8 to 3.4 (m, CH of C_3H_7), 4.20 (s, C_5H_5); 4.42 (m, 1H) and 4.25 (m, 2H) (C_5H_3)	[38]
*30	$CH(CH_3)_2$	$CH_2C(CH_3)_3$	Ih m.p. 54 to 55°	[7, 8]
31	$C(CH_3)_3$	C_3H_7-n	Is b.p. 158 to 162°/2 Torr	[6]
32	$C(CH_3)_3$	$CH_2CH(CH_3)_2$	Is (m with the 1,1′-isomer) for reduction see p. 191	[7]
33	$C(CH_3)_3$	$C(CH_3)_3$	Ip (25) deep orange colored solid, m.p. 107 to 108° (pentane) ^{1}H NMR: 1.17 (s, C_4H_9), 1.25 (s, C_4H_9), 4.08 (s, C_5H_5); 4.27 (m, 1H) and 4.71 (m, 2H)(C_5H_3) for reduction see p. 191	[38]
34	$C(CH_3)_3$	$CH_2C(CH_3)_3$	Iq m.p. 110° for reduction see p. 191	[6]
35	$C(CH_3)_3$	$(CH_2)_2C_6H_5$	Is m.p. 81 to 82°	[6]
36	$CH_2CH(CH_3)-$ $CH_2C(CH_3)_3$	$CH_2C(CH_3)_3$	Ir for reduction see p. 191	[7]

References on pp. 203/4

37	$CH_2C_6H_5$	$CH_2C(CH_3)_3$	I (small amount) m.p. 111 to 112° for "Applications" see p. 191	[8]
38	$CH{=}CHC_6H_5$	C_6H_5	II (10) m.p. 119 to 120°	[44]
39	$CH{=}CHC_6H_5$	$C_6H_4CH_3$-3	II (17) m.p. 110°	[44]
40	$CH{=}CHC_6H_5$	$C_6H_3(CH_3)_2$-3,4	II (17) m.p. 155 to 156°	[44]
41	$CH{=}CHC_6H_5$	$C_6H_3(CH_3)_2$-3,5	II (17) m.p. 141 to 142°	[44]
42	$CH{=}CHC_6H_5$	C_6H_4Cl-3	II (22) m.p. 110 to 111° IR (CCl_4): 1661.5 (νC=O) for s-cis COR′	[26, 44]
43	$CH{=}CHC_6H_5$	C_6H_4Br-3	II (13) m.p. 156 to 157° IR (CCl_4): 1662 (νC=O) for s-cis COR′	[26, 44]
44	$CH{=}CHC_6H_5$	$C_6H_3Cl_2$-3,4	II (14) m.p. 156 to 157°	[44]
45	$CH{=}CHC_6H_5$	$C_6H_4CF_3$-3	II (12) m.p. 123 to 125° IR (CCl_4): 1662.5 (νC=O) for s-cis COR′	[26, 44]
46	$CH{=}CHC_6H_5$	$C_6H_3(CF_3$-3)Cl-4	II (19) m.p. 145 to 146°	[44]
47	$CH{=}CHC_6H_5$	$C_6H_4OCH_3$-4	II (26) m.p. 111 to 113° IR (CCl_4): 1660.5 (νC=O) for s-cis COR′	[26, 44]
*48	C_6H_5	CH_3	Ih (21), Ii (51m/37%) oil (oxime: m.p. 170°) ^{1}H NMR ($CDCl_3$): 2.03 (s, CH_3), 4.15 (s, C_5H_5); 4.52 (m, 1H) and 4.83 (d, 2H)(C_5H_3); 7.4 to 8.1 (m, C_6H_5) ^{1}H NMR (CCl_4): 2.04 (s, CH_3)	[21, 27, 31]

References on pp. 203/4

Table 11 (continued)

No.	$C_5H_5FeC_5H_3(CO{-}R'{-}1)R^2{-}3$ R′ of COR′	R²	method of preparation (yield in %, "m" indicates isomeric mixture, number following solidus is fraction present of the 1,3-compound), properties, and remarks	Ref.
*49	C_6H_5	$C(CH_3)_3$	Ik (84 m) (m = 1/2 mixture of No. 49 and its 1,1′-isomer) viscous red liquid (m) ^{1}H NMR ($OS(CD_3)_2$): 4 to 5 (cyclopentadienyl protons), 7 to 8 (C_6H_5) (m) IR ($OS(CH_3)_2$): 1638 (νC=O) (m) mass spectrum (rel. intensity in %, ion); m/e = 346 (100, M^+), 331 (15, $[M-CH_3]^+$), 316 (1.5, $[M-2CH_3]^+$), 3.1 (0.3, $[M-3CH_3]^+$), 289 (0.5, $[M-C_4H_9]^+$), 226 (43, $[M-C_4H_9-C_5H_3]^+$), 197 (8, $[M-C_3H_8-C_6H_5CO]^+$), 186 (3, $[FeC_{10}H_{10}]^+$), 137 (29, $[M-C_6H_5CO-C_3H_4-C_5H_4]^+$), 131 (16, $[M-C_6H_5-O-C_4H_9-C_5H_5]^+$), 63 (2, $C_5H_3^+$), 56 (30, Fe), 28 (53, CO^+) (m)	[33]
*50	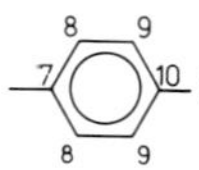 CH_3	CH_3	Ii (63 m/35.7%) ^{13}C NMR ($CDCl_3$): 14.9 (0.3 C, CH_3), 70.8 (C_5H_5), 77.5 (C-1), 88.8 (C-3); 128.1 (0.6 C) and 128.2 (1.0 C) and 128.4 (0.4 C) and 128.8 (2.0 C) (C-8,8,9,9); 137.2 (C-7), 141.7 (C-10), 198.3 or 198.9 (C=O) mass spectrum: m/e = 318 $[M]^+$, 224, 202, 199, 121, 119, 91, 56	[48]
51	C_6H_{11}-cyclo	$CH_2C(CH_3)_3$	Ir m.p. 100 to 105° for reduction and "Applications" see p. 191	[7, 8]
*52	C_6H_4Cl-2	CH_3	Io (29 related to unrecovered $C_5H_5FeC_5H_4{-}CH_3$) m.p. 98 to 101° ^{1}H NMR ($CDCl_3$): 2.04 (s, CH_3-3) IR (CH_2Cl_2): 1650 (νC=O)	[41]
53	C_6H_4Cl-2	$CH_2C(CH_3)_3$	Ir m.p. 113 to 114° for "Applications" see p. 191	[8]
*54	C_6H_4Cl-2	C_6H_5	Ig (76 m/24%)	[41]
supplement:				
55	$CH{=}CHC_6H_5$	$C_6H_4CH_3$-4	IR (CCl_4): 1661 (νC=O) for s-cis COR′	[26]

* Further information:

$C_5H_5FeC_5H_3(CO–CH_3\text{-}1)CH_3\text{-}3$ (Table **11**, No. **1**) is isolated from the isomer mixture obtained by Methods Ia or b (p. 189) by column chromatography on neutral [16] Al_2O_3 in benzene/hexane [9] with pentane/ether (1/1) [16] or benzene/ether [9] as eluent, see also [39]. From the 1,2-isomer No. 1 can be separated by means of high pressure liquid-liquid chromatography on columns filled with Carbowax 400 on Corasil, using 10% CH_2Cl_2 in isooctane as eluent [34]. The three isomers cannot be distinguished polarographically, since they all have the same potential for the reversible one electron oxidation to the ferrocenium ion [13].

The singlet of the CH_3 protons on C-3 is shifted in CCl_4 by 0.05 ppm towards lower field vs. that of $C_5H_5FeC_5H_4–CH_3$, in benzene by 0.01 ppm towards higher field [21]. X-ray diffraction patterns on powder samples with a counter tube diffractometer give somewhat differing values for a sample levelled-off in the sample holder and for an unlevelled sample; these deviate significantly from those obtained with a photographic camera (sample in a glass cylinder) [17].

The reaction with CH_3Li in ether yields $C_5H_5FeC_5H_3(C(CH_3)_2OH\text{-}1)CH_3\text{-}3$ [36, 37]. Reduction with $LiAlH_4$ is treated on p. 191.

Optically Active Forms. (+)(1S)-$C_5H_5FeC_5H_3(CO–CH_3\text{-}1)CH_3\text{-}3$ is prepared from the acid chloride of (+)-$C_5H_5FeC_5H_3(CO_2H\text{-}1)CH_3\text{-}3$ (absolute configuration (1S) [24]) by reaction with $Cd(CH_3)_2$ in ether in an 84% yield (referred to the acid). It melts at 49 to 52°C and has a specific rotation $[\alpha]_D^{20} = +39.0°$ (C_6H_6, c = 0.68) and a ν(C=O) band in CCl_4 at 1665 cm^{-1}. The electronic spectrum in C_2H_5OH has absorptions at $\lambda_{max}(\varepsilon)$ = 227 (12560), 268 (6290), 335 (1050), 372 sh (478), 455 (492) nm. The optical rotatory dispersion (C_2H_5OH, c ≈ 0.1): $[M]_{490} = +1490°$, $[M]_{453} = -50°$, $[M]_{448} = +50°$, $[M]_{406} = -760°$, $[M]_{399} = -735°$, $[M]_{365} = -4100°$ [19]. The absolute configuration is given by the preparation method [24].

Reduction with $LiAlH_4$ yields the mixture (1S,6SR)-$C_5H_5FeC_5H_3(CHOH–CH_3\text{-}1)\text{-}CH_3\text{-}3$ (IIa + IIIa with R** = R² = CH_3, R* = H), cf. p. 191. Chloroformylation with the Vilsmeier complex of $(CH_3)_2NCHO$ and $POCl_3$ leads to (1S)-$C_5H_5FeC_5H_3(C(Cl)=CHCHO\text{-}1)CH_3\text{-}3$ [19].

$C_5H_5FeC_5H_3(CO–CH_3\text{-}1)C_2H_5\text{-}3$ (Table **11**, No. **2**) is also obtained by the acetylation of $C_5H_5FeC_5H_4–C_2H_5$ with the silicoanhydride of acetic acid in the presence of $SnCl_4$ (mole ratio 1/1/2), by reacting for 4 h $C_5H_5FeC_5H_4–C_2H_5$ and $SnCl_4$ at the boiling point with the mixture formed by boiling $SiCl_4$ for four hours with glacial acetic acid in benzene. After the usual processing $FeC_{10}H_8(COCH_3)C_2H_5$ is separated chromatographically from $FeC_{10}H_7(COCH_3)_2C_2H_5$ in 2/1 petroleum ether/benzene on an Al_2O_3 column by elution with benzene [2].

No. 2 is won free of contamination from the positional isomer mixture by column chromatography on SiO_2 with 1/1 hexane/ether [47]. The separation also is successful by column chromatography on Al_2O_3 with 1/1 pentane/ether [16]. Twenty-one fractions are eluted from Al_2O_3 with benzene, of which the fractions 12 to 20 contain No. 2 [14]. No. 2 is eluted from Al_2O_3 with 1/9 $CHCl_3/C_6H_6$ following the elution of the 1,2- and 1,1′-isomers with C_6H_6 [1]. Only the 1,2-isomer (R_f = 0.48) can be separated by preparative thin layer chromatography on Al_2O_3 plates with 1/6 ether/benzene, since due to the similarity of the absorption properties No. 2 is eluted mixed with the 1,1′-isomer (R_f = 0.32). The mixture can be obtained free of the 1,2-isomer by repeated rechromatographing [40]. The separation by gas-liquid chromatography on poly-(ethylene glycol) M-20 or neopentylglycol adipate is described in [18]. For the latter the relative retained volumes at 150°C (referred to 1 for that of $C_5H_5FeC_5H_4–CN$) for the 1,2-, 1,1′-, and 1,3-isomers are 1.33, 1.76, and 2.02, respectively [18]. Conversion into the semicarbazones and their fractional crystallization from 2/1 C_2H_5OH/H_2O enables only the separation of $H_2NCONHN=C(CH_3)–C_5H_4FeC_5H_4–C_2H_5$ from $C_5H_5FeC_5H_3(C(CH_3)=NNHCONH_2\text{-}1)C_2H_5\text{-}2$ and -3 [2]. If the mixture of No. 2 and the 1,1′-isomer is oxidized with MnO_2 to $FeC_{10}H_8(COCH_3)_2$, then

References on pp. 203/4

the subsequent thin layer chromatography yields a $C_5H_5FeC_5H_3(CO{-}CH_3)_2$-1,3 contaminated with 10% $Fe(C_5H_4{-}CO{-}CH_3)_2$ [40]. The determination of the ratio of the isomers without their preceding separation is effected by ^{1}H NMR spectroscopy in CCl_4 after addition of $Eu(dpm)_3$ (dpm = 2,2,6,6-tetramethyl-3,5-heptamedionate, see Formula II, p. 114) [32].

In contrast to the 1,2-isomer, for No. 2 neither at −60°C nor at room temperature is it possible to identify with certainty a magnetic nonequivalence for the CH_2 protons of the C_2H_5 group, in the ^{1}H NMR spectrum in $CDCl_3$ [23].

Activated MnO_2 in boiling methylcyclohexane [12], CH_2Cl_2 [47], or $CHCl_3$ [40] forms $C_5H_5FeC_5H_3(CO{-}CH_3)_2$-1,3 [12, 23, 40, 43, 47]. The 2,4-dinitrophenylhydrazone **$C_5H_5FeC_5H_3(C(CH_3){=}NNHC_6H_3(NO_2)_2$-1)$C_2H_5$-3** crystallizes from CH_3CO_2H/C_2H_5OH as small, dark violet rods that melt at 179 to 180°C [1, 4]. Clemmensen reduction and reduction with $LiAlH_4/AlCl_3$ or with $LiAlH_4$ are covered on p. 191.

$C_5H_5FeC_5H_3(CO{-}CH_3$-1)$CH(CH_3)_2$-3 (Table **11**, No. **3**) is more difficult than Nos. 1 and 2 to isolate from the isomer mixture. After the chromatographic separation of the 1,2-isomer on a neutral Al_2O_3 column with 9/1 pentane/ether, the main band containing No. 3 and the 1,1′-isomer is divided into ten fractions by use of pentane/ether mixtures with increasing ether content. Of these, the last two yield No. 3 [16].

$C_5H_5FeC_5H_3(CO{-}CH_3$-1)$C(CH_3)_3$-3 (Table **11**, No. **4**) is also formed by the acid-catalyzed hydration at 50°C of $C_5H_5FeC_5H_3(C{\equiv}CH$-1)$C(CH_3)_3$-3 in C_2H_5OH plus 0.75% H_2O, 0.05 M in H_2SO_4 [45]. According to [49] the Friedel-Crafts acetylation of $C_5H_5FeC_5H_4{-}C(CH_3)_3$ with CH_3COCl yields No. 4 together with its 1,2-isomer and Fe $(C_5H_4{-}CO{-}CH_2C(CH_3)_3)_2$.

Isolation from the isomer mixture has been attempted by the procedures described for No. 3; despite testing of a multitude of columns, No. 4 nevertheless could not be separated from its 1,1′-isomer [16]. Reaction with CH_3Li in ether yields $C_5H_5FeC_5H_3(C(CH_3)_2OH$-1)$C(CH_3)_3$-3 [37]. $C_5H_5FeC_5H_3(C(Cl){=}CHCHO$-1)$C(CH_3)_3$-3 and $C_5H_5FeC_5H_3(C(Cl){=}CH_2$-1)$C(CH_3)_3$-3 are obtained with the Vilsmeier reagent from $(CH_3)_2NCHO$ and $POCl_3$ [49].

$C_5H_5FeC_5H_3(CO{-}CH_3$-1)$CH_2C(CH_3)_3$-3 (Table **11**, No. **5**) is won from the mother liquor remaining after crystallizing the 1,1′-isomer out of petroleum ether solution, by column chromatography on Al_2O_3 [8]. For "Applications" see p. 191.

(E)-$C_5H_5FeC_5H_3(CO{-}CH_3$-1)$CH_2CH{=}CHC_6H_5$-3 (Table **11**, No. **6**). Preparation by Method I c (p. 189) but with a doubled excess of $(CH_3CO)_2O$ and BF_3 fails just like that with $CH_3COCl/AlCl_3$ [36]. Addition of the solution of No. 6 in CH_2Cl_2 to $[(C_6H_5)_3C]BF_4$ forms with a dark brown coloration the carbenium ion **$[C_5H_5FeC_5H_3(CO{-}CH_3$-1)$CHCHCHC_6H_5-3]^+$** (IV), which reacts instantaneously with $NaOCH_3/CH_3OH$ to form in a quantitative yield the isomeric ethers No. 21 and 22 in a 36/64 ratio.

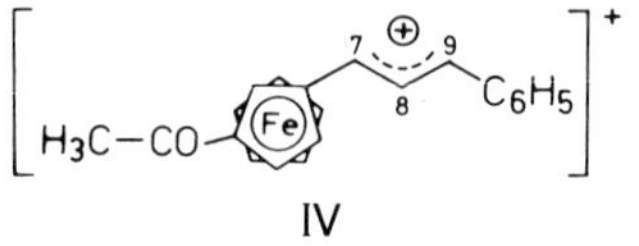

IV

The fact that the kinetically controlled addition of CH_3CO^- occurs more often on C-9 with the formation of No. 22 than on C-7 with formation of No. 21 may be traced back to the acetyl group, which reduces the ability of the ferrocenyl group to delocalize positive charge. The isolation of the mixture of Nos. 21 and 22 from the reaction product takes place by column chromatography on Al_2O_3 (previously exposed 24 h to the atmosphere) with 4/1 petroleum ether/ether following that of $(C_6H_5)_3CH$ (with petroleum ether) and unreacted No. 6 [36].

$C_5H_5FeC_5H_3(CO{-}CH_3\text{-}1)C_6H_5\text{-}3$ (Table **11**, No. **7**) is collected in the main fraction over a period of two weeks together with the 1,1′-isomer from the isomer mixture obtained by Method Im (p. 190) by column chromatography on Al_2O_3 in Skellysolve B beginning after eight weeks. From this fraction it is separated by repeated rechromatography on Al_2O_3 with ether as the eluent [5]. No. 7 has been separated from the 1,2- but not the 1,1′-isomer by column chromatography on acid-washed Al_2O_3 of activity grade 1 with ether as eluent [10]. Separation of the 1,2-isomer by high pressure liquid-liquid chromatography at room temperature with isooctane containing 10 to 20% CH_2Cl_2 as eluent has been effected on a column filled with Carbowax 400 on Corasil [34]. For reaction with $C_6H_5CHO/NaOH$ to form No. 38 see Method II.

$C_5H_5FeC_5H_3(CO{-}CH_3\text{-}1)OCH_3\text{-}3$ (Table **11**, No. **20**). While Method If (p. 189) yields only the mixture of the three positional isomers, Method Ih gives in addition to these $C_5H_5FeC_5H_3(OCH_3)_2$-1,2 and -1,3. On chromatographing the reaction product in C_6H_6 on an alumina column these and $C_5H_5FeC_5H_4{-}OCH_3$ elute out first with benzene and then with 1/1 benzene/ether the mixture of the $FeC_{10}H_8(COCH_3)OCH_3$. No. 20 is not separable from this mixture even by multiple rechromatographing. Even after conversion of the mixture into $FeC_{10}H_8(C_2H_5)OCH_3$ by a Clemmensen reduction or with $LiAlH_4/AlCl_3$ the 1,3-isomer can to be sure be separated from the 1,2-isomer, but not from the 1,1′-isomer [20].

(E)-$C_5H_5FeC_5H_3(CO{-}CH_3\text{-}1)CH(CH{=}CHC_6H_5)OCH_3\text{-}3$ and **(E)-$C_5H_5FeC_5H_3(CO{-}CH_3\text{-}1)CH{=}CHCH{-}(C_6H_5)OCH_3\text{-}3$** (Table **11**, Nos. **21** and **22**) can be enriched to 80% in each case, if the 36/64 mixture of Nos. 21 and 22 obtained from No. 6 (see p. 200) is subjected to preparative thin layer chromatography on SiO_2 with 9/1 petroleum ether/ether as eluent, No. 22 in the first, No. 21 in the last fraction. Both compounds decompose on standing, do not isomerize in the presence of $NaOCH_3$ under the preparation conditions, and are reduced by $NaBH_4$ on the CH_3CO group to the CH_3CHOH group [36].

$C_5H_5FeC_5H_3(CO{-}CH_3\text{-}1)CHO\text{-}3$ (Table **11**, No. **23**) forms in traces [28] as the sole homoannular disubstitution product [35] of the acetylation in the dark of $C_5H_5FeC_5H_4{-}CHO$ in CH_2Cl_2 with a threefold molar quantity of the $CH_3COCl/AlCl_3$ complex (1 h in an ice-bath and 3 h at room temperature). The isolation takes place chromatographically on a deactivated Al_2O_3 column with 1/1 n-hexane/ether, during which No. 23 elutes after $C_5H_5FeC_5H_4{-}CHO$ and before the 1,1′-isomer (see p. 133) [28].

$C_5H_5FeC_5H_3(CO{-}C_2H_5\text{-}1)CH_3\text{-}3$ (Table **11**, No. **24**) is separated from the isomer mixture prepared following Method Ih (p. 189) by column chromatography on Al_2O_3 with 10/1 benzene/ether following the elution of the 1,2- and 1,1′-isomers. $NH_2OH \cdot HCl$ and KOH in ethanol solution yield the oxime [27, 31].

$C_5H_5FeC_5H_3(CO{-}C_2H_5\text{-}1)C_4H_9\text{-}n\text{-}3$ (Table **11**, No. **25**) is oxidized by atmospheric oxygen in $(CH_3)_2SO$ in the presence of excess $KOC(CH_3)_3$ to the stable radical anion **$[C_5H_5FeC_5H_3(CO{=}C(CH_3)O\text{-}1)C_4H_9\text{-}n\text{-}3]^{\dot{-}}$** (V with $R^2 = n\text{-}C_4H_9$). This reacts further with excess O_2 to form the likewise paramagnetic dianion VI. The monoanion possesses a g value of 2.00706 and a hyperfine splitting coupling constant for the CH_3 protons of $A_H = 4.20$ G and for the H-2,5 protons of 0.50 G. This compound group is remarkable for its characteristically slight H/D exchange rate for the α alkyl protons in $(CD_3)_2SO$. The time for the complete formation of $[C_5H_5FeC_5H_3(CO{=}C(CD_3)O\text{-}1)C_4H_9\text{-}n\text{-}3]^{\dot{-}}$ at room temperature is 10 h [25].

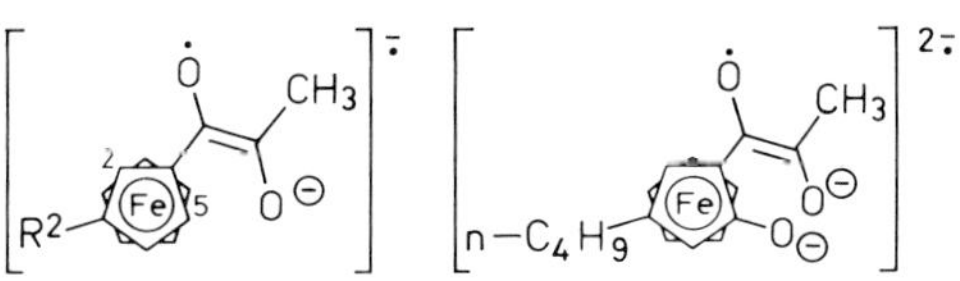

V VI

References on pp. 203/4

$C_5H_5FeC_5H_3(CO{-}C_2H_5\text{-}1)C(CH_3)_3\text{-}3$ (Table **11**, No. **26**). A mixture of isomers, consisting primarily of No. 26 and its 1,1′-isomer, is obtained by addition of $CH_2{=}CHCOCl$ in CH_2Cl_2 to a suspension of $C_5H_5FeC_5H_4{-}C(CH_3)_3/AlCl_3$ (molar ratio 1/1/1) in CH_2Cl_2 at −78°C with subsequent stirring overnight at this temperature. After the usual processing a 16% yield is recovered, based on the reacted $C_5H_5FeC_5H_4{-}C(CH_3)_3$. In the column chromatography the unreacted $C_5H_5FeC_5H_4{-}C(CH_3)_3$ is recovered (80%) first with petroleum ether. Then with 3/1 petroleum ether/ether are eluted a mixture of VII, consisting of at least three isomers, in 12% yield referred to the reacted $C_5H_5FeC_5H_4{-}C(CH_3)_3$, the mixture of No. 26 and its 1,1′-isomer, and finally VIIIa and b in 61% yield. The mixture is a viscous oil. Its ^1H NMR spectrum in $CDCl_3$ exhibits the following shifts (in ppm): $\delta = 1.05$ to 1.35 (m, $C(CH_3)_3 + CH_3$ of C_2H_5), 2.75 (br. q, $COCH_2$), 4.0 to 4.2 (m), 4.20 (s), 4.4 to 4.6 (m), and 4.7 to 4.9 (m) ($C_5H_5 + C_5H_3 + C_5H_4$) [42].

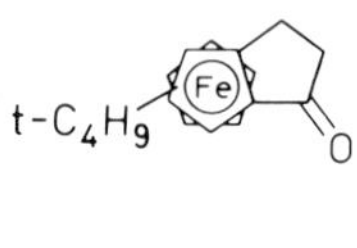

VII

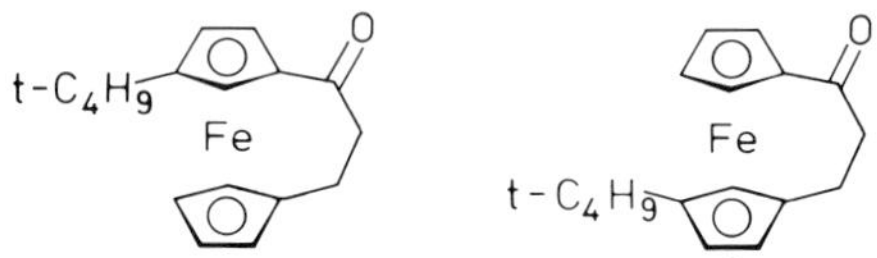

a b

VIII

$C_5H_5FeC_5H_3(CO{-}C_2H_5\text{-}1)(C_6H_4Cl\text{-}4)\text{-}3$ (Table **11**, No. **27**) is obtained mixed with the 1,2-isomer by the Friedel-Crafts reaction for 4.5 h of $C_5H_5FeC_5H_4{-}C_6H_4Cl\text{-}4$ with $(C_2H_5CO)_2O$ and $AlCl_3$ (molar ratio 1/1.3/1) in CH_2Cl_2. The unreacted $C_5H_5FeC_5H_4{-}C_6H_4Cl\text{-}4$ is separated by column chromatography on SiO_2 with Skelly B, and the 1,1′-isomer is precipitated by selective recrystallization. In the subsequent column chromatography of the resulting mixture on SiO_2, No. 27 is eluted in 19% yield with 7/93 benzene/cyclohexane before the 1,2-isomer [30].

A 1 s exposure of No. 27 to atmospheric oxygen in $(CH_3)_2SO$ in the presence of a three- to fourfold excess of $KOC(CH_3)_3$ forms the paramagnetic radical anion **$[C_5H_5FeC_5H_3(CO{=}C(CH_3)O\text{-}1)(C_6H_4Cl\text{-}4)\text{-}3]^{\dot{-}}$** (V with $R^2 = 4\text{-}ClC_6H_4$). This anion has $g = 2.00680$ and the hyperfine splitting coupling constants A_H of 4.20 G for the CH_3 protons and 0.50 G for the H-2,5 protons [30].

$C_5H_5FeC_5H_3(CO{-}C_3H_7\text{-}n\text{-}1)CH_3\text{-}3$ (Table **11**, No. **28**) is obtained as an 84/16 mixture with the 1,1′-isomer when the reaction product won by Method Ii (p. 189) is enriched in No. 28 and its 1,1′-isomer by chromatographic separation of $C_5H_5FeC_5H_4{-}CH_3$ and the main part of the 1,2-isomer on an acid-washed Al_2O_3 column with hexane and 2/1 hexane/benzene, respectively, as eluents. Chromatography is then repeated on neutral Al_2O_3 with 2/1 CH_2Cl_2/ether [29]. For reduction see p. 191.

$C_5H_5FeC_5H_3(CO{-}CH(CH_3)_2\text{-}1)C(CH_3)_3\text{-}3$ (Table **11**, No. **29**). The mixture obtained by Method Ii of No. 29 and the 1,1′-isomer is resolved into its constituents by repeated column chromatography on Al_2O_3 and thin layer chromatography on SiO_2 [38]. No. 29 reacts with CH_3Li to form $C_5H_5FeC_5H_3(C(CH_3)(CH(CH_3)_2)OH\text{-}1)C(CH_3)_3\text{-}3$ [38].

$C_5H_5FeC_5H_3(CO{-}CH(CH_3)_2\text{-}1)CH_2C(CH_3)_3\text{-}3$ (Table **11**, No. **30**) is separated from the 1,1′-isomer by chromatography on an Al_2O_3 column in petroleum ether [8]. For reduction see p. 191.

$C_5H_5FeC_5H_3(CO{-}C_6H_5\text{-}1)CH_3\text{-}3$ (Table **11**, No. **48**) could be separated, to be sure, from the 1,2- but not from the 1,1′-isomer by preparative thin layer chromatography on SiO_2 with benzene/petroleum ether of the isomer mixture produced by Method Ii [21]. The isomer mixture obtained by Method Ih is according to [31] separable chromatographically on an Al_2O_3 column with 10/1 benzene/ether as eluent. In the ^1H NMR spectrum in CCl_4 the signal of the CH_3 protons is shifted by 0.09 ppm toward lower field with respect to that for $C_5H_5FeC_5H_4{-}CH_3$ [21]. KBH_4

reduces No. 48 in a nonstereoselective reaction to form a 1/1 mixture of (1SR, 6SR)- and (1SR, 6RS)-$C_5H_5FeC_5H_3(CHOHC_6H_5\text{-}1)CH_3\text{-}3$ (see Formulas II and III (p. 191), respectively, with $R^{**} = C_6H_5$, $R^2 = CH_3$, $R^* = H$) [31]. The oxime $C_5H_5FeC_5H_3(C(C_6H_5){=}NOH\text{-}1)CH_3\text{-}3$ is obtained with $NH_2OH \cdot HCl$ and KOH in C_2H_5OH [31].

$C_5H_5FeC_5H_3(CO\text{–}C_6H_5\text{-}1)C(CH_3)_3\text{-}3$ (Table **11**, No. **49**). Irradiation of the 2/1 mixture of No. 49 and its 1,1′-isomer in moist, oxygen-free $(CH_3)_2SO$ with UV of $\lambda > 280$ nm causes photolysis with formation of $[\pi\text{-}C_5H_5Fe(OS(CH_3)_2)_n]^+$, $[\pi\text{-}(CH_3)_3C\text{–}C_5H_4Fe(OS(CH_3)_2)_2]^+$, $[C_6H_5CO_2]^-$, C_5H_6, and $C_5H_5\text{–}C(CH_3)_3$ [33].

$C_5H_5FeC_5H_3(CO\text{–}(C_6H_4CH_3\text{-}4)\text{-}1)CH_3\text{-}3$ (Table **11**, No. **50**) is obtained in a 2/2.6/1 molar ratio as a red, syrupy mixture with its 1,1′- and 1,2-isomers by chromatography on dry SiO_2 columns with 6/1 hexane/$CH_3CO_2C_2H_5$ of the product from Method I i (p. 189) after prior separation of the ditoluoyl by-products. The ^1H NMR spectrum of this mixture ($CDCl_3$; enumeration see Table 11, p. 198) is δ (in ppm) = 1.83 (s, CH_3-1′ of 1,1′-isomer), 2.04 (s, CH_3-3 of No. 50), 2.35 (s, CH_3-2 of 1,2-isomer), 2.42 (s, CH_3-10), 4.0 to 4.2 (m, 4.4H; at δ = 4.05 s, cyclopentadienyl protons of $C_5H_4CH_3$ and at δ = 4.11 s, for C_5H_5), 4.48 (m, 2H, cyclopentadienyl protons in β position to CO), 4.81 (m, 2H, cyclopentadienyl protons in α position to CO), 7.1 to 7.9 (m, C_6H_4). The mixture can be resolved by gas-liquid chromatography into the three isomers on a column filled with 3% OV-225 on Chromosorb W-HP. Retention times for the 1,2-, 1,1′-, and 1,3-isomers are, respectively, 6.7, 8.6, and 9.3 min for a 225°C column temperature and 9.1, 11.8, and 12.9 min for 215°C. If the syrup is recrystallized from 1/10 $CHCl_3$/hexane, one obtains a brass-colored powder melting at 80 to 85°C that consists only of No. 50 and the 1,1′-isomer in a 1/1.4 ratio [48].

$C_5H_5FeC_5H_3(CO\text{–}(C_6H_4Cl\text{-}2)\text{-}1)CH_3\text{-}3$ (Table **11**, No. **52**). The mixture of No. 52 and the 1,1′-isomer, produced by Method I o, can be satisfactorily separated by slow crystallization of the 1,1′-isomer from a concentrated solution of the mixture in 4/1 n-hexane/$CHCl_3$ [41]. Preparative high pressure liquid chromatography yields – to be sure in only 2 mg quantities – good results, while multiple chromatography on long Al_2O_3 columns with petroleum ether/CH_2Cl_2/ether leads to only a partial separation [41]. $KOC(CH_3)_3$ and a little H_2O in $CH_3OCH_2CH_2OCH_3$ react with No. 52 to produce better than 80% $C_5H_5FeC_5H_3(CO_2H\text{-}1)CH_3\text{-}3$. The presence of possible traces of by-product $C_5H_5FeC_5H_4\text{–}CH_3$ was not established [41].

$C_5H_5FeC_5H_3(CO\text{–}(C_6H_4Cl\text{-}2)\text{-}1)C_6H_5\text{-}3$ (Table **11**, No. **54**). Only the 1,2-isomer can be separated chromatographically from the isomer mixture produced by Method I g. The isomer ratio in the raw product is determined by converting it with $KOC(CH_3)_3$ in $H_2O/CH_3OCH_2CH_2OCH_3$ into the isomer mixture $FeC_{10}H_8(CO_2H)C_6H_5$, which after esterification is converted to $FeC_{10}H_8(C_6H_5)CH_3$ and as such is analyzed by NMR spectroscopy [41].

References:

[1] M. Rosenblum (Diss. Harvard Univ. 1953, p. 140). – [2] A. N. Nesmeyanov, N. A. Volkenau (Dokl. Akad. Nauk SSSR **111** [1956] 605/8; Proc. Acad. Sci. USSR Chem. Sect. **106/111** [1956] 689/91). – [3] M. Rosenblum (Chem. Ind. [London] **1958** 953). – [4] M. Rosenblum, R. B. Woodward (J. Am. Chem. Soc. **80** [1958] 5443/9). – [5] M. Rosenblum (J. Am. Chem. Soc. **81** [1959] 4530/6).

[6] R. J. Stephenson, Imperial Chemical Industries Ltd. (Brit. 864197 [1959/61]; C.A. **55** [1961] 17647). – [7] L. A. Day, Imperial Chemical Industries Ltd. (Brit. 864198 [1959/61]; C.A. **55** [1961] 17647). – [8] S. Birtwell, Imperial Chemical Industries Ltd. (Ger. 1116663 [1960/62]; Brit. 861834 [1959/61]; C.A. **1961** 16565). – [9] E. A. Hill, J. H. Richards (J. Am. Chem. Soc. **83** [1961] 4216/21). – [10] M. Rosenblum, W. G. Howells (J. Am. Chem. Soc. **84** [1962] 1167/72).

[11] R. A. Benkeser, Y. Nagai, J. Hooz (Bull. Chem. Soc. Japan **36** [1963] 482/3). – [12] A. F. Ellis (Diss. Univ. Illinois 1963, pp. 1/152 from Diss. Abstr. **24** [1963] 510). – [13] J. Tirouflet, E. Laviron, R. Dabard, J. Komenda (Bull. Soc. Chim. France **1963** 857/62). – [14] R. A. Benkeser, Y. Nagai, J. Hooz (J. Am. Chem. Soc. **86** [1964] 3742/6). – [15] M. I. Levenberg, J. H. Richards (J. Am. Chem. Soc. **86** [1964] 2634/7).

[16] Y. Nagai, J. Hooz, R. A. Benkeser (Bull. Chem. Soc. Japan **37** [1964] 53/60). – [17] R. Dabard, D. Weigel (Bull. Soc. Chim. France **1966** 390/6). – [18] A. N. Nesmeyanov [Nessmejanow], L. P. Yur'eva [Jurjewa], N. S. Kochetkova, S. V. Vitt [Witt] (Izv. Akad. Nauk SSSR Ser. Khim. **1966** 560/2; Bull. Acad. Sci. USSR Div. Chem. Sci. **1966** 530/2; Abhandl. Deut. Akad. Wiss. Berlin Kl. Chem. Geol. Biol. **1966** No. 2, pp. E265/E268). – [19] G. Haller, K. Schlögl (Monatsh. Chem. **98** [1967] 603/18). – [20] S. McVey, I. G. Morrison, P. L. Pauson (J. Chem. Soc. C **1967** 1847/50).

[21] K. Schlögl, H. Falk, G. Haller (Monatsh. Chem. **98** [1967] 82/94). – [22] D. E. Bublitz (J. Organometal. Chem. **16** [1969] 149/61). – [23] H. Egger, A. Nikiforov (Monatsh. Chem. **100** [1969] 242/53). – [24] H. Falk, O. Hofer, K. Schlögl (Monatsh. Chem. **100** [1969] 624/48, 636). – [25] J. J. McDonnell (Tetrahedron Letters **1969** 2039/42).

[26] A. Perjéssy, Š. Toma (Collection Czech. Chem. Commun. **48** [1983] 1635/40). – [27] C. Moïse, J. Tirouflet, D. Sautrey (Compt. Rend. C **271** [1970] 951/4). – [28] M. Sato, M. Koga, I. Motoyama, K. Hata (Bull. Chem. Soc. Japan **43** [1970] 1142/7). – [29] J. W. Huffman, J. F. Cope (J. Org. Chem. **36** [1971] 4068/72). – [30] J. J. McDonnell, D. J. Pochopien (J. Org. Chem. **36** [1971] 2092/8).

[31] C. Moïse, D. Sautrey, J. Tirouflet (Bull. Soc. Chim. France **1971** 4562/76). – [32] J. Paul, K. Schlögl, W. Silhan (Monatsh. Chem. **103** [1972] 243/54). – [33] L. H. Ali, A. Cox, T. J. Kemp (J. Chem. Soc. Dalton Trans. **1973** 1468/75). – [34] R. Eberhardt, H. Lehner, K. Schlögl (Monatsh. Chem. **104** [1973] 1409/20). – [35] E. V. Leonova, N. S. Kochetkova, E. I. Fedin, P. V. Petrovskii (Ref. Dokl. Soobshch. 11th Mendeleevsk. S'ezd Obshch. Prikl. Khim., Alma-Ata 1974 [1975], Vol. 2, pp. 98/9; C.A. **84** [1976] No. 164987).

[36] T. D. Turbitt, W. E. Watts (J. Chem. Soc. Perkin Trans. II **1974** 185/9). – [37] T. D. Turbitt, W. E. Watts (J. Chem. Soc. Perkin Trans. II **1974** 189/95). – [38] T. S. Abram, W. E. Watts (J. Chem. Soc. Perkin Trans. I **1975** 116/21). – [39] F. Jursík (J. Chromatog. Libr. **3** [1975] 1087/114, 1108). – [40] A. N. Nesmeyanov, E. V. Leonova, N. S. Kochetkova, A. I. Malkova, A. G. Makarovskaya (J. Organometal. Chem. **96** [1975] 275/8).

[41] V. Rapić, K. Schlögl, B. Steinitz (J. Organometal. Chem. **94** [1975] 87/98). – [42] W. Crawford, T. D. Turbitt, W. E. Watts (J. Organometal. Chem. **105** [1976] 341/8). – [43] S. L. Sosin, B. A. Antipova, T. M. Yezhova, L. V. Dzhashi, V. V. Korshak (Vysokomol. Soedin. A **18** [1976] 34/9; Polym. Sci. [USSR] **18** [1976] 40/6). – [44] Š. Toma, A. Maholányiová, E. Solčániová (Acta Fac. Rerum Nat. Univ. Comenianae Chim. No. 24 [1976] 53/80). – [45] T. S. Abram, W. Crawford, A. C. Knipe, W. E. Watts (Proc. Roy. Irish Acad. Sect. B **77** [1977] 317/21).

[46] Š. Toma, A. Maholanyiová, E. Solčaniová (Collection Czech. Chem. Commun. **42** [1977] 1013/26). – [47] M. Hisatome, O. Tachikawa, M. Sasho, K. Yamakawa (J. Organometal. Chem. **217** [1981] C17/C20). – [48] B. E. Maryanoff (J. Org. Chem. **46** [1981] 70/4). – [49] L. I. Tolstykh, L. Errais (Tr. Mosk. Inst. Neftekhim. Gazov. Prom. **158** [1981] 144/9 from C.A. **99** [1983] No. 5775).

5.2.3.5.3.3 Ketones of the Type $C_5H_5FeC_5H_3(A–CO–R'-1)R^2$-3 (A–CO–R' = R[1])

The compounds listed in Table 12 are prepared by the following methods:

Method I. The Gomberg-Bachmann-Hey reaction of ferrocene with $[4\text{-}CH_3COC_6H_4N_2]_2SO_4$ in CH_3CO_2H/H_2O for 12 h at 10 to 15°C [1]. On column chromatography of the raw product in $CHCl_3/C_6H_6$ on Al_2O_3 elution with 4/1 benzene/Skellysolve B yields in succession ferrocene, $C_5H_5FeC_5H_4–C_6H_4COCH_3$-4, an unidentified colorless compound melting at 121 to 123°C, the 1,2- and 1,1'-isomers, and finally No. 1 [1].

Method II. Friedel-Crafts acylation of $C_5H_5FeC_5H_4–R^1$:

a) CH_3COCl and $AlCl_3$ in CH_2Cl_2 at room temperature are added dropwise to $C_5H_5FeC_5H_4–R^1$ (molar ratio 3/6/2) in CH_2Cl_2, and the reaction is continued another 4 h. No. 2 is separated from its 1,1'-isomer chromatographically on an SiO_2 column with 3/97 $CH_3CO_2C_2H_5/C_6H_6$ [3].

b) $AlCl_3$ is added in portions at 0 to 5°C to $C_5H_5FeC_5H_4–(CH_2)_n–CO–C_6H_4OCH_3$-4 (n = 2 to 9) and $C_6H_5CH{=}CHCOCl$, and the reaction is continued for 4 h at room temperature. Column chromatography on SiO_2 with 1/19 to 1/10 $CH_3CO_2C_2H_5/C_6H_6$ gives (7 to 10%) yields of the 1,2-isomer. The 1,3-isomer is isolable only in the mixture (indicated by "m") with the 1,1'-isomer (total yield 20 to 25%), from which it can be separated neither chromatographically nor by fractional crystallization [3].

c) $C_5H_5FeC_5H_4–(CH_2)_n–CO–C_6H_4OCH_3$-4 in CH_2Cl_2 at 0 to 5°C is added to a solution of $C_6H_5CH{=}CHCOCl$ and $AlCl_3$ (molar ratio 1/5/10) in CH_2Cl_2, and reacted for n = 3 to 6 for 1 h at 35°C, for n = 7 to 9 for 2 to 3 h at room temperature. Separation as for Method IIb gives in addition to a 5 to 7% yield of the 1,2-isomer the mixture "m" of the 1,3- and 1,1'-isomers in a total yield of 10 to 15% plus $C_6H_5CH{=}CHCO–C_5H_4FeC_5H_3(CO–CH{=}CHC_6H_5\text{-}1)(CH_2)_n–CO(C_6H_4OCH_3\text{-}4)$-2 and -3 in yields of 7 to 10% and 10 to 15%, respectively [3].

Method III. Condensation of $C_5H_5FeC_5H_3(CHO\text{-}1)R^2$-3 at room temperature:

a) With $CH_3COC_6H_5$ and NaOH (mole ratio 1/1/10) in 2/1 C_2H_5OH/H_2O (reaction time 12 h). Starting not with the pure compound but with a mixture of its positional isomers yields an isomer mixture that is separable by column chromatography on SiO_2 with benzene eluent, the elution sequence being 1,2-, 1,1'-, and 1,3-isomers [5].

b) With indane-1,3-dione (see Formula I) in benzene in the presence of a few drops of $(C_2H_5)_3N$ (reaction time 12 h). Starting with the mixture of the positional isomers yields an isomer mixture that can be separated as for Method IIIa. The elution sequence is here, however, 1,2-, 1,3-, and the 1,1'-isomer [5].

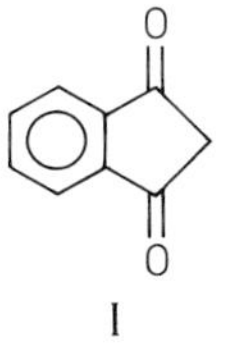

I

Properties. With the exception of No. 1 all compounds exhibit a plane of chirality, see "Ferrocene" 7, 1980, pp. 1/4 (English), pp. 4/7 (German), since they bear on one cyclopentadienyl ring two different substituents. However, they have been prepared only in the racemic form.

References on p. 209

A linear correlation corresponding to a modified and extended Seth-Paul-Van Duyse equation has been observed between the νC=O vibration in CCl_4 and the $X^+(R^{A,B})$ constants of the structural fragments of 154 compounds with ferrocene structures of the general formula R^A–CO–R^B, among them Nos. 9, 11 to 14, 17, with $R^A = C_5H_5FeC_5H_3(R^2\text{-}3)CH{=}CH\text{–}$ and $R^B = C_6H_5$. The $X^+(R^{A,B})$ used have been confirmed by the recently derived empirical relationship for calculation of the substituent constants and transmission factors [4].

Table 12

Ketones of the Type $C_5H_5FeC_5H_3(A\text{–}CO\text{–}R'\text{-}1)R^2\text{-}3$.

Further information for the number preceded by an asterisk is given at the end of the table, p. 209.

For abbreviations and dimensions see p. VIII.

No.	$C_5H_5FeC_5H_3(R^1\text{-}1)R^2\text{-}3$ $R^1 = A\text{–}CO\text{–}R'$	R^2	method of preparation (yield in %, "m" indicates mixture of the 1,3-compound with its 1,1′-isomer), properties, and remarks	Ref.
compound with $R^1 = R^2$:				
1	$C_6H_4\text{–}CO\text{–}CH_3\text{-}4$		I orange needles, m.p. 202 to 204° (CH_2Cl_2/petroleum ether) IR ($CHCl_3$): 895.2, 903.3, 1001, 1106.2 UV (C_2H_5OH/H_2O, 95/5): $\lambda_{max}(\varepsilon) = 300$ (31000), 407 (7520), 480sh (2630)	[1,2]
compounds with $R^1 \neq R^2$:				
2	$(CH_2)_2CO\text{–}CH_3$	$COCH_3$	IIa (13.5) deep red oil IR (Nujol): 1005 and 1110 (cyclopentadienyl), 1115 ($\nu_sC(3)\text{–}C_{acetyl}$), 1680 and 1728 (νC=O)	[3]
3	$(CH_2)_n\text{–}CO\text{–}C_6H_4OCH_3\text{-}4$ n = 2 to 6	$COCH{=}CHC_6H_5$	IIb, IIc	[6]
4	$(CH_2)_7\text{–}CO\text{–}C_6H_4OCH_3\text{-}4$	$COCH_3$	see No. 5, p. 209 red-orange oil (m) ^{1}H NMR: see figure in [6]	[6]
*5	$(CH_2)_7\text{–}CO\text{–}C_6H_4OCH_3\text{-}4$	$COCH{=}CHC_6H_5$	IIb (19.2m), IIc m.p. 92 to 102° (C_6H_6/petroleum ether) (m) ^{1}H NMR: see figure in [6]	[6]
6	$(CH_2)_n\text{–}CO\text{–}C_6H_4OCH_3\text{-}4$ n = 8,9	$COCH{=}CHC_6H_5$	IIb, IIc	[6]

References on p. 209

Table 12 (continued)

No.	$C_5H_5FeC_5H_3(R^1\text{-}1)R^2\text{-}3$ $R^1 = A\text{–}CO\text{–}R'$	R^2	method of preparation (yield in %, "m" indicates mixture of the 1,3-compound with its 1,1′-isomer), properties, and remarks	Ref.
7	$CH{=}CH\text{–}CO\text{–}C_6H_5$	C_6H_5	IIIa (15) m.p. 87 to 89° ^{1}H NMR ($CDCl_3$): 4.088 (s, C_5H_5); 4.775 (m, H-4), 4.938 (m, H-5), 5.075 (m, H-2) (C_5H_3)	[5]
8	$CH{=}CH\text{–}CO\text{–}C_6H_5$	$C_6H_4CH_3$-4	IIIa (26) m.p. 182 to 183°	[5]
9	$CH{=}CH\text{–}CO\text{–}C_6H_5$	$C_6H_4CH_3$-3	IIIa (28) m.p. 98 to 100° IR (CCl_4): 1665.5 (νC=O for s-cis COR^A), 1643.5 (νC=O for s-trans COR^A)	[4, 5]
10	$CH{=}CH\text{–}CO\text{–}C_6H_5$	$C_6H_3(CH_3)_2$-3, 4	IIIa (21) m.p. 115 to 116°	[5]
11	$CH{=}CH\text{–}CO\text{–}C_6H_5$	C_6H_4Cl-4	IIIa (40) m.p. 196 to 198° IR (CCl_4): 1666.5 (νC=O for s-cis COR^A), 1644.5 (νC=O for s-trans COR^A)	[4, 5]
12	$CH{=}CH\text{–}CO\text{–}C_6H_5$	C_6H_4Cl-3	IIIa (23) m.p. 101 to 102° IR (CCl_4): 1666 (νC=O for s-cis COR^A), 1644.5 (νC=O for s-trans COR^A)	[4, 5]
13	$CH{=}CH\text{–}CO\text{–}C_6H_5$	C_6H_4Br-4	IIIa (17) m.p. 189 to 191° IR (CCl_4): 1666.5 (νC=O for s-cis COR^A), 1644.5 (νC=O for s-trans COR^A)	[4, 5]
14	$CH{=}CH\text{–}CO\text{–}C_6H_5$	C_6H_4Br-3	IIIa (20) m.p. 105 to 108° IR (CCl_4): 1667 (νC=O for s-cis COR^A), 1645.5 (νC=O for s-trans COR^A)	[4, 5]
15	$CH{=}CH\text{–}CO\text{–}C_6H_5$	$C_6H_3Cl_2$-3, 4	IIIa (18) m.p. 157 to 158°	[5]
16	$CH{=}CH\text{–}CO\text{–}C_6H_5$	$C_6H_4CF_3$-3	IIIa (21) m.p. 164 to 165°	[5]
17	$CH{=}CH\text{–}CO\text{–}C_6H_5$	$C_6H_4OCH_3$-4	IIIa (23) m.p. 116 to 118° IR (CCl_4): 1664.5 (νC=O for s-cis COR^A), 1642 (νC=O for s-trans COR^A)	[4, 5]

References on p. 209

Table 12 (continued)

No.	$C_5H_5FeC_5H_3(R^1\text{-}1)R^2\text{-}3$ $R^1 = A\text{–}CO\text{–}R'$	R^2	method of preparation (yield in %, "m" indicates mixture of the 1,3-compound with its 1,1′-isomer), properties, and remarks	Ref.
18	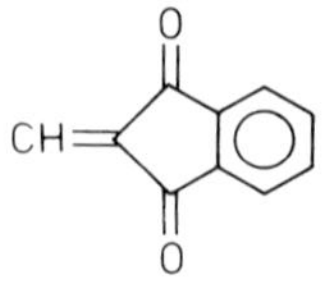	C_6H_5	IIIb (29) m.p. 209 to 210° ^{1}H NMR ($CDCl_3$): 4.088 (s, C_5H_5); 5.325 (m, H-4), 5.488 (m, H-5), 6.033 (m, H-2) (C_5H_3)	[5]
19	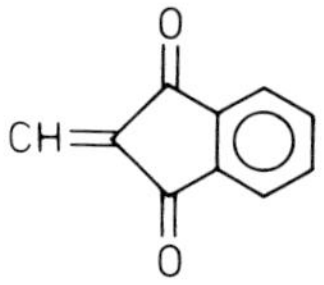	$C_6H_4CH_3$-4	IIIb (27) m.p. 192 to 193°	[5]
20	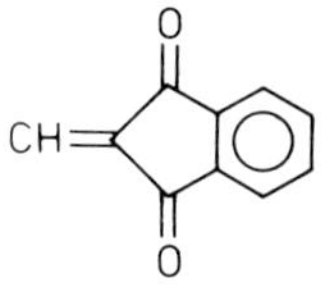	$C_6H_4CH_3$-3	IIIb (40) m.p. 204 to 206°	[5]
21	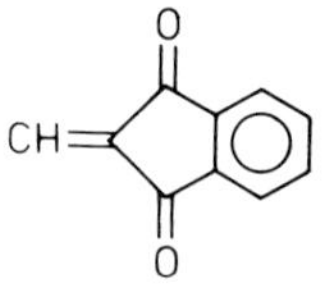	$C_6H_3(CH_3)_2$-3,4	IIIb (40) m.p. 210 to 212°	[5]
22	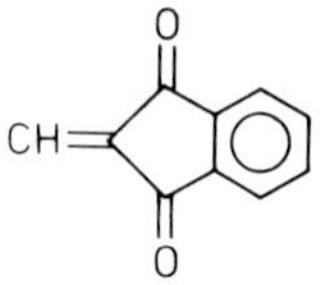	$C_6H_3(CH_3)_2$-3,5	IIIb (34) m.p. 172 to 174°	[5]
23	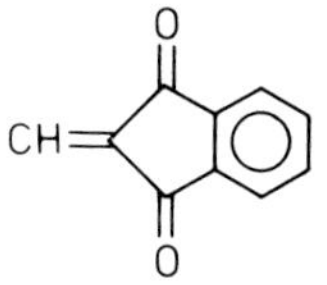	C_6H_4Cl-4	IIIb (29) m.p. 188 to 189°	[5]
24	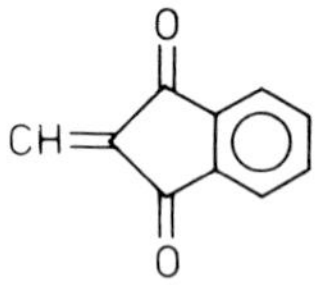	C_6H_4Cl-3	IIIb (17) m.p. 119 to 121°	[5]
25	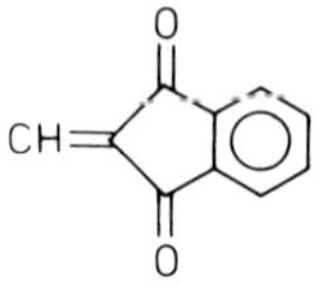	C_6H_4Br-4	IIIb (16) m.p. 208 to 209°	[5]

References on p. 209

Table 12 (continued)

No.	$C_5H_5FeC_5H_3(R^1\text{-}1)R^2\text{-}3$ $R^1 = A\text{–}CO\text{–}R'$	R^2	method of preparation (yield in %, "m" indicates mixture of the 1,3-compound with its 1,1′-isomer), properties, and remarks	Ref.
26	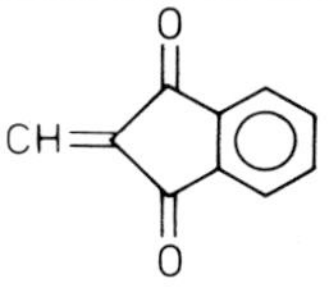	$C_6H_4Br\text{-}3$	IIIb (30) m.p. 174 to 175°	[5]
27	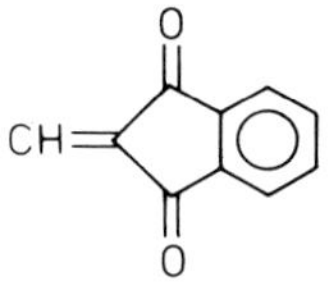	$C_6H_3Cl_2\text{-}3,4$	IIIb (28) m.p. 205 to 208°	[5]
28	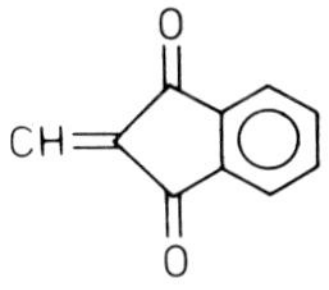	$C_6H_4CF_3\text{-}3$	IIIb (21) m.p. 249 to 251°	[5]
29	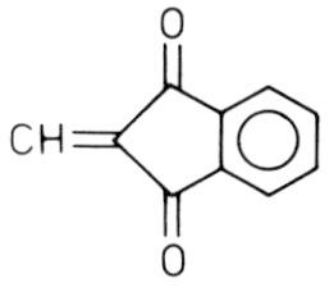	$C_6H_3(CF_3\text{-}3)Cl\text{-}4$	IIIb (5) m.p. 204 to 205°	[5]
30	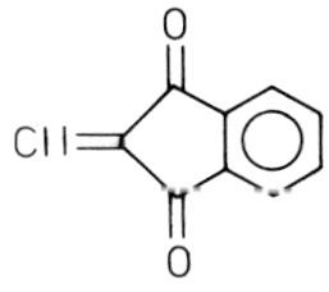	$C_6H_4OCH_3\text{-}4$	IIIb (35) m.p. 210 to 211°	[5]

* Further information:

$C_5H_5FeC_5H_3([CH_2]_7\text{–}CO\text{–}(C_6H_4OCH_3\text{-}4)\text{-}1)CO\text{–}CH{=}CHC_6H_5\text{-}3$ (Table **12**, No. **5**) mixed with its 1,1′-isomer gives on a 2 h refluxing with NaOH in C_2H_5OH by a reverse Claisen-Schmidt reaction No. 4 plus $4\text{-}CH_3OC_6H_4\text{–}CO\text{–}[CH_2]_7\text{–}C_5H_4FeC_5H_4\text{–}COCH_3$ (total yield 32%), which are separated from the mixture with the starting compound (recovery 30%) by column chromatography on SiO_2 with 1/20 to 1/10 $CH_3CO_2C_2H_5/C_6H_6$ [6].

References:

[1] M. Rosenblum, W. G. Howells, A. K. Banerjee, C. Bennett (J. Am. Chem. Soc. **84** [1962] 2726/32). – [2] M. Rosenblum, J. O. Santer, W. G. Howells (J. Am. Chem. Soc. **85** [1963] 1450/8). – [3] Š. Toma, M. Sališová (J. Organometal. Chem. **57** [1973] 191/8). – [4] A. Perjéssy, Š. Toma (Collection Czech. Chem. Commun. **48** [1983] 1635/40). – [5] Š. Toma, A. Maholányiová, E. Solčaniová (Acta Fac. Rerum Nat. Univ. Comenianae Chim. No. 24 [1976] 53/80).

[6] M. Sališová, E. Solčaniová, Š. Toma (Acta Fac. Rerum Nat. Univ. Comenianae Chim. No. 28 [1980] 103/26)

5.2.3.6 Carboxylic Acids and Their Salts

5.2.3.6.1 1,1′-Compounds

5.2.3.6.1.1 1,1′-Dicarboxyferrocene $Fe(C_5H_4–CO_2H)_2$

Preparation and Formation

From $Fe(C_5H_4–CO–CH_3)_2$. Oxidation with excess NaOCl solution gives a practically quantitative yield when the reaction is conducted in ~1/1 $H_2O/CH_3O(CH_2)_2O(CH_2)_2OCH_3$ beginning at 0 to 5°C with a slow increase over 3 h to room temperature [37] or in H_2O/CH_3OH after 30 min reaction at 5°C [22], 8 to 15°C [25] and subsequent increase within 30 min to 50°C [22, 25]. In contrast, reaction for 4 h at 43°C in H_2O/CH_3OH gives a mixture of $Fe(C_5H_4–CO_2H)_2$ plus $HO_2C–C_5H_4FeC_5H_4–CO–CH_3$ (joint yield 82%). The yield of the dicarboxylic acid drops with extended reaction times and increased NaOCl excess, without cessation of the formation of the monocarboxylic acid [2]. See p. 212 regarding the separation of the two acids.

Oxidation of solid $Fe(C_5H_4–CO–CH_3)_2$ with commercial 5.25% aqueous NaOCl solution in the dark at 44 to 50°C for 5 h yields the dicarboxylic acid free from contamination by its methyl ester, in yields of 75% [48], 77% [42], 80% [27]. Reaction with aqueous KOCl solution for 3 h at 50°C gives a 90 to 92% yield [26, 29, 31].

Oxidation with excess alkaline NaOBr solution proceeds optimally in H_2O/dioxane (2/1 [32], 1/1 [9]) between 0 and 5°C [32], 5 and 8°C [9], for a reaction time of 1 h. The yield of the raw product is 91% [32], of the purified compound 53% [9]. Oxidation in a purely aqueous medium at 100°C is described by [23].

Yields of 55 to 60% have been achieved with NaOI. $Fe(C_5H_4–CO–CH_3)_2$ in CH_3OH is mixed dropwise simultaneously at room temperature with a 5 N NaOH solution and an aqueous I_2/KI solution [1, 2, 10] or at 50°C with a solution of I_2 in CH_3OH and one of NaOH in 4/3 CH_3OH/H_2O over a 1 h period [22].

No yield above 7.5% of theoretical has been obtained with iodine in pyridine at 30 to 95°C [9].

Oxidation by molecular oxygen, passed for 24 h through a solution of $Fe(C_5H_4–CO–CH_3)_2$ in $OP(N(CH_3)_2)_3$ at 80°C in the presence of a 20-fold molar quantity of $KOC(CH_3)_3$, gives the dicarboxylic acid in a 17% yield along with $C_5H_5FeC_5H_4–CO_2H$ (34%) and ferrocene (26%) [36]; separation of this mixture is covered on p. 212.

From $Fe(C_5H_4–M)_2$, M = Li, Na, MgBr. Carboxylation of $Fe(C_5H_4–M)_2$ with subsequent hydrolysis yields $Fe(C_5H_4–CO_2H)_2$ in nearly quantitative amounts. Since, however, in general the $Fe(C_5H_4–M)_2$ is not previously isolated from the reaction mixture in which it is produced but this mixture as a whole is reacted, often with a more or less large content of $C_5H_5FeC_5H_4–M$, depending on the preparation conditions, the yields referred to the starting compound for the $Fe(C_5H_4–M)_2$ can be considerably reduced.

Maximum yields of 98% referred to ferrocene along with 1.8% of $C_5H_5FeC_5H_4–CO_2H$ have been obtained when the reaction solution gotten by lithiation of ferrocene with $n\text{-}C_4H_9Li \cdot (CH_3)_2N(CH_2)_2N(CH_3)_2$ in hexane at room temperature for 6.5 h is subsequently poured into a slurry of dry ice in diethyl ether [30]. The compound I gotten from ferrocene, $n\text{-}C_4H_9Li$, and $(CH_3)_2N(CH_2)_2N(CH_3)(CH_2)_2N(CH_3)_2$ in hexane in 50 to 60% yield gives in the corresponding reaction with CO_2 a product free from $C_5H_5FeC_5H_4–CO_2H$ [39], as does also the $Fe(C_5H_4–Li)_2$ prepared by reaction of $Fe(C_5H_4–HgCl)_2$ with C_2H_5Li (yield of $Fe(C_5H_4–CO_2H)_2$ 44% referred to $Fe(C_5H_4–HgCl)_2$) [18].

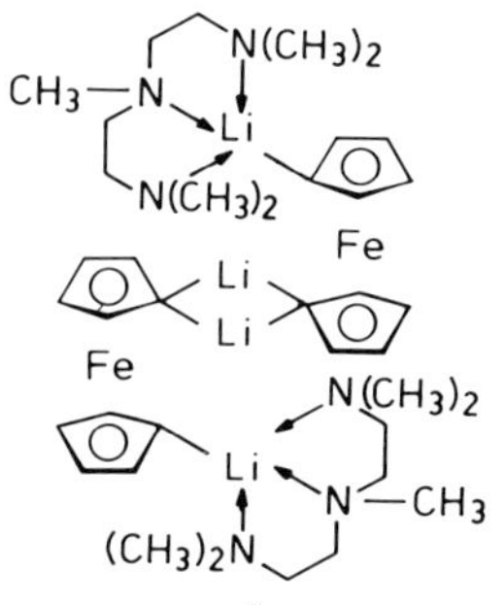

I

In contrast, when the reaction product obtained by lithiation of ferrocene with n-C_4H_9Li in tetrahydrofuran/ether, in tetrahydrofuran/hexane, or in ether is employed, $C_5H_5FeC_5H_4$–CO_2H forms in considerable quantities. In reaction media containing tetrahydrofuran the yield of $Fe(C_5H_4$–$CO_2H)_2$ is around 30 to 40%, that of $C_5H_5FeC_5H_4$–CO_2H between 35 and 43%, in each case referred to ferrocene, see for example [6, 12, 14, 46]. In ether the fraction of $C_5H_5FeC_5H_4$–CO_2H in the reaction product rises to 50 to 70% with a declining total yield, see for example [4, 5, 13, 21]. Carboxylation of a solid mixture of $Fe(C_5H_4$–$Li)_2$ and $C_5H_5FeC_5H_4$–Li with CO_2 has been effected in an autoclave at 10 atm, with subsequent column chromatographic separation on Al_2O_3 [20].

$Fe(C_5H_4$–$Na)_2$ is likewise predominantly carboxylated with dry ice, see for example [3, 8, 11]. For reaction with aqueous Na_2CO_3 see [16]. An 85 to 90% yield of the dicarboxylic acid along with 15 to 10% of the monocarboxylic acid (yields referred to the starting ferrocene) is obtained using the mixture of $Fe(C_5H_4$–$Na)_2$ and $C_5H_5FeC_5H_4$–Na formed by metallation of ferrocene with n-$C_5H_{11}Na$ in the presence of $(CH_3)_2N(CH_2)_2N(CH_3)_2$ [38]. Yields of 55 to 70% referred to ferrocene are obtained with ferrocene reacted with n-$C_5H_{11}Na$ in hydrocarbons such as pentane [3, 18] or octane [11]; 42% is gotten with a mixture obtained with C_6H_5Na in C_6H_6 [8]. According to [38] the $FeC_{10}H_{10-n}Na_n$ prepared in the absence of $(CH_3)_2N(CH_2)_2N(CH_3)_2$ is obtained not only in distinctly lower total yield but also possesses besides the products with n = 2 (60% fraction) and 1 (20%) some 20% with n = 3 to 6; see also [18].

Preparation from $Fe(C_5H_4$–$MgBr)_2$ with dry ice in ether [15, 18] gives a yield of 59% referred to the $Fe(C_5H_4$–$Br)_2$ when the Mg compound is prepared from the dibromoferrocene with Mg powder in the presence of CH_3I in tetrahydrofuran [15]. A 30% yield (based on the reacted ferrocene) is gotten when the Mg compound is obtained from $Fe(C_5H_4$–$Na)_2$ with $MgBr_2$ and Mg in $Br(CH_2)_2Br$ [18].

From Other Ferrocene Derivatives. By cleavage of $Fe(C_5H_4$–CO–C_6H_4Cl-2$)_2$ with $KOC(CH_3)_3$ and traces of H_2O in $CH_3OCH_2CH_2OCH_3$ with boiling for 10 to 20 min; yield 85 to 95% [47]. By oxidation with O_2 of $Fe(C_5H_4$–$CHO)_2$ or $Fe(C_5H_4$–$CH_2OH)_2$ dissolved in $OP(N(CH_3)_2)_3$ in the presence of a 20-fold molar quantity of $KOC(CH_3)_3$, in practically quantitative yield (86, 83%, respectively, of pure product) [36].

- From HO_2C–$C_5H_4FeC_5H_4$–CO–CH_3 with iodine in pyridine at 100°C [7].
- By saponification of $Fe(C_5H_4$–CO–S–$CH_3)_2$ with $KOH/H_2O/C_2H_5OH$ [24].
- By saponification of $Fe(C_5H_4$–$CN)_2$ with boiling $NaOH/H_2O/C_2H_5OH$ (34% yield) [19, 33].
- By alkaline hydrolysis of $Fe(C_5H_4$–$CO_2C_2H_5)_2$ with NaOH (molar ratio 1/1) in boiling H_2O/C_2H_5OH, in 16% yield after 30 min along with a 55% yield of $C_2H_5O_2C$–$C_5H_4FeC_5H_4$–CO_2H with a 20% recovery of the starting compound [17].
- From $K_3[Fe(Fe(C_5H_4$–$CO_2)_2)_3]$ in protogenic solvents such as CH_3CO_2H or in amphiprotic solvents such as H_2O [41].

References on pp. 212/3

Purification and Separation

Purification is effected by recrystallization from glacial acetic acid [9, 12, 28] in the dark [28], from glacial acetic acid/ethanol [29], 1/1 glacial acetic acid/dioxane [9], dioxane [9, 25], $CH_3CO_2C_2H_5$ [22], C_2H_5OH [47], 95% C_2H_5OH [3], or CH_3OH/H_2O [2, 47], or by sublimation at 180 to 200°C and 0.5 Torr [3]. However, it has been stated that the compound cannot be satisfactorily recrystallized due to the liability of its solutions to decompose. Reprecipitating the acid with dilute hydrochloric acid after dissolving it in an $NaHCO_3$ solution is recommended [42].

$C_5H_5FeC_5H_4–CO_2H$ accumulates as the most frequent by-product of the preparation. It can be separated from $Fe(C_5H_4–CO_2H)_2$ by fractional crystallization from glacial acetic acid [4, 21] or by extraction with boiling toluene [46], with hot [36] benzene [30, 36], or with ether [13] in an Soxhlet apparatus [12, 31]. In the separation by thin layer chromatography on SiO_2 gel with 15/1/3 toluene/petroleum ether/acetone the relative R_f values (referred to that for ferrocene as 1) are 0.08 for $Fe(C_5H_4–CO_2H)_2$ and 0.26 for $C_5H_5FeC_5H_4–CO_2H$ [40]. At a mixture ratio of 15/1/1 or with 15/2 petroleum ether/acetone or 30/1 petroleum ether/ethanol as the developer, the separation is equally possible [44]. It succeeds by column chromatography on Al_2O_3 using 4/4/1/1 C_2H_5OH/C_4H_9OH/conc. NH_4OH/H_2O [20]. High pressure liquid chromatography on SiO_2 with 1/3 ether/CH_3OH as the mobile phase is reported in [34].

$Fe(C_5H_4–CO_2H)_2$ can be determined in a mixture of the mono- and dicarboxylic acids by use of the IR absorptions at 1002 or 1108.6 cm^{-1} [43].

The separation of $HO_2C–C_5H_4FeC_5H_4–CO–CH_3$ takes place by its extraction with hot benzene [7, 29] or after conversion of the two acids into their methyl esters by column chromatography on Al_2O_3 with benzene [2].

$C_2H_5O_2C–C_5H_4FeC_5H_4–CO_2H$ is removed from the mixture of the two acids by extraction with diethyl ether [17].

The R_f values for $Fe(C_5H_4–CO_2H)_2$ in thin layer chromatography on microcrystalline cellulose are 0.00 in pentane or toluene and 0.42 in H_2O. R_f rises with the increasing dielectric constants of acetone/methanol mixtures [35].

Analysis

Its high oxidation potential impedes the determination of $Fe(C_5H_4–CO_2H)_2$ by potentiometric titration with $Cu(ClO_4)_2$ in $CH_3CN/HClO_4/H_2O$ [45].

References:

[1] R. B. Woodward, M. Rosenblum, M. C. Whiting (J. Am. Chem. Soc. **74** [1952] 3458/9). – [2] M. Rosenblum (Diss. Harvard Univ. 1953, pp. 114/5). – [3] P. J. Graham, E. I. DuPont de Nemours & Co. (U.S. 2835686 [1953/58]; C.A. **1958** 16366). – [4] R. A. Benkeser, D. Goggin, G. Schroll (J. Am. Chem. Soc. **76** [1954] 4025/6). – [5] A. N. Nesmeyanov, E. G. Perevalova, R. V. Golovnya, O. A. Nesmeyanova (Dokl. Akad. Nauk SSSR **97** [1954] 459/61).

[6] D. W. Mayo, P. D. Shaw, M. Rausch (Chem. Ind. [London] **1957** 1388/9). – [7] N. A. Nesmeyanov, O. A. Reutov (Dokl. Akad. Nauk SSSR **115** [1957] 518/21; Proc. Acad. Sci. USSR Chem. Sect. **112/117** [1957] 763/6). – [8] A. N. Nesmeyanov, E. G. Perevalova, Z. A. Beinoravichute (Dokl. Akad. Nauk SSSR **112** [1957] 439/40). – [9] W. Göller (Diss. Stuttgart T.H. 1958). – [10] K. L. Motz (Diss. Univ. Illinois 1958, pp. 1/130, 73; Diss. Abstr. **19** [1958] 958).

[11] A. N. Nesmeyanov, E. G. Perevalova, Z. A. Beinoravichute, I. L. Malina (Dokl. Akad. Nauk SSSR **120** [1958] 1263/6; Proc. Acad. Sci. USSR Chem. Sect. **118/123** [1958] 499/502). – [12] M. Rausch, M. Vogel, H. Rosenberg, D. Mayo, P. Shaw (WADC-TR-57-62-II [1958] 1/18). – [13] R. A. Benkeser, L. W. Hall (AD-215025 [1959]). – [14] M. R. Rausch, E. O. Fischer, H. Grubert (J. Am. Chem. Soc. **82** [1960] 76/82). – [15] H. Shechter, J. F. Helling (J. Org. Chem. **26** [1961] 1034/7).

[16] M. Okawara, Y. Takemoto, H. Kitaoka, E. Haruki, E. Imoto (Kogyo Kagaku Zasshi **65** [1962] 685/8 from C.A. **58** [1963] 577). – [17] R. L. Schaaf (J. Org. Chem. **27** [1962] 107/11). – [18] D. Seyferth, H. P. Hofmann, R. Burton, J. F. Helling (Inorg. Chem. **1** [1962] 227/31; AD-264343 [1961] 1/17; AD-283933 [1962] 1/37, 27/31). – [19] A. N. Nesmeyanov, E. G. Perevalova, L. P. Yur'eva, K. I. Grandberg (Izv. Akad. Nauk SSSR Otd. Khim. Nauk **1963** 1377/80; Bull. Acad. Sci. USSR Div. Chem. Sci. **1963** 1254/6). – [20] K. Schlögl, M. Fried (Monatsh. Chem. **94** [1963] 537/43).

[21] J. Tirouflet, E. Laviron, R. Dabard, J. Komenda (Bull. Soc. Chim. France **1963** 857/62). – [22] K. Yamakawa, H. Ochi, K. Arakawa (Chem. Pharm. Bull. [Tokyo] **11** [1963] 905/11). – [23] C. Rai, Union Oil Co. of California (U.S. 3461287 [1963/69]; C.A. **71** [1969] No. 81541). – [24] D. E. Bublitz, G. H. Harris (J. Organometal. Chem. **4** [1965] 404/8). – [25] J. T. Suh, Colgate-Palmolive Co. (U.S. 3382267 [1965/68] 1/5; C.A. **69** [1968] No. 59378).

[26] I. A. Golubeva, T. P. Vishnyakova (Neftekhim. Pererab. Nefti Gaza Temat. Analit. Obs. Ts. NIIT Neftekhim. **1966** 36/42). – [27] P. Petrovitch (Double-Liaison No. 133 [1966] 1093/109). – [28] H. Valot (personal communication from [27]). – [29] I. A. Golubeva, T. P. Vishnyakova, N. N. Bul'on (Tr. Mosk. Inst. Neftekhim. Gazov. Prom. No. 72 [1967] 61/7). – [30] M. D. Rausch, D. J. Ciappenelli (J. Organometal. Chem. **10** [1967] 127/36).

[31] E. B. Sokolova, M. P. Shebanova, L. A. Datskevich (Sin. Fiz. Khim. Polim. No. 6 [1970] 69/72; C.A. **73** [1970] No. 131397). – [32] A. Sonoda, I. Moritani (J. Organometal. Chem. **26** [1971] 133/40). – [33] V. I. Boev, A.V. Dombrovskii (Zh. Obshch. Khim. **46** [1976] 626/8; J. Gen. Chem. [USSR] **46** [1976] 625/6). – [34] D. T. Haworth, T. Liu (J. Chromatog. Sci. **14** [1976] 519/20). – [35] D. T. Haworth, T.-C. Liu (Separ. Sci. **11** [1976] 329/34).

[36] G. Schmitt, S. Özman (J. Org. Chem. **41** [1976] 3331/2). – [37] V. P. Tverdokhlebov, I. V. Tselinskii, B. V. Gidaspov, G. Yu. Chikisheva (Zh. Org. Khim. **12** [1976] 2335/40; J. Org. Chem. [USSR) **12** [1976] 2268/72). – [38] E. W. Post, T. F. Crimmins (J. Organometal. Chem. **161** [1978] C17/C19). – [39] M. Walczak, K. Walczak, R. Mink, M. D. Rausch, G. Stucky (J. Am. Chem. Soc. **100** [1978] 6382/8). – [40] X. Yang, Z. Xie, F. Ni, S. Shi (Fundam. Res. Organometal. Chem. Proc. 1st China-Japan-U.S. Trilateral Semin. Organometal. Chem., Beijing 1980 [1982], pp. 765/9; C.A. **97** [1982] No. 103654).

[41] T. Nomura (Bull. Chem. Soc. Japan **56** [1983] 2937/40). – [42] F. W. Knobloch, W. H. Rauscher (J. Polym. Sci. **54** [1961] 651/6). – [43] E. F. Wolfarth (Anal. Chem. **30** [1958] 185/6). – [44] K. Tanikawa, K. Arakawa (Bunseki Kagaku **20** [1971] 278/81). – [45] P. F. Quirk, B. Kratochvil (Anal. Chem. **42** [1970] 535/6).

[46] R. L. Schaaf, C. T. Lenk (J. Chem. Eng. Data **9** [1964] 103). – [47] V. Rapić, K. Schlögl, B. Steinitz (J. Organometal. Chem. **94** [1975] 87/98). – [48] K. Gonsalves, L. Zhan-ru, M. D. Rausch (J. Am. Chem. Soc. **106** [1984] 3862/3).

Physical and Structural Properties

$Fe(C_5H_4–CO_2H)_2$ crystallizes as fine [21] needles [5] or prisms [17], whose color is reported from yellow [17, 28] through red [5] and dark red [3, 29] to brown [6], in the majority of cases, however, as yellow-orange to a brilliant orange-red [2, 4, 7, 12, 13, 17, 19, 20, 23, 30, 42].

Inconsistent and in part contradictory data exist regarding the behavior of the compound on heating. This is possibly explained by the fact that according to [5, 20] it decomposes without melting at a decomposition temperature which depends strongly on the heating rate; for a heating time of $\geqq$8 h this is as low as 150°C [5]. Decomposition takes place with discoloration at temperatures (in °C) between 225 and 230 [5], >240 [24], from 242 to 247 [13], not below 250 [4, 23], at ~250 [2, 31], not until 265 [16], not below 280 [45], slowly between 265 and 300 [6]. In

References on pp. 219/20

contrast sublimation without decomposition has been observed above 230°C [1]. Melting with sublimation is reported for 215 to 235 [3], 240 to 250 [18]; melting with decomposition for >230 [47], 240 [17, 30], ca. 246 [42], not below 250 [4, 15, 23], 250 [34, 37], between 250 and 260°C [46]. The melting point: 250 [7, 19], >250 [29], >265 [14], not below 280°C [45]. Under N_2 up to 350°C neither visible decomposition nor visible melting are observed [46]. Sublimation at 0.5 Torr occurs between 180 and 240°C [3].

^{57}Fe Mössbauer Spectrum

The isomer shifts (referred to stainless steel) at 330 and 80 K are $\delta = 0.43 \pm 0.05$ and 0.48 ± 0.05 mm/s, respectively. The quadrupole splitting is $\Delta = 2.16 \pm 0.05$ mm/s in each case [11]. At 78 K $\delta = 0.49 \pm 0.01$ mm/s (referred to a metallic iron standard) and $\Delta = 2.16 \pm 0.01$ mm/s [35]. The relative probability of a recoil-free transition f' (referred to that for ferrocene as 1 at 300 K) is at 300 K 1.4 ± 0.14 and at 80 K 5.5 ± 0.55. The intensity ratio I_π/I_σ of the quadrupole doublet peaks is 1.5 ± 0.15 at 300 K and 1.16 ± 0.11 at 80 K [11]. The apparent correlation between the quadrupole splitting and the oxidation potential is treated in [43], the relationship between the Mössbauer data and the molecular structure in [35, 36].

IR Spectrum

The most important bands (in cm^{-1}), their intensities, and their assignments (for abbreviations see p. VIII) are:

	paraffin oil	paraffin oil	KBr	KBr	KBr	KBr	assignment
						3120 }	ν(CH)
						2860 }	
				2625		2625 vw	ν(OH)
				2544		2545 vw	ν(OH)
	1700 s			1692 }	1690	1720 s, sh }	ν_a(C=O)
	1686 s }		1669	1672 }		1680 vs, br. }	
	1660 s }						
	1470 vs					1492 s	ν_s(CO) + δ(OH)
					1403	1405 m	ν(CC)
						1374 w	
	1382 s						
	1218 w		1298			1300 s	δ(OH) + ν_s(CO)
	1170 s					1166 m	ν_a(CC)
	1054 w						
	1030 m						
		1025 sh				1025 m	δ(CH)
	918 m		914 br.	919		914 m, br.	γ(OH)
	878 w					823 m, br.	
						775 vw	
						742 w	
Ref.	[9]	[26]	[33]	[10]	[17]	[42]	[42]

Figures displaying IR spectra can be found in [8, 20].

Electronic Absorption Spectrum. Magnetic Circular Dichroism

Absorption maxima λ_{max} in nm and their extinction coefficients ε in $L \cdot mol^{-1} \cdot cm^{-1}$ (log ε) at room temperature for various solvents and their assignments:

solvent	A_{2g} ($e_{2g} \rightarrow e^*_{2g}$)		A_{1u} ($e_{2g} \rightarrow e_{2u}$)		charge transfer to CO_2H		A_{2u} ($e_{2g} \rightarrow e_{2u}$)		Ref. [44]
	λ_{max}	ε	λ_{max}	ε	λ_{max}	ε	λ_{max}	ε	
CH_2Cl_2	454	252	345	384	312	1110	267	5250	[44]
$(CH_3)_2SO$	450.4	(2.46)	349.6	(2.60)	309.6	(3.13)	—	—	[42]
CH_3OH	447	313	—	—	—	—	—	—	[32]
C_2H_5OH	445	300	—	—	—	—	255	9900	[17]
C_2H_5OH/ H_2O (95/5)	448.4	(2.47)	349.6	(2.58)	308	(3.12)	—	—	[42]
H_2O	452.5	—	—	—	309.6	sh	—	—	[42]

In the region of the first band (A_{2g}) an S-shaped MCD band is observed with peaks in CH_3OH at $\lambda_1 = 485$ and $\lambda_2 = 417$ nm ($\lambda_{\varepsilon=0} = 440$ nm), which are partially overlapped by a Gaussian MCD curve with $\lambda_{max} = 357$ nm [32].

X-ray Photoelectron Spectroscopy

Fe $2p_{3/2}$ has been determined as 708.6 eV referred to C_{1s} with 285.0 eV [38, 39].

Crystal and Molecular Structure

The compound crystallizes in two different modifications, one triclinic, which is obtained by recrystallization from glacial acetic acid at room temperature with slow evaporation, and one monoclinic, which deposits only from hot solutions at temperatures ≧333 K. In both modifications $Z = 4$ and the unit cell contains two crystallographically independent dimers, which for their part are each linked by hydrogen bonds between the carboxyl groups. The location of the acidic protons constitutes the main difference between the two modifications. While in the triclinic form (I) the protons are localized firmly on the associated oxygen atoms and the bonding distances are distinctly different for C=O and C–O, in the monoclinic form (II) the protons assume no fixed location but are found between the two O atoms, to each of which they are 50% assigned, with almost equal carbon-oxygen distances [41].

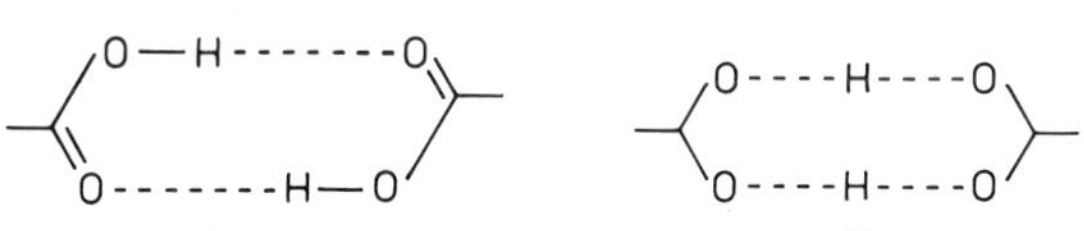

I II

Triclinic Form. Crystal data determined by neutron diffraction ($\lambda = 1.1302$ Å) at 78 K on single crystals (Method I) and by X-ray diffraction (Mo K_α, $\lambda = 0.71069$ Å) (Method II) [41] (space group $P\bar{1}$-C_i^1 (No. 2)):

method	a (Å)	b (Å)	c (Å)	α	β	γ	V (Å^3)	D (g/cm^3)
I	7.424(2)	7.809(2)	18.118(4)	98.79(2)°	91.07(2)°	97.21(3)°	1029.1(5)	1.769
II	7.477(1)	7.895(1)	18.235(3)	99.04(1)°	91.35(3)°	97.41(2)°	1053.0(3)	1.729

References on pp. 219/20

The structure is given by **Fig. 7**. $C_5H_4CO_2H$ forms four ligands A, B, C, and D which are somewhat different in bond lengths and angles (values obtained by Method I, see **Figs. 8** and **9**, respectively) and which yield the two types of molecule, A–Fe–B and C–Fe–D. The average of the two cases is an Fe–carbon distance of 2.049(2) Å, an Fe–cyclopentadienyl ring distance of 1.649(1) Å. The rings are nearly planar and the H atoms 2 to 5 (except H-5 of D) are displaced

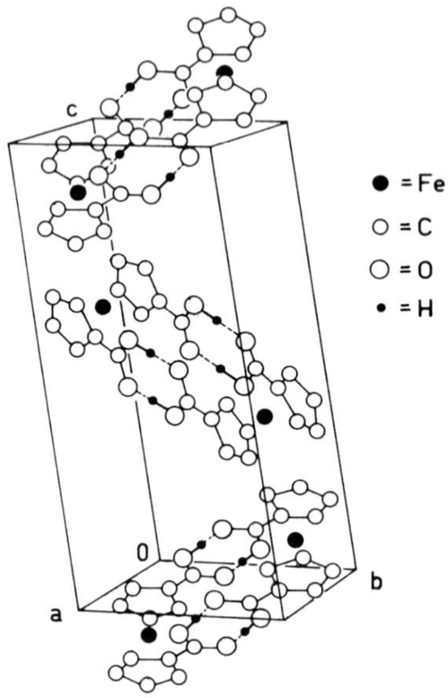

Fig. 7. $Fe(C_5H_4{-}CO_2H)_2$ in triclinic modification [41].

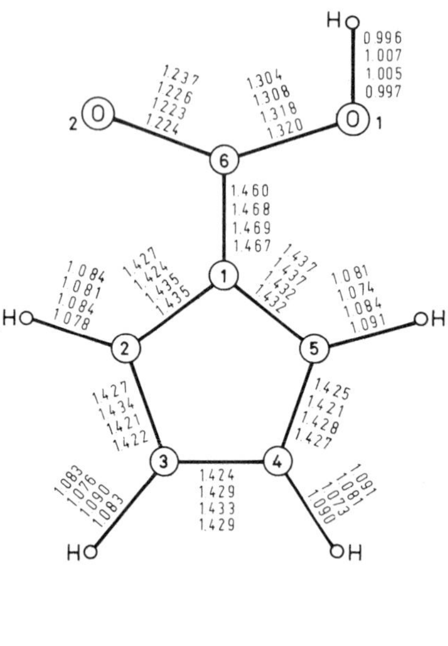

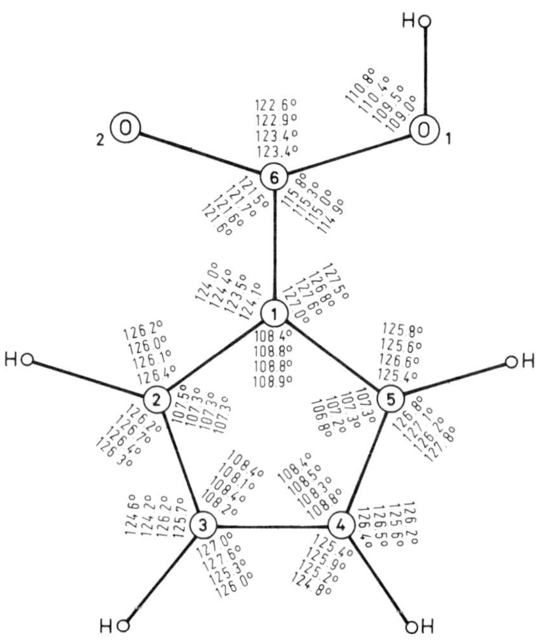

Fig. 8. Bond distances (in Å) in $Fe(C_5H_4{-}CO_2H)_2$ [41].

Fig. 9. Bond angles in $Fe(C_5H_4{-}CO_2H)_2$ [41].

References on pp. 219/20

away from the ring planes toward the Fe. The molecules exhibit a practically completely eclipsed configuration; the torsion angle (angle C_n–ring center 1–ring center 2–C_n with n = 1 to 5) amounts on the average for A–Fe–B by Method I to 1.58°, by Method II to 1.24°, for C–Fe–D by Method I to −0.76°, by Method II −0.48°. The angle defined by ring center 1–Fe–ring center 2 deviates distinctly from 180°C, because the Fe on the average is displaced by 0.013 Å in the direction of the C-1 atom. The ring planes are slightly inclined, with an enlarged opening between the C-1 atoms. In each case two molecules form a dimer held together by hydrogen bonding between the carboxylate groups (see I). Neighboring dimers are bound by a weak C–H .... O=C interaction, in which H-2 and H-5 function as donors. This interaction occurs noticeably only between molecules of the same type. For the hydrogen bonding geometry of A–Fe–B see **Fig. 10**, of C–Fe–D **Fig. 11**. The interaction between the C–Fe–D molecules is somewhat weaker than that between the A–Fe–B molecules [41].

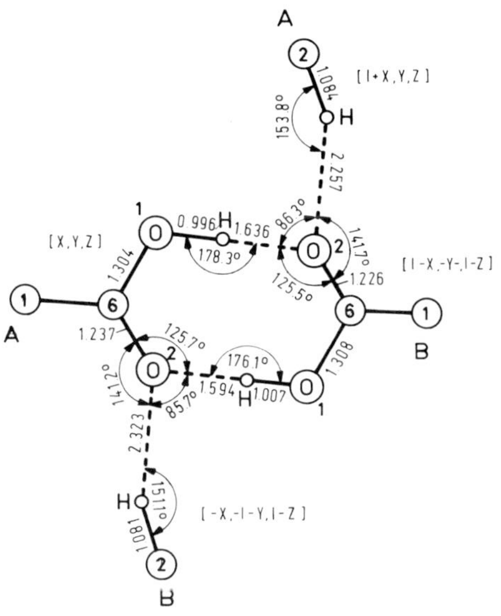

Fig. 10. Hydrogen bonding geometry of A–Fe–B (bond distances in Å) [41].

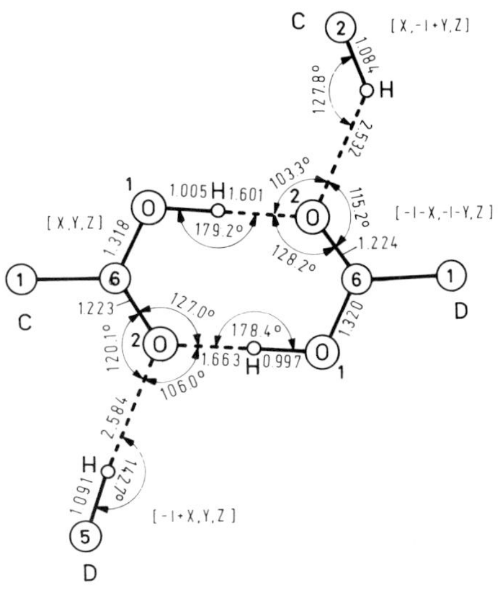

Fig. 11. Hydrogen bonding geometry of C–Fe–D [41].

A comparison of salient molecular dimensions with $C_5H_5FeC_5H_4–CO_2H$ shows for the monocarboxylic acid with 1.407(15) Å a noticeably shorter C(1)–C(6) distance and with 1.261 Å a longer C=O distance, although the latter is at the limit of statistical significance. The greater difference may be due to only a slight possibility of electronic perturbations resulting from interactions between neighboring dimers in $Fe(C_5H_4–CO_2H)_2$ or to perturbations attributable to the presence of an 'extra' carboxylate group on the second cyclopentadienyl ring in the diacid [48].

Monoclinic Form. $Fe(C_5H_4–CO_2H)_2$ crystallizes from hot glacial acetic acid as laths elongated along the b axis [25], in space group $P2_1/c-C^5_{2h}$ (No. 14) [25, 41]. X-ray structural determinations (Mo K_α, $\lambda = 0.71069$ Å) at 78 K [40] and at 298 K on single crystals [25] give the crystal data:

T in K	a(Å)	b(Å)	c(Å)	α	β	γ	V(Å³)	D(g/cm³)
78	8.330(2)	8.750(1)	14.123(2)	90.0°	90.76(3)°	90.0°	1029.3(3)	1.769
298	8.403(10)	8.910(10)	14.192(18)	90.0°	90.41(8)°	90.0°	1062(2)	1.714

References on pp. 219/20

The density measured by flotation in CCl_4/CH_2I_2 is 1.713 g/cm³ [25]. The enumeration of the atoms, bond distances in Å, and the packing of the molecules in the unit cell are given in **Fig. 12**, the bond angles in **Fig. 13** [25]. **Fig. 14** shows the crystal structure of the monoclinic modification [41]. Here also the molecule possesses an almost completely eclipsed configuration (torsion angle on average 1°30′ [41], 1°38′ [25]) and planar cyclopentadienyl rings whose planes include an angle of 1°17′. The planes of the likewise planar carboxyl groups form with the plane of the adjacent C_5H_4 ring an angle of 0°50′ for C(6)–O(1)–O(2) and of 4°40′ for C(6′)–O(1′)–O(2′)

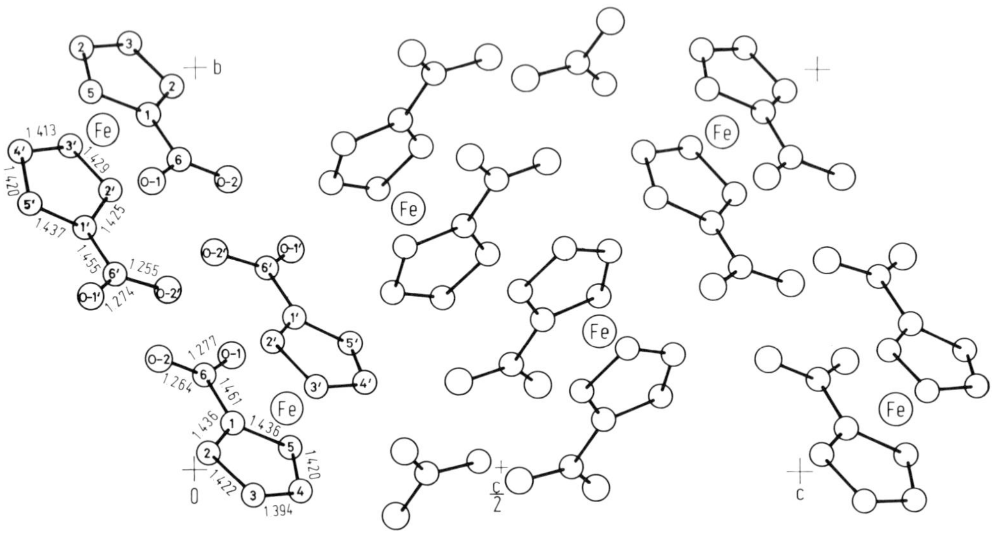

Fig. 12. A view down the a axis illustrating the atom numbering and the packing of the dimer in the unit cell for $Fe(C_5H_4{-}CO_2H)_2$ in the monoclinic modification from [25].

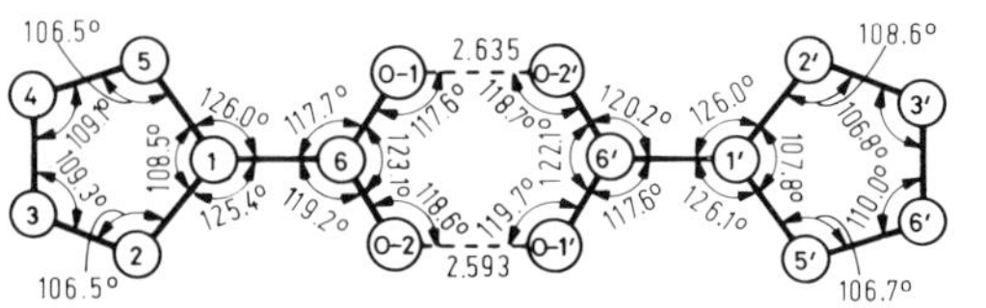

Fig. 13. Bond angles in the dimer unit of $Fe(C_5H_4{-}CO_2H)_2$ in the monoclinic modification from [25].

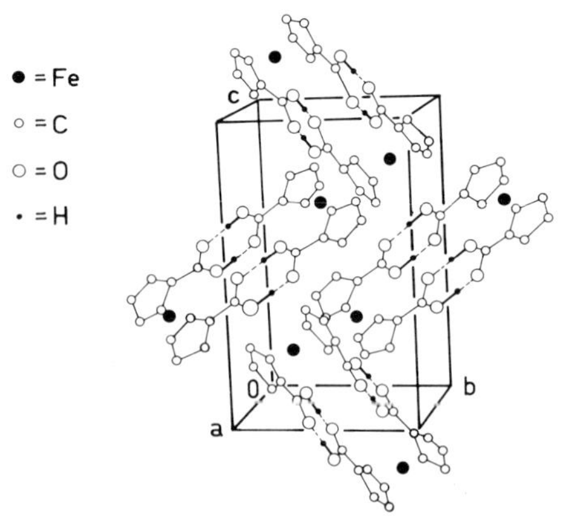

Fig. 14. Crystal structure of $Fe(C_5H_4{-}CO_2H)_2$ in the monoclinic modification from [41].

[25]. In the dimer linked by hydrogen bonding (see II) the four C–O distances are not very different from each other [25]. Both the oxygen of the C=O and that of the HO group can serve as acceptors. The H···O distances are equal in both cases and amount to between 2.25 and 2.6 Å [41].

Both these structures and those found for $Fe(C_5H_4–CO–CH_3)_2$ refute the supposition that ferrocene possesses preferentially a stepped configuration [27]. Also, calculations by Dashevskii and coworkers are untenable, giving for $Fe(C_5H_4–CO_2H)_2$ the conformations III or IV [22].

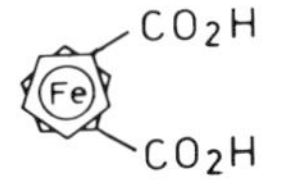

III

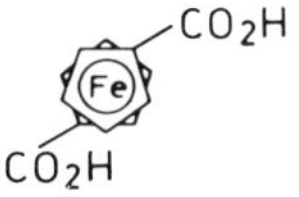

IV

References:

[1] R. B. Woodward, M. Rosenblum, M. C. Whiting (J. Am. Chem. Soc. **74** [1952] 3458/9). – [2] M. Rosenblum (Diss. Harvard Univ. 1953, pp. 114/5). – [3] P. J. Graham, E. I. DuPont de Nemours & Co. (U.S. 2835686 [1953/58]; C.A. **1958** 16366). – [4] A. N. Nesmeyanov, E. G. Perevalova, R. V. Golovnya, O. A. Nesmeyanova (Dokl. Akad. Nauk SSSR **97** [1954] 459/61). – [5] W. Göller (Diss. Stuttgart T.H. 1958).

[6] K. L. Motz (Diss. Univ. Illinois 1958, pp. 1/130; Diss. Abstr. **19** [1958] 958). – [7] M. Rausch, M. Vogel, H. Rosenberg, D. Mayo, P. Shaw (WADC-TR-57-62-II [1958] 1/18). – [8] E. F. Wolfarth (Anal. Chem. **30** [1958] 185/6). – [9] L. A. Kazitsyna, B. V. Lokshin, A. N. Nesmeyanov (Dokl. Akad. Nauk SSSR **127** [1959] 333/6; Proc. Acad. Sci. USSR Chem. Sect. **124/129** [1959] 531/4). – [10] M. R. Rausch, E. O. Fischer, H. Grubert (J. Am. Chem. Soc. **82** [1960] 76/82).

[11] R. A. Stukan, S. P. Gubin, A. N. Nesmeyanov, V. I. Gol'danskii, E. F. Makarov (Teor. Eksperim. Khim. **2** [1966] 805/11; Theor. Exptl. Chem. [USSR] **2** [1966] 581/4). – [12] F. W. Knobloch, W. H. Rauscher (J. Polym. Sci. **54** [1961] 651/6). – [13] M. Okawara, Y. Takemoto, H. Kitaoka, E. Haruki, E. Imoto (Kogyo Kagaku Zasshi **65** [1962] 685/8 from C.A. **58** [1963] 577). – [14] R. L. Schaaf (J. Org. Chem. **27** [1962] 107/11). – [15] D. Seyferth, H. P. Hofmann, R. Burton, J. F. Helling (Inorg. Chem. **1** [1962] 227/31; AD-264343 [1961] 1/17; AD-283933 [1962] 1/37, 27/31).

[16] J. Tirouflet, E. Laviron, R. Dabard, J. Komenda (Bull. Soc. Chim. France **1963** 857/62). – [17] K. Yamakawa, H. Ochi, K. Arakawa (Chem. Pharm. Bull. [Tokyo] **11** [1963] 905/11). – [18] C. Rai, Union Oil Co. of California (U.S. 3461287 [1963/69]; C.A. **71** [1969] No. 81541). – [19] Colgate-Palmolive Co. (U.S. 3382267 [1965/68] 1/5; C.A. **69** [1968] No. 59378). – [20] P. Petrovitch (Double-Liaison No. 133 [1966] 1093/109).

[21] H. Valot (personal communication from [20]). – [22] G. Dashevskii, L. Kaluskii, A. I. Kitaigorodskii (Roszniki Chem. **41** [1967] 559/66). – [23] I. A. Golubeva, T. P. Vishnyakova, N. N. Bul'on (Tr. Mosk. Inst. Neftekhim. Gazov. Prom. No. 72 [1967] 61/7). – [24] M. D. Rausch, D. J. Ciappenelli (J. Organometal. Chem. **10** [1967] 127/36). – [25] G. J. Palenik (Inorg. Chem. **8** [1969] 2744/9).

[26] A. Perjéssy, Š. Toma (Chem. Zvesti **23** [1969] 533/9). – [27] G. J. Palenik (Inorg. Chem. **9** [1970] 2424/30). – [28] E. B. Sokolova, M. P. Shebanova, L. A. Datskevich (Sin. Fiz. Khim. Polim. No. 6 [1970] 69/72; C.A. **73** [1970] No. 131397). – [29] A. Sonoda, I. Moritani (J. Organometal. Chem. **26** [1971] 133/40). – [30] K. Tanikawa, K. Arakawa (Bunseki Kagaku **20** [1971] 278/81).

[31] H. Imai (Bull. Soc. Chem. Japan **45** [1972] 1264/5). – [32] D. Nielson, D. Boone, H. Eyring (J. Phys. Chem. **76** [1972] 511/5). – [33] P. A. Vigato, U. Casellato, D. A. Clemente, G. Bandoli, M. Vidali (J. Inorg. Nucl. Chem. **35** [1973] 4131/9). – [34] K. Yamakawa, M. Hisatome (J. Organometal. Chem. **52** [1973] 407/24). – [35] M. L. Good, J. Buttone, D. Foyt (Ann. N.Y. Acad. Sci. **239** [1974] 193/207).

[36] R. A. Abramovitch, J. L. Atwood, M. L. Good, B. A. Lampert (Inorg. Chem. **14** [1975] 3085/9). – [37] G. Schmitt, S. Özman (J. Org. Chem. **41** [1976] 3331/2). – [38] A. B. Fischer, M. S. Wrighton, M. Umana, R. W. Murray (J. Am. Chem. Soc. **101** [1979] 3442/6). – [39] A. B. Fischer, M. S. Wrighton, M. Umana, R. W. Murray (AD-AO66631 [1979] 1/27; C. A. **91** [1981] No. 201771). – [40] F. Takusagawa, T. F. Koetzle (Abstr. Am. Cryst. Assoc. Winter Meeting, Honolulu 1979, PB 36, p. 91 from [41]).

[41] F. Takusagawa, T. F. Koetzle (Acta Cryst. B **35** [1979] 2888/96). – [42] T. Nomura (Bull. Chem. Soc. Japan **56** [1983] 2937/40). – [43] R. M. G. Roberts, J. Silver (J. Organometal. Chem. **263** [1984] 235/41). – [44] O. Travers, R. Rossi, V. Carassiti (Ann. Univ. Ferrara Sez. V **3** [1974] 103/10). – [45] H. Shechter, J. F. Helling (J. Org. Chem. **26** [1961] 1034/7).

[46] R. L. Schaaf, C. T. Lenk (J. Chem. Eng. Data **9** [1964] 103). – [47] V. Rapić, K. Schlögl, B. Steinitz (J. Organometal. Chem. **94** [1975] 87/98). – [48] F. A. Cotton, A. H. Ried (Acta Cryst. C **41** [1985] 686/8).

Solubility. Chemical Behavior

The reactions of the acid in aqueous alkali hydroxides are treated in connection with the corresponding alkali salts.

Solubility. Dissociation

$Fe(C_5H_4{-}CO_2H)_2$ is well soluble only in aqueous alkali hydroxide solutions [6], glacial acetic acid [4, 6], formic acid, or dioxane [6]. It is slightly soluble in alcohols [4, 6]. In almost all other organic solvents it is at best difficultly soluble, or practically insoluble; 100 mL boiling toluene, for example, dissolve 30 mg of the acid [6]. Similarly poor is the solubility in benzene [6, 30, 61], xylene [6], or ether [6, 14, 30]. The solutions decompose rapidly [10]. The very low solubility in C_6H_6 did not allow examination of the dielectric spectrum for obtaining information about the conformation and the energy barrier to internal rotation [61].

The compound dissociates in two steps. In 2/1 C_2H_5OH/H_2O the dissociation constants $K_1=3.1\times10^{-7}$ and $K_2=2.7\times10^{-8}$ at 27°C and an ionic strength <0.01 [1, 2]. The relatively slight difference between the two values indicates that the two carboxyl groups do not influence each other [6]. In 19/11 C_2H_5OH/H_2O $pK_a=6.22$ [8]. In 1/1 CH_3OH/H_2O $pK_a=6.80$ at 25°C [18].

Mass Spectrum

On fragmentation by electron impact the molecular ion (m/e 274) has the highest relative intensity (taken as 100%). It decomposes mainly according to scheme in I. The sequence m/e $274\rightarrow256\rightarrow212$ is confirmed by metastable peaks at m/e 239.2 and 175.6. The fragment m/e 230 could transform into the fragment ions $[FeC_5H_4{-}CO_2H]^+$ (m/e 165, 39%) and $[HOFeC_5H_5]^+$ (m/e 138, 25%), as [22] reports [42]. The binuclear ion $[Fe_2(C_5H_4{-}CO_2H)_2]^+$ forms by reaction of the fragmentation product Fe^+ with a neutral molecule, at 50 V, 150°C [55].

Photochemical Reactions

In absolute ethanol, methanol, cyclohexane, or CH_2Cl_2, the compound photooxidizes only on irradiation with 254 nm light. It behaves essentially like $C_5H_5FeC_5H_4{-}CO_2H$, but exhibits a threefold greater reaction rate [46]. In $(CH_3)_2SO$ irradiation by a high pressure Hg lamp causes the absorptions at 718, 1155, 1264, and 1700 cm^{-1} to be replaced by ones at 795, 1580, 1650, and 2323 cm^{-1} [43].

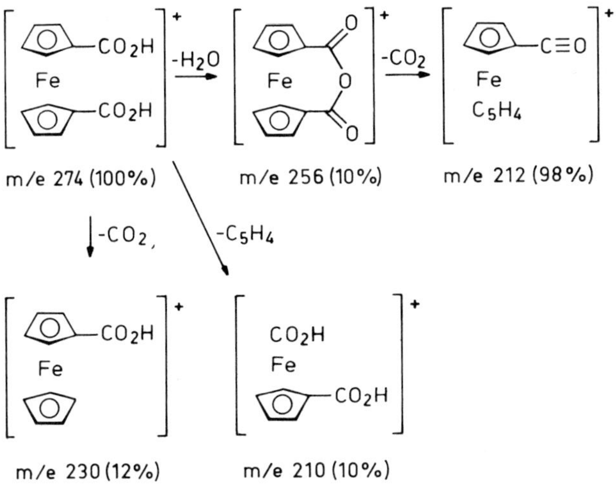

I [42]

Electrochemical Behavior

Reversible **oxidation** to the ferrocenium ion per $Fe(C_5H_4{-}CO_2H)_2 - e^- \rightleftharpoons [Fe(C_5H_4{-}CO_2H)_2]^+$ in CH_3CN (0.2 M $LiClO_4$) at 25°C gives by chronopotentiometry a quarter-wave potential of $E_{\tau/4}$ = 0.768 V (vs. SCE, the saturated calomel electrode). It is higher by 0.453 V than that of ferrocene [21]. The formal oxidation potential determined by cyclic voltammetry in CH_3CN (0.1 M $[(C_2H_5)_4N]ClO_4$) is at room temperature E° = 0.838 V (vs. NaCl-saturated calomel electrode) [50], 0.69 V (vs. SCE) [51], in potassium phosphate buffer of pH 7.0 at 25°C E° = 0.420 V (vs. SCE) [64]. The formal oxidation potential for the compound immobilized by covalent bonding on a Pt/PtO surface is E° = 0.757 V (vs. NaCl-saturated CE) [50], 0.76 ± 0.010 V (vs. SCE) [51], differing only little from these values. The potential has been determined oscillopolarographically in 1/1 C_2H_5OH/H_2O (1 M $HClO_4$) on a Pt electrode [12, 17, 20]. Because $E°_{\tau/4}$ is governed by the level of the HOMO orbital in the ferrocene which is generally accepted as being the e_{2g} level, $E°_{\tau/4}$ would be expected to be related to σ_p rather than to σ_p^+ [58]. Regarding the effect exerted by CO_2H on the redox potential of ferrocene, see [12, 17, 20, 21].

The electrochemical **reduction** forms by successive one-electron transitions the radical anion $\mathbf{[Fe(C_5H_4{-}CO_2H)_2]^{-\cdot}}$ and the dianion $\mathbf{[Fe(C_5H_4{-}CO_2H)_2]^{2-}}$. The half-wave potentials measured polarographically in $(CH_3)_2NCHO$/0.1 M $[(C_4H_9)_4N]PF_6$ are $E_{1/2}$ = −2.11 and −2.60 V (vs. SCE). When the compound is electrolyzed in this solution at a controlled potential (−>2.11 or −>2.60 V), Fe and $[C_5H_4CO_2H]^-$ form with the consumption of two electrons [52].

Other Reactions

$Fe(C_5H_4{-}CO_2H)_2$ on **heating** in air decomposes with an exothermic peak at 246°C [57]. The compound adsorbed on microspheres of SiO_2 forms in an inert atmosphere at 350°C Fe^{III} silicate, at 550°C the Fe^{II} silicate, and at 800°C Fe_2SiO_4 [29].

Hydriding over Raney nickel in cyclohexane for 5 h at 265°C and 200 atm H_2 yields 37% C_5H_6 and small quantities of ferrocene. In alcohol solution $C_5H_5CO_2H$ (14% yield) and its ethyl ester (17%) were obtained [59]. **Bromine** in boiling CCl_4 likewise decomposes the ferrocene core and forms a Br-containing carboxylic acid, but no pentabromocyclopentane, evidence that the two CO_2H groups are located on different cyclopentadienyl rings [59].

References on pp. 224/5

Ligand exchange up to 70% takes place in the presence of $AlCl_3$ (molar ratio 1/4) in boiling mesitylene. The ions $[HO_2C–C_5H_4FeC_6H_3(CH_3)_3]^+$ and $[Fe(C_6H_3(CH_3)_3)_2]^{2+}$ form by the simple and double exchange, respectively. In contrast only very little exchange occurs in boiling benzene [47].

Decarboxylation is observed neither on heating with Cu powder and basic Cu carbonate nor in quinoline solution [2].

Cu^{II} is not able to cause **oxidation** of $Fe(C_5H_4–CO_2H)_2$ [38]. The compound evolves a gas with H_2SO_4, which must be CO_2 because the ferrocenium ion $[C_5H_5FeC_5H_4–CO_2H]^+$ forms as shown by the ESR spectrum of the solution ($g_{\parallel}=3.58$, $g_{\perp}=1.76$) [37].

Reduction with $LiAlH_4$ gives $Fe(C_5H_4–CH_2OH)_2$ quantitatively [13, 44, 62], that with $LiAlH_4$/$AlCl_3$ gives $Fe(C_5H_4–CH_3)_2$ [24].

Salt and Complex Formation. On treatment of the dry acid with concentrated alkalies the alkali salts deposit in a crystalline form [7]. Introduction of NH_3 into the solution of the compound in $(CH_3)_2NCHO$ yields the diammonium salt [27, 30]. K_2CO_3 forms $K_2[Fe(C_5H_4–CO_2)_2]\cdot xH_2O$, $Ba(OH)_2\cdot 8H_2O$ gives $Ba[Fe(C_5H_4–CO_2)_2]\cdot H_2O$ [57].

In absolute ethanol or ether $Fe(C_5H_4–CO_2H)_2$ interacts with HgX_2 (X = Cl, Br, I, SCN) with a distinct increase in the solubility of the HgX_2, which possibly indicates a reaction between the two compounds. The electronic absorption spectrum is in all cases that of $Fe(C_5H_4–CO_2H)_2$ superimposed on that of the HgX_2 [45].

Esterification with formation of $Fe(C_5H_4–CO_2R)_2$ ($R=CH_3, C_2H_5$) takes place in absolute CH_3OH [2, 6, 13, 18, 25, 34, 39, 41, 63] or C_2H_5OH [14], respectively, under catalysis by mineral acids [3], conc. H_2SO_4 [14, 34, 39, 41, 48, 63], gaseous HCl [6, 13, 18, 25], hydrochloric acid [2, 11], or $BF_3\cdot O(C_2H_5)_2$ [36] on refluxing for several hours [6, 13, 14, 18, 25, 36, 39, 41]. $CH_3O_2C–C_5H_4FeC_5H_4–CO_2H$ forms when $Fe(C_5H_4–CO_2H)_2$ is saturated with HCl in absolute CH_3OH at -8 to -3°C [18] or the suspension of the acid in methanol in the presence of several drops of conc. H_2SO_4 is diluted with CH_3OH with warming on a water bath just until a clear solution is formed [34]. In contrast, esterification with C_2H_5OH did not succeed on introduction of HCl gas or in the presence of small amounts of conc. H_2SO_4, p-toluenesulfonic acid, or hydrochloric acid by azeotropic distilling off of the water of reaction with toluene or benzene. Likewise, no ester formation under these conditions was observed with excess $n-C_4H_9OH$ or $HOCH_2CH_2OH$ [6].

Reaction with CH_2N_2 in ether/CH_3OH [61] or ether gives $Fe(C_5H_4–CO_2CH_3)_2$ in high yields [2, 4, 5, 16, 35], that with diphenyl sulfite in pyridine in the dark $Fe(C_5H_4–CO_2C_6H_5)_2$ [23]. $Fe(C_5H_4–CO_2H)_2$ on reaction with hydroquinone in the presence of $(CF_3CO)_2O$ in benzene/CH_3CN at room temperature forms $Fe(C_5H_4–CO_2C_6H_4OH-4)_2$ and II [56].

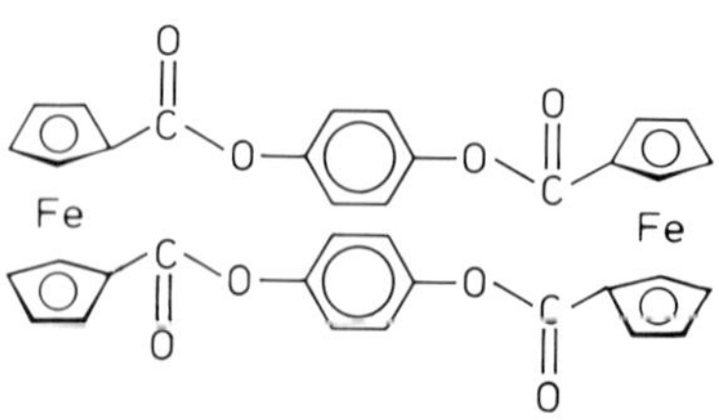

II

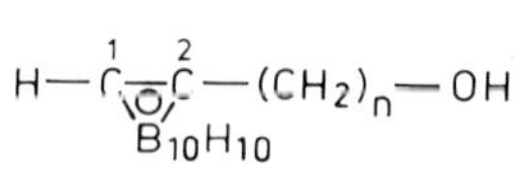

III

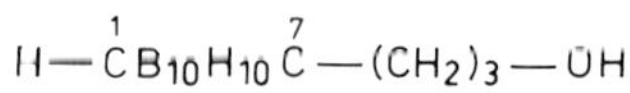

IV

The 1,2- or 1,7-dicarba-closo-dodecacarborane derivatives III (n=1) and IV give in boiling benzene or toluene, respectively, in the presence of catalytic quantities of n-butyl titanate the corresponding monoesters $HO_2C-C_5H_4FeC_5H_4-CO_2(CH_2)_nC_2B_{10}H_{11}$-1,2 (n=1) or -1,7 (n=3). With III (n=1, 2, 4 to 8) in a 1/2 molar ratio the diesters $Fe(C_5H_4-CO_2(CH_2)_nC_2B_{10}H_{11}\text{-}1,2)_2$ are obtained. For the latter reaction xylene can also be used as the solvent instead of benzene or toluene and instead of butyl titanate also another titanate with a lower alkyl group [26].

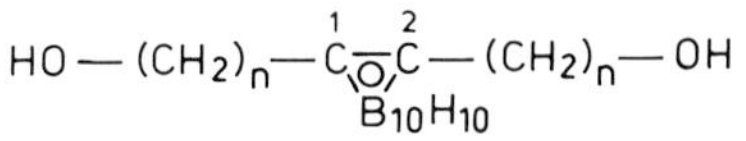

V

$HOCH_2-\overset{1}{C}B_{10}H_{10}\overset{7}{C}-CH_2OH$

VI

Extended refluxing of equimolar quantities of $Fe(C_5H_4-CO_2H)_2$ and V (n=1, 4, 6, or 8) in excess solvent (toluene for n=1, 4, or 6; xylene for n=8) in the presence of catalytic amounts of p-toluenesulfonic acid (for n=1), H_2SO_4 (for n=6 or 8), or amyl titanate (for n=4) forms $\text{-}[O(CH_2)_n-CB_{10}H_{10}C(CH_2)_nO_2C-C_5H_4FeC_5H_4-CO]_x\text{-}$. The corresponding reaction with VI in excess toluene yields $\text{-}[OCH_2-CB_{10}H_{10}C-CH_2O_2C-C_5H_4FeC_5H_4-CO]_x\text{-}$ after 40 h refluxing in the presence of amyl titanate [26]. Diols such as $HO(CH_2)_mOH$ (for example, m=4 or 6) and the acid in a 1/1 molar ratio in contrast gave no polyester formation, since preparation was impossible for either homogeneous melts of the two components or – due to the low solubility of the acids in alcohols – solutions of equimolar quantities. Also, reaction in dioxane with azeotropic distilling off of the water formed failed due to the thermal instability of the acid [6]; see also [31]. Cf. "Applications", p. 225.

The **acid chloride** $Fe(C_5H_4-COCl)_2$ forms in good yields [9] on refluxing with PCl_3 [9, 27, 28, 30, 53] in excess [9, 27, 30] in the dark [28] without indications of decomposition. The reaction also proceeds in benzene at 50 to 60°C [7] with addition of pyridine [53]. It further forms in the reaction with PCl_5 [9, 13, 54] in benzene [13, 54] at 30 to 50°C [13] or with $(COCl)_2$ [60] in CH_2Cl_2 in the presence of pyridine in the dark, first at room, then at boiling temperatures, in good yield [10, 28] or with $SOCl_2$ [9, 40] in benzene [53]. The latter reaction is accompanied by an extensive formation of resin [53].

Other Reactions. Heating with $(CH_3CO)_2O$ in N_2 with distilling off of the CH_3CO_2H formed gives even under the mildest reaction conditions not the mixed anhydride but $CH_3-CO[-O_2C-C_5H_4FeC_5H_4-CO-]_nO_2CCH_3$ in quantitative yield. At a reaction temperature of 60°C, it is obtained as a soluble mobile brown oil, at 70°C as a soluble viscous dark brown oil, and at 170°C as an insoluble hard brittle solid [15]. Heating a 1/1 mixture of $Fe(C_5H_4-CO_2H)_2$ and $1,3\text{-}C_6H_4(CO_2H)_2$ in $(CH_3CO)_2O$ to 70°C yields a dark brown oil containing no dicarboxylic acid, which converts on warming for 48 h at 40°C and 10^{-2} Torr into a dark brown mobile resin soluble in CH_2Cl_2 [15].

Condensation experiments with $N_2H_4 \cdot H_2SO_4$ in fuming H_2SO_4 or polyphosphoric acid at 25 to 30°C failed due to protolytic cleavage [33]. Also, the attempted high temperature polycondensations in $C_6H_5NO_2$ at 180°C with N_2H_4 or $H_2NNHCO-A_n-CONHNH_2$ (n=0 or 1, A=$(CH_2)_4$ or $1,3\text{-}C_6H_4$) have been fruitless, and the acid is unchanged after 5 h heating [32]. Condensation with VII in the presence of polyphosphoric acid gives VIII [19].

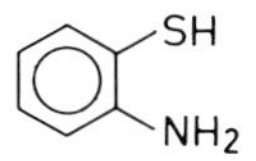

VII

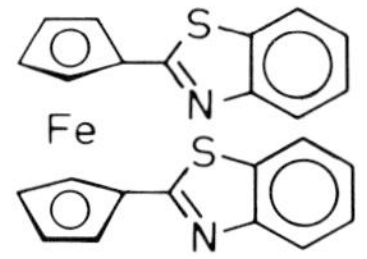

VIII

Microbiological Activity. $Fe(C_5H_4-CO_2H)_2$ possesses in high concentrations bactericidal effects, in moderate concentrations bacteriostatic properties [49].

References on pp. 224/5

References:

[1] R. B. Woodward, M. Rosenblum, M. C. Whiting (J. Am. Chem. Soc. **74** [1952] 3458/9). – [2] M. Rosenblum (Diss. Harvard Univ. 1953). – [3] R. A. Benkeser, D. Goggin, G. Schroll (J. Am. Chem. Soc. **76** [1954] 4025/6). – [4] A. N. Nesmeyanov, E. G. Perevalova, R. V. Golovnya, O. A. Nesmeyanova (Dokl. Akad. Nauk SSSR **97** [1954] 459/61). – [5] N. A. Nesmeyanov, O. A. Reutov (Dokl. Akad. Nauk SSSR **115** [1957] 518/21).

[6] W. Göller (Diss. Stuttgart T. H. 1958). – [7] N. A. Nesmeyanov, O. A. Reutov (Dokl. Akad. Nauk SSSR **120** [1958] 1267/70; Proc. Acad. Sci. USSR Chem. Sect. **118/123** [1958] 503/6). – [8] R. A. Benkeser, L. W. Hall Jr. (AD-215025 [1959] 1/9; AFOSR-TN-59-450 [1959]; PB-140776 [1959] 1/9; U.S. Govt. Res. Rept. **32** [1959] 165). – [9] S. I. Goldberg (J. Org. Chem. **25** [1960] 482/3). – [10] F. W. Knobloch, W. H. Rauscher (J. Polym. Sci. **54** [1961] 651/6).

[11] H. Shechter, J. F. Helling (J. Org. Chem. **26** [1961] 1034/7). – [12] J. Komenda, J. Tirouflet (Compt. Rend. **254** [1962] 3093/5). – [13] M. Okawara, Y. Takemoto, H. Kitaoka, E. Haruki, E. Imoto (Kogyo Kagaku Zasshi **65** [1962] 685/8 from C.A. **58** [1963] 577). – [14] R. L. Schaaf (J. Org. Chem. **27** [1962] 107/11). – [15] W. A. Lee (RAE-TN-CPM-5 [1963] 1/10; X-63-13082 [1963] 1/10).

[16] A. N. Nesmeyanov, E. G. Perevalova, L. P. Yur'eva, K. I. Grandberg (Izv. Akad. Nauk SSSR Otd. Khim. Nauk **1963** 1377/80; Bull. Acad. Sci. USSR Div. Chem. Sci. **1963** 1254/6). – [17] J. Tirouflet, E. Laviron, R. Dabard, J. Komenda (Bull. Soc. Chim. France **1963** 857/62). – [18] K. Yamakawa, H. Ochi, K. Arakawa (Chem. Pharm. Bull. [Tokyo] **11** [1963] 905/11). – [19] C. Rai, Union Oil Co. of California (U.S. 3461287 [1963/69]; C.A. **71** [1969] No. 81541). – [20] J. Komenda (Chem. Zvesti **18** [1964] 378/84).

[21] W. F. Little, C. N. Reilley, J. D. Johnson, A. P. Sanders (J. Am. Chem. Soc. **86** [1964] 1382/6). – [22] A. Mandelbaum, M. Cais (Tetrahedron Letters **1964** 3847/52). – [23] L. Plummer, C. S. Marvel (J. Polym. Sci. A **2** [1964] 2559/69). – [24] K. Schlögl, M. Fried (Monatsh. Chem. **95** [1964] 558/75). – [25] J. T. Suh, Colgate-Palmolive Co. (U.S. 3382267 [1965/68] 1/5; C. A. **69** [1968] No. 59378).

[26] M. M. Fein, N. Mayes, Thiokol Chemical Corp. (U.S. 3414596 [1965/68]; C.A. **70** [1969] No. 37909). – [27] I. A. Golubeva, T. P. Vishnyakova (Neftekhim. Pererab. Nefti Gaza Temat. Anal. Obs. Ts. NIIT Neft. **1966** 36/42). – [28] P. Petrovitch (Double-Liaison No. 133 [1966] 1013/109). – [29] C. F. Cook, P. R. Gray, H. M. Barton (ORNL-IIC-10-Vol. 1 [1967] 130/41; C.A. **69** [1968] No. 46289). – [30] I. A. Golubeva, T. P. Vishnyakova, N. N. Bul'on (Tr. Mosk. Inst. Neftekhim. Gazov. Prom. No. 72 [1967] 61/7).

[31] H.-J. Lorkowski (Fortschr. Chem. Forsch. **9** [1967] 207/53, 209/41). – [32] H.-J. Lorkowski, R. Pannier, A. Wende (J. Prakt. Chem. **35** [1967] 149/58). – [33] E. W. Neuse, H. Rosenberg (J. Polym. Sci. A **6** [1968] 1567/82). – [34] K. Schlögl, H. Soukup (Monatsh. Chem. **99** [1968] 927/46). – [35] G. P. Chalykh, E. B. Sokolova, O. D. Shalygina, N. V. Noskova (Tr. Mosk. Khim. Tekhnol. Inst. No. 61 **1969** 164).

[36] T. P. Vishnyakova, I. A. Golubeva, L. M. Timofeeva (Zh. Obshch. Khim. **39** [1969] 2534/6; J. Gen. Chem. [USSR] **39** [1969] 2475/6). – [37] R. Prins, A. R. Korswagen (J. Organometal. Chem. **25** [1970] C74/C76). – [38] P. F. Quirk, B. Kratochvil (Anal. Chem. **42** [1970] 535/6). – [39] E. B. Sokolova, M. P. Shebanova, L. A. Datskevich (Sin. Fiz. Khim. Polim. No. 6 **1970** 69/72). – [40] Yu. A. Krylov, A. A. Usvyatsov, V. P. Ivanov, E. G. Rozantsev (Izv. Akad. Nauk SSSR Khim. **1971** 2296/8; Bull. Acad. Sci. USSR Div. Chem. Sci. **1971** 2167/9).

[41] A. Sonoda, I. Moritani (J. Organometal. Chem. **26** [1971] 133/40). – [42] H. Imai (Bull. Soc. Chem. Japan **45** [1972] 1264/5). – [43] L. H. Ali, A. Cox, T. J. Kemp (J. Chem. Soc. Dalton Trans. **1973** 1468/75). – [44] K. Yamakawa, M. Hisatome (J. Organometal. Chem. **52** [1973] 407/24). – [45] K. R. Mann, W. H. Morrison, D. N. Hendrickson (Inorg. Chem. **13** [1974] 1180/5).

[46] O. Traverso, R. Rossi, V. Carassiti (Ann. Univ. Ferrara Sez. V **3** [1974] 103/10). – [47] D. Astruc, R. Dabard (Tetrahedron **32** [1976] 245/9). – [48] V. P. Tverdokhlebov, I. V. Tselinskii, B. V. Gidaspov, G. Yu. Chikisheva (Zh. Org. Khim. **12** [1976] 2335/40; J. Org. Chem. [USSR] **12** [1976] 2268/72). – [49] Š. Toma, P. Foltinová, E. Zikova (Acta Fac. Rerum Nat. Univ. Comenianae Chim. No. 25 [1977] 9/19). – [50] J. R. Lenhard, R. W. Murray (J. Am. Chem. Soc. **100** [1978] 7870/5).

[51] J. R. Lenhard, R. Rocklin, H. Abruna, K. Willman, K. Kuo, R. Nowak, R. W. Murray (J. Am. Chem. Soc. **100** [1978] 5213/5). – [52] N. El Murr, A. Chaloyard, E. Laviron (Nouv. J. Chim. **2** [1978] 15/7). – [53] A. M. Shevchik, Yu. P. Losev, Ya. M. Paushkin, R. V. Chibrikova, Zh. I. Zaremba (Vestsi Akad. Navuk Belarusk. SSR Ser. Khim. Navuk **1978** 114/6). – [54] V. P. Red'ko, V. K. Skubin, D. F. Kutepov, V. V. Korshak, V. G. Mochalov (Vysokomol. Soedin. B **22** [1980] 222/4). – [55] D. V. Zagorevskii, Yu. S. Nekrasov, G. A. Nurgalieva (J. Organometal. Chem. **194** [1980] 77/81).

[56] A. Kasahara, T. Izumi, K. Saito (Bull. Chem. Soc. Japan **56** [1983] 2865/6). – [57] T. Nomura (Bull. Chem. Soc. Japan **56** [1983] 2937/40). – [58] R. M. G. Roberts, J. Silver (J. Organometal. Chem. **263** [1984] 235/41). – [59] A. N. Nesmeyanov, E. G. Perevalova, R. V. Golovnya, T. V. Nikitina, N. A. Simukova (Izv. Akad. Nauk SSSR Otd. Khim. Nauk **1956** 739/41; Bull. Acad. Sci. USSR Div. Chem. Sci. **1956** 749/51). – [60] A. J. Sicree, H. Rosenberg (unpublished data from R. L. Schaaf, C. T. Lenk, J. Chem. Eng. Data **9** [1964] 103).

[61] E. Jakusek, H. A. Kołoziej, M. Rozpenk, S. Sorriso (J. Mol. Struct. **79** [1982] 341/4). – [62] A. Ratajczak, B. Czech, L. Drobek (Syn. Reactiv. Inorg. Metal-Org. Chem. **12** [1982] 557/63). – [63] K. Gonsalves, L. Zhan-ru, M. D. Rausch (J. Am. Chem. Soc. **106** [1984] 3862/3). – [64] I. J. Higgins, H. A. O. Hill, E. V. Plotkin (Eur. Appl. 78636 [1983] 1/28).

Catalysis. Applications

Addition of 1 to 3 wt% $Fe(C_5H_4–CO_2H)_2$ to furfuryl alcohol yields using HCl [8] or another acidic catalyst [8, 15] at 60 to 70°C a polymeric product, whose pyrolytic decomposition in the temperature range from 500 to 2000°C in an inert atmosphere leads to amorphous glassy carbons containing dispersed Fe [8, 15]. The polyesters prepared from $HOCH_2CH_2OH$ and dicarboxylic acids or their anhydrides, such as fumaric acid and phthalic or maleic anhydride, in the presence of 0.01 to 10 wt% $Fe(C_5H_4–CO_2H)_2$ are even in the cold hardenable with peroxide catalysts into transparent products which find application as adhesives [6]. For the hardening catalyzed by benzoyl peroxide of styrene-containing polyesters made from $HOCH_2CH_2OH$ and fumaric acid, the optimum concentration of $Fe(C_5H_4–CO_2H)_2$ is 0.002 to 0.003% [9, 10]. Polyesters are prepared from $HO(CH_2)_4OH$ with adipic acid + $Fe(C_5H_4–CO_2H)_2$ in the molar ratio 1/(0.90 to 0.98)/(0.10 to 0.02) [17].

With $Fe(C_5H_4–CO_2H)_2$ modified polyester fibers of $HOCH_2CH_2OH$ and terephthalic acid possess an increased photostability and oxidative thermostability. Their crystallinity on orientational elongation is higher and the degree of orientation is lower than that of pure poly(ethylene terephthalate) fibers. So the use of the former is suggested for low shrinkage fiber preparation [18].

Electrodes of n-type Si derivatized with $Fe(C_5H_4–CO_2H)_2$ show photoresponses which vanish for frequencies above several hundred cycles [16]. The compound covalently immobilized on PtO has a redox potential that differs only around 0.007 V from that of the free acid [14]. An amperometric $Fe(C_5H_4–CO_2H)_2$ mediated enzyme electrode can determine glucose in undiluted whole blood or plasma samples; $[Fe(C_5H_4–CO_2H)_2]^+$ serves as a mediator of electron transfer between immobilized glucose oxidase and a graphite electrode [19], see also [20].

References on p. 226

Added in small quantities the compound increases the thermal oxidation stability of polyorganosiloxanes [1, 5] and also the UV stability of organic coatings (e.g., ones based on melamine [3]) on surfaces exposed to irradiation in space [4]. To be sure, the low solubility of $Fe(C_5H_4{-}CO_2H)_2$ in common solvents is disadvantageous for its effective application for these purposes [2].

Additions of 0.01 wt% to a positive photoresist made from a methacrylchloride/methylmethacrylate polymer raises its electron radiation sensitivity by fivefold [11]; 0.05% raises that of glycidylmethacrylate/methylmethacrylate by 6.25-fold [12].

Treatment of raw coal with the acid and CO or N_2 at 280 or 275°C, respectively, increases the magnetic susceptibility of certain impurities enough that they can be removed magnetically [13].

The acid is used together with an Fe chelating agent for preparation of photosensitive copying paper [7].

References:

[1] W. A. Piccoli, Dow Corning Corp. (U.S. 2979482 [1956/61]; C.A. **1961** 18176). – [2] J. J. Mattice (Off. Dig. Fed. Soc. Paint. Technol. **34** [1962] 602/21). – [3] R. M. Van Vliet (U.S. 3287314 [1962/66]; C.A. **66** [1967] No. 29600). – [4] R. G. Schmitt, R. C. Hirt (J. Appl. Polym. Sci. **7** [1963] 1565/80). – [5] G. S. Tubyanskaya, R. I. Kobzova, E. M. Oparina, V. A. Zaitsev, A. A. Egorova (Plasticheskie Massy **1967** No. 4, pp. 37/8; Soviet Plast. **1967** No. 4, pp. 39/40).

[6] V. P. Red'ko, V. K. Skubin, D. F. Kutepov, V. V. Korshak (U.S.S.R. 394397 [1971/73]; C.A. **81** [1974] No. 64289). – [7] T. Goto, K. Kojima, S. Kusakata, Ricoh Co., Ltd. (Japan. Kokai 74-01224 [1972/74] 1/4 from C. A. **81** [1974] No. 8386). – [8] R. Kammereck, M. Nakamizo, P. L. Walker (Extend. Abstr. Program Biennal Conf. Carbon **11** [1973] 116/7). – [9] D. F. Kutepov, V. P. Red'ko, V. K. Skubin (Tr. Mosk. Khim. Tekhnol. Inst. No. 74 [1973] 123/5 from C.A. **82** [1975] No. 43765). – [10] D. F. Kutepov, V. P. Red'ko, V. K. Skubin, V. G. Mochalov (Tr. Mosk. Khim. Tekhnol. Inst. No. 74 [1973] 126/9 from C.A. **82** [1975] No. 43963).

[11] M. Miyamura, Tokyo Shibaura Electric Co., Ltd. (Japan. Kokai 77-146219 [1976/77] 1/3 from C.A. **88** [1978] No. 201078). – [12] M. Miyamura, Tokyo Shibaura Electric Co., Ltd. (Japan. Kokai 77-146223 [1976/77] 1/3 from C.A. **88** [1978] No. 201074). – [13] J. K. Kindig, R. L. Turner, Hazen Research Inc. (U.S. 4098584 [1977/78]; C.A. **89** [1978] No. 217845). – [14] J. R. Lenhard, R. Rocklin, H. Abruna, K. Willman, K. Kuo, R. Nowak, R. W. Murray (J. Am. Chem. Soc. **100** [1978] 5213/5). – [15] L. N. Mulay, D. W. Collins, A. W. Thompson, P. L. Walker (J. Organometal. Chem. **178** [1979] 217/26).

[16] A. J. Blake, F. R. Mayers, A. G. Osborne, D. R. Rosseinsky (J. Chem. Soc. Dalton Trans. **1982** 2379/83). – [17] W. Göller (Diss. Stuttgart T.H. 1958). – [18] A. I. Volozhin, L. Yu. Verkhovodka, S. S. Gusev, P. N. Zernov, V. A. Levdanskii, K. A. Bogdanov (Preprinty 2nd Mezhdunar. Simp. Khim. Voloknam, Kalinin 1977, Vol. 2, pp. 21/6 from C.A. **88** [1978] No. 122522). – [19] A. E. G. Cass, G. Davis, G. D. Francis, H. A. O. Hill, W. J. Aston, et al. (Anal. Chem. **56** [1984] 667/71). – [20] I. J. Higgins, H. A. O. Hill, E. V. Plotkin (Eur. Appl. 78636 [1983] 1/28, 17).

5.2.3.6.1.2 Ions, Salts, and Complexes Derived from $Fe(C_5H_4{-}CO_2H)_2$

Ions

$[Fe(C_5H_4{-}CO_2H)_2]^{\cdot}$ and $[Fe(C_5H_4{-}CO_2H)_2]^{2-}$ are treated on p. 221 in connection with the electrochemical behavior of the acid.

Ferrocenium Ion $[Fe(C_5H_4{-}CO_2H)_2]^+$. Electrochemical preparation is covered on p. 221. Its instability, due to the electron-withdrawing effect of the substituents, is so large that no reproducible ESR spectrum can be obtained [12]. It decomposes in strongly acidic solutions per $3[Fe(C_5H_4{-}CO_2H)_2]^+ \rightarrow 2Fe(C_5H_4{-}CO_2H)_2 + Fe^{3+} + 2C_5H_4CO_2H^0$. The rate constant k is dependent on the temperature and on the acidic strength H_0 of the solution. At 25°C $k = 0.28 \times 10^{-2}$ for $H_0 = -2$ and $1.12 \times 10^{-3}\ s^{-1}$ for $H_0 = -3$ [21]. The rate constant for the oxidation of glucose oxidase in 0.1 M aqueous $KH_2PO_4/HClO_4$ at pH 7 and 25°C is $0.26 \times 10^5\ L \cdot mol^{-1} \cdot s^{-1}$ [28], see also [29].

Salts and Complexes

Alkali Metal Salts

Salts with the alkali metal cations are obtained in a crystalline form by treatment of solid $Fe(C_5H_4{-}CO_2H)_2$ with concentrated alkalies and are purified by washing with small amounts of CH_3OH [4]. They are easily soluble in H_2O [3, 4] and difficultly soluble in strong alkalies [4]. $AgNO_3$ precipitates the Ag salt nearly quantitatively from their aqueous solutions [3].

$Na_2[Fe(C_5H_4{-}CO_2)_2]$ is a yellow, crystalline [5] compound, which is also obtained by dissolving the acid in $NaHCO_3/H_2O$ [8]. Its IR spectrum exhibits absorptions at 1390 and 1554 cm^{-1}, which are assigned to the symmetric and the asymmetric CO_2 vibrations, respectively [16].

Hydrogenation in H_2O for 1 h over Raney nickel at 180°C and for 2.5 h at 265°C at an initial pressure of 122 atm and a working pressure of 260 atm gives $C_5H_5CO_2H$ in 65% yield [2]. With cold saturated $MgCl_2$, $BaCl_2$, or $Pb(O_2CCH_3)_2$ aqueous solutions the corresponding salts precipitate [4]. No decarboxylation occurs on heating with Na_2O [1]. Addition of an aqueous solution of the salt to one of $UO_2(NO_3)_2$ (molar ratio 1/1) precipitates the red $[UO_2[Fe(C_5H_4{-}CO_2)_2] \cdot 2H_2O]_n$, for which the polymeric structure I is assumed [16].

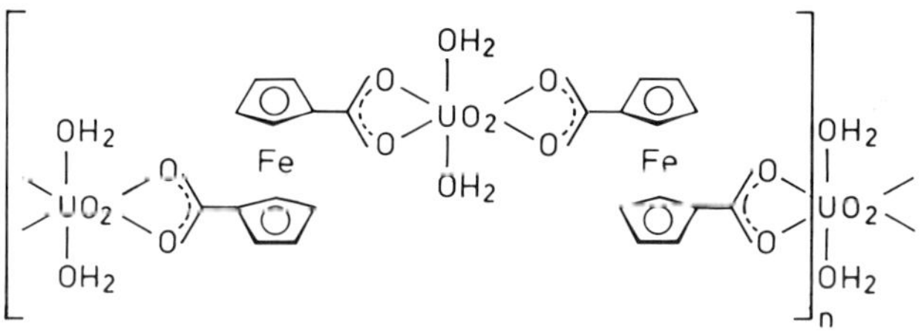

I

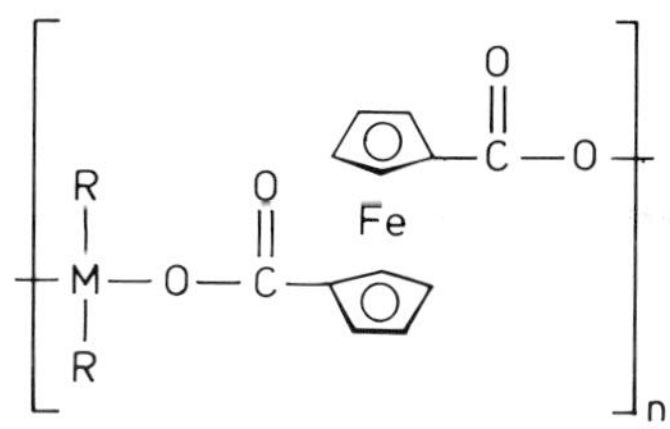

II

Interfacial condensation of aqueous solutions of $Na_2[Fe(C_5H_4{-}CO_2)_2]$ with the solution of R_2MX_2 in $CHCl_3$ in a 1/1 molar ratio produces oligomers of Type II with n = 2 to 5 in the following yields, referred to R_2MX_2 [18]:

compound	M Si	Ge	Sn
$(CH_3)_2MCl_2$	21%	49%	86%
$(CH_3)_2MBr_2$	—	—	90%
$(C_2H_5)_2MCl_2$	—	41%	79%
$(C_2H_5)_2MBr_2$	—	—	85%
$(n\text{-}C_8H_{17})_2MCl_2$	—	—	13%
$(t\text{-}C_4H_9)_2MCl_2$	—	—	10%

compound	M Si	Ge	Sn
$(t\text{-}C_4H_9)_2MBr_2$	—	—	11%
$(t\text{-}C_4H_9)_2MI_2$	—	—	18%
$(C_6H_5CH_2)_2MCl_2$	—	—	21%
$(C_6H_5)_2MCl_2$	21%	21%	81%
$(C_6H_5)(CH_2{=}CH)MCl_2$	37%	—	—

References on pp. 230/1

The yields appear to decrease in the sequence Sn>Ge>Si, I>Br>Cl, and $CH_3 > C_2H_5 >$ t-C_4H_9 [18]. The compounds II degrade so rapidly in air, sometimes explosively, that they are possibly useful as solid propellant fuels [25]. The interfacial polycondensation with R_2SnX_2 (R = CH_3, C_2H_5, t-C_4H_9, n-C_8H_{17}; X = Cl, Br, I) is also treated by [15, 17]. A kinetics investigation of the interfacial condensation with $(C_5H_5)_2ZrCl_2$ [13, 14] reveals a first order reaction referred to $(C_5H_5)_2ZrCl_2$ and zero order with respect to $[Fe(C_5H_4{-}CO_2)_2]^{2-}$; the overall reaction is pseudo first order. A modified Langmuir adsorption isotherm model, in which the anions represent the species adsorbed on the aqueous boundary surface, reproduces the experimental results [14], see also [27]. Reaction with $(C_6H_5)_3SbCl_2$ in CCl_4 gives a 40% yield of III with an average molecular weight of 70000 [19, 22].

Addition of IV in benzene to a rapidly stirred solution of the salt in dilute sodium hydroxide gives in a short reaction time the insoluble V (yield after 20 min 18.0%). It has semiconductor properties. The lack of solubility indicates a certain cross-linking via the third Cl atom [20], see also [23].

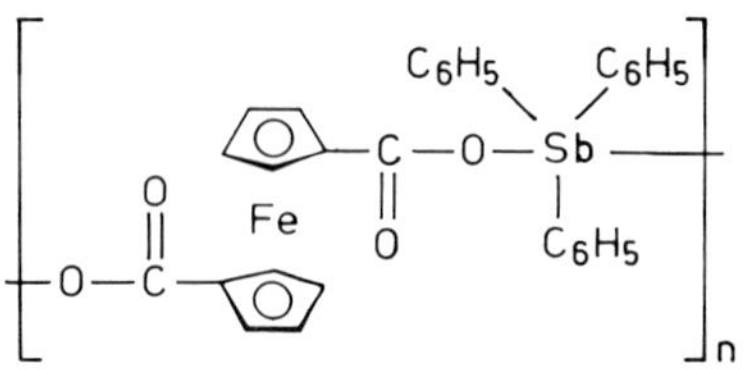

III

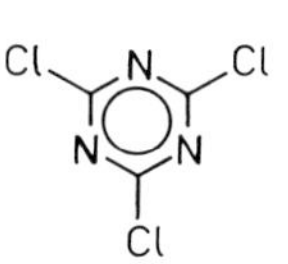

IV

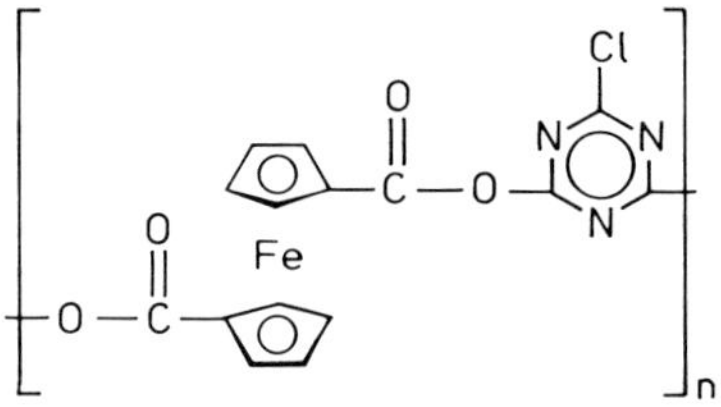

V

$K_2[Fe(C_5H_4{-}CO_2)_2] \cdot xH_2O$ is likewise obtained from the acid and K_2CO_3 [26]. The most important bands of its IR spectrum (KBr) in cm^{-1}, their intensities, and their assignments (abbreviations on p. VIII) are given in Table 13 [26].

Table 13

$K_2[Fe(C_5H_4{-}CO_2)_2] \cdot xH_2O$	$Ba[Fe(C_5H_4{-}CO_2)_2] \cdot H_2O$	assignment
3380 m, br.	3380 m, br.	$\nu(H_2O)$
3070 w	3090 w	$\nu(CH)$
1575 vs, br.	1555 vs, br. 1525 vs, br.	$\nu_a(OCO)$
1465 s	1475 s	$\nu_s(OCO)$
1387 s 1362 s 1340 m, sh	1388 s 1358 m 1346 m	$\nu_s(OCO)$
1187 w 1176 w	1183 w	$\nu_a(CC)$
1035 w 1010 w	1028 m	$\delta(CH)$
808 m 784 m	808 m 780 w, sh	$\delta(OCO) + \gamma(CH)$

In KOH/H_2O irradiation by a 150 watt incandescent lamp causes photolysis, sunlight more rapidly, with formation of $Fe(OH)_3$ and $(C_5H_5CO_2K)_2$ [6]. Digestion with $Ba[Fe(C_5H_4{-}CO_2)_2]$ and $Fe_2(SO_4)_3$ in H_2O on a water bath forms the chelate complex $K_3[Fe(Fe(C_5H_4{-}CO_2)_2)_3]$ [26]:

$$Fe_2(SO_4)_3 \cdot n\,H_2O + 3\,Ba[Fe(C_5H_4{-}CO_2)_2] \rightarrow Fe_2[Fe(C_5H_4{-}CO_2)_2]_3 + 3\,BaSO_4$$
$$Fe_2[Fe(C_5H_4{-}CO_2)_2]_3 + 3\,K_2[Fe(C_5H_4{-}CO_2)_2] \rightarrow 2\,K_3[Fe(Fe(C_5H_4{-}CO_2)_2)_3].$$

Other Salts

$[NH_4]_2[Fe(C_5H_4{-}CO_2)_2]$ forms by the reaction of NH_4OH with VI [4] and practically quantitatively on passing NH_3 at room temperature into a solution of $Fe(C_5H_4{-}CO_2H)_2$ in $(CH_3)_2NCHO$ [7, 10] up to saturation [10]. The water-soluble salt [3] does not melt, but turns dark at 230°C [10]. $AgNO_3$ precipitates the Ag salt from the aqueous solution [3].

Polycondensation in an autoclave with exclusion of O_2 and use of $ZnCl_2$ (molar ratio 1/2) as a catalyst gives after 5 h reaction at 200°C VII in 48.4% yield [9]; see also [24, p. 1337].

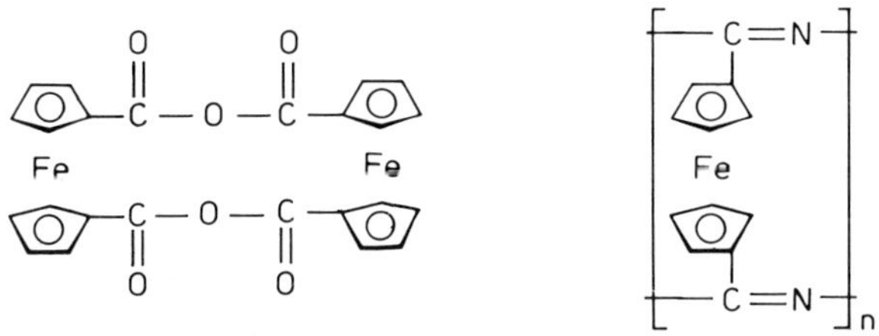

VI VII

Soaking PVC fibers, thermally stabilized by 10 min at 230°C, for 5 to 10 min in a 20 to 30% solution of the salt makes these stable against boiling water after a 20 to 30 min heat treatment at 130°C [11].

The **Mg salt** is treated in connection with $Na_2[Fe(C_5H_4{-}CO_2)_2]$, p. 227.

$Ba[Fe(C_5H_4{-}CO_2)_2] \cdot H_2O$ is obtained from $Ba(OH)_2 \cdot 8\,H_2O$ and $Fe(C_5H_4{-}CO_2H)_2$ [26] or by addition of a saturated $BaCl_2$ solution to an aqueous solution of $Na_2[Fe(C_5H_4{-}CO_2)_2]$ [4]. The most important IR bands and their intensities and assignments are included in Table 13. The reaction with $Fe_2(SO_4)_3$ and $K_2[Fe(C_5H_4{-}CO_2)_2]$ is covered with the latter compound, see above.

The **Pb salt** precipitates on mixing aqueous solutions of $Fe(C_5H_4{-}CO_2H)_2$ and $Pb(O_2CCH_3)_2$, each 0.01 M [4].

For the **UO_2 salt** see $Na_2[Fe(C_5H_4CO_2)_2]$, p. 227.

The **Cu^{II} salt** is a yellow-green substance practically insoluble in H_2O [4].

The **Ag salt** precipitates on addition of $AgNO_3$ to aqueous solutions of the Li, Na, K, or NH_4 salts [3].

Complexes

$K_3[Fe^{III}(Fe(C_5H_4{-}CO_2)_2)_3]$ (see Formula VIII) precipitates nearly quantitatively on 3 h digestion of $Fe_2(SO_4)_3$, $Ba[Fe(C_5H_4CO_2)_2]$, and $K_2[Fe(C_5H_4{-}CO_2)_2]$ in H_2O on a water bath, as yellow-brown single crystals, which are purified by washing with 50% C_2H_5OH. Attempts to recrystallize the chelate complex from H_2O or $(CH_3)_2NCHO$ failed. In the latter a weak decomposition was observed after one week. In amphiprotic solvents, as for example H_2O, or protogenic solvents, such as CH_3CO_2H, $Fe(C_5H_4{-}CO_2H)_2$ forms [26].

References on pp. 230/1

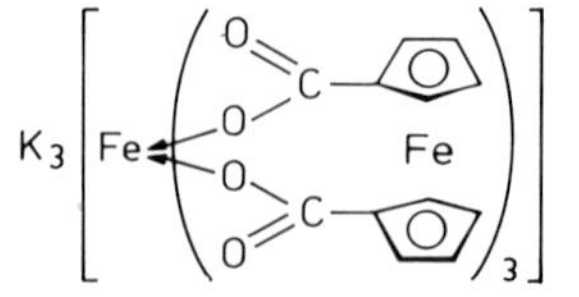

VIII

IR bands (KBr) in cm^{-1}, their intensities (for abbreviations see p. VIII), and assignments [26]:

3100 vw, br.	$\nu(CH)$	1280 vs, br.	$\nu_s(CO) + \delta(OCO)$
1677 vs	$\nu_a(C\overset{\cdots}{-}O)$	1160 m	
1586 m 1542 m	$\nu_a(OCO)$	1025 w 905 vw	$\delta(CH)$
1465 s, br.	$\nu_s(OCO)$	816 w 775 w	$\delta(OCO)$
1380 m 1353 w 1325 w	$\nu_s(OCO)$	724 w, br.	

The electronic absorption spectrum displays the following absorptions λ_{max} (log ε) as a function of the solvent [26]:

solvent	λ_{max} in nm		
$(CH_3)_2SO$	309.6 sh (3.63)	349 sh (3.25)	494.6 sh (2.95)
C_2H_5OH/H_2O (95/5)	309.6 sh	349.7 sh	448
H_2O	309.6 sh	—	494.6 sh

The compound is stable in air and insoluble in most organic solvents. It decomposes at 284°C. The thermal decomposition process takes place in two steps, as shown by the exothermic peaks at 284 and 339°C on heating in air, at 289 and 343°C on heating in N_2. Completion of the first step is connected with a 12% weight loss, of the second stage with one of 49.8%. The partial cleavage of the Fe–O bond is followed by sublimation of ferrocene or its derivative. The residue remaining at 360°C is K_2CO_3 [26].

References:

[1] M. Rosenblum (Diss. Harvard Univ. 1953). – [2] A. N. Nesmeyanov, E. G. Perevalova, R. V. Golovnya, T. V. Nikitina, N. A. Simukova (Izv. Akad. Nauk SSSR Otd. Khim. Nauk **1956** 739/41; Bull. Acad. Sci. USSR Div. Chem. Sci. **1956** 749/51). – [3] W. Göller (Diss. Stuttgart T. H. 1958). – [4] N. A. Nesmeyanov, O. A. Reutov (Dokl. Akad. Nauk SSSR **120** [1958] 1267/70; Proc. Acad. Sci. USSR Chem. Sect. **118/123** [1958] 503/6). – [5] K. Yamakawa, H. Ochi, K. Arakawa (Chem. Pharm. Bull. [Tokyo] **11** [1963] 905/11).

[6] A. N. Nesmeyanov, V. A. Sazonova, V. I. Romanenko, G. P. Zol'nikova (Izv. Akad. Nauk SSSR Ser. Khim. **1965** 1694/5; Bull. Acad. Sci. USSR Div. Chem. Sci. **1965** 1660/1). – [7] I. A. Golubeva, T. P. Vishnyakova (Neftekhim. Pererab. Nefti Gaza Temat. Analit. Obs. Ts. NIIT Neftekhim. **1966** 36/42). – [8] P. Petrovitch (Double-Liaison No. 133 [1966] 1093/109). – [9] T. P. Vishnyakova, I. A. Golubeva, Ya. M. Paushkin (Vysokomol. Soedin. **8** [1966] 181/5). – [10] I. A. Golubeva, T. P. Vishnyakova, N. N. Bul'on (Tr. Mosk. Inst. Neftekhim. Gazov. Prom. No. 72 [1967] 61/7).

[11] G. I. Bochkov, I. I. Borshcheva, L. A. Vol'f, Yu. K. Kirilenko, E. P. Okladnikova, et al. (U.S.S.R. 348650 [1969/72]). – [12] R. Prins (Mol. Phys. **19** [1970] 603/20). – [13] C. E. Carraher,

J. T. Reimer (J. Polym. Sci. Polym. Chem. Ed. **10** [1972] 3367/72). – [14] C. E. Carraher, J. T. Reimer (Polymer **13** [1972] 153/6). – [15] C. E. Carraher, P. J. Lessek (Am. Chem. Soc. Div. Org. Coatings Plastics Chem. Papers **33** [1973] 420/6).

[16] P. A. Vigato, U. Casellato, D. A. Clemente, G. Bandoli, M. Vidali (J. Inorg. Nucl. Chem. **35** [1973] 4131/9). – [17] C. E. Carraher, P. J. Lessek (Angew. Makromol. Chem. **38** [1974] 57/66). – [18] C. E. Carraher, S. Jorgensen, P. J. Lessek (J. Appl. Polym. Sci. **20** [1976] 2255/8). – [19] C. E. Carraher, H. S. Blaxall, J. A. Schröder, W. Venable (Org. Coat. Plast. Chem. **39** [1977] 549/54). – [20] C. E. Carraher, R. J. Foust (Polym. Preprints Am. Chem. Soc. Div. Polym. Chem. **19** [1978] 523/7).

[21] J. Holeček, K. Handlíř, J. Klikorka (Proc. Conf. Coord. Chem. **7** [1978] 65/70). – [22] C. E. Carraher (Polym. Sci. Technol. **14** [1981] 215/26). – [23] J. E. Sheats (J. Macromol. Sci. Chem. A **15** [1981] 1173/99). – [24] C. E. Carraher (J. Macromol. Sci. Chem. A **17** [1982] 1293/356). – [25] C. E. Carraher, J. Schröder (unpublished results from [24]).

[26] T. Nomura (Bull. Chem. Soc. Japan **56** [1983] 2937/40). – [27] C. E. Carraher (Interfacial Syn. **2** [1977] 367/416, 402/5). – [28] A. E. G. Cass, G. Davis, G. D. Francis, H. A. O. Hill, W. J. Aston et al. (Anal. Chem. **56** [1984] 667/71). – [29] I. J. Higgins, H. A. O. Hill, E. V. Plotkin (Eur. Appl. 78636 [1983] 1/28).

5.2.3.6.1.3 Carboxylic Acids of the Type $HO_2C–C_5H_4FeC_5H_4–R$

For compounds of the type $HO_2C–C_5H_4FeC_5H_4–A–CO_2H$ see Section 5.2.3.6.1.4, pp. 241/9.

The compounds compiled in Table 14 have been prepared by the following methods and by those reported under "Further information", pp. 237/9:

Method I. Alkaline saponification at the boiling point:

a) Of $NC–C_5H_4FeC_5H_4–R$ with KOH in C_2H_5OH/H_2O [8, 16, 34] or n-C_4H_9OH [20, 27]. Boiling is continued for 6 ($R=CH_3$) [27], 8 (R = Cl, Br) [16], or 15 h (R = I). In the latter reaction in addition to No. 10 is obtained a 19% yield of $H_2N–CO–C_5H_4FeC_5H_4–I$. This becomes the main product (77% yield) with only 14.6% of No. 10 if the reaction is stopped after only 4 h. Rather than No. 10 the amide precipitates in $KOH/H_2O/C_2H_5OH$ [34].

b) Of $H_2N–CO–C_5H_4FeC_5H_4–R$ with KOH in n-C_4H_9OH [20].

c) Of $(C_6H_5)_2N–CO–C_5H_4FeC_5H_4–R$ with KOH/C_2H_5OH [5, 7, 36]; 2/1 10% ethanolic KOH/C_6H_6 is used with $(C_6H_5)_2N–CO–C_5H_4FeC_5H_4–CHOHC_6H_5$ [7].

d) Of $CH_3O_2C–C_5H_4FeC_5H_4–R$ with 20% NaOH in H_2O [2, 18], with 15% KOH in CH_3OH [30].

Method II. The mixture of $Li–C_5H_4FeC_5H_4–R$ and $Li–C_5H_4FeC_5H_3(R)Li$ obtained by lithiation of $C_5H_5FeC_5H_4–R$ with n-C_4H_9Li in 1/2 tetrahydrofuran/ether is poured into a slurry of dry ice in ether. For separation of the acids their mixture is esterified with CH_2N_2 in ether and chromatographed in petroleum ether on an Al_2O_3 column. The monoester elutes with 1/1 petroleum ether/benzene before the diester. The esters are then saponified with 25% KOH solution in C_2H_5OH/H_2O. The yields reported refer to $C_5H_5FeC_5H_4–R$ [26].

Method III. Clemmensen reduction of $HO_2C–C_5H_4FeC_5H_4–CO–R'$ (Nos. 14 to 17, 19) in glacial acetic acid with amalgamated Zn and concentrated hydrochloric acid [2, 18] with addition of toluene [3].

References on pp. 239/40

Method IV. Oxidation of $CH_3CO–C_5H_4FeC_5H_4–R$:

a) With I_2 in pyridine at room temperature (No. 14) [13], with heating on a water bath (No. 9) [9].

b) With NaOCl in CH_3OH [1].

Method V. Friedel-Crafts acylation:

a) Of $C_5H_5FeC_5H_4–CO_2H$ with RCl and $AlCl_3$ in CCl_4 [2].

b) Of $C_5H_5FeC_5H_4–CO_2CH_3$ with RCl or R_2O and $AlCl_3$ in CH_2Cl_2 and subsequent saponification of the ester [3].

Method VI. Cleavage of 2-$ClC_6H_4–CO–C_5H_4FeC_5H_4–R$ with $KOC(CH_3)_3$ and traces of H_2O in $CH_3OCH_2CH_2OCH_3$:

a) With boiling for 10 to 20 min [42].

b) With prolonged heating. An unresolved mixture of 2-$ClC_6H_4–CO–C_5H_4FeC_5H_4–C_6H_5$ and $C_5H_5FeC_5H_3(CO–(C_6H_4Cl$-2)-1)$C_6H_5$-3 gave only a mixture of No. 22 with its 1,3-isomer; separation was not attempted [42].

The compounds are **purified** by recrystallization from hexane (Nos. 2, 3, 5 to 7) [3, 21], cyclohexane (No. 6) [7], benzene (Nos. 10, 11, 14, 18, 19, 21) [2, 5, 7, 34, 40], benzene/hexane (Nos. 12, 13, 15, 16) [3, 9, 16], benzene/heptane (No. 20) [40], CH_3OH/H_2O (Nos. 1, 9, 14) [3, 16, 27, 42], C_2H_5OH/H_2O (Nos. 9, 17) [9, 21], ether (Nos. 1, 16) [20, 36], ligroin/CH_3OH (No. 5) [2], ligroin/C_2H_5OH (No. 2) [2], CCl_4/CH_3OH (No. 16) [2], or $CHCl_3$/hexane (No. 19) [3].

Physical Properties. For description of the NMR spectra the cyclopentadienyl rings are numbered as in Formula I.

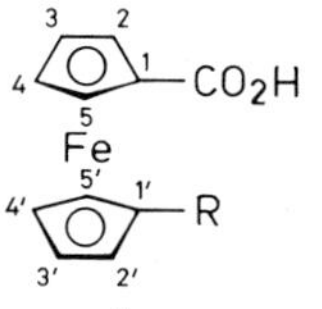

I

In the IR spectrum of the compounds in solution the heteroannular substituent R shifts the νC=O band vs. that of $C_5H_5FeC_5H_4–CO_2H$ to higher wave numbers when R is electron withdrawing, to lower when it is electron releasing [4].

Chemical Behavior

The reversible **redox reaction** to the ferrocenium ion per $HO_2C–C_5H_4FeC_5H_4–R \rightleftharpoons [HO_2C–C_5H_4FeC_5H_4–R]^+ + e^-$ yields the following quarter-wave potentials $E_{\tau/4}$ by chronopotentiometry in CH_3CN (0.2 M $LiClO_4$) and formal redox potentials E by potentiometric titration with $K_2Cr_2O_7$ in 98% CH_3CO_2H/dilute $HClO_4$ (75/25 vol%, 0.066 M $HClO_4$). The last column gives the differences $\Delta E = |E - E_{calc}|$ between the measured values E and the formal potentials E_{calc}, calculated under the assumption of the additivity of the effects of the substituents (in %):

No.	R	$E_{\tau/4}$ in mV vs. SCE	E in mV vs. normal H_2 electrode	$\frac{\lvert E-E_{calc}\rvert}{E}$
2	C_2H_5	497	455 ± 2	0.4
3	n-C_3H_7	—	453 ± 2	0.4
5	n-C_4H_9	—	451 ± 2	0.9
Ref.		[15]	[12, 17]	[12, 17]

No.	R	$E_{\tau/4}$ in mV vs. SCE	E in mV vs. normal H_2 electrode	$\frac{\vert E-E_{calc}\vert}{E}$
6	$CH_2C_6H_5$	535	—	—
13	OCH_3	—	410±2	2.5
14	$COCH_3$	781	—	—
19	COC_6H_5	783	—	—
Ref.		[15]	[12,17]	[12,17]

For Nos. 2 and 11 see also "Further information", p. 238.

A linear dependence exists between $E_{\tau/4}$ and $\Sigma\sigma_p$ (σ_p=para σ Hammett constant) for a large number of heteroannular disubstituted ferrocene derivatives, which is also well fulfilled even when CO_2H is not one of the substituents [15]. For the correlation between E and $\Sigma\sigma^*$ (σ^* = Taft constant) for a number of heteroannular disubstituted ferrocene derivatives, among them No. 2, see [25].

The **acidity** of the compounds is affected by R. It is greater than that for $C_5H_5FeC_5H_4$–CO_2H when R is electron withdrawing, smaller when electron releasing [2, 4, 35]. The following table contains a compilation of electrometrically determined pK_a values:

No.	R	C_2H_5OH/H_2O 50/50 vol %			C_2H_5OH/H_2O 63/37 vol%	C_2H_5OH/H_2O 67/33 vol%	CH_3OH/H_2O 68/32 vol%	H_2O (1 M KCl)
		25±1°C	24°C	20°C	—	24°C	20°C	—
2	C_2H_5	6.30	6.34±0.01	—	6.77	—	6.43	4.7±0.1
3	C_3H_7	6.35	—	—	6.81	—	—	—
5	C_4H_9	6.33	—	—	6.84	—	6.50	—
6	$CH_2C_6H_5$	6.26	6.25±0.01	—	6.74	—	—	—
	H	6.09	6.11±0.01	—	—	—	6.29	—
8	Cl	5.90	—	5.90	—	—	—	—
9	Br	5.90	—	—	—	—	—	
11	$CHOHC_6H_5$	—	5.84±0.01	—	—	—	—	—
13	OCH_3	6.07	—	6.07	—	—	—	—
14	$COCH_3$	5.59	5.76±0.02	—	6.03	6.30	5.91	—
15	COC_2H_5	5.67	—	—	6.10	—	—	—
16	COC_3H_7	5.66	—	—	—	—	5.95	
19	COC_6H_5	5.63	5.90±0.04	—	6.08	—	—	—
Ref.		[14]	[6, 7]	[31]	[3]	[13]	[2]	[32]

For the correlation of pK_a with the constants of the substituents, see "Ferrocene" 3, 1978, pp. 58/9, as well as [3].

The ferrocenium ion **$[HO_2C–C_5H_4FeC_5H_4–C_2H_5]^+$** is a stronger acid than the parent compound $HO_2C–C_5H_4FeC_5H_4–C_2H_5$ (see above for preparation of the ion); $pK_a = 2.0 \pm 0.4$ in H_2O (1 M KCl) [32].

Esterification of the acids to $CH_3O_2C–C_5H_4FeC_5H_4–R$ with R = CH_3 [20, 27, 42], Cl [16], Br [9, 16], O_2CCH_3 [9], OCH_3 [16], $COCH_3$ [1, 10], $COCH_2C_6H_5$ [21] takes place with CH_2N_2 in ether/CH_3OH [42] or ether at room temperature in practically quantitative yield. The esters with R = $COCH_3$ [1] or $COCH{=}CHC_6H_5$ [40] are also obtained in boiling methanol with 5% hydrochloric acid or several drops of concentrated H_2SO_4 as a catalyst, respectively, likewise for R = $COCH_3$ in HCl-saturated absolute methanol in an ice bath [13].

The esterification rates of the acids No. 2, 6, 11, 14, and 19 with $(C_6H_5)_2CN_2$ in toluene at 30°C have been studied under the conditions of a second order reaction. This order is valid for all compounds up to the reaction of the first n% with the following rate constants k; the subsequent deviation is possibly due to a dimerization of the acids in toluene [6].

No.	R	k in $L \cdot mol^{-1} \cdot min^{-1}$	n %
2	C_2H_5	0.0566 ± 0.0010	29 and 30
6	$CH_2C_6H_5$	0.0595 ± 0.0002	18 and 17
11	$CHOHC_6H_5$	10.04 ± 0.14	33 and 36
14	$COCH_3$	0.9110 ± 0.0048	58 and 47
19	COC_6H_5	0.7547 ± 0.0013	52 and 49

With the exception of No. 11, which reacts significantly more rapidly, the esterification rate parallels the acidic strength. The deviation of No. 11 cannot be due to an inductive effect alone, but is possibly attributable to an intramolecular hydrogen bond which hinders the rotation of the CO_2H group and thus eases the attack of the $(C_6H_5)_2CN_2$ molecule on the OH (see II) [6]. A plot of log K vs. σ_p (para σ Hammett constant) for R yields a straight line [6].

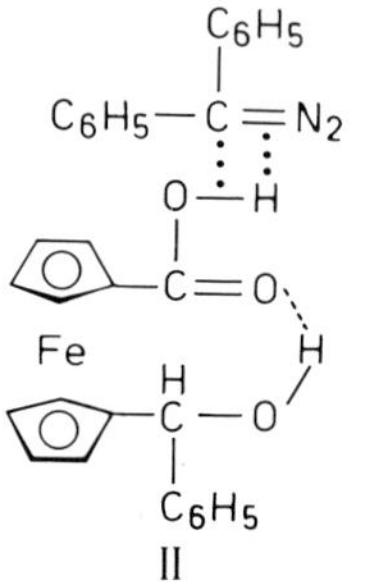

II

Clemmensen **reduction** of the keto acids (Nos. 14 to 17, 19) (see Method III) converts the CO–R′ groups into CH_2R' = R (Nos. 2, 3, 5 to 7) [2, 3, 18, 21].

Table 14

Carboxylic Acids of the Type $HO_2C{-}C_5H_4FeC_5H_4{-}R$.
Further information for numbers preceded by an asterisk is given at the end of the table, pp. 237/9.
For abbreviations and dimensions see p. VIII.
R_f values are normally for descending paper chromatography with $n\text{-}C_4H_9OH/C_2H_5OH$/conc. NH_4OH/H_2O in the volume ratio 4/4/1/1.

No.	$HO_2C{-}C_5H_4FeC_5H_4{-}R$ R	method of preparation (yield in %), properties, and remarks	Ref.
*1	CH_3	Ia, Ib (73), Id (95), II (13.8), VIa (85 to 95), see "Further information", p. 237 m.p. 139 to 140°, 148 to 151° (CH_3OH/H_2O), 148 to 150° (ether), 152 to 154° ^1H NMR (CCl_4): 1.93 (CH_3) IR: 1673 ($\nu C{=}O$) R_f: 0.62	[20, 26 to 30, 42]

Table 14 (continued)

No.	$HO_2C–C_5H_4FeC_5H_4–R$ R	method of preparation (yield in %), properties, and remarks	Ref.
*2	C_2H_5	Ia, II (13), III (89, 70, 25 to 40) m.p. 74.5 to 76.5°, 75.5 to 77.5° (ligroin/C_2H_5OH), 76.5 to 77.5°, 77.5 to 78.5° (hexane), 81.5 to 82° IR (vaseline oil): 811(s), 911(s), 935(m), 1024(s), 1163(s), 1461(vs), 1663(vs; νC=O) IR ($CHCl_3$): 1672 (νC=O) IR (CCl_4): 1682 (νC=O) IR: 1685 to 1690 (νC=O)	[2 to 4, 7, 8, 18, 26]
3	C_3H_7-n	II (14), III (86, 25 to 40) m.p. 93 to 94°, 94.5 to 95°, 95 to 97° (hexane) ^{1}H NMR ($CDCl_3$ or CCl_4): 4.15 (t, H-2′ to 5′), 4.47 (t, H-3, 4), 4.80 (t, H-2, 5)	[3, 18, 24, 26]
4	$CH(CH_3)_2$	Ib R_f: 0.72	[20]
5	C_4H_9-n	III (88, 74, 25 to 40) m.p. 73 to 74.5° (ligroin/CH_3OH), 77 to 78°, 78.5 to 79.5° (hexane) IR (vaseline oil): 817(vs), 866(m) 916(s), 927(s), 1026(s), 1109(w), 1162(s), 1188(vs), 1212(w), 1376(s), 1397(m), 1462(vs), 1666(vs; νC=O) IR ($CHCl_3$): 1672 (νC=O) IR (CCl_4): 1682 (νC=O)	[2 to 4, 18]
6	$CH_2C_6H_5$	III (87, 25 to 40), see No. 11, p. 238 m.p. 151 to 152.5° dec. (cyclohexane), 152 to 153.5°, 154 to 155° dec. (hexane)	[3, 7, 18]
*7	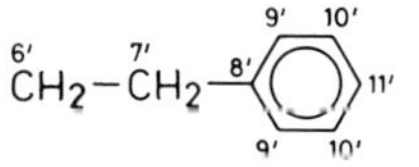	III (70) golden color, m.p. 122° (hexane) ^{1}H NMR ($CDCl_3$): 2.68 to 2.73 (CH_2CH_2), 4.05 (H-2′ to 5′), 4.4 (H-3, 4), 4.75 (H-2, 5), 7.14 (C_6H_5) ^{13}C NMR ($CDCl_3$): 30.2 (C-6′), 37.0 (C-7′), 69.1 to 69.8 (C-3′, 4′, C-2′, 5′), 70.9 (C-2, 5), 72.3 (C-3, 4), 89.8 (C-1, 1′), 125.4 (C-11′), 127.9 (C-9′, 10′), 141.2 (C-8′), 177.8 (C-6)	[21, 38]
8	Cl	Ia (78) m.p. 174 to 176° dec. ^{1}H NMR ($CDCl_3$ or CCl_4): 4.16 (t, -H-3′, 4′), 4.52 (t, H-2′, 5′, 3, 4), 4.97 (t, H-2, 5)	[16, 24]
*9	Br	Ia (89), IVa (41) m.p. 154 to 158° dec. (C_2H_5OH/H_2O), 163.5 to 164.5° (in N_2, CH_3OH/H_2O)	[9, 16]
10	I	Ia (74) red-orange crystals (C_6H_6), m.p. 138.5° dec.	[34]
*11	$CHOH–C_6H_5$	Ic (81 raw product) m.p. 129.5 to 130° (C_6H_6)	[7]

References on pp. 239/40

Table 14 (continued)

No.	$HO_2C{-}C_5H_4FeC_5H_4{-}R$ R	method of preparation (yield in %), properties, and remarks	Ref.
*12	O_2CCH_3	see No. 9, p. 238 yellow crystals (C_6H_6/hexane) m.p. 126 to 130° dec.	[9]
13	OCH_3	see No. 12, p. 238 m.p. 139.5 to 140.5° (C_6H_6/hexane)	[16]
*14	$COCH_3$	Ic (86), Id, IVa (33), IVb, Va, Vb, see No. 18, p. 239 red needles, scarlet prisms m.p. 153 to 155° (C_6H_6), 155 to 156°, 155 to 157° (C_6H_6), 155.5 to 156.5°, 159.5 to 161.0° (CH_3OH/H_2O) ^{1}H NMR ($CDCl_3$): 2.44 (s, CH_3), 4.48 (t, H-3, 4), 4.58 (t, H-3′, 4′), 4.86 (t, H-2, 2′, 5, 5′), 10.45 (s, CO_2H) ^{1}H NMR ($CDCl_3$ or CCl_4): 4.40 (t, H-3, 4), 4.51 (t, H-3′, 4′), 4.81 (t, H-2, 2′, 5, 5′) IR (vaseline oil); 826 (vs), 852 (vs), 894 (m), 910 (vw), 970 (m), 1021 (s), 1090 (s), 1064 (w), 1144 (vs), 1210 (w), 1252 (vs), 1290 (vs), 1382 (vs), 1466 (vs), 1644 (vs), 1690 (vw), 1722 (vs) IR (vaseline oil): 913, 962, 1025, 1034, 1111, 1161, 1271, 1349, 1370, 1445, 1471, 1667, 1704 IR (KBr): 1637 (νC=O of $COCH_3$), 1725 (νC=O of CO_2H) IR ($CHCl_3$): 1663 sh, 1682 UV (C_2H_5OH): $\lambda_{max}(\varepsilon) = 227$ (14200), 455 (420) R_f: 0 (on SiO_2 with n-hexane or toluene)	[1 to 5, 7, 13, 18, 24, 37, 40]
15	COC_2H_5	Id, Vb m.p. 130 to 130.5°, 133 to 134° (C_6H_6/hexane)	[3, 18]
*16	$CO\overset{7'}{C}H_2\overset{8'}{C}H_2CH_3$	Ic (76), Id, Va, Vb orange crystals, m.p. 114 to 115° (CCl_4 and CH_3OH), 117.5 to 118.5°, 118.5 to 119.5° (C_6H_6/hexane) ^{1}H NMR ($CDCl_3$ or CCl_4): 1.01 (t, CH_3), 1.78 (m, CH_2-8′), 2.78 (t, CH_2-7′), 4.47 (t, H-3, 4), 4.53 (t, H-3′, 4′), 4.83 (t, H-2, 2′, 5, 5′)	[2, 3, 18, 24, 36]
*17	$COCH_2C_6H_5$	see "Further information", p. 239 orange-red to red crystals, m.p. 194 to 199°, 208°, 210° (C_2H_5OH) ^{1}H NMR ($OS(CD_3)_2$): 4.05 (s, CH_2), 4.43 (t, H-3, 4), 4.59 (t, H-3′, 4′), 4.70 (t, H-2, 5), 4.84 (t, H-2′, 5′), 7.29 (apparent s, C_6H_5)	[11, 21, 23, 40]

References on pp. 239/40

Table 14 (continued)

No.	$HO_2C{-}C_5H_4FeC_5H_4{-}R$ R	method of preparation (yield in %), properties, and remarks	Ref.
*18	$COCH{=}CHC_6H_5$	see "Further information", p. 239 deep red crystals, m.p. 165 to 170° (C_6H_6), 167 to 168° ^1H NMR ($CDCl_3$): 4.48 (t, H-3,4), 4.63 (t, H-3',4'), 4.86 (t, H-2,5), 4.95 (t, H-2',5'), 7.15 (1H) and 7.77 (1H) (CH=CH, AB pattern; J(A,B) = 15.5), 7.2 to 7.67 (m, $C_6H_5 + CO_2H$) UV (CH_3OH): $\lambda_{max}(\varepsilon) = 220$ (2.36×10^4), 2.58 (1.44×10^4), 307 (2.56×10^4), 4.85 (1.54×10^3)	[40]
*19	COC_6H_5	Ic (74), Va, Vb (31), VIb red crystals, m.p. 125 to 126° (C_6H_6), 128 to 129° ($CHCl_3$/hexane) IR: 912, 952, 1026, 1047, 1111, 1143, 1163, 1279, 1332, 1368, 1389, 1441, 1471, 1570, 1592, 1634, 1672, 1704 IR: 1045 (m), 1303 (s), 1400 (s), 1430 (s), 1456 (s), 1620 (m), 1645 (s, br.), 1700 (m), 3090 (m), 3410 (m), 3450 (m, br.)	[3, 5, 7, 18, 43]
*20	$COCOCH_3$	see "Further information", p. 239 m.p. 114 to 116° (C_6H_6/n-heptane) ^1H NMR ($CDCl_3$): 2.51 (s, CH_3), 4.50 (t, H-3,4), 4.74 (t, H-3',4'), 4.89 (t, H-2,5), 5.15 (t, H-2',5'), 9.25 (br.s, CO_2H)	[40]
*21	$COCOC_6H_5$	see "Further information", p. 239 m.p. 170 to 180° (C_6H_6) ^1H NMR ($CDCl_3$): 4.55 (t, H-3,4), 4.71 (t, H-3',4'), 4.94 (m, H-2,2',5,5'), 7.2 to 8.17 (m, C_6H_5), 8.6 (br.s, CO_2H)	[40]
supplement:			
22	C_6H_5	VIb only as mixture with its 1,3-isomer	[42]

* Further information:

$HO_2C{-}C_5H_4FeC_5H_4{-}CH_3$ (Table **14**, No. **1**) can also be gotten from $Fe(C_5H_4{-}CH_3)_2$ in $OP(N(CH_3)_2)_3$ in the presence of $KOC(CH_3)_3$ by oxidation with O_2 at 80°C after acidification with HCl, in 12.5% yield [39]. If the isomer mixture of $FeC_{10}H_8(CH_3)CHO$ is oxidized with AgO in a 10% solution of KOH in 1/1 tetrahydrofuran/H_2O at room temperature, one obtains after acidification with H_3PO_4 a total yield of ca. 20% for $FeC_{10}H_8(CH_3)CO_2H$. The composition of the mixture of 1,1'-, 1,2-, and 1,3-isomers, determined by NMR spectroscopy after esterification with CH_2N_2, is shifted from 41/15/44 to 36/6/58 due to steric effects which impede the oxidation of the CHO groups in an α position with respect to the CH_3 [33].

References on pp. 239/40

No. 1 is separated from $C_5H_5FeC_5H_3(CH_3\text{-}1)CO_2H\text{-}2$ or -3 by esterifying the isomer mixture with CH_2N_2 in 5/1 petroleum ether/benzene and chromatographing on Al_2O_3 over the total column length, dividing up the mixture into 20 to 25 fractions. The elution sequence is 1,2-, 1,1'-, and 1,3-isomer. However, no sharp separation is achieved. Here also a distinct shift of the ratios occurs in favor of the 1,3-isomer, which is due to greater ease of saponification of the 1,1'-and 1,2-isomers and the consequent higher absorbability on the alkaline column [27].

The singlet of the CH_3 protons in CCl_4 is shifted versus that of $C_5H_5FeC_5H_4\text{–}CH_3$ by 0.02 ppm toward higher field [28].

In the mass spectrum appear the fragment ions $[FeC_5H_4\text{–}CO_2H]^+$ and $[FeC_5H_4\text{–}CH_3]^+$ in a 1/6.7 ratio, which indicates the strengthening of the ligand–Fe bonding by electron donor substituents [19]. $ClCO\text{–}C_5H_4FeC_5H_4\text{–}CH_3$ forms with PCl_3 in benzene [29].

$HO_2C\text{–}C_5H_4FeC_5H_4\text{–}C_2H_5$ (Table **14**, No. **2**). In H_2O (1 M KCl) the formal redox potential (see p. 232) is $E = 453 \pm 2$ mV vs. normal H_2 electrode. Calculation for $[C_2H_5\text{–}C_5H_4FeC_5H_4\text{–}CO_2]^- \rightleftharpoons [C_2H_5\text{–}C_5H_4FeC_5H_4\text{–}CO_2] + e^-$ gives $E = 429 \pm 16$ mV [32].

$HO_2C\text{–}C_5H_4FeC_5H_4\text{–}(CH_2)_2C_6H_5$ (Table **14**, No. **7**) cyclizes neither in the presence of polyphosphoric acid nor $(CF_3CO)_2O$. In contrast, III forms by reaction with ClCOCOCl in CH_2Cl_2 with a subsequent intramolecular Friedel-Crafts condensation using $AlCl_3$ [21].

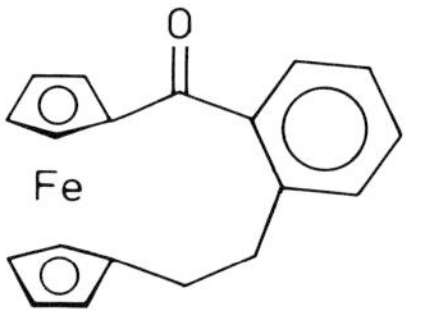

III

$HO_2C\text{–}C_5H_4FeC_5H_4\text{–}Br$ (Table **14**, No. **9**) reacts with $Cu(O_2CCH_3)_2$ in boiling C_2H_5OH/H_2O with complete replacement of the Br by O_2CCH_3 to form No. 12. The reaction is finished within 15 min, 60% yield. The increase of the mobility of the Br by the CO_2H group can be explained by the formation of a Cu–Br coordination compound in the Cu salt initially formed. The related increase in the polarization of the C–Br bond favors the substitution. Esterification of the CO_2H group reduces the mobility of the Br considerably [9].

$HO_2C\text{–}C_5H_4FeC_5H_4\text{–}CHOH\text{–}C_6H_5$ (Table **14**, No. **11**). In CH_3CN (0.1 M $[(C_2H_5)_4N]ClO_4$) the formal redox potential (cf. pp. 232/3) is 636 mV (vs. an NaCl-saturated calomel electrode) by cyclic voltammetry. $E = 592$ mV for the compound immobilized by covalent bonding on a Pt/PtO surface [41]. Catalytic hydration in 20/1 $CH_3CO_2C_2H_5/CH_3CO_2H$ over Pd gives No. 6 in 92% yield (raw product) [7].

$HO_2C\text{–}C_5H_4FeC_5H_4\text{–}O_2CCH_3$ (Table **14**, No. **12**) is more soluble in H_2O than No. 9 [9]. Reaction with $(CH_3O)_2SO_2$ in boiling C_2H_5OH produces No. 13 in 88% yield [16].

$HO_2C\text{–}C_5H_4FeC_5H_4\text{–}COCH_3$ (Table **14**, No. **14**). Partial saponification occurs on condensation of its methyl ester in boiling xylene with $KOC(C_6H_5)_3$, so that in addition to IV (R' = H, 78% yield) No. 14 (7% yield) also results from the chromatographic processing on SiO_2 columns with 3/1 benzene/$CH_3CO_2C_2H_5$ as eluent [40].

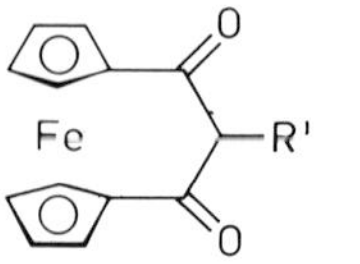

IV

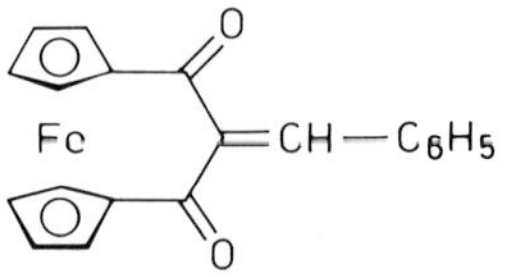

V

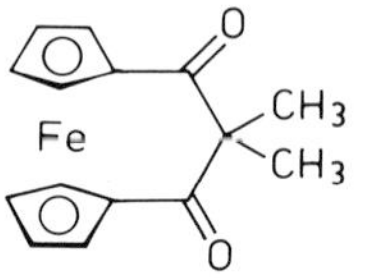

VI

The compound is moderately soluble in H_2O and well soluble in bases or concentrated acids [5]. It is not oxidized at the lower temperatures [13]. At 100°C it forms $Fe(C_5H_4{-}CO_2H)_2$ with I_2 in pyridine [2]. The reaction with ClCOCOCl in CH_2Cl_2 yields $ClCO{-}C_5H_4FeC_5H_4{-}COCH_3$ [1]. That with $ClCO_2CH_3$, $(C_2H_5)_3N$, and subsequently NaN_3 in acetone at 0°C gives $N_3CO{-}C_5H_4FeC_5H_4{-}COCH_3$ [22]. Regarding condensation with C_6H_5CHO, see "Further information" for No. 18, below.

$HO_2C{-}C_5H_4FeC_5H_4{-}COC_3H_7$-n (Table **14**, No. **16**) is reduced by $LiAlH_4/AlCl_3$ in ether to $CH_3{-}C_5H_4FeC_5H_4{-}C_4H_9$ [36].

$HO_2C{-}C_5H_4FeC_5H_4{-}COCH_2C_6H_5$ (Table **14**, No. **17**) is prepared by addition of a boiling mixture of $Fe(C_5H_4{-}COCH_2C_6H_5)_2$ and $OC(OC_2H_5)_2$ (molar ratio 1/4) in benzene to a boiling suspension of $NaNH_2$ in benzene. The yield is 60% relative to the reacted diketone [11, 21]. For the formation and separation of the products resulting from the cyclization of its methyl ester, see "Further information" for No. 21, below.

X-ray diffraction patterns for powder samples with a counter diffractometer give identical values for a smoothed and an unsmoothed sample in a holder. The values measured with a photographic camera (sample in a glass cylinder) agree with them [23].

$HO_2C{-}C_5H_4FeC_5H_4{-}COCH{=}CHC_6H_5$ (Table **14**, No. **18**) is obtained by condensation of No. 14 with C_6H_5CHO in methanol with 10% NaOH as catalyst in 5 h at room temperature in a 60% yield. The separation takes place by column chromatography on SiO_2 with 3/1 C_6H_6/$CH_3CO_2C_2H_5$ as eluent. If instead of No. 14 its methyl ester serves as the starting compound, the column chromatography yields a large number of bands, from which only two compounds were isolated, namely No. 18 (42% yield) and its methyl ester (25%). Condensation of IV (R' = H, p. 238) with C_6H_5CHO and 10% NaOH in C_2H_5OH at 50 to 65°C produces No. 18 (50%) and No. 14 (28%), which are likewise separated by column chromatography. If the latter reaction is conducted with $(C_2H_5)_3N$ as the catalyst, only traces of Nos. 18 and 14 form in addition to the main product V (66% yield) [40].

The absorption at $\lambda_{max} = 485$ nm in the electronic spectrum points to conjugation between the C_5H_4 ring and the $=CHC_6H_5$ [40].

$HO_2C{-}C_5H_4FeC_5H_4{-}COC_6H_5$ (Table **14**, No. **19**) is soluble easily in bases, moderately in concentrated hydrochloric acid, and in H_2SO_4 with a blue coloration [5].

$HO_2C{-}C_5H_4FeC_5H_4{-}COCOR'$ (Table **14**, Nos. **20** and **21** with R' = CH_3, C_6H_5). These compounds form very easily on oxidation of IV (p. 238) by atmospheric oxygen. They also result as by-products in the cyclization of $CH_3O_2C{-}C_5H_4FeC_5H_4{-}COCH_2R'$ with $KOC(C_6H_5)_3$ in boiling xylene to form IV. The separation of IV (R' = CH_3) and No. 20 takes place chromatographically, that of IV (R' = C_6H_5), No. 17, and No. 21 by fractional crystallization. No. 20 also forms on methylation of IV (R' = H) with CH_3I in the presence of K_2CO_3 in anhydrous acetone along with IV (R' = CH_3) and VI. These products are separated by column chromatography on SiO_2 with (90 to 95)/(10 to 5) $C_6H_6/CH_3CO_2C_2H_5$ [40].

References:

[1] M. Rosenblum (Diss. Harvard Univ. 1953, p. 118). – [2] N. A. Nesmeyanov, O. A. Reutov (Dokl. Akad. Nauk SSSR **115** [1957] 518/21; Proc. Acad. Sci. USSR Chem. Sect. **112/117** [1957] 763/6). – [3] R. A. Benkeser, L. W. Hall (AD-215025 [1959] 1/7). – [4] L. A. Kazitsyna, B. V. Lokshin, A. N. Nesmeyanov (Dokl. Akad. Nauk SSSR **127** [1959] 333/6; Proc. Acad. Sci. USSR Chem. Sect. **124/129** [1959] 531/4). – [5] W. F. Little, R. Eisenthal (J. Am. Chem. Soc. **82** [1960] 1577/80).

[6] W. F. Little, R. Eisenthal (J. Am. Chem. Soc. **83** [1961] 4936/9). – [7] W. F. Little, R. Eisenthal (J. Org. Chem. **26** [1961] 3609/10). – [8] A. N. Nesmeyanov, V. A. Sazonova, V. N. Drozd (Dokl. Akad. Nauk SSSR **137** [1961] 102/5; Proc. Acad. Sci. USSR Chem. Sect. **136/141** [1961] 272/4). – [9] A. N. Nesmeyanov, V. A. Sazonova, V. N. Drozd (Izv. Akad. Nauk SSSR Otd. Khim. Nauk **1962** 45/7; Bull. Acad. Sci. USSR Div. Chem. Sci. **1962** 38/9). – [10] K. Schlögl, M. Peterlik, H. Seiler (Monatsh. Chem. **93** [1962] 1309/27).

[11] B. Gautheron, R. Dabard (Bull. Soc. Chim. France **1963** 2009). – [12] E. G. Perevalova, S. P. Gubin (Tr. Konf. Probl. Primen. Korrel. Uravn. Org. Khim., Tartu 1962 [1963], Vol. 2, pp. 26/39; C.A. **63** [1965] 7873). – [13] K. Yamakawa, H. Ochi, K. Arakawa (Chem. Pharm. Bull. [Tokyo] **11** [1963] 905/11). – [14] S. P. Gubin, K. I. Grandberg, E. G. Perevalova, A. N. Nesmeyanov (Dokl. Akad. Nauk SSSR **159** [1964] 1075/8; Dokl. Chem. Proc. Acad. Sci. USSR **154/159** [1964] 1314/6). – [15] W. F. Little, C. N. Reilley, J. D. Johnson, A. P. Sanders (J. Am. Chem. Soc. **86** [1964] 1382/6).

[16] A. N. Nesmeyanov, E. G. Perevalova, K. I. Grandberg (Izv. Akad. Nauk SSSR Ser. Khim. **1964** 1903/5; Bull. Acad. Sci. USSR Div. Chem. Sci. **1964** 1811/3). – [17] E. G. Perevalova, S. P. Gubin, S. A. Smirnova, A. N. Nesmeyanov (Dokl. Akad. Nauk SSSR **155** [1964] 857/60; Dokl. Chem. Proc. Acad. Sci. USSR **154/159** [1964] 328/31). – [18] E. G. Perevalova, M. D. Reshetova, K. I. Grandberg, A. N. Nesmeyanov (Izv. Akad. Nauk SSSR Ser. Khim. **1964** 1901/3; Bull. Acad. Sci. USSR Div. Chem. Sci. **1964** 1808/10). – [19] K. L. Rinehart (Abstr. 16th Southeastern Regional Meeting Am. Chem. Soc., Charleston, West Virginia, 1964, Paper 120 from D. W. Slocum, C. R. Ernst, Advan. Organometal. Chem. **10** [1972] 79/114, 97). – [20] K. Schlögl, M. Fried (Monatsh. Chem. **95** [1964] 558/75).

[21] B. Gautheron (Diss. Univ. Dijon 1965, pp. 1/100, 58). – [22] A. N. Nesmeyanov, V. N. Drozd, V. A. Sazonova (Izv. Akad. Nauk SSSR Ser. Khim. **1965** 1205/8; Bull. Acad. Sci. USSR Div. Chem. Sci. **1965** 1169/71). – [23] R. Dabard, D. Weigel (Bull. Soc. Chim. France **1966** 390/6). – [24] G. G. Dvoryantseva, S. L. Portnova, Yu. N. Sheinker, L. P. Yur'eva, A. N. Nesmeyanov (Dokl. Akad. Nauk SSSR **169** [1966] 1083/6; Dokl. Chem. Proc. Acad. Sci. USSR **166/171** [1966] 773/5). – [25] S. P. Gubin (Izv. Akad. Nauk SSSR Ser. Khim. **1966** 1551/8; Bull. Acad. Sci. USSR Div. Chem. Sci. **1966** 1495/501).

[26] A. N. Nesmeyanov, E. G. Perevalova, V. D. Tyurin, S. P. Gubin (Izv. Akad. Nauk SSSR Ser. Khim. **1966** 1938/43; Bull. Acad. Sci. USSR Div. Chem. Sci. **1966** 1874/8). – [27] H. Falk, G. Haller, K. Schlögl (Monatsh. Chem. **98** [1967] 592/602). – [28] K. Schlögl, H. Falk, G. Haller (Monatsh. Chem. **98** [1967] 82/94). – [29] K. Schlögl, H. Soukup (Monatsh. Chem. **99** [1968] 927/46). – [30] H. Falk, O. Hofer, K. Schlögl (Monatsh. Chem. **100** [1969] 624/48).

[31] A. N. Nesmeyanov, T. V. Baukova, K. I. Grandberg, Yu. A. Ustynyuk, S. P. Gubin, E. G. Perevalova (Izv. Akad. Nauk SSSR Ser. Khim. **1969** 721/3; Bull. Acad. Sci. USSR Div. Chem. Sci. **1969** 658/9). – [32] A. A. Pendin, P. K. Leont'evskaya, T. I. L'vova, B. P. Nikol'skij (Dokl. Akad. Nauk SSSR **189** [1969] 115/8; Dokl. Phys. Chem. Proc. Acad. Sci. USSR **184/189** [1969] 717/20). – [33] K. Schlögl, M. Walser (Monatsh. Chem. **100** [1969] 840/3). – [34] M. Sato, M. Koga, I. Motoyama, K. Hata (Bull. Chem. Soc. Japan **43** [1970] 1142/7). – [35] M. Cais (Jerusalem Symp. Quantum Chem. Biochem. **3** [1971] 96/113).

[36] J. W. Huffman, J. F. Cope (J. Org. Chem. **36** [1971] 4068/72). – [37] K. Tanikawa, K. Arakawa (Bunseki Kagaku **20** [1971] 278/81). – [38] M. L. Martin, J. Tirouflet, B. Gautheron (J. Organometal. Chem. **97** [1975] 261/73). – [39] G. Schmitt, S. Oezman (Tetrahedron Letters **1975** 3689/90). – [40] M. Sališová, Š. Toma, E. Solčaniová (J. Organometal. Chem. **132** [1977] 419/27).

[41] J. R. Lenhard, R. W. Murray (J. Am. Chem. Soc. **100** [1978] 7870/5). – [42] V. Rapic, K. Schlögl, B. Steinitz (J. Organometal. Chem. **94** [1975] 87/98). – [43] M. J. Golightly (AD-A 124405 [1982] 1/55, 35; C.A. **99** [1983] No. 88334).

5.2.3.6.1.4 Carboxylic Acids of the Type $HO_2C–A–C_5H_4FeC_5H_4–R^2$ ($A–CO_2H = R^1$)

See also 5.2.3.7.1.4 for cases in which only the ester but not its free acid has been isolated.

The acids collected in Table 15 are prepared by the following methods:

Method I. Alkaline saponification

a) of $Fe(C_5H_4–CH_2CN)_2$ with boiling $NaOH/H_2O/C_2H_5OH$ [37, 44, 51],

b) of I with 2 N NaOH solution at the boiling point [8],

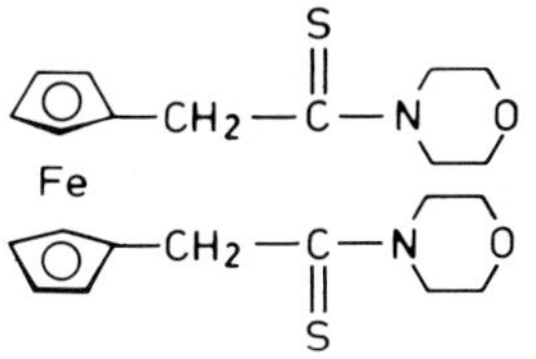

I

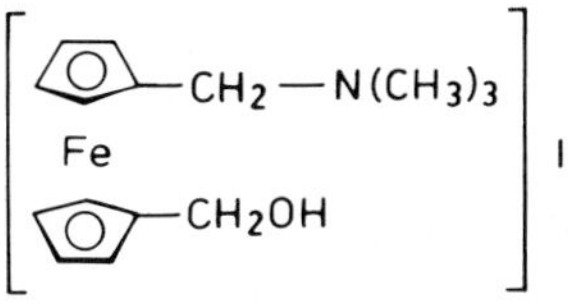

II

c) of $R'O_2C–A–C_5H_4FeC_5H_4–R^2$ (for preparation see 5.2.3.7.1.4, pp. 307/22; $R' = CH_3$ for Nos. 3, 11, 26, 28 to 31, and C_2H_5 for Nos. 2, 3, 10) with NaOH in H_2O [5], C_2H_5OH/H_2O [15, 17, 18, 20, 23, 48] at room temperature [15] or with KOH in C_2H_5OH [30, 33, 43] at room [30,33] or boiling [43] temperatures,

d) of $2\text{-}CH_3O_2CC_6H_4–C_5H_4FeC_5H_4–CH_2CN$ [44].

Method II. Reaction of II with $NaO_2CCH_2CO_2C_2H_5$, which yields the compound No. 2 instead of the expected $HO_2C(CH_2)_2–C_5H_4FeC_5H_4–CH_2OH$ [35].

Method III. Reaction of ferrocenium ion with the diazonium ion $[RN_2]^+$:

a) In a mole ratio >1 in dilute H_2SO_4 at room temperature. The precipitated $FeC_{10}H_8R_2$ has a sharp melting point, so that it possibly consists only of No. 7, free of positional isomer by-products [4].

b) In the mole ratio 1/4 first at 0°C then at room temperature, for 12 h. After separation of $FeC_{10}H_{10}$ from the precipitated $FeC_{10}H_{10-n}R_n$ (n = 0, 1, 2, 4, 5) by dissolving the mixture in 10% potassium hydroxide, the mixture of $C_5H_5FeC_5H_4–R$ and $FeC_{10}H_8R_2$ is extracted into benzene from the remaining $FeC_{10}H_{10-n}R_n$ (n = 1, 2, 4, 5) which is precipitated by acid. $FeC_{10}H_{10-n}R_n$ (n = 1, 2) (total yield 10%) has a decomposition temperature of 214 to 216°C and contains $FeC_{10}H_8R_2$ and $C_5H_5FeC_5H_4–R$ in the ratio 93/7 [27], see also [11].

Method IV. Reaction of $Fe(C_5H_4–OH)_2$ with a 100% excess of $ClCH_2CO_2H$ in 10% potassium hydroxide at the boiling point. The free acid on acidification with 10% H_2SO_4 and shaking with ether is only slowly obtained from the $\mathbf{Fe(C_5H_4–O–CH_2CO_2K)_2}$, which is difficultly soluble in water [13, 14].

Method V. Friedel-Crafts acylation:

a) Of ferrocene in CH_2Cl_2 with $AlCl_3$ and III (n = 2 for No. 11)

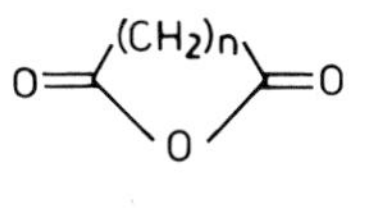

III

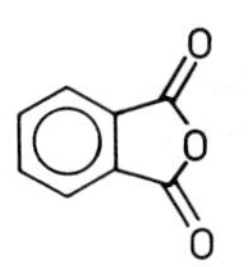

IV

in the mole ratio 1/4/2 without an external supply of heat [7, 9], with III (n = 3 for No. 12, 8 for No. 13) or terephthalic acid (for No. 14) [26].

b) Of ferrocene in CS_2 with $AlCl_3$ and III (n = 2 for No. 11) [5] or IV (mole ratio 1/2/2) for No. 15, at the boiling point [1, 2].

c) Of $C_5H_5FeC_5H_4{-}C_2H_5$ with $AlCl_3$ and III (n = 2). For separation of the mixture of the three positionally isomeric ketoacids, it is esterified with CH_2N_2 and subsequently separated by column chromatography on Al_2O_3 with benzene/petroleum ether into three main fractions, which are immediately saponified and purified by repeated fractional crystallization [22, 25, 35].

Method VI. Condensation of $Fe(C_5H_4{-}CO{-}CH_3)_2$ with $C_2H_5O_2CCO_2C_2H_5$ in boiling $C_2H_5OH/C_2H_5O^-$ and precipitation from the filtrate (in H_2O) with 20% hydrochloric acid [36].

Method VII. By Knoevenagel condensation of $OHC{-}C_5H_4FeC_5H_4{-}R^2$ with malonic acid (mole ratio 1/2) in pyridine with addition of small quantities of piperidine on a boiling water bath [20].

Method VIII. Claisen condensation of $OHC{-}C_5H_4FeC_5H_4{-}CH_3$ with CH_3COCO_2H in methanolic KOH at 30°C [33].

Purification of the acids takes place by recrystallization from the solvents given in parentheses following the melting points in the table.

Chemical Behavior

Reduction. The following table lists the starting materials and end products. It is evident that the ketoacids can be converted into the normal acids with amalgamated Zn and concentrated hydrochloric acid by the Clemmensen procedure with the simultaneous reduction of any $COCH_3$ (No. 30) or CH=CH groups (No. 33) to C_2H_5 and CH_2CH_2, respectively. During the catalytic hydrogenation of keto to normal acids and of CH=CH to CH_2CH_2 groups, the $COCH_3$ groups are not reduced. For the reduction of Nos. 1 and 11, see also "Further information", pp. 247 and 248.

Reduction Reactions

starting compound No.	process	end product No.	yield in %	Ref.
10	Pt/H_2 in glacial acetic acid	2	—	[17]
11	Pt/H_2 in glacial acetic acid	3	—	[15, 17]
	Clemmensen in glacial acetic acid	3	77	[5]
	Clemmensen in $C_6H_6/C_2H_5OH/H_2O$/conc. HCl (1/1/1/1)	$Fe(C_5H_4(CH_2)_2{-}CO_2C_2H_5)_2$	—	[10]
12	Pt/H_2 in glacial acetic acid	4	67	[15,17,26]
13	Pt/H_2 in glacial acetic acid	5	—	[26]
14	Pt/H_2 in glacial acetic acid	6	—	[26]
24	Pt/H_2 in C_2H_5OH	18	—	[34]
25	$Pd/C/H_2$ in C_2H_5OH	20	~100	[20]
28	Clemmensen in toluene	21	52 to 58	[32, 33]
29	Clemmensen in 4/1 CH_3CO_2H/H_2O	22	—	[28,30,35]
30	Clemmensen in 2/1 CH_3CO_2H/H_2O	22	57	[30]
32	Pt/H_2 in glacial acetic acid	23	60	[15]
33	Clemmensen in toluene	21	—	[33]

Cyclization. In general compounds with the $(CH_2)_2CO_2H$ group cyclize with heteroannular ring closure, those with the $(CH_2)_3CO_2H$ with homoannular closure. Thus, No. 2 gives with $(CF_3CO)_2O$ in CCl_4 in the dark at 0°C a mixture of V and VI ($R^2 = (CH_2)_2CO_2H$), in which the latter distinctly predominates [16, 20, 23, 43, 48]. No. 18 similarly gives V and VI ($R^2 = CH_3$)

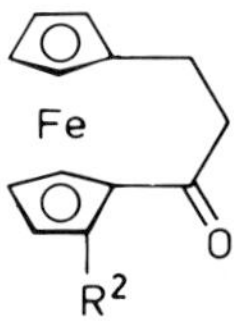

V

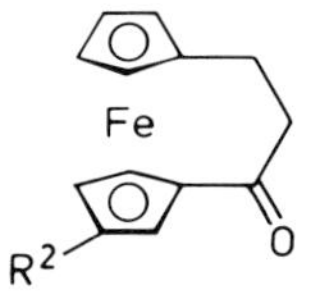

VI

in approximately a 1/4 ratio [34]. The corresponding reaction of No. 3 leads in contrast to the 1,2-ferrocenophane derivatives VII and VIII [18, 40, 50]. In addition to these two polyphosphoric acid at 60°C gives a third compound, white, melting at 135°C, which probably is IX [25], see also

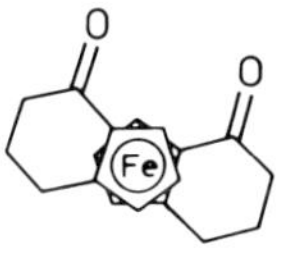

rac-form
VII

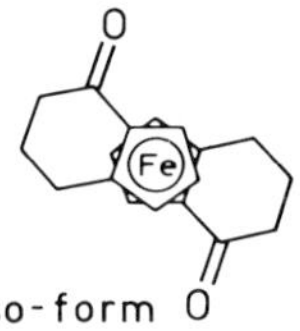

meso-form
VIII

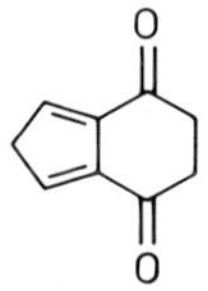

IX

[5, 12]. No. 15 forms with polyphosphoric acid at 130°C a substance, probably X, whose chlorination product appears to form during the action of $TiCl_4$ on No. 15 [3]. Refluxing of Nos. 21 or 22 with $(CF_3CO)_2O$ in CH_2Cl_2 yields XI with $R^2 = CH_3$ [32, 33, 49] or C_2H_5 [28, 30, 35], respectively. Unproven is the conjecture of [21] that during the acid/catalyzed cyclization of

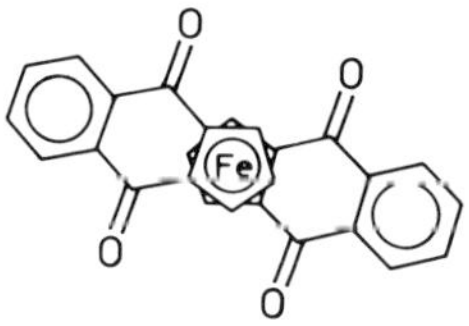

X

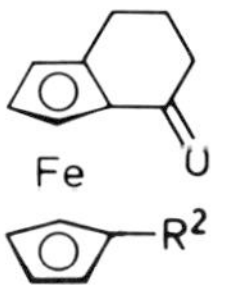

XI

No. 22 a heteroannular bridged ketone can also be formed in addition to XI. No. 20 cannot be brought to homoannular ring closure either with $(CF_3CO)_2O$ or by the Friedel-Crafts reaction; the formation of V or VI with $R^2 = COCH_3$ is not to be expected due to the presence of two CO groups on one cyclopentadienyl ring [20].

Additional Reactions. Esterification of the acids with CH_2N_2 in ether to form the dimethyl esters has been carried out with Nos. 1 [8, 37], 2 [15], 4 [15], 15 [2], and 23 [15]. Azeotropic distillation of No. 11 with C_2H_5OH, conc. H_2SO_4, and toluene yields the diethyl ester, yield 70 [9], 94% [7].

Reaction with PCl_3 forms the acid chlorides; corresponding reactions have been conducted with the compounds No. 2 [19, 20, 43], 3 [41], and 26 [44].

The reaction of the dicarboxylic acids No. 1 and 27 with $(cyclo\text{-}C_6H_{11}N)_2C$ in CH_2Cl_2 leads to $(cyclo\text{-}C_6H_{11}NH)_2CO$ and to the intramolecular anhydrides XII [42, 44 to 46] and XIII [44 to 46], respectively. No. 27 shows with respect to this reaction such an extreme readiness that it is complete after 1 h at 0°C. No. 1 requires 6 h at 25°C [44].

References on pp. 249/50

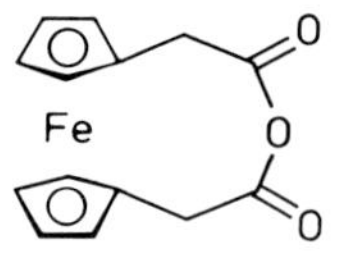

XII

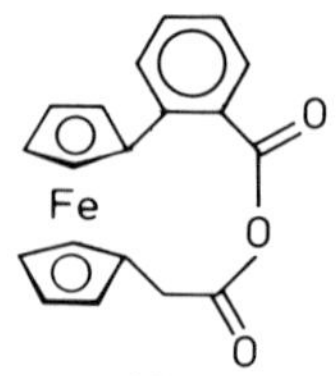

XIII

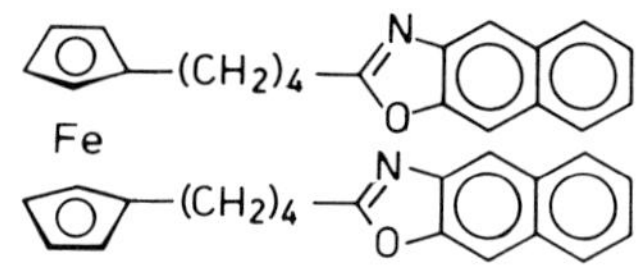

XIV

The action of polyphosphoric acid condenses No. 4 with 2-amino-3-hydroxy-naphthalene to form XIV, No. 5 with 2-amino-3-hydroxy-anthracene to form XV, and No. 6 with 2,3-diamino-naphthalene to form XVI [26].

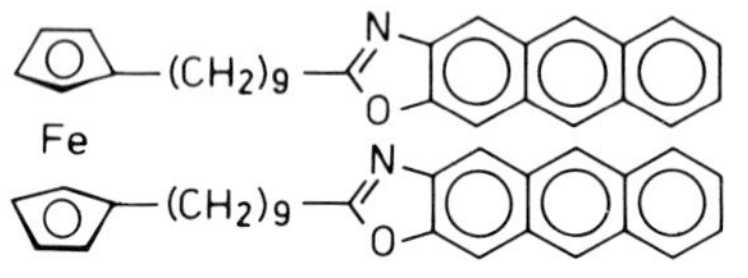

XV

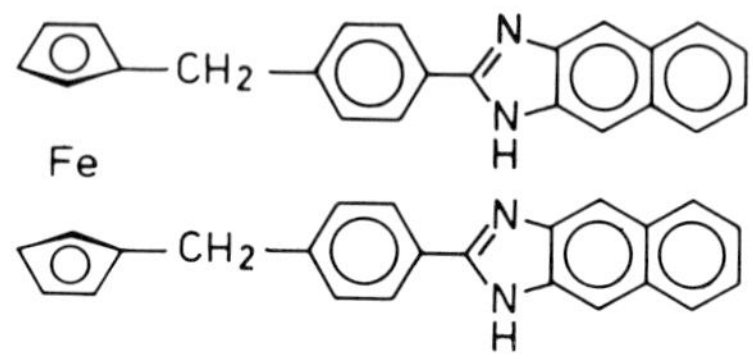

XVI

Refluxing Nos. 2 or 3 in CH_2Cl_2 for 18 h with XVII (mole ratio 1/1) in the presence of $[2\text{-ClC}_5H_4NCH_3]$ I (C_5H_5N = pyridine) and $(n\text{-}C_4H_9)_3N$ under conditions of high dilution leads to the douple and triple layered ferrocenoporphyrinophanes XVIII (yield 10% for n = 2, 22% for n = 3) and XIX (yield 0.8% for n = 2, 1.3% for n = 3). Under the same mild conditions of "Mukaiyama's method" No. 3 reacts with XVIII (n = 3) to form XIX (n = 3) in a 10% yield [52].

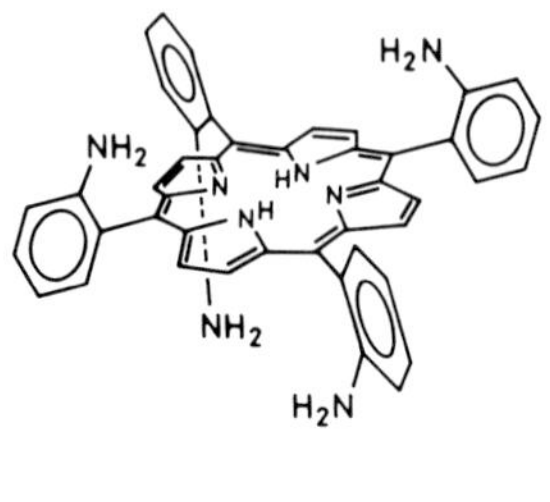

XVII

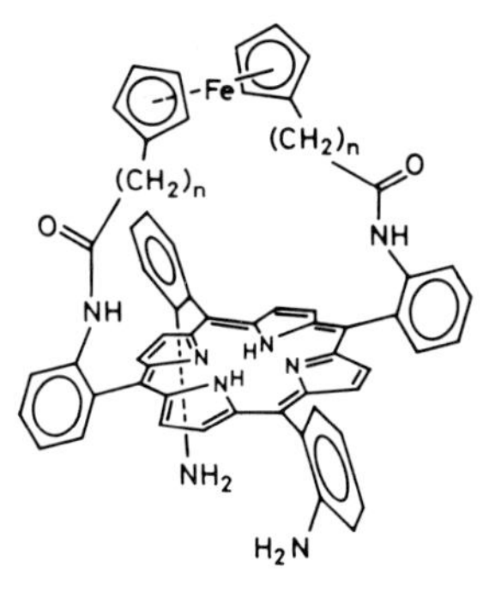

XVIII

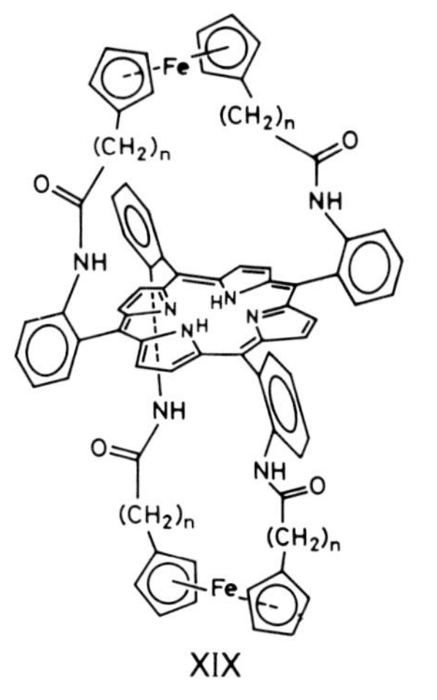

XIX

Table 15

Carboxylic Acids of the Type HO_2C–A–$C_5H_4FeC_5H_4$–R^2.

Further information for numbers preceded by an asterisk is given at the end of the table, pp. 247/9.

For abbreviations and dimensions see p. VIII.

R_f values are for descending paper chromatography with n-C_4H_9OH/C_2H_5OH/conc. NH_4OH/H_2O in the volume ratio 4/4/1/1 as the mobile phase.

No.	R^1–$C_5H_4FeC_5H_4$–R^2 R^1 = A–CO_2H	R^2	method of preparation (yield in %), properties, and remarks	Ref.
compounds with $R^1 = R^2$:				
*1	CH_2CO_2H		Ia (70, 72), Ib (13) fine yellow plates (ether/pentane) darkened below 137° m.p. 140 to 143° (sintered, dec.), 165° (sintered) and 175 to 178° (dec.) IR (Nujol): 664, 775, 825, 930, 945, 1023, 1041, 1161, 1212, 1240, 1295, 1329, 1405, 1418, 1693, 1713 (νC=O) IR (Nujol): between 1705 and 1712 (νC=O) R_f: 0.27	[8, 15, 37, 44, 51]
2	$(CH_2)_2CO_2H$		Ic (73, 74, 75 to 80), II, see "Chemical Behavior" of No. 10, p. 242 yellow solid, m.p. 135 to 137°, 140 to 141°, 142 to 145° (C_2H_5OH/H_2O) IR (KBr): 1710 (νC=O) R_f: 0.38, 0.40	[15, 17, 20, 23, 35, 43]
3	$(CH_2)_3CO_2H$		Ic (<100), see "Chemical Behavior" of No. 11, p. 242 bright yellow crystals, m.p. 109 to 110° dec. (ether/hexane), 109.5 to 110.5° (petroleum ether/ C_6H_6, 1/1), 110 to 113° (petroleum ether/C_6H_6) R_f: 0.48	[5, 10, 15, 17 18]
4	$(CH_2)_4CO_2H$		see "Chemical Behavior" of No. 12, p. 242 m.p. 151 to 155° ($CH_3CO_2C_2H_5$/ petroleum ether) R_f: 0.53	[15, 17, 26]
5	$(CH_2)_9CO_2H$		see "Chemical Behavior" of No. 13, p. 242	[26]
6	$CH_2C_6H_4CO_2H$-4		see "Chemical Behavior" of No. 14, p. 242	[26]
*7	$C_6H_4CO_2H$-2		IIIa (15) pale yellow crystals, m.p. 195° (glacial acetic acid)	[4]

Table 15 (continued)

No.	$R^1–C_5H_4FeC_5H_4–R^2$ $R^1 = A–CO_2H$	R^2	method of preparation (yield in %), properties, and remarks	Ref.
*8		$C_6H_3(OH\text{-}3)CO_2H\text{-}4$	IIIb	[11, 27]
9		$O–CH_2CO_2H$	IV (76) m.p. 168.5 to 169.5° (H_2O), stable in air	[13, 14]
10		$CO–CH_2CO_2H$	Ic	[17]
*11		$CO(CH_2)_2CO_2H$	Ic (78), Va (32, 38), Vb (18), see "Further information", p. 248 lustrous orange to orange-yellow crystals, decomposes at 162 to 164° with darkening, m.p. 164 to 166° dec., 165° dec. (CH_3OH/H_2O, 50/50), 167 to 169° dec. (CH_3OH/H_2O) R_f: 0.23	[5, 7, 9, 15, 17]
12		$CO(CH_2)_3CO_2H$	Ic (70), Va m.p. 190 to 191° dec. (C_2H_5OH) R_f: 0.27	[15, 17, 26]
13		$CO(CH_2)_8CO_2H$	Va	[26]
14		$CO–C_6H_4CO_2H\text{-}4$	Va	[26]
*15		$CO–C_6H_4CO_2H\text{-}2$	Vb (8) fine amorphous brownish powder (dimethylester, m.p. 144 to 144.5°)	[1, 2]
16		$CO–CH_2CO–CO_2H$	VI (52) violet crystals, m.p. 150 to 153° (C_2H_5OH) IR (KBr): 930 (carboxyl + enol OH), 1620 (enol CO), 1740 (carboxyl C=O), 2500 to 3100 (dimer's carboxyl C=O)	[36]
compounds with $R^1 \neq R^2$:				
*17	CH_2CO_2H	$COC_6H_4CH_3\text{-}4$	see "Further information", p. 248	[47]
18	$(CH_2)_2CO_2H$	CH_3	see "Chemical Behavior" of No. 24, p. 242	[34]
*19	$(CH_2)_2CO_2H$	C_2H_5	see "Further information", p. 248 R_f: 0.75	[15]
20	$(CH_2)_2CO_2H$	$COCH_3$	see "Chemical Behavior" of No. 25, p. 242 m.p. 65 to 70° (CH_2Cl_2/petroleum ether) R_f: 0.50	[20]
21	$(CH_2)_3CO_2H$	CH_3	see "Chemical Behavior" of Nos. 28, 33, p. 242, and No. 28, p. 249 yellow solid, m.p. 70 to 71° (pentane)	[32, 33]

References on pp. 249/50

Table 15 (continued)

No.	R^1–$C_5H_4FeC_5H_4$–R^2 R^1 = A–CO_2H	R^2	method of preparation (yield in %), properties, and remarks	Ref.
22	$(CH_2)_3CO_2H$	C_2H_5	see "Chemical Behavior" of Nos. 29, 30, p. 242 yellow crystals, m.p. 45 to 46° (pentane) IR ($CHCl_3$): 905	[28, 30, 35]
23	$(CH_2)_4CO_2H$	$(CH_2)_3CO_2H$	see "Chemical Behavior" of No. 32, p. 242 m.p. 84 to 86° (C_6H_6/petroleum ether) R_f: 0.51	[15, 17]
24	CH=CH–CO_2H	CH_3	VII	[34]
25	CH=CH–CO_2H	$COCH_3$	VII (64) m.p. 158 to 160° dec. R_f: 0.49	[20]
26	$C_6H_4CO_2H$-2	CH_2O–CH_3	Ic	[44]
27	$C_6H_4CO_2H$-2	CH_2CO_2H	Id	[44]
*28	$CO(CH_2)_2CO_2H$	CH_3	Ic orange solid, m.p. 101 to 102° (pentane/ether, 2/1)	[32, 33]
29	$CO(CH_2)_2CO_2H$	C_2H_5	Ic, Vc m.p. 94 to 96° (hexane) ^{1}H NMR ($CDCl_3$): 1.16 (t, CH_3), 2.35 (m, CH_2 of C_2H_5), 2.8 to 3.4 (m, CH_2CH_2), 4.19 (s, C_5H_4–R^2); 4.52 (m, H-3,4) and 4.83 (m, H-2,5) (C_5H_4–R^1) IR ($CHCl_3$): 800, 907	[22, 25, 28, 30, 31, 35]
*30	$CO(CH_2)_2CO_2H$	$COCH_3$	Ic (66) m.p. 182° (CH_3CO_2H)	[29, 30]
31	$CO(CH_2)_2CO_2H$	$(CH_2)_3CO_2H$	Ic (65) m.p. 120 to 123° (C_6H_6/petroleum ether) R_f: 0.33	[15]
32	$CO(CH_2)_3CO_2H$	$CO(CH_2)_2CO_2H$	Ic (54) m.p. 142 to 149° dec. (CH_3OH/H_2O) R_f: 0.25	[15]
33	CH=CH–CO–CO_2H	CH_3	VIII violet paste (unpurified product)	[33]

* Further information:

$Fe(C_5H_4$–$CH_2CO_2H)_2$ (Table **15**, No. **1**) is considerably unstable in hot tetrahydrofuran, CH_3OH, or $CHCl_3$ [37]. With $LiAlH_4$ in tetrahydrofuran it is reduced to $Fe(C_5H_4$–$(CH_2)_2OH)_2$ [51]. The Fe^{2+} of No. 1 acts as a polymerization accelerator in the synthesis of an unsaturated polyester from $HOCH_2CH_2OH$, fumaric acid, and No. 1 at 200°C [38]. A mixture of this polyester with styrene can be hardened with benzoyl peroxide even at room temperature; the optimal concentration of No. 1 in the polyester is 0.002 to 0.003% [38, 39].

$Fe(C_5H_4–C_6H_4CO_2H\text{-}2)_2$ (Table **15**, No. **7**) dissolves in 96% H_2SO_4 with a green coloration, changing to red [4].

$Fe(C_5H_4–C_6H_3(OH\text{-}3)CO_2H\text{-}4)_2$ (Table **15**, No. **8**) may be obtained as the distinctly predominating main component in the mixture of the three positional isomers. The mixture is diamagnetic, possesses at 20°C a specific conductivity of $2.0\times10^{-8}\Omega^{-1}\cdot cm^{-1}$, and is converted by boiling in NH_3 solution into the NH_4 salt, which yields with $AgNO_3$ the chelate complex **$Ag_2FeC_{24}H_{16}O_6\cdot 2H_2O$.** The mixture of the acids dissolved in CH_3OH gives with $Cu(O_2CCH_3)_2$ in a 64% yield the chelate complex **$Cu_2FeC_{24}H_{16}O_6\cdot 6H_2O$** [27], see also [11].

$Fe(C_5H_4–CO(CH_2)_2CO_2H)_2$ (Table **15**, No. **11**) is according to [9] obtained by Method Va but not by Method Vb (p. 242). In contrast [5] succeeded in preparation by Method Vb in an 18% yield. No. 11 dissolves easily in benzene, alcohol, acetone, or dioxane, moderately in CCl_4, and difficultly in petroleum ether [5]. Its oscillopolarogram (in $C_2H_5OH/HClO_4$ on a flowing Hg electrode) differs distinctly from that of $C_5H_5FeC_5H_4–CO(CH_2)_2CO_2H$, so that in this way each is detectable as an impurity in the other [24].

In contradiction to the results of [7], according to which reduction attempts by both the Wolff-Kishner and Clemmensen reactions failed, No. 11 has been reduced to No. 3 via the latter method, cf. p. 242, by [5, 10]. Catalytic hydrogenation in C_2H_5OH over RuO_2 with 165 atm H_2 at 85°C goes only to the hydroxy group, as established by the isolation of the dilactone XX [7].

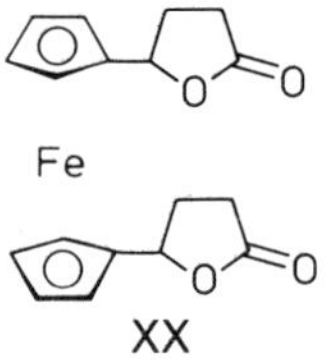

XX

For the catalytic hydrogenation with Pt/H_2 see p. 242. Attempts at cyclization miscarried with both conc. H_2SO_4 and polyphosphoric acid [5].

$Fe(C_5H_4–COC_6H_4CO_2H\text{-}2)_2$ (Table **15**, No. **15**) decomposes on attempts to cyclize it with H_2SO_4 [2]. Additions of small quantities of No. 15 enhance the thermal stability of dimethylpolysiloxanes [6].

$HO_2CCH_2–C_5H_4FeC_5H_4–COC_6H_4CH_3\text{-}4$ (Table **15**, No. **17**). The Friedel-Crafts acylation of $C_5H_5FeC_5H_4–CH_2CO_2CH_3$ with $4\text{-}CH_3C_6H_4COCl/AlCl_3$ (molar ratio 1/1/1) in CH_2Cl_2 with subsequent alkaline saponification and purification via the Ca salts yields the Ca salt of No. 17 mixed with its 1,2- and 1,3-isomers in the ratio 10/7/3. Data for the mixture: ^1H NMR spectrum (1/3 $OS(CD_3)_2/CDCl_3$): $\delta=2.40$ (s, CH_3), 2.9 to 3.5 (m, 3.2 H, $H_2O+CH_2CO_2$), 3.9 to 4.9 (m, $C_{10}H_8$), 7.21 (m, 2H) and 7.74 (m, 2H) (C_6H_4) ppm. IR (KBr): 1607 and 1634 (νC=O), 3428 (OH) cm^{-1}. Electronic spectrum (CH_3OH): $\lambda_{max}(\varepsilon)=256\,(30290)$, 281 sh (20925), 358 (3500), 484 (1700) nm. Studies of the biological activity of the mixture gave neither antiarthritic activity in a rat adjuvant arthritis model nor prostaglandin synthetase inhibitory activity in a human blood aggregation assay, nor an effect on the human platelet serotinin release reaction [47].

$HO_2C(CH_2)_2–C_5H_4FeC_5H_4–C_2H_5$ (Table **15**, No. **19**) is won in addition to No. 2, as the more soluble acid, from the mother liquor of the latter. Both compounds are obtained by the reaction of $Fe(C_5H_4–COCH_3)_2$ with $(C_2H_5O)_2CO$ to form Fe $(C_5H_4–COCH_2CO_2C_2H_5)_2$ followed by catalytic hydrogenation over PtO_2 in acetic acid, with subsequent saponification of the raw product following Method Ic (p. 241). No. 19 cannot be crystallized [15].

$HO_2C(CH_2)_2CO$–$C_5H_4FeC_5H_4$–CH_3 (Table **15**, No. **28**). Method Vc (p. 242) starting with $C_5H_5FeC_5H_4$–CH_3 produces $FeC_{10}H_8(CO(CH_2)_2CO_2CH_3)CH_3$ as a 51/16/33 mixture of the 1,1′-, 1,2-, and 1,3-isomers, from which only the 1,2-isomer can be separated by preparative thin layer chromatography on SiO_2 with 30/1 C_6H_6/C_2H_5OH. The mixture of No. 28 and its 1,3-isomer remaining after saponification melts at 94 to 106°C; its Clemmensen reduction in 2/2/1 $C_6H_6/CH_3OH/H_2O$ yields No. 21 mixed with $C_5H_5FeC_5H_3(CH_3\text{-}3)(CH_2)_3CO_2H\text{-}1$ [49].

$HO_2C(CH_2)_2CO$–$C_5H_4FeC_5H_4$–$COCH_3$ (Table **15**, No. **30**). X-ray diffraction patterns recorded with a counting tube diffractometer on powder samples give for a sample carefully smoothed into a carrier and an unsmoothed sample somewhat differing values. The latter value agrees with that measured in a photographic camera for a sample in a glass cylinder [29].

References:

[1] R. B. Woodward, M. Rosenblum, M. C. Whiting (J. Am. Chem. Soc. **74** [1952] 3458/9). – [2] M. Rosenblum (Diss. Harvard Univ. 1953, p. 112). – [3] W. B. Hardy, E. Klingsberg, American Cyanamid Co. (U.S. 2900401 [1954/59]; C.A. **1960** 569). – [4] V. Weinmayr (J. Am. Chem. Soc. **77** [1955] 3012/4). – [5] A. N. Nesmeyanov, N. A. Volkenau, V. D. Vilchevskaya (Dokl. Akad. Nauk SSSR **111** [1956] 362/4; Proc. Acad. Sci. USSR Chem. Sect. **106/111** [1956] 675/7).

[6] W. A. Piccoli, Dow Corning Corp. (U.S. 2979482 [1956/61]; C.A. **1960** 18176). – [7] P. J. Graham, R. V. Lindsey, G. W. Parshall, M. L. Peterson, G. M. Whitman (J. Am. Chem. Soc. **79** [1957] 3416/20). – [8] K. L. Rinehart, R. J. Curby, P. E. Sokol (J. Am. Chem. Soc. **79** [1957] 3420/4). – [9] W. Göller (Diss. Stuttgart T.H. 1958). – [10] J. B. Thomson (Chem. Ind. [London] **1959** 1122).

[11] A. A. Berlin, T. V. Kostroma (U.S.S.R. 129018 [1959/60]; C.A. **1961** 3612). – [12] K. Arakawa, K. Yamakawa (Kagaku No Ryoiki **14** [1960] 632/44). – [13] A. N. Nesmejanow, W. A. Ssasonowa, V. N. Drosd (Chem. Ber. **93** [1960] 2717/29). – [14] A. N. Nesmeyanov, V. A. Sazonova, V. N. Drozd, L. A. Nikonova (Dokl. Akad. Nauk SSSR **133** [1960] 126/9; Proc. Acad. Sci. USSR Chem. Sect. **130/135** [1960] 751/4). – [15] K. Schlögl, H. Seiler (Monatsh. Chem. **91** [1960] 79/91).

[16] K. Schlögl, H. Seiler (Tetrahedron Letters **1960** No. 7, pp. 4/8). – [17] K. Schlögl, H. Seiler (Angew. Chem. **72** [1960] 38). – [18] K. L. Rinehart, R. J. Curby, D. H. Gustafson, K. G. Harrison, R. E. Bozak, D. E. Bublitz (J. Am. Chem. Soc. **84** [1962] 3263/9). – [19] K. Schlögl, M. Peterlik (Tetrahedron Letters **1962** 573/6). – [20] K. Schlögl, M. Peterlik, H. Seiler (Monatsh. Chem. **93** [1962] 1309/27).

[21] R. L. Dotson, M. D. Barnett (7th Rept. Res. Petrol. Res. Fund., 1963, p. 36 from W. E. Watts, Organometal. Chem. Rev. **2** [1967] 231/54, 238). – [22] J. P. Monin, G. Tainturier, R. Dabard, J. Tirouflet (Bull. Soc. Chim. France **1963** 667/8). – [23] K. L. Rinehart, D. E. Bublitz, D. H. Gustafson (J. Am. Chem. Soc. **85** [1963] 970/82). – [24] J. Tirouflet, R. Dabard, E. Laviron (Bull. Soc. Chim. France **1963** 1655/60). – [25] J. Tirouflet, J.-P. Monin, G. Tainturier, R. Dabard (Compt. Rend. **256** [1963] 433/5).

[26] C. Rai, Union Oil Corp. of California (U.S. 3222373 [1963/65]; C.A. **64** [1966] 5139). – [27] A. A. Berlin, T. V. Kostroma (Vysokomol. Soedin. Geterots. Vysokomol. Soedin. **1964** 53/8; C.A. **61** [1964] 9591). – [28] J. Tirouflet, G. Tainturier (Tetrahedron Letters **1965** 4177/9). – [29] R. Dabard, D. Weigel (Bull. Soc. Chim. France **1966** 390/6). – [30] G. Tainturier, J. Tirouflet (Bull. Soc. Chim. France **1966** 595/600).

[31] J. Tirouflet, G. Tainturier, H. Singer (Bull. Soc. Chim. France **1966** 2565/70). – [32] R. Dabard, P. Dixneuf (Compt. Rend. C **265** [1967] 324/7). – [33] P. Dixneuf, R. Dabard (Bull. Soc. Chim. France **1969** 2164/71). – [34] H. Falk, O. Hofer, K. Schlögl (Monatsh. Chem. **100** [1969] 624/48). – [35] J. Tirouflet (Colloq. Intern. Centre Natl. Rech. Sci. [Paris] No. 191 [1970] 267/78).

[36] M. Laćan, R. Šarac-Arneri (Croat. Chem. Acta **43** [1971] 215/9). – [37] A. Sonoda, I. Moritani (J. Organometal. Chem. **26** [1971] 133/40). – [38] D. F. Kutepov, V. P. Red'ko, V. K. Skubin (Tr. Mosk. Khim. Tekhnol. Inst. No. 74 [1973] 123/5 from C.A. **82** [1975] No. 43765). – [39] D. F. Kutepov, V. P. Red'ko, V. K. Skubin, V. G. Mochalov (Tr. Mosk. Khim. Tekhnol. Inst. No. 74 [1973] 126/9 from C.A. **82** [1975] No. 43963). – [40] K. Yamakawa, M. Hisatome (J. Organometal. Chem. **52** [1973] 407/24).

[41] A. N. Nesmeyanov, M. I. Rybinskaya, G. B. Shul'pin, A. A. Progrebnyak (J. Organometal. Chem. **92** [1975] 341/67). – [42] E. I. Edwards, R. Epton, G. Marr (J. Organometal. Chem. **122** [1976] C49/C53). – [43] M. Hisatome, T. Sakamoto, K. Yamakawa (J. Organometal. Chem. **107** [1976] 87/101). – [44] E. I. Edwards, R. Epton, G. Marr, G. K. Rogers. K. J. Thompson (Chem. Soc. [London] Spec. Publ. No. 28 [1977] 92/100). – [45] E. I. Edwards, R. Epton, G. Marr (J. Organometal. Chem. **168** [1979] 259/72).

[46] E. R. Epton, G. Marr (J. Organometal. Chem. **168** [1979] 259/72). – [47] B. E. Maryanoff, S. L. Keeley, F. J. Persico (J. Med. Chem. **26** [1983] 226/9). – [48] R. L. Schaaf, K. L. Rinehart (WADD-TR-61-108-III; AD-403929 [1963] 1/80, 73). – [49] G. Haller, K. Schlögl (Monatsh. Chem. **98** [1967] 2044/57). – [50] K. Arakawa, K. Yamakawa (Kagaku No Ryoiki **14** [1960] 632/44).

[51] K. Gonsalves, L. Zhan-ru, M. D. Rausch (J. Am. Chem. Soc. **106** [1984] 3862/3). – [52] M. Hisatome, S.-I. Takano, K. Yamakawa (Tetrahedron Letters **26** [1985] 2347/50).

5.2.3.6.2 1,2-Compounds

5.2.3.6.2.1 1,2-Dicarboxyferrocene $C_5H_5FeC_5H_3(CO_2H)_2$-1,2

Preparation. The compound is prepared by the oxidation of I in dioxane with aqueous NaOBr solution at 0 to 5°C (reaction time 4 h, yield 56%) [4] or of $C_5H_5FeC_5H_3(COCH_3)_2$-1,2 with NaOCl in H_2O with addition of a minimal volume of CH_3OH for attainment of a homogeneous solution [1]. The reaction of $C_5H_5FeC_5H_3Li_2$-1,2 with Dry Ice proceeds practically quantitatively. The starting compound is not isolated, but is previously won from the isomer mixture of II and III (see "Ferrocene" 6, 1977, p. 212) in 9/4 ether/benzene [2, 3] or from $C_5H_5FeC_5H_3$ $(HgCl)_2$-1,2 in ether [3] by reaction with n-C_4H_9Li in hexane at room temperature; the reaction mixture is poured onto the crushed Dry Ice. The yield is between 30 and 60% referred to II and III. The product obtained from $C_5H_5FeC_5H_3(HgCl)_2$-1,2 is contaminated with $C_5H_5FeC_5H_4$–CO_2H [3]. The acid is purified by recrystallization from $CHCl_3$ [3] or benzene/petroleum ether [4].

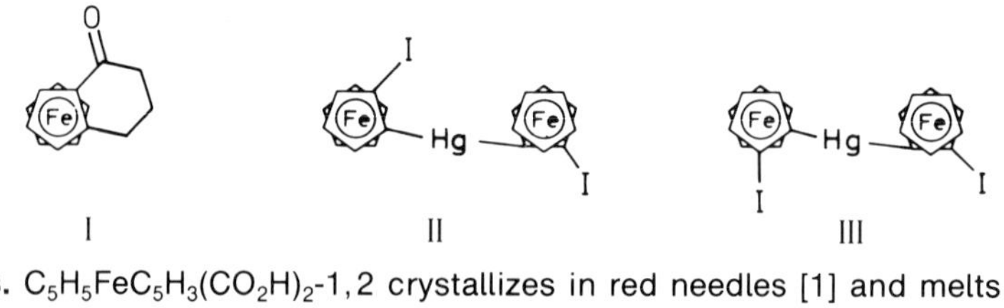

I II III

Properties. $C_5H_5FeC_5H_3(CO_2H)_2$-1,2 crystallizes in red needles [1] and melts under N_2 in a sealed tube at 195 to 198°C with foaming [3], in air with decomposition at 202 to 204 [4], 205 to 207 [2, 3], 206 to 206.5°C [1]. ^{1}H NMR spectrum (acetone): $\delta = 4.40$ (s, C_5H_5), 4.81 [2] (4.89 [3]) (t, H-4), 5.23 (d, H-3,5; $J \approx 2.6$ [2], 2.8 [3] Hz) [2, 3]. IR (KBr): 1600, 1690, 2900 cm^{-1} [3]; ($CHCl_3$): 1000, 1104, 1631, 1721, 2617, 2994 cm^{-1} [1].

References on p. 251

Chemical Behavior. The compound both as a solid and in solution must be protected from light, since otherwise photochemical decomposition occurs within a short time [3]. The dissociation constants $pK_1 = 3.64$ and $pK_2 \geqq 12$ have been determined by potentiometric titration in 2/1 C_2H_5OH/H_2O [1]. Although the compound reacts very sensitively to the usual chemical reagents, it can easily be converted into its anhydride IV with $(cyclo\text{-}C_6H_{11}N)_2C$ in acetone [1 to 3]. Quantum mechanical calculations by the EHT method indicate that the compound should be unstable in acidic solutions and be decomposed both thermochemically and photochemically [5].

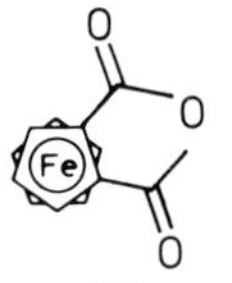

IV

References:

[1] J. H. Richards, T. J. Curphey (Chem. Ind: [London] **1956** 1456). – [2] M. D. Rausch (Pure Appl. Chem. **30** [1972] 523/38). – [3] P. V. Roling, M. D. Rausch (J. Org. Chem. **39** [1974] 1420/4). – [4] N. A. Maier, S. A. Shklyar, Yu. A. Ol'dekop (Zh. Obshch. Khim. **50** [1980] 2626/7). – [5] P. Zahradnik, Š. Toma (Acta Fac. Rerum Nat. Univ. Comenianae Chim. No. 29 [1981] 1/7).

5.2.3.6.2.2 Carboxylic Acids of the Type $C_5H_5FeC_5H_3(CO_2H\text{-}1)R\text{-}2$

All compounds listed in Table 16 possess on one cyclopentadienyl ring two different substituents and thus a plane of chirality; for the centrochiral R,S nomenclature used see "Ferrocene" 7, 1980, pp. 4/7 [German], pp. 1/4 [English translation]. Nos. 3 and 12 have in addition in the asymmetrically substituted C atom of R a chirality center. Nevertheless, optically active forms have been prepared only for Nos. 1, 4, and 7 (see "Further information", pp. 255, 256, 257). All other data refer to the racemic forms or the diastereoisomeric mixtures.

The compounds have been prepared by the following methods and those discussed under "Further information":

Method I. Alkaline saponification:

a) Of $C_5H_5FeC_5H_3(CN\text{-}1)R\text{-}2$ with KOH in n-C_4H_9OH [11,22] or with NaOH in 2/3 C_2H_5OH/H_2O [4, 5].

If one starts with the product obtained by cyaniding of $C_5H_5FeC_5H_4\text{–}C_2H_5$ with liquid HCN in tetrahydrofuran in the presence of $FeCl_3$, No. 2 is obtained in an isomer mixture (m) predominantly with the 1,3-isomer [4, 5]; a separation was not carried out.

The product acquired by the reaction of $[C_5H_5FeC_5H_4\text{–}CH_2N(CH_3)_3]I$ with NaCN is not $C_5H_5FeC_5H_3(CN\text{-}1)CH_3\text{-}2$ as reported in [2] but $C_5H_5FeC_5H_4\text{–}CH_2CN$, according to a more recent investigation by the same authors [3].

b) Of $C_5H_5FeC_5H_3(CO\text{–}NH_2\text{-}1)R\text{-}2$ with KOH in n-C_4H_9OH [7].

c) Of $C_5H_5FeC_5H_3(CO\text{–}N(CH_3)C_2H_5\text{-}1)R\text{-}2$ with KOH in 19/1 C_2H_5OH/H_2O [28].

Method II. Cleavage of $C_5H_5FeC_5H_3(CO(C_6H_4Cl\text{-}2)\text{-}1)R\text{-}2$ with $KOC(CH_3)_3$ and traces of H_2O in $CH_3OCH_2CH_2OCH_3$ at the boiling point, 10 to 20 min for $R = CH_3$, longer for $R = C_6H_5$. In the latter case $C_5H_5FeC_5H_4\text{–}C_6H_5$ forms in 10% yield along with No. 4 [30].

References on p. 258

Method III. Oxidation of $C_5H_5FeC_5H_3(COCH_3\text{-}1)R\text{-}2$ with NaOCl
- a) in refluxing 10/1 CH_3OH/H_2O [22],
- b) in absolute CH_3OH. Chromatography on SiO_2 yields No. 13 as the main product [1].

Method IV. Oxidation of $C_5H_5FeC_5H_3(CHO\text{-}1)R\text{-}2$
- a) with AgO in a mixture of 5 M aqueous KOH and tetrahydrofuran at 20°C [19],
- b) with active MnO_2 in glacial acetic acid in the presence of KCN and subsequent saponification of the ester [19].

Method V. The product obtained by lithiation of $C_5H_5FeC_5H_4\text{–}R$ with n-C_4H_9Li in ether [6, 14] or ether/hexane [8, 18, 32] is poured onto a slurry of Dry Ice in ether [6, 8, 14, 18, 32]. The yields given here are relative to $C_5H_5FeC_5H_4\text{–}R$. No. 6 is prepared by starting with I instead of $C_5H_5FeC_5H_4\text{–}OH$ and decomposing in C_2H_5OH with 2M HCl the product obtained after lithiation and carboxylation [31, 34].

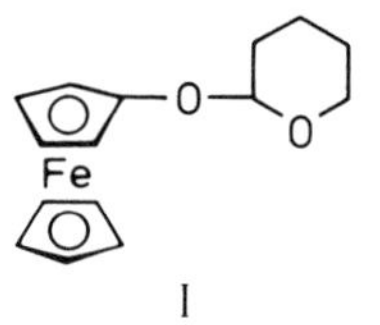

I

Table 16

Carboxylic Acids of the Type $C_5H_5FeC_5H_3(CO_2H\text{-}1)R\text{-}2$.

Further information for numbers preceded by an asterisk is given at the end of the table, pp. 254/8.

For abbreviations and dimensions see p. VIII.

R_f values are for descending paper chromatography with n-C_4H_9OH/C_2H_5OH/conc. NH_4OH/H_2O in the volume ratio 4/4/1/1 as the mobile phase.

No.	$C_5H_5FeC_5H_3$–(CO_2H-1)R-2 R	method of preparation (yield in %), properties, and remarks	Ref.
*1	CH_3	Ia (4), Ib (10), Ic (35), II (85 to 95), see "Further information", p. 254 m.p. 154 to 158° (CH_3OH/H_2O), 155 to 156° (after sublimation), 156 to 159° (CH_3OH/H_2O), 158 to 160° (CH_2Cl_2/ether), 160° (C_2H_5OH/H_2O) ^{1}H NMR ($CDCl_3$): 2.29 (s, CH_3), 4.17 (s, C_5H_5), 4.33 (m, 2.4H), 4.79 (m, 1.5H) ^{1}H NMR (CCl_4): 2.30 (s, CH_3), 4.18 (s, C_5H_5), 4.36 (m, 2H), 4.80 (m, 1H), see "Further information", p. 254 IR (Nujol): 1000, 1120, 1240, 1280, 1670 (νC=O), 3300 (OH) R_f: 0.65	[7, 10, 11, 15, 27, 28, 30]
2	C_2H_5	Ia (61m) m.p. 113 to 114° (cyclohexane) (m)	[4, 5]

Table 16 (continued)

No.	$C_5H_5FeC_5H_3$–(CO_2H-1)R-2 R	method of preparation (yield in %), properties, and remarks	Ref.
3	$CH(CH_3)_2$	Ib (≈10) R_f: 0.75	[7]
*4	C_6H_5	Ia (13), II (70), IIIa (43), see "Further information", p. 256 m.p. 218 to 220° (C_2H_5OH or CH_3OH/H_2O), 220 to 222° dec. (CH_3OH) mass spectrum: m/e = 306 $[M]^+$, 168, 138, 73, 56	[22, 30]
*5	Cl	V (31 (raw product), 18) yellow solid (raw product), m.p. 170 to 171° (hexane), 170 to 172° (hexane) ^{1}H NMR: 3.81 (m, 1H), 4.05 (s, C_5H_5), 4.43 (m, 1H), 4.68 (m, 1H) IR: 1000, 1111	[8, 18]
*6	OH	V (78 relative to I) deep red crystals, m.p. 160° dec. (ether/petroleum ether) IR (KBr): 1650 (νC=O), 3380 (νOH) mass spectrum: m/e = 246 $[M]^+$	[31, 34]
*7	CH_2OH	see No. 1, p. 254	[10]
*8	$C(C_6H_5)_2OH$	V (73) red crystals (ether), m.p. 144 to 146° (with effervescence) IR: 1000 and 1111 (C_5H_5), 1681 (νC=O), 3448 (νOH) mol. wt: 418 (by Rast method)	[6]
*9	O_2CCH_3	see No. 6, p. 257 orange needles, m.p. 136° (petroleum ether) ^{1}H NMR (CCl_4): 2.22 (s, CH_3), 4.30 (s) and 4.60 (m) ($C_5H_5+C_5H_3$), 11.48 (s, OH) mass spectrum: m/e = 288 $[M]^+$	[31, 34]
*10	OCH_3	V (69) red crystals (absolute C_6H_6), m.p. 150 to 152° (evacuated capillary) IR (paraffin oil): see figure in [14]	[14, 18]
11	$CH_2OC_2H_5$	see No. 1, p. 254	[10]
12	$CH(CH_3)OCH_3$	V (1.7) m.p. 202° dec.	[32]
13	$COCH_3$	IIIb orange-red needles, m.p. 105.0 to 106.0° IR ($CHCl_3$): 1002, 1109, 1639, 1667, 1724, 2660, 2941	[1]

References on p. 258

* Further information:

$C_5H_5FeC_5H_3(CO_2H\text{-}1)CH_3\text{-}2$ (Table **16**, No. **1**) can also be prepared from II.

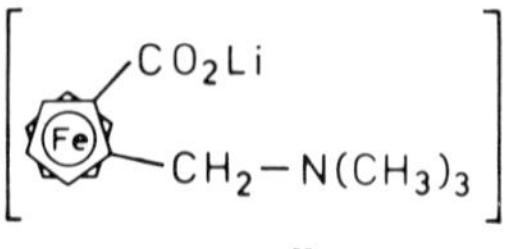

II

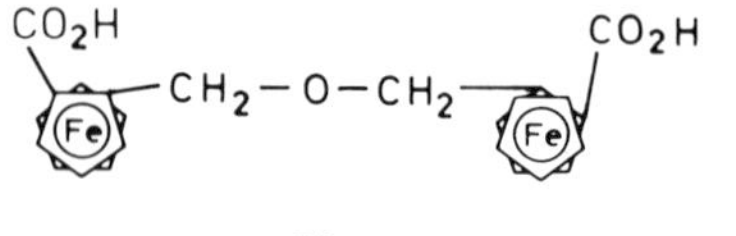

III

The aqueous solution of II is added to a sodium amalgam in portions with cooling and then heated for 8 h on a water bath after cessation of the exothermic reaction. The usual processing gives a mixture of three acids as the raw product. Almost pure No. 1 is won by recrystallization in CH_3OH/H_2O. A total yield of 48% (referred to $C_5H_5FeC_5H_4\text{–}CH_2N(CH_3)_2$) is obtained including that recovered by preparative thin layer chromatography of the mother liquor on SiO_2 with 10/1 C_6H_6/C_2H_5OH. The acids remaining in the mother liquor are No. 7 and III. On the other hand, if one decomposes II in hot C_2H_5OH rapidly with Na, then in addition to No. 1 (yield 15% referred to $C_5H_5FeC_5H_4\text{–}CH_2N(CH_3)_2$) compound No. 11 forms as the main product [10]. Instead of II the compound $[C_5H_5FeC_5H_3(CO_2CH_3\text{-}1)CH_2N(CH_3)_3\text{-}2]I$ can also be reduced with Na amalgam following Emde's procedure [37]. Preparation by Method IV (p. 252) from the 15/44/41 mixture of the 1,2-, 1, 3-, and 1,1′-isomers of $FeC_{10}H_8(CHO)CH_3$ for steric reasons strongly shifts the isomer ratio at the expense of No. 1, with Method IVa to 6, 58, and 36%, respectively, and with Method IVb to 8, 65, and 27% [19]. Separation of the isomers is covered below.

The **deuterated** product $C_5H_5FeC_5H_3(CO_2H\text{-}1)CH_2D\text{-}2$ melts at 154 to 156°C (CH_3OH/H_2O); ^{1}H NMR ($CDCl_3$): $\delta = 2.31$ (CH_2D) ppm; molecular weight 240 by titrimetry; mass spectrum m/e = 245 $[M]^+$. It is prepared from II, which is dissolved in D_2O after evaporation with absolute C_6H_6 in vacuum, and then reduced with Na amalgam with exclusion of moisture. For isolation the acid mixture is esterified with CH_2N_2 in petroleum ether and then chromatographed on an Al_2O_3 column in petroleum ether. $C_5H_5FeC_5H_3(CO_2CH_3\text{-}1)CH_2D\text{-}2$ is eluted with ether and subsequently distilled in a bulb tube at 0.3 Torr on an air bath. After saponification with KOH/CH_3OH the yield amounts to 32 to 40% referred to $C_5H_5FeC_5H_4\text{–}CH_2N(CH_3)_2$ [16].

Separation of the positional isomers is effected by esterification of the mixture of $FeC_{10}H_8(CO_2H)CH_3$ with CH_2N_2 and chromatographing it in ca. 5/1 petroleum/benzene on alumina over the total column length. It is divided into 20 to 25 fractions in the elution sequence 1,2- before 1,1′- and 1,3-isomers, but nevertheless a sharp separation is not achieved, and 20% of the amount of ester introduced remains irreversibly adsorbed. Subsequent saponification with KOH in CH_3OH yields the acids. A distinct shift of the fraction of the total yield in favor of the 1,3-isomer is attributed to the easier saponification of the 1,2- and 1,1′-isomers and the consequent stronger absorbability on the alkaline column. Separation via the $4\text{-}BrC_6H_4CH_2$ ester offers no advantages [11].

Properties. It is conjectured from intensity integration that the CO_2H proton, which cannot be recognized in the ^{1}H NMR spectrum, is broadly spread between the two multiplets with $\delta = 4.33$ and 4.79 ppm [28]. The signal of the CH_3 proton is shifted around 0.35 ppm to lower field versus that of $C_5H_5FeC_5H_4\text{–}CH_3$ in CCl_4 [15].

Chemical Reactions. CH_2N_2 in ether gives the methyl ester [7, 10, 11], PCl_3 in absolute benzene $C_5H_5FeC_5H_3(COCl\text{-}1)CH_3\text{-}2$ [10]. Resolution of the racemic mixture is discussed under "Optically Active Forms".

$C_5H_5FeC_5H_3(CO_2Li\text{-}1)CH_3\text{-}2$ formed from II (p. 254) with Na amalgam yields on refluxing with R′Li in a 1/3 molar ratio, $R'=CH_3$, C_2H_5, C_6H_5, after hydrolysis the ketones $C_5H_5FeC_5H_3(CO\text{–}R'\text{-}1)CH_3\text{-}2$. With $R'=C_6H_5$ in addition an unidentified red oil and $C_5H_5FeC_5H_3(C(C_6H_5)_2OH\text{-}1)CH_3\text{-}2$ are obtained [23, 27).

Optically Active Forms. (+)(1S)-$C_5H_5FeC_5H_3(CO_2H\text{-}1)CH_3\text{-}2$ (see Formula IV) is won from the racemic mixture by resolution with (−)-$C_6H_5CH(CH_3)NH_2$, on whose addition (+)-$C_5H_5FeC_5H_3(CO_2NH_3\text{–}CH(CH_3)C_6H_5\text{-}1)CH_3\text{-}2$ spontaneously crystallizes out of ether solution. This yields with H_3PO_4 (+)-No. 1 after multiple recrystallization from ethanol/ether [10, 27]. (+)-No. 1 is liberated by the kinetic racemate splitting of $(C_5H_5FeC_5H_3(CH_3\text{-}2)C(O)\text{-}1)_2O$ with (−)-$C_6H_5CH(CH_3)NH_2$ in pyridine. After 60 h at 0°C the chemical yield amounts to around 90%, the optical to around 6%. The kinetic resolution with (−)-menthol in pyridine yields after 10 d at 20°C a chemical yield of ca. 16% and an optical yield of ca. 20% (both values are uncertain, since the menthol ester of No. 1 is only difficultly separated from the unreacted anhydride) [10, 38]. (+)-No. 1 is obtained from the methiodide of (+)-$C_5H_5FeC_5H_3(CO_2CH_3\text{-}1)CH_2N(CH_3)_2\text{-}2$ via the Emde degradation with Na amalgam, in a 71% yield, and chromatographically purified on SiO_2 [35, 36].

(+)-No. 1 sublimes above 90°C and decomposes at 158°C [10]. Specific rotation values (extrapolated to p = 100% when not otherwise specified): In C_2H_5OH $[\alpha]_D^{20}=+37.5°$ (c = 1, p ≈ 71.5%) [36], +52° ± 2° (c = 0.3 to 1.2) [10, 12], 53.0° ± (0.2 to 0.9)° (c = 0.1 to 0.6) [33]; $[\alpha]_{570}^{20}=+37.1°$ (c = 0.1 to 0.6); $[\alpha]_{546}^{20}=+42.7°$ (c = 0.1 to 0.6) [33]. In C_6H_6 $[\alpha]_D^{20}=+34° \pm 1.5°$ (c = 1) [10], +25.8° ± (0.2 to 0.9)° (c = 0.1 to 0.6); $[\alpha]_{570}^{20}=+8.5° \pm (0.2\ \text{to}\ 0.9)°$ (c = 0.1 to 0.6); $[\alpha]_{546}^{20}=-30.0° \pm (0.2\ \text{to}\ 0.9)°$ (c = 0.1 to 0.6) [33]. ^1H NMR spectrum ($CHCl_3$): δ = 2.30 (CH_3), 4.25 (C_5H_5), 4.42 and 4.87 (C_5H_3), 7.95 (CO_2H) ppm [36]. Electronic spectrum in C_2H_5OH: $\lambda_{max}(\varepsilon)$ = 439(189), 348 sh (226), 302(720), 259(3360) nm [12]. Optical rotatory dispersion (C_2H_5OH, c = 0.1): $[M]_D^{20}=+130°$ [33], $[M]_{540}^{20}=+25°$, $[M]_{460}^{20}=+1090°$, $[M]_{419}^{20}=-34°$, $[M]_{357}^{20}=+960°$ [12]. Circular dichroism (C_2H_5OH, c ≈ 0.1): $\Delta\varepsilon_{501}=-0.20$, $\Delta\varepsilon_{432}=+0.44$, $\Delta\varepsilon_{384}=+0.07$ [33]. A band analysis of the circular dichroism curve by a computer curve fitting program shows it to arise from the superposition of three bands with the following rotational strengths R in $erg \cdot cm^3 \cdot 10^{40}$: +2.11 (439 nm), −1.02 (479 nm), and −0.30 (510 nm) [17]. An empirical relationship between the sum of the absolute values of the rotational and the dipole strengths indicates the transition to be electrically and magnetically dipole forbidden; a dissymmetrically distorted, inherently symmetric chromophor is present [24].

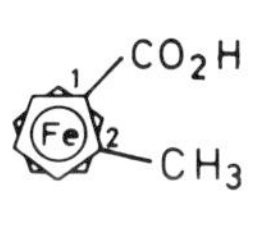

(+)(1S)-No. 1
IV

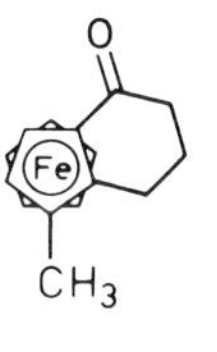

(+)-V

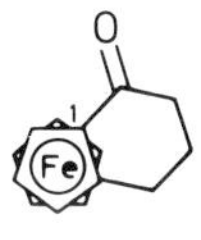

(+)(1S)-VI

The absolute configuration (1S) is derived from (+)(1S)-$C_5H_5FeC_5H_3(CHO\text{-}1)CH_3\text{-}2$ by chemical correlation [10] as well by comparison of the optical rotatory dispersion curve of (+)-V derived from (+)-No. 1 with that of (+)-VI with the known absolute configuration (1S) [13], see also [38].

While no racemization occurs in CH_3NO_2 at 368.2 K in the presence of ascorbic acid as an antioxidant and an equimolar amount of $HClO_4$, it takes place quantitatively under otherwise equal conditions under the oxidizing influence of a small quantity of HNO_3, with a rate constant

References on p. 258

$k = 8 \times 10^{-3}\ s^{-1}$ [25]. Apparently the acquisition of a positive charge effects a loosening of the Fe–cyclopentadienyl bond and thus enables the racemization [20].

Reaction with PCl_3 leads to the acid chloride [9, 12, 33], esterification with CH_2N_2 in ether to (−)(1S)-$C_5H_5FeC_5H_3(CO_2CH_3\text{-}1)CH_3\text{-}2$ [12, 33], and reduction with $LiAlD_4/AlCl_3$ in ether to (−)(1S)-$C_5H_5FeC_5H_3(CD_3\text{-}1)CH_3\text{-}2$ [24].

For the anion (measured in 0.1 N ethanolic $NaOC_2H_5$ solution) the specific rotation values are, for c = 0.1 to 0.6, values ±(0.2 to 0.9)°: $[\alpha]^{20}_{589} = +54.3°$, $[\alpha]^{20}_{546} = +58.5°$. $[M]^{20}_{589} = +133°$. Circular dichroism (c ≈ 0.1) $\Delta\varepsilon_{489} = +0.20$, $\Delta\varepsilon_{438} = -0.03$, $\Delta\varepsilon_{383} = +0.16$ [33].

(+)(1S)-$C_5H_5FeC_5H_3(CO_2Li\text{-}1)CH_3\text{-}2$ forms (−)(1S)-$C_5H_5FeC_5H_3(CO\text{–}R'\text{-}1)CH_3\text{-}2$ on reaction with R'Li (R' = CH_3, C_2H_5) and gives (+)(1S)-$C_5H_5FeC_5H_3(CO\text{–}C_6H_5\text{-}1)CH_3\text{-}2$ with C_6H_5Li [27].

(−)(1R)-$C_5H_5FeC_5H_3(CO_2H\text{-}1)CH_3\text{-}2$ is won by cleavage of the racemic mixture with optically active $C_6H_5CH(CH_3)NH_2$ [27]. For the circular dichroism curve see [27, p. 4570]. The optical purity has been determined by the isotope dilution process with $C_5H_5FeC_5H_3(CO_2H\text{-}1)CH_2D\text{-}2$ as tracer [16], by NMR spectroscopy on the diastereomeric acid amides prepared with (−)-$C_6H_5CH(CH_3)NH_2$ [16] as well as by separation by means of high pressure liquid chromatography of the diastereomeric amides obtained with (−)-α-$C_{10}H_7CH(CH_3)NH_2$ [29]. The resulting specific rotation values extrapolated to p = 100% are $[\alpha]_D = -55.6° \pm 1°$, $-54° \pm 3°$ [16], and $-53° \pm 1.6°$, respectively [29].

(+)(1R)-$C_5H_5FeC_5H_3(COCH_3\text{-}1)CH_3\text{-}2$ is obtained from (−)(1R)-No. 1 via the acid chloride [21] or from the Li salt by reaction with CH_3Li. The reaction of the salt with R'Li (R' = C_2H_5, C_6H_5) yields (+)(1R)-$C_5H_5FeC_5H_3(COC_2H_5\text{-}1)CH_3\text{-}2$ and (−)(1R)-$C_5H_5FeC_5H_3(COC_6H_5\text{-}1)CH_3\text{-}2$, respectively [27].

$C_5H_5FeC_5H_3(CO_2H\text{-}1)C_6H_5\text{-}2$ (Table **16**, No. **4**). Method IVa (p. 252) completely fails to work for preparation of No. 4, while Method IVb gives only very small yields. Also Method V (p. 252) does not yield pure No. 4, since the yield after separation of the isomer mixture is very slight [22]. Separation of No. 4 from the mixture of the three positional isomers takes place after esterification with CH_2N_2, by preparative thin layer chromatography on SiO_2 in C_6H_6 with subsequent saponification with KOH in CH_3OH [22]. For resolution of the racemate with cinchonidine (preferential crystallization with (−)-No. 4) and cinchonine (preferential crystallization with (+)-No. 4), see the following sections.

Optically Active Forms. (−)(1S)-$C_5H_5FeC_5H_3(CO_2H\text{-}1)C_6H_5\text{-}2$ is won from the racemic mixture by evaporation of it with cinchonidine in $CHCl_3$ followed by twice recrystallizing the salt with 5/1 ethanol/ether. The optical purity is 74% from NMR of the diastereoisomeric $C_5H_5FeC_5H_3(CONHCH(CH_3)C_6H_5\text{-}1)C_6H_5\text{-}2$ obtained from (−)-No. 4 and (−)-$C_6H_5CH(CH_3)NH_2$. The resulting rotation value extrapolated to p = 100% is $[\alpha]^{20}_D = -130° \pm 13°$ (C_2H_5OH, c ≈ 0.2 to 0.3). The optical purity derived by chemical correlation of (+)-No. 4 with (+)(1S)-No. 1 yields $[\alpha]_D = 135°$ [22]. This agrees well with the value $[\alpha]^{20}_D = -136° \pm 4.1°$ yielded by the enantiomeric purity, measured on the diastereoisomeric acid amide, obtained by separation via high pressure liquid chromatography of the product with (−)-α-$C_{10}H_7CH(CH_3)NH_2$ [29].

Cyclization with polyphosphoric acid leads to (−)(1S)-VII [22].

(+)(1R)-$C_5H_5FeC_5H_3(CO_2H\text{-}1)C_6H_5\text{-}2$ (see VIII). The acid liberated from the mother liquor of the first salt crystallization (see previous paragraph) is similarly reacted with cinchonine, the salt likewise subjected to a twofold recrystallization, and the liberated acid purified by preparative thin layer chromatography on SiO_2 in C_6H_6; optical purity p = 71.5% [22].

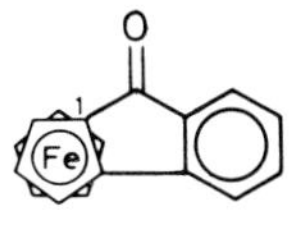

(−)(1S)-VII

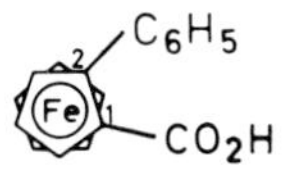

(+)(1R)-No. 4
VIII

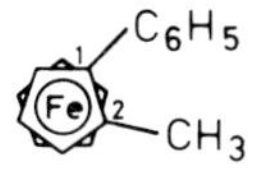

(+)(1S)-IX

Electronic spectrum (C_2H_5OH): $\lambda_{max}(\varepsilon) = 435$ (343) nm. Circular dichroism (C_2H_5OH, c = 0.02 to 0.06; p = 100%): $\Delta\varepsilon_{435} = -1.42$, $\Delta\varepsilon_{495} = +0.73$. A band analysis using a computer curve fitting program shows the circular dichroism curve to consist of a superposition of three bands with the following rotatory strengths R in erg·cm^3·10^{40}: −4.2 (435 nm), −0.8 (445 nm), and +2.6 (510 nm). The absolute configuration (1R, 2S) is yielded by the chemical correlation with (+)-No. 1 of the assured configuration (1S) via the (+)(1S)-IX [22], see also [26].

The compound is esterified with CH_2N_2 to form (+)(1R)-$C_5H_5FeC_5H_3(CO_2CH_3\text{-}1)C_6H_5\text{-}2$, is reduced with $LiAlH_4/AlCl_3$ to (+)(1S)-IX, and is converted by reaction of its acid chloride with $Cd(CH_3)_2$ into (+)(1R)-$C_5H_5FeC_5H_3(COCH_3\text{-}1)C_6H_5\text{-}2$ [22].

$C_5H_5FeC_5H_3(CO_2H\text{-}1)Cl\text{-}2$ (Table **16**, No. **5**) is purified by recrystallization from hexane [8] or by conversion to the methyl ester with column chromatography on Al_2O_3 with 1/1 heptane/benzene as eluent and subsequent saponification with 20% sodium hydroxide [18]. In 1/1 C_2H_5OH/H_2O at 20°C $pK_a = 6.19$ [18]. Unpurified the compound decomposes rapidly in air [8].

$C_5H_5FeC_5H_3(CO_2H\text{-}1)OH\text{-}2$ (Table **16**, No. **6**), the ferrocene analogue of salicylic acid, can be handled in air and recrystallized without taking special precautions. This high stability in comparison to the other hydroxyferrocene derivatives is probably due to both the electron withdrawing effect of the CO_2H group and to the formation of an intramolecular hydrogen bond between the two substituents (cf. the IR spectrum in the table). CH_2N_2 in ether forms the methyl ester. Reaction with $(CH_3CO)_2O$ (molar ratio ca. 1/2) in pyridine first for 10 min at 60 to 70°C, then 30 min at room temperature, gives No. 9, the analogue of aspirin, in a 64% yield [31, 34].

$C_5H_5FeC_5H_3(CO_2H\text{-}1)CH_2OH\text{-}2$ (Table **16**, No. **7**). **Optically Active Forms.** Saponification of (−)(1S)-$C_5H_5FeC_5H_3(CO_2CH_3\text{-}1)CH_2OH\text{-}2$ gives (1S)-$C_5H_5FeC_5H_3(CO_2H\text{-}1)CH_2OH\text{-}2$. Cyclization to the γ-lactone failed [39].

$C_5H_5FeC_5H_3(CO_2H\text{-}1)C(C_6H_5)_2OH\text{-}2$ (Table **16**, No. **8**) converts into the lactone X on vigorous shaking with hydrochloric acid or on heating to >146°C, which is evidence for the 1,2-positions of the substituents. With NaOH or aqueous $NaHCO_3$ solution it immediately forms the yellow salt, insoluble in ether, **$C_5H_5FeC_5H_3(CO_2Na\text{-}1)C(C_6H_5)_2OH\text{-}2$** [6].

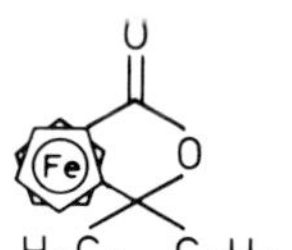

X

$C_5H_5FeC_5H_3(CO_2H\text{-}1)O_2CCH_3\text{-}2$ (Table **16**, No. **9**). The carrageen paw test of this aspirin analogous compound on rats gave little indication of an inflammation arresting action. Intraperitoneal injection of a high dose (200 mg/kg) killed all experimental animals, one of 50 mg/kg still 20%, with the remaining animals exhibiting severe health damage. On oral application the toxicity is quite significantly less, however an anti-inflammatory activity was not determinable [34].

References on p. 258

$C_5H_5FeC_5H_3(CO_2H\text{-}1)OCH_3\text{-}2$ (Table **16**, No. **10**) precipitates as a red oil that can be crystallized by cooling and trituration [14]. The formal redox potential is $E = -382$ mV (vs. standard calomel electrode) at 25 ± 1°C from potentiometric titration in 3/1 $CH_3CO_2H/HClO_4$ (0.033 M $HClO_4$). This value deviates considerably with respect to the additively calculated $E = -420$ mV [18]. In 1/1 C_2H_5OH/H_2O at 20°C the $pK_a = 6.39$ [18].

References:

[1] J. H. Richards, T. J. Curphey (Chem. Ind. [London] **1956** 1456). – [2] C. R. Hauser, J. K. Lindsay, D. Lednicer, C. E. Cain (J. Org. Chem. **22** [1957] 717/8). – [3] D. Lednicer, J. K. Lindsay, C. R. Hauser (J. Org. Chem. **23** [1958] 653/5). – [4] A. N. Nesmeyanov [Nesmejanow], E. G. Perevalova [Perewalowa], L. P. Yur'eva [Jurjewa] (Chem. Ber. **93** [1960] 2729/35). – [5] E. G. Perevalova, L. P. Yur'eva, Yu. I. Baukov (Dokl. Akad. Nauk SSSR **135** [1960] 1402/5; Proc. Acad. Sci. USSR Chem. Sect. **130/135** [1960] 1455/8).

[6] R. A. Benkeser, W. P. Fitzgerald, M. S. Melzer (J. Org. Chem. **26** [1961] 2569/71). – [7] K. Schlögl, M. Fried (Monatsh. Chem. **95** [1964] 558/75). – [8] J. W. Huffman, L. H. Keith, R. L. Asbury (J. Org. Chem. **30** [1965] 1600/4). – [9] H. Falk, K. Schlögl (Tetrahedron **22** [1966] 3047/53). – [10] H. Falk, K. Schlögl, W. Steyrer (Monatsh. Chem. **97** [1966] 1029/44).

[11] H. Falk, G. Haller, K. Schlögl (Monatsh. Chem. **98** [1967] 592/602). – [12] G. Haller, K. Schlögl (Monatsh. Chem. **98** [1967] 603/18). – [13] G. Haller, K. Schlögl (Monatsh. Chem. **98** [1967] 2044/57). – [14] A. N. Nesmeyanov, T. V. Baukova, K. I. Grandberg (Izv. Akad. Nauk SSSR Ser. Khim. **1967** 1867/9; Bull. Acad. Sci. USSR Div. Chem. Sci. **1967** 1796/8). – [15] K. Schlögl, H. Falk, G. Haller (Monatsh. Chem. **98** [1967] 82/94).

[16] P. Reich-Rohrwig, K. Schlögl (Monatsh. Chem. **99** [1968] 1752/63). – [17] H. Falk, O. Hofer (Monatsh. Chem. **100** [1969] 1499/506). – [18] A. N. Nesmeyanov, T. V. Baukova, K. I. Grandberg, Yu. A. Ustynyuk, S. P. Gubin, E. G. Perevalova (Izv. Akad. Nauk SSSR Ser. Khim. **1969** 721/3; Bull. Acad. Sci. USSR Div. Chem. Sci. **1969** 658/9). – [19] K. Schlögl, M. Walser (Monatsh. Chem. **100** [1969] 840/3). – [20] K. Bauer, H. Falk, H. Lehner, K. Schlögl, U. Wagner (Monatsh. Chem. **101** [1970] 941/3).

[21] H. Falk, H. Lehner, K. Schlögl (Monatsh. Chem. **101** [1970] 967/75). – [22] H. Lehner, K. Schlögl (Monatsh. Chem. **101** [1970] 895/911). – [23] C. Moïse, J. Tirouflet, D. Sautrey (Compt. Rend. C **271** [1970] 951/4). – [24] H. Falk, H. Lehner (Tetrahedron **27** [1971] 2279/91). – [25] H. Falk, H. Lehner, J. Paul, U. Wagner (J. Organometal. Chem. **28** [1971] 115/24).

[26] H. Lehner, K. Schlögl (Monatsh. Chem. **102** [1971] 277/91). – [27] C. Moïse, D. Sautrey, J. Tirouflet (Bull. Soc. Chim. France **1971** 4562/76). – [28] D. W. Slocum, F. E. Stonemark (J. Org. Chem. **38** [1973] 1677/81). – [29] R. Eberhardt, C. Glotzmann, H. Lehner, K. Schlögl (Tetrahedron Letters **1974** 4365/8). – [30] V. Rapić, K. Schlögl, B. Steinitz (J. Organometal. Chem. **94** [1975] 87/98).

[31] R. Epton, G. Marr, G. K. Rogers (J. Organometal. Chem. **110** [1976] C42/C44). – [32] D. W. Slocum, B. P. Koonsvitsky (J. Org. Chem. **41** [1976] 3664/8). – [33] V. Rapić, K. Schlögl, B. Steinitz (Monatsh. Chem. **108** [1977] 767/80). – [34] R. Epton, G. Marr, G. K. Rogers (J. Organometal. Chem. **150** [1978] 93/100). – [35] V. I. Sokolov, L. L. Troitskaya (9th Intern. Conf. Organometal. Chem., Dijon 1979, Abstr. Papers B 36).

[36] V. I. Sokolov, L. L. Troitskaya, O. A. Reutov (J. Organometal. Chem. **182** [1979] 537/46). – [37] I. Omae (Yuki Gosei Kagaku Kyokaishi **40** [1982] 147/57). – [38] H. Falk, K. Schlögl (Monatsh. Chem. **99** [1968] 578/87). – [39] T. Aratani, T. Gonda, H. Nozaki (Tetrahedron **26** [1970] 5453/64).

5.2.3.6.2.3 Carboxylic Acids of the Type $C_5H_5FeC_5H_3(A{-}CO_2H\text{-}1)R^2\text{-}2$ ($A{-}CO_2H = R^1$)

Of the compounds collected in Table 17 Nos. 9 to 25 with $R^1 \neq R^2$ possess a plane of chirality (for the centrochiral R,S nomenclature used for the ferrocene chirality see "Ferrocene" 7, 1980, pp. 4/7 [German], pp. 1/4 [English translation]). Of these the compounds No. 19 to 21, 24, and 25 possess in addition in the asymmetrically substituted C-6 atom of R^1 a center of chirality, so that they can occur in the form of the diastereoisomers a and b, which themselves are racemic forms. The compounds No. 3 and 4, with the constitutionally equivalent but chiral substituents, should each form one racemic and two meso forms.

The product $FeC_{10}H_8(C_6H_4{-}CO_2H\text{-}2)_2$ obtained at room temperature from the ferrocenium ion with the diazonium ion $[2\text{-}HO_2CC_6H_4N_2]^+$ in dilute H_2SO_4 [1] appears to consist completely of the 1,1′-isomer (see 5.2.3.6.1.4, Table 15, No. 7, p. 245).

The compounds are prepared by the following methods.

Method I. Alkaline saponification:

a) Of $C_5H_5FeC_5H_3(CH_2CN)_2$-1,2 with KOH in boiling C_2H_5OH/H_2O [24].

b) As for Method Ia, but of $C_5H_5FeC_5H_3(CH_2{-}CR'(CO_2C_2H_5)_2)_2$-1,2 (see 5.2.3.7.2.2, p. 335, for its preparation) with R′ = H for No. 6, CH_3 for No. 7, C_6H_5 for No. 8 [20].

c) As for Method Ia, but with $C_5H_5FeC_5H_3(CH_2CN\text{-}1)R^2$-2 (boiling 10 h for No. 10, 36 h for No. 11) [9, 24].

d) As for Method Ic, but after 30 min boiling decomposition with $HOCH_2CH_2OH$ and subsequent distillation off of the solvent up to a boiling temperature of 100°C [2].

e) Of $C_5H_5FeC_5H_3(CO(CH_2)_2CO_2CH_3\text{-}1)R^2$-2 (for preparation see 5.2.3.7.2.2, p. 337) with KOH (molar ratio 1/2) in C_2H_5OH without heating [13, 17].

f) Of (1RS, 6SR)-$C_5H_5FeC_5H_3(\overset{6}{C}H(C_6H_5)CH(CO_2C_2H_5)_2\text{-}1)CH_3$-2 with KOH in boiling ethanol [22].

g) As for Method If, but of the (1RS, 6RS) diastereoisomer [22].

h) Of $C_5H_5FeC_5H_3(CH(C_6H_5)CH(CO_2C_2H_5)_2\text{-}1)CH_2CH(CO_2C_2H_5)_2$-2 (m.p. 76°C) with KOH in boiling ethanol/water [20].

i) As for Method Ih, but of the diastereoisomer with m.p. 80°C [20].

Method II. Knoevenagel condensation of $C_5H_5FeC_5H_3(CHO\text{-}1)R^2$-2 with malonic acid (mole ratio 1/2) in pyridine and small quantities of piperidine with warming on the water bath [4].

Method III. a) Conversion of the acid chloride obtained from compound No. 19a by the Arndt Eistert method first into the diazoketone with CH_2N_2 (molar ratio 1/2) in ether, then warming this to 170°C in 1/1 collidine/$C_6H_5CH_2OH$. After extraction in ether and boiling down the extract, the remaining oil is purified on an Al_2O_3 chromatographic column with benzene and subsequently saponified [22].

b) As for Method IIIa, but with the acid chloride obtained from compound No. 19b [22].

References on p. 266

Method IV. Friedel-Crafts acylation of $C_5H_5FeC_5H_4$–R^2 with succinic acid anhydride and $AlCl_3$ (mole ratio about 1/1/2) in CH_2Cl_2 at room temperature [14].

a) The 16/51/33 mixture of No. 22 and its 1,1'- and 1,3-isomers ($R^2 = CH_3$), obtained in 68% yield, is converted into the methyl ester, and the $C_5H_5FeC_5H_3(CO(CH_2)_2CO_2CH_3\text{-}1)CH_3\text{-}2$ is separated on SiO_2 by thin layer chromatography with 30/1 benzene/ethanol and saponified with KOH/CH_3OH [14].

b) The isomer mixture obtained with $R^2 = C_2H_5$ is separated after esterification with CH_2N_2 by column chromatography on Al_2O_3 with benzene/petroleum ether and purified after saponification by numerous recrystallizations [5, 6].

Purification of the compounds is in general by recrystallization from the solvents shown in parentheses in Table 17 and by reprecipitation with acids after dissolution in alkalies.

Chemical Behavior

The acids are converted into the methyl esters with CH_2N_2 in ether (No. 1) or ether/methanol (No. 10) [24].

Dissolution in PCl_3 at 0°C and warming slowly over 20 h to room temperature forms from Nos. 19a and b the corresponding acid chlorides [22]. No. 10 for this purpose is stirred for 90 min at 70 to 80°C with PCl_3 and traces of pyridine in benzene [24].

The CO groups of the keto acids can be reduced to CH_2 groups by the Clemmensen reaction with amalgamated zinc and concentrated hydrochloric acid with several hours heating. The conversion of No. 23 to No. 15 is carried out in aqueous acetic acid [10, 11, 21], that of No. 22 to No. 14 occurs in toluene [13, 17] or C_6H_6/CH_3OH. In the latter case a mixture forms of $C_5H_5FeC_5H_3([CH_2]_3CO_2H\text{-}1)CH_3\text{-}2$ and $C_5H_5FeC_5H_3([CH_2]_3CO_2CH_3\text{-}1)CH_3\text{-}2$, which after saponification gives No. 14 in a 47% yield [14].

The $CR'(CO_2H)_2$ groups decarboxylate on heating the compounds with Cu powder in quinoline to 180°C in an easy S_N1 reaction to form CHR'–CO_2H [20]. Nos. 6, 7, and 8 are thus converted to Nos. 2, 3, and 4, respectively [20], and Nos. 24a, b, 25a, b into Nos. 19a, b, 20a, and b [22].

Propionic and butyric acid derivatives cyclize intramolecularly to ketones under the action of $(CF_3CO)_2O$ in CH_2Cl_2. For steric reasons ring closure of the former compounds proceeds preferentially heteroannularly, that of the latter homoannularly. Details are mentioned under "Further information" for Nos. 2, 13 to 15, 19a, b, 21a, and b, pp. 264/5.

Table 17

Carboxylic Acids of the Type $C_5H_5FeC_5H_3(A{-}CO_2H\text{-}1)R^2\text{-}2$ ($A{-}CO_2H = R^1$).

Further information for numbers preceded by an asterisk is given at the end of the table, pp. 264/5.
For abbreviations and dimensions see p. VIII.

No.	$C_5H_5FeC_5H_3(R^1\text{-}1)R^2\text{-}2$ $R^1 = A{-}CO_2H$ R^2	method of preparation (yield in %), properties, and remarks	Ref.
compounds with $R^1 = R^2$:			
*1	CH_2CO_2H	Ia (82) yellow solid, does not melt below 250° but decomposes slowly from 160°	[24]
*2	$(CH_2)_2CO_2H$	see "Chemical Behavior" of No. 6, p. 260, and No. 18, p. 265 yellow leaflets, m.p. 186° (CH_3OH/H_2O) ^1H NMR (CD_3OD): around 2.58 ($[CH_2]_2$ protons), around 4.06 ($C_5H_5 + C_5H_3$)	[4, 15, 19, 20]
*3	$CH_2CH(CH_3)CO_2H$	see "Chemical Behavior" of No. 7, p. 260 yellow leaflets, m.p. 200° (C_2H_5OH/H_2O)	[20]
*4	$CH_2CH(C_6H_5)CO_2H$	see "Chemical Behavior" of No. 8, p. 260 crystals, m.p. 230° (C_2H_5OH/H_2O)	[20]
5	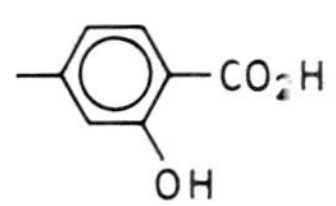	a minor constituent in the mixture of the positional isomers that consists predominantly of the 1,1′-isomer for details see 5.2.3.6.1.4 (Table 15, No. 8, p. 246)	[3, 7]
6	$CH_2CH(CO_2H)_2$	Ib solid, melts at 230° with decarboxylation, too unstable to be analyzed or recrystallized	[15, 19, 20]
7	$CH_2C(CH_3)(CO_2H)_2$	Ib solid, m.p. 230° dec.	[20]
8	$CH_2C(C_6H_5)(CO_2H)_2$	Ib yellow solid	[20]

References on p. 266

Table 17 (continued)

No.	$C_5H_5FeC_5H_3(R^1\text{-}1)R^2\text{-}2$ $R^1 = A\text{–}CO_2H$	R^2	method of preparation (yield in %), properties, and remarks	Ref.
compounds with $R^1 \neq R^2$:				
9	CH_2CO_2H	C_2H_5	Id (70) stout amber rod-like crystals, m.p. 123 to 125°, 124 to 125° (C_6H_6/hexane)	[2]
10	CH_2CO_2H	$(CH_2)_2C_6H_5$	Ic (81) oil	[23, 24]
*11	CH_2CO_2H	$C(C_6H_5)_2OH$	Ic (71) m.p. 150 to 151° (C_2H_5OH/H_2O) IR (Nujol): 897, 930, 1005, 1105, 1709 (νC=O), 3521 (νO–H)	[8, 9]
*12	CH_2CO_2H	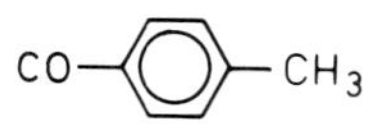	see "Further information", p. 264	[26]
*13	$(CH_2)_2CO_2H$	CH_3	see No. 16, p. 265	[18]
*14	$(CH_2)_3CO_2H$	CH_3	see "Chemical Behavior" of No. 22, p. 260 yellow solid, m.p. 69° (pentane)	[13, 14, 17]
*15	$(CH_2)_3CO_2H$	C_2H_5	see "Chemical Behavior" of No. 23, p. 260 m.p. 68 to 69° (pentane) IR ($CHCl_3$): 878, 1000, 1104	[10, 11, 21]
*16	$CH{=}CHCO_2H$	CH_3	see "Further information", p. 265	[18]
*17	$CH{=}CHCO_2H$	OH	see "Further information", p. 265 IR: 1660 (νC=O), 3400 (νO–H)	[25]
*18	$CH{=}CHCO_2H$	$(CH_2)_2CO_2H$	see "Further information", p. 265	[4]
*19a	$\overset{6}{C}H(C_6H_5)CH_2CO_2H$	CH_3	see "Chemical Behavior" of No. 24a, p. 260 m.p. 148° (C_2H_5OH/H_2O) ^1H NMR (CD_3COCD_3): 2.08 (s, CH_3), 2.75 (d, CH_2; J = 7.5), 3.78 (s, C_5H_5 + 1H of C_5H_3), 3.90 (m, 1H of C_5H_3), 4.00 (m, 1H of C_5H_3), 4.25 (t, CH), 7.35 (m, C_6H_5)	[22]

References on p. 266

*19b	$\overset{\smile}{C}H_C(C_6H_5)$–$CH_AH_B$–$CO_2H$ 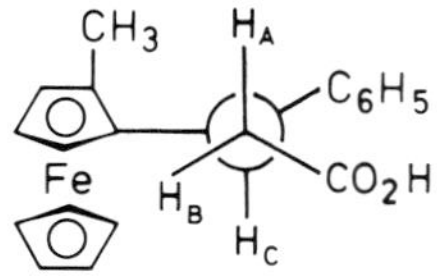	CH_3	see "Chemical Behavior" of No. 24b, p. 260 m.p. 188° (C_2H_5OH/H_2O) ^{1}H NMR (CD_3COCD_3): 1.75 (s, CH_3), 2.78 (dd, H_A; J(A,B) = 15), 3.38 (dd, H_B; J(B,C) = 3.5), 4.10 (s, C_5H_5 + 2H of C_5H_3), 4.18 (m, 1H of C_5H_3), 4.38 (dd, H_C; J(A,C) = 10.5), 7.12 (m, C_6H_5)	[22]
20a	$\overset{6}{C}H(C_6H_5)CH_2CO_2H$	$(CH_2)_2CO_2H$	see "Chemical Behavior" of No. 25a, p. 260 m.p. 188° (ether/hexane)	[16, 19, 20]
20b	$\overset{6}{C}H(C_6H_5)CH_2CO_2H$	$(CH_2)_2CO_2H$	see "Chemical Behavior" of No. 25b, p. 260 m.p. 185° (ether/hexane)	[16, 19, 20]
*21a	$\overset{6}{C}H(C_6H_5)CH_2CH_2CO_2H$	CH_3	IIIa	[22]
*21b	$\overset{6}{C}H(C_6H_5)CH_2CH_2CO_2H$	CH_3	IIIb	[22]
*22	$CO(CH_2)_2CO_2H$	CH_3	Ie, IVa (<10) orange solid, m.p. 138 to 144°, 146° (pentane/ether, 2/1)	[13, 14, 17]
23	$CO(CH_2)_2CO_2H$	C_2H_5	Ie, IVb m.p. 120° (hexane) ^{1}H NMR ($CDCl_3$): 1.17 (t, CH_3), 2.5 to 3.2 (m, CH_2 of C_2H_5 + $(CH_2)_2$), 4.19 (s, C_5H_5); 4.38 (m, H-3 or H-4), 4.48 (m, H-4 or H-3), and 4.7 (m, H-5) (C_5H_3) IR ($CHCl_3$): 912, 1001, 1106	[5, 6, 10 to 12]
24a	$\overset{6}{C}H(C_6H_5)CH(CO_2H)_2$	CH_3	If	[22]
24b	$\overset{6}{C}H(C_6H_5)CH(CO_2H)_2$	CH_3	Ig	[22]
25a	$\overset{6}{C}H(C_6H_5)CH(CO_2H)_2$	$CH_2CH(CO_2H)_2$	Ih yellow, fluffy solid (ether/hexane)	[20]
25b	$\overset{6}{C}H(C_6H_5)CH(CO_2H)_2$	$CH_2CH(CO_2H)_2$	Ii yellow solid (ether/hexane), m.p. 220° with decarboxylation	[20]

References on p. 266

* Further information:

$C_5H_5FeC_5H_3(CH_2CO_2H)_2$-1,2 (Table **17**, No. **1**). The attempt at Friedel-Crafts acylation of ferrocene with No. 1 failed [24].

$C_5H_5FeC_5H_3(CH_2CH_2CO_2H)_2$-1,2 (Table **17**, No. **2**) cyclizes to I with $(CF_3CO)_2O$ [15, 20].

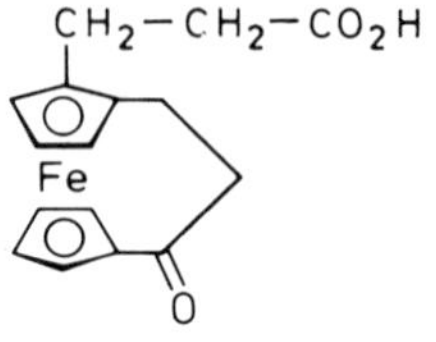

I

$C_5H_5FeC_5H_3(CH_2CHR'CO_2H)_2$-1,2 (Table **17**, Nos. **3** and **4** with $R' = CH_3$, C_6H_5). Systematic attempts, with use of numerous eluents, failed in each case to separate by thin layer chromatography the products after conversion into their ethyl esters. There was in no case any evidence for the presence of several diastereomers [20].

$C_5H_5FeC_5H_3(CH_2CO_2H$-1$)C(C_6H_5)_2OH$-2 (Table **17**, No. **11**) quite surprisingly yields no lactone under the usual conditions [9].

$C_5H_5FeC_5H_3(CH_2CO_2H$-1$)CO(C_6H_4CH_3$-4$)$-2 (Table **17**, No. **12**). The Ca salt of No. 12 is obtained in a 7/10/3 mixture with the Ca salts of its 1,1'- and 1,3-isomers by Friedel-Crafts acylation of $C_5H_5FeC_5H_4$–CH_2CO_2H with 4-$CH_3C_6H_4COCl/AlCl_3$ (mole ratio 1/1/1) in CH_2Cl_2, subsequent alkaline saponification, and purification of $FeC_{10}H_8(CH_2CO_2H)CO(C_6H_4CH_3$-4) via the Ca salts. Data for the mixture are: ^{1}H NMR spectrum (1/3 $OS(CD_3)_2/CDCl_3$), $\delta = 2.40$ (s, CH_3), 2.9 to 3.5 (m, 3.2H, $H_2O + CH_2CO_2$), 3.9 to 4.9 (m, $FeC_{10}H_8$), 7.21 (m, 2H), and 7.74 (m, 2H) (C_6H_4) ppm; IR (KBr), 1607 and 1634 (νC=O), 3428 (OH) cm^{-1}; electronic spectrum (CH_3OH), $\lambda_{max}(\varepsilon) = 256$ (30290), 281 sh (20925), 358 (3500), 484 (1700) nm. Studies of the biological activity of the mixture gave neither antiarthritic activity in a rat adjuvant arthritis model nor prostaglandin synthetase inhibitory activity in a human blood aggregation assay nor an effect on the human platelet serotonin release reaction [26].

$C_5H_5FeC_5H_3(CH_2CH_2CO_2H-1)CH_3$-2 (Table **17**, No. **13**). **Optically Active Forms.** (1S)-No. 13 (see Formula II) does represent a not very stable compound; it was therefore immediately cyclized (without measurement of the specific rotation) with $(CF_3CO)_2O$ to form (+)(1S)-III [18].

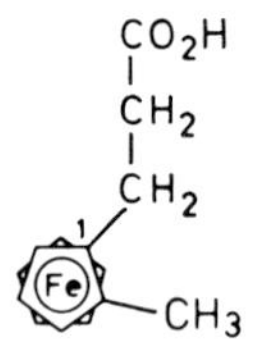

(1S)-No. 13
II

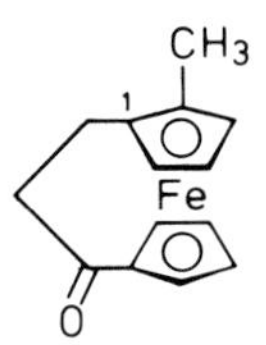

(+)(1S)-III

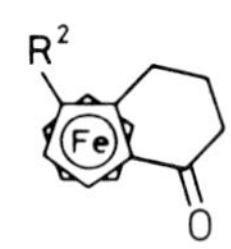

IV

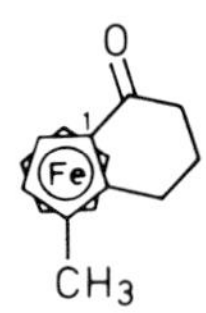

(+)(1S)-V

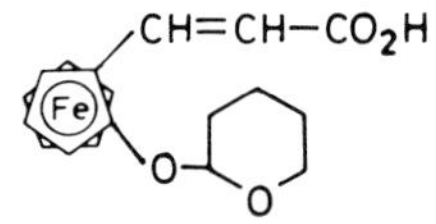

VI

$C_5H_5FeC_5H_3(CH_2CH_2CH_2CO_2H-1)CH_3$-2 (Table **17**, No. **14**) cyclizes to IV ($R^2 = CH_3$) under the action of $(CF_3CO)_2O$ [13, 14, 17].

Optically Active Forms. (1S)-$C_5H_5FeC_5H_3(CH_2CH_2CH_2CO_2H-1)CH_3$-2 is obtained from (−)(1S)-No. 22, see p. 265, as a rather oxidation-sensitive compound. It was therefore immediately (without measurement of the specific rotation) converted with $(CF_3CO)_2O$ to (+)(1S)-V, a chemical correlation which confirms the absolute configuration (1S) [14].

References on p. 266

$C_5H_5FeC_5H_3(CH_2CH_2CH_2CO_2H\text{-}1)C_2H_5\text{-}2$ (Table **17**, No. **15**) cyclizes with $(CF_3CO)_2O$ to IV ($R^2 = C_2H_5$) [10, 11, 21].

$C_5H_5FeC_5H_3(CH{=}CHCO_2H\text{-}1)CH_3\text{-}2$ (Table **17**, No. **16**). **Optically Active Forms.** (1S)-No. 16 is prepared from (+)(1S)-$C_5H_5FeC_5H_3(CHO\text{-}1)CH_3$-2 by Method II (p. 259) and is catalytically hydrogenated in C_2H_5OH to (1S)-No. 13 [18].

$C_5H_5FeC_5H_3(CH{=}CHCO_2H\text{-}1)OH\text{-}2$ (Table **17**, No. **17**) is obtained from VI via an HCl-catalyzed hydrolysis. It can be handled in air and is recrystallized without taking special precautions. This high stability compared to that of other hydroxyferrocene derivatives is probably due both to the electron withdrawing action of the CO_2H group and also to the formation of an intramolecular hydrogen bond between the two substituents (cf. IR spectrum in Table 17) [25].

$C_5H_5FeC_5H_3(CH{=}CHCO_2H\text{-}1)(CH_2)_2CO_2H\text{-}2$ (Table **17**, No. **18**). The product obtained by the Vilsmeier formylation of $C_5H_5FeC_5H_4\text{–}(CH_2)_2CO_2CH_3$, which probably consists of much 1,3- and little 1,2-isomer, is reacted following Method II (p. 259) and the residue remaining after distilling off of the pyridine in vacuum is taken up in $CHCl_3$. On subsequent extraction with 1 N NaOH saponification of the ester group occurs simultaneously. Concentrated HCl precipitates out of the alkaline solution in 65% yield the mixture "m" of the dicarboxylic acids, m.p. 147 to 150°C (CH_3OH/H_2O), that correspondingly consists of little No. 18 and much of its 1,3-isomer. Catalytic hydrogenation of the mixture in C_2H_5OH over Pd/C yields quantitatively little No. 2 mixed with much $C_5H_5FeC_5H_3(CH_2CH_2CO_2H)_2$-1,3 [4].

(1RS, 6SR)-$C_5H_5FeC_5H_3(\overset{6}{C}H(C_6H_5)CH_2CO_2H\text{-}1)CH_3\text{-}2$ (Table **17**, No. **19a**) gives the ketone VII (m.p. 228°C) on cyclization with $(CF_3CO)_2O$, along with small quantities of the ketone VIII [22].

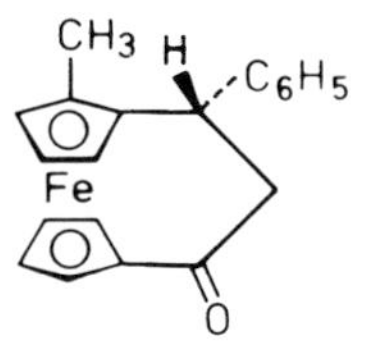

VII

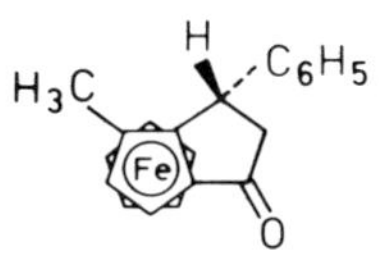

VIII

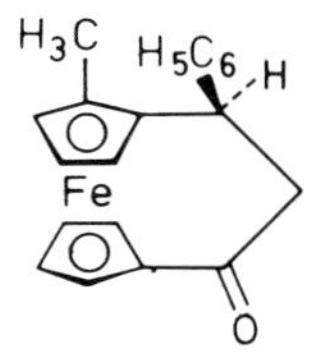

IX

(1RS 6RS)-$C_5H_5FeC_5H_3(\overset{6}{C}H(C_6H_5)CH_2CO_2H\text{-}1)CH_3\text{-}2$ (Table **17**, No. **19b**) in contrast yields with $(CF_3CO)_2O$ only one ketone, namely IX (m.p. 114°C), diasterereoisomeric to VII [22].

(1RS, 6SR)- and (1RS, 6RS)-$C_5H_5FeC_5H_3(\overset{6}{C}H(C_6H_5)(CH_2)_2CO_2H\text{-}1)CH_3\text{-}2$ (Table **17**, Nos. **21a** and **21b**) are cyclized with $(CF_3CO)_2O$ to the diastereomeric ketones X (m.p. 163°C) and XI (m.p. 146°C), respectively [22].

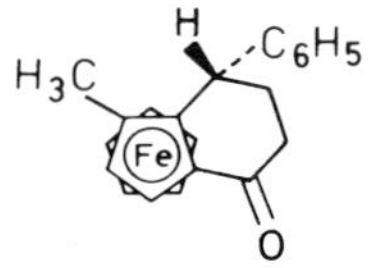

X

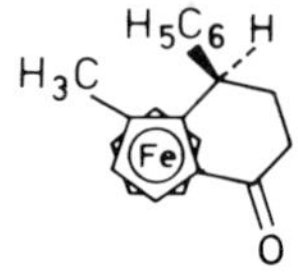

XI

$C_5H_5FeC_5H_3(CO(CH_2)_2CO_2H\text{-}1)CH_3\text{-}2$ (Table **17**, No. **22**). **Optically Active Forms.** The (–)(1S)-$C_5H_5FeC_5H_3(CO(CH_2)_2CO_2H\text{-}1)CH_3$-2 precipitates in 57% yield on refluxing with KOH in C_2H_5OH/H_2O (1S)-$C_5H_5FeC_5H_3(CO(CH_2)_2CN\text{-}1)CH_3$-2. It melts at 133 to 139°C and has a specific rotation $[\alpha]_D^{20} = -432°$ (C_6H_6, c = 0.94, p = 100%). It is reduced by the Clemmensen reaction with amalgamated Zn and concentrated HCl in C_6H_6/CH_3OH to a mixture of (1S)-No. 14 and its methyl ester, which after saponification yields the free acid in a 47% yield [14].

References on p. 266

References:

[1] V. Weinmayr (J. Am. Chem. Soc. **77** [1955] 3012/4). – [2] D. Lednicer, C. R. Hauser (J. Org. Chem. **24** [1959] 43/6). – [3] A. A. Berlin, T. V. Kostroma (U.S.S.R. 129 018 [1959/60]; C.A. **1961** 3612). – [4] K. Schlögl, M. Peterlik, H. Seiler (Monatsh. Chem. **93** [1962] 1309/27). – [5] J.-P. Monin, G. Tainturier, R. Dabard, J. Tirouflet (Bull. Soc. Chim. France **1963** 667/8).

[6] J. Tirouflet, J.-P. Monin, G. Tainturier, R. Dabard (Compt. Rend. **256** [1963] 433/5). – [7] A. A. Berlin, T. V. Kostroma (Vysokomol. Soedin. Geterotsepnye Vysokomol. Soedin. **1964** 53/8; C.A. **61** [1964] 9591). – [8] D. W. Slocum, B. W. Rockett, C. R. Hauser (Chem. Ind. [London] **1964** 1831/2). – [9] D. W. Slocum, B. W. Rockett, C. R. Hauser (J. Am. Chem. Soc. **87** [1965] 1241/6). – [10] J. Tirouflet, G. Tainturier (Tetrahedron Letters **1965** 4177/9).

[11] G. Tainturier, J. Tirouflet (Bull. Soc. Chim. France **1966** 595/600). – [12] J. Tirouflet, G. Tainturier, H. Singer (Bull. Soc. Chim. France **1966** 2565/70). – [13] R. Dabard, P. Dixneuf (Compt. Rend. C **265** [1967] 324/7). – [14] G. Haller, K. Schlögl (Monatsh. Chem. **98** [1967] 2044/57). – [15] C. Moïse, J. Tirouflet (Compt. Rend. C **265** [1967] 457/9).

[16] C. Moïse, J. Tirouflet (Compt. Rend. C **267** [1968] 414/6). – [17] P. Dixneuf, R. Dabard (Bull. Soc. Chim. France **1969** 2164/71). – [18] H. Falk, O. Hofer, K. Schlögl (Monatsh. Chem. **100** [1969] 624/48). – [19] C. Moïse, J. Tirouflet (Proc. 4th Intern. Conf. Organometal. Chem., Bristol 1969, p. G3). – [20] C. Moïse, J. Tirouflet (Bull. Soc. Chim. France **1970** 2656/65).

[21] J. Tirouflet (Colloq. Intern. Centre Natl. Rech. Sci. [Paris] No. 191 [1970] 267/78). – [22] C. Moïse, Y. Mugnier, J. Tirouflet (J. Organometal. Chem. **51** [1973] 343/52). – [23] M. Benedikt, K. Schlögl (8th Intern. Conf. Organometal. Chem., Kyoto 1977, p. 23). – [24] M. Benedikt, K. Schlögl (Monatsh. Chem. **109** [1978] 805/22). – [25] R. Epton, G. Marr, G. K. Rogers (J. Organometal. Chem. **150** [1978] 93/100).

[26] B. E. Maryanoff, S. L. Keeley, F. J. Persico (J. Med. Chem. **26** [1983] 226/9).

5.2.3.6.3 1,3-Compounds

5.2.3.6.3.1 Carboxylic Acids of the Type $C_5H_5FeC_5H_3(CO_2H\text{-}1)R\text{-}3$

Of the acids collected in Table 18, Nos. 1 to 4 with two different substituents on one cyclopentadienyl ring possess a plane of chirality (for the centrochiral R,S nomenclature used for the ferrocene chirality see "Ferrocene" 7, 1980, pp. 4/7 [German], pp. 1/4 [English translation]). Nevertheless, optically active forms have been prepared only for No. 1 (see "Further information", p. 268). In contrast to Nos. 1 to 4, No. 5, $C_5H_5FeC_5H_3(CO_2H)_2\text{-}1,3$, is achiral.

The compounds are prepared by the following methods:

Method I. Alkaline saponification at the boiling point:

a) Of the isomeric mixture $FeC_{10}H_8(CN)R$ with 19% KOH in n-C_4H_9OH [5] or with NaOH in C_2H_5OH/H_2O [1, 2]. The yields quoted refer to the starting isomer mixture.
Starting with the product obtained by the cyaniding of $C_5H_5FeC_5H_4\text{–}C_2H_5$ in tetrahydrofuran with liquid HCN in the presence of $FeCl_3$ yields predominantly No. 2 mixed with its 1,2-isomer [1, 2]. No separation was conducted, the data refer to the mixture (m).

b) Of $C_5H_5FeC_5H_3(CONH_2\text{-}1)R\text{-}3$ with 15% KOH in n-C_4H_9OH [3].

Method II. Cleavage of $C_5H_5FeC_5H_3(CO(C_6H_4Cl\text{-}2)\text{-}1)R\text{-}3$ with $KOC(CH_3)_3$ and traces of H_2O in boiling $CH_3OCH_2CH_2OCH_3$ (10 to 20 min for $R=CH_3$, longer for $R=C_6H_5$). In the latter case the reaction was started with $C_5H_5FeC_5H_3(CO(C_6H_4Cl\text{-}2)\text{-}1)C_6H_5\text{-}3$ mixed with its 1,1'-isomer and correspondingly a mixture of No. 4 with $HO_2C\text{–}C_5H_4FeC_5H_4\text{–}C_6H_5$ obtained; no separation was attempted [15].

References on p. 269

Method III. Oxidation:

a) Of the isomer mixture $FeC_{10}H_8(CHO)CH_3$ with AgO in KOH/tetrahydrofuran/H_2O; the yield of $FeC_{10}H_8(CO_2H)CH_3$ amounts to 20%. Steric effects shift the ratio of the 1,3-, 1,1′-, and 1,2-isomers from 44/41/15 in the aldehyde mixture in favor of No. 1 to 58/36/6 in the acid mixture [11]. Isomer separation is covered under "Further information" for No. 1, see below.

b) Of $C_5H_5FeC_5H_3(COCH_3)_2$-1,3 in dioxane with NaOBr solution at 0 to 5°C [17].

Table 18

Carboxylic Acids of the Type $C_5H_5FeC_5H_3(CO_2H\text{-}1)R\text{-}3$.
Further information for numbers preceded by an asterisk is given at the end of the table, pp. 267/8.
For abbreviations and dimensions see p. VIII.
R_f values are for descending paper chromatography with n-C_4H_9OH/C_2H_5OH/conc.NH_4OH/H_2O in the volume ratio 4/4/1/1 as mobile phase.

No.	$C_5H_5FeC_5H_3(CO_2H\text{-}1)R\text{-}3$ R	method of preparation (yield in %), properties, and remarks	Ref.
*1	CH_3	Ia (33), Ib (≈10), II (85 to 95), IIIa m.p. 159 to 161° (CH_3OH/H_2O) ^{1}H NMR(CCl_4): 2.03 (s, CH_3) R_f: 0.60	[3, 5, 8, 11, 15]
2	C_2H_5	Ia (61m) m.p. 113 to 114° (cyclohexane) (m)	[1, 2]
3	$CH(CH_3)_2$	Ib (10) R_f: 0.69	[3]
4	C_6H_5	II	[15]
*5	CO_2H	IIIb (50) orange-red crystals, m.p. 228° dec. IR (KBr): 1000, 1100, 1675 (νC=O), 3100 mass spectrum: m/e = 274 $[M]^+$	[17]

* Further information:

$C_5H_5FeC_5H_3(CO_2H\text{-}1)CH_3\text{-}3$ (Table **18**, No. **1**). The positional isomers are separated by esterifying the mixture of $FeC_{10}H_8(CO_2H)CH_3$ with CH_2N_2 and then chromatographing it in ca. 5/1 petroleum ether/benzene over the total length of an Al_2O_3 column with subdivision into 20 to 25 fractions (elution sequence 1,2-, 1,1′-, and 1,3-isomers); however, a sharp separation was not achieved. Nevertheless, a distinct shift of the fraction of the total yield toward the 1,3-isomer was observed, which is attributed to the easier saponification of the 1,2- and 1,1′-isomers and the consequent stronger absorptivity on the alkaline column. Of the inserted quantity of ester, 20% remains irreversibly adsorbed. Subsequent saponification with KOH in CH_3OH yields the acids. A separation via the 4-$BrC_6H_4CH_2$ ester proceeds similarly and offers no advantages [5].

The NMR signal of the CH_3 protons in CCl_4 is shifted by 0.08 ppm toward lower field versus that of $C_5H_5FeC_5H_4$–CH_3 [8].

References on p. 269

CH_2N_2 in ether [3, 5] or ether/methanol [15] forms the methyl ester [3, 5, 15], $LiAlH_4/AlCl_3$ the $C_5H_5FeC_5H_3(CH_3)_2$-1,3 [3]. Cleavage of the racemate is treated in the following paragraph.

Optically Active Forms. (+)(1S)-$C_5H_5FeC_5H_3(CO_2H$-1)CH_3-3 (see Formula I) is won from the racemate by splitting with (−)-$C_6H_5CH(CH_3)NH_2$; on its addition the (+)-$C_5H_5FeC_5H_3(CO_2NH_3CH(CH_3)C_6H_5$-1)$CH_3$-3 precipitates from ether solution at 0°C. This yields (+)-No. 1 after multiple recrystallization from ethanol and then suspension in ether and repeated shaking with dilute H_3PO_4 [6, 16]. The kinetic racemate resolution of $(C_5H_5FeC_5H_3(CH_3$-3)C(O)-1$)_2O$ with (−)-$C_6H_5CH(CH_3)NH_2$ in absolute pyridine at 0°C liberates (+)-No. 1 in an optical yield of 0.9% [9].

(+)-No. 1 melts at 146 to 149°C [6, 9] and exhibits the following specific rotation values at 20°C and p = 100%: In C_2H_5OH $[\alpha]_{589} = +33° \pm 2°$ (c = 0.3 to 1.0) [9], +33.0° (c = 0.7 to 1.2) [6], 33.0° ± (0.2 to 0.9)° (c = 0.1 to 0.6); $[\alpha]_{570} = +46.3° \pm (0.2$ to $0.9)°$ (c = 0.1 to 0.6); $[\alpha]_{546} = +53.5° \pm (0.2$ to $0.9)°$ (c = 0.1 to 0.6) [16]. In benzene (c = 0.1 to 0.6; ± (0.2 to 0.9)° error limits), $[\alpha]_{589} = +37.0°$, $[\alpha]_{570} = +39.4°$, $[\alpha]_{546} = +60.0°$ [16]. Electronic spectrum (C_2H_5OH): $\lambda_{max}(\varepsilon)$ = 439 (229), 346sh (285), 303 (696), 262 (3940) nm [6]. Optical rotatory dispersion (C_2H_5OH, c = 0.1, 20°C): $[M]_{589} = +81°$ [16], $[M]_{502} = +252°$, $[M]_{462} = +65°$, $[M]_{445} = +130°$, $[M]_{377} = -46°$ (see the curve in the paper) [6]; in benzene (c = 0.1 to 0.6), $[M]^{20}_{589} = +91°$ [16]. Circular dichroism (C_2H_5OH, c ≈ 0.1): $\Delta\varepsilon_{535} = -0.01$, $\Delta\varepsilon_{478} = +0.07$, $\Delta\varepsilon_{461} = +0.06$, $\Delta\varepsilon_{415} = +0.16$ [16]. An empirical relationship between the rotatory and dipole strengths of the ferrocene bands indicates an electrically and magnetically dipole-forbidden transition [12].

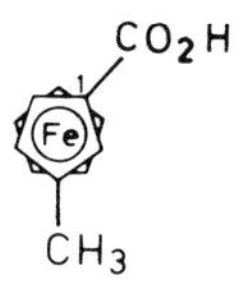

(+)(1S)-No. 1
I

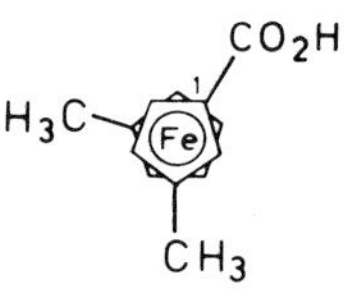

(+)(1S)-II

The absolute configuration of (1S) derived by chemical correlations and optical comparison is confirmed by conversion of the compound into (+)(1S)-II for whose enantiomer the absolute configuration has been determined by [4] by X-rays [7, 10, 13].

CH_2N_2 esterifies the acid to (+)(1S)-$C_5H_5FeC_5H_3(CO_2CH_3$-1)CH_3-3 [6, 13, 16]. PCl_3 in C_6H_6 gives the acid chloride [16], which forms (+)(1S)-$C_5H_5FeC_5H_3(COCH_3$-1)CH_3-3 on further reaction with $Cd(CH_3)_2$ [6].

The anion $[C_5H_5FeC_5H_3(CO_2$-1)CH_3-3$]^-$ exhibits the following values in 0.1 N ethanolic $NaOC_2H_5$ solution at 20°C: specific rotation (c = 0.1 to 0.6; ± (0.2 to ± 0.9)° error limits) $[\alpha]_{589} = +19.8°$, $[\alpha]_{546} = +33.7°$; $[M]_{589} = +49°$; circular dichroism (c ≈ 0.1) $\Delta\varepsilon_{444} = +0.15$, $\Delta\varepsilon_{420} = +0.13$ [16].

$C_5H_5FeC_5H_3(CO_2H)_2$-1,3 (Table **18**, No. **5**). The following chemical shifts have been calculated for the protons of the substituted cyclopentadienyl ring for a solution in acetone: δ = 5.05 (d, 2H; J = 1 to 1.5 Hz), 5.40 (t, 1H) ppm [14].

Refluxing a suspension of No. 5 in CH_3OH after addition of concentrated H_2SO_4 forms $C_5H_5FeC_5H_3(CO_2CH_3)_2$-1,3 in 90% yield after 28 h [17].

References on p. 269

References:

[1] A. N. Nesmejanow [Nesmeyanov], E. G. Perewalowa [Perevalova], L. P. Jurjewa [Yur'eva] (Chem. Ber. **93** [1960] 2729/35). – [2] E. G. Perevalova, L. P. Yur'eva, Yu. I. Baukov (Dokl. Akad. Nauk SSSR **135** [1960] 1402/5; Proc. Acad. Sci. USSR Chem. Sect. **130/135** [1960] 1455/8). – [3] K. Schlögl, M. Fried (Monatsh. Chem. **95** [1964] 558/75). – [4] O. L. Carter, A. T. McPhail, G. A. Sim (J. Chem. Soc. A **1967** 365/73). – [5] H. Falk, G. Haller, K. Schlögl (Monatsh. Chem. **98** [1967] 592/602).

[6] G. Haller, K. Schlögl (Monatsh. Chem. **98** [1967] 603/18). – [7] G. Haller, K. Schlögl (Monatsh. Chem. **98** [1967] 2044/57). – [8] K. Schlögl, H. Falk, G. Haller (Monatsh. Chem. **98** [1967] 82/94). – [9] H. Falk, K. Schlögl (Monatsh. Chem. **99** [1968] 578/87). – [10] H. Falk, O. Hofer, K. Schlögl (Monatsh. Chem. **100** [1969] 624/48).

[11] K. Schlögl, M. Walser (Monatsh. Chem. **100** [1969] 840/3). – [12] H. Falk, H. Lehner (Tetrahedron **27** [1971] 2279/91). – [13] H. Falk, K. Schlögl (Monatsh. Chem. **102** [1971] 33/6). – [14] P. V. Roling, M. D. Rausch (J. Org. Chem. **39** [1974] 1420/4). – [15] V. Rapić, K. Schlögl, B. Steinitz (J. Organometal. Chem. **94** [1975] 87/98).

[16] V. Rapić, K. Schlögl, B. Steinitz (Monatsh. Chem. **108** [1977] 767/80). – [17] A. Kasahara, T. Izumi, Y. Yoshida, I. Shimizu (Bull. Chem. Soc. Japan **55** [1982] 1901/6).

5.2.3.6.3.2 Carboxylic Acids of the Type $C_5H_5FeC_5H_3(A\text{–}CO_2H\text{-}1)R^2\text{-}3$ ($A\text{–}CO_2H = R^1$)

Of the compounds compiled in Table 19, all of those with $R^1 \neq R^2$ (Nos. 4 to 13) possess a chirality plane (for the centrochiral R,S nomenclature used for the ferrocene chirality see "Ferrocene", 1980, pp. 4/7 [German], pp. 1/4 [English translation]). Optically active forms have been prepared of the compounds No. 5, 6, 8, and 11 (see "Further information", pp. 272/3).

The product $FeC_{10}H_8(C_6H_4\text{–}CO_2H\text{-}2)_2$ obtained in dilute H_2SO_4 from the ferrocenium ion with the diazonium ion $[2\text{-}HO_2CC_6H_4N_2]^+$ appears to consist completely of the 1,1′-isomer (see 5.2.3.6.1.4, Table 15, No. 7, p. 245).

For preparation the following methods are used:

Method I. Knoevenagel condensation of $C_5H_5FeC_5H_3(CHO\text{-}1)R^2\text{-}3$ with malonic acid (molar ratio 1/2) in pyridine and small quantities of piperidine with heating on the water bath [3].
No. 9 is prepared from the product, probably a mixture of much 1,3- and a little 1,2-isomer, of the Vilsmeier reaction of $C_5H_5FeC_5H_4\text{–}(CH_2)_2CO_2CH_3$ with $C_6H_5N(CH_3)CHO$ and $POCl_3$ in CH_2Cl_2. The pyridine is distilled off and the residue is taken up in $CHCl_3$. On subsequent extraction of the solution with 1 N NaOH simultaneous saponification of the ester group takes place. Concentrated HCl precipitates No. 9 from the alkaline solution [3].

Method II. Alkaline saponification in C_2H_5OH:

a) Of I with 6 N aqueous NaOH solution at room temperature in the dark, for 24 h [15].

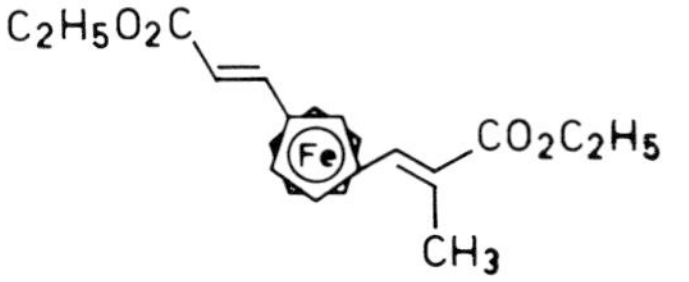

I

b) Of $C_5H_5FeC_5H_3(CO(CH_2)_2CO_2CH_3\text{-}1)R^2\text{-}3$, cold, with excess KOH [10, 12], mole ratio 1/2 [7, 8].

References on pp. 273/4

Method III. Friedel-Crafts succinoylation with succinic acid anhydride and $AlCl_3$:

a) Of $C_5H_5FeC_5H_4$–$(CH_2)_3CO_2CH_3$; hydrogenolysis and saponification subsequently follow [17].

b) Of $C_5H_5FeC_5H_4$–C_2H_5 in CH_2Cl_2. The resulting mixture of the three positionally isomeric keto acids is immediately esterified with CH_2N_2 and separated into three main fractions by chromatography on an Al_2O_3 column with benzene/petroleum ether. After saponification these fractions are purified by numerous fractional crystallizations [4, 5].

Method IV. Condensation of $C_5H_5FeC_5H_3(CHO\text{-}1)CH_3\text{-}3$ with pyruvic acid in CH_3OH with KOH at 30°C. The K salt precipitated on cooling is dissolved in water and No. 13 liberated with 20% H_3PO_4 with simultaneous extraction with ether [10, 12].

Method V. Claisen condensation of $C_5H_5FeC_5H_3(CO\text{–}CH_3)_2\text{-}1,3$ with $(C_2H_5O)_2CO$/NaH; hydrogenolysis and saponification subsequently follow [17].

Purification of the compounds is effected by recrystallization from the solvents given in parentheses in the table as well as by precipitation after dissolution in alkalies by acids with shaking in ether.

Chemical Behavior. The CO groups of the keto acids are converted by the Clemmensen reaction into CH_2 groups. Amalgamated Zn and concentrated HCl at the boiling points reduce No. 12 to No. 7 in 4/1 CH_3CO_2H/H_2O [7, 8, 14], No. 11 to No. 6 (yield 52 to 57%) in toluene [10, 12] (see "Further information" for the behavior of (+)-No. 11, p. 273), and also No. 13 to No. 6 in toluene, with simultaneous hydrogenation of the double bond, in a 55% yield (relative to the starting aldehyde for No. 13) [12].

Catalytic hydrogenation of the $CH{=}CHCO_2H$ group in C_2H_5OH over Pd/C [2, 3] or PtO_2 [13] yields the $CH_2CH_2CO_2H$ group. No. 1 is prepared nearly quantitatively from No. 9 by the first procedure [2, 3], for the latter see "Further information" for No. 8, p. 273.

The acids with the $(CH_2)_2CO_2H$ and $(CH_2)_3CO_2H$ groups are convertible with $(CF_3CO)_2O$ in CH_2Cl_2 into the ketones with intramolecular ring closure. This takes place for steric reasons preferentially heteroannularly for $(CH_2)_2CO_2H$, homoannularly for $(CH_2)_3CO_2H$. Details are given under "Further information" for Nos. 1 and 5 to 7, pp. 272/3.

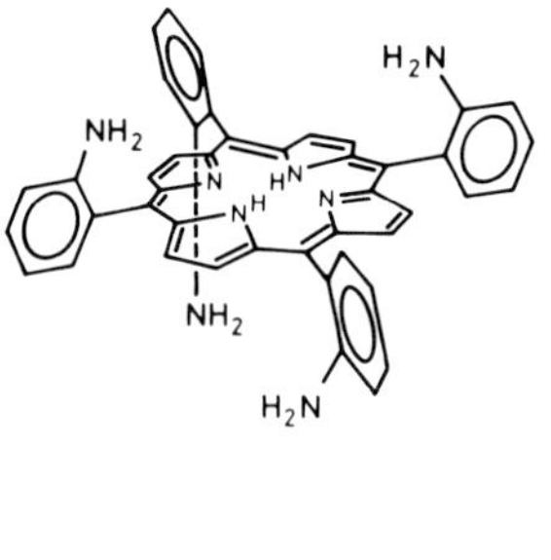

II

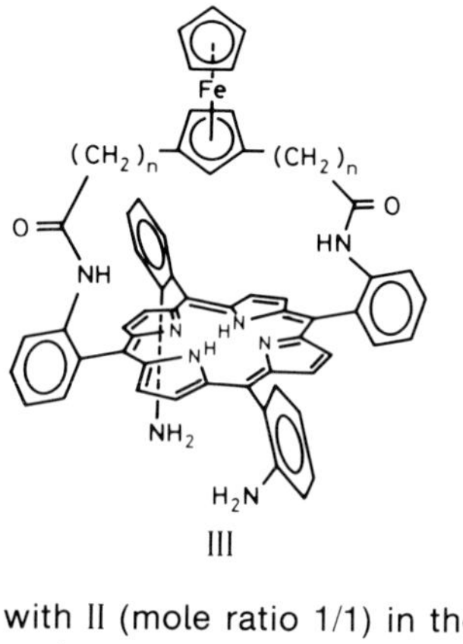

III

Refluxing the compounds No. 1 or 2 in CH_2Cl_2 with II (mole ratio 1/1) in the presence of $[2\text{-}ClC_5H_4NCH_3]I$ (C_5H_5N = pyridine) and $(n\text{-}C_4H_9)_3N$ under conditions of high dilution leads to the double layered ferrocenoporphyrinophanes III with n = 2 (yield 4.7%) and 3 (yield 7.2%), respectively. Dibridged phanes have not yet been isolated [17].

Table 19

Carboxylic Acids of the Type $C_5H_5FeC_5H_3(A–CO_2H\text{-}1)R^2\text{-}3$ ($A–CO_2H = R^1$).

Further information for numbers preceded by an asterisk is given at the end of the table, pp. 272/3.

For abbreviations and dimensions see p. VIII.

R_f values are for descending paper chromatography with n-C_4H_9OH/C_2H_5OH/conc. NH_4OH/H_2O in the volume ratio 4/4/1/1 as mobile phase.

No.	$C_5H_5FeC_5H_3(R^1\text{-}1)R^2\text{-}3$ $R^1 = A–CO_2H$	R^2	method of preparation (yield in %), properties, and remarks	Ref.
compounds with $R^1 = R^2$:				
*1	$(CH_2)_2CO_2H$		V, see "Chemical Behavior" for No. 9, p. 270 m.p. 113 to 114° R_f: 0.38	[2, 3, 17]
2	$(CH_2)_3CO_2H$		IIIa m.p. 99 to 101°	[17]
3	$C_6H_3(OH\text{-}3)CO_2H\text{-}4$		minor constituent in mixture of positional isomers consisting mostly of the 1,1′-isomer; for details see 5.2.3.6.1.4 (Table 15, No. 8, p. 246)	[1, 6]
compounds with $R^1 \neq R^2$:				
*4	CH_2CO_2H	$COC_6H_4CH_3\text{-}4$	see "Further information", p. 272	[16]
*5	$(CH_2)_2CO_2H$	CH_3	see No. 8, p. 273	[13]
*6	$(CH_2)_3CO_2H$	CH_3	see "Chemical Behavior" for Nos. 11 and 13, p. 270 yellow solid, m.p. 67 to 68° (pentane)	[10, 12]
*7	$(CH_2)_3CO_2H$	C_2H_5	see "Chemical Behavior" for No. 12, p. 270 oil IR ($CHCl_3$): 906(?), 915, 1000, 1104	[7, 8]
*8	$CH{=}CHCO_2H$	CH_3	see "Further information", p. 273	[13]
9	$CH{=}CHCO_2H$	$(CH_2)_2CO_2H$	I (65) m.p. 147 to 150° (CH_3OH/H_2O) R_f: 0.35	[2, 3]
*10	$CH{=}C(CH_3)CO_2H$-(E)	$CH{=}CHCO_2H$-(E)	IIa (84) red crystals, m.p. 217.5 to 218° (CH_2Cl_2) ^1H NMR ($CDCl_3$): 2.08 (s, CH_3), 4.19 (s, C_5H_5), 4.77 (s, 2H of C_5H_3), 4.90 (s, 1H of C_5H_3), 6.14 (d, 1H; J=15), 7.61 (s, 1H), 7.72 (d, 1H; J–15) IR (KBr): 1622, 1678 mass spectrum (CI): m/e = 341	[15]

References on pp. 273/4

Table 19 (continued)

No.	$C_5H_5FeC_5H_3(R^1\text{-}1)R^2\text{-}3$ $R^1 = A\text{–}CO_2H$	R^2	method of preparation (yield in %), properties, and remarks	Ref.
*11	$CO(CH_2)_2CO_2H$	CH_3	IIb orange solid, m.p. 131° (pentane/ether, 2/1)	[10, 12]
12	$CO(CH_2)_2CO_2H$	C_2H_5	IIb, IIIb m.p. 106 to 108°, 108 to 110° (hexane) ^1H NMR ($CDCl_3$): 1.2 (t, CH_3), 2.42 (q, CH_2 of C_2H_5), 2.8 to 3.3 (m, $(CH_2)_2$), 4.29 (s, C_5H_5); 4.57 (m, H-4) and 4.88 (m-H-2,5) (C_5H_3) IR ($CHCl_3$): 888, 1002, 1106	[4, 5, 7 to 9, 14]
13	$CH{=}CHCOCO_2H$	CH_3	IV violet paste (unpurified product)	[10, 12]

* Further information:

$C_5H_5FeC_5H_3(CH_2CH_2CO_2H)_2$-1,3 (Table **19**, No. **1**) cyclizes to IV under the action of $(CF_3CO)_2O$ [2, 3].

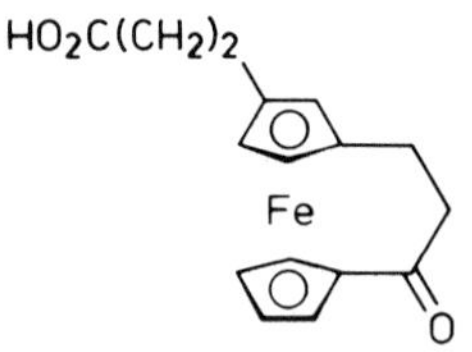

IV

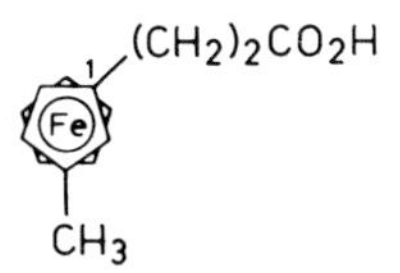

(1S)-No. 5

V

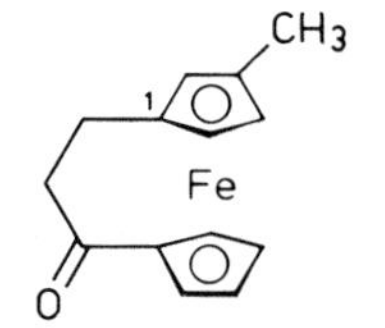

(−)(1S)-VI

$C_5H_5FeC_5H_3(CH_2CO_2H\text{-}1)CO(C_6H_4CH_3\text{-}4)$-3 (Table **19**, No. **4**). Friedel-Crafts acylation of $C_5H_5FeC_5H_4\text{–}CH_2CO_2CH_3$ with $4\text{-}CH_3C_6H_4COCl/AlCl_3$ (molar ratio 1/1/1) in CH_2Cl_2, subsequent alkaline saponification, and purification of $FeC_{10}H_8(CH_2CO_2H)CO(C_6H_4CH_3\text{-}4)$ via the Ca salts yield the Ca salt of No. 4 mixed with its 1,1′- and 1,2-isomers in a 3/10/7 ratio. Data for this mixture: ^1H NMR spectrum (1/3 $OS(CD_3)_2/CDCl_3$), $\delta = 2.40$ (s, CH_3), 2.9 to 3.5 (m, 3.2H, $H_2O + CH_2CO_2$), 3.9 to 4.9 (m, $FeC_{10}H_8$), 7.21 (m, 2H) and 7.74 (m, 2H) (C_6H_4) ppm. IR (KBr), 1607 and 1634 (νC=O), 3428 (νOH) cm^{-1}. Electronic spectrum (CH_3OH), $\lambda_{max}(\varepsilon) = 256\,(30290)$, 281sh (20925), 358 (3500), 484 (1700) nm. Investigations of the biological activity of the mixture gave neither antiarthritic activity in a rat adjuvant arthritis model nor prostaglandin synthetase inhibitory activity in a human blood aggregation assay nor an effect on the human platelet serotonin release reaction [16].

$C_5H_5FeC_5H_3(CH_2CH_2CO_2H\text{-}1)CH_3$-3 (Table **19**, No. **5**). **Optically Active Forms.** (1S)-No. 5 (see Formula V) is not very stable and is therefore immediately (without measurement of the specific rotation value) converted with $(CF_3CO_2)_2O$ into the (−)(1S)-VI [13].

$C_5H_5FeC_5H_3(CH_2CH_2CH_2CO_2H\text{-}1)CH_3$-3 (Table **19**, No. **6**) yields on ring closure with $(CF_3CO)_2O$ the products VII and VIII with $R^2 = CH_3$ [12].

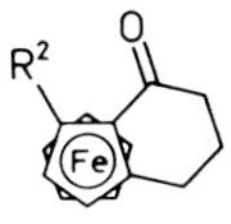

VII

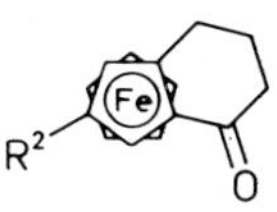

VIII

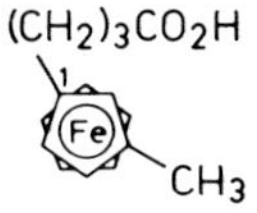

(1S)-No. 6

IX

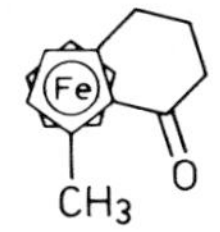

(−)-X

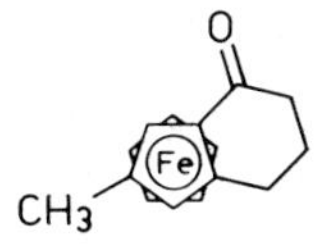

(+)-XI

Optically Active Forms. (1S)-$C_5H_5FeC_5H_3(CH_2CH_2CH_2CO_2H\text{-}1)CH_3\text{-}3$ (see Formula IX) is prepared from (+)-No. 11 (see below) and with regard to its sensitivity is immediately (without measurement of the specific rotation value) converted with $(CF_3CO_2)_2O$. This forms a mixture of the ketones (−)-X in 51% yield and of (+)-XI in 30.6% yield, which is separable by preparative thin layer chromatography with 15/1 C_6H_6/C_2H_5OH [11].

$C_5H_5FeC_5H_3(CH_2CH_2CH_2CO_2H\text{-}1)C_2H_5\text{-}3$ (Table **19**, No. **7**) cyclizes with $(CF_3CO)_2O$ to the mixture of VII and VIII with $R^2 = C_2H_5$, which is separable by chromatography on an Al_2O_3 column with ether [7, 8, 14].

$C_5H_5FeC_5H_3(CH{=}CHCO_2H\text{-}1)CH_3\text{-}3$ (Table **19**, No. **8**). **Optically Active Forms.** (1S)-No. 8 is prepared by Method I from (−)(1S)-XII and is hydrogenated in C_2H_5OH over PtO_2 catalytically to form (1S)-No. 5 [13].

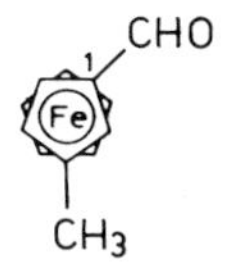

(−)(1S)-XII

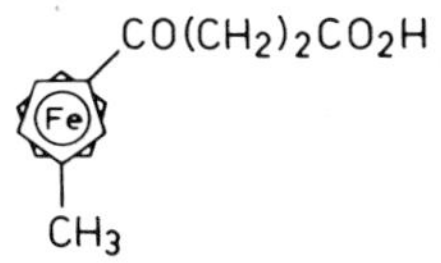

(+)(1S)-No. 11

XIII

(E,E)-$C_5H_5FeC_5H_3(CH{=}C(CH_3)CO_2H\text{-}1)CH{=}CHCO_2H\text{-}3$ (Table **19**, No. **10**) reacts with 4-$O_2NC_6H_4OH$ in CH_2Cl_2 under the action of dicyclohexylcarbodiimide and small quantities of 4-dimethylaminopyridine to form the di(4-nitrophenyl) ester [15].

$C_5H_5FeC_5H_3(CO(CH_2)_2CO_2H\text{-}1)CH_3\text{-}3$ (Table **19**, No. **11**) can be separated from its 1,1′-isomer by fractional crystallization from 2/1 pentane/ether [10, 12].

Optically Active Forms. (+)(1S)-$C_5H_5FeC_5H_3(CO(CH_2)_2CO_2H\text{-}1)CH_3\text{-}3$ (see Formula XIII) is obtained from (+)-$C_5H_5FeC_5H_3(CO(CH_2)_2CN\text{-}1)CH_3\text{-}3$ by refluxing with 6/1 C_2H_5OH/H_2O, in 57% yield as an oil. Specific rotation $[\alpha]_D^{20} = +29.6°$ (C_6H_6, c = 0.5 for p = 100%). Its reduction at the boiling point with amalgamated Zn and 1/1 conc. HCl/H_2O in 1/1 C_6H_6/CH_3OH gives (1S)-No. 6 in an 80% yield. The reaction goes considerably more slowly than that of the corresponding 1,2-isomer [11].

References:

[1] A. A. Berlin, T. V. Kostroma (U.S.S.R. 129018 [1959/60]; C.A. **1961** 3612). – [2] K. Schlögl, H. Seiler (Tetrahedron Letters **1960** No. 7, p. 4/8). – [3] K. Schlögl, M. Peterlik, H. Seiler (Monatsh. Chem. **93** [1962] 1309/27). – [4] J.-P. Monin, G. Tainturier, R. Dabard, J. Tirouflet (Bull. Soc. Chim. France **1963** 667/8). – [5] J. Tirouflet, J.-P. Monin, G. Tainturier, R. Dabard (Compt. Rend. **256** [1963] 433/5).

[6] A. A. Berlin, T. V. Kostroma (Vysokomol. Soedin. Geterotsepnye Soedin. **1964** 53/8; C.A. **61** [1964] 9591). – [7] J. Tirouflet, G. Tainturier (Tetrahedron Letters **1965** 4177/9). – [8] G. Tain-

turier, J. Tirouflet (Bull. Soc. Chim. France **1966** 595/600). – [9] J. Tirouflet, G. Tainturier, H. Singer (Bull. Soc. Chim. France **1966** 2565/70). – [10] R. Dabard, P. Dixneuf (Compt. Rend. C **265** [1967] 324/7).

[11] G. Haller, K. Schlögl (Monatsh. Chem. **98** [1967] 2044/57). – [12] P. Dixneuf, R. Dabard (Bull. Soc. Chim. France **1969** 2164/71). – [13] H. Falk, O. Hofer, K. Schlögl (Monatsh. Chem. **100** [1969] 624/48). – [14] J. Tirouflet (Colloq. Intern. Centre Natl. Rech. Sci. [Paris] No. 191 [1970] 267/78). – [15] R. Breslow, G. Trainor, A. Ueno (J. Am. Chem. Soc. **105** [1983] 2739/44).

[16] B. E. Maryanoff, S. L. Keeley, F. J. Persico (J. Med. Chem. **26** [1983] 226/9). – [17] M. Hisatome, S.-I. Takano, K. Yamakawa (Tetrahedron Letters **26** [1985] 2347/50).

5.2.3.7 Carboxylic Acid Esters

5.2.3.7.1 1,1′-Compounds

5.2.3.7.1.1 1,1′-Di(methoxycarbonyl)ferrocene $Fe(C_5H_4{-}CO_2CH_3)_2$

Preparation and Formation

$\mathbf{Fe(C_5H_4{-}CO_2H)_2}$ yields the title compound on boiling with CH_3OH in the presence of a small quantity of a mineral acid as a catalyst, such as concentrated H_2SO_4 [2, 19, 24] (90% yield after 24 to 25 h reaction) [20, 31] or hydrochloric acid [1, 7] (72% yield) [8]. Refluxing of the solid acid in absolute CH_3OH saturated with HCl is recommended [10, 13]; the yield after 2 h reaction is 82% [10]. When only a few bubbles of HCl are introduced into the CH_3OH, 60% of theory forms after 7 h refluxing [5]. $BF_3 \cdot O(C_2H_5)_2$ can also serve as the catalyst [18]. Esterification with an ethereal solution of CH_2N_2 [1, 3, 4, 9] in CH_3OH [30] at room temperature gives a practically quantitative yield [16].

Crystalline $\mathbf{Fe(C_5H_4{-}COCl)_2}$ reacts with anhydrous CH_3OH, first cold, then at the boiling point, to give an 81 to 82% yield [17, 29], see also [14].

From Cyclopentadienide. The reaction of $NaC_5H_4CO_2CH_3$ with $FeCl_2$ in tetrahydrofuran at room temperature gives $Fe(C_5H_4{-}CO_2CH_3)_2$ in yields of 21% [15], of 29% when the $NaC_5H_4CO_2CH_3$ is prepared in situ from NaC_5H_5 and $ClCO_2CH_3$ [6]. It can be increased to 54% if the reaction of $FeCl_2$ with $C_5H_5CO_2CH_3$ in $(C_2H_5)_2NH$ (molar ratio 1/2/4) and ether is conducted first at 0°C for 1 h, then at room temperature for 12 h. The yield is raised to 58% in a variant in which 2/1 $(C_2H_5)_2NH$ and n-C_4H_9Li after 30 min refluxing with $FeCl_2$ in tetrahydrofuran are stirred for 3 h at room temperature and then reacted with $C_5H_5CO_2CH_3$ first at the boiling point, then at room temperature [15].

In these processes the compound is separated chromatographically on an Al_2O_3 column with C_6H_6 [6, 15].

From Other Compounds. The mixture of $Fe(C_5H_4{-}Na)_2$ and $C_5H_5FeC_5H_4{-}Na$ obtained by metallation of ferrocene with n-$C_5H_{11}Na$ (mole ratio 1/1) reacts with CO_2 and CH_2N_2 to a product consisting of 70% $Fe(C_5H_4{-}CO_2CH_3)_2$ and 30% $C_5H_5FeC_5H_4{-}CO_2CH_3$. Initial metallation by n-C_4H_9Li gives 41 and 59%, respectively [11]. For separation see p. 275.

Carbonylation of $Fe(C_5H_4{-}HgCl)_2$ in CH_3OH with CO at 50 atm and 70°C with Li_2PdCl_4 catalyst forms the ester in 37% yield together with I (8%) and II (29%), from which it is separated by column chromatography on Al_2O_3 with C_6H_6. Elution with C_6H_6 afforded the ester after I, further elution with $CHCl_3$ gave II [22].

References on pp. 275/6

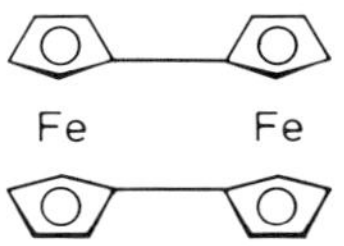

I

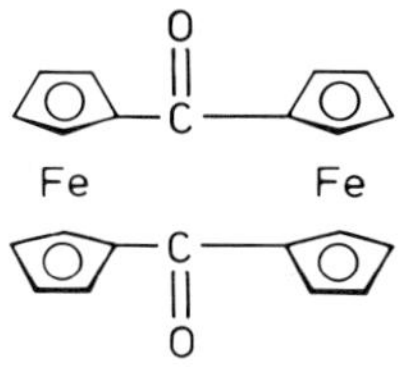

II

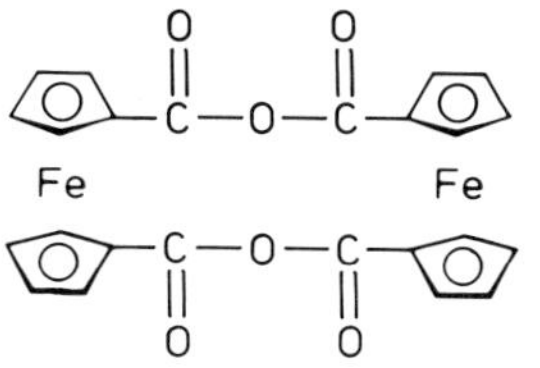

III

A solution of ferrocene and oxalic acid in dry CH_3OH is electrolyzed between Pt electrodes at 3.8×10^{-2} A/cm^2 and 25 V at 15°C. After passage of 0.4 F the solution is reduced with $Na_2S_2O_3$ and extracted with benzene. The extract is evaporated and chromatographed on a thin Al_2O_3 layer. Elution with petroleum ether gives a 65% recovery of the ferrocene inserted, then benzene a 20% yield of $C_5H_5FeC_5H_4$–CO_2CH_3, and finally a 4% yield of the title compound (yield based on reacted ferrocene) (unidentified ferrocene-containing products are subsequently eluted with aqueous KOH solution) [26, 27].

Lengthy heating of III with CH_3OH in the presence of HCl [28] forms $Fe(C_5H_4$–$CO_2CH_3)_2$, as does the photodecomposition of $C_5H_5FeC_5H_4$–CO_2CH_3 in $(CH_3)_2SO$. The chromatographic processing of the residue with 19/1 benzene/ether forms seven bands, that with $R_f = 0.25$ gives the compound in 7% yield [23].

^{59}Fe-labelled $Fe(C_5H_4$–$CO_2CH_3)_2$ is obtained in 0.8% yield on heating of $C_5H_5{}^{59}FeC_5H_4$–$COCH_3$ and $C_5H_5FeC_5H_4$–CO_2CH_3 (molar ratio 1/1) in the presence of catalytic quantities of $FeCl_3$ in a sealed tube for 1 h at 180°C. Identification and separation from $^{59}Fe(C_5H_4$–$COCH_3)_2$ and also from $C_5H_5{}^{59}FeC_5H_4$–CO_2CH_3 take place by thin layer chromatography on SiO_2 in 5/1 C_6H_6/$CH_3CO_2C_2H_5$ and in 8/1 C_6H_6/CH_3OH, respectively [25].

Purification. Separation

Purification is conducted chromatographically on an activated Al_2O_3 column with ether eluent [29] or by recrystallization from hexane [19, 29], cyclohexane [1], petroleum ether [10], CH_3OH [4, 5, 7, 20], hexane/ethanol [9], or petroleum ether/ethanol [3].

$Fe(C_5H_4$–$CO_2CH_3)_2$ is separable from $C_5H_5FeC_5H_4$–CO_2CH_3 by column chromatography on Al_2O_3 with petroleum ether. After development of two bands the monoester is eluted with benzene, the diester with ether [11]. The free acid is removed with 5 to 10% aqueous Na_2CO_3 solution from a mixture with the diester [19].

The separation from $Fe(C_5H_4$–$COCH_3)_2$ can be conducted by thin layer chromatography on SiO_2 with 5/1 C_6H_6/$CH_3CO_2C_2H_5$ [25] or 15/1/1 toluene/petroleum ether/acetone [21] as well as by gas chromatography at 170 to 200°C on a column filled with 2.5% Apiezon L on Chromosorb W [12]. The two latter processes are applicable also in the case of the simultaneous presence of $Fe(C_5H_4$–$CH_2OH)_2$ and ferrocene [12, 21]. The compound can be separated by thin layer chromatography on SiO_2 with 8/1 benzene/methanol from $C_5H_5FeC_5H_4$–$COCH_3$ [25].

For thin layer chromatography on SiO_2 gel plates: $R_f = 0$ (petroleum ether or n-hexane), 0.06 (toluene) [21].

References:

[1] M. Rosenblum (Diss. Harvard Univ. 1953, pp. 116/8). – [2] R. A. Benkeser, D. Goggin, G. Schroll (J. Am. Chem. Soc. **76** [1954] 4025/6). – [3] A. N. Nesmeyanov, E. G. Perevalova, R. V. Golovnya, O. A. Nesmeyanova (Dokl. Akad. Nauk SSSR **97** [1954] 459/61). – [4] A. N. Nesmeyanov, O. A. Reutov (Dokl. Akad. Nauk SSSR **115** [1957] 518/21; Proc. Acad. Sci. USSR Chem. Sect. **112/117** [1957] 763/6). – [5] W. Göller (Diss. Stuttgart T.H. 1958).

[6] J. M. Osgerby, P. L. Pauson (J. Chem. Soc. **1961** 4604/9). – [7] H. Shechter, J. F. Helling (J. Org. Chem. **26** [1961] 1034/7). – [8] M. Okawara, Y. Takemoto, H. Kitaoka, E. Haruki, I. Imoto (Kogyo Kagaku Zasshi **65** [1962] 685/8 from C.A. **58** [1963] 577). – [9] A. N. Nesmeyanov, E. G. Perevalova, L. P. Yur'eva, K. I. Grandberg (Izv. Akad. Nauk SSSR Otd. Khim. Nauk **1963** 1377/80; Bull. Acad. Sci. USSR Div. Chem. Sci. **1963** 1254/6). – [10] K. Yamakawa, H. Ochi, K. Arakawa (Chem. Pharm. Bull. [Tokyo] **11** [1963] 905/11).

[11] R. A. Benkeser, J. L. Bach (J. Am. Chem. Soc. **86** [1964] 890/5). – [12] K. Tanikawa, K. Arakawa (Chem. Pharm. Bull. [Tokyo] **13** [1965] 926/31). – [13] J. T. Suh, Colgate-Palmolive Co. (U.S. 3382267 [1965/68] 1/5; C.A. **69** [1968] No. 59378). – [14] P. Petrovitch (Double Liaison No. 133 [1966] 1093/109). – [15] P. L. Pauson, M. A. Sandhu, W. E. Watts (J. Chem. Soc. C **1968** 860/3).

[16] G. P. Chalykh, E. B. Sokolova, O. D. Shalygina, N. V. Noskova (Tr. Mosk. Khim. Tekhnol. Inst. No. 61 [1969] 164). – [17] H. Valot (Bull. Soc. Chim. France **1969** 1972/5). – [18] T. P. Vishnyakova, I. A. Golubeva, L. M. Timofeeva (Zh. Obshch. Khim. **39** [1969] 2534/6; J. Gen. Chem. [USSR] **39** [1969] 2475/6). – [19] E. B. Sokolova, M. P. Shebanova, L. A. Datskevich (Sin. Fiz. Khim. Polim. No. 6 [1970] 69/72; C.A. **73** [1970] No. 131 397). – [20] A. Sonoda, I. Moritani (J. Organometal. Chem. **26** [1971] 133/40).

[21] K. Tanikawa, K. Arakawa (Bunseki Kagaku **20** [1971] 278/81). – [22] A. Kasahara, T. Izumi, S. Ohnishi (Bull. Chem. Soc. Japan **45** [1972] 951/2). – [23] L. H. Ali, A. Cox, T. J. Kemp (J. Chem. Soc. Dalton Trans. **1973** 1468/75). – [24] V. P. Tverdokhlebov, I. V. Tselinskii, B. V. Gidaspov, G. Yu. Chikisheva (Zh. Org. Khim. **12** [1976] 2335/40; J. Org. Chem. [USSR] **12** [1976] 2268/72). – [25] E. A. Stadlbauer, E. Nipper, M. Wenzel (J. Labelled Compounds Radiopharm. **13** [1977] 491/508).

[26] L. I. Denisovich, M. G. Peterleitner, S. P. Gubin (Izv. Akad. Nauk SSSR Ser. Khim. **1978** 465/8; Bull. Acad. Sci. USSR Div. Chem. Sci. **1978** 403/6). – [27] S. P. Gubin, L. I. Denisovich, N. V. Zakurin, M. G. Peterleitner (J. Organometal. Chem. **146** [1978] 267/78). – [28] N. A. Nesmeyanov, O. A. Reutov (Dokl. Akad. Nauk SSSR **120** [1958] 1267/70; Proc. Acad. Sci. USSR Chem. Sect. **118/123** [1958] 503/6). – [29] V. P. Red'ko, V. K. Skubin, D. F. Kutepov, V. V. Korshak, V. G. Mochalov (Vysokomol. Soedin B **22** [1980] 222/4). – [30] V. Rapić, K. Schlögl, B. Steinitz (J. Organometal. Chem. **94** [1975] 87/98).

[31] K. Gonsalves, L. Zhan-ru, M. D. Rausch (J. Am. Chem. Soc. **106** [1984] 3862/3).

Properties

The compound forms yellow-orange [2, 17] to orange-red [21] needle-like [7, 17] crystals [21] or flakes [2]; according to [32], brown crystals.

The melting point (recrystallization solvent in parentheses) most often given in the literature is 114 to 115°C [1, 3, 8, 15, 27], (benzene) [34], (methanol) [6], (petroleum ether/ethanol) [4], (hexane/ethanol) [14], (hexane) [39]. Only one value lies higher, namely 116.5 to 117.5°C [2]. Lower values, in °C, are: 114 [23, 26], 113.5 to 114.5 [16, 44], 113 to 114 (methanol) [32], 112 to 115 (petroleum ether) [17, 33], 112 to 114 [12, 42], 112 to 113 (methanol) [7, 11], 112 to 112.5 [30], 111.5 to 113 [21], 111 to 113 (hexane) [31], 108 to 109 [26], 107 to 109 (ligroin) [10].

Nuclear Magnetic Resonance Spectra. The ^{1}H NMR spectrum exhibits the following signals, δ in ppm:

solvent	CH_3 protons	C_5H_4 protons		Ref.
		H-3,4	H-2,5	
$CDCl_3$	3.78s	4.35t	4.77t	[34]
$CDCl_3$	3.80s	4.47t; J=2Hz	4.82t; J=2Hz	[26]

References on pp. 278/9

solvent	CH_3 protons	C_5H_4 protons H-3,4	H-2,5	Ref.
CCl_4	3.80 s	4.31 t	4.76 t	[37, 38]
CCl_4	—	4.34*) t	4.74*) t	[19]
CCl_4	—	4.40 t	4.80 t	[18]
$(CD_3)_2SO$	3.74 s	4.55 t	4.80 t	[36]

*) Values relative to the shift of ferrocene, $\delta = 4.05$ ppm.

The introduction of the CO_2CH_3 substituent onto the unsubstituted cyclopentadienyl ring of $C_5H_5FeC_5H_4$–CO_2CH_3 influences the location of the proton signals in the substituted ring only insignificantly. Only for substituents with a weak acceptor action is the additivity rule valid for calculation of the ring proton signals [19].

^{57}Fe NMR spectrum ($CDCl_3$): $\delta = 379.9 \pm 2.5$ [40], 379.8 [43] ppm (vs. ferrocene as an internal standard; compound with the natural abundance of ^{57}Fe) [40, 43].

IR spectra (figures in the papers [12, 16, 39]). The solid compound and its solutions yield for the (νC=O) band the following values, in cm^{-1}:

solvent	(Vaseline)	(Nujol)		(KBr)	heptane	$CHCl_3$
band	1651 and 1707	1709	1700	1704	1733	1712
Ref.	[9]	[26]	[32]	[21, 34]	[13]	[9]

solvent	CCl_4			dioxane	$(CH_3)_2SO$	?
band	1725	1722	1730	1725	1713	1750
Ref.	[13]	[24]	[9]	[13]	[36]	[37, 38]

For half-widths and integral intensities see [13, 24]. Additional bands (for explanation of the intensities see p. VIII) are in Nujol at 1144 [26], 1145, 1195, 1285, and 1375 [32]; in Vaseline oil at 823(m), 838(s), 896(s), 913(s), 968(m), 1031(m), 1064(m), 1146(vs), 1194(m), 1282(vs), and 1464(vs) [9]; in $(CH_3)_2SO$ at 960, 1145, 1192, 1284, 1468, 2944, and 3078 [36] cm^{-1}. In accordance with Rosenblum's rule there are no bands around 1000 and 1111 cm^{-1} [5]. For a relationship between (νC=O) bands in CCl_4 and the X^+ ($R^{A,B}$) constants of structural fragments for 154 compounds with ferrocene structure of the general formula R^A CO R^B, among them $Fe(C_5H_4$–$CO_2CH_3)_2$ with $R^A = C_5H_4FeC_5H_4$–CO_2CH_3 and $R^B = OCH_3$, in the form of a modified and extended Seth-Paul-Van Duyse equation, see [41].

Electronic Absorption Spectrum. Magnetic Circular Dichroism. Absorption maxima λ_{max} in nm, their extinction coefficients ε in $L \cdot mol^{-1} \cdot cm^{-1}$ at room temperature for various solvents, and their assignments:

solvent	A_{2g} $e_{2g} \rightarrow e^*_{2g}$		A_{1u} $e_{2g} \rightarrow e_{2u}$		charge transfer to CO_2CH_3		Ref.
	λ	ε	λ*)	ε	λ	ε	
isooctane	449	220	342	280	300	1200	[20, 29]
n-heptane	449	250	342	300	—	—	[22]
cyclohexane	445	275	—	—	—	—	[35]
benzene	448	260	—	—	—	—	[28]
ethanol	446	340	—	—	—	—	[17]
ethanol	—	—	—	—	304.8	1100	[25]

λ*) is the wavelength of the midpoint of the inflection.

References on pp. 278/9

solvent	A_{2u} $e_{2g} \rightarrow e_{2u}$		E_{1u} $e_{2g} \rightarrow e_{2u}$		E_{1u} $e_{2g} \rightarrow e_{2u}$		Ref.
	λ	ε	λ	ε	λ	ε	
isooctane	255	8000	218	37000	193	26500	[20, 29]
n-heptane	255	6800	218	25000	—	—	[22]
cyclohexane	—	—	—	—	—	—	[35]
benzene	—	—	—	—	—	—	[28]
ethanol	277	10000	—	—	—	—	[17]
ethanol	—	—	—	—	—	—	[25]

The oscillator strengths f in isooctane for λ_{max} of 449, 255, 218, and 193 nm amount to $f \cdot 10^3 = 3.5$, 173, 415, and 660, respectively [20, 29]. The introduction of a second CO_2CH_3 group into $C_5H_5FeC_5H_4$–CO_2CH_3 effects a bathochromic displacement of the A_{2g}, A_{1u}, and A_{2u} transitions by 1000, 200, and 300 nm, respectively, and a hypsochromic shift of the charge transfer transition by 1200 nm [22].

In the region of the A_{2g} transition is observed an S-shaped magnetic circular dichroism band with peaks in cyclohexane at $\lambda_1 = 480$ and $\lambda_2 = 415$ nm ($\lambda_{\varepsilon=0} = 440$ nm), which is partially overlapped by a Gaussian MCD curve with $\lambda = 355$ nm [35]. In benzene the peaks lie at $\lambda_1 = 485$, $\lambda_2 = 415$ nm (A/D ≈ −1.17) [28].

Crystal Structure. $Fe(C_5H_4$–$CO_2CH_3)_2$ crystallizes in the monoclinic space group A 2/a-C^6_{2h} (No. 15) with the lattice parameters a = 13.34, b = 5.96, c = 32.67 Å, β = 104°20′. The density calculated with Z = 8 is 1.594 g/cm³; the density measured is 1.589 g/cm³ [44].

References:

[1] R. B. Woodward, M. Rosenblum, M. C. Whiting (J. Am. Chem. Soc. **74** [1952] 3458/9). – [2] M. Rosenblum (Diss. Harvard Univ. 1953, pp. 116/8). – [3] R. A. Benkeser, D. Goggin, G. Schroll (J. Am. Chem. Soc. **76** [1954] 4025/6). – [4] A. N. Nesmeyanov, E. G. Perevalova, R. V. Golovnya, O. A. Nesmeyanova (Dokl. Akad. Nauk SSSR **97** [1954] 459/61). – [5] A. N. Nesmeyanov, L. A. Kazitsyna, B. V. Lokshin, I. I. Kritskaya (Dokl. Akad. Nauk SSSR **117** [1957] 433/6; Proc. Acad. Sci. USSR Chem. Sect. **112/117** [1957] 1033/6).

[6] N. A. Nesmeyanov, O. A. Reutov (Dokl. Akad. Nauk SSSR **115** [1957] 518/21; Proc. Acad. Sci. USSR Chem. Sect. **112/117** [1957] 763/6). – [7] W. Göller (Diss. Stuttgart T.H. 1958). – [8] A. N. Nesmeyanov, E. G. Perevalova, Z. A. Beinoravichute, I. L. Malgina (Dokl. Akad. Nauk SSSR **120** [1958] 1263/6; Proc. Acad. Sci. USSR Chem. Sect. **118/123** [1958] 499/502). – [9] L. A. Kazitsyna, B. V. Lokshin, A. N. Nesmeyanov (Dokl. Akad. Nauk SSSR **127** [1959] 333/6; Proc. Acad. Sci. USSR Chem. Sect. **124/129** [1959] 531/4). – [10] J. M. Osgerby, P. L. Pauson (J. Chem. Soc. **1961** 4604/9).

[11] H. Shechter, J. F. Helling (J. Org. Chem. **26** [1961] 1034/7). – [12] M. Okawara, Y. Takemoto, H. Kitaoka, E. Haruki, I. Imoto (Kogyo Kagaku Zasshi **65** [1962] 685/8 from C.A. **58** [1963] 577). – [13] G. G. Dvoryantseva, Yu. N. Sheinker (Izv. Akad. Nauk SSSR Otd. Khim. Nauk **1963** 924/7; Bull. Acad. Sci. USSR Div. Chem. Sci. **1963** 832/5). – [14] A. N. Nesmeyanov, E. G. Perevalova, L. P. Yur'eva, K. I. Grandberg (Izv. Akad. Nauk SSSR Otd. Khim. Nauk **1963** 1377/80; Bull. Acad. Sci. USSR Div. Chem. Sci. **1963** 1254/6). – [15] A. D. Petrov, E. B. Sokolova, G. P. Bakunchik (Dokl. Akad. Nauk SSSR **148** [1963] 598/600; Proc. Acad. Sci. USSR Chem. Sect. **148/153** [1963] 77/9).

[16] R. L. Schaaf, K. L. Rinehart (WADD-TR-61-108-Pt. III [1965] 1/80, 49; AD-403929 [1965] 1/80, 49). – [17] K. Yamakawa, H. Ochi, K. Arakawa (Chem. Pharm. Bull. [Tokyo] **11** [1963] 905/11). – [18] R. A. Benkeser, J. L. Bach (J. Am. Chem. Soc. **86** [1964] 890/5). – [19] G. G. Dvoryantseva, S. L. Portnova, K. I. Grandberg, S. P. Gubin, Yu. N. Sheinker (Dokl. Akad. Nauk SSSR **160** [1965] 1075/8; Dokl. Chem. Proc. Acad. Sci. USSR **160/165** [1965] 155/8). – [20] A. N. Nesmeyanov, B. M. Yavorskii, G. B. Zaslavskaya, N. S. Kochetkova (Dokl. Akad. Nauk SSSR **160** [1965] 837/40; Dokl. Chem. Proc. Acad. Sci. USSR **160/165** [1965] 131/4).

[21] J. T. Suh, Colgate-Palmolive Co. (U.S. 3382267 [1965/68] 1/5; C.A. **69** [1968] No. 59378). – [22] K. I. Grandberg, S. P. Gubin (Izv. Akad. Nauk SSSR Ser. Khim. **1966** 551/4; Bull. Acad. Sci. USSR Div. Chem. Sci. **1966** 519/21). – [23] P. Petrovitch (Double Liaison No. 133 [1966] 1093/109). – [24] G. G. Dvoryantseva, M. I. Struchkova, Yu. N. Sheinker, K. I. Grandberg, S. P. Gubin (Izv. Akad. Nauk SSSR Ser. Khim. **1967** 945/7; Bull. Acad. Sci. USSR Div. Chem. Sci. **1967** 919/20). – [25] S. P. Gubin, K. I. Grandberg, I. P. Shepilov (Izv. Akad. Nauk SSSR Ser. Khim. **1967** 1149/51; Bull. Acad. Sci. USSR Div. Chem. Sci. **1967** 1114/6).

[26] P. L. Pauson, M. A. Sandhu, W. E. Watts (J. Chem. Soc. C **1968** 860/3). – [27] G. P. Chalykh, E. B. Sokolova, O. D. Dhalygina, N. V. Noskova (Tr. Mosk. Khim. Tekhnol. Inst. No. 61 [1969] 164; C.A. **73** [1970] No. 25627). – [28] H. Falk (Monatsh. Chem. **100** [1969] 411/8). – [29] A. N. Nesmeyanov, G. B. Zaslavskaya, B. M. Yavorskii, N. S. Kochetkova, N. P. Gambaryan (Opt. Spektrosk. **26** [1969] 945/9; Opt. Spectrosc. [USSR] **26** [1969] 512/5). – [30] H. Valot (Bull. Soc. Chim. France **1969** 1972/5).

[31] E. B. Sokolova, M. P. Shebanova, L. A. Datskevich (Sin. Fiz. Khim. Polim. No. 6 [1970] 69/72; C.A. **73** [1970] No. 131397). – [32] A. Sonoda, I. Moritani (J. Organometal. Chem. **26** [1971] 133/40). – [33] K. Tanikawa, K. Arakawa (Bunseki Kagaku **20** [1971] 278/81). – [34] A. Kasahara, T. Izumi, S. Ohnishi (Bull. Chem. Soc. Japan **45** [1972] 951/2). – [35] D. Nielson, D. Boone, H. Eyring (J. Phys. Chem. **76** [1972] 511/5).

[36] L. H. Ali, A. Cox, T. J. Kemp (J. Chem. Soc. Dalton Trans. **1973** 1468/75). – [37] L. I. Denisovich, M. G. Peterleitner, S. P. Gubin (Izv. Akad. Nauk SSSR Ser. Khim. **1978** 465/8; Bull. Acad. Sci. USSR Div. Chem. Sci. **1978** 403/6). – [38] S. P. Gubin, L. I. Denisovich, N. V. Zakurin, M. G. Peterleitner (J. Organometal. Chem. **146** [1978] 267/78). – [39] V. P. Red'ko, V. K. Skubin, D. F. Kutepov, V. V. Korshak, V. G. Mochalov (Vysokomol. Soedin B **22** [1980] 222/4). – [40] E. Haslinger, W. Robien, K. Schlögl, W. Weissensteiner (J. Organometal. Chem. **218** [1981] C11/C14).

[41] A. Perjéssy, Š. Toma (Collection Czech. Chem. Commun. **48** [1983] 1635/46). – [42] V. Rapić, K. Schlögl, B. Steinitz (J. Organometal. Chem. **94** [1975] 87/98). – [43] E. Haslinger, K. Kooi, W. Robien, K. Schlögl (Monatsh. Chem. **114** [1983] 495/9). – [44] D. W. Fischer (Acta Cryst. **17** [1964] 619).

Solubility. Chemical Reactions

The compound is well **soluble** in organic solvents [2].

The molecular ion peak is observed in the **mass spectrum** (m/e = 302) [29, 36, 37]. The spectrum is characterized by little fragmentation [40].

Electrochemical Behavior

The compound undergoes a one-electron reversible oxidation with $E^\circ = 0.88$ V vs. a saturated calomel electrode (SCE) to the ferrocenium ion $\mathbf{[Fe(C_5H_4{-}CO_2CH_3)_2]^+}$ measured in CH_3CN (0.10 M $HClO_4$) at room temperature by cyclic voltammetry. It acts as a mediator according

References on pp. 281/2

to $2[Fe(C_5H_4\text{–}CO_2CH_3)_2]^+ + [(RO)_2PS_2]_2Zn \rightarrow 2Fe(C_5H_4\text{–}CO_2CH_3)_2 + Zn^{2+} + [(RO)_2PS_2]_2$ in the electrochemical oxidation of $[(RO)_2PS_2]_2Zn$ with $R = C_2H_5$, $CH(CH_3)_2$ to the corresponding disulfides in high yields in CH_3CN at a glassy carbon electrode at 0.90 V (SCE). In the absence of $Fe(C_5H_4\text{–}CO_2CH_3)_2$, the constant potential oxidation of $[(RO)_2PS_2]_2Zn$ occurs with a large overpotential [42].

$Fe(C_5H_4\text{–}CO_2CH_3)_2$ is reduced to the radical anion $\mathbf{[Fe(C_5H_4\text{–}CO_2CH_3)_2]^{\overline{\cdot}}}$ in an irreversible one-electron reaction. The polarographic half-wave potential $E_{1/2} = -2.18$ V vs. SCE, in $(CH_3)_2NCHO/0.1$ M $[(n\text{-}C_4H_9)_4N]PF_6$ at room temperature. Electrolysis of the compound in this solution at < -2.18 V on a mercury pool electrode forms Fe and $[C_5H_4\text{–}CO_2CH_3]^-$ with consumption of 2 F; this solution has an oxidation potential of $E_{1/2} = -0.18$ V vs. SCE. A comparison of the two processes yields the following reaction mechanism, according to which the acceptance of the first electron with formation of the radical anion is a slow irreversible reaction (half-life several minutes) and then acceptance of the second electron follows:

$$Fe(C_5H_4\text{–}CO_2CH_3)_2 + e^- \rightarrow [Fe(C_5H_4\text{–}CO_2CH_3)_2]^{\overline{\cdot}}$$
$$[Fe(C_5H_4\text{–}CO_2CH_3)_2]^{\overline{\cdot}} \xrightarrow{\text{slow}} CH_3O_2C\text{–}C_5H_4Fe^{\cdot} + [CH_3O_2C\text{–}C_5H_4]^-$$
$$2\,CH_3O_2C\text{–}C_5H_4Fe^{\cdot} \xrightarrow{(CH_3)_2NCHO} [CH_3O_2C\text{–}C_5H_4Fe(OHCN(CH_3)_2)_2]_2$$
$$[CH_3O_2C\text{–}C_5H_4Fe(OHCN(CH_3)_2)_2]_2 \xrightarrow{\text{slow}} 2[CH_3O_2C\text{–}C_5H_4]^- + Fe^{2+} + Fe^0$$
$$Fe^{2+} + 2e^- \rightarrow Fe^0$$
$$\overline{\qquad\qquad\qquad\qquad\qquad\qquad\qquad\qquad\qquad\qquad}$$
$$Fe(C_5H_4\text{–}CO_2CH_3)_2 + 2e^- \rightarrow 2[CH_3O_2C\text{–}C_5H_4]^- + Fe^0$$

The resulting $[CH_3O_2CC_5H_4]^-$ is very stable and can be immediately converted again with $FeCl_3$ quantitatively back to the starting compound [38, 39].

Other Reactions

Addition of 10^{-3} to 10^{-4} M $Fe(C_5H_4\text{–}CO_2CH_3)_2$ suppresses the photoreduction of fluorenone by $(C_2H_5)_3N$ in cyclohexane on irradiation at 404 nm. The rate constant for the **deactivation** of the triplet state of fluorenone by the ester is $k = 5.3 \times 10^9$ L·mol^{-1}·s^{-1} (precision ±20%). It is assumed that the deactivation takes place via a triplet-triplet energy transfer [30], with the energy of the excited triplet state for $Fe(C_5H_4\text{–}CO_2CH_3)_2$ lying somewhat below that of the fluorenone (53.3 kcal/mol [27]) [34].

In the presence of $AlCl_3$ (mole ratio 1/4) in boiling mesitylene a 70% **ligand exchange** occurs to the ions $[CH_3O_2C\text{–}C_5H_4FeC_6H_3(CH_3)_3]^+$ and $[Fe(C_6H_3(CH_3)_3)_2]^{2+}$ arising through simple and double exchange. In contrast only very little exchange occurs in boiling benzene [32].

Sulfonation is limited in comparison to that of ferrocene by the electrophilic groups; to be sure it does occur with SO_3 in ClH_2CCH_2Cl, but it does not in dioxane. The resulting $CH_3O_2C\text{–}C_5H_4FeC_5H_3(CO_2CH_3)SO_3H$ precipitates largely as its Fe salt, whose formation is made possible by partial decomposition of the ester [9]. The arylation with $[4\text{-}O_2NC_6H_4N_2]Cl$ in glacial acetic acid is accompanied by considerable resin formation and partial decomposition of the ferrocene compound. The yield of the mixture of $CH_3O_2C\text{–}C_5H_4FeC_5H_3(CO_2CH_3\text{-}1)(C_6H_4NO_2\text{-}4)\text{-}2$ and -3 is less than 2% [10].

Reduction with $LiAlH_4$ in ether [4, 15, 24, 41], tetrahydrofuran [16, 33], or ether/C_6H_6 [28] at the boiling point gives $Fe(C_5H_4\text{–}CH_2OH)_2$ [4, 6, 15, 16, 24, 33, 41], in high yields [4, 11, 28, 41]. If the reaction is conducted in the mole ratio 1/0.76 for 5 min at room temperature, it forms a 23% yield of $CH_3CO_2C\text{–}C_5H_4FeC_5H_4\text{–}CH_2OH$ along with an 18% yield of $Fe(C_5H_4\text{–}CH_2OH)_2$ and a 45% recovery of $Fe(C_5H_4\text{–}CO_2CH_3)_2$ [21]. $LiAlH_4/AlX_3$ (X = Cl [7, 14], Br [4]) reduces the compound in boiling ether to $Fe(C_5H_4\text{–}CH_3)_2$ [4, 7, 14], and likewise $NaBH_4/AlCl_3$ in $CH_3OCH_2CH_2OCH_3$ at 70°C [31].

References on pp. 281/2

Half saponification with NaOH (mole ratio 1/1 to 1.2/1) in 30/1 CH_3OH/H_2O gives $CH_3O_2C{-}C_5H_4FeC_5H_4{-}CO_2H$ [3, 19].

With Metal Compounds. The Grignard reagents R'MgX (R' = C_6H_5 or 3-$CF_3C_6H_4$, X = Br; R' = $(CH_3)_3Si(CH_2)_3$, X = Cl) form in boiling ether on subsequent hydrolysis the tertiary alcohols $Fe(C_5H_4{-}C(OH)R'_2)_2$ with R' = C_6H_5 [4], $C_6H_4CF_3$-3 [20], or $(CH_2)_3Si(CH_3)_3$ [12]. In anomalous contrast – probably for steric reasons – the reaction with $(CH_3)_3SiMgCl$ leads to $(CH_3)_3SiCH_2CO{-}C_5H_4FeC_5H_4{-}CO_2CH_3$ [12]. $LiC_6H_4N(CH_3)_2$-4 yields correspondingly $Fe(C_5H_4{-}COH(C_6H_4N(CH_3)_2\text{-}4)_2)_2$, the analogue in ferrocene chemistry to the color base of terephthalic green [22].

Condensation Reactions. Transesterification to $Fe(C_5H_4{-}CO_2(CH_2)_2OH)_2$ takes place on heating with $HOCH_2CH_2OH$ to 170°C in the presence of a catalyst of KOH and $Zn(O_2CCH_3)_2$ [26]. Polyester formation to I occurs on heating of $Fe(C_5H_4{-}CO_2CH_3)_2$ and $HOCH_2CH_2OH$ with $Ca(O_2CCH_3)_2 \cdot H_2O$ and small quantities of Sb_2O_3 in a polymerization tube, first to 175 to 178°C and then to 283°C, at 30 to 8 Torr [13]. Compound I with a low molecular weight was obtained by [8]. No transesterification with $HOCH_2CH_2OH$ was observed with Na as catalyst on heating in either toluene or absolute ethanol. An attempted polyester formation with aliphatic dialcohols in the melt with addition of sodium also failed [5].

The compound reacts with a large excess of N_2H_4 in alcohol to form $Fe(C_5H_4{-}CONHNH_2)_2$ [1, 17, 18, 40]. No sort of polyamide formation occurs in attempts with a series of diamines [11]. $NH_2OH \cdot HCl$ in the presence of KOH forms the K salt of the dihydroxamic acid $Fe(C_5H_4{-}CONHOH)_2$ [25].

The attempt to form II by means of the Claisen condensation with $CH_3CO_2CH_3$ was unsuccessful [35]. $NaNH_2$ in liquid NH_3 constitutes an effective condensation agent for the reaction with CH_3CN to form $Fe(C_5H_4{-}COCH_2CN)_2$ [23].

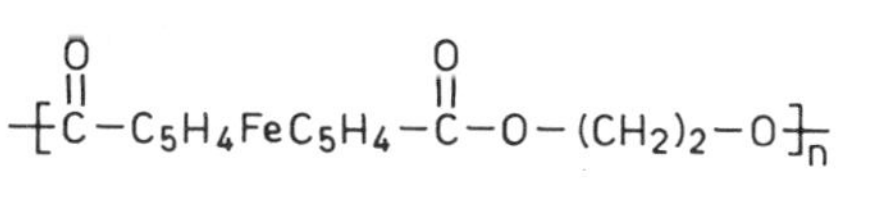

I

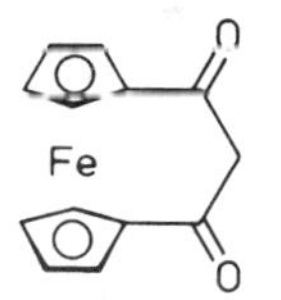

II

References:

[1] M. Rosenblum (Diss. Harvard Univ. 1953, p. 116/8). – [2] A. N. Nesmeyanov, E. G. Perevalova, R. V. Golovnya, O. A. Nesmeyanova (Dokl. Akad. Nauk SSSR **97** [1954] 459/61). – [3] N. A. Nesmeyanov, O. A. Reutov (Dokl. Akad. Nauk SSSR **115** [1957] 518/21; Proc. Acad. Sci. USSR Chem. Sect. **112/117** [1957] 763/6). – [4] A. N. Nesmeyanov, E. G. Perevalova, Z. A. Beinoravichute (Dokl. Akad. Nauk SSSR **112** [1957] 439/40). – [5] W. Göller (Diss. Stuttgart T.H. 1958).

[6] K. L. Rinehart, A. K. Frerichs, P. A. Kittle, L. F. Westman, D. H. Gustafson, R. L. Pruett, J. E. McMahon (J. Am. Chem. Soc. **82** [1960] 4111/2). – [7] E. A. Hill, J. H. Richards (J. Am. Chem. Soc. **83** [1961] 4216/21). – [8] F. W. Knobloch, W. H. Rauscher (J. Polym. Sci. **54** [1961] 651/6). – [9] N. A. Nesmeyanov, B. N. Strunin (Dokl. Akad. Nauk SSSR **137** [1961] 106/8; Proc. Acad. Sci. USSR Chem. Sect. **136/141** [1961] 275/6). – [10] E. G. Perevalova, N. A. Simukova, T. V. Nikitina, P. D. Reshetov, A. N. Nesmeyanov (Izv. Akad. Nauk SSSR Otd. Khim. Nauk **1961** 77/83; Bull. Acad. Sci. USSR Div. Chem. Sci. **1961** 67/71).

[11] M. Okawara, Y. Takemoto, H. Kitaoka, E. Haruki, I. Imoto (Kogyo Kagaku Zasshi **65** [1962] 685/8 from C. A. **58** [1963] 577). – [12] A. D. Petrov, E. B. Sokolova, G. P. Bakunchik (Dokl. Akad. Nauk SSSR **148** [1963] 598/600; Dokl. Chem. Proc. Acad. Sci. USSR **148/153** [1963] 77/9). – [13] K. Yamakawa, H. Ochi, K. Arakawa (Chem. Pharm. Bull. [Tokyo] **11** [1963] 905/11). – [14] R. A. Benkeser, J. L. Bach (J. Am. Chem. Soc. **86** [1964] 890/5). – [15] J. T. Suh, Colgate-Palmolive Co. (U.S. 3382267 [1965/68] 1/5; C.A. **69** [1968] No. 59378).

[16] A. N. Nesmeyanov, S. S. Churbanov, Yu. A. Ustynyuk, E. G. Perevalova (Izv. Akad. Nauk SSSR Ser. Khim. **1966** 1648/50; Bull. Acad. Sci. USSR Div. Chem. Sci. **1966** 1591/2). – [17] H.-J. Lorkowski, R. Pannier, A. Wende (J. Prakt. Chem. **35** [1967] 149/58). – [18] E. W. Neuse, H. Rosenberg (J. Polym. Sci. Polym. Chem. Ed. **6** [1968] 1567/82). – [19] T. H. Barr, H. L. Lentzner, W. E. Watts (Tetrahedron **25** [1969] 6001/13). – [20] G. P. Chalykh, E. B. Sokolova, O. D. Shalygina, N. V. Noskova (Tr. Mosk. Khim. Tekhnol. Inst. No. 61 [1969] 164).

[21] H. Falk, O. Hofer, K. Schlögl (Monatsh. Chem. **100** [1969] 624/48). – [22] G. Hallas, G. A. Hirst (J. Soc. Dyers Colourists **85** [1969] 310/2). – [23] A. A. Koridze, T. P. Vishnyakova (Zh. Obshch. Khim. **39** [1969] 922/3; J. Gen. Chem. [USSR] **39** [1969] 886). – [24] H. Valot (Bull. Soc. Chim. France **1969** 1972/5). – [25] T. P. Vishnyakova, I. A. Golubeva. L. M. Timofeeva (Zh. Obshch. Khim **39** [1969] 2534/6; J. Gen. Chem. [USSR] **39** [1969] 2475/6).

[26] E. B. Sokolova, M. P. Shebanova, L. A. Datskevich (Sin. Fiz. Khim. Polim. No. 6 [1970] 69/72; C.A. **73** [1970] No. 131397). – [27] P. Engel, B. Monroe (in: J. N. Pitts, G. S. Hammond, W. A. Noyes, Advan. Photochem. **8** [1971] 245). – [28] A. Sonoda, I. Moritani (J. Organometal. Chem. **26** [1971] 133/40). – [29] L. H. Ali, A. Cox, T. J. Kemp (J. Chem. Soc. Dalton Trans. **1973** 1468/75). – [30] A. Gilbert, J. M. Kelly, E. Koerner von Gustorf (Mol. Photochem. **6** [1974] 225/30).

[31] A. N. Nesmeyanov, G. B. Shul'pin, M. V. Tolstaya, M. I. Rybinskaya (U.S.S.R. 555114 [1975/77]; C.A. **87** [1977] No. 135954). – [32] D. Astruc, R. Dabard (Tetrahedron **32** [1976] 245/9). – [33] V. P. Tverdokhlebov, I. V. Tselinskii, B. V. Gidaspov, G. Yu. Chikisheva (Zh. Org. Khim. **12** [1976] 2335/40; J. Org. Chem. [USSR] **12** [1976] 2268/72). – [34] C. R. Bock, E. A. Koerner von Gustorf (Advan. Photochem. **10** [1977] 221/310, 277). – [35] M. Sališová, Š. Toma, E. Solčaniová (J. Organometal. Chem. **132** [1977] 419/27).

[36] L. I. Denisovich, M. G. Peterleitner, S. P. Gubin (Izv. Akad. Nauk SSSR Ser. Khim. **1978** 465/8; Bull. Acad. Sci. USSR Div. Chem. Sci. **1978** 403/6). – [37] S. P. Gubin, L. I. Denisovich, N. V. Zakurin, M. G. Peterleitner (J. Organometal. Chem. **146** [1978] 267/78). – [38] N. El Murr, A. Chaloyard, E. Laviron (Nouv. J. Chim. **2** [1978] No. 1, pp. 15/7). – [39] A. Chaloyard, N. El Murr (Inorg. Chem. **19** [1980] 3217/20). – [40] V. Rapić, K. Schlögl, B. Steinitz (J. Organometal. Chem. **94** [1975] 87/98).

[41] K. Gonsalves, L. Zhan-ru, M. D. Rausch (J. Am. Chem. Soc. **106** [1984] 3862/3). – [42] R. L. Blankespoor (Inorg. Chem. **24** [1985] 1126/8).

Applications

Polyesters prepared from $HOCH_2CH_2OH$ and dicarboxylic acids or their anhydrides such as fumaric acid and phthalic or maleic acid anhydrides in the presence of 0.01 to 10 wt% $Fe(C_5H_4–CO_2CH_3)_2$ are low temperature hardenable with peroxide to transparent products, which are used as adhesives [3]. Addition of 0.001 to 0.01 mol% of the ferrocene compound catalytically accelerates the polyester formation from $HOCH_2CH_2OH$ and fumaric acid in the presence of KOH at 200°C, with the ferrocene residue incorporated into the polymer chain [11].

Polyethyleneterephthalates $-[4-O_2C–C_6H_4–CO_2(CH_2)_2–]_m–[O_2C–C_5H_4FeC_5H_4–CO_2(CH_2)_2–]_n$ (m = 0.995 to 0.999, n = 0.005 to 0.001, m + n = 1) modified with $Fe(C_5H_4–CO_2CH_3)_2$ with an improved UV-light resistance in comparison to the compound with n = 0 (m = 1), are prepared in two stages by transesterifying the mixture of $Fe(C_5H_4–CO_2CH_3)_2$ and dimethylterephthalate with $HOCH_2CH_2OH$ and subsequent polycondensation of the bisglycolesters in vacuum at 270 to 280°C [10]. Mn acetate plus Sb_2O_3 is used as catalyst for the transesterification stage [4]. Because of the catalytic activity of $Fe(C_5H_4–CO_2CH_3)_2$ an additional catalyst is not necessary [10]. Added in small quantities, $Fe(C_5H_4–CO_2CH_3)_2$ raises the thermooxidative stability of polyorganosiloxanes [1]. It is used as a photostabilizer for textile fibers based on polyvinylalcohols [2] and for polypropylene, to which it is added in quantities of 0.5 to 1.5% in solvents such as acetone, heptane, or mixtures [9].

An increase of the electron radiation sensitivity of positive working radiation resists made of copolymers of methacrylic chloride/methylmethacrylate [6] or glycidylmethacrylate/methylmethacrylate [5] is effected by additions in the range of 0.01 to 20 wt% [5, 6].

$^{59}Fe(C_5H_4–CO_2CH_3)_2$ is used for diagnostic purposes in nuclear medicine [7].

References:

[1] W. A. Piccoli, Dow Corning Corp. (U.S. 2979482 [1956/61]; C.A. **1961** 18 176). – [2] E. A. Kalennikov, G. I. Bochkov, Yu. K. Krylenko, D. A. Kaushanskii (Khim. Volokna **1964** No. 2, p. 27 from [8]). – [3] V. P. Red'ko, V. K. Skubin, D. F. Kutepov, V. V. Korshak (U.S.S.R. 394397 [1971/73]; C.A. **81** [1974] No. 64289). – [4] A. I. Volozhin, L. Yu. Verkhovodka, E. T. Krut'ko, E. A. Kalennikov, Ya. M. Paushkin, E. M. Aizenshtein, L. P. Repina, K. A. Bogdanov (U.S.S.R. 403702 [1972/73]; C.A. **81** [1974] No. 121385). – [5] M. Miyamura, Tokyo Shibaura Electric Co., Ltd. (Japan. Kokai 77-146223 [1976/77] from C.A. **88** [1978] No. 201074).

[6] M. Miyamura, Tokyo Shibaura Electric Co., Ltd. (Japan. Kokai 77-146219 [1976/77] from C.A. **88** [1978] No. 201078). – [7] E. A. Stadlbauer, E. Nipper, M. Wenzel (J. Labelled Compounds Radiopharm. **13** [1977] 491/508). – [8] N. S. Kochetkova, Yu. K. Krynkina (Usp. Khim. **47** [1978] 934/48; Russ. Chem. Rev. **47** [1978] 486/94, 486). – [9] L. Dauksaite, Z. Kaikariene, V. D. Vil'chevskaya, N. S. Kochetkova, Yu. A. Ershov (U.S.S.R. 732322 [1973/80]; C.A. **93** [1980] No. 133370). – [10] A. I. Volozhin, L. Yu. Verkhovodka, A. A. Rozmyslova, Ya. M. Paushkin (Vysokomol. Soedin. B **19** [1977] 593/4).

[11] V. P. Red'ko, V. K. Skubin, D. F. Kutepov, V. V. Korshak, V. G. Mochalov (Vysokomol. Soedin. B **22** [1980] 222/4).

5.2.3.7.1.2 Carboxylic Acid Esters of the Type $R'O_2C–C_5H_4FeC_5H_4–CO_2R''$

The ester with $R' = R'' = CH_3$ is treated in 5.2.3.7.1.1, pp. 274/83.

The compounds listed in Table 20 are usual prepared by the following methods:

Method I. Reaction of $Fe(C_5H_4–COCl)_2$

a) with ROH [30] in C_6H_6 at room temperature [29],

b) with ROH in boiling ether after addition of freshly roasted K_2CO_3 [20],

c) with ROH in CH_3CN/C_6H_6 under reflux for 70 h [31],

d) with $HOCH_2–A_x–CH_2OH$ (x = 0 for No. 8; x = 1, A = $(CH_2OCH_2)_n$ with n = 1 to 3 for Nos. 9 to 11, A = 1,3-C_6H_4 for No. 12) in refluxing dry C_6H_6 (68 h) for Nos. 8 to 11, in refluxing 4/5 C_6H_6/CH_3CN in the presence of $(C_2H_5)_3N$ (40 h) for No. 12. The ferrocenophane I is produced in 22 to 28% yield along with Nos. 8 to 11. Compounds I and II (with x = 1 and A = 1,3-C_6H_4) are obtained in 2 and 4% yields, respectively, along with No. 12. The reaction products are separated by silica gel column chromatography. Nos. 8 to 11 are eluted with $CHCl_3$ after elution of I with C_6H_6. No. 12 is eluted with 9/1 $CHCl_3/CH_3OH$ after the elution of I with $CHCl_3$ and of II with 9/1 $CHCl_3$/ether [33].

References on pp. 289/90

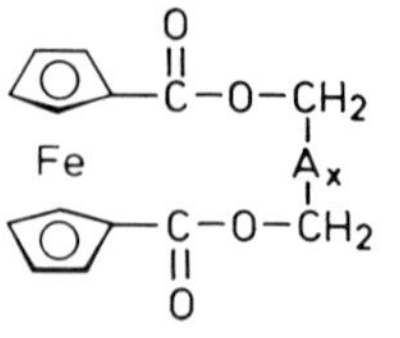

I

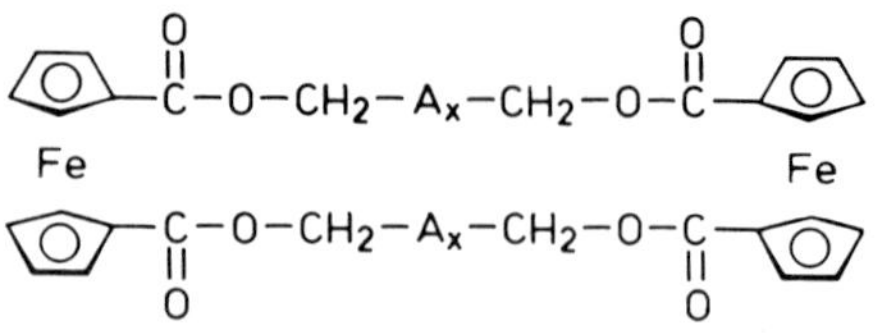

II

Method II. Reaction of $Fe(C_5H_4{-}CO_2H)_2$

a) with $(C_6H_5O)_2SO$ in pyridine at 65°C in the dark [12],

b) with 1,4-$(HO)_2C_6H_4$ in 6/5 CH_3CN/C_6H_6 at room temperature in the presence of $(CF_3CO)_2O$ [31],

c) with CH_3OH by heating on the water bath in the presence of small quantities of concentrated H_2SO_4 and dilution with CH_3OH until a clear solution forms [17],

d) with CH_3OH by saturation of the solution with dry HCl at −8 to −13°C [9],

e) with absolute C_2H_5OH in the presence of concentrated H_2SO_4 at the boiling point [4, 6]. This forms Nos. 1 and 16, and also $C_5H_5FeC_5H_4{-}CO_2C_2H_5$ when starting from a mixture of $Fe(C_5H_4{-}CO_2H)_2$ and $C_5H_5FeC_5H_4{-}CO_2H$ [6]. After evaporation in vacuum [6] or dilution with water [4] all three compounds are extracted from the reaction mixture into ether, and then from this extract No. 16 is extracted with aqueous $NaHCO_3$ solution [4, 6]. For separation of $C_5H_5FeC_5H_4{-}CO_2C_2H_5$ and No. 1, see "Further information" for the latter, p. 287.

Method III. Reaction for 20 h of $Fe(C_5H_4{-}CO_2Ag)_2$ with boiling C_2H_5Br [2].

Method IV. Reaction of $Fe(C_5H_4{-}CO_2CH_3)_2$ (see previous section):

a) With NaOH (mole ratio 1/1) in 30/1 [1] or 14/1 [18] CH_3OH/H_2O at the boiling point for 24 h [18] or 50 min [1].

b) With $HOCH_2CH_2OH$ in the presence of $KOH/Zn(O_2CCH_3)_2$ (weight ratio 1/1) catalyst with heating to 170°C. Unreacted starting compound is extracted from the resulting dark, viscous mass using hot cyclohexane [21].

Purification is effected – when not otherwise specified under "Further information" – by recrystallization from the solvents given in parentheses in Table 20 or by vacuum distillation.

Applications. Additions of 0.001 to 0.1 mol% of Nos. 1 or 4 catalytically accelerate polyester formation from $HOCH_2CH_2OH$ and fumaric acid in the presence of KOH at 200°C, with partial transesterification and incorporation of the ferrocene residue into the polymer chain [30]. Polyvinylbutyrate films are stabilized by additions of 0.1 to 5 wt% of No. 1 against deterioration under the action of UV irradiation. The greater the content of No. 1 the greater the stability, but unfortunately also the darker the film [7]. The effect of the esters of 1,1′-ferrocenedicarboxylic acids on the UV stability of high pressure polyethylene films is treated in [15].

An increase of the electron radiation stability of positive working radiation resists made from copolymers of methacrylic chloride/methylmethacrylate [25] or glycidylmethacrylate/methylmethacrylate [24] is effected by additions of the compounds No. 1 to 3 in the range from 0.01 to 20 wt% [24, 25].

Table 20

Carboxylic Acid Esters of the Type $R'O_2C–C_5H_4FeC_5H_4–CO_2R''$.

Further information for numbers preceded by an asterisk is given at the end of the table, pp. 286/9.

For abbreviations and dimensions see p. VIII.

R_f values are for thin layer chromatography.

No.	$R'O_2C–C_5H_4FeC_5H_4–CO_2R''$ R' R''	method of preparation (yield in %), properties, and remarks	Ref.
compounds with R' = R'':			
*1	C_2H_5	I (70), IIe (18 to 21 referred to $FeC_{10}H_{10}$), III (90 raw product) red crystals, m.p. 39 to 40° (pentane), 39 to 41° (C_2H_5OH), 41.5 to 42° (hexane), b.p. 158°/0.4 Torr, 174 to 175°/3 Torr n_D^{20}: 1.5680 IR see figure in [8, 30] UV see "Further information", p. 287 R_f: 0 (petroleum ether or n-hexane), 0.06 (toluene) (on SiO_2 gel)	[2, 6, 8, 22, 30]
2	C_3H_7	for "Applications" see p. 284	[24, 25]
3	C_4H_9	for "Applications" see p. 284	[24, 25]
4	$CH_2CH{=}CH_2$	I (74) b.p. 182°/4 Torr n_D^{20}: 1.5765 IR plot in figure in [30]	[30]
*5	$CH_2C{\equiv}CH$	Ia, Ib (74) bright orange crystals, m.p. 91 to 92° (hexane), 103 to 104° IR (Vaseline oil): 1130 to 1280 (νC–O), 2130 (νC≡C), 3300 (ν ≡C–H) IR (KBr): 1130 to 1260 (νC–O), 1715 (νC=O), 3270 (ν ≡C–H) R_f: 0.5 (with C_6H_6 on Al_2O_3)	[20, 29]
*6	$CH_2C_6H_5$	—	[23]
*7	C_6H_5	IIa red needles, m.p. 142.5 to 143° (CH_3OH)	[12]
*8	CH_2CH_2OH	Id (51), IVb (60) red, viscous oil	[21, 33]
*9	$CH_2CH_2OCH_2CH_2OH$	Id (59) red oil	[33]
*10	$CH_2(CH_2OCH_2)_2CH_2OH$	Id (53) red oil	[33]

References on pp. 289/90

Table 20 (continued)

No.	$R'O_2C-C_5H_4FeC_5H_4-CO_2R''$ R'	R''	method of preparation (yield in %), properties, and remarks	Ref.
*11	$CH_2(CH_2OCH_2)_3CH_2OH$		Id (55) red oil	[33]
*12	$CH_2C_6H_4CH_2OH$-3		Id (58) red oil	[33]
*13	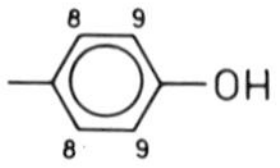		Ic (38), IIb (25) orange crystals (C_2H_5OH), m.p. 226 to 228° dec. ^1H NMR (acetone-d_6): 4.68 (t, H-3,4 of C_5H_4), 5.01 (t, H-2,5 of C_5H_4), 6.83 (d, H-9 of C_6H_4; J = 9), 7.05 (d, H-8 of C_6H_4; J = 9), 8.49 (br. s, OH) IR (KBr): 805 (C_5H_4), 865 (C_6H_4), 1690 (C=O), 3100 (C_5H_4), 3370 (OH) mass spectrum: m/e = 458 $[M]^+$	[31]
*14	$C_{10}H_{19}$-l ($C_{10}H_{19}$ = menthyl)		preparation method not reported	[28]
compounds with R' ≠ R'':				
*15	CH_3	H	IIc (52), IId (32), IVa (71.25) yellow needles, m.p. 147 to 148° (C_6H_6/petroleum ether 1/1), 147.5 to 149.5°, 148 to 150° (ether), 149 to 150° ^1H NMR ($CDCl_3$): 3.89 (s, CH_3), 4.54 (t) and 4.96 (t) (C_5H_4), 7.91 (s, OH) IR ($CHCl_3$): 1678 and 1696 to 1708 (diffuse) (νC=O) IR (Vaseline oil): 830, 899, 917, 972, 1040, 1154, 1165, 1290 (vs), 1410, 1484, 1686 (s), 1722 (s) IR (Nujol): 1667 and 1709 (νC=O)	[1, 5, 9, 17, 18]
*16	C_2H_5	H	IIe (18; 24 based on reacted $Fe(C_5H_4-CO_2H)_2$), see No. 1, p. 287 m.p. 144 to 145° (toluene), 145 to 146° (C_2H_5OH/H_2O) IR (Nujol) see figure in [8] UV: $\lambda_{max}(\log \varepsilon)$ = 448 (2.48), 255 (3.98), 218 (4.54)	[4, 6, 8]

* Further information:

$Fe(C_5H_4-CO_2C_2H_5)_2$ (Table **20**, No. **1**). The attempt to prepare No. 1 from ferrocene and $ClCO_2C_2H_5$ in the presence of $AlCl_3$ in CH_2Cl_2 at −35°C failed due to the decomposition of $ClCO_2C_2H_5$ by $AlCl_3$ [6].

No. 1 is purified by column chromatography on activated Al_2O_3 with ether as eluent, by vacuum distillation at 174 to 175°C/3 Torr, and by recrystallization from pentane [30], hexane [6, 8], or a little C_2H_5OH [2]. The separation of $C_5H_5FeC_5H_4{-}CO_2C_2H_5$ can be conducted chromatographically on Al_2O_3 columns, with elution of No. 1 with ether following elution of the monoester with petroleum ether/benzene, or by fractional distillation, with No. 1 distilling off at 158°C/0.4 Torr after the monoester at (114 to 129°C)/(0.3 to 0.45) Torr [6]. For separation of $Fe(C_5H_4{-}CO_2H)_2$ see Method IIe (p. 284) and below.

Absorption maxima (λ_{max} in nm, extinction coefficient ε (log ε) in $L\cdot mol^{-1}\cdot cm^{-1}$) of the electronic spectrum at room temperature and their assignments:

solvent	A_{2g} $e_{2g}\rightarrow e^*_{2g}$		A_{1u} $e_{2g}\rightarrow e_{2u}$		charge transfer to $CO_2C_2H_5$		A_{2u} $e_{2g}\rightarrow e_{2u}$		E_{1u} $e_{2g}\rightarrow e_{2u}$		E_{1u} $e_{2g}\rightarrow e_{2u}$		Ref.
	λ	ε	λ*⁾	ε	λ*⁾	ε	λ	ε	λ	ε	λ	ε	
isooctane	449	220	342	280	300	1200	255	8000	218	37000	193	27000	[13,19]
?	450	(2.40)	—	—	—	—	256	(3.94)	218	(4.59)	—	—	[6]

λ*⁾ is the wavelength of the midpoint of inflection.

The oscillator strengths of the bands in isooctane at 449, 255, 218, and 193 nm amount to $f\cdot10^3 = 3.5$, 173, 415, and 660, respectively [13, 19].

Oxidation to the ferrocenium ion $\mathbf{[Fe(C_5H_4{-}CO_2C_2H_5)_2]^+}$ is made more difficult by the electron withdrawing action of the substituents. The cation decomposes even in strongly acid solution per $3[Fe(C_5H_4{-}CO_2C_2H_5)_2]^+ \rightarrow 2Fe(C_5H_4{-}CO_2C_2H_5)_2 + Fe^{3+} + 2C_5H_4CO_2C_2H_5$. The rate constant k is dependent on the temperature and the acid strength H_o of the solution: at 25°C k (in s^{-1}) $= 5.62\times10^{-3}$ for $H_o = -1$, 2.48×10^{-3} for $H_o = -2$, and 9.48×10^{-4} for $H_o = -3$ [27].

Under N_2 slow decomposition takes place around 345°C [6]. Saponification with NaOH (mole ratio 1/1) in 1/1 C_2H_5OH/H_2O at the boiling point gives after 30 min with 20% recovery of No. 1 the compounds No. 16 in 55% yield and $Fe(C_5H_4{-}CO_2H)_2$ in 16% yield. No. 1 is extracted from the diluted alkaline reaction solution with ether, and subsequently acidification precipitates the mixture of No. 16 and $Fe(C_5H_4{-}CO_2H)_2$, from which after drying No. 16 is extracted with ether [6].

$\mathbf{Fe(C_5H_4{-}CO_2CH_2C{\equiv}CH)_2}$ (Table **20**, No. **5**) is separated by column chromatography on Al_2O_3 and recrystallized from hexane [20]. Its decomposition begins at 156°C and attains its maximum at 380°C [29]. Mixing the solution in absolute ethanol with CuI in NH_4OH at room temperature precipitates out $\mathbf{Fe(C_5H_4{-}CO_2CH_2C{\equiv}CCu)_2}$ in 27% yield (m.p. 115 to 118°C, IR bands in Vaseline oil at 1130 to 1280 cm^{-1}) [20].

$\mathbf{Fe(C_5H_4{-}CO_2CH_2C_6H_5)_2}$ (Table **20**, No. **6**) occurs in the smectic B, liquid crystal phase of BBOA (III) as a suspension of crystallites, but not as a homogeneous solution of individual molecules. According to ^{57}Fe γ-spectroscopic studies the change of the total intensity of the Mössbauer spectrum with the angle between the magnetic field and the γ-ray gives a lattice contribution of $\varepsilon_L = 0.18$ to the nuclear vibration ellipsoid [23]. The ratio f(298 K)/f(77 K) = 0.32 for the temperature dependence of the recoil free fraction f for a 1% solution of No. 6 is higher than expected. It seems that No. 6 possesses a relatively large alignment in BBOA [32].

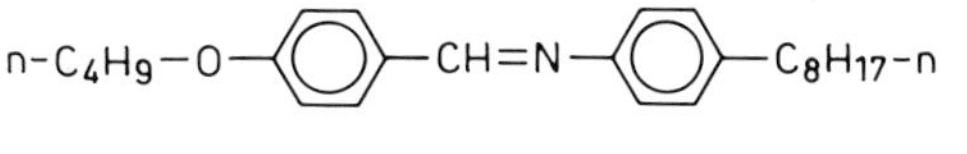

III

References on pp. 289/90

$Fe(C_5H_4\text{-}CO_2C_6H_5)_2$ (Table **20**, No. **7**) yields the polybenzimidazol IV on heating with 3,3′-diaminobenzidine in the complete absence of O_2 at 0.1 Torr, initially at 220 to 280°C and finally at 400°C [11]. Under normal pressure the reaction sets in at about 210 to 215°C [12].

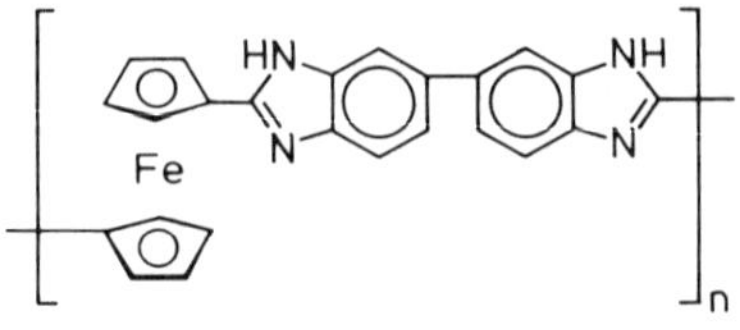

IV

$Fe(C_5H_4\text{-}CO_2(CH_2)_2OH)_2$ (Table **20**, No. **8**). For formation as an intermediate see "Applications" of $Fe(C_5H_4\text{-}CO_2CH_3)_2$, pp. 282/3.

II (x=0) (see p. 284) is obtained with $Fe(C_5H_4\text{-}COCl)_2$ in refluxing C_6H_6 [33]. No. 8 is converted to $Fe(C_5H_4\text{-}CO_2(CH_2)_2O_2CNH(CH_2)_6NCO)_2$ on reaction with $OCN(CH_2)_6NCO$ in a 1/2 mole ratio in boiling C_6H_5Cl. At a 1/1 mole ratio V precipitates as a very viscous, dark red polymeric product soluble in acetone. If No. 8 is warmed with m-toluylenediisocyanate to 80°C, then after 5 min foaming occurs, and a red, brittle mass forms on cooling. Its solution in acetone poured onto a glass plate yields after 3 d a transparent, light orange film of VI, which is only with difficulty separated from the glass [21].

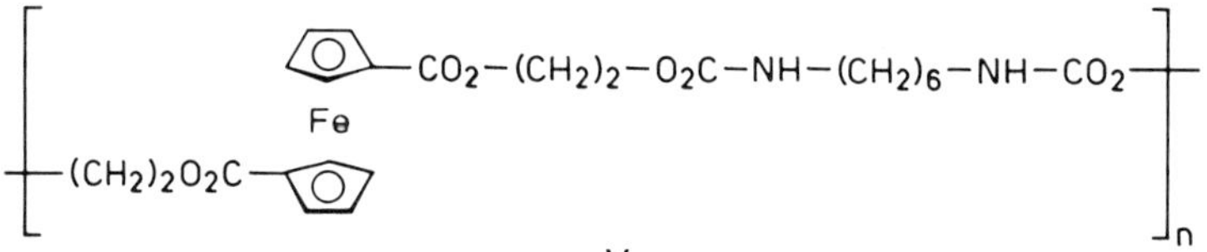

V

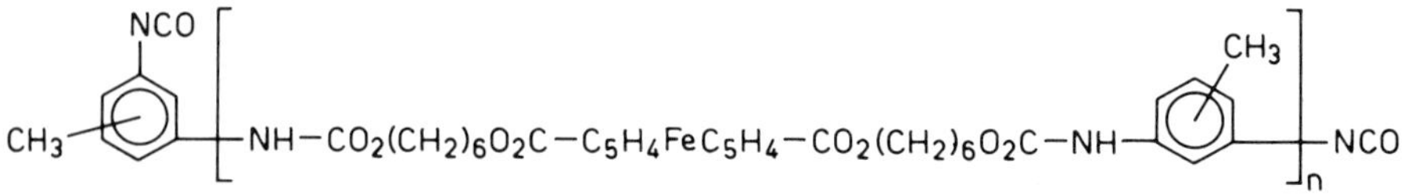

VI

$Fe(C_5H_4\text{-}CO_2CH_2(CH_2OCH_2)_nCH_2OH)_2$ (Table **20**, Nos. **9** to **11** with n=1 to 3) reacts with $Fe(C_5H_4\text{-}COCl)_2$ in refluxing C_6H_6 to form II (see p. 284) with x=1 and $A=(CH_2OCH_2)_n$ [33].

$Fe(C_5H_4\text{-}CO_2CH_2C_6H_4CH_2OH\text{-}3)_2$ (Table **20**, No. **12**) gives II (see p. 284) with x=1 and A=1,3-C_6H_4 with $Fe(C_5H_4\text{-}COCl)_2$ in boiling benzene [33].

$Fe(C_5H_4\text{-}CO_2C_6H_4OH\text{-}4)_2$ (Table **20**, No. **13**) in chromatography on SiO_2 columns with 5/1 $CHCl_3$/acetone elutes before VII, the by-product resulting in 1 and 11% yields, respectively, in the preparations by both Methods Ic and IIb (pp. 283 and 284) [5]. Reaction with $Fe(C_5H_4\text{-}COCl)_2$ following Method Ic gives a 6% yield of VII [31].

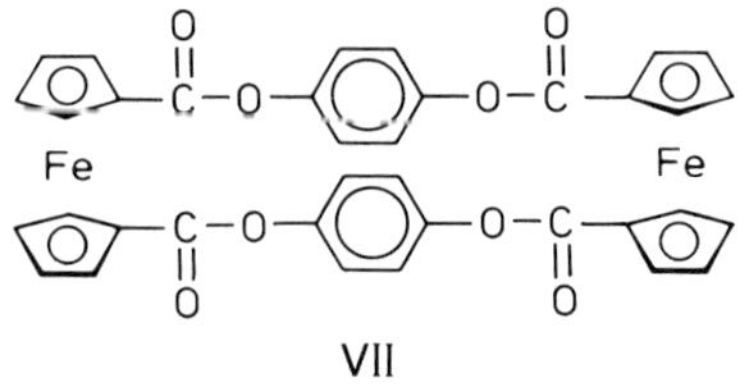

VII

$Fe(C_5H_4–CO_2C_{10}H_{19}$-l$)_2$ (Table **20**, No. **14**). Studies of the atropisomerism show that at room temperature the free rotation of the cyclopentadienyl rings with respect to each other is not hindered. The chromatogram and the ^{13}C NMR spectrum both indicate that only a single product exists [28].

$CH_3O_2C–C_5H_4FeC_5H_4–CO_2H$ (Table **20**, No. **15**) results during the preparation together with $Fe(C_5H_4–CO_2CH_3)_2$ and $Fe(C_5H_4–CO_2H)_2$. From the alkaline solution of this mixture is extracted first $Fe(C_5H_4–CO_2CH_3)_2$ with ether and subsequently after acidification No. 15 [18]. The main amount of the dicarboxylic acid remains behind undissolved. Quantitatively the separation of No. 15 from $Fe(C_5H_4–CO_2H)_2$ succeeds only through preparative thin layer chromatography on SiO_2 with 5/1 C_6H_6/C_2H_5OH [17].

The half-wave potential in 9/1 C_2H_5OH/H_2O (0.01 M $NaClO_4$, 0.01 M $HClO_4$) at 25±1°C from polarography is $E_{1/2}=0.85\pm0.01$ V versus the saturated calomel electrode for the electrochemical oxidation to the ferrocenium ion **$[CH_3O_2C–C_5H_4FeC_5H_4–CO_2H]^+$**. The value 0.90 V is calculated on the basis of the additivity of the polar substituent effects [26].

The acidity of No. 15 is greater than that of $C_5H_5FeC_5H_4–CO_2H$. In CH_3OH/H_2O (68/32 wt%) $pK_a=6.08$ at 20°C [1], in C_2H_5OH/H_2O (1/1 v/v) at 25±1°C $pK_a=5.81$ [10]. For the correlation of pK_a with the substituent constants of R for a series of compounds of the type $HO_2C–C_5H_4FeC_5H_4–R$, among them No. 15, see "Ferrocene" 3, 1978, pp. 58/9, and for the transmission of the electron effects see [16]. The introduction of CO_2CH_3 into $C_5H_5FeC_5H_4–CO_2H$ causes a slight bathochromic shift of its νC=O bands, like that of CO_2H in $C_5H_5FeC_5H_4–CO_2CH_3$ [5].

Refluxing with PCl_3 in C_6H_6 gives $ClCO–C_5H_4FeC_5H_4–CO_2CH_3$ [3, 17, 18].

$C_2H_5O_2C–C_5H_4FeC_5H_4–CO_2H$ (Table **20**, No. **16**) possesses in C_2H_5OH/H_2O (63/37) a pK_a value of 6.25 and thus constitutes a stronger acid than $C_5H_5FeC_5H_4–CO_2H$ [4]. It reacts with the neocarborane VIII (n=4,5) in benzene in the presence of catalytic amounts of n-butyl titanate on refluxing to form the half-ester $(C_2H_{11}B_{10})–(CH_2)_n–O_2C–C_5H_4FeC_5H_4–CO_2H$ [14].

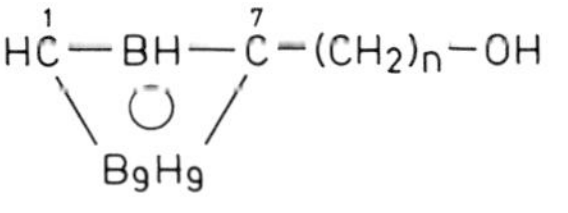

VIII

References:

[1] N. A. Nesmeyanov, O. A. Reutov (Dokl. Akad. Nauk SSSR **115** [1957] 518/21; Proc. Acad. Sci. USSR Chem. Sect. **112/117** [1957] 763/6). – [2] W. Göller (Diss. Stuttgart T.H. 1958). – [3] N. A. Nesmeyanov, O. A. Reutov (Dokl. Akad. Nauk SSSR **120** [1958] 1267/70; Proc. Acad. Sci. USSR Chem. Sect. **118/123** [1958] 503/6). – [4] R. A. Benkeser, L. W. Hall Jr. (AFOSR-TN-59-450 [1959] 1/9; AD-215025 [1959] 1/9; PB-140776 [1959] 1/9; U.S. Govt. Res. Rept. **32** [1959] 165). – [5] L. A. Kazitsyna, B. V. Lokshin, A. N. Nesmeyanov (Dokl. Akad. Nauk SSSR **127** [1959] 333/6; Proc. Acad. Sci. USSR Chem. Sect. **124/129** [1959] 531/4).

[6] R. L. Schaaf (J. Org. Chem. **27** [1962] 107/11). – [7] R. M. Van Vliet (U.S. 3287314 [1962/66]; C.A. **66** [1967] No. 29600). – [8] R. L. Schaaf, K. L. Rinehart (WADD-TR-61-108-II [1961] 1/80; AD-403929 [1961/65] 1/80, 49). – [9] K. Yamakawa, H. Ochi, K. Arakawa (Chem. Pharm. Bull. [Tokyo] **11** [1963] 905/11). – [10] S. P. Gubin, K. I. Grandberg, E. G. Perevalova, A. N. Nesmeyanov (Dokl. Akad. Nauk SSSR **159** [1964] 1075/8; Dokl. Chem. Proc. Acad. Sci. USSR **154/159** [1964] 1314/6).

[11] C. S. Marvel (SPE [Soc. Plast. Eng.] J. **20** [1964] 220/6). – [12] L. Plummer, C. S. Marvel (J. Polym. Sci. A **2** [1964] 2559/69). – [13] A. N. Nesmeyanov, B. M. Yavorskii, G. B. Zaslavskaya, N. S. Kochetkova (Dokl. Akad. Nauk SSSR **160** [1965] 837/40; Dokl. Chem. Proc. Acad. Sci. USSR **160/165** [1965] 131/4). – [14] M. M. Fein, N. Mayes,Thiokol Chemical Corp. (U.S. 3414596 [1965/68]; C.A. **70** [1969] No. 37909). – [15] Z. Beinoravichute (Mater. Vses. Simp. Vopr. Proizvod. Primen. Trub Detalei Truboprovodov Polietilena, Vilnius 1966, pp. 125/9; C.A. **68** [1968] No. 30546).

[16] S. P. Gubin (Izv. Akad. Nauk SSSR Ser. Khim. **1966** 1551/8; Bull. Acad. Sci. USSR Div. Chem. Sci. **1966** 1495/501). – [17] K. Schlögl, H. Soukup (Monatsh. Chem. **99** [1968] 927/46). – [18] T. H. Barr, H. L. Lentzner, W. E. Watts (Tetrahedron **25** [1969] 6001/13). – [19] A. N. Nesmeyanov, G. B. Zaslavskaya, B. M. Yavorskii, N. S. Kochetkova, N. P. Gambaryan (Opt. Spektrosk. **26** [1969] 945/9; Opt. Spectrosc. [USSR] **26** [1969] 512/5). – [20] V. G. Shershun, B. S. Sakharov, T. P. Vishnyakova, Ya. M. Paushkin (Tr. Mosk. Inst. Neftekhim. Gazov. Prom. No. 3 [1969] 48/54).

[21] E. B. Sokolova, M. P. Shebanova, L. A. Datskevich (Sin. Fiz. Khim. Polim. **1970** No. 6, pp. 69/72). – [22] K. Tanikawa, K. Arakawa (Bunseki Kagaku **20** [1971] 278/81). – [23] V. O. Aimiuwa (Diss. Kent State Univ. 1976 from Diss. Abstr. Intern. B **37** [1976] 2351). – [24] M. Miyamura, Tokyo Shibaura Electric Co., Ltd. (Japan. Kokai 77-146223 [1976/77] from C.A. **88** [1978] No. 201074). – [25] M. Miyamura, Tokyo Shibaura Electric Co., Ltd. (Japan. Kokai 77-146219 [1976/77] from C.A. **88** [1978] No. 201078).

[26] M. M. Sabbatini, E. Cesarotti (Inorg. Chim. Acta **24** [1977] L9/L10). – [27] J. Holeček, K. Handlíř, J. Klikorka (Proc. Conf. Coord. Chem. **7** [1978] 65/70). – [28] B. Nickel, A. Rassat (Tetrahedron Letters **1978** 633/6). – [29] A. M. Shevchik, Yu. P. Losev, Ya. M. Paushkin, R. V. Chibrikova, Z. I. Zaremba (Vestsi Akad. Navuk Belarusk. SSR Ser. Khim. Navuk **1978** No. 4, pp. 114/6; C.A. **89** [1978] No. 163719). – [30] V. P. Red'ko, V. K. Skubin, D. F. Kutepov, V. V. Korshak, V. G. Mochalov (Vysokomol. Soedin. B **22** [1980] 222/4).

[31] A. Kasahara, T. Izumi, K. Saito (Bull. Chem. Soc. Japan **56** [1983] 2865/6). – [32] V. O. Aimiuwu, D. L. Uhrich (Mol. Cryst. Liq. Cryst. **43** [1977] 295/312). – [33] T. Izumi, T. Tezuka, S. Yusa, A. Kasahara (Bull. Chem. Soc. Japan **57** [1984] 2435/9).

5.2.3.7.1.3 Carboxylic Acid Esters of the Type $R'O_2C–C_5H_4FeC_5H_4–R^2$ ($R'O_2C = R^1$)

The compounds collected in Table 21 are prepared by the following methods and by those described under "Further information", pp. 302/5:

Method I. Esterification of $HO_2C–C_5H_4FeC_5H_4–R^2$

a) with CH_2N_2 in ether [1, 8, 15, 21, 23, 34],

b) in boiling CH_3OH for 7 h [7] or overnight [48] with addition of 5% [1] or conc. hydrochloric acid [48],

c) in absolute CH_3OH saturated with dry HCl on an ice bath [12],

d) in boiling CH_3OH with addition of several drops of conc. H_2SO_4 [44],

e) with $(C_6H_5)_2CN_2$ in toluene at room temperature [6].

The products obtained by Methods Id or Ie are purified by column chromatography on SiO_2 [44] or Florisil [6], respectively.

Method II. Clemmensen reduction with amalgamated Zn and concentrated hydrochloric acid of $CH_3O_2C–C_5H_4FeC_5H_4–COC_nH_{2n+1}$ (n = 0, 1, 2, and 3 for Nos. 1, 2, 3, and 4, respectively) or of $CH_3O_2C–C_5H_4FeC_5H_4–COC_6H_5$ for No. 5

a) in CH_3CO_2H [16, 37],

b) in 1/1/1 $C_6H_6/CH_3OH/H_2O$ at the boiling point with subsequent purification by thin layer chromatography on SiO_2 with C_6H_6/petroleum ether [27, 29].

Method III. Reaction of $Fe(C_5H_4–CO_2CH_3)_2$:

a) With $LiAlH_4$ (mole ratio 1/0.75) in ether for 5 min at room temperature, followed by thin layer chromatography of the raw product to give in addition to a 23% yield of No. 9 an 18% yield of $Fe(C_5H_4–CH_2OH)_2$ and 45% of the unreacted starting compound [29].

b) With $CF_3(CH_2)_2MgCl$ (mole ratio 1/29) in boiling ether, forming in addition to Nos. 11 and 22 also $Fe(C_5H_4–C(OH)(CH_2CH_2CF_3)_2)_2$ (27.5% yield) and $CF_3(CH_2)_2CO–C_5H_4FeC_5H_4–C(OH)(CH_2CH_2CF_3)_2$ (31% yield). All four products are separated by chromatography with petroleum ether, benzene, and ethanol as eluents on an Al_2O_3 column protected from light [28].

Method IV. Reaction of $CH_3O_2C–C_5H_4FeC_5H_4–COCl$ in tetrahydrofuran with $LiAlH(OC(CH_3)_3)_3$ initially at −75°C with gradual warming to room temperature. No. 15 is purified by preparative thin layer chromatography on SiO_2 in C_6H_6 [27].

Method V. Friedel-Crafts acylation of $C_5H_5FeC_5H_4–R^1$:

a) With R^2Cl and $AlCl_3$ (mole ratio 1/(2 to 6)/(4 to 7.4) for Nos. 16 to 18, 21 [3, 16, 43] or 1/3.5/2 for No. 19 [44]) in CCl_4 with cooling at room temperature, for 1 h [3, 16, 43].

b) As for Method Va, but for 3 to 6 h in CH_2Cl_2 with the mole ratio 1/(>1)/(ca. 3) [4, 26, 45].

c) With $(CH_3CO)_2O$ and $BF_3 \cdot O(C_2H_5)_2$ in CH_2Cl_2 for 6 h at room temperature [43].

The product obtained by Method Va after the usual processing is chromatographed in 1/1 petroleum ether/benzene on an Al_2O_3 column and the compound eluted with ether [16]. No. 19 is isolated from the reaction product by column chromatography on Al_2O_3 with 9/1 $C_6H_6/CH_3CO_2C_2H_5$ [44].

Method VI. Photoethoxycarboxylation of $C_5H_5FeC_5H_4–R^2$ in 1/1 C_2H_5OH/CCl_4 with irradiation at 365 nm for 4 h at room temperature. The reaction product is extracted with CH_2Cl_2 or ether, the solvent evaporated off, and the residue subjected to thin layer or column chromatography for separation of the starting compound. A second thin layer chromatographing isolates the compound from the mixture with its 1,2- and 1,3-positional isomers. Nos. 24 and 28 are exceptions; in the former case $C_5H_5FeC_5H_4–C_6H_4CO_2C_2H_5$-4 also forms in 7.7% yield, in the latter there is no yield of $C_5H_5FeC_5H_3(CO_2C_2H_5\text{-}1)OCH_3$-2. The yields quoted for Method VI refer in each case to unrecovered $C_5H_5FeC_5H_4–R^2$ [42].

Purification of the compounds is in general by recrystallization from the solvents given in parentheses following the melting points shown in Table 21 and/or by chromatography. For details see under the Methods above or "Further information" following the table.

Properties

The ^{1}**H NMR spectra** in general display the two sets of triplets H-2,5 and H-3,4 for the R^1–C_5H_4 and for the R^2–C_5H_4 protons, whose spin-spin splitting belongs to the A_2B_2 type. The influence of the substituents on the proton screening in their own cyclopentadienyl ring is greater than on that in the other. The chemical shifts in CCl_4 calculated by the additivity rule from the spectra for $C_5H_5FeC_5H_4$–R^1 and $C_5H_5FeC_5H_4$–R^2 agree well with the measured values; the differences for Nos. 7, 14, and 17 are at most 0.07 ppm [19].

The **IR spectra** of CH_3O_2C–$C_5H_4FeC_5H_4$–R^2 undergo frequency changes in the solid state caused by dipole-dipole interactions. In solution the νC=O bands shift to higher values, the more so the more electrophilic the R^2. Thus, the frequency is smaller with $R^2 = C_2H_5$ (No. 2) or C_4H_9 (No. 4) and larger with $R^2 = COCH_3$ (No. 16) or COC_3H_7-n (No. 18) than that for $R^2 = H$ (i.e., for $C_5H_5FeC_5H_4$–CO_2CH_3) [5]. A linear correlation corresponding to a modified and extended Seth-Paul-Van Duyse equation has been observed between the νC=O vibration in CCl_4 and the $X^+(R^{A,B})$ constants of the structural fragments for 154 compounds with ferrocene structures of the general formula R^A–CO–R^B, among them the compounds No. 2 to 5, 7, 16 to 18, and 21, with $R^A = C_5H_4FeC_5H_4$–R^2 and $R^B = OCH_3$. The $X^+(R^{A,B})$ used have been confirmed by the recently derived empirical relationship for calculation of the substituent constants and transmission factors [47]. The integrated intensity of the νC=O bands of R^1 constitutes a measure of the transannular effect of R^2 [22]. With increasing electron acceptor strength for R^2 the integral intensity A (measured in CCl_4) drops from 4.1×10^4 for $R^2 = H$ through 3.9×10^4 for No. 18 with $R^2 = COC_3H_7$-n to 3.8×10^4 for No. 17 with $R^2 = COC_2H_5$, all in $L^2 \cdot mol^{-1} \cdot cm^{-2}$ [9]. The intensity change ΔA (referred to that for $R^2 = H$) measured in CCl_4 on Nos. 2 to 5, 7, 14, 16 to 18, and 21 is correlated with the induction constants σ_i and the conjugation constants σ_c^0 of R^2 via $\Delta A \cdot 10^{-4} = -1.38\,\sigma_c^0 - 0.15\,\sigma_i$. A considerable deviation from this equation was found only for No. 14 ($R^2 = OCH_3$). The effect of R^2 on the integral intensity of the νC=O bands of CH_3O_2C–$C_5H_4FeC_5H_4$–R^2 is about half as large as in CH_3O_2C–C_6H_4–R^2-4 [22].

The νC=O band of R^2 of the keto acid esters, such as Nos. 16 and 18, is shifted to lower frequencies versus that of the aliphatic CO compounds, which is attributed to the conjugation of the CO group with the cyclopentadienyl ring [2]. The νC–H bands of the cyclopentadienyl rings lie in the region of 3080 to 3120 cm^{-1}. They are subdivided into two components for H-2,5 and H-3,4 protons. The substituents R^1 and R^2 affect their location only a little but in contrast their intensity significantly [10].

The substituents exercise an additive effect on the location of the "ferrocene band" around λ = 440 nm, in the **electronic absorption spectrum**, but not on those of the remaining bands, as shown by studies of the compound No. 5, 7, 14, 16, 18, and 21 in n-heptane solution. Introduction of the substituents $R^2 = CH_2C_6H_5$, Br, $COCH_3$, COC_3H_7, and also CN and CO_2CH_3 into the C_5H_5 ring of $C_5H_5FeC_5H_4$–CO_2CH_3 as a rule displaces its band at $\lambda_{max} = 263$ nm hypsochromically. Nos. 14 ($R^2 = OCH_3$) and 21 ($R^2 = COC_6H_5$) are exceptions, with bathochromic shifts of 1600 and 700 (sh) cm^{-1}, respectively. The compounds No. 5, 7, 14, 16, 18, and 21 exhibit in n-heptane in the 300 to 320 nm region (ε ≈ 1000) an additional band, the charge transfer band to CO_2CH_3 [20]. In C_2H_5OH this does not appear for Nos. 7 and 14, apparently due to complete masking by the bands around 260 nm [24].

Chemical Behavior

The following **quarter-wave potentials** $E_{\tau/4}$ versus the saturated calomel electrode (SCE) have been measured in CH_3CN (0.2 M $LiClO_4$) at 25°C by chronopotentiometry, for the reversible oxidation of the compounds No. 32 to 36 to the ferrocenium ion per

$(C_6H_5)_2CHO_2C–C_5H_4FeC_5H_4–R^2 - e^- \rightleftharpoons [(C_6H_5)_2CHO_2C–C_5H_4FeC_5H_4–R^2]^+$; they lie by $\Delta E_{\tau/4}$ above those of ferrocene [14]:

No.	R^2	$E_{\tau/4}$ in V (vs. SCE)	$\Delta E_{\tau/4}$ in V
32	C_2H_5	0.519	0.204
33	$CH_2C_6H_5$	0.535	0.220
34	$CHOH–C_6H_5$	0.557	0.242
35	$CO–CH_3$	0.808	0.493
36	$CO–C_6H_5$	0.783	0.468

A well fulfilled linear dependence exists between $E_{\tau/4}$ and the sum of the para σ Hammett constants of the substituents R^1 and R^2 [14].

Alkaline **saponification** of the esters gives the free acids in high yields. It is conducted for $CH_3O_2C–C_5H_4FeC_5H_4–R^2$ with 15% methanolic KOH for $R^2 = CH_3$ (No. 1) [23, 29], see also [27], or with 20% NaOH for $R^2 = COCH_3$ (No. 16) [3, 16], COC_2H_5 (No. 17) [16], COC_3H_7 (No. 18) [3, 16], and COC_6H_5 (No. 21) [16], see also [4].

Vigorous **reduction** with $LiAlH_4/AlCl_3$ in ether [13, 34] or tetrahydrofuran [23] converts the CH_3O_2C into the CH_3 group in good yields, forming $Fe(C_5H_4–CH_3)_2$ from No. 1 [13, 23] and $I–C_5H_4FeC_5H_4–CH_3$ from No. 8. The fact that in the latter case the yield is so small may be attributed to the decomposition of No. 8 by the reducing agent [34].

With $LiAlH_4$ alone $HOCH_2–C_5H_4FeC_5H_4–R^2$ is obtained when $R^2 = CH_3$ [29], C_2H_5, $n\text{-}C_3H_7$, $n\text{-}C_4H_9$ [16], $CH_2C_6H_5$ [16, 48]. The action of $LiAlH_4$ on $CH_3O_2C–C_5H_4FeC_5H_4–COCH_3$ (No. 16), which primarily occurs at the $COCH_3$, is treated under "Further information" for No. 16, p. 304.

The keto esters $CH_3O_2C–C_5H_4FeC_5H_4–CO–R''$ with $R'' = C_nH_{2n+1}$ (n = 0 to 3) (Nos. 15 to 18) or C_6H_5 (No. 21) easily undergo Clemmensen reduction (see Method II, p. 291) in good yields to form $CH_3O_2C–C_5H_4FeC_5H_4–CH_2–R''$ (Nos. 1 to 5) [16, 27, 37]. For the ionic hydrogenation of No. 16 ($R'' = CH_3$), see its "Further information", p. 304.

The **internal Claisen condensation** of the keto esters $CH_3O_2C–C_5H_4FeC_5H_4–COCH_2R'''$ with $R''' = H$, CH_3, C_6H_5 (Nos. 16, 17, 19) leads to I.

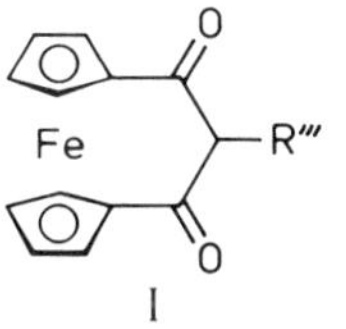

I

This cyclization requires a high activation energy. It occurs with $KOC(C_6H_5)_3$ as a catalyst in boiling xylene (yield 78% for $R''' = H$, 60% for CH_3, and 20% for C_6H_5), but not at room temperature. $KOC(CH_3)_3$ possesses a somewhat lesser effectiveness as a catalyst, giving a 50% yield for $R''' = H$ in boiling xylene. Along with the cyclization occurs partial saponification and, due to the high sensitivity to oxygen of I with $R''' = CH_3$ or C_6H_5, its oxidation, so that the by-products include $HO_2C–C_5H_4FeC_5H_4–COCH_2R'''$ (7% yield for $R''' = H$, 44% for C_6H_5) and $HO_2C–C_5H_4FeC_5H_4–COCOR'''$ (yield 4% for $R''' = CH_3$, 11% for $R''' = C_6H_5$) [44].

Applications

The photoimaging composition which contains 4 to 5 wt% Nos. 16, 29, or 31 in addition to 13 to 20% CBr_4 and 1.2 to 2.0% tanning additive requires a lower developing temperature and a shorter development time compared to other ferrocene-based compositions [46].

Table 21

Carboxylic Acid Esters of the Type $R'O_2C–C_5H_4FeC_5H_4–R^2$.

Further information for numbers preceded by an asterisk is given at the end of the table, pp. 302/5. For abbreviations and dimensions see p. VIII.

No.	$R^1–C_5H_4FeC_5H_4–R^2$ $R^1 = CO_2R'$	R^2	method of preparation (yield in %), properties, and remarks	Ref.
*1	CO_2CH_3	CH_3	Ia, IIb (96), see "Further information", p. 302 m.p. 36 to 38° ^{1}H NMR(CCl_4): 1.89 (s, C–CH_3), 3.99 (s, $C_5H_4–R^2$); 4.22 (s, H-3,4) and 4.61 (s, H-2,5)($C_5H_4–R^1$) IR: 1710 IR (CCl_4): ca. 1720 (νC=O)	[13, 21, 23, 25, 27, 29, 32]
2	CO_2CH_3	C_2H_5	IIa (74), see Nos. 10, p. 303, and 16, p. 304 b.p. 115 to 117°/1.5 Torr IR: 1714 (νC=O) IR (Vaseline oil): 814(m), 876(m), 900(m), 970(s), 1028(m), 1142(vs), 1192(m), 1280(vs), 1378(m), 1464(vs), 1514(w), 1654(sh), 1718(vs, νC=O) IR (CCl_4): 1718 or 1721 (νC=O) IR ($CHCl_3$): 1707 (νC=O) n_D^{20}: 1.5911	[5, 16, 21, 22, 37, 38, 47]
3	CO_2CH_3	C_3H_7-n	IIa (95) b.p. 124 to 126°/2 Torr IR: 1718 (νC=O) IR (CCl_4): 1721 (νC=O) n_D^{20}: 1.5817	[16, 21, 22, 47]
4	CO_2CH_3	C_4H_9-n	IIa (88) b.p. 143 to 144°/2 Torr IR (Vaseline oil): 800(s), 822(s), 894(s), 920(m), 958(s), 964(s), 1020(s), 1022(s), 1184(s), 1190(s), 1264(s), 1280(s), 1370(s), 1454(s), 1460(s), 1696 to 1712 (vs, νC=O)	[5, 16, 22, 47]

			IR (CCl_4): 1704 to 1712 (νC=O) IR (CCl_4): 1721 (νC=O) IR ($CHCl_3$): 1696 to 1704 (νC=O) n_D^{20}: 1.5739	
5	CO_2CH_3	$CH_2C_6H_5$	Ia (74), see No. 21, p. 305 reddish oil, m.p. 57.5 to 58°, b.p. 198 to 200°/2.5 Torr ^{1}H NMR (CCl_4): 3.98 (C_5H_4–R^2); 4.17 (t, H-3,4) and 4.61 (t, H-2,5)(C_5H_4–R^1) IR (CCl_4): 1726 (νC=O) IR: 1140(s), 1193(s), 1256(s), 1280(s), 1300(s), 1376(m), 1438(m), 1465(m), 1496(m), 1593(m), 1750(s, br.), 2865(s), 2948(m), 3020(w), 3080(w) UV (n-heptane): $\lambda_{max}(\varepsilon)$ = 444(240), 340sh (360), 261sh (6000), 214 (33000) UV (C_2H_5OH): λ_{max} (ε) = 302 (1200)	[16, 19, 20, 22, 24, 47, 48]
6	CO_2CH_3	Cl	Ia m.p. 53 to 54°	[15]
*7	CO_2CH_3	Br	Ia (96) m.p. 68 to 69.5 (hexane) ^{1}H NMR (CCl_4): 4.07 (H-3,4 of C_5H_4–R^2), 4.33 (H-3,4 of C_5H_4–R^1 and H-2,5 of C_5H_4–R^2), 4.75 (H-2,5 of C_5H_4–R^1) IR (CCl_4): 1723(νC=O) UV (n-heptane): $\lambda_{max}(\varepsilon)$ = 443(210), 340sh (200), 252sh (5000)	[7, 15, 19, 20, 22, 24, 47]
*8	CO_2CH_3	I	Ia (76), see "Further information", p. 303 orange needles, m.p. 71 to 72° (CH_3OH/H_2O), 71 to 72.5° (n-hexane) ^{1}H NMR ($CDCl_3$): 3.86 (s, CH_3), 4.2 (t, H-3,4 of C_5H_4–R^2), 4.42 (m, H-3,4 of C_5H_4–R^1 and H-2,5 of C_5H_4–R^2), 4.83 (t, H-2,5 of C_5H_4–R^1) IR (KBr): 1025, 1140, 1190, 1275, 1380, 1460, 1715(νC=O)	[34, 36]

Table 21 (continued)

No.	R^1–$C_5H_4FeC_5H_4$–R^2 $R^1 = CO_2R'$	R^2	method of preparation (yield in %), properties, and remarks	Ref.
*9	CO_2CH_3	CH_2OH	IIIa (23), see "Further information", p. 303 oil ^{1}H NMR ($CDCl_3$): 2.30 (OH), 3.82 (CH_3), 4.23 to 4.44 ($CH_2 + 6C_5H_4$ protons), 4.82 ($2C_5H_4$ protons) IR (CCl_4): 1720 (νC=O), 3500 (O–H ··· O), 3620 (free O–H)	[29, 41]
*10	CO_2CH_3	$CHOHCH_3$	see No. 16, p. 304, and "Further information", p. 303 yellow oil, m.p. 55.5 to 56° (hexane), 57.5 to 58.5° (hexane) IR: 1720 (νC=O), 3500 (br., coordinated OH) IR (CCl_4): 1705 and 1724 (νC=O), 3508 (O–H ··· O), 3552 (O–H ··· Fe), 3605 (free O–H) IR (CCl_4): 3512, 3550 sh IR (C_6H_6): 3510 ± 13 (br.), 3550 sh	[8, 18, 26, 31]
11	CO_2CH_3	$COH(CH_2CH_2CF_3)_2$	IIIb (9) m.p. 74° IR: 1710 (νC=O), 3480 (νO–H)	[28]
12	CO_2CH_3	O_2CCH_3	Ia (quantitative), see No. 7, p. 302 m.p. 46 to 47.5° (hexane) partially decomposes during chromatography on Al_2O_3	[7]
*13	CO_2CH_3	$CH(CH_3)O_2CCH_3$	see No. 10, p. 303 oil IR: 1720 and 1725 sh (νC=O) n_D^{25}: 1.5721	[26]
14	CO_2CH_3	OCH_3	Ia m.p. 37.5 to 39° ^{1}H NMR (CCl_4): 3.82 (H-2,5 of C_5H_4–R^2), 4.05 (H-3,4 of C_5H_4–R^2), 4.33 (H-3,4 of C_5H_4–R^1), 4.76 (H-2,5 of C_5H_4–R^1) IR (CCl_4): 1718 (νC=O) UV (n-heptane): $\lambda_{max}(\varepsilon)$ = 446 (250), 342 sh (390), 275 (4400), 214 (27000)	[15, 19, 20, 22, 24, 47]

15	CO_2CH_3	CHO	IV (46), see No. 9, p. 303 m.p. 81 to 83° (petroleum ether), 81 to 84° IR (CCl_4): 1690 and 1760 (νC=O) IR (CCl_4): 1692 and 1750 (νC=O), 2760	[27, 29]
*16	CO_2CH_3	$COCH_3$	Ia (82), Ib, Ic (85), Va (80, 58, 35, 30), Vb (92), Vc (66), see "Further information", p. 304 dark red to orange crystals, orange-red prisms, m.p. 91 to 94° (ether), 92.5 to 94.5° (ligroin), 98 to 99° (hexane), 101 to 101.7° (heptane), 101 to 102° (heptane or C_6H_6/petroleum ether), 102 to 102.5° (hexane), 102 to 103° (cyclohexane) ^{1}H NMR (C_6H_6): 2.0 (s, $COCH_3$), 3.43 (s, CO_2CH_3); 3.82 (t, 2H), 3.95 (t, 2H), 4.45 (t, 2H), 4.57 (t, 2H) (C_5H_4–R^1 + C_5H_4–R^2) ^{1}H NMR (CCl_4): 4.20 (m) and 4.60 (m) (C_5H_4–R^1 + C_5H_4–R^2) IR (Vaseline oil): 822 (s), 840 (vs), 883 (m), 894 (s), 906 (vw), 921 (w), 940 (vw), 974 (vs), 1035 (m), 1052 (m), 1072 (s), 1120 (s), 1154 (vs), 1288 (vs), 1466 (s), 1663 (s, νC=O), 1718 (s, νC=O) IR ($CHCl_3$): 1676 (νC=O of R^2), 1716 (νC=O of R^1), 3096 to 3114 (νC–H of C_5H_4) IR (CCl_4): 1678 or 1680 (νC=O of R^2), 1720 or 1722 or 1726 (νC=O of R^1), see figure in [1] UV (n-heptane): λ_{max} (ε) = 449 (250), 342sh (300), 255 (6800), 218 (25000) UV (C_2H_5OH): λ_{max}(ε) = 455 (430), 311 (1300), 277 (14000) mass spectrum: m/e = 286 $[M]^+$, 194 $[CH_3COC_5H_4FeOCH_3]^+$ see "Applications", p. 293	[1 to 5, 8, 10, 12, 16, 20, 22, 24, 26, 43, 47]
*17	CO_2CH_3	COC_2H_5	Va (88), Vb m.p. 64°, 64 to 65° (hexane) ^{1}H NMR ($CDCl_3$): 1.18 (t, CH_3 of C_2H_5), 2.73 (m, CH_2), 3.80 (s, CO_2CH_3); 4.38 (t, 2H), 4.48 (t, 2H), and 4.79 (t, 4H) (C_5H_4–R^1 + C_5H_4–R^2) ^{1}H NMR (CCl_4): 4.36 (H-3,4 of C_5H_4–R^1), 4.45 (H-3,4 of C_5H_4–R^2), 4.77 (H-2,5 of C_5H_4–R^1 and C_5H_4–R^2) IR (CCl_4): 1685 (νC=O of R^2), 1725 (νC=O of R^1) IR ($CHCl_3$): 3098 to 3116 (C–H of C_5H_4)	[4, 9, 10, 16, 22, 35, 47]

References on pp. 305/7

Table 21 (continued)

No.	$R^1–C_5H_4FeC_5H_4–R^2$ $R^1 = CO_2R'$	R^2	method of preparation (yield in %), properties, and remarks	Ref.
18	CO_2CH_3	COC_3H_7-n	Va (80, 45), Vb m.p. 54.5 to 55.5° (ligroin), 57.5 to 58.5° (hexane) IR (Vaseline oil): 830(m), 842(m), 849(vw), 876(m), 885(vs), 904(m), 972(m), 1034(vs), 1064(m), 1077(m), 1153(vs), 1278(vs), 1344(w), 1400(w), 1460(vs), 1663 (vs, νC=O of R^2), 1716 (vs, νC=O of R^1) IR ($CHCl_3$): 1676 (νC=O of R^2), 1715 (νC=O of R^1), 3100 to 3118 (νC–H of C_5H_4) IR (CCl_4): 1679 or 1685 (νC=O of R^2), 1718 or 1725 (νC=O of R^1) UV (n-heptane): $\lambda_{max}(\varepsilon) = 454(310)$, 352sh (500), 258 (8600), 221(22000) UV (C_2H_5OH): $\lambda_{max}(\varepsilon) = 305(1400)$	[2 to 5, 9, 10, 16, 20, 22, 24, 47]
19	CO_2CH_3	$COCH_2C_6H_5$	Ia (~100), Va (70) fine orange needles, m.p. 123 to 125° (C_6H_6), 125° (C_2H_5OH) ^1H NMR ($OS(CD_3)_2$): 3.72 (s, CH_3), 4.03 (s, CH_2), 4.48 (t, H-3,4 of $C_5H_4–R^1$), 4.60 (t, H-3,4 of $C_5H_4–R^2$), 4.77 (t, H-2,5 of $C_5H_4–R^1$), 4.88 (t, H-2,5 of $C_5H_4–R^2$), 7.29 (apparent s, C_6H_5)	[17, 44]
20	CO_2CH_3	$COCH{=}CHC_6H_5$	Id (95), see No. 16, p. 304 deep red crystals, No. 20 exhibits dimorphism with m.p. of 65 to 70° and 125 to 130° (C_6H_6/petroleum ether) ^1H NMR ($CDCl_3$): 3.76 (s, CH_3), 4.40 (t, H-3,4 of $C_5H_4–R^1$), 4.56 (t, H-3,4 of $C_5H_4–R^2$), 4.82 (t, H-2,5 of $C_5H_4–R^1$), 4.91 (t, H-2,5 of $C_5H_4–R^2$), 7.10 and 7.78 (A Bq, CH=CH; J(A, B) = 15.6), 7.22 to 7.62 (m, C_6H_5)	[44]
*21	CO_2CH_3	COC_6H_5	Va (85), Vb (51), see "Further information", p. 305 dark, sweet smelling oil (raw product), m.p. 79.5 to 80.5° (hexane), 81 to 83° (hexane) ^1H NMR (CCl_4): 4.35 (H-3,4 of $C_5H_4–R^1$), 4.48 (H-3,4 of $C_5H_4–R^2$), 4.72 (H-2,5 of $C_5H_4–R^1$), 4.85 (H-2,5 of $C_5H_4–R^2$)	[4, 16, 19, 20, 22, 47, 48]

No.	R^1	R^2	method (yield in %), properties and remarks	Ref.
			IR (CCl_4): 1726 (νC=O of R^1) UV (n-heptane): $\lambda_{max}(\varepsilon)$ = 463 (500), 365 sh (1000), 268 sh (8300), 241 (11000)	
22	CO_2CH_3	$CO(CH_2)_2CF_3$	IIIb (13) m.p. 110° IR: 1670 (νC=O of R^2), 1725 (νC=O of R^1)	[28]
23	$CO_2C_2H_5$	C_2H_5	VI (41) red oily substance ^{1}H NMR (CCl_4): 1.12 (t, CH_3 of R^2), 1.32 (t, CH_3 of R^1), 2.30 (q, CH_2 of R^2), 3.98 (s, H-2 to 5 of C_5H_4–R^2), 4.15 to 4.24 (CH_2 of R^1 and H-3,4 of C_5H_4–R^1), 4.61 (t, H-2,5 of C_5H_4–R^1) IR (neat): 890, 900, 1270 (νC–O), 1710 (νC=O), 2920, 2960, 3080 mass spectrum (70 eV): m/e = 286 $[M]^+$	[42]
24	$CO_2C_2H_5$	C_6H_5	VI (21) yellow-orange solid ^{1}H NMR (CCl_4): 1.26 (t, CH_3), 4.01 (q, CH_2), 4.14 (t, H-3,4 of C_5H_4–R^2), 4.25 (t, H-3,4 of C_5H_4–R^1), 4.55 (t, H-2,5 of C_5H_4–R^1 and C_5H_4–R^2), 7.1 to 7.5 (m, C_6H_5) IR (KBr): 880, 920, 1280 (νC–O), 1600, 1685 (νC=O), 3050, 3090 mass spectrum (70 eV): m/e = 334 $[M]^+$	[42]
25	$CO_2C_2H_5$	Cl	VI (4) yellow oily substance ^{1}H NMR (CCl_4): 1.35 (t, CH_3), 4.03 (t, H-3,4 of C_5H_4–R^2), 4.24 (q, CH_2), 4.35 (t, H-3,4 of $C_5H_4R^1$ + H-2,5 of C_5H_4–R^2), 4.77 (t, H-2,5 of C_5H_4–R^1) IR (KBr): 880, 915, 1090, 1280 (νC–O), 1710 (νC=O), 3120 mass spectrum (70 eV): m/e = 292 $[M]^+$	[42]
26	$CO_2C_2H_5$	Br	VI (11) yellow oily substance ^{1}H NMR (CCl_4): 1.38 (t, CH_3), 4.07 (t, H-3,4 of C_5H_4–R^2), 4.23 (q, CH_2), 4.35 (t, H-3,4 of C_5H_4–R^1 + H-2,5 of C_5H_4–R^2), 4.73 (t, H-2,5 of C_5H_4–R^1)	[42]

References on pp. 305/7

Table 21 (continued)

No.	$R^1–C_5H_4FeC_5H_4–R^2$ $R^1 = CO_2R'$	R^2	method of preparation (yield in %), properties, and remarks	Ref.
			IR (KBr): 865, 910, 1270 (νC–O), 1705 (νC=O), 3100 mass spectrum (70 eV): m/e = 337 $[M]^+$	
27	$CO_2C_2H_5$	I	VI (4) orange-red oily substance ^{1}H NMR (CCl_4): 1.31 (t, CH_3), 4.12 (t, H-3,4 of $C_5H_4–R^2$), 4.22 (q, CH_2), 4.27 (t, H-2,5 of $C_5H_4–R^2$), 4.34 (t, H-3,4 of $C_5H_4–R^1$), 4.70 (t, H-2,5 of $C_5H_4–R^1$) IR (KBr): 860, 905, 1270 (νC–O), 1700 (νC=O), 3100 mass spectrum (70 eV): m/e = 384 $[M]^+$	[42]
28	$CO_2C_2H_5$	OCH_3	VI (24) yellow oily substance ^{1}H NMR (CCl_4): 1.33 (t, CH_3 of R^1), 3.55 (s, CH_3 of R^2), 3.74 (t, H-3,4 of $C_5H_4–R^2$), 3.97 (t, H-2,5 of $C_5H_4–R^2$), 4.19 (q, CH_2), 4.26 (t, H-3,4 of $C_5H_4–R^1$), 4.71 (t, H-2,5 of $C_5H_4–R^1$) IR (KBr): 865, 920, 1100, 1280 (νC–O), 1710 (νC=O), 2860, 2920, 2960, 3100 mass spectrum (70 eV): m/e = 288 $[M]^+$	[42]
*29	$CO_2C_2H_5$	$COCH_3$	IR (CCl_4): 1671.3 (νC=O of R^2), see p. 114 "Applications" see p. 293	[30]
*30	$CO_2C_2H_5$	$COCH_2CH_2Cl$	Vb (49) deep red oil ^{1}H NMR ($CDCl_3$): 1.4 (t, CH_3), 4.32 (q, $COCH_2$ of R^2), 3.20 (t, CH_2Cl), 3.90 (t, CH_2 of C_2H_5); 4.55 (m, 4H) and 4.82 (m, 4H) ($C_5H_4–R^1 + C_5H_4–R^2$)	[45]
31	$CO_2C_3H_7$-n	$COCH_3$	"Applications" see p. 293	[46]
32	$CO_2CH(C_6H_5)_2$	C_2H_5	Ie (91) m.p. 81.5° (hexane)	[6]

33	$CO_2CH(C_6H_5)_2$	$CH_2C_6H_5$	Ie (50) m.p. 87 to 87.5° (cyclohexane/hexane)	[6]
34	$CO_2CH(C_6H_5)_2$	$CHOHC_6H_5$	Ie (95) m.p. 88 to 89° (hexane)	[6]
35	$CO_2CH(C_6H_5)_2$	$COCH_3$	Ie (67) m.p. 103 to 104.5° (hexane)	[6]
36	$CO_2CH(C_6H_5)_2$	COC_6H_5	Ie (94) m.p. 133 to 133.5° (cyclohexane)	[6]
*37	$CO_2CH_2C_6H_4Br$-4	CH_3	see "Further information", p. 305	[23]

References on pp. 305/7

* Further information:

$CH_3O_2C–C_5H_4FeC_5H_4–CH_3$ (Table **21**, No. **1**) forms mixed with its 1,2- and 1,3-isomers on oxidation of $FeC_{10}H_8(CHO)CH_3$ (fraction of the 1,1′-compound 41%) with active, pulverized MnO_2 in the presence of KCN and glacial acetic acid in CH_3OH at 20°C. A 46% total yield of $FeC_{10}H_8(CO_2CH_3)CH_3$ results after purification by preparative thin layer chromatography on SiO_2 with benzene, with the fraction of No. 1 being 36% [32]. For separation of the positional isomers see below. Metallation of $C_5H_5FeC_5H_4–CH_3$ in decane with $n\text{-}C_5H_{11}M$ (M = Na, K) in a 1/1 mole ratio yields a mixture of $FeC_{10}H_8MCH_3$ and $FeC_{10}H_7M_2CH_3$. Reaction first with CO_2 and then CH_2N_2 in ether gives a mixture of $FeC_{10}H_8(CO_2CH_3)CH_3$ and $FeC_{10}H_7(CO_2CH_3)_2CH_3$, which is separated by column chromatography on Al_2O_3 with development by petroleum ether. Benzene elutes the mixture of the monoester isomers, consisting of No. 1 and its 1,3-isomer in a ca. 2.4/1 ratio, and ether that of the diesters [13].

A sharp separation of No. 1 and its two positional isomers is not possible, even when the mixture is subdivided into 20 to 25 fractions by chromatography over the total length of an Al_2O_3 column in petroleum ether/benzene (elution sequence 1,2-, 1,1′-, 1,3-isomers). A distinct shift of the fraction occurs in favor of the 1,3-isomer, which is attributed to the easier saponificability of the 1,1′- and 1,2-isomers and the consequent higher adsorbability on the alkaline column: 20% of the amount of ester introduced remains in each case irreversibly adsorbed [23]. Also, vapor phase chromatography failed to work for the separation of the 1,1′- and 1,3-isomers, in spite of use of eight different column fillings. Vapor phase chromatography on Craig polyester succinate at 228°C and 20 psi (1.4 atm) He pressure gave the two compounds in a single peak (5.5 min retention time) [13]. No. 1 is separated from $CH_3O_2C–C_5H_4FeC_5H_3(CO_2CH_3)CH_3$ by column chromatography on Al_2O_3 with petroleum ether, with subsequent elution of the diester with a 1/1 mixture of petroleum ether/benzene [21] (see also above). From $Fe(C_5H_4–CH_3)_2$ and $Fe(C_5H_4–CO_2CH_3)_2$ the separation is effectuated by column chromatography on SiO_2, where No. 1 is eluted with C_6H_6 after the hydrocarbon and $Fe(C_5H_4–CO_2CH_3)_2$ is subsequently eluted with 10/1 C_6H_6/C_2H_5OH [29].

The singlet of the CH_3 protons of R^2 is shifted by 0.06 ppm to higher field vs. that of $C_5H_5FeC_5H_4–CH_3$ in CCl_4 [25].

$CH_3O_2C–C_5H_4FeC_5H_4–Br$ (Table **21**, No. **7**). Reacting a boiling solution of No. 7 in ethanol with a hot, aqueous $Cu(O_2CCH_3)_2$ solution forms No. 12 in a 16% yield after 30 min on a boiling water bath. The ether extract from the diluted aqueous reaction solution is evaporated to remove the ether and the residue chromatographed on Al_2O_3. Benzene elutes unreacted No. 7 in a 45% recovery, ether then elutes No. 12. No. 12 partially decomposes on the Al_2O_3 column [7]. The Ullmann coupling of No. 7 with Cu is hindered by the CH_3O_2C group. At 125°C a satisfactory yield of II occurs only after reaction for 16 to 24 h [11], see "Ferrocene" 6, 1977, Table 3, No. 15, p. 20.

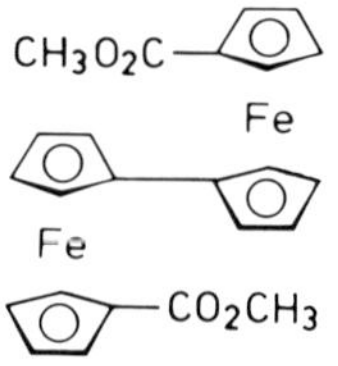

II

$CH_3O_2C–C_5H_4FeC_5H_4–I$ (Table **21**, No. **8**) is obtained in 90% yield by reaction of $CH_3O_2C–C_5H_4FeC_5H_4–HgCl$ with a saturated solution of I_2 in CH_2Cl_2 [36]. Ullmann coupling with activated Cu bronze gives a 43% yield of II. A 40% yield of $CH_3O_2C–C_5H_4FeC_5H_4–CN$ forms CuCN in N-methylpyrrolidone solution at 110°C [36].

$CH_3O_2C–C_5H_4FeC_5H_4–CH_2OH$ (Table **21**, No. **9**). No. 9 is formed in 15% yield by treating first with CO_2 and then with CH_2N_2 the product of lithiation of $C_5H_5FeC_5H_4–CH_2OCH_3$ in ether with 1.6 M n-C_4H_9Li in hexane for 3 h at 25°C. It is unclear whether the CH_3OCH_2 group was cleaved during carbonation, acidification, or chromatography of the product. It should be noted that this spontaneous hydrolysis is a result of the inherent instability of this particular compound relative to its methyl ester under reaction or isolation conditions and does not constitute a general method of cleavage of ferrocenyl methyl ethers [41].

No. 9 is oxidized to No. 15 by active MnO_2 in boiling $CHCl_3$ for 3 h. The No. 15 is isolated by preparative thin layer chromatography on SiO_2 in C_6H_6 from the raw product, which still contains $Fe(C_5H_4–CHO)_2$ and $Fe(C_5H_4–CO_2CH_3)_2$ [29].

$CH_3O_2C–C_5H_4FeC_5H_4–CHOHCH_3$ (Table **21**, No. **10**). $NC–C_5H_4FeC_5H_4–CHOHCH_3$ is saponified with 20% KOH and extracted with ether. After evaporating off the ether the remaining acid is immediately esterified by Method Ia (p. 290), giving a 64% yield referred to the nitrile [18]. Ether formation is avoided by not drying No. 10 after extraction in CH_2Cl_2 over $MgSO_4$ and chromatographing only on neutral or basic Al_2O_3 that is deactivated with 5 wt% H_2O [26].

The carbenium ion **$[CH_3O_2C–C_5H_4FeC_5H_4–CHCH_3]^+$** forms in $HClO_4$. In the UV $\lambda_{max} = 253$ nm in 60% $HClO_4$. The degree of dissociation is 12 to 90% in 30 to 44% acid. In 60% $HClO_4$ $pK_{R^+} = -5.27 \pm 0.30$, measured spectroscopically at 253 nm [33]; see also "Further information" for No. 13, below. Ionic hydrogenation with $(C_2H_5)_3SiH$ to No. 2 also proceeds via the carbenium ion. It takes place in glacial acetic acid solution only after addition of CF_3CO_2H. At a 1/5/10 molar ratio of No. 10/silane/CF_3CO_2H the yield amounts to 85% after 3 h at 20°C. In contrast to the ionic hydrogenation of $X–C_5H_4FeC_5H_4–COCH_3$ (cf. No. 16), for the series X = H, Cl, CO_2CH_3, CN in $X–C_5H_4FeC_5H_4–CHOHCH_3$, the observed rate constants decrease with the increasing acceptor strength of X. This is attributed to the increasingly hindered formation of $[X–C_5H_4FeC_5H_4–CHCH_3]^+$, which is not compensated for even by its growing reactivity with respect to H^- [38].

Mixing with $(CH_3CO)_2O$ and pyridine at room temperature forms No. 13, whose raw product remains behind after distilling off of the pyridine and excess anhydride at 0.5 Torr up to 50°C [26].

$CH_3O_2C–C_5H_4FeC_5H_4–CH(CH_3)O_2CCH_3$ (Table **21**, No. **13**) is purified by distillation of the raw product taken up in CH_2Cl_2. The following rate constants have been measured as a function of temperature for the hydrolysis proceeding via the S_N1 mechanism in 4/1 acetone/water [26]:

t in °C	35.00	45.00	63.50
$k \cdot 10^5$ in s^{-1} ...	0.438 ± 0.019	1.39 ± 0.02	10.0 ± 0.5

At 45°C $k_{rel} = 0.0176$ referred to the rate constant of 1.000 for $C_5H_5FeC_5H_4–CH(CH_3)O_2CCH_3$ at this temperature. The relative activation parameters based on 0.0 ± 0.57 kcal/mol and 0.00 $cal \cdot mol^{-1} \cdot K^{-1}$ for the enthalpy and entropy of $C_5H_5FeC_5H_4–CH(CH_3)O_2CCH_3$ are $\Delta H^{\ddagger}_{rel} = +2.33 \pm 1.13$ kcal/mol and $\Delta S^{\ddagger}_{rel} = +1.5$ $cal \cdot mol^{-1} \cdot K^{-1}$. From the relationship of k_{rel} and the quarter-wave potential $\Delta E_{\tau/4}$ (relative to that for $C_5H_5FeC_5H_5$) for the $X–C_5H_4FeC_5H_4–CH(CH_3)O_2CCH_3$, X = CH_3, Cl, Br, $COCH_3$, CO_2CH_3, or CN, it is evident that the hydrolysis is predominantly affected by the inductive and field effects of X, while resonance

References on pp. 305/7

effects – if any at all – are of subordinate importance [26]. A linear relationship exists between the logarithm of the solvolysis constants of these compounds and the pK_{R^+} values of $[X–C_5H_4FeC_5H_4–CH(CH_3)]^+$ (for X = CH_3O_2C see "Further information" for No. 10, p. 303) [33].

$CH_3O_2C–C_5H_4FeC_5H_4–COCH_3$ (Table **21**, No. **16**) also forms within a minute on acidification of the aqueous solution of $[CH_3O_2C–C_5H_4FeC_5H_4–C(CH_3)=NNHCOCH_2N(CH_3)_3]Cl$ with concentrated HCl. It is extracted from the solution with $CHCl_3$ [1].

Since the CH_3O_2C group impedes acetylation in general, but quite especially in its own cyclopentadienyl ring, Method V gives primarily No. 16 and only to a slight extent its 1,2-isomer. The separation of No. 16 from the raw product therefore in general takes place only via recrystallization from CH_3OH [3], heptane [26, 43], or hexane [43]. Preparative thin layer chromatography of the mother liquor on SiO_2 with 1/1 petroleum ether/ether yields $C_5H_5FeC_5H_4–CO_2CH_3$ and the 1,2-isomer (11% yield by Method Vc, p. 291, 0.7% by Method Va, p. 291), eluting ahead of a residual amount of No. 16 [43]. The raw product is separated by chromatography on Al_2O_3 columns in 1/1 petroleum ether/benzene with elution of No. 16 with ether [16]. No. 16 can be separated from $Fe(C_5H_4–CO_2CH_3)_2$ by column chromatography on Al_2O_3 in benzene, with the diester eluting with benzene before the monoester [1].

In $CDCl_3$ the NMR signals of the $COCH_3$ protons are shifted vs. those of $C_5H_5FeC_5H_4–COCH_3$ by $\Delta\delta = 1.4$ Hz to higher field. This value does not lie on the straight line that forms when $\Delta\delta$ for $X–C_5H_4FeC_5H_4–COCH_3$ (X = C_2H_5, H, Cl, CN) is plotted against the para σ Hammet constants for the X [34].

The ionic hydrogenation under mild conditions with $(C_2H_5)_3SiH$ and CF_3CO_2H (molar ratio 1/2.5/5) at 20°C leads to No. 2, in 65% yield after 2 h reaction. The rate-determining step is the addition of the H^- to the protonated CO group, whose positive charge is strongly localized on the C atom by the electron acceptor activity of the CH_3O_2C group. Within the series X = H, Cl, CO_2CH_3, CN in $X–C_5H_4FeC_5H_4–COCH_3$, the rate constant for hydrogenation increases with the increasing acceptor strength of X (cf. in contrast "Further information" for No. 10, p. 303) [37].

$NaBH_4$ powder is added within 2 to 5 min to the methanol solution of No. 16 and the reaction is continued for 30 min. After hydrolysis, No. 10 is extracted with CH_2Cl_2 in 56% yield [26]. No. 10 is also obtained, in a 16% yield after 1 h, with $LiAlH_4$ under mild conditions (mole ratio 1/1 in ether at room temperature). It is separated from the simultaneously forming III (yield 83%) by chromatography on Al_2O_3 columns with benzene containing increasing quantities of ether. The vigorous reduction with $LiAlH_4$ in a 1/2 mole ratio in boiling tetrahydrofuran gives $CH_3CHOH–C_5H_4FeC_5H_4–CH_2OH$, which very easily dehydrates to III [8].

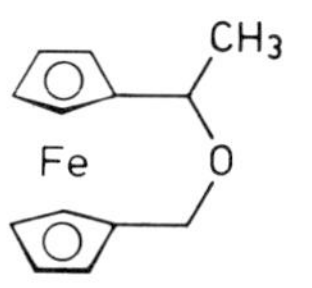

III

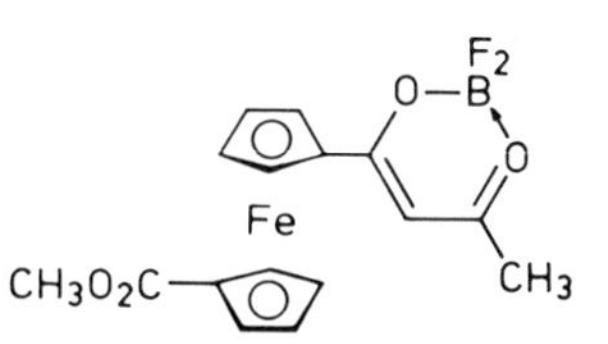

IV

No. 16 condenses with C_6H_5CHO in CH_3OH under the action of 10% NaOH at room temperature to form No. 20 in 25% yield. Simultaneously a 42% yield forms of the free acid $HO_2C–C_5H_4FeC_5H_4–COCH=CHC_6H_5$. The two products after the usual processing are isolated successively by chromatography of the reaction product [44]. The acetylation reagents $(CH_3CO)_2O/BF_3 \cdot O(C_2H_5)_2$ (cf. Method Vc, p. 291) produce at room temperature IV. This reaction characteristic for enolizing ketones had previously not been observed in the ferrocene series. It only occurs when the second cyclopentadienyl ring carries a substituent with electron acceptor properties [43]. Boiling with Girard's reagent in CH_3OH with addition of glacial acetic acid precipitates $[CH_3O_2C{-}C_5H_4FeC_5H_4{-}C(CH_3){=}NNHCOCH_2N(CH_3)_3]Cl$, which is soluble in 0.1 M NaOH [1].

$CH_3O_2C{-}C_5H_4FeC_5H_4{-}COC_2H_5$ (Table **21**, No. **17**) is oxidized to the stable paramagnetic radical anion **$[CH_3O_2C{-}C_5H_4FeC_5H_4{-}C(O){=}C(CH_3)O]^{\dot{-}}$** (see Formula V) in dry $(CH_3)_2SO$ under the action of a three- to fourfold excess of $KOC(CH_3)_3$ by small quantities of O_2 (one second action of atmospheric oxygen).

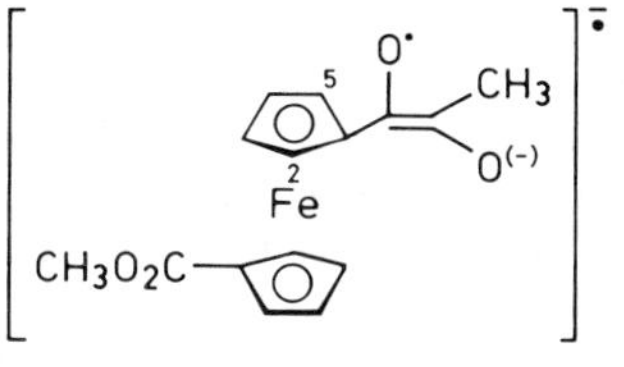

V

The ESR g value is 2.00728. The hfs coupling constant with the $C{-}CH_3$ protons is $A_H = 3.82$ G, with the cyclopentadienyl protons H-2,5 is 0.55 G. Protons of the $CH_3O_2CC_5H_4$ ring do not interact with the unpaired spin, nevertheless polar effects of its substituents transfer via the ring [35].

$CH_3O_2C{-}C_5H_4FeC_5H_4{-}COC_6H_5$ (Table **21**, No. **21**) forms quantitatively by action of bases on $[CH_3O_2C{-}C_5H_4FeC_5H_4{-}C(C_6H_5){=}NH_2]PF_6$ in H_2O/C_2H_5OH [39] and yields the oxime with $NH_2OH \cdot HCl$ in $C_2H_5OH/H_2O/KOH$ [40]. No. 5 is obtained by introducing gaseous HCl into a flask containing No. 21 and granular zinc in ether at 0°C. After removing the solvent from the ether layer the residue is chromatographed on an SiO_2 column with 1/1 C_6H_6/CH_2Cl_2 [48].

$C_2H_5O_2C{-}C_5H_4FeC_5H_4{-}COCH_3$ (Table **21**, No. **29**) does not form on reaction by Method VI (p. 291) of $C_5H_5FeC_5H_4{-}COCH_3$ [42].

$C_2H_5O_2C{-}C_5H_4FeC_5H_4{-}COCH_2CH_2Cl$ (Table **21**, No. **30**) is isolated from the reaction product obtained by Method Vb (p. 291) by column chromatography on SiO_2 with $C_6H_6/CH_3CO_2C_2H_5$ as eluent. Crystallization attempts led to HCl cleavage and polymerization [45]. It possesses in low concentrations bacteriostatic, in higher concentrations bactericidal effects, but is on the whole less effective than $ClCH_2CO{-}C_5H_4FeC_5H_4{-}COCH_3$ [45].

$4\text{-}BrC_6H_4CH_2O_2C{-}C_5H_4FeC_5H_4{-}CH_3$ (Table **21**, No. **37**). The isomer mixture won from $FeC_{10}H_8(COCl)CH_3$ with $4\text{-}BrC_6H_4CH_2OH$, composed of the $FeC_{10}H_8(CO_2CH_2C_6H_4Br\text{-}4)CH_3$, can be separated by column chromatography on Al_2O_3 only little better (somewhat higher separability of the 1,1'- and 1,3- isomers) than $FeC_{10}H_8(CO_2CH_3)CH_3$ (see "Further information" for No. 1, p. 302) [23].

References:

[1] M. Rosenblum (Diss. Harvard Univ. 1953). – [2] A. N. Nesmeyanov, L. A. Kazitsyna, B. V. Lokshin, I. I. Kritskaya (Dokl. Akad. Nauk SSSR **117** [1957] 433/6; Proc. Acad. Sci. USSR Chem. Sect. **114/117** [1957] 1033/6). – [3] N. A. Nesmeyanov, O. A. Reutov (Dokl. Akad. Nauk SSSR **115** [1957] 518/21; Proc. Acad. Sci. USSR Chem. Sect. **112/117** [1957] 763/6). – [4] R. A. Benkeser, L. W. Hall Jr. (AFOSR-TN-59-450 [1959] 1/9; AD-215025 [1959] 1/9; PB-140776 [1959] 1/9; U.S. Govt. Res. Rept. **32** [1959] 165). – [5] L. A. Kazitsyna, B. V. Lokshin, A. N. Nesmeyanov (Dokl.

Akad. Nauk SSSR **127** [1959] 333/6; Proc. Acad. Sci. USSR Chem. Sect. **124/129** [1959] 531/4).

[6] W. F. Little, R. Eisenthal (J. Am. Chem. Soc. **83** [1961] 4936/9). – [7] A. N. Nesmeyanov, V. A. Sazonova, V. N. Drozd (Izv. Akad. Nauk SSSR Otd. Khim. Nauk **1962** 45/7; Bull. Acad. Sci. USSR Div. Chem. Sci. **1962** 38/9). – [8] K. Schlögl, M. Peterlik, H. Seiler (Monatsh. Chem. **93** [1962] 1309/27). – [9] G. G. Dvoryantseva, Yu. N. Sheinker (Izv. Akad. Nauk SSSR Otd. Khim. Nauk **1963** 924/7; Bull. Acad. Sci. USSR Div. Chem. Sci. **1963** 832/5). – [10] G. G. Dvoryantseva, M. I. Struchkova, Yu. N. Sheinker (Dokl. Akad. Nauk SSSR **152** [1963] 617/20; Dokl. Chem. Proc. Acad. Sci. USSR Chem. Sect. **148/153** [1963] 740/3).

[11] A. N. Nesmeyanov, V. N. Drozd, V. A. Sazonova, V. I. Romanenko, A. K. Prokof'ev, L. A. Nikonova (Izv. Akad. Nauk SSSR Otd. Khim. Nauk **1963** 667/74; Bull. Acad. Sci. USSR Div. Chem. Sci. **1963** 597/603). – [12] K. Yamakawa, H. Ochi, K. Arakawa (Chem. Pharm. Bull. [Tokyo] **11** [1963] 905/11). – [13] R. A. Benkeser, J. L. Bach (J. Am. Chem. Soc. **86** [1964] 890/5). – [14] W. F. Little, C. N. Reilley, J. D. Johnson, A. P. Sanders (J. Am. Chem. Soc. **86** [1964] 1382/6). – [15] A. N. Nesmeyanov, E. G. Perevalova, K. I. Grandberg (Izv. Akad. Nauk SSSR Ser. Khim. **1964** 1903/5; Bull. Acad. Sci. USSR Div. Chem. Sci. **1964** 1811/3).

[16] E. G. Perevalova, M. D. Reshetova, K. I. Grandberg, A. N. Nesmeyanov (Izv. Akad. Nauk SSSR Ser. Khim. **1964** 1901/3; Bull. Acad. Sci. USSR Div. Chem. Sci. **1964** 1808/10). – [17] B. Gautheron (Diss. Dijon 1965, pp. 1/100, 59). – [18] M. D. Reshetova, L. M. Yarysheva, E. G. Perevalova, A. N. Nesmeyanov (Izv. Akad. Nauk SSSR Ser. Khim. **1965** 2196/8; Bull. Acad. Sci. USSR Div. Chem. Sci. **1965** 2160/1). – [19] G. G. Dvoryantseva, S. L. Portnova, Yu. N. Sheinker, L. P. Yur'eva, A. N. Nesmeyanov (Dokl. Akad. Nauk SSSR **169** [1966] 1083/6; Dokl. Chem. Proc. Acad. Sci. USSR **166/171** [1966] 773/5). – [20] K. I. Grandberg, S. P. Gubin (Izv. Akad. Nauk SSSR Ser. Khim. **1966** 551/4; Bull. Acad. Sci. USSR Div. Chem. Sci. **1966** 519/21).

[21] A. N. Nesmeyanov, E. G. Perevalova, V. D. Tyurin, S. P. Gubin (Izv. Akad. Nauk SSSR Ser. Khim. **1966** 1938/43; Bull. Acad. Sci. USSR Div. Chem. Sci. **1966** 1874/8). – [22] G. G. Dvoryantseva, M. I. Struchkova, Yu. N. Sheinker, K. I. Grandberg, S. P. Gubin (Izv. Akad. Nauk SSSR Ser. Khim. **1967** 945/7; Bull. Acad. Sci. USSR Div. Chem. Sci. **1967** 919/20). – [23] H. Falk, G. Haller, K. Schlögl (Monatsh. Chem. **98** [1967] 592/602). – [24] S. P. Gubin, K. I. Grandberg, I. P. Shepilov (Izv. Akad. Nauk SSSR Ser. Khim. **1967** 1149/51; Bull. Acad. Sci. USSR Div. Chem. Sci. **1967** 1114/6). – [25] K. Schlögl, H. Falk, G. Haller (Monatsh. Chem. **98** [1967] 82/94).

[26] D. W. Hall, E. A. Hill, J. H. Richards (J. Am. Chem. Soc. **90** [1968] 4972/6). – [27] K. Schlögl, H. Soukup (Monatsh. Chem. **99** [1968] 927/46). – [28] G. P. Chalykh, E. B. Sokolova, O. D. Shalygina, N. V. Noskova (Tr. Mosk. Khim. Tekhnol. Inst. No. 61 **1969** 164; C.A. **73** [1970] No. 25627). – [29] H. Falk, O. Hofer, K. Schlögl (Monatsh. Chem. **100** [1969] 624/48). – [30] W. F. Little, H. T. Busby (Proc. 4th Intern. Conf. Organometal. Chem., Bristol 1969, Abstr. G 2).

[31] A. N. Nesmeyanov, M. D. Reshetova, E. G. Perevalova (Izv. Akad. Nauk SSSR Ser. Khim. **1969** 1939/44; Bull. Acad. Sci. USSR Div. Chem. Sci. **1969** 1795/9). – [32] K. Schlögl, M. Walser (Monatsh. Chem. **100** [1969] 840/3). – [33] A. N. Nesmeyanov, L. I. Kazakova, M. D. Reshetova, L. A. Kazitsina, E. G. Perevalova (Izv. Akad. Nauk SSSR Ser. Khim. **1970** 2804/6; Bull. Acad. Sci. USSR Div. Chem. Sci. **1970** 2639/41). – [34] M. Sato, I. Motoyama, K. Hata (Bull. Chem. Soc. Japan **43** [1970] 1142/7). – [35] J. J. McDonnell, D. J. Pochopien (J. Org. Chem. **36** [1971] 2092/8).

[36] R. F. Kovar, M. D. Rausch (J. Organometal. Chem. **35** [1972] 351/66). – [37] L. I. Kazakova, N. M. Loim, Z. N. Parnes (Zh. Obshch. Khim. **43** [1973] 1577/80; J. Gen. Chem. [USSR] **43** [1973] 1561/3). – [38] L. I. Kazakova, N. M. Loim, E. G. Perevalova, Z. N. Parnes (Zh. Obshch. Khim. **43** [1973] 2306/10; J. Gen. Chem. [USSR] **43** [1973] 2294/7). – [39] H. Patin, D. Mourot

(Compt. Rend. C **281** [1975] 737/40). – [40] D. Mourot, H. Patin (J. Organometal. Chem. **114** [1976] 89/97).

[41] D. W. Slocum, B. P. Koonsvitsky (J. Org. Chem. **41** [1976] 3664/8). – [42] T. Akiyama, T. Kitamura, T. Kato, H. Watanabe, T. Serizawa, A. Sugimori (Bull. Chem. Soc. Japan **50** [1977] 1137/41). – [43] M. N. Nefedova, V. N. Setkina, V. A. Kataev, T. V. Fisun, D. N. Kursanov (Dokl. Akad. Nauk SSSR **237** [1977] 352/5; Dokl. Chem. Proc. Acad. Sci. USSR **232/237** [1977] 662/5). – [44] M. Sališová, Š. Toma, E. Solčaniová (J. Organometal. Chem. **132** [1977] 419/27). – [45] Š. Toma, P. Foltínová, E. Zíkova (Acta Fac. Rerum Nat. Univ. Comenianae Chim. **25** [1977] 9/19).

[46] A. N. Nesmeyanov, L. V. Balabanova, M. A. Shvekhgeimer et al. (U.S.S.R. 1029125 [1977/83] from C.A. **99** [1983] No. 149643). – [47] A. Perjéssy, Š. Toma (Collection Czech. Chem. Commun. **48** [1983] 1635/46). – [48] M. J. Golightly (AD-A-124405 [1982] 1/55, 35, 36).

5.2.3.7.1.4 Carboxylic Acid Esters of the Type $R'O_2C{-}A{-}C_5H_4FeC_5H_4{-}R^2$ ($R'O_2C{-}A = R^1$)

The compounds assembled in Table 22 are usually prepared by the following methods:

Method I. Esterification of the free acid

a) with CH_2N_2 in ether [23] or CH_3OH/ether [10],

b) with C_2H_5OH in the presence of small amounts of concentrated H_2SO_4 with azeotropic distillation of the H_2O with toluene [5, 7].

Method II. Friedel-Crafts alkylation of ferrocene with $CH_3O_2C(CH_2)_nCl$ (≙ RCl; n = 1 to 3, 5) in heptane in the presence of $AlCl_3$ as catalyst and Al powder as reductant. The isomer mixture (m) $FeC_{10}H_8R_2$ forms in addition to $C_5H_5FeC_5H_4{-}R$. The reaction product is subdivided into three bands (ferrocene, $C_5H_5FeC_5H_4{-}R$, and m) by preparative thin layer chromatography on SiO_2 with 75/5/30/8 $C_6H_6/CH_3CO_2C_4H_9$/heptane/$(CH_3)_2SO$. Each band is separately removed and extracted with CH_3OH and ether [35, 36].

a) For 4 h at 80 to 95°C [35].

b) Reactant molar ratio 1/>1/<2/>2, initially at 68°C, finally at >100°C for 3 h [36].

c) As for Method IIb, but with 1/2/<3/>2, for 6 h [36].

d) As for Method IIb, but for 6 h at 95 to 98°C [36].

Method III. Reaction of Li_2PdCl_4 in C_2H_5OH with $Fe(C_5H_4{-}HgCl)_2$ and $HCR''{=}CHCO_2C_2H_5$ (R'' = H for No. 8, CH_3 for No. 9) or $H_2C{=}C(CH_3)CO_2CH_3$ (for No. 10) for 12 h at room temperature. After filtering off the Pd and distilling off the C_2H_5OH the remaining raw product is chromatographed on an Al_2O_3 column with benzene. I appears first, in 8% yield along with Nos. 8 or 10, 10% with No. 9, and subsequently the desired compound, which can also be eluted with $CHCl_3$ [24].

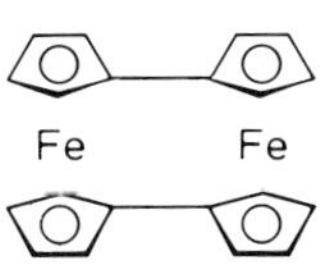

I

References on pp. 322/3

Method IV. Gomberg arylation of ferrocene in CH_3CO_2H with a sulfuric acid aryl diazonium sulfate solution at room temperature. The compound is isolated from the $CHCl_3$ extract of the reaction product by column chromatography on SiO_2 with benzene after elution of ferrocene and $C_5H_5FeC_5H_4$–R [42] or with 4/1 hexane/ether after elution of $C_5H_5FeC_5H_4$–R^2 [43]:

a) Aryl = $C_6H_4CO_2CH_3$-2 for No. 12, $C_6H_4CO_2C_2H_5$-4 for No. 13 [42].

b) Aryl = $C_6H_4CO_2H$-3 for No. 11 [37], 4-C_6H_4–$C_6H_4CO_2H$-3 for No. 14, 4-C_6H_4–$C_6H_4CO_2H$-2 for No. 15 [41], and $C_{10}H_6CO_2H$-5 to 7 for Nos. 46 to 48 [43]. The reaction product is esterified in boiling CH_3OH or C_2H_5OH by addition of a few drops of concentrated H_2SO_4 before the extraction with $CHCl_3$ [37, 41, 43].

c) As for Method IVa, but of $C_5H_5FeC_5H_4$–$C_6H_4CO_2C_2H_5$-4; aryl = $C_{10}H_6CO_2C_2H_5$-4 [43].

Method V. Condensation of $Fe(C_5H_4–COCH_3)_2$:

a) With $(C_2H_5O)_2CO$ under the action of NaH in benzene, initially at room temperature, then at the boiling point [16, 27]. The raw product is chromatographed on an SiO_2 column with 10/1 $C_6H_6/CH_3CO_2C_2H_5$ [27].

b) As for Method Va, but with $NaNH_2$ in liquid NH_3/ether at room temperature. A considerable amount of urethane forms along with No. 18 [10].

c) As for Method Vb, but for suppression of the urethane formation the NH_3 is replaced by ether after conversion of $Fe(C_5H_4–COCH_3)_2$ with $NaNH_2$ in liquid NH_3 into the disodium salt, and only then is $(C_2H_5O)_2CO$ in ether added dropwise [15].

d) With $C_2H_5O_2CCO_2C_2H_5$ in $C_2H_5OH/C_2H_5O^-$, first with cooling, then at room temperature, for 2 d reaction time [22].

Method VI. Friedel-Crafts acylation of ferrocene with R″COCl and $AlCl_3$, if not otherwise stated first under cooling, then at the boiling point:

a) With R″ = $(CH_2)_2CO_2CH_3$ for No. 19 [3], $CH{=}CHCO_2CH_3$ for No. 28 in the mole ratio 1/2/4 in CS_2. The product resulting after the usual processing is chromatographed on an Al_2O_3 column in C_6H_6, and No. 28 is eluted with benzene after elution of ferrocene with heptane [17].

b) With R″ = $(CH_2)_nCO_2R'$ (n = 2 to 8) in the mole ratio 1/>2/>4 in CH_2Cl_2 [10, 38]. The compounds occur as by-products of $C_5H_5FeC_5H_4$–$CO(CH_2)_nCO_2R'$, their yield increases with rising n. Ferrocene and the monoacylation product are separated by column chromatography on SiO_2 with 9/1 to 17/3 $C_6H_6/CH_3CO_2C_2H_5$ as eluent [38].

c) With R″ = $(CH_2)_4COCl$ in the mole ratio 1/1.4/2.5 in CH_2Cl_2, but at room temperature for 20 h. After the usual processing the acid fraction is extracted with 10% Na_2CO_3 solution from the ether extract of the reaction product, and esterified with CH_2N_2 (No. 44) or C_2H_5OH/HCl (Nos. 23, 45) after acidification. The final preparative thin layer chromatography gives II in 5% yield and No. 44

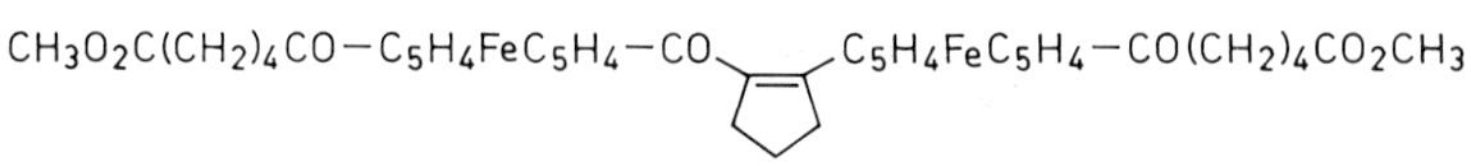

II

References on pp. 322/3

or Nos. 45 and 23, respectively. The main product consists of the polymer $H[C_5H_4FeC_5H_4-CO(CH_2)_4CO]_n-OH$ [31].

Method VII. Friedel-Crafts acylation of $C_5H_5FeC_5H_4-CH_2CO_2CH_3$ with 4-$CH_3C_6H_4COCl$ and $AlCl_3$ (mole ratio 1/1/1) in CH_2Cl_2 at room temperature. No. 31 mixed with its two positional isomers (m) is isolated by dry column chromatography on SiO_2 with 2/1 hexane/$CH_3CO_2C_2H_5$ [39].

Method VIII. Friedel-Crafts acylation of $C_5H_5FeC_5H_4-R^2$:

a) With $CH_3O_2C(CH_2)_nCOCl$ and $AlCl_3$ in CH_2Cl_2. For n = 2, in the mole ratios 1/0.75/1.1 for No. 39, 1/1/1 for No. 40, and 1/1/4 for No. 41. For n = 3, 1/1/3 for No. 43. At room temperature only for Nos. 39 to 41, previously at 50°C for No. 43 [10, 18, 21]. Nos. 39 and 40 are obtained mixed with their two positional isomers. The separation of excess $C_5H_5FeC_5H_4-R^2$, small amounts of diacylation products, and the 1,2-isomers goes smoothly by column chromatography on deactivated Al_2O_3 with benzene and petroleum ether. More difficult is the separation from the 1,3-isomer. While it is possible for No. 40 by repeated preparative column chromatography on Al_2O_3 in benzene [18], No. 39 and its 1,3-isomer are not separable chromatographically. Only after saponification of the mixture with KOH in ethanol can the free acids be separated by fractional crystallization [21]. No. 41 is isolated by crystallizing the reaction product after dissolving out $C_5H_5FeC_5H_4-COCH_3$ with petroleum ether [18].

b) With III and $AlCl_3$ in the mole ratio 1/1/1.8 in CH_2Cl_2. The acid mixture precipitated with 50% H_3PO_4 after the usual processing is extracted with ether and esterified with CH_2N_2. After column chromatographic separation from the 1,2-isomer on SiO_2 with 30/1 C_6H_6/C_2H_5OH there remains behind the mixture of No. 39 and its 1,3-isomer, that were not further separated [20].

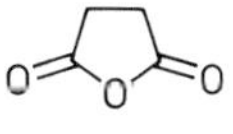

III

The compounds are in general isolated chromatographically (for details see above in connection with the preparation methods) and are **purified** by distillation or recrystallization.

Chemical Behavior

Saponification. NaOH [3, 10, 11, 15, 16] or KOH [18, 20, 21] in C_2H_5OH/H_2O [10, 11, 15, 16, 18, 21, 22] or CH_3OH [20] saponify the esters in almost quantitative yield to the anions of the free acids. The reaction has been conducted with the compounds No. 3 [10, 11, 15, 22], 4 [10], 5 [8, 18], 19 [3, 10], 21 [10], 38 [29], 39 [20, 21], 40 [18, 19], 41 [18], 42 and 43 [10].

Reduction. Starting compounds and end products for the reduction of the **keto carboxylic acids** are shown in the following table. The reduction can be effected by the Clemmensen technique using amalgamated zinc and concentrated hydrochloric acid or by catalytic hydrogenation over PtO_2 in glacial acetic acid at room temperature.

starting compound No.	procedure	reduction product No.	yield in %	Ref.
18	Clemmensen reduction in C_6H_6/C_2H_5OH/conc. HCl (10/3/8) at 50°C	3*)	60	[27]
	with H_2/PtO_2 (10 to 15 atm)	3	87 to 94, 59	[10, 11, 15, 16]
19	Clemmensen in CH_3OH	4		[13]
	with H_2/PtO_2	4	~100	[10, 11]
	with H_2/PtO_2 with premature interruption after taking up of 2 mol H_2/mol No. 19	42*)	33	[10]
20	Clemmensen in boiling 5/6/3/3 $C_6H_6/C_2H_5OH/H_2O$/conc. HCl	5		[14]

*) See also the following section.

The more rapidly migrating reduced esters can be separated by column chromatography on Al_2O_3 in benzene from the unreacted keto esters remaining at the beginning of the column [14]. The product obtained by the Clemmensen method from No. 18 and won by subsequent chromatography on SiO_2 is contaminated with small quantities of C=C-containing compounds and is therefore hydrogenated with H_2 at 4 atm for 20 h over Pd/C in 5/1 C_2H_5OH/C_6H_6. No. 3 is isolated from the reaction mixture in 60% yield by chromatography on SiO_2 columns with 20/1 $C_6H_6/CH_3CO_2C_2H_5$ [27]. The raw substance obtained after the take up of 2 mol H_2/mol No. 19 and the usual processing was separated chromatographically on Al_2O_3, yielding first No. 4 on elution with petroleum ether containing increasing amounts of benzene, and later No. 42 on elution with ether [10]. For the reduction of the keto acid ester No. 18 with $NaBH_4$, see its "Further information", p. 320.

The **aryl carboxylic acid esters** No. 11 to 15, and 46 to 48, $Fe(C_5H_4\text{–}A\text{–}CO_2R')_2$ with A = C_6H_4, $C_6H_4C_6H_4$-4, or $C_{10}H_6$-5 to 7, can be reduced with $LiAlH_4$ in ether to the corresponding hydroxymethyl compounds $Fe(C_5H_4\text{–}A\text{–}CH_2OH)_2$ [37, 41, 42, 43]; No. 49 yields 4-$HOCH_2C_{10}H_6\text{–}C_5H_4FeC_5H_4\text{–}C_6H_4CH_2OH$-4 [43].

Acyloine Ring Closure. The bismethyl esters No. 1, 2, 4, 6, and 32 are able to react in boiling xylene at great dilution with finely dispersed Na under conditions of absolute oxygen exclusion with vigorous stirring to form the acyloines IV, with n = m = 1 for No. 1, n = m = 2 for No. 2, n = m = 3 for No. 4, n = m = 4 for No. 6, and n = 3, m = 4 for No. 32. The yields amount to 20 to 75%, the higher the greater the n and m [10], see also [9, 11].

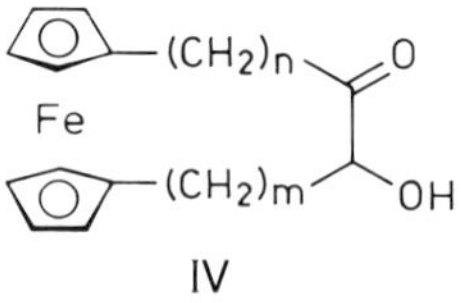

IV

Dehydration. The hydroxy carboxylic acid esters $Fe(C_5H_4\text{–}C(R'')OHCH_2CO_2C_2H_5)_2$ with R″ = H (No. 16) and CH_3 (No. 17) are dehydrated on boiling with $POCl_3$ in CH_2Cl_2 to form $Fe(C_5H_4\text{–}C(R'')\text{=}CHCO_2C_2H_5)_2$. The yield of No. 8 (R″ = H) is 46% raw product (referred to $Fe(C_5H_4\text{–}COCH_2CN)_2$, the starting compound for No. 16) [26] and is 62% for No. 9 (R″ = CH_3) (referred to $Fe(C_5H_4\text{–}COCH_3)_2$, the starting compound for No. 17) [25].

Table 22

Carboxylic Acid Esters of the Type $R'O_2C–A–C_5H_4FeC_5H_4–R^2$ ($R'O_2C–A = R^1$).

Further information for numbers preceded by an asterisk is given at the end of the table, pp. 320/2.
For abbreviations and dimensions see p. VIII.

No.	$R^1–C_5H_4FeC_5H_4–R^2$ $R^1 = A–CO_2R'$	R^2	method of preparation (yield in %; "m" indicates isomeric mixture, number following solidus is fraction present of 1,1'-compound), properties, and remarks	Ref.
compounds with $R^1 = R^2$:				
*1	$CH_2CO_2CH_3$		Ia (94, 75), IIa (4 m) reddish brown oil, brown liquid, b.p. 140 to 145°/0.01 Torr ^{1}H NMR (CCl_4): 3.67 (s, CH_3), 3.98 (CH_2 + 2H of C_5H_4), 4.04 (2H of C_5H_4) IR (neat): 812, 830, 923, 1012, 1041, 1145, 1200, 1231, 1265, 1315, 1435, 1460, 1740 (νC=O), 2980, 3130	[6, 23, 35]
2	$(CH_2)_2CO_2CH_3$		Ia, IIb (22 m) deep red oil (m) IR (KBr): 820, 824, 1005 to 1029, and 3100 (C_5H_4); 890 to 896, 900, 915, 965, and 1285 (C_5H_3); 1107, 1123, and 1140 (C_5H_5); 1450, 1585, and 2865 to 2880 (CH_2); 2935 and 2965 (CH_3); 1675 (νC=O) (m) mol. wt: 341 from dates of hydrolysis (m)	[10, 36]
3	$(CH_2)_2CO_2C_2H_5$		see "Chemical Behavior" for No. 18, pp. 309/10 yellow to orange oil IR (neat): 1735 (νC=O)	[10, 11, 15, 16, 27]
*4	$(CH_2)_3CO_2CH_3$		Ia, IIc (42 m), see "Further information", p. 320, and "Chemical Behavior" for No. 19, pp. 309/10 viscous yellow to deep orange oil, m.p. 16.5 to 17.5°, b.p. 148°/10^{-4} Torr	[3, 10, 11, 36]

References on 322/3

Table 22 (continued)

No.	R^1–$C_5H_4FeC_5H_4$–R^2 R^1 = A–CO_2R'	R^2	method of preparation (yield in %; "m" indicates isomeric mixture, number following solidus is fraction present of 1,1′-compound), properties, and remarks	Ref.
*5	$(CH_2)_3CO_2C_2H_5$		see "Further information", p. 320, and "Chemical Behavior" for No. 20, pp. 309/10 orange oil (raw product)	[8, 14]
*6	$(CH_2)_4CO_2CH_3$		Ia	[10, 11]
7	$(CH_2)_5CO_2CH_3$		IId (18m) viscous deep red oil (m)	[36]
*8	$\overset{6}{C}H{=}\overset{7}{C}H$– $CO_2C_2H_5$		III (28), see "Further information", p. 320, and "Chemical Behavior" for No. 16, p. 310 brown crystals, m.p. 83 to 84°, 84 to 85° (C_6H_6), 98.5 to 99.5° (hexane) ^{1}H NMR ($CDCl_3$): 1.33 (t, CH_3), 4.33 (d, CH_2), 4.55 (s, C_5H_4), 6.01 (d, H-7), 7.52 (d, H-6) ^{1}H NMR (CCl_4): 1.29 (t, CH_3), 4.17 (q, CH_2); 4.34 (t) and 4.42 (t) (C_5H_4); 5.89 (d, H-7; J(H-7, 6) = 16.0), 7.33 (d, H-6) IR: 497, 566, 639, 730, 765, 822, 848, 874, 963, 1035, 1043, 1164, 1194, 1245, 1302, 1365, 1397, 1478, 1637, 1721, 2901, 2933, 3078, 3088 IR (KBr): 1618, 1710	[24, 26, 30]
*9	$C(CH_3){=}CH$– $CO_2C_2H_5$		III (25) see "Chemical Behavior" for No. 17, p. 310 brown-orange crystals, m.p. 57°, 65 to 66° ^{1}H NMR ($CDCl_3$): 1.26 (t, CH_3 of C_2H_5), 1.98 (d, =CCH_3), 4.20 (q, CH_2 of C_2H_5); 4.38 (m, 2H) and 4.53 (m, 2H)(C_5H_4); 6.04 (q, =CH) ^{13}C NMR (CH_2Cl_2): 14.3 and 17.0 (CH_3), 59.5 (CH_2); 68.3 (C-2,5), 72.1 (C-3,4) and 86.5 (C-1) (C_5H_4); 102.5 and 154.3 (C=C), 166.5 (C=O) IR (KBr): 1620, 1711	[24, 25]

10	$CH{=}C(CH_3)$–CO_2CH_3	III (31) m.p. 96 to 97° ^{1}H NMR ($CDCl_3$): 2.01 (d, CCH_3), 3.77 (s, CO_2CH_3); 4.37 (m, 2H), 4.48 (m, 2H)(C_5H_4); 7.45 (q, =CH) IR (KBr): 1624, 1710	[24]
11	$C_6H_4CO_2CH_3$-3	IVb (20) orange-yellow needles, m.p. 113 to 114° ^{1}H NMR ($CDCl_3$): 3.91 (s, CH_3), 4.26 (t, H-3,4 of C_5H_4), 4.55 (t, H-2,5 of C_5H_4), 7.10 to 7.75 (m, C_6H_4) IR (KBr): 800 (C_6H_4-1,3), 805 (C_5H_4), 860 (C_6H_4-1,3), 1720 (νC=O), 3100 (C_5H_4)	[37]
12	$C_6H_4CO_2CH_3$-2	IVa (23) reddish crystals, m.p. 92 to 94° ^{1}H NMR ($CDCl_3$): 3.69 (s, CH_3), 4.19 (t, H-3,4 of C_5H_4), 4.33 (t, H-2,5 of C_5H_4), 7.18 to 7.60 (m, C_6H_4) IR (KBr): 735 and 760 (C_6H_4-1,2), 805 (C_5H_4), 1500 and 1600 (C_6H_4-1,2), 1720 (νC=O), 3100 (C_5H_4) mass spectrum: m/e = 454 $[M]^+$	[42]
13	$C_6H_4CO_2C_2H_5$-4	IVa (26) reddish yellow crystals, m.p. 130 to 131° ^{1}H NMR ($CDCl_3$): 1.36 (t, CH_3), 4.24 (t, H-3,4 of C_5H_4), 4.37 (q, CH_2), 4.49 (t, H-2,5 of C_5H_4); 7.22 (d, 2H) and 7.80 (d, 2H)(C_6H_4) IR (KBr): 805 (C_5H_4), 820 and 1600 (C_6H_4-1,4), 1720 (νC=O), 3100 (C_5H_4) mass spectrum: m/e = 482 $[M]^+$	[42]
14	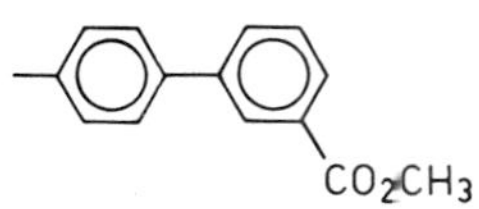	IVb (35) reddish yellow crystals, m.p. 153 to 155° ^{1}H NMR ($CDCl_3$): 3.87 (s, CH_3), 4.08 (m, C_5H_4), 7.21 to 8.12 (m, C_6H_4–C_6H_4) IR (KBr): 805 (C_5H_4); 835, 880, 1580, and 1600 (C_6H_4-1,4 and -1,3); 1720 (νC=O) mass spectrum: m/e = 606 $[M]^+$	[41]

References on pp. 322/3

Table 22 (continued)

No.	R^1–$C_5H_4FeC_5H_4$–R^2 R^1 = A–CO_2R'	R^2	method of preparation (yield in %; "m" indicates isomeric mixture, number following solidus is fraction present of 1,1'-compound), properties, and remarks	Ref.
15	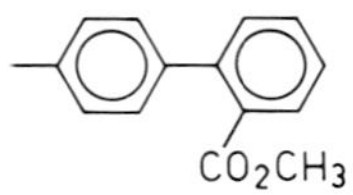		IVb (18) reddish yellow crystals, m.p. 62 to 63° ^{1}H NMR ($CDCl_3$): 3.91 (s, CH_3), 4.22 (t, H-3,4 of C_5H_4), 4.51 (t, H-2,5 of C_5H_4), 7.11 to 7.85 (m, C_6H_4–C_6H_4) IR (KBr): 730 (C_6H_4-1,4 or -1,2), 810 (C_5H_4); 820, 845, 1500, and 1600 (C_6H_4-1,4 and -1,2); 1720 (νC=O)	[41]
16	$CHOHCH_2CO_2C_2H_5$		see No. 18, p. 320	[26]
*17	$C(CH_3)OHCH_2CO_2C_2H_5$		see "Further information", p. 320 yellow-orange oil (raw product)	[25]
*18	$COCH_2CO_2C_2H_5$		Va (70 to 74 raw product, 64), Vb (34), Vc (90 raw product), see "Further information", p. 320 viscous red oil (raw product)	[10, 11, 16, 27]
*19	$CO(CH_2)_2CO_2CH_3$		VIa (27), VIb (33) orange needles, m.p. 98° (C_6H_6/petroleum ether, 1/1), 102 to 105° (C_6H_6/petroleum ether)	[3, 10]
*20	$CO(CH_2)_2CO_2C_2H_5$		Ib (94, 70) shiny orange-brown platelets, m.p. 134 to 136° (C_2H_5OH)	[5, 7]
21	$CO(CH_2)_3CO_2CH_3$		VIb (70 raw product) red oil which could not be crystallized	[10]
22	$\overset{6}{C}O\overset{7}{C}H_2\overset{8}{C}H_2\overset{9}{C}H_2$– $CO_2C_2H_5$		VIb (17) m.p. 25° (petroleum ether) ^{1}H NMR ($CDCl_3$): 1.27 (t, CH_3), 2.00 (quint, CH_2-8), 2.43 (t, CH_2-7), 2.73 (t, CH_2-9), 4.14 (q, CH_2 of C_2H_5), 4.49 (t, H-3,4 of C_5H_4), 4.78 (t, H-2,5 of C_5H_4)	[38]

No.	R	preparation, properties	Ref.
23	$\overset{6}{C}O\overset{7}{C}H_2(CH_2)_2\overset{10}{C}H_2$–$CO_2C_2H_5$	VIb (20), VIc (11) oil, crystals, m.p. 27 to 29° (ether/petroleum ether) ^{1}H NMR ($CDCl_3$): 1.25 (t, CH_3), 1.4 to 2.0 (m, $(CH_2)_2$), 2.36 (t, CH_2-7), 2.66 (t, CH_2-10), 4.13 (q, CH_2 of C_2H_5), 4.47 (t, H-3,4 of C_5H_4), 4.76 (t, H-2,5 of C_5H_4) IR (CCl_4): 1670 and 1733 (νC=O)	[31, 38]
24	$CO\overset{7}{C}H_2(CH_2)_3\overset{11}{C}H_2$–$CO_2C_2H_5$	VIb (30) oil ^{1}H NMR ($CDCl_3$): 1.25 (t, CH_3), 1.1 to 2.0 (m, $(CH_2)_3$), 2.33 (t, CH_2-7), 2.65 (t, CH_2-11), 4.13 (q, CH_2 of C_2H_5), 4.47 (t, H-3,4 of C_5H_4), 4.75 (t, H-2,5 of C_5H_4)	[38]
25	$CO\overset{7}{C}H_2(CH_2)_4\overset{12}{C}H_2$–$CO_2C_2H_5$	VIb (23) m.p. 29 to 32° (C_6H_6/petroleum ether) ^{1}H NMR ($CDCl_3$): 1.23 (t, CH_3), 1.1 to 1.9 (m, $(CH_2)_4$), 2.29 (t, CH_2-7), 2.63 (t, CH_2-12), 4.10 (q, CH_2 of C_2H_5), 4.46 (t, H-3,4 of C_5H_4), 4.75 (t, H-2,5 of C_5H_4)	[38]
26	$CO\overset{7}{C}H_2(CH_2)_5\overset{13}{C}H_2$–$CO_2C_2H_5$	VIb (35) m.p. 33 to 44° (ether/petroleum ether) ^{1}H NMR ($CDCl_3$): 1.25 (t, CH_3), 1.1 to 1.9 (m, $(CH_2)_5$), 2.30 (t, CH_2-7), 2.63 (t, CH_2-13), 4.11 (q, CH_2 of C_2H_5), 4.46 (t, H-3,4 of C_5H_4), 4.75 (t, H-2,5 of C_5H_4)	[38]
27	$CO\overset{7}{C}H_2(CH_2)_6\overset{14}{C}H_2$–$CO_2C_2H_5$	VIb (40) m.p. 54 to 55° (acetone/petroleum ether) ^{1}H NMR ($CDCl_3$): 1.25 (t, CH_3), 1.1 to 1.9 (m, $(CH_2)_6$), 2.29 (t, CH_2-7), 2.64 (t, CH_2-14), 4.13 (q, CH_2 of C_2H_5), 4.49 (t, H-3,4 of C_5H_4), 4.78 (t, H-2,5 of C_5H_4)	[38]
28	$COCH{=}CHCO_2CH_3$	VIa (22 related to ferrocene reacted) dark violet substance, m.p. 164° (petroleum ether/C_6H_6, 1/1) soluble in usual organic solvents and in bases	[17]
*29	$COC_6H_4CO_2CH_3$-2	Ia red needles, m p. 144.0 to 144.5° (CH_3OH/H_2O) IR ($CHCl_3$): 1661 (νC=O of ketone), 1724 (νC=O of ester), see figure in [2] UV ($CHCl_3$): λ_{max}(log ε) = 480 (3.04), 355 (3.19), 275 (4.05), 234 (4.29)	[1, 2]

References on pp. 322/3

Table 22 (continued)

No.	R^1–$C_5H_4FeC_5H_4$–R^2 R^1 = A–CO_2R'	R^2	method of preparation (yield in %; "m" indicates isomeric mixture, number following solidus is fraction present of 1,1′-compound), properties, and remarks	Ref.
30	$COCH_2COCO_2C_2H_5$		Vd (52) deep violet crystals (absolute C_2H_5OH), m.p. 111 to 115°, above 95° with carbonization ^{1}H NMR ($CDCl_3$): 1.4 (t, CH_3), 4.4 (q, CH_2 of C_2H_5), 6.2 (OH, enol-form), 6.5 (=CH) IR (KBr): 940 (w, enol OH), 1610 (s, enol CO), 1740 (s, νC=O of ester)	[22]
compounds with $R^1 \neq R^2$:				
*31	$\overset{6}{C}H_2\overset{7}{C}O_2\overset{8}{C}H_3$	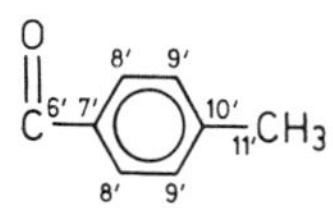	VII (76 m/50%) photosensitive deep red oil (m) ^{13}C NMR ($CDCl_3$): δ = 21.6 (C-11′), 34.2 (0.5 C, C-6), 51.8 (C-8), 70.1 (C-3,4 of C_5H_4–R^1), 70.9 (C-2,5 of C_5H_4–R^1), 71.1 (d*⁾), 72.2 (C-2′,5′ of C_5H_4–R^2), 72.7 (d*⁾), 73.1 (C-3′,4′ of C_5H_4–R^2), 74.7 (d*⁾), 82.1 (0.5 C, C-1 of C_5H_4–R^1); 128.2 (1.4 C), 128.4 (0.6 C), 128.7 (0.6 C), and 128.8 (1.4 C) (C-8′,9′); 136.9 (C-7′), 142.1 (C-10′), 171.1 (0.5 C, C-7), ~199.5 (C-6′). C-1′ of C_5H_4–R^2 is not discernible or indeterminate ^{13}C NMR (CD_2Cl_2): δ = 33.7 (0.5 C, C-6), 69.7 (C-3,4 of C_5H_4–R^1), 70.6 (C-2,5 of C_5H_4–R^1), 71.7 (C-2′,5′ of C_5H_4–R^2), 72.8 (C-3′,4′ of C_5H_4–R^2), 78.7 (C-1′ of C_5H_4–R^2), 81.9 (0.5 C, C-1 of C_5H_4–R^1), 170.7 or 171.4 (C-7), 197.5 (C-6′) mass spectrum m/e = 317 $[M - CO_2CH_3]^+$, 184, 119, 91, 65; no $[M]^+$ *⁾ unassigned minor peaks	[39]
32	$(CH_2)_4CO_2CH_3$	$(CH_2)_3CO_2CH_3$	Ia	[10, 11]
*33	$C_6H_4CO_2C_2H_5$-4	CHO	see "Further information", p. 321 ^{1}H NMR ($CDCl_3$): 1.37 (t, CH_3), 4.39 (m, H-3,4 of C_5H_4–R^1 and C_5H_4–R^2), 4.58 (t, H-2,5 of C_5H_4–R^1), 4.75 (t, H-2,5 of C_5H_4–R^2), 7.0 to 8.0 (m, C_6H_4), 9.78 (s, CHO)	[28, 32]

*34	$C_6H_4CO_2C_2H_5$-4	$COCH_3$	see "Further information", p. 321 ^{1}H NMR ($CDCl_3$): 1.37 (t, CH_3 of C_2H_5), 2.07 (s, $COCH_3$); 4.30(m) and 4.35(m) (H-3,4 of C_5H_4–R^1 and C_5H_4–R^2); 4.36 (q, CH_2 of C_2H_5); 4.56(t) and 4.66(t) (H-2,5 of C_5H_4–R^1 and C_5H_4–R^2); 7.2 to 8.1 (m, C_6H_4)	[28, 32]
35	$C_6H_4CO_2C_2H_5$-4	$COCH{=}CHC_6H_5$	see No. 34, pp. 321/2 m.p. 128 to 130° IR (CCl_4): 1662.9 (νC=O of R^2) for s-cis	[28, 32, 40]
36	$C_6H_4CO_2C_2H_5$-4	$CH{=}CHCOC_6H_5$	see No. 33, p. 321 m.p. 160 to 165°	[28]
37	$C_6H_4CO_2C_2H_5$-4	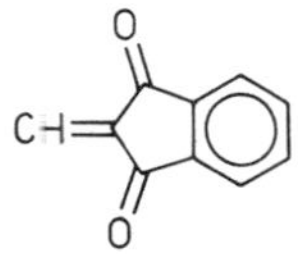	see No. 33, p. 321 m.p. 156 to 158°	[28]
*38	$C_6H_4CO_2CH_3$-2	CH_2OCH_3	see "Further information", p. 322	[29]
39	$CO(CH_2)_2CO_2CH_3$	CH_3	VIIIa, VIIIb (No. 39 + 1,3-isomer: 60/51%) red oil, b.p. 176 to 179°/0.05 Torr (No. 39 + 1,3-isomer) ^{1}H NMR ($CDCl_3$): 1.91(CH_3 of R^2)	[20, 21]
40	$CO(CH_2)_2CO_2CH_3$	C_2H_5	VIIIa red oil	[18, 19]
41	$CO(CH_2)_2CO_2CH_3$	$COCH_3$	VIIIa red crystals, m.p. 71 to 72° (hexane or petroleum ether/CH_2Cl_2, 3/1)	[18]
42	$CO(CH_2)_2CO_2CH_3$	$(CH_2)_3CO_2CH_3$	see "Chemical Behavior" for No. 19, p. 310 crystallizable oil, m.p. 40 to 65° (raw product)	[10]
43	$CO(CH_2)_3CO_2CH_3$	$CO(CH_2)_2CO_2CH_3$	VIIIa (65) red, viscous oil, contaminated with $C_5H_5FeC_5H_4$–$CO(CH_2)_2CO_2CH_3$, not distillable	[10]

References on pp. 322/3

Table 22 (continued)

No.	$R^1-C_5H_4FeC_5H_4-R^2$ $R^1 = A-CO_2R'$	R^2	method of preparation (yield in %; "m" indicates isomeric mixture, number following solidus is fraction present of 1,1'-compound), properties, and remarks	Ref.
44	$\overset{8}{C}O\overset{8}{C}H_2\overset{9}{C}H_2\overset{10}{C}H_2-$ $\overset{11}{C}H_2CO_2CH_3$	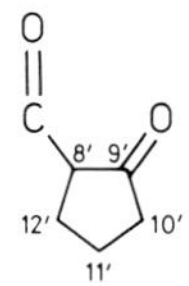	VIc (30) m.p. 94 to 96° (C_2H_5OH) ^{1}H NMR ($CDCl_3$): 1.75 (m, CH_2-9,10,11',12'), 2.38 (m, CH_2-11), 2.71 (m, CH_2-8,10'), 4.15 (m, CH-8'), 4.50 (m, H-3,4 of $C_5H_4-R^1$ and $C_5H_4-R^2$), 4.80 (m, H-2,5 of $C_5H_4-R^1$ and $C_5H_4-R^2$) IR (KBr): 1672 (νC=O), 1740 (νC=O of CO_2CH_3 and C=O-9') UV (C_2H_5OH): λ_{max} (ε) = 282 (1900), 222 (15000)	[31]
45	$CO\overset{8}{C}H_2\overset{9}{C}H_2\overset{10}{C}H_2-$ $\overset{11}{C}H_2CO_2C_2H_5$	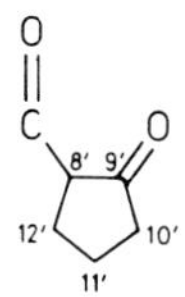	VIc (30) m.p. 72 to 73° (C_2H_5OH) ^{1}H NMR ($CDCl_3$): 1.72 (m, CH_2-9,10,11',12'), 2.35 (m, CH_2-11), 2.70 (m, CH_2-8,10'), 4.50 (m, H-3,4 of $C_5H_4-R^1$ and $C_5H_4-R^2$), 4.80 (m, H-2,5 of $C_5H_4-R^1$ and $C_5H_4-R^2$) IR (CCl_4): 1675 (νC=O), 1735 (νC=O of $CO_2C_2H_5$ and C=O-9') mass spectrum: m/e = 452 $[M]^+$	[31]
supplement:				
compounds with $R^1 = R^2$:				
46	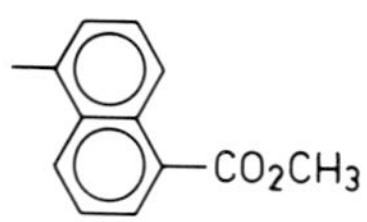		IVb (14.2) reddish crystals, m.p. 156 to 158° ^{1}H NMR ($CDCl_3$): 3.91 (s, CH_3); 4.42 (m, H-3,4) and 4.58 (m, H-2,5) (C_5H_4); 7.32 to 8.50 (m, $C_{10}H_6$) IR (KBr): 770, 790, 820, 1500, and 1600 ($C_{10}H_6$); 1720 (CO_2CH_3); 805 and 3100 (C_5H_4) mass spectrum: m/e = 554 $[M]^+$	[43]
47	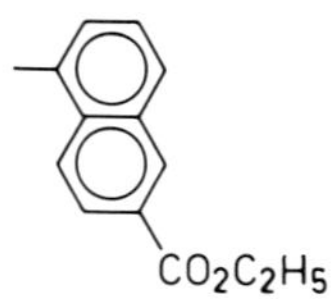		IVb (8) reddish crystals, m.p. 130 to 132° ^{1}H NMR ($CDCl_3$): 1.38 (t, CH_3), 4.40 (t, H-3,4 of C_5H_4), 4.44 (q, CH_2), 4.61 (t, H-2,5 of C_5H_4), 7.20 to 8.49 ($C_{10}H_6$)	[43]

References on pp. 322/3

IR (KBr): 770, 780, 1500, and 1600 ($C_{10}H_6$); 1720 ($CO_2C_2H_5$); 810 and 3100 (C_5H_4)
mass spectrum: m/e = 582 $[M]^+$

48 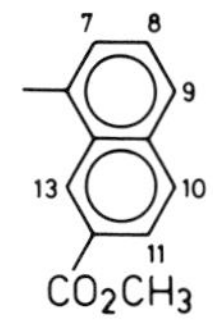

IV b (6.2) [43]
reddish orange crystals, m.p. 152 to 153°
^{1}H NMR ($CDCl_3$): 3.96 (s, CH_3); 4.50 (t, H-3, 4) and 4.90 (t, H-2, 5) (C_5H_4); 7.11 to 7.93 (m, H-7 to 11), 9.57 (br. s, H-13)
IR (KBr): 740, 835, 1500, and 1600 ($C_{10}H_6$); 1720 (CO_2CH_3); 800 and 3100 (C_5H_4)
mass spectrum: m/e = 554 $[M]^+$

compound with $R^1 \neq R^2$:

49 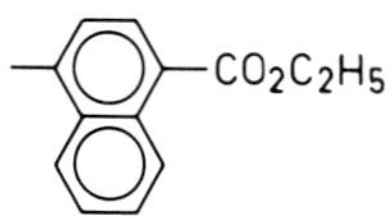 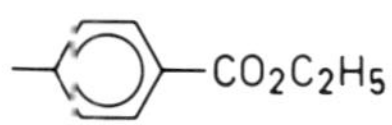

IV c (15.3) [43]
reddish oil
^{1}H NMR ($CDCl_3$): 1.37 (t, CH_3), 1.44 (t, CH_3), 4.31 (t, H-3, 4 of C_5H_4–R^2), 4.38 (t, H-3, 4 of C_5H_4–R^1), 4.41 (m, CH_2), 4.57 (t, H-2, 5 of C_5H_4–R^2), 4.69 (t, H-2.5 of C_5H_4–R^1), 7.34 to 8.94 ($C_{10}H_6 + C_6H_4$)
IR ($CHCl_3$): 730, 780, 1500, and 1600 ($C_{10}H_6$ and C_6H_4); 1720 ($CO_2C_2H_5$); 805 and 3100 (C_5H_4)
mass spectrum: m/e = 532 $[M]^+$

References on pp. 322/3

* Further information:

$Fe(C_5H_4{-}CH_2CO_2CH_3)_2$ (Table **22**, No. **1**) is eluted from a Celite column with benzene [6]. Dieckmann cyclization to V takes place easily in ether under the action of $NaC(C_6H_5)_3$ [13, 23].

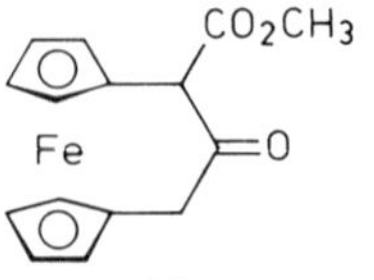

V

$Fe(C_5H_4{-}(CH_2)_3CO_2CH_3)_2$ (Table **22**, No. **4**). A solution of ferrocene and $CH_3O_2C(CH_2)_3CO_2H$ (40% neutralized with KOH) is electrolyzed in CH_3OH between Pt electrodes at 2.3×10^{-2} A/cm^2 and 3 V at 20°C. After passage of 0.3 F it is reduced with $Na_2S_2O_3$. Subsequent extraction in C_6H_6 and evaporation of the extract yields a raw product from which the mixture of the methoxycarbonylpropyl ferrocenes is obtained by column chromatography on SiO_2 with benzene. This mixture is then separated by thin layer chromatography on SiO_2, with isolation of $FeC_{10}H_8([CH_2]_3CO_2CH_3)_2$ in 1.6% yield referred to the starting ferrocene (IR bands at 1000, 1100, 1450, 1750 cm^{-1}; mass spectrum: m/e = 386 $[M]^+$), after $C_5H_5FeC_5H_4{-}(CH_2)_3CO_2CH_3$ (7%) and before $FeC_{10}H_7([CH_2]_3CO_2CH_3)_3$ (1.8%). Due to the similarity of the R_f values, the components are not fully separable [33, 34]. No. 4 is distillable at an air bath temperature of 200 to 210°C at 0.2 Torr [10]. For its thin layer chromatographic identification and separation for example from $C_5H_5FeC_5H_4{-}(CH_2)_3CO_2CH_3$ and ferrocene on SiO_2 in benzene, see [12].

$Fe(C_5H_4{-}(CH_2)_3CO_2C_2H_5)_2$ (Table **22**, No. **5**) also forms during the Clemmensen reduction of $Fe(C_5H_4{-}CO(CH_2)_2CO_2H)_2$ (see Table 15, No. 11, p. 246) with amalgamated Zn in boiling 1/1/1/1 $C_6H_6/C_2H_5OH/H_2O$/conc. HCl [8].

$Fe(C_5H_4{-}(CH_2)_4CO_2CH_3)_2$ (Table **22**, No. **6**) is distillable at 0.3 Torr at an air bath temperature of 210 to 215°C [10]. Use of thin layer chromatography for identification and separation, for example, along with $C_5H_5FeC_5H_4{-}(CH_2)_4CO_2CH_3$ and ferrocene on SiO_2 in C_6H_6 is treated in [12].

$Fe(C_5H_4{-}CH{=}CHCO_2C_2H_5)_2$ (Table **22**, No. **8**) forms on heating $Fe(C_5H_4{-}I)_2$, $(C_2H_5)_3N$, $(C_6H_5)_3P$, $Pd(O_2CCH_3)_2$, and $CH_2{=}CHCO_2C_2H_5$ in CH_3CN in a sealed tube for 8 h at 100°C, and is subsequently separated in the ether phase on addition of water and ether, and subjected to column chromatography on Al_2O_3, with a 62% yield [30].

$Fe(C_5H_4{-}C(CH_3){=}CHCO_2C_2H_5)_2$ (Table **22**, No. **9**) in concentrated H_2SO_4 displays the following chemical shifts in the ^{13}C NMR spectrum, which for the most part can not be assigned with certainty: δ = 13.9, 14.9, 20.1, and 27.0 (CH_3), 47.0, 68.9, 79.7, 80.8, 81.5, 82.4, 85.3, 95.4, 98.0, 103.1, 111.9, 146.8, 168.1, 179.8 ppm. Cis-trans isomerization is facilitated by protonation on the carbonyl oxygen to form the carbenium ion [25].

$Fe(C_5H_4{-}C(CH_3)OHCH_2CO_2C_2H_5)_2$ (Table **22**, No. **17**) is formed by the Reformatzki reaction on boiling a mixture of $Fe(C_5H_4{-}COCH_3)_2$, $BrCH_2CO_2C_2H_5$, and Zn dust in dry benzene [25].

$Fe(C_5H_4{-}COCH_2CO_2C_2H_5)_2$ (Table **22**, No. **18**) also forms on introduction of dry HCl into a suspension of $Fe(C_5H_4{-}COCH_2CN)_2$ in absolute ethanol. It is reduced to No. 16 by dropwise addition of a solution of $NaBH_4$ in 19/1 ethanol/water at 50°C [26].

$Fe(C_5H_4{-}CO(CH_2)_2CO_2CH_3)_2$ (Table **22**, No. **19**) does not cyclize on heating with polyphosphoric acid [3].

$Fe(C_5H_4{-}CO(CH_2)_2CO_2C_2H_5)_2$ (Table **22**, No. **20**) cannot be prepared by Method VIb (p. 308), which yields only the monoacylation product on use of excess $C_2H_5O_2C(CH_2)_2COCl$ [38]. Catalytic hydrogenation over Raney Ni at 75°C with 135 atm H_2 gives VI [5].

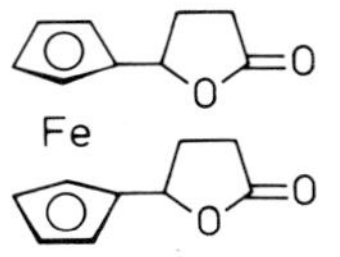

VI

$Fe(C_5H_4{-}COC_6H_4CO_2CH_3\text{-}2)_2$ (Table **22**, No. **29**) increases the thermal stability of dimethylpolysiloxane when added in small amounts [4].

$CH_3O_2CCH_2{-}C_5H_4FeC_5H_4{-}COC_6H_4CH_3\text{-}4$ (Table **22**, No. **31**). A mixture of No. 31 and its two positional isomers is given by Method VII (p. 309) or by esterification of $FeC_{10}H_8(CH_2CO_2H){-}COC_6H_4CH_3\text{-}4$ by Method Ia (p. 307). ^1H NMR spectrum ($CDCl_3$): $\delta = 2.43$ (s, CH_3 of R^2), 3.24 (s, 0.5 H, CH_2 of No. 31), 3.42 (s, 0.2 H, CH_2 of 1,3-isomer), 3.67 (s, CO_2CH_3), 3.73 (s, 0.3 H, CH_2 of 1,2-isomer), 4.1 to 4.3 (m, 5 H, $C_5H_5 + C_5H_4{-}R^1$), 4.5 to 4.7 (m, 2 H, H-3,4 of $C_5H_4{-}R^2$), 4.8 to 5.0 (m, 1.4 H, H-2,5 of $C_5H_4{-}R^2$), 7.1 to 8.0 (m, C_6H_4) ppm. The mixture is separable by gas chromatography at a column temperature of 250°C with 3% OV-225 on Chromosorb W HP. No. 31 appears ($R_f = 12.5$ min) after the 1,2-isomer ($R_f = 9.3$ min) and before the 1,3-isomer ($R_f = 13.5$ min). Thin layer chromatography on SiO_2 with 2/1 hexane/$CH_3CO_2C_2H_5$ gives only two spots, a small cinnabar red one at $R_f = 0.47$ (1,2-isomer) and a large red one at $R_f = 0.39$ (1,1′-and 1,3-isomers) [39].

$4\text{-}C_2H_5O_2CC_6H_4{-}C_5H_4FeC_5H_4{-}CHO$ (Table **22**, No. **33**). The Vilsmeier-Haack formylation of $C_5H_5FeC_5H_4{-}C_6H_4CO_2C_2H_5\text{-}4$ with $C_6H_5N(CH_3)CHO$ and $POCl_3$ in the absence of solvents at 65 to 70°C for 3 h yields No. 33 mixed with its two positional isomers, as a spontaneously decomposing oil [32] in a total yield of 86%, whose fraction of No. 33 is 60%. This fraction is determined after chromatographic enrichment, from the ^1H NMR spectrum, under the assumption that the elution takes place in the sequence 1,2-, 1,1′-, and 1,3-isomer and that the cyclopentadienyl protons in the α position with respect to the electron-withdrawing substituents are less screened than those in the β position. A complete separation of No. 33 does not succeed by column chromatography on SiO_2 with C_6H_6. Nevertheless, separation experiments with very dilute solutions on Silufol 254 showed that the isomers are separable [28, 32].

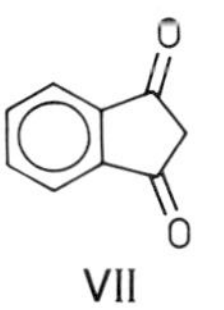

VII

Condensation with $CH_3COC_6H_5$ and NaOH (mole ratio 1/1/10) on 2/1 ethanol/water or with VII in benzene in the presence of several drops of $(C_2H_5)_3N$ (12 h reaction) forms $FeC_{10}H_8(C_6H_4CO_2C_2H_5\text{-}4)CH{=}CHCOC_6H_5$ (No. 36 + positional isomers) or $FeC_{10}H_8(C_6H_4CO_2C_2H_5\text{-}4){-}CH{=}C(CO)_2C_6H_4$ (No. 37 + positional isomers), respectively. Both isomer mixtures are separable by column chromatography on SiO_2 with benzene as eluent. The elution sequence is 1,2-, 1,1′-, and 1,3-isomer for the first mixture, 1,2, 1,3, and 1,1′ for the latter. The yield of No. 36 is 28%, of No. 37 80% [28].

$4\text{-}C_2H_5O_2CC_6H_4{-}C_5H_4FeC_5H_4{-}COCH_3$ (Table **22**, No. **34**). Friedel-Crafts acetylation of $C_5H_5FeC_5H_4{-}C_6H_4CO_2C_2H_5\text{-}4$ with $(CH_3CO)_2O$ under the action of 85% H_3PO_4 (mole ratio 1/20/4) at 90 to 95°C, with the anhydride being at the same time the reaction medium, yields an oil after column chromatographic processing on SiO_2 with benzene as eluent, in 46% relative yield. This consists 73% of No. 34 and 27% of its 1,3-isomer, which are not separable chromatographically. The content of the 1,2-isomer is practically zero [28, 32]. Condensation of the mixture with C_6H_5CHO in C_2H_5OH with addition of NaOH leads to $FeC_{10}H_8(C_6H_4CO_2C_2H_5\text{-}4)COCH{=}CHC_6H_5$, from which No. 35 can be separated by column chromatography on SiO_2 with benzene in 42% yield [28].

2-$CH_3O_2CC_6H_4$–$C_5H_4FeC_5H_4$–CH_2OCH_3 (Table **22**, No. **38**) is obtained by solvolysis of [2-$CH_3O_2CC_6H_4$–$C_5H_4FeC_5H_4$–$CH_2N(CH_3)_3$]I in CH_3OH [29].

References:

[1] R. B. Woodward, M. Rosenblum, M. C. Whiting (J. Am. Chem. Soc. **74** [1952] 3458/9). – [2] M. Rosenblum (Diss. Harvard Univ. 1953, p. 112). – [3] A. N. Nesmeyanov, N. A. Vol'kenau, V. D. Vil'chevskaya (Dokl. Akad. Nauk SSSR **111** [1956] 362/4; Proc. Acad. Sci. USSR Chem. Sect. **106/111** [1956] 675/7). – [4] W. A. Piccoli, Dow Corning Corp. (U.S. 2979482 [1956/61]; C. A. **1961** 18176). – [5] P. J. Graham, R. V. Lindsey, G. W. Parshall, M. L. Peterson, G. M. Whitman (J. Am. Chem. Soc. **79** [1957] 3416/20).

[6] K. L. Rinehart, R. J. Curby, P. E. Sokol (J. Am. Chem. Soc. **79** [1957] 3420/4). – [7] W. Göller (Diss. Stuttgart T.H. 1958). – [8] J. B. Thomson (Chem. Ind. [London] **1959** 1122). – [9] K. Arakawa, K. Yamakawa (Kagaku No Ryoiki **14** [1960] 632/44). – [10] K. Schlögl, H. Seiler (Monatsh. Chem. **91** [1960] 79/91).

[11] K. Schlögl, H. Seiler (Angew. Chem. **72** [1960] 38; 17th Intern. Conf. Pure Appl. Chem., München 1959, Vol. 1, Abstr. A 148). – [12] K. Schlögl, H. Pelousek, A. Mohar (Monatsh. Chem. **92** [1961] 533/41). – [13] W. Mock, J. H. Richards (J. Org. Chem. **27** [1962] 4050/1). – [14] K. L. Rinehart, R. J. Curby, D. H. Gustafson, K. G. Harrison, R. E. Bozak, D. E. Bublitz (J. Am. Chem. Soc. **84** [1962] 3263/9). – [15] K. Schlögl, M. Peterlik, H. Seiler (Monatsh. Chem. **93** [1962] 1309/27).

[16] K. L. Rinehart, D. E. Bublitz, D. H. Gustafson (J. Am. Chem. Soc. **85** [1963] 970/82). – [17] A. N. Nesmeyanov, V. D. Vil'chevskaya, N. S. Kochetkova (Dokl. Akad. Nauk SSSR **152** [1963] 627/8; Dokl. Chem. Proc. Acad. Sci. USSR **148/153** [1963] 750/1). – [18] G. Tainturier, J. Tirouflet (Bull. Soc. Chim. France **1966** 595/600). – [19] J. Tirouflet, G. Tainturier (Tetrahedron Letters **1965** 4177/9). – [20] G. Haller, K. Schlögl (Monatsh. Chem. **98** [1967] 2044/57).

[21] P. Dixneuf, R. Dabard (Bull. Soc. Chim. France **1969** 2164/71). – [22] M. Laćan, R. Šarac-Arneri (Croat. Chem. Acta **43** [1971] 215/9). – [23] A. Sonoda, I. Moritani (J. Organometal. Chem. **26** [1971] 133/40). – [24] A. Kasahara, T. Izumi, G. Saito, M. Yodono, R. Saito, Y. Goto (Bull. Chem. Soc. Japan **45** [1972] 895/900). – [25] A. N. Nesmeyanov, G. B. Shul'pin, M. I. Rybinskaya, P. V. Petrovskii (Dokl. Akad. Nauk SSSR **215** [1974] 599/602; Dokl. Chem. Proc. Acad. Sci. USSR **214/219** [1974] 188/91).

[26] A. N. Nesmeyanov, M. I. Rybinskaya, G. B. Shul'pin, A. A. Pogrebnyak (J. Organometal. Chem. **92** [1975] 341/67). – [27] M. Hisatome, T. Sakamoto, K. Yamakawa (J. Organometal. Chem. **107** [1976] 87/101). – [28] Š. Toma, A. Maholányiová, E. Solčániová (Acta Fac. Rerum Nat. Univ. Comenianae Chim. **24** [1976] 53/80). – [29] E. I. Edwards, R. Epton, G. Marr, G. K. Rogers, K. J. Thompson (Chem. Soc. [London] Spec. Publ. No. 28 [1977] 92/100). – [30] A. Kasahara, T. Izumi, M. Maemura (Bull. Chem. Soc. Japan **50** [1977] 1021/2).

[31] M. Laćan, V. Rapić, A. Brbot-Šaranović (Croat. Chem. Acta **49** [1977] 857/67). – [32] Š. Toma, A. Maholanyiová, E. Solčaniová (Collection Czech. Chem. Commun. **42** [1977] 1013/26). – [33] L. I. Denisovich, M. G. Peterleitner, S. P. Gubin (Izv. Akad. Nauk SSSR Ser. Khim. **1978** 465/8; Bull. Acad. Sci. USSR Div. Chem. Sci. **1978** 403/6). – [34] S. P. Gubin, L. I. Denisovich, N. V. Zakurin, M. G. Peterleitner (J. Organometal. Chem. **146** [1978] 267/78). – [35] N. A. Maier, S. A. Shklyar (Zh. Obshch. Khim. **48** [1978] 702/3; J. Gen. Chem. [USSR] **48** [1978] 645/6).

[36] N. A. Maier, S. A. Shklyar, Yu. A. Ol'dekop (Zh. Obshch. Khim. **49** [1979] 2518/21; J. Gen. Chem. [USSR] **49** [1979] 2223/5). – [37] A. Kasahara, T. Izumi, H. Umezawa (Chem. Letters **1980**

1039/42). – [38] M. Salisová, E. Solčaniová, Š. Toma (Acta Fac. Rerum Nat. Univ. Comenianae Chim. **28** [1980] 103/26). – [39] B. E. Maryanoff (J. Org. Chem. **46** [1981] 70/4). – [40] A. Perjéssy, Š. Toma (Collection Czech. Chem. Commun. **48** [1983] 1635/46).

[41] I. Shimizu, Y. Kamei, T. Tezuka, T. Izumi, A. Kasahara (Bull. Chem. Soc. Japan **56** [1983] 192/8). – [42] I. Shimizu, H. Umezawa, T. Kanno, T. Izumi, A. Kasahara (Bull. Chem. Soc. Japan **56** [1983] 2023/8). – [43] A. Kasahara, T. Izumi, I. Shimizu, T. Oikawa, H. Umezawa, M. Murakami, O. Watanabe (Bull. Chem. Soc. Japan **58** [1985] 1560/9).

5.2.3.7.2 1,2-Compounds

5.2.3.7.2.1 Carboxylic Acid Esters of the Type $C_5H_5FeC_5H_3(CO_2R'\text{-}1)R^2\text{-}2$ ($CO_2R' = R^1$)

Compounds of the type $C_5H_5FeC_5H_3(CO_2R'\text{-}1)A\text{–}CO_2R''$ are treated in 5.2.3.7.2.2, pp. 332/45.

All compounds in Table 23 possess with two different substituents on one cyclopentadienyl ring a plane of chirality; for the "ferrocene chirality" and the centrochiral R, S nomenclature used throughout see "Ferrocene" 7, 1980, pp. 1/4 (English translation) and 4/7 (German). Optically active forms of the compounds No. 1, 2, 5, 9, 12, 13, and 15 are described under "Further information", pp. 327/31.

The compounds are prepared by the methods following and those noted under "Further information":

Method I. Esterification of $C_5H_5FeC_5H_3(CO_2H\text{-}1)R^2\text{-}2$

a) with CH_2N_2 in ether [2, 8],

b) with $4\text{-}BrC_6H_4COCH_2Br$ in hot absolute ethanol at pH 8 [5].

The compounds are isolated by column chromatography on Al_2O_3 with C_6H_6 (No. 22) [5] or 1/1 C_6H_6/heptane (No. 8 migrates here as the only band) [8].

Method II. Photoethoxycarbonylation of $C_5H_5FeC_5H_4\text{–}R^2$ in 1/1 C_2H_5OH/CCl_4 with irradiation at 365 nm, for 4 h, at room temperature. The extract of the reaction product with CH_2Cl_2 or ether is evaporated and the residue subjected to separation from $C_5H_5FeC_5H_4\text{–}R^2$ by thin layer or column chromatography. A second thin layer chromatographing isolates the compound from the mixture with its 1,1'- and 1,3-positional isomers. No. 16 is an exception, with $C_5H_5FeC_5H_4\text{–}C_6H_4CO_2C_2H_5\text{-}4$ precipitating as a fourth product during its preparation. The yields given for Method II refer in each case to unrecovered $C_5H_5FeC_5H_4\text{–}R^2$ [18].

Table 23

Carboxylic Acid Esters of the Type $C_5H_5FeC_5H_3(CO_2R'\text{-}1)R^2\text{-}2$

Further information for numbers preceded by an asterisk is given at the end of the table, pp. 327/31.

For abbreviations and dimensions see p. VIII.

No.	$C_5H_5FeC_5H_3(R^1\text{-}1)R^2\text{-}2$ $R^1 = CO_2R'$	R^2	method of preparation (yield in %), properties, and remarks	Ref.
*1	CO_2CH_3	CH_3	Ia, see "Further information", p. 327 oil, b.p. 80 to 90°/0.005 Torr ^1H NMR (CCl_4): 2.24 (s, CH_3 of R^2), 4.04 (m, C_5H_5+2H of C_5H_3), 4.61 (s, 1H of C_5H_3) IR (CCl_4): ca. 1720 (νC=O)	[2, 6]

References on p. 331

Table 23 (continued)

No.	$C_5H_5FeC_5H_3(R^1\text{-}1)R^2\text{-}2$ $R^1 = CO_2R'$	R^2	method of preparation (yield in %), properties, and remarks	Ref.
*2	CO_2CH_3	C_6H_5	for (+)(1R)-No. 2 see "Further information", p. 328	[26]
*3	CO_2CH_3	Cl	Ia, see "Further information", p. 328 b.p. 140 to 150°/1 Torr ^{1}H NMR (CCl_4): 3.79 (CH_3), 4.20 (C_5H_5); 4.17(1H), 4.52(1H), and 4.64(1H)(C_5H_3) IR: 1003, 1105, 1150 (νC–O), 1193, 1260, 1280, 1712 (νC=O)	[8, 11]
*4	CO_2CH_3	I	see "Further information", p. 328 dark red oil, which solidified below 0° ^{1}H NMR ($CDCl_3$): 3.87 (s, CH_3), 4.26 (s, C_5H_5); 4.47 (m, H-4), 4.74 (m, H-3), and 4.85 (m, H-5)(ABC, C_5H_3) IR (neat): 1000, 1110, 1150, 1190, 1260, 1275, 1450, 1725, 2915	[12]
*5	CO_2CH_3	OH	Ia (59), see "Further information", p. 329 orange crystals, m.p. 46° (ether/light petroleum ether) IR: 1682 (νC=O), 3460 (νO–H) mass spectrum: m/e = 260 $[M]^+$	[21]
*6	CO_2CH_3	CH_2OH	see "Further information", p. 329	[17]
7	CO_2CH_3	$CH_2O_2CC_6H_5$	for (−)(1S)-No. 7 see (−)(1S)-No. 6, p. 329	[10]
8	CO_2CH_3	OCH_3	Ia, see No. 5, p. 329 orange crystals, m.p. 39° (light petroleum ether) ^{1}H NMR (CCl_4): 3.65 (s, CH_3), 3.75 (s, CH_3), 3.90 (m, 2H of C_5H_3), 4.10 (s, C_5H_5), 4.40 (m, 1H of C_5H_3) ^{1}H NMR (CCl_4): 3.67 (s, CH_3 of R^2), 3.74 (s, CH_3 of R^1), 3.85 (t, H-4 of C_5H_3; J (H-4,3) = J (H-4,5) ≈ 2.6), 4.11 (s, C_5H_5 + H-3 of C_5H_3), 4.40 (dd, H-5 of C_5H_3; J (H-5,3) = 1.5) mass spectrum: m/e = 274 $[M]^+$	[5, 8, 16, 21]
*9	CO_2CH_3	CHO	for (−)(1S)-No. 9 see "Further information", p. 329	[13, 20]

Table 23 (continued)

No.	$C_5H_5FeC_5H_3(R^1\text{-}1)R^2\text{-}2$ $R^1 = CO_2R'$	R^2	method of preparation (yield in %), properties, and remarks	Ref.
*10	CO_2CH_3	$COCH_3$	see "Further information", p. 330 dark yellow oil crystallizing slowly upon storage in the cold, m.p. 39° ^{1}H NMR (C_6H_6): 2.37 (s, CH_3 of R^2), 3.48 (s, CH_3 of R^1), 3.97 (s, C_5H_5), 4.11 (t, H-4 of C_5H_3; J = 3), 4.69 (m, H-3,5 of C_5H_3) IR (CCl_4): 1008, 1110, 1680 (νC=O of R^2), 1720 (νC=O of R^1) mass spectrum: m/e = 286 $[M]^+$, 221 $[M-C_5H_5]^+$, 152 $[C_5H_5FeOCH_3]^+$	[19]
*11	CO_2CH_3	CO_2H	see "Further information", p. 330 m.p. 103 to 103.5°	[23]
*12	CO_2CH_3	$(CH_2)_2CO_2H$	for (−)(1S)-No. 12 see "Further information", p. 330	[13]
*13	CO_2CH_3	$CH{=}CHCO_2H$	for (−)(1S)-No. 13 see "Further information", p. 330	[13]
14	$CO_2C_2H_5$	C_2H_5	II (11) dark red liquid ^{1}H NMR (CCl_4): 1.13 (t, CH_3 of R^2), 1.34 (t, CH_3 of R^1), 2.45 to 2.93 (m, CH_2 of R^2), 4.0 (s, C_5H_5), 4.20 (q, CH_2 of R^1), 4.09 to 4.26 (H-3,4 of C_5H_3), 4.60 (t, H-5 of C_5H_3) IR (neat): 920, 1000, 1100, 2920, 2960, and 3100 ($C_5H_5FeC_5H_3C_2H_5$-2), 1290 (νC–O), 1705 (νC=O) mass spectrum; m/e = 286 $[M]^+$	[18]
*15	$CO_2C_2H_5$	$CH{=}CH_2$	for "Optically Active Forms" see "Further information", p. 330	[22]
16	$CO_2C_2H_5$	C_6H_5	II (2) yellow oily substance IR (KBr): 950, 997, 1105, 1600, 3050, and 3100 ($C_5H_5FeC_5H_3C_6H_5$-2), 1240 (νC–O), 1705 (νC=O) mass spectrum: m/e = 334 $[M]^+$	[18]
17	$CO_2C_2H_5$	Cl	II (0.2) yellow oily substance	[18]

References on p. 331

Table 23 (continued)

No.	$C_5H_5FeC_5H_3(R^1\text{-}1)R^2\text{-}2$ $R^1 = CO_2R'$	R^2	method of preparation (yield in %), properties, and remarks	Ref.
			^{1}H NMR (CCl_4): 1.33 (t, CH_3), 4.10 (s, C_5H_5), 4.16 to 4.32 (CH_2 + H-3,4 of C_5H_3), 4.68 (t, H-5 of C_5H_3) IR (KBr): 915, 1000, 1100, 3020, and 3100 ($C_5H_5FeC_5H_3Cl$-2), 1280 (νC–O), 1710 (νC=O) mass spectrum: m/e = 292 $[M]^+$	
18	$CO_2C_2H_5$	Br	II (4) yellow oily substance ^{1}H NMR (CCl_4): 1.32 (t, CH_3), 4.11 (s, C_5H_5), 4.15 to 4.30 (CH_2 + H-3,4 of C_5H_3), 4.63 (t, H-5 of C_5H_3) IR (KBr): 905, 1000, 1095, and 3100 ($C_5H_5FeC_5H_3Br$-2), 1265 (νC–O), 1700 (νC=O) mass spectrum: m/e = 337 $[M]^+$	[18]
19	$CO_2C_2H_5$	I	II (1.6) orange-red oily substance ^{1}H NMR (CCl_4): 1.33 (t, CH_3), 4.23 (q, CH_2), 4.24 (t, H-4 of C_5H_3), 4.68 (t, H-3,5 of C_5H_3) IR (KBr): 930(sh), 960, 1000(sh), 1100, and 3060 ($C_5H_5FeC_5H_3I$-2), 1270 (νC–O), 1700 (νC=O) mass spectrum: m/e = 384 $[M]^+$	[18]
*20	$CO_2C_2H_5$	CHO	see "Further information", p. 330 m.p. 66° (ether/n-hexane) ^{1}H NMR ($CDCl_3$): 1.39 (t, CH_3; J = 7), 4.32 (s, C_5H_5), 4.36 (q, CH_2), 4.79 (t, H-4 of C_5H_3; J = 3), 5.17 (d, H-3,5 of C_5H_3), 10.69 (s, CHO) IR (KBr): 1660 (νC=O of R^2), 1710 (νC=O of R^1) UV: λ_{max} (log ε) = 445 (2.72), 362 (3.09), 334 (3.02), 254sh (3.87), 222 (4.33) mass spectrum: m/e (relative intensity in % referred to ^{56}Fe) = 286 ($[M]^+$, 70), 258 (10), 230 (17), 214 (23), 186 (15), 121 (45), 56 (100)	[25]
*21	$CO_2CH_2C_6H_4Br$-4	CH_3	see "Further information", p. 331	[2]
22	$CO_2CH_2COC_6H_4Br$-4	OCH_3	Ib (81) m.p. 135 to 136° (absolute C_2H_5OH)	[5]

References on p. 331

* Further information:

$C_5H_5FeC_5H_3(CO_2CH_3$-1)CH_3-2 (Table **23**, No. **1**) forms mixed with the 1,1′- and 1,3-isomers on oxidation of $FeC_{10}H_8(CHO)CH_3$ (fraction of 1,2-isomer 15%) with active, pulverized MnO_2 in the presence of KCN and glacial acetic acid in CH_3OH at 20°C. The total yield of the mixture purified by preparative thin layer chromatography on SiO_2 with benzene amounts to 46%, 6% of the total yield being the fraction of No. 1 [9].

A sharp separation of No. 1 and its positional isomers is not possible, even when the mixture is chromatographed in 3/1 petroleum ether/benzene over the total length of an Al_2O_3 column which is then divided into 20 to 25 fractions, and the mixed fractions rechromatographed (elution sequence 1,2-, 1,1′-, and 1,3-isomer). In part due to the easier saponificability of No. 1 and the 1,1′-isomer and the consequent higher absorbability the fraction in the mixed product shifts from 15/41/44 for the 1,2-, 1,1′-, and 1,3-isomers in the aldehyde mixture to 8/27/65 in the isolated esters. Of the inserted amount of ester mixture 20% remains in each case irreversibly adsorbed [9]. High pressure liquid-liquid chromatography with Carbowax 400 on a Corasil-filled column successfully separates No. 1 from the 1,3-isomer in both 1/9 and 1/4 CH_2Cl_2/iso-octane at 116 to 155 atm [14].

The ^1H NMR singlet of the CH_3 protons of R^2 is displaced versus that of $C_5H_5FeC_5H_4$–CH_3 in CCl_4 by 0.29 ppm toward lower field [6]. Saponification with methanolic KOH gives the free acid, reduction with $LiAlH_4/AlCl_3$ the $C_5H_5FeC_5H_3(CH_3)_2$-1,2 [2].

Optically Active Forms. (−)(1S)-$C_5H_5FeC_5H_3(CO_2CH_3$-1)CH_3-2 (see Formula I) is obtained by Method Ia from (+)(1S)-$C_5H_5FeC_5H_3(CO_2H$-1)CH_3-2 and isolated by preparative thin layer chromatography on SiO_2 in 30/1 benzene/ethanol, in 92 to 93% yield [1, 4, 20]. Heating the aqueous solution of the methiodide of (+)(1S, 8S)-II with Na amalgam and esterifying with CH_2N_2 the acid products extracted in ether gives a 24% yield referred to (+)(1S, 8S)-II [7, 10], see also [15]. The isolation takes place by column chromatography on Al_2O_3 via elution with hexane/benzene [10]. Winning from (+)(1R)-III by lithiation with n-C_4H_9Li in hexane/tetrahydrofuran at 0°C and subsequent carboxylation likewise leads via the $C_5H_5FeC_5H_3(CO_2Li$-1)CH_3-2, which is treated as in the preceding; the yield is 16% referred to (+)(1R)-III [10].

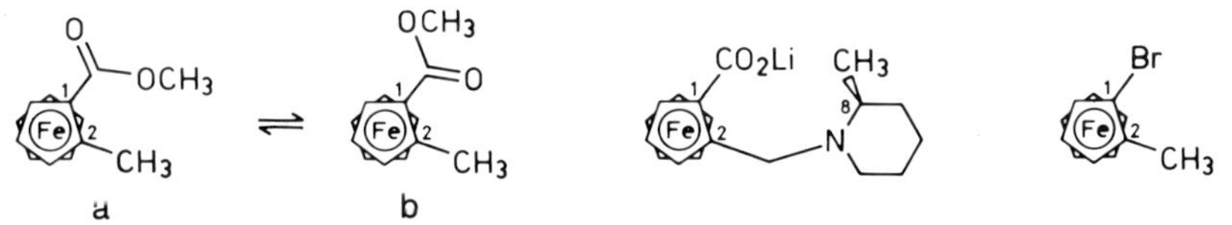

(−)(1S)-No. 1

I

(+)(1S, 8S)-II

(+)(1R)-III

Red [10] oil [1, 10, 20], boiling at 120°C (bath temperature) at 1.5 Torr [7, 10]. Specific rotation (20°C, extrapolated to p = 100% unless otherwise stated): In benzene $[\alpha]_D = -24.5° \pm (0.2$ to $0.9)°$ (c = 0.1 to 0.6) [20], −11.1° ± 0.2° (c = 2.6, p = 98%) [1], −11° (c = 0.3 to 1.2) [4], −10.0° (c = 1.9, p = 94%) [7, 10]; $[\alpha]_{570} = -44.7° \pm (0.2$ to $0.9)°$ (c = 0.1 to 0.6); $[\alpha]_{546} = -83.8° \pm (0.2$ to $0.9)°$ (c = 0.1 to 0.6) [20]. In C_2H_5OH (c = 0.1 to 0.6, values ± (0.2 to 0.9)°) $[\alpha]_D = -0.5°$, $[\alpha]_{570} = -14.2°$, $[\alpha]_{546} = 42.8°$ [20].

^1H NMR spectrum: In $CDCl_3$ δ = 2.31 (s) and 3.85 (s) (CH_3 and CO_2CH_3), 4.17 (s, C_5H_5), 4.30 (m, 2H), and 4.74 (m, 1H) (C_5H_3) ppm [20]. In CCl_4 δ = 2.25 (s) and 3.75 (s) (CH_3 and CO_2CH_3), 4.05 (s, C_5H_5), 4.15 (m, 2H), and 4.60 (m, 1H) (C_5H_3) ppm [10].

References on p. 331

IR spectrum (neat): 820, 1000, 1030, 1090, 1210, 1440, 1720 cm^{-1} [10]; (CH_2Cl_2): 1715 cm^{-1} [20].

Electronic spectrum (C_2H_5OH): $\lambda_{max}(\varepsilon) = 440(200)$, 346(292), 305(620), 259(3930) nm [4]. Optical rotatory dispersion (C_2H_5OH, $c \approx 0.1$, 20°C): $[M]_{635} = +67°$ [4], $[M]_D = -1.3°$ [20], $[M]_{534} = -160°$, $[M]_{460} = +3050°$, $[M]_{416} = +110°$, $[M]_{367} = +1930°$ [4]. In C_6H_6 $[M]_D^{20} = -63°$ (c = 0.1 to 0.6) [20]. Circular dichroism (C_2H_5OH, $c \approx 0.1$): $\Delta\varepsilon_{497} = -0.22$, $\Delta\varepsilon_{435} = +0.36$, $\Delta\varepsilon_{386} = +0.05$; ($C_6H_6$, $c \approx 0.1$): $\Delta\varepsilon_{494} = -0.26$, $\Delta\varepsilon_{431} = +0.35$, $\Delta\varepsilon_{384} = +0.06$ [20].

The absolute configuration is derived from the correlation with (+)-$C_5H_5FeC_5H_3(CO_2H\text{-}1)CH_3\text{-}2$ of the known absolute configuration (1S) [4]. Due to the comparable sizes of OCH_3 and =O the conformations a) and b) for (−)(1S)-No. 1 (see Formula I) are equally preferred. The Gibbs free energy difference between the two forms is zero at 283 K, as derived from the temperature dependence of the molar amplitude of the Cotton effect in the region of the ferrocene bands for 283 to 343 K [3].

$LiAlH_4$ in ether forms (+)(1S)-$C_5H_5FeC_5H_3(CH_2OH\text{-}1)CH_3\text{-}2$ [1, 4]. The oxidation with acidic instead of the usually used alkaline MnO_2 in $ClCH_2CH_2Cl$ at 110°C leads to (−)(1S)-No. 9 in 62% yield, which is separated by preparative thin layer chromatography on SiO_2 with 20/1 C_6H_6/C_2H_5OH [20].

$C_5H_5FeC_5H_3(CO_2CH_3\text{-}1)C_6H_5\text{-}2$ (Table **23**, No. **2**). **Optically Active Forms.** (+)(1R)-No. 2 is prepared from (+)(1R)-$C_5H_5FeC_5H_3(CO_2H\text{-}1)C_6H_5\text{-}2$ by Method Ia (p. 323) and purified by preparative thin layer chromatography on SiO_2 with C_6H_6; yield 91%. Specific rotation (C_2H_5OH, $c \approx 0.2$ to 0.3; p = 100%): $[\alpha]_D^{20} = +100°$. Electronic spectrum in C_2H_5OH: $\lambda_{max}(\varepsilon)$ – 435 (368) nm. Circular dichroism (C_2H_5OH, $c \approx 0.02$ to 0.06; p = 100%): $\Delta\varepsilon_{435} = -1.57$, $\Delta\varepsilon_{495} = +0.71$. According to a band analysis with the aid of a curve fitting computer program the circular dichroism curve results from three overlapping bands with the rotatory strengths in $erg \cdot cm^3 \cdot 10^{40}$: −5.3 (435 nm), −0.1 (445 nm), and +2.4 (510 nm).

The reduction with $LiAlH_4$ leads to (+)(1R)-$C_5H_5FeC_5H_3(CH_2OH\text{-}1)C_6H_5\text{-}2$, that with $LiAlH_4$/$AlCl_3$ to (+)(1S)-IV. Saponification with methanolic KOH gives back the starting acid [26].

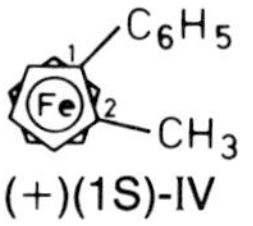

(+)(1S)-IV

$C_5H_5FeC_5H_3(CO_2CH_3\text{-}1)Cl\text{-}2$ (Table **23**, No. **3**). Lithiation of $C_5H_5FeC_5H_4\text{–}Cl$ with n-C_4H_9Li, carboxylation, and esterification per Method Ia (p. 323) forms No. 3 in 36% yield along with a 4% yield of $CH_3O_2C\text{–}C_5H_4FeC_5H_3(CO_2CH_3\text{-}1)Cl\text{-}2$ [11]. It is isolated by chromatography on Al_2O_3 columns in 1/1 heptane/benzene. Saponification with 20% NaOH gives the free acid [8].

$C_5H_5FeC_5H_3(CO_2CH_3\text{-}1)I\text{-}2$ (Table **23**, No. **4**) forms from $C_5H_5FeC_5H_3(CO_2CH_3\text{-}1)HgCl\text{-}2$ on reaction in CH_2Cl_2 with a saturated solution of I_2 in CH_2Cl_2 with separation by preparative thin layer chromatography on SiO_2 in CH_2Cl_2, in an 86% yield [12]. On reduction with $LiAlH_4$ in ether the C–I bond is split and $C_5H_5FeC_5H_4\text{–}CH_2OH$ is obtained as the sole product. $NaBH_4$ and $AlCl_3$ (mole ratio 1/7.4/<2) in $CH_3OCH_2CH_2OCH_3$ form in contrast $C_5H_5FeC_5H_3(CH_2OH\text{-}1)I\text{-}2$ (36% yield) and $C_5H_5FeC_5H_3(CH_3\text{-}1)I\text{-}2$ (62%). The reaction with CuCN in N-methylpyrrolidone solution leads to $C_5H_5FeC_5H_3(CO_2CH_3\text{-}1)CN\text{-}2$ in 41% yield. The mixed Ullmann coupling reaction with $C_5H_5FeC_5H_4\text{–}I$ under the action of Cu bronze yields V (55%), VI, and VII [12].

References on p. 331

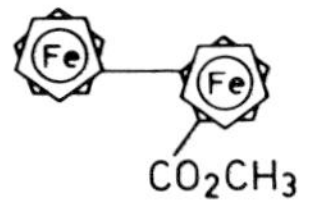

V

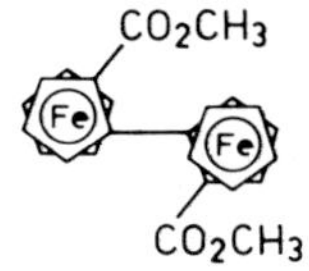

VI

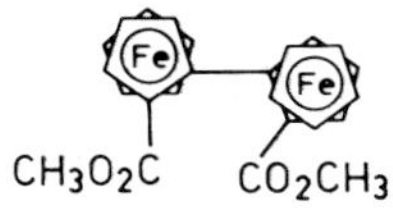

VII

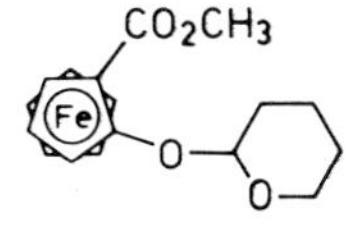

VIII

$C_5H_5FeC_5H_3(CO_2CH_3\text{-}1)OH\text{-}2$ (Table **23**, No. **5**) is obtained in a 75% yield on hydrolysis, catalyzed by 2M HCl, of VIII in C_2H_5OH at room temperature [21]. No. 5 can be handled in air and recrystallized without special precautions. This unusual stability compared with that of other hydroxy compounds of ferrocene is probably due to the electron withdrawing effect of the CO_2CH_3 substituent and the formation of an intramolecular hydrogen bond. Reaction with CH_2N_2 in ether forms No. 8, which is isolated from the oil remaining after evaporation of the solvent by column chromatography on Al_2O_3 with petroleum ether as eluent, in 21% yield [21].

$C_5H_5FeC_5H_3(CO_2CH_3\text{-}1)CH_2OH\text{-}2$ (Table **23**, No. **6**) possibly forms along with the 1,1′-isomer (cf. Table 21, No. 9, p. 296) on carboxylation and subsequent esterification by Method Ia (p. 323) of the product resulting from lithiation of $C_5H_5FeC_5H_4\text{–}CH_2OCH_3$. The IR spectrum was taken of the isolated compound, but it proved too unstable for further studies [17].

Optically Active Forms. (–)(1S)-$C_5H_5FeC_5H_3(CO_2CH_3\text{-}1)CH_2OH\text{-}2$ is won from the methiodide of (+)(1S, 8S)-II (see p. 327) by heating with aqueous KOH (5%, 26 h [10]; 10%, 25 h [13]) and esterification of the acid products extracted with ether, in accordance with Method Ia (p. 323), in a yield of 19 [10] or 23% [13] (in each case referred to (+)(8S)-$C_5H_5FeC_5H_4\text{–}CH_2NC_5H_9\text{–}CH_3\text{-}8$) [10, 13] and eluted as a red oil with ether from an Al_2O_3 chromatographic column [10]. Specific rotation in $CHCl_3$: $[\alpha]_D^{20} = -22°$ (c = 0.81, p = 94%) [10], −22.5° (c = 5.5) [13]. ^1H NMR spectrum (CCl_4): δ = 3.87 (s, CH_3), 4.27 (s) and 4.30 to 4.83 (m)($C_5H_5 + C_5H_3CH_2$) ppm [10]. IR spectrum (neat): 780, 820, 1000, 1095, 1210, 1290, 1460, 1710, 3450 cm^{-1} [10]. For the optical rotatory dispersion and circular dichroism curves (C_2H_5OH), see [13, p. 2885].

Active MnO_2 oxidizes the compound on 2 h refluxing in $CHCl_3$ to (–)(1S)-No. 9; 99% yield [13]. The benzoate (1S)-No. 7 is obtained in the usual way; it melts at 132°C (CH_3OH) [10].

$C_5H_5FeC_5H_3(CO_2CH_3\text{-}1)CHO\text{-}2$ (Table **23**, No. **9**). **Optically Active Forms.** Preparation of (–)(1S)-No. 9, an oil [20], from (–)(1S)-No. 1 or (–)(1S)-No. 6 is described under "Further information", p. 328 and above, respectively. Specific rotation values (± (0.2 to 0.9)°, c = 0.1 to 0.6, 20°C, p = 100%): In C_2H_5OH $[\alpha]_D = -759°$; $[\alpha]_{570} = -919°$; $[\alpha]_{546} = -1218°$. In C_6H_6 $[\alpha]_D = -696°$; $[\alpha]_{570} = -866°$; $[\alpha]_{546} = -1212°$ [20]. A distinctly lower value of $[\alpha]_D^{20} = -384°$ (c = 0.60, C_2H_5OH) is reported in [13]. ^1H NMR spectrum ($CDCl_3$): δ = 3.93 (s, CH_3), 4.39 (s, C_5H_5), 4.84 (m, 1H), 5.19 (m, 1H), and 5.23 (m, 1H)(C_5H_3), 10.6 (s, CHO) [20]; 3.88 (s, CH_3), 4.33 (s, C_5H_5), 4.78 (m, 1H), and 5.15 (m, 2H)(C_5H_3), 10.68 (s, CHO) [13] ppm. IR spectrum (neat): 750, 805, 825, 1005, 1035, 1080, 1150, 1160, 1190, 1220, 1275, 1290, 1670, 1710 cm^{-1} [13]; in CH_2Cl_2: 1655 and 1695 (νC=O) cm^{-1} [20]. Optical rotatory dispersion: $[M]_D^{20} = -2065°$ (C_2H_5OH), −1894° (C_6H_6) [20]. Circular dichroism (c ≈ 0.1): $\Delta\varepsilon_{478} = -1.84$ (C_2H_5OH), $\Delta\varepsilon_{476} = -2.14$ (C_6H_6) [20].

(−)(1S)-No. 9 yields in the usual way **(1S)-$C_5H_5FeC_5H_3(CO_2CH_3\text{-}1)CH{=}NNHCONH_2\text{-}2$**, the semicarbazone melting at 210°C with decomposition [13]. The Knoevenagel-Doebner condensation with $HO_2CCH_2CO_2H$ in boiling pyridine with addition of piperidine, yields (−)(1S)-No. 13 in a 22% yield (referred to (−)(1S)-No. 6) [13]. (1S)-$C_5H_5FeC_5H_3(CO_2CH_3\text{-}1)CN\text{-}2$ can be obtained in a single reaction vessel process – without isolation of the intermediate products – in which (−)(1S)-No. 9 in pyridine is reacted first with aqueous $NH_2OH \cdot HCl$ solution and then after 1 h

References on p. 331

stirring with $CuSO_4 \cdot 5H_2O$, $(C_2H_5)_3N$, and finally dicyclohexylcarbodiimide in CH_2Cl_2. In contrast, attempts at oxidation with AgO in tetrahydrofuran/KOH to form (1S)-No. 11 were unsuccessful [20].

$C_5H_5FeC_5H_3(CO_2CH_3\text{-}1)COCH_3\text{-}2$ (Table **23**, No. **10**) is prepared by Friedel-Crafts acetylation of $C_5H_5FeC_5H_4\text{–}CO_2CH_3$ with $(CH_3CO)_2O$ and $BF_3 \cdot O(C_2H_5)_2$ in CH_2Cl_2 at room temperature in an 11% yield along with a 66% yield of the 1,1′-isomer. The 1,2-isomer is isolated from the mother liquor of the 1,1′-isomer, after dissolution of the raw product in hot hexane, by preparative thin layer chromatography on SiO_2 with petroleum ether/ether, following $C_5H_5FeC_5H_4\text{–}CO_2CH_3$ and preceding the residual quantity of the 1,1′-isomer. On acetylation with $CH_3COCl/AlCl_3$ in CCl_4 the yield is 0.7% (35% 1,1′-isomer) [19]. Quantum mechanical calculations based on the extended Hückel theory indicate that No. 10 should decompose, especially in acid solution, with formation of IX and $[FeC_5H_5]^+$ [24].

$C_5H_5FeC_5H_3(CO_2CH_3\text{-}1)CO_2H\text{-}2$ (Table **23**, No. **11**) is obtained by oxidation of X with KOCl in CH_3OH at 50 to 60°C in a 63 to 64% yield, after thin layer chromatographic isolation on SiO_2 with 75/5/70/8 $C_6H_6/CH_3CO_2C_4H_9$/heptane/$(CH_3)_2NCHO$ as eluent [23]. Preparation of (1S)-No. 11 from (−)(1S)-No. 9 with AgO in tetrahydrofuran/KOH was unsuccessful [20].

$C_5H_5FeC_5H_3(CO_2CH_3\text{-}1)CH_2CH_2CO_2H\text{-}2$ (Table **23**, No. **12**). **Optically Active Forms.** (−)(1S)-No. 12 is obtained from (−)(1S)-No. 13 (see below) in the form of fine, yellow-orange crystals. The specific rotation (c=4.5, C_2H_5OH, 20°C) is $[\alpha]_D = 0°$, $[\alpha]_{545} = -25.0°$. Esterification with CH_2N_2 gives (−)(1S)-$C_5H_5FeC_5H_3(CO_2CH_3\text{-}1)CH_2CH_2CO_2CH_3\text{-}2$ [13].

$C_5H_5FeC_5H_3(CO_2CH_3\text{-}1)CH{=}CHCO_2H\text{-}2$ (Table **23**, No. **13**). **Optically Active Forms.** (−)(1S)-No. 13, with $[\alpha]_D^{20} = -520°$, c=0.10, acetone, is prepared from (−)(1S)-No. 9 (see there). It is esterified with CH_2N_2 to (−)(1S)-$C_5H_5FeC_5H_3(CO_2CH_3\text{-}1)CH{=}CHCO_2CH_3\text{-}2$ or is hydrogenated in C_2H_5OH catalytically over PtO_2 at room temperature in an H_2 atmosphere to (−)(1S)-No. 12, in 90% yield. The latter compound is purified chromatographically on activated Al_2O_3 in C_2H_5OH [13].

$C_5H_5FeC_5H_3(CO_2C_2H_5\text{-}1)CH{=}CH_2\text{-}2$ (Table **23**, No. **15**). **Optically Active Forms.** (−)(1S)-$C_5H_5FeC_5H_3(CO_2C_2H_5\text{-}1)CH{=}CH_2\text{-}2$ is prepared from (−)(1S, 6S)-XI by heating with CH_3I. It is purified by column chromatography on SiO_2 with benzene; yield 35% [22].

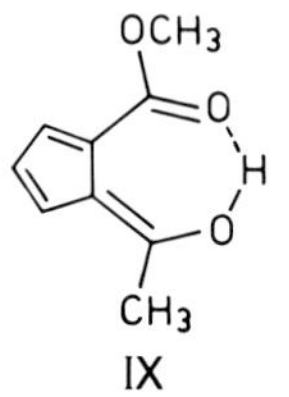

IX

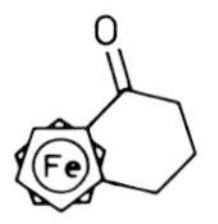

X

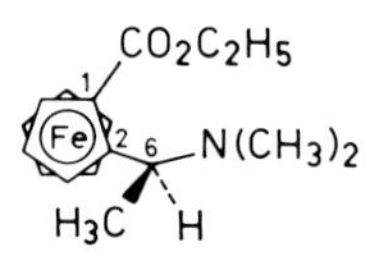

(−)(1S, 6S)-XI

The reddish viscous oil exhibits a specific rotation of $[\alpha]_D^{20} = -175.5°$ (c=0.182, $CHCl_3$). ^1H NMR spectrum: δ=1.28 (t, CH_3), 4.27 (q, CH_2 of C_2H_5); 4.13, 4.41, and 4.75 (m, $C_5H_5 + C_5H_3$); 5.15 and 5.44 (m, $=CH_2$), 7.23 (m, CH=) ppm. IR spectrum: 900, 995, and 1410 ($CH{=}CH_2$); 910, 1000, and 1100 ($C_5H_5FeC_5H_3$-1,2); 1710 (νC=O) cm^{-1}. Mass spectrum: m/e=284 $[M]^+$ [22].

(+)(1R)-$C_5H_5FeC_5H_3(CO_2C_2H_5\text{-}1)CH{=}CH_2\text{-}2$ is analogously obtained from the enantiomeric (+)(1R, 6R)-XI in a 40% yield as a reddish viscous oil with $[\alpha]_D^{20} = +179.8°$ (c=0.26, $CHCl_3$). ^1H NMR, IR, and mass spectra exhibit the same values as for (−)(1S)-No. 15 [22].

$C_5H_5FeC_5H_3(CO_2C_2H_5\text{-}1)CHO\text{-}2$ (Table **23**, No. **20**). For preparation a solution of XII and XIII (molar ratio 1/1.5) in CH_2Cl_2 is irradiated for 14 h with a 300 W sun lamp and reacted after

References on p. 331

cooling with 1/1 C_2H_5OH/2 N NaOH. No. 20 is won from the concentrated organic layer as the first band on chromatography on a short Al_2O_3 column and is recrystallized from ether/ n-hexane, yield 67% [25].

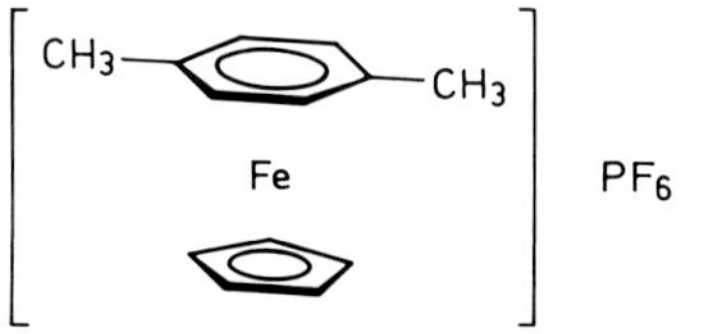

XII

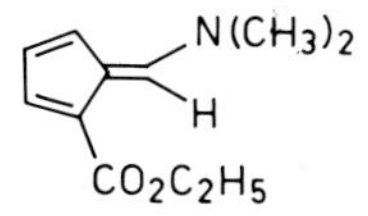

XIII

$C_5H_5FeC_5H_3(CO_2CH_2(C_6H_4Br\text{-}4)\text{-}1)CH_3\text{-}2$ (Table **23**, No. **21**). The isomer mixture obtained from $FeC_{10}H_8(COCl)CH_3$ with 4-$BrC_6H_4CH_2OH$ of $FeC_{10}H_8(CO_2CH_2C_6H_4Br\text{-}4)CH_3$ can be separated chromatographically on Al_2O_3 columns only little better than $FeC_{10}H_8(CO_2CH_3)CH_3$ (see "Further information" for No. 1, p. 327), so that it offers practically no advantage with respect to the latter compound for the separation of the isomers of the free acid [2].

References:

[1] H. Falk, K. Schlögl, W. Steyrer (Monatsh. Chem. **97** [1966] 1029/44). – [2] H. Falk, G. Haller, K. Schlögl (Monatsh. Chem. **98** [1967] 592/602). – [3] H. Falk, G. Haller, K. Schlögl (Monatsh. Chem. **98** [1967] 2058/69). – [4] G. Haller, K. Schlögl (Monatsh. Chem. **98** [1967] 603/18). – [5] A. N. Nesmeyanov, T. V. Baukova, K. I. Grandberg (Izv. Akad. Nauk SSSR Ser. Khim. **1967** 1867/9; Bull. Acad. Sci. USSR Div. Chem. Sci. **1967** 1796/8).

[6] K. Schlögl, H. Falk, G. Haller (Monatsh. Chem. **98** [1967] 82/94). – [7] T. Aratani, T. Gonda, H. Nozaki (Tetrahedron Letters **1969** 2265/8). – [8] A. N. Nesmeyanov, T. V. Baukova, K. I. Grandberg, Yu. A. Ustynyuk, S. P. Gubin, E. G. Perevalova (Izv. Akad. Nauk SSSR Ser. Khim. **1969** 721/3; Bull. Acad. Sci. USSR Div. Chem. Sci. **1969** 658/9). – [9] K. Schlögl, M. Walser (Monatsh. Chem. **100** [1969] 840/3). – [10] T. Aratani, T. Gonda, H. Nozaki (Tetrahedron **26** [1970] 5453/64).

[11] A. Sonoda, I. Moritani (Nippon Kagaku Zasshi **91** [1970] 566/71 [Eng. Abstr. p. A 33]). – [12] R. F. Kovar, M. D. Rausch (J. Organometal. Chem. **35** [1972] 351/66). – [13] T. Shirafuji, A. Odaira, Y. Yamamoto, H. Nozaki (Bull. Chem. Soc. Japan **45** [1972] 2884/7). – [14] R. Eberhardt, H. Lehner, K. Schlögl (Monatsh. Chem. **104** [1973] 1409/20). – [15] S. I. Goldberg, W. D. Bailey (J. Am. Chem. Soc. **96** [1974] 6381/7).

[16] R. Epton, G. Marr, G. K. Rogers (J. Organometal. Chem. **110** [1976] C42/C44). – [17] D. W. Slocum, B. P. Koonsvitsky (J. Org. Chem. **41** [1976] 3664/8). – [18] T. Akiyama, T. Kitamura, T. Kato, H. Watanabe, T. Serizawa, A. Sugimori (Bull. Chem. Soc. Japan **50** [1977] 1137/41). – [19] M. N. Nefedova, V. N. Setkina, V. A. Kataev, T. V. Fisun, D. N. Kursanov (Dokl. Akad. Nauk SSSR **237** [1977] 352/5; Dokl. Chem. Proc. Acad. Sci. USSR **232/237** [1977] 662/5). – [20] V. Rapič, K. Schlögl, B. Steinitz (Monatsh. Chem. **108** [1977] 767/80).

[21] R. Epton, G. Marr, G. K. Rogers (J. Organometal. Chem. **150** [1978] 93/100). – [22] A. Kasahara, T. Izumi, H. Watanabe (Bull. Chem. Soc. Japan **52** [1979] 957/8). – [23] N. A. Maier, S. A. Shklyar, Yu. A. Ol'dekop (Zh. Obshch. Khim. **50** [1980] 2626/7; C.A. **94** [1981] No. 121685). – [24] P. Zahradník, Š. Toma (Acta Fac. Rerum Nat. Univ. Comenianae Chim. **29** [1981] 1/7). – [25] P. Bickert, B. Hildebrandt, K. Hafner (Organometallics **3** [1984] 653/7).

[26] H. Lehner, K. Schlögl (Monatsh. Chem. **101** [1970] 895/911).

5.2.3.7.2.2 Carboxylic Acid Esters of the Type $C_5H_5FeC_5H_3(A{-}CO_2R'{-}1)R^2{-}2$
($A{-}CO_2R'{-}1 = R^1$)

The compounds compiled in Table 24 with $R^1 \neq R^2$, Nos. 13 to 27, possess a plane of chirality. The centrochiral R,S nomenclature used for them is described in "Ferrocene" 7, 1980, pp. 1/4 (English) and 4/7 (German). In addition compounds No. 18, 20, 21, 26, and 27 have a chirality center, so that they occur in diastereoisomeric forms.

Compounds No. 4 and 7 to 9 occupy a special position. Their substituents are to be sure constitutionally equal, but are themselves chiral. The compounds, thus, can occur in the isomeric forms A, B, and C (see I). A and B are achiral and pseudoasymmetric, with the configurationally opposed substituents, enantiotopic with respect to each other. The racemic form C is composed of the enantiomers a and b, whose substituents are in each case configurationally equal and are diastereotopic with respect to each other. The meso forms A and B are diastereomeric both with each other and with C.

a b

A B C

I

While all isomers could be isolated of Nos. 4 and 8 (see "Further information", pp. 338/42), chromatography with a multitude of eluents by [9] failed to resolve Nos. 7 and 9.

The preparation of the compounds is conducted by the following methods and by those described under "Further information", pp. 338/45:

Method I. Esterification of $C_5H_5FeC_5H_3(A{-}CO_2H{-}1)R^2{-}2$

a) with CH_2N_2 in ether, ether/methanol [17], or 1/1/1 ether/C_6H_6/H_2O [12],

b) with C_2H_5OH [9].

Starting with the isomeric mixture $FeC_{10}H_8(CO(CH_2)_2CO_2H)CH_3$ Method Ia yields No. 24 together with its 1,1'- and 1,3-isomers [7]. For separation see "Further information", p. 344.

Method II. Friedel-Crafts alkylation of ferrocene with $CH_3O_2C(CH_2)_nCl$ ($\triangleq$ RCl, $n = 1$ to 3) in heptane in the presence of $AlCl_3$ as catalyst and Al powder as reductant. This forms in addition to $C_5H_5FeC_5H_4{-}R$ the $FeC_{10}H_8R_2$, as a mixture (m) of the 1,2-, 1,1'-, and 1,3-isomers. Preparative thin layer chromatography on SiO_2 with 75/5/30/8 C_6H_6/$CH_3CO_2C_4H_9$/C_7H_{16}/$(CH_3)_2SO$ separates the reaction product into three bands ($C_5H_5FeC_5H_5$, $C_5H_5FeC_5H_4{-}R$, and m), each of which is separately removed and extracted with methanol and ether [20, 21].

a) For 4 h at 80 to 95°C [20].

b) Reaction in molar ratio 1/>1/<2/>2 first at 68°C and then for 3 h at >100°C [21].

c) As for Method II b, but with molar ratio 1/2/<3/>2 for 6 h [21].

d) As for Method II b, but for 6 h at 95 to 98°C [21].

Method III. Refluxing for 5 to 7 h in xylene a substituted ferrocenylcarbinol with $NaCR''(CO_2C_2H_5)_2$ ($R''=H$ for Nos. 10, 26, 27; CH_3 for No. 11; C_6H_5 for No. 12), followed by hydrolysis and extraction with ether. The residue remaining after evaporation of the solvent is purified by chromatography on Al_2O_3 columns with anhydrous benzene [5, 9, 13] or 9/1 benzene/ether as eluent [9].

a) Starting compound is II [9].

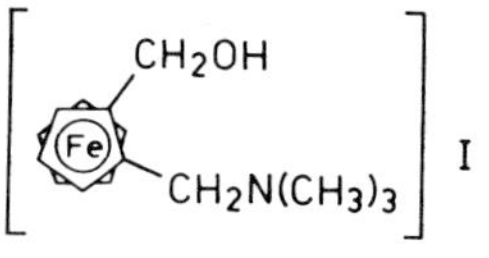

II

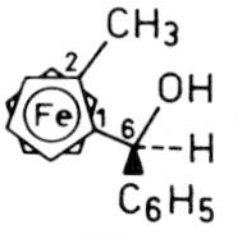

III

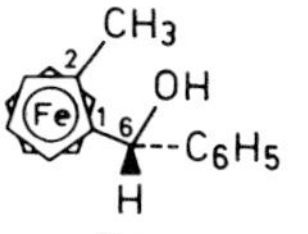

IV

b) Starting compound is (1RS, 6SR)-$C_5H_5FeC_5H_3(CHOHC_6H_5\text{-}1)CH_3\text{-}2$ (see Formula III, which shows one enantiomer of the racemate) [13].

c) Starting compound is (1RS, 6RS)-$C_5H_5FeC_5H_3(CHOHC_6H_5\text{-}1)CH_3\text{-}2$ (see Formula IV, which shows one enantiomer of the racemate) [13].

d) Starting compound is II, with reaction in xylene instead of boiling ethanol [5, 9].

Method IV. Friedel-Crafts acylation of $C_5H_5FeC_5H_4\text{–}CH_2CO_2CH_3$ with 4-$CH_3C_6H_4COCl$ and $AlCl_3$ (molar ratio 1/1/1) in CH_2Cl_2 at room temperature. No. 14 is isolated in a mixture (m) with its 1,1'- and 1,3-isomers by dry column chromatography on SiO_2 with 2/1 hexane/$CH_3CO_2C_2H_5$ [22].

Method V. Vilsmeier-Haack formylation of $C_5H_5FeC_5H_4\text{–}R^1$ with $C_6H_5N(CH_3)CHO$ and $POCl_3$

a) in CH_2Cl_2 at room temperature in the dark [1],

b) in 3d at 65 to 70°C without solvent with subsequent chromatographic purification on SiO_2 columns with C_6H_6 [15, 16].

While No. 23 could not be separated by this chromatographic procedure from the mixture (m) with its two positional isomers obtained by Method Vb, thin layer chromatography of very dilute solutions on Silufol 254 nevertheless showed that the isomers are separable [15, 16].

Method VI. Acylation of $C_5H_5FeC_5H_4\text{–}R^2$ with $CH_3O_2C(CH_2)_2COCl$ and $AlCl_3$ (mole ratio 1/0.75/>1 [7]; 1/1/1 [3]) in CH_2Cl_2 at room temperature [2, 3, 7]. Column chromatography on Al_2O_3 in C_6H_6 yields No. 25 after $C_5H_5FeC_5H_4\text{–}C_2H_5$ and before the mixture of its 1,1'- and 1,3-isomers [3]. Such chromatography separates No. 24 from $C_5H_5FeC_5H_4\text{–}CH_3$ and $FeC_{10}H_7(CO(CH_2)_2CO_2CH_3)_2CH_3$ only as a mixture with its two positional isomers. For separation of the isomer mixture see "Further information" for No. 24, p. 344.

Chemical Behavior

The esters can be smoothly saponified. The reaction is usually effected with potassium hydroxide in ethanol (see for example Nos. 3 [5], 7, 9 to 12 [9], 24 [7], 25 [3], and 26 [13]) or in methanol (No. 24) [4].

References on p. 345

Table 24

Carboxylic Acid Esters of the Type $C_5H_5FeC_5H_3(A–CO_2R'\text{-}1)R^2\text{-}2$.

Further information for numbers preceded by an asterisk is given at the end of the table, pp. 338/45.
For abbreviations and dimensions see p. VIII.

No.	$C_5H_5FeC_5H_3(R^1\text{-}1)R^2\text{-}2$ $R^1 = A–CO_2R'$	R^2	method of preparation (yield in %; "m" indicates isomeric mixture, number following solidus is fraction present of 1,2-compound), properties, and remarks	Ref.
compounds with $R^1 = R^2$:				
1	$CH_2–CO_2CH_3$		Ia, IIa (4m) m.p. 67 to 72° ^{1}H NMR ($CDCl_3$): 3.37 (s, CH_2), 3.67 (s, CH_3), 4.05 (s, C_5H_5), 4.05 to 4.30 (m, C_5H_3) R_f: 0.8 (thin layer chromatography on Al_2O_3 with 10/1 C_6H_6/C_2H_5OH)	[17, 20]
2	$(CH_2)_2–CO_2CH_3$		IIb (22m) deep red oil (m) IR (KBr): 820, 824, 1005 to 1029, and 3100 (C_5H_4); 890 to 896, 900, 915, 965, and 1285 (C_5H_3); 1107, 1123, and 1140 (C_5H_5); 1450, 1585, and 2865 to 2880 (CH_2); 2935 and 2965 (CH_3); 1675 (νC=O) (m) mol. wt: 341 from hydrolysis data (m)	[21]
3	$(CH_2)_2–CO_2C_2H_5$		Ib m.p. 50° ^{1}H NMR ($CDCl_3$): 1.25 (t, CH_3), 2.60 ($[CH_2]_2$), 4.01 (main peak of $C_5H_5FeC_5H_3$), 4.16 (q, CH_2 of C_2H_5; J = 7)	[5, 9]
*4	$(CH_2)_2–CO_2C_6H_9(C_3H_7\text{-i-}10)CH_3\text{-}13$		"Diastereoisomeric and Optically Active Forms" see "Further information", pp. 338/9, and p. 332	[10, 14]
*5	$(CH_2)_3–CO_2CH_3$		IIc (42m)	[21]

6	$(CH_2)_5$–CO_2CH_3		IId (18 m) deep red, viscous oil (m)	[21]
7	$CH_2CH(CH_3)$–$CO_2C_2H_5$		Ib see pp. 332 and 333	[9]
*8	CH=CH–$CO_2C_6H_9(C_3H_7$-i-10)CH_3-13		"Diastereoisomeric and Optically Active Forms" see "Further information", pp. 339/42, and p. 332	[10, 14]
9	$CH_2CH(C_6H_5)$–$CO_2C_2H_5$		Ib see pp. 332 and 333	[9]
10	$CH_2CH(CO_2C_2H_5)_2$		IIId (80) viscous oil, which carbonizes during distillation	[5, 9]
11	$CH_2C(CH_3)(CO_2C_2H_5)_2$		IIIa (60) orange oil, which crystallizes easily, m.p. 62° (pentane)	[9]
12	$CH_2C(C_6H_5)(CO_2C_2H_5)_2$		IIIa (34) m.p. 98° (ether/hexane)	[9]

compounds with $R^1 \neq R^2$:

13	CH_2–CO_2CH_3	$CH_2CH_2C_6H_5$	Ia oil ^{1}H NMR ($CDCl_3$): 2.73 ("s", $(CH_2)_2$ of R^2), 3.32 (s, CH_2 of R^1), 3.67 (s, CH_3), 4.03 (s, C_5H_5), 3.9 to 4.3 (m, C_5H_3), 7.20 (s, C_6H_5) R_f: 0.85 (thin layer chromatography on Al_2O_3 with 20/1 C_6H_6/C_2H_5OH)	[17]
*14	$\overset{6}{C}H_2$–$\overset{7}{C}O_2\overset{8}{C}H_3$	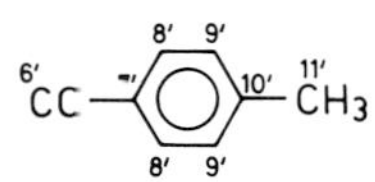	IV (76 m/30%) photosensitive deep red oil (m) ^{13}C NMR ($CDCl_3$): δ = 21.6 (C-11′), 34.6 (or 35.1?) (C-6), 51.8 (C-8), 70.9 (C-1′ to 5′), 71.1 (d*⁾), 72.7 (d*⁾), 74.1 (d*⁾), ~84.7 (C-1); 128.2 (1.4C), 128.4 (0.6C), 128.7 (0.6C), and 128.8 (1.4C) (C-8′, 9′); 136.9 (C-7′), 142.1 (C-10′), ~171.5 (or 170.9?) (C-7), ~201 (C-6′); C-2 to 5 are indeterminate or not discernible	[22]

*⁾ unassigned minor peaks

References on p. 345

Table 24 (continued)

No.	$C_5H_5FeC_5H_3(R^1\text{-}1)R^2\text{-}2$ $R^1 = A\text{–}CO_2R'$	R^2	method of preparation (yield in %; "m" indicates isomeric mixture, number following solidus is fraction present of 1,2-compound), properties, and remarks	Ref.
			^{13}C NMR (CD_2Cl_2): $\delta = 34.7$ (or 34.2?) (C-6), 70.6 (C-1′ to 5′), 77.1 (C-2), 84.1 (or 84.0?) (C-1), 170.7 (or 171.4?) (C-7), ~199 (C-6′); C-3 to 5 are indeterminate or not discernible mass spectrum: m/e = 317 $[M-CO_2CH_3]^+$, 184, 119, 91, 65; no $[M]^+$	
15	$(CH_2)_2\text{–}CO_2CH_3$	CHO	Va (65 m), the product consists mainly of the 1,3-isomer only contaminated with No. 15	[1]
*16	$(CH_2)_2\text{–}CO_2CH_3$	CO_2CH_3	for "Optically Active Forms" see "Further information", p. 342	[12]
*17	$(CH_2)_6\text{–}CO_2C_2H_5$	C_8H_{17}-n	see "Further information", p. 342	[23 to 25]
*18	$(CH_2)_6\text{–}CO_2C_2H_5$	$CH{=}CHCH(OH)C_5H_{11}$-n	for "Optically Active Forms" see "Further information", p. 343	[23, 24]
*19	$CH{=}CH\text{–}CO_2CH_3$	CO_2CH_3	for "Optically Active Forms" see "Further information", p. 343	[12]
*20	$CH{=}CH\text{–}CO_2C_6H_9(C_3H_7\text{-i-}10)CH_3\text{-}13$	CH_2OH	for "Optically Active Forms" see "Further information", p. 342	[10, 14]
*21	$CH{=}CH\text{–}CO_2C_6H_9(C_3H_7\text{-i-}10)CH_3\text{-}13$	CHO	for "Optically Active Forms" see "Further information", p. 344	[11, 14]
*22	$CH{=}CH(CH_2)_4\text{–}CO_2C_2H_5$	$CH{=}CHCOC_5H_{11}$-n	for "Optically Active Forms" see "Further information", p. 344	[23]
23	$C_6H_4\text{–}CO_2C_2H_5\text{-}4$	CHO	Vb (86 m/13%) oil (m) ^1H NMR ($CDCl_3$): 1.37 (t, CH_3), 4.20 (s, C_5H_5), 7.0 to 8.0 (m, C_6H_4), 10.18 (s, CHO)	[15, 16]

*24	$CO(CH_2)_2–CO_2CH_3$	CH_3	Ia (88 m), VI oil; yellow solid, m.p. 71° (pentane) ^1H NMR ($CDCl_3$): 2.31 (CH_3 of R^2) IR (CCl_4): 1680 and 1745 (νC=O)	[4, 7]
25	$CO(CH_2)_2–CO_2CH_3$	C_2H_5	VI crystals, m.p. 50 to 52° (pentane)	[2, 3]
*26	$CH(C_6H_5)CH(CO_2C_2H_5)_2$	CH_3	for "Diastereoisomeric Forms" see "Further information", p. 344	[13]
*27	$CH(C_6H_5)CH(CO_2C_2H_5)_2$	$CH_2CH(CO_2C_2H_5)_2$	for "Diastereoisomeric Forms" see "Further information", p. 345	[6, 8, 9]

References on p. 345

* Further information:

$C_5H_5FeC_5H_3([CH_2]_2–CO_2C_6H_9(CH(CH_3)_2\text{-}10)CH_3\text{-}13)_2\text{-}1,2$ (Table **24**, No. **4**). Mass spectrum: m/e = 606 $[M]^+$ [14].

Diastereoisomeric and Optically Active Forms. (9S,10R,13S,9′R,10′S,13′R)-$C_5H_5FeC_5H_3$–($[CH_2]_2$–$CO_2C_6H_9(CH(CH_3)_2$-10)CH_3-13$)_2$-1,2 (No. 4 A, see V) is obtained from No. 8 A (see p. 340) as yellow crystals melting at 116 to 118°C [10, 14]. The compound is achiral and exhibits in the ^{1}H NMR spectrum ($CDCl_3$) the following shifts: δ = 0.75, 0.82, 0.88, and 0.95 (m, CH_3 of menthyl); 1.0 to 2.1 (m, 18H, H-10 to 15 and 10′ to 15′ of menthyl), 2.55 (m, CH_2-6, 7, 6′, 7′), 3.95 (apparent t, C_5H_3), 4.04 (s, C_5H_5), 4.97 (apparent sext, H-9,9′ of menthyl) ppm. Due to the enantiotopic relationship between the configurationally opposed substituents three different signals (a, b, c) are to be expected for the six CH_3 groups, each as a doublet. IR spectrum ($CHCl_3$): 1000, 1100, and 3095 (ferrocenyl C–H); 1455, 2875, 2935, and 2965 (aliphatic C–H); 1725 (νC=O) cm^{-1}. Electronic absorption spectrum (C_2H_5OH): $\lambda_{max}(\varepsilon)$ = 210 (21000), 320 sh (90), 440 (80) nm. The configuration (2r) is given by the relationship to No. 8 A [14].

(9R,10S,13R,9′S,10′R,13′S)-$C_5H_5FeC_5H_3([CH_2]_2$–$CO_2C_6H_9(CH(CH_3)_2$-10)CH_3-13$)_2$-1,2 (No. 4 B, see V) is prepared from No. 8 B (see p. 340) as a viscous yellow oil [10, 14]. The achiral compound, diastereoisomeric to Nos. 4 A and 4 C, displays the following spectra: ^{1}H NMR ($CDCl_3$): δ = 0.73, 0.79, 0.84, 0.88, 0.92, and 0.94 (CH_3 of menthyl); 1.0 to 2.1 (m, H-10 to 15 and 10′ to 15′ of menthyl), 2.55 (m, CH_2-6, 7, 6′, 7′), 3.95 (perturbed d, C_5H_3), 4.04 (s, C_5H_5), 4.97 (sext, H-9,9′ of menthyl) ppm. In contrast to No. 4 A all three doublets a, b, c of the CH_3 protons are resolved here. IR ($CHCl_3$): 1000, 1100, and 3095 (ferrocenyl C–H); 1455, 2875, 2935, and 2965 (aliphatic C–H); 1725 (νC=O) cm^{-1}. Electronic absorption spectrum (C_2H_5OH): $\lambda_{max}(\varepsilon)$ = 211 (24000), 322 sh (80), 410 (80) nm. The relationship with No. 8 B gives the absolute configuration (1s) [14].

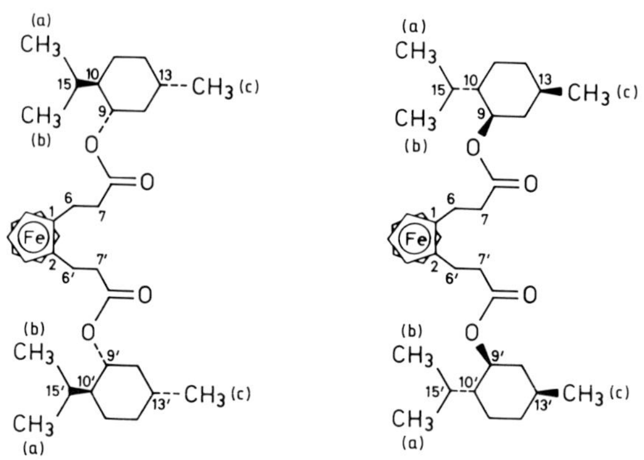

V

(+)(9S,10R,13S,9′S,10′R,13′S)-$C_5H_5FeC_5H_3([CH_2]_2$–$CO_2C_6H_9(CH(CH_3)_2$-10)CH_3-13$)_2$-1,2 (No. 4C a, see VI) is obtained from (+)-No. 8C a (see p. 341). It is a viscous yellow oil with a specific rotation (c = 0.450, C_2H_5OH) of $[\alpha]_D^{22}$ = +56.3° ± 1.0° [10, 14], $[\alpha]_{546}^{22}$ = +61.8° ± 1.3°. The ^{1}H NMR, IR, and electronic absorption spectra coincide with those of its enantiomer (−)-No. 4C b [14], see below.

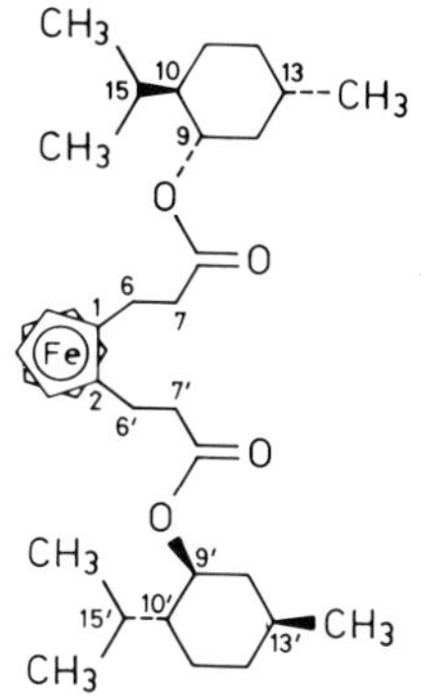

(+)-No. 4 C a

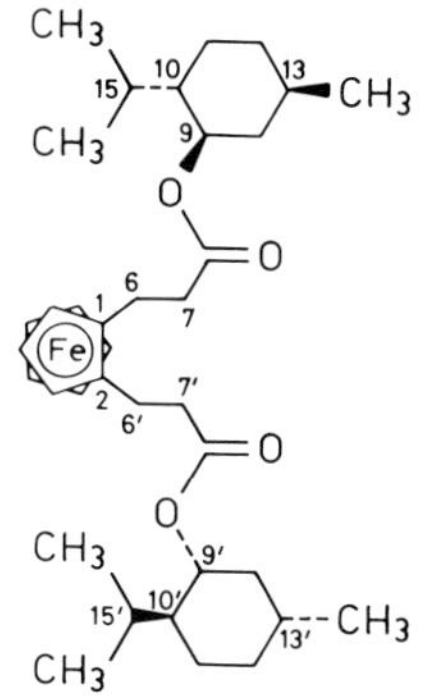

(−)-No. 4 C b

VI

(−)(9 R,10 S,13 R,9′R,10′S,13′R)-$C_5H_5FeC_5H_3([CH_2]_2–CO_2C_6H_9(CH(CH_3)_2$-10)$CH_3-13)_2$-1,2 (No. 4 C b, see VI) is obtained from (−)-No. 8 C b (see p. 342) as a viscous yellow oil. Specific rotation (c = 0.50, C_2H_5OH): $[\alpha]_D^{22} = -54.2° \pm 0.9°$ [10, 14], $[\alpha]_{546}^{22} = -61.9° \pm 0.7°$. ^{1}H NMR spectrum ($CDCl_3$): δ = 0.73, 0.75, 0.79, 0.82, 0.85, 0.88, 0.90, and 0.95 (CH_3 of menthyl); 1.0 to 2.1 (m, H-10 to 15 and 10′ to 15′ of menthyl), 2.55 (m, CH_2-6, 7, 6′, 7′), 3.97 (perturbed d, C_5H_3), 4.04 (s, C_5H_5), 4.72 (sext, H-9, 9′ of menthyl) ppm. The constitutionally and configurationally equivalent substituents are diastereotopic, so that the protons of each CH_3 group should show different signals, in each case as a doublet. Even if not all 12 lines are resolved, the higher complexity is still distinctly comparable with the spectra of Nos. 4 A and 4 B. IR spectrum ($CHCl_3$): 1000, 1100, and 3095 (ferrocenyl C–H); 1455, 2875, 2935, and 2965 (aliphatic C–H); 1725 (νC=O) cm^{-1} [14].

$C_5H_5FeC_5H_3([CH_2]_3–CO_2CH_3)_2$-1,2 (Table **24**, No. **5**). Electrolysis of a solution of ferrocene and $CH_3O_2C(CH_2)_3CO_2H$ (40% neutralized with KOH) at 20°C in CH_3OH between Pt electrodes at 2.3×10^{-2} A/cm² and 3 V followed after passage of 0.3 F by reduction with $Na_2S_2O_3$, extraction into C_6H_6, and evaporation of the extract, yields a raw product from which the mixture of the methoxycarbonylpropylferrocenes is separated by column chromatography on SiO_2 with C_6H_6. The separation of this mixture is conducted by thin layer chromatography on SiO_2 with isolation of $FeC_{10}H_8([CH_2]_3CO_2CH_3)_2$ in 1.6% yield relative to the unrecovered ferrocene (IR: 1000, 1100, 1450, 1750 cm^{-1}; mass spectrum: m/e = 386 $[M]^+$) after $C_5H_5FeC_5H_4–(CH_2)_3CO_2CH_3$ (7%) and before $FeC_{10}H_7([CH_2]_3CO_2CH_3)_3$ (1.8%). Due to the similarity of the R_f values the components are not completely separable [18, 19]. No attempt was made to isolate No. 5 from the mixed $FeC_{10}H_8([CH_2]_3CO_2CH_3)_2$ obtained either by this preparation or by Method IIc, p. 332.

(E, E)-$C_5H_5FeC_5H_3(CH=CH–CO_2C_6H_9(CH(CH_3)_2$-10)$CH_3-13)_2$-1,2 (Table **24**, No. **8**). Mass spectrum: m/e = 602 $[M]^+$ [14].

Diastereoisomeric and Optically Active Forms. (9 S,10 R,13 S, 9′R,10′S,13′R)-$C_5H_5FeC_5H_3$–($CH=CH–CO_2C_6H_9(CH(CH_3)_2$-10)$CH_3-13)_2$-1,2 (No. 8 A, see VII) is obtained together with its diastereoisomer No. 8 B from the diastereoisomer mixture of (+)(E)(1 S, 9 R,10 S,13 R)-No. 21 and (−)(E)(1 R, 9 R,10 S,13 R)-No. 21 (see p. 344). Purification of the mixture by chromatography on an Al_2O_3 column with 1/3 ether/hexane gives a red oil, which is not separable by thin layer chromatography on SiO_2 with 2/3 ether/hexane but is separable by repeated fractional crystallization from hot pentane [10, 14].

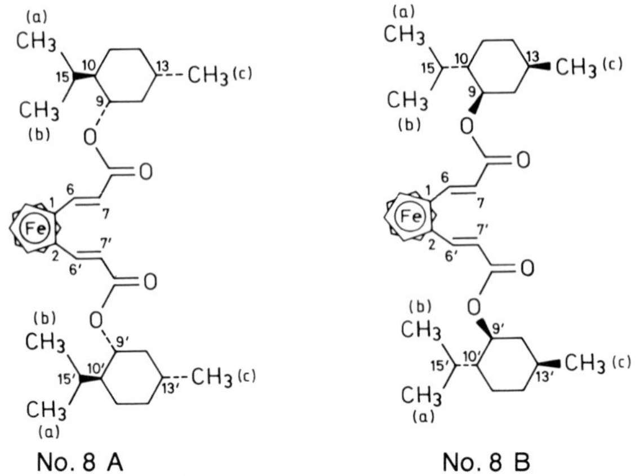

No. 8 A No. 8 B

VII

Initially No. 8 A deposits as fluffy orange crystals, which melt at 173.5 to 174.5°C (pentane) [10, 14]. The achiral compound exhibits the following shifts in the ^{1}H NMR spectrum ($CDCl_3$): $\delta = 0.80$, 0.87, 0.91, 0.92, and 0.99 (CH_3 of menthyl); 1.0 to 2.2 (m, H-10 to 15 and 10′ to 15′ of menthyl), 4.14 (s, C_5H_5), 4.62 and 4.72 (m, H-9, 9′ of menthyl and C_5H_3), 6.12 (d) and 7.75 (d) (CH=CH; J = 16 Hz) ppm. Not all CH_3 signals are resolved. Due to the enantiotopic relationship between the constitutionally equivalent, configurationally opposed substituents, three signals (a, b, c), each a doublet, are to be expected for the protons of the six CH_3 groups. IR spectrum ($CHCl_3$): 1000, 1100, and 3100 (ferrocenyl C–H); 1450, 2870, 2930, and 2960 (aliphatic C–H); 1280 and 1695 (conjugated ester), 1635 (C=C) cm^{-1}. Electronic absorption spectrum (C_2H_5OH): $\lambda_{max}(\varepsilon) = 210$ (19000), 284 (26000), 367 (3200), 477 (2300) nm. The relationship to No. 8 B gives the configuration (2r) [14].

Catalytic hydrogenation in C_2H_5OH over PtO_2 at atmospheric pressure and room temperature yields No. 4 A. The color change from red-orange to yellow indicates the end of the reaction. The residue remaining after evaporation in vacuum is chromatographed on Al_2O_3 and No. 4 A eluted with 1/3 ether/pentane [10, 14].

(9R,10S,13R,9′S,10′R,13′S)-$C_5H_5FeC_5H_3(CH{=}CH{-}CO_2C_6H_9(CH(CH_3)_2\text{-}10)CH_3\text{-}13)_2\text{-}1,2$ (No. 8 B, see VII). The achiral compound, diastereoisomeric to No. 8 A, is separated from its mother liquor as red plates by concentrating the solution and recrystallization from pentane [10, 14]. Reaction of only (+)(E)(1S,9R,10S,13R)-No. 21 forms exclusively No. 8 B [10, 14]; the yield is 26% referred to (+)(1S,9R,10S,13R,7′S)-VIII, the starting compound for (+)-No. 21 [11, 14]. No. 8 B melts at 146 to 148°C (pentane) [10, 14], 147 to 149°C [14]. ^{1}H NMR spectrum ($CDCl_3$): $\delta = 0.81$, 0.88, 0.93, and 1.0 (CH_3 of menthyl); 1.0 to 2.2 (m, H-10 to 15 and 10′ to 15′ of menthyl), 4.15 (s, C_5H_5), 4.62 and 4.75 (m, H-9, 9′ of menthyl and C_5H_3), 6.11 (d) and 7.74 (d) (CH=CH; J = 16 Hz) ppm. IR spectrum ($CHCl_3$): 1000, 1100, and 3100 (ferrocenyl C–H); 1450, 2870, 2930, and 2960 (aliphatic C–H); 1280 and 1695 (conjugated ester), 1635 (C=C) cm^{-1}. Electronic absorption spectrum (C_2H_5OH): $\lambda_{max}(\varepsilon) = 209$ (21000), 283 (24000), 358 (5000), 475 (1900) nm. $R_f = 0.75$ (thin layer chromatography on SiO_2 with 2/3 ether/hexane) [14]. The relationship of No. 8 B to (+)(1S,9R,10S,13R,7′S)-VIII and the correlation with (−)-$C_5H_5FeC_5H_3(CO_2CH_3\text{-}1)CH_3\text{-}2$ of the known absolute configuration (1S) yields for No. 8 B the configuration (1s) [11].

Catalytic hydrogenation as for No. 8 A yields No. 4 B [10, 14].

References on p. 345

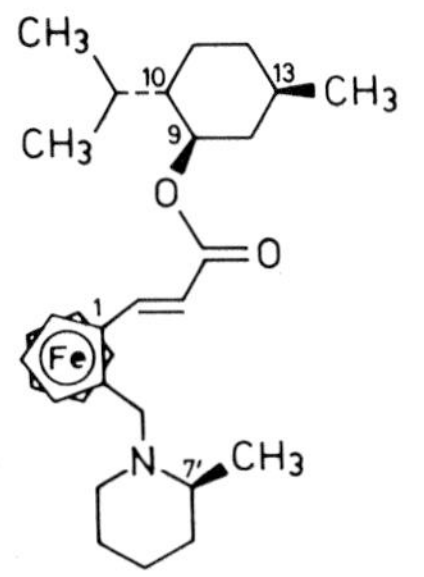

(+)(1S,9R,10S,13R,7′S)-VIII

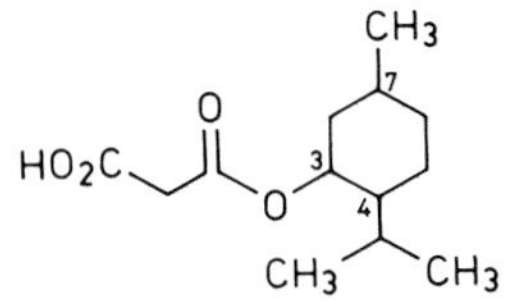

IX

(+)(9S,10R,13S,9′S,10′R,13′S)-$C_5H_5FeC_5H_3(CH{=}CH{-}CO_2C_6H_9(CH(CH_3)_2\text{-}10)CH_3\text{-}13)_2$-1,2 (No. 8 Ca, see X) is prepared at 100°C by a modified Knoevenagel condensation of $C_5H_5FeC_5H_3(CHO)_2$-1,2 with (+)(3S,4R,7S)-IX (mole ratio 1/3) in pyridine with addition of piperidine [10, 14] in 40% yield. Most of it is crystallized out from boiling pentane, the residual quantity is isolated from the mother liquor by chromatography on an Al_2O_3 column [14].

The chiral compound, enantiomeric to (−)-No. 8 Cb, diastereomeric to Nos. 8 A and 8 B, melts at 144 to 146°C and has a specific rotation of $[\alpha]_D^{22} = +70.0° \pm 1.2°$ (c = 0.25, C_2H_5OH) [10, 14]. The ^1H NMR, IR, and electronic absorption spectra are identical with those of (−)-No. 8 Cb [14], see below.

Catalytic hydrogenation as for No. 8 A leads to (+)-No. 4 C [10, 14].

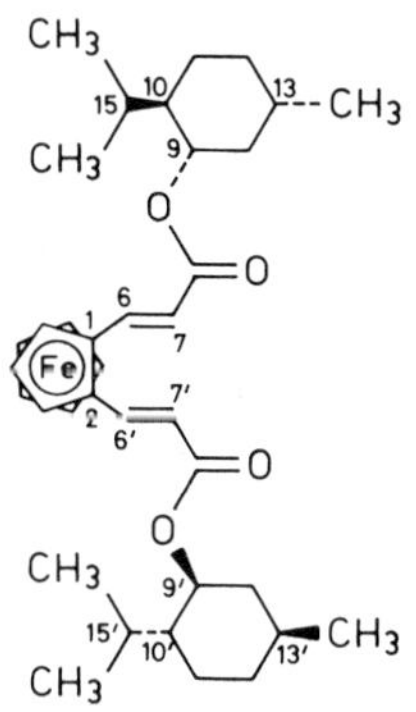

(+)-No. 8 C a

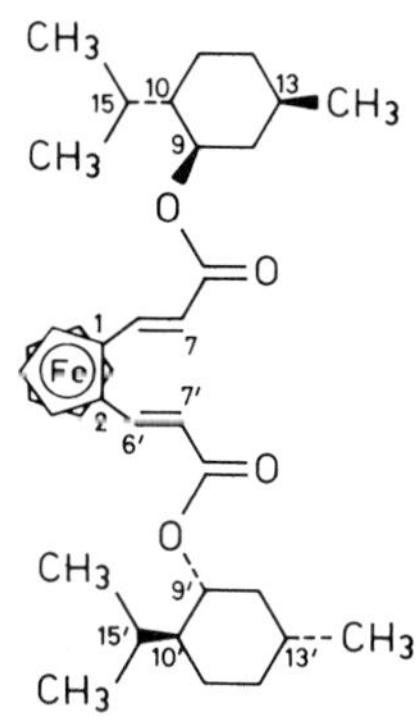

(−)-No. 8 C b

X

(−)(9R,10S,13R,9′R,10′S,13′R)-$C_5H_5FeC_5H_3(CH{=}CH{-}CO_2C_6H_9(CH(CH_3)_2\text{-}10)CH_3\text{-}13)_2$-1,2 (No. 8Cb, see X) is prepared in a 48% yield as for (+)-No. 8C but with (−)(3R,4S,7R)-IX [10, 14].

The chiral, crystalline compound melts at 144 to 146°C (pentane) and has a specific rotation of $[\alpha]_D^{22} = -72.3° \pm 2.0°$ (c = 0.250, C_2H_5OH) [10, 14]. ^1H NMR spectrum ($CDCl_3$): δ = 0.80, 0.87, 0.92, and 0.99 (CH_3 of menthyl); 1.0 to 2.2 (m, H-10 to 15 and 10′ to 15′ of menthyl), 4.14 (s, C_5H_5), 4.62 and 4.76 (m, H-9, 9′ of menthyl and C_5H_3), 6.12 (d) and 7.44 (dd)(CH=CH; J = 16 Hz) ppm. Since the configurationally equivalent substituents are diastereotopic with respect to each other, the spectrum when completely resolved should exhibit for each CH_3 group a different line doublet. IR spectrum ($CHCl_3$): 1000, 1100, and 3095 (ferrocenyl C–H); 1455, 2870, 2925, and 2960 (aliphatic C–H); 1270 and 1695 (conjugated ester), 1630 (C=C) cm^{-1}. Electronic absorption spectrum (C_2H_5OH): $\lambda_{max}(\varepsilon)$ = 209 (19000), 285 (24000), 360 (3300), 477 (2000) nm [14].

References on p. 345

(−)-No. 8 C is catalytically hydrogenated as for No. 8 A to (−)-No. 4 C [10, 14].

$C_5H_5FeC_5H_3(CH_2$–CO_2CH_3-1)CO($C_6H_4CH_3$-4)-2 (Table **24**, No. **14**). A mixture (m) of No. 14 with its 1,1′- and 1,3-positional isomers is won by Method IV (p. 333) or by esterification by Method Ia (p. 332) of $FeC_{10}H_8(CH_2CO_2H)COC_6H_4CH_3$-4. It is separable at 250°C by gas chromatography on a column filled with 3% OV-225 on Chromosorb W HP. No. 14 appears first (R_f = 9.3 min). No. 14 can also be separated by thin layer chromatography (R_f = 0.47) on SiO_2 with 2/1 hexane/$CH_3CO_2C_2H_5$ from the mixture of the 1,1′- and 1,3-isomers (R_f = 0.39). ^{1}H NMR spectrum ($CDCl_3$) of (m): δ = 2.43 (s, CH_3 of R^2), 3.24 (s, 0.5 H, CH_2 of 1,1′-isomer), 3.42 (s, 0.2 H, CH_2 of 1,3-isomer), 3.67 (s, CO_2CH_3), 3.73 (s, 0.3 H, CH_2 of No. 14), 4.1 to 4.3 (m, 5 H, $C_5H_5 + C_5H_4$–R^1), 4.5 to 4.7 (m, 2 H, H-3,4 of C_5H_4–R^2), 4.8 to 5.0 (m, 1.4 H, H-2,5 of C_5H_4–R^2), 7.1 to 8.0 (m, C_6H_4) ppm [22].

$C_5H_5FeC_5H_3([CH_2]_2$–CO_2CH_3-1)CO_2CH_3-2 (Table **24**, No. **16**). **Optically Active Forms.** (−)(1R)-No. 16 (see XI) is prepared from (−)-$C_5H_5FeC_5H_3([CH_2]_2CO_2H$-1)$CO_2CH_3$-2 by Method Ia (p. 332) and isolated in 90% yield by column chromatography on activated Al_2O_3 with C_6H_6 as eluent. The fine, yellow-orange crystals melt at 66.5 to 68°C (C_2H_5OH). Specific rotation $[\alpha]^{22}_D = -10°$ (c = 0.56, C_2H_5OH). ^{1}H NMR spectrum ($CDCl_3$): δ = 2.35 to 2.65 and 2.85 to 3.30 (m, $(CH_2)_2$), 3.15 (s) and 3.79 (s) (CH_3 of R^1 and R^2), 4.13 (s, C_5H_5), 4.28 (m, 2 H) and 4.73 (m, 1 H) (C_5H_3) ppm. IR spectrum (Nujol): 780, 818, 1030, 1105, 1138, 1175, 1193, 1215, 1257, 1290, 1305, 1715, 1730 cm^{-1}. Optical rotatory dispersion (c = 0.3 mg/L C_2H_5OH): $[M]_{524} = -270°$, $[M]_{494} = 0°$, $[M]_{454} = +540°$, $[M]_{418} = 0°$, $[M]_{400} = -210°$, $[M]_{384} = 0°$, $[M]_{364} = +270°$, $[M]_{350} = 0°$, $[M]_{296} = -620°$, $[M]_{260} = 0°$. The circular dichroism curve is plotted in the paper. The absolute configuration (1R, 2S) is derived from the similarity of the optical rotatory dispersion and circular dichroism curves to those of (−)-XII of the known configuration (1S) [12].

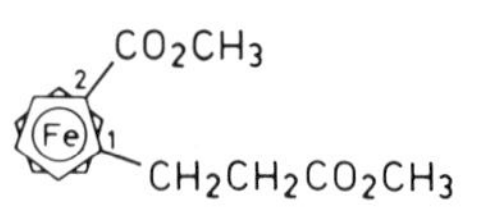

(−)(1R, 2S)-No. 16

XI

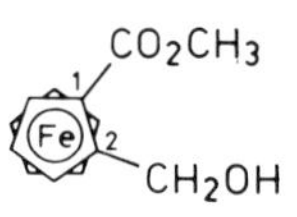

(−)(1S)-XII

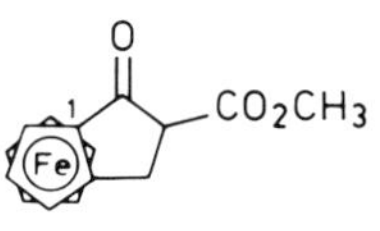

(−)(1S)-XIII

The compound undergoes Dieckmann cyclization to (−)(1S)-XIII with NaH in tetrahydrofuran [12].

$C_5H_5FeC_5H_3(\overset{6}{C}H_2[CH_2]_4\overset{11}{C}H_2$–$CO_2C_2H_5$-1)$C_8H_{17}$-n-2 (Table **24**, No. **17**). Hydrogenolysis in 10% $HClO_4/CH_3CO_2H$ with 10% Pd/C as catalyst (ratio 1/2) at an H_2 pressure of 1 atm and a temperature ≦ 55°C gives cyclopentane and its derivatives $C_5H_8([CH_2]_6CO_2C_2H_5$-1)$C_8H_{17}$-n-2 and $C_5H_8([CH_2]_6CO_2H$-1)C_8H_{17}-n-2 [25].

Optically Active Forms. (+)(1S)-$C_5H_5FeC_5H_3([CH_2]_6$–$CO_2C_2H_5$-1)C_8H_{17}-n-2, the analog of prostanoic acid, is obtained by ionic hydrogenation of (−)(1S)-XIV with $(C_2H_5)_3SiH$ in CF_3CO_2H at

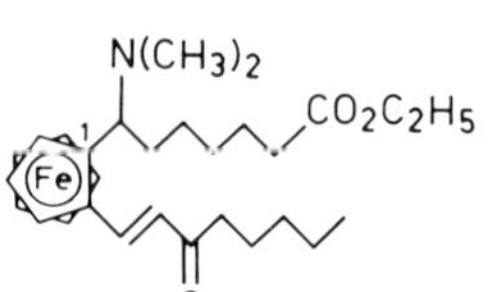

(−)(1S)-XIV

20°C and subsequent chromatography in 67% yield as an oil with a specific rotation of $[\alpha]_{578} = +0.176°$ (c = 11.3, C_2H_5OH) [23, 24]. ^1H NMR spectrum ($CDCl_3$): δ = 0.88 and 1.26 (CH_3 of R^1 and R^2), 2.25 and 2.35 (CH_2-6 to 11), 3.92 (C_5H_3), 3.98 (C_5H_5), 4.12 (CH_2 of C_2H_5) ppm [24].

$C_5H_5FeC_5H_3([CH_2]_6$–$CO_2C_2H_5$-1)CH=CH$\overset{8'}{C}$HOH–C_5H_{11}-n-2 (Table **24**, No. **18**). **Optically Active Forms.** Reduction of the methiodide of (−)(1S)-XIV with $NaBH_4$ in 5/1 C_2H_5OH/H_2O forms the diastereomer mixture (−)(1S,8′RS)-No. 18, in a yield after chromatographic purification on SiO_2 with 5/1 C_6H_6/ether as eluent of 33 [24] or 37% [23]. The oily product possesses a specific rotation in C_2H_5OH of $[\alpha]_{578} = -170°$ (c = 3.1) [24], −114.5° (c = 2.6) [23]. The ^1H NMR spectrum (see figure on p. 16 in [24]) indicates a ratio other than 1/1 for the two diastereomers in the mixture and that the asymmetric induction on the C-8′ amounts to 38 ± 2% [24]. IR spectrum: 1660 (νC=C), 1730 (νC=O), 3400 (νO–H) cm^{-1} [23].

$C_5H_5FeC_5H_3$(CH=CH–CO_2CH_3-1)CO_2CH_3-2 (Table **24**, No. **19**). **Optically Active Forms.** (−)(1R)-No. 19 is gotten by esterification of (−)(1R)-$C_5H_5FeC_5H_3$(CH=CH–CO_2H-1)CO_2CH_3-2 by Method Ia (p. 332) and purified chromatographically on activated alumina columns with benzene. The dark red needles melt at 92.5 to 94°C (C_2H_5OH). Specific rotation: $[\alpha]_D^{20} = -324°$ (c = 0.56, C_2H_5OH). ^1H NMR spectrum ($CDCl_3$): δ = 3.80(s) and 3.85(s) (CH_3 of R^1 and R^2), 4.20 (s, C_5H_5); 4.60(m), 4.80(m), and 5.00(m) (C_5H_3); 6.15(d) and 8.37(d) (CH=CH) ppm. IR spectrum (KBr): 812, 850, 940, 992, 1015, 1030, 1083, 1155, 1190, 1220, 1250, 1273, 1627, 1695, 1710 cm^{-1}. Optical rotatory dispersion (c = 0.7 mg/L C_2H_5OH): $[M]_{520} = -2280°$, $[M]_{490} = 0°$, $[M]_{446} = +4560°$, $[M]_{384} = 0°$, $[M]_{368} = -1930°$, $[M]_{324} = 0°$, $[M]_{300} = -300°$. Circular dichroism: see figure in [12]. The absolute configuration (1R, 2S) is derived from the similarity of the optical rotatory dispersion and circular dichroism curves to those of (−)(1S)-XII [12].

(E)-$C_5H_5FeC_5H_3$(CH=CH–$CO_2C_6H_9(CH(CH_3)_2$-10)(CH_3-13)-1)CH_2OH-2 (Table **24**, No. **20**). **Optically Active Forms.** The modified Knoevenagel condensation of $C_5H_5FeC_5H_3$(CHO-1)CH_2OH-2 with (−)(3R,4S,7R)-IX, p. 341, at 100°C in pyridine with addition of piperidine and subsequent chromatographic purification on Al_2O_3 columns with 49/1 ether/methanol gives a 59% yield of the mixture of the diastereomers (+)(1S,9R,10S,13R)- and (−)(1R,9R,10S,13R)-No. 20 (see XV) [10, 14]. The mixture is not separable by chromatography (R_f = 0.20 on SiO_2 in 2/3 ether/hexane for both diastereomers). It exhibits the following spectra: ^1H NMR ($CDCl_3$), δ = 0.75, 0.86, and 0.96 (CH_3 of menthyl); 1.0 to 2.0 (m, H-10 to 15 of menthyl), 3.42 (OH), 4.10 (s, C_5H_5), 4.45 (m, H-9 of menthyl, CH_2, and C_5H_3), 6.03(d) and 7.60(d) (CH=CH; J = 16 Hz) ppm. IR (neat), 1000, 1100, and 3095 (ferrocenyl C–H); 1450, 2880, 2930, and 2960 (aliphatic C–H); 1270 and 1700 (conjugated ester), 1630 (C=C), 3420 (O–H) cm^{-1}. Mass spectrum: m/e = 424 $[M]^+$ [14].

Mild oxidation with activated MnO_2 in $CHCl_3$ at room temperature for 30 min followed by evaporation of the filtrate gives in 86% yield a diastereoisomer mixture (+)(1S,9R,10S,13R)- and (−)(1R,9R,10S,13R)-No. 21 (see XV) [10, 14].

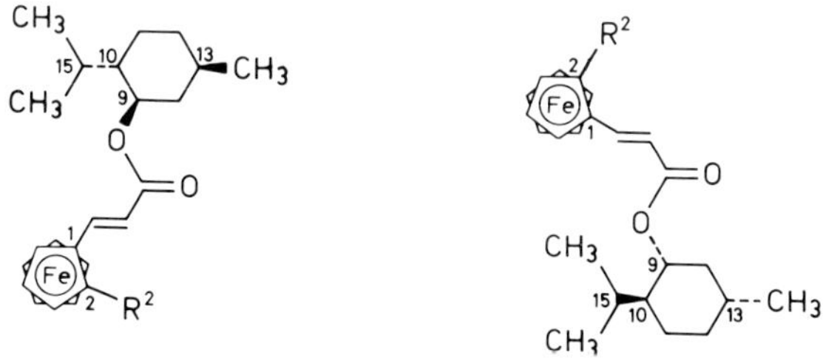

(+)(1S,9R,10S,13R)-No. 20	$R^2 = CH_2OH$	(−)(1R,9R,10S,13R)-No. 20
(+)(1S,9R,10S,13R)-No. 21	R^2 = CHO	(−)(1R,9R,10S,13R)-No. 21

XV

References on p. 345

(E)-$C_5H_5FeC_5H_3(CH{=}CH{-}CO_2C_6H_9(CH(CH_3)_2\text{-}10)(CH_3\text{-}13)\text{-}1)CHO\text{-}2$ (Table **24**, No. **21**). **Optically Active Forms.** (+)(1S,9R,10S,13R)-No. 21 (formula in XV) (see also below) and (−)(1R,9R,10S,13R)-No. 21 (see XV) are obtained as a mixture (see preceding paragraph) in the form of an unstable red oil. $R_f = 0.37$ for both isomers on SiO_2 with 2/3 ether/hexane. Spectra of the mixture: ^1H NMR ($CDCl_3$), δ = 0.75, 0.87, and 0.98 (CH_3 of menthyl); 4.22 (s, C_5H_5), 4.75 (t) and 4.92 (d) (C_5H_3), 6.26 (d) and 7.93 (dd) (CH=CH; J = 16 Hz), 10.5 (s, CHO) ppm. IR (neat), 995, 1100, and 3100 (ferrocenyl C–H); 1450, 2875, 2930, and 2960 (aliphatic C–H); 1670 and 2725 (conjugated CHO), 1270 and 1705 (conjugated ester), 1630 (C=C) cm^{-1}. Mass spectrum: m/e = 422 $[M]^+$ [14]. The modified Knoevenagel condensation of the mixture with (+)(3S,4R,7S)-IX (see p. 341) in pyridine/piperidine at 100°C for 1 h forms a mixture of the pseudoasymmetric compounds No. 8 A and 8 B [14].

(+)(1S,9R,10S,13R)-No. 21 alone forms on oxidation of (+)(1S,9R,10S,13R,7'S)-VIII (see p. 341) with activated MnO_2 for 30 h in $CHCl_3$. Due to its instability on Al_2O_3 no attempt has been made to separate it from the unreacted starting compound. It reacts with (+)(3S,4R,7S)-IX to form No. 8 B [11, 14].

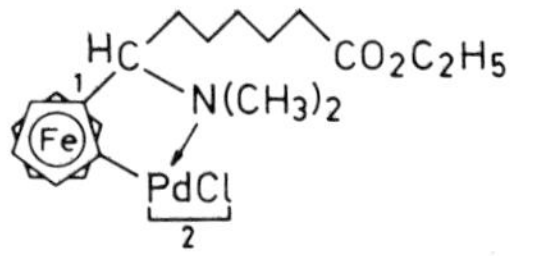

(−)(1S)-XVI

$C_5H_5FeC_5H_3(CH{=}CH(CH_2)_4{-}CO_2C_2H_5\text{-}1)CH{=}CHCOC_5H_{11}\text{-}n\text{-}2$ (Table **24**, No. **22**). **Optically Active Forms.** (−)(1S)-No. 22 is prepared from (−)(1S)-XVI by refluxing for 5 h with n-$C_5H_{11}COCH{=}CH_2$ in toluene and subsequent addition of $(C_2H_5)_3N$. It is isolated from the evaporated filtrate by chromatography on Silufol in a 10% yield, ahead of the 10% yield of (−)(1S)-XIV. The specific rotation $[\alpha]_{578} = -184°$ (c = 0.85, C_2H_5OH) [23].

$C_5H_5FeC_5H_3(CO(CH_2)_2{-}CO_2CH_3\text{-}1)CH_3\text{-}2$ (Table **24**, No. **24**) is separated from the mixture with its 1,1'-, and 1,3-positional isomers by column chromatography on deactivated Al_2O_3 with benzene and petroleum ether ($R_f = 0.55$) [7] or by preparative thin layer chromatography on SiO_2 with 30/1 C_6H_6/C_2H_5OH [4].

$C_5H_5FeC_5H_3(\overset{6}{C}H(C_6H_5)\overset{7}{C}H(CO_2C_2H_5)_2\text{-}1)CH_3\text{-}2$ (Table **24**, No. **26**). **Diastereoisomeric Forms.** (1RS,6SR)-No. 26 is obtained by Method IIIb (p. 333) in 90% yield as a crystalline compound melting at 78°C (pentane). ^1H NMR spectrum ($CDCl_3$): δ = 0.94 (t, CH_3 of C_2H_5), 1.00 (t, CH_3 of C_2H_5), 2.03 (s, CH_3 of R^2), 3.5 (s, C_5H_5), 3.7 to 4.1 (m, CH_2 of C_2H_5 and C_5H_3), 3.90 (d) and 4.30 (d) (CH-6,7; J = 10.5 Hz), 7.3 (m, C_6H_5) ppm. The conformation presented in Newman's style, see XVII, is deduced from the spectrum [13].

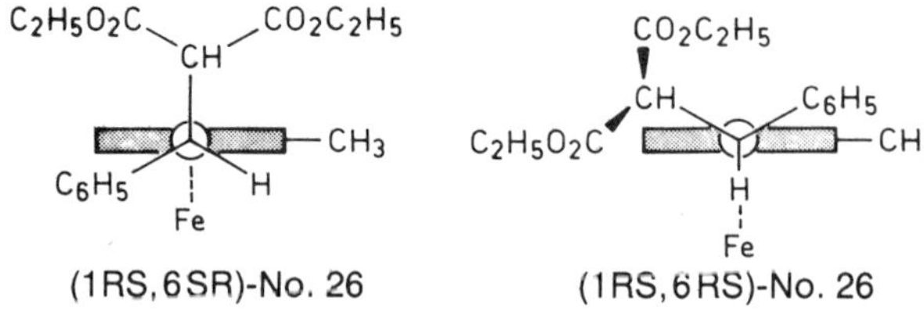

(1RS,6SR)-No. 26 (1RS,6RS)-No. 26

XVII

(1RS,6RS)-No. 26 is obtained by Method IIIc (p. 333) in 30% yield. After recrystallization from pentane the compound melts at 75°C and has the following ^1H NMR spectrum ($CDCl_3$):

$\delta = 0.87$ (t, CH_3 of C_2H_5), 1.30 (t, CH_3 of C_2H_5), 1.63 (s, CH_3 of R^2), 3.72 (q, CH_2 of C_2H_5), 4.02 (s, $C_5H_5 + 2H$ of C_5H_3), 4.32 (q, CH_2 of C_2H_5), 4.35 (m, 1H of C_5H_3), 4.30 (d) and 4.75 (d) (CH-6, 7; J = 5 Hz), 7.05 (s, C_6H_5) ppm. The Newman conformation derived from this spectrum is shown in XVII [13].

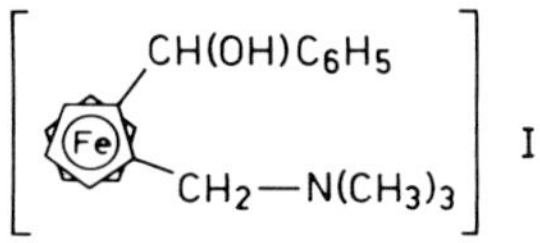

XVIII

$C_5H_5FeC_5H_3(CH(C_6H_5)CH(CO_2C_2H_5)_2$-1)$CH_2CH(CO_2C_2H_5)_2$-2 (Table **24**, No. **27**). The diastereoisomer forms a and b arise by Method III during the reaction of $NaCH(CO_2C_2H_5)_2$ in absolute ethanol with pulverized XVIII at the boiling point. They are separable by repeated column chromatography on SiO_2 with 1/2 ether/petroleum ether as eluent; for form a $R_f = 0.45$, for b $R_f = 0.37$. The reaction proceeds stereoselectively, and in fact little a (4.8% yield, m.p. 76°C) and much b (25.4% yield, m.p. 80°C) (both recrystallized from ether/hexane) are obtained from XVIII (m.p. 220°C). In contrast the form of XVIII melting at 210°C yields much a (22.5% yield) and only little b (6.4% yield) [6, 8, 9].

No. 27a yields on saponification with potassium hydroxide in ethanol the free acid as a fluffy, yellow product, No. 27b the acid melting at 220°C with decarboxylation [9].

References:

[1] K. Schlögl, M. Peterlik, H. Seiler (Monatsh. Chem. **93** [1962] 1309/27). – [2] J. Tirouflet, G. Tainturier (Tetrahedron Letters **1965** 4177/9). – [3] G. Tainturier, J. Tirouflet (Bull. Soc. Chim. France **1966** 595/600). – [4] G. Haller, K. Schlögl (Monatsh. Chem. **98** [1967] 2044/57). – [5] C. Moïse, J. Tirouflet (Compt. Rend. C **265** [1967] 457/9).

[6] C. Moïse, J. Tirouflet (Compt. Rend. C **267** [1968] 414/6). – [7] P. Dixneuf, R. Dabard (Bull. Soc. Chim. France **1969** 2164/71) – [8] C. Moïse, J. Tirouflet (Proc. 4th Intern. Conf. Organometal. Chem., Bristol 1969, Abstr. G 3). – [9] C. Moïse, J. Tirouflet (Bull. Soc. Chim. France **1970** 2656/65). – [10] S. I. Goldberg, W. D. Bailey (J. Am. Chem. Soc. **93** [1971] 1046/7).

[11] S. I. Goldberg, W. D. Bailey (Tetrahedron Letters **1971** 4087/90). – [12] T. Shirafuji, A. Odaira, Y. Yamamoto, H. Nozaki (Bull. Chem. Soc. Japan **45** [1972] 2884/7). – [13] C. Moïse, Y. Mugnier, J. Tirouflet (J. Organometal. Chem. **51** [1973] 343/52). – [14] S. I. Goldberg, W. D. Bailey (J. Am. Chem. Soc. **96** [1974] 6381/7). – [15] Š. Toma, A. Maholányiová, E. Solčániová (Acta Fac. Rerum Nat. Univ. Comenianae Chim. **24** [1976] 53/80).

[16] Š. Toma, A. Maholányiová, E. Solčániová (Collection Czech. Chem. Commun. **42** [1977] 1013/26). – [17] M. Benedikt, K. Schlögl (Monatsh. Chem. **109** [1978] 805/22). – [18] L. I. Denisovich, M. G. Peterleitner, S. P. Gubin (Izv. Akad. Nauk SSSR Ser. Khim. **1978** 465/8; Bull. Acad. Sci. USSR Div. Chem. Sci. **1978** 403/6). – [19] S. P. Gubin, L. I. Denisovich, N. V. Zakurin, M. G. Peterleitner (J. Organometal. Chem. **146** [1978] 267/78). – [20] N. A. Maier, S. A. Shklyar (Zh. Obshch. Khim. **48** [1978] 702/3; J. Gen. Chem. [USSR] **48** [1978] 645/6).

[21] N. A. Maier, S. A. Shklyar, Yu. A. Ol'dekop (Zh. Obshch. Khim. **49** [1979] 2518/21; J. Gen. Chem. [USSR] **49** [1979] 2223/5). – [22] B. E. Maryanoff (J. Org. Chem. **46** [1981] 70/4). – [23] V. I. Sokolov, L. L. Troitskaya, N. S. Khrushcheva (J. Organometal. Chem. **250** [1983] 439/46). – [24] V. I. Sokolov, N. S. Khrushcheva, L. L. Troitskaya, O. A. Reutov (Dokl. Akad. Nauk SSSR **274** [1984] 342/7; Dokl. Chem. Proc. Acad. Sci. USSR **274/279** [1984] 15/9). – [25] V. I. Sokolov, L. L. Troitskaya, N. S. Khrushcheva, O. A. Reutov (Dokl. Akad. Nauk SSSR **281** [1985] 861/3).

5.2.3.7.3 1,3-Compounds

5.2.3.7.3.1 Carboxylic Acid Esters of the Type $C_5H_5FeC_5H_3(CO_2R'\text{-}1)R^2\text{-}3$ ($CO_2R' = R^1$)

All compounds of Table 25 with $R^1 \neq R^2$ (Nos. 2 to 11) possess a plane of chirality. The centrochiral R, S nomenclature used for them is described in "Ferrocene" 7, 1980, pp. 1/4 (English) and 4/7 (German). Optically active forms have been obtained only of compounds No. 2 and 3, described below under "Further information", pp. 349 and 350.

The compounds are prepared by the following methods and by those given under "Further information":

Method I. Esterification of $C_5H_5FeC_5H_3(CO_2H\text{-}1)R^2\text{-}3$ with CH_2N_2 in ether or ether/methanol [2, 3, 8, 10].

Method II. Photoethoxycarbonylation of $C_5H_5FeC_5H_4\text{–}R^2$ in 1/1 C_2H_5OH/CCl_4 with irradiation at 365 nm for 4 h at room temperature. The extract of the reaction product obtained with CH_2Cl_2 or ether is evaporated, and the residue is subjected to thin layer or column chromatography for the separation of $C_5H_5FeC_5H_4\text{–}R^2$. A second thin layer chromatographing isolates the compound from the mixture with its 1,1'- and 1,2-positional isomers. No. 5 is an exception, with $C_5H_5FeC_5H_4\text{–}C_6H_4CO_2C_2H_5\text{-}4$ forming as a fourth product. The yields cited for Method II are based in each case on the amount of unrecovered $C_5H_5FeC_5H_4\text{–}R^2$ [9].

Table 25
Carboxylic Acid Esters of the Type $C_5H_5FeC_5H_3(CO_2R'\text{-}1)R^2\text{-}3$.
Further information for numbers preceded by an asterisk is given at the end of the table, pp. 348/50.
For abbreviations and dimensions see. p. VIII.

No.	$C_5H_5FeC_5H_3(R^1\text{-}1)R^2\text{-}3$ $R^1 = CO_2R'$	R^2	method of preparation (yield in %), properties, and remarks	Ref.
compound with $R^1 = R^2$:				
*1	CO_2CH_3		see "Further information", p. 348 m.p. 155 to 157° (C_6H_6/hexane), 160 to 161° ^1H NMR ($CDCl_3$): 3.81 (s, CH_3), 4.95 (d, H-4,5 of C_5H_3; J = 1.5), 5.40 (t, H-2 of C_5H_3; J = 1.5) IR (KBr): 1000, 1100, 1710 (νC=O), 3100 mass spectrum: m/e = 302 $[M]^+$	[11, 12]
compounds with $R^1 \neq R^2$:				
*2	CO_2CH_3	CH_3	I, see "Further information", p. 349 m.p. 64 to 65° ^1H NMR (CCl_4): 2.01 (s, CH_3 of R^2), 4.06 (s, C_5H_5); 4.15 (m, 1H) and 4.64 (m, 2H) (C_5H_3) IR (CCl_4): 1720 (νC=O)	[2, 4, 8]

Table 25 (continued)

No.	$C_5H_5FeC_5H_3(R^1\text{-}1)R^2\text{-}3$ $R^1 = CO_2R'$	R^2	method of preparation (yield in %), properties, and remarks	Ref.
*3	CO_2CH_3	CHO	for (−)(1S)-No. 3 see "Further information", p. 350	[10]
4	$CO_2C_2H_5$	C_2H_5	II (36) dark red, oily substance ^{1}H NMR (CCl_4): 1.19 (t, CH_3 of R^2), 1.32 (t, CH_3 of R^1), 2.35 (q, CH_2 of R^2), 4.05 (s, C_5H_5), 4.2 to 4.26 (q and t, CH_2 of R^1 + H-4 of C_5H_3), 4.62 (t, H-2,5 of C_5H_3) IR (neat): 910, 930, 1000, 1100, 2920, 2960, and 3100 ($C_5H_5FeC_5H_3$–C_2H_5-1,3); 1285 (νC–O), 1705 (νC=O) mass spectrum: m/e = 286 $[M]^+$	[9]
5	$CO_2C_2H_5$	C_6H_5	II (12) yellow-orange solid ^{1}H NMR (CCl_4): 1.36 (t, CH_3), 4.01 (s, C_5H_5), 4.24 (q, CH_2), 4.74 (t, H-4 of C_5H_3), 4.83 (t, H-5 of C_5H_3), 5.18 (t, H-2 of C_5H_3), 7.05 to 7.50 (m, C_6H_5) IR (KBr): 890, 925, 1000, 1100, 1600, 3050, and 3100 ($C_5H_5FeC_5H_3$–C_6H_5-1,3); 1240 (νC–O), 1705 (νC=O) mass spectrum: m/e = 334 $[M]^+$	[9]
6	$CO_2C_2H_5$	Cl	II (1) yellow oily substance ^{1}H NMR (CCl_4): 1.31 (t, CH_3), 4.21 (q, CH_2), 4.23 (s, C_5H_5); 4.54 (t, H-4), 4.65 (t, H-5), and 4.97 (t, H-2) (C_5H_3) IR (KBr): 900, 905, 1000, 1100, and 3100 ($C_5H_5FeC_5H_3$–Cl-1,3); 1280 (νC–O), 1710 (νC=O) mass spectrum: m/e = 292 $[M]^+$	[9]
7	$CO_2C_2H_5$	Br	II (3) yellow oily substance ^{1}H NMR (CCl_4): 1.31 (t, CH_3), 4.15 to 4.30 (q, CH_2), 4.20 (s, C_5H_5); 4.53 (t, H-4), 4.67 (t, H-5), and 4.97 (t, H-2) (C_5H_3) IR (KBr): 875, 1000, 1100, and 3110 ($C_5H_5FeC_5H_3$–Br-1,3); 1275 (νC–O), 1700 (νC=O) mass spectrum: m/e = 337 $[M]^+$	[9]

References on p. 350

Table 25 (continued)

No.	$C_5H_5FeC_5H_3(R^1\text{-}1)R^2\text{-}3$ $R^1 = CO_2R'$	R^2	method of preparation (yield in %), properties, and remarks	Ref.
8	$CO_2C_2H_5$	I	II (1.4) red-orange, oily substance ^1H NMR (CCl_4): 1.24 (t, CH_3), 4.12 (s, C_5H_5), 4.19 (q, CH_2); 4.52 (t, H-4), 4.68 (t, H-5), and 4.98 (t, H-2) (C_5H_3) IR (KBr): 860, 915, 990, 1100, and 3080 ($C_5H_5FeC_5H_3$–I-1,3); 1265 (νC–O), 1710 (νC=O) mass spectrum: m/e = 384 $[M]^+$	[9]
9	$CO_2C_2H_5$	OCH_3	II (8.6) yellow liquid ^1H NMR (CCl_4): 1.32 (t, CH_3 of R^1), 3.60 (s, CH_3 of R^2), 4.10 (s, C_5H_5), 4.19 (q, CH_2); 4.23 (q, H-4), 4.45 (q, H-5), and 4.58 (t, H-2) (C_5H_3) IR (KBr): 920, 940, 1000sh, 1100, 2860, 2920, 2960, and 3100 ($C_5H_5FeC_5H_3$–OCH_3-1,3); 1300 (νC–O of ester), 1700 (νC=O) mass spectrum: m/e = 228 $[M]^+$	[9]
*10	$CO_2C_2H_5$	CHO	see "Further information", p. 350 m.p. 66° (ether/n-hexane) ^1H NMR ($CDCl_3$): 1.38 (t, CH_3; J = 7), 4.32 (s, C_5H_5), 4.34 (q, CH_2; J = 7), 4.98 (dd, H-5 of C_5H_3; ^3J = 3, ^4J = 1.2), 5.19 (dd, H-4 of C_5H_3; ^3J = 3, ^4J = 1.2), 5.45 (t, H-2 of C_5H_3; J = 1.2), 10.00 (s, CHO) IR (KBr): 1660 and 1710 (νC=O) UV (n-hexane): λ_{max} (log ε) = 219 (4.42), 252sh (3.84), 275 (3.66), 335 (3.05), 357 (3.11), 458 (2.52) mass spectrum: m/e (relative intensity in % referred to ^{56}Fe) = 286 ($[M]^+$, 100), 258 (38), 214 (20), 186 (15), 121 (50), 56 (92)	[13]
*11	$CO_2CH_2C_6H_4Br\text{-}4$	CH_3	see "Further information", p. 350	[2]

* Further information:

$C_5H_5FeC_5H_3(CO_2CH_3)_2$-1,3 (Table **25**, No. **1**), is prepared in practically quantitative yield from $C_5H_5FeC_5H_3(COCH_3)_2$-1,3 by oxidation with NaOCl in 1/2 C_2H_5OH/CH_2Cl_2 and subsequent saponification with CH_2N_2 [11]. Esterification of $C_5H_5FeC_5H_3(CO_2H)_2$-1,3 by refluxing for 28 h in CH_3OH in the presence of a few drops of concentrated H_2SO_4 as catalyst gives a 90% yield. The compound is purified by column chromatography on SiO_2 with C_6H_6 [12]. It is reduced to $C_5H_5FeC_5H_3(CH_2OH)_2$-1,3 by $LiAlH_4$ in ether [11, 12].

References on p. 350

$C_5H_5FeC_5H_3(CO_2CH_3\text{-}1)CH_3\text{-}3$ (Table **25**, No. **2**) forms mixed with its 1,1′- and 1,2-isomers on oxidation of $FeC_{10}H_8(CHO)CH_3$ containing a 44% fraction of the 1,3-isomer with acidic, active MnO_2 in the presence of KCN and glacial acetic acid in methanol at 20°C. The total yield of the product purified by preparative thin layer chromatography on SiO_2 with C_6H_6 amounts to 46%, of which the fraction of No. 2 is 58% [5]. The shift in favor of No. 2 of the isomer ratio versus that in the starting compound is discussed below.

The mixture of $FeC_{10}H_8MCH_3$ and $FeC_{10}H_7M_2CH_3$ from metallation of $C_5H_5FeC_5H_4\text{–}CH_3$ with n-$C_5H_{11}M$ (M = Na or K) in a 1/1 molar ratio in decane reacts first with CO_2 and subsequently with CH_2N_2 to form a mixture of $FeC_{10}H_8(CO_2CH_3)CH_3$ and $FeC_{10}H_7(CO_2CH_3)_2CH_3$ in a 1/>2 molar ratio. This is separated by column chromatography on Al_2O_3 with development by petroleum ether. Benzene elutes the isomer mixture of monoesters consisting of No. 2 and its 1,1′-isomer in a ratio of ca. 1/2.4, and ether the mixture of the diesters [1].

A sharp separation of No. 2 and its positional isomers, especially the 1,1′-compound, is not possible by column chromatography on Al_2O_3, even when the mixture is developed with 5/1 petroleum ether over the whole column length and is subdivided into 20 to 25 fractions (elution sequence 1,2-, 1,1′-, and 1,3-isomers) and the mixed fraction is rechromatographed. Due to the easier saponificability of the 1,1′- and 1,2-compounds and the consequent higher absorbability the isomer ratio in the isolated esters is shifted in favor of No. 2 from 44/41/15 (1,3/1,1′/1,2-isomer) in the aldehyde mixture to 65/27/8 for No. 2 and its 1,1′- and 1,2-isomers, respectively. Of the quantity of ester introduced 20% remains irreversibly absorbed [2]. Vapor phase chromatography also failed for separation of the 1,1′-isomer in spite of use of eight different column packings. The two isomers both possess a retention time of 5.5 min on Craig polyester succinate at 228°C and 20 psi (0.14 MPa) He pressure [1]. The separation from the 1,2-isomer succeeds with high pressure liquid-liquid chromatography on a column filled with Carbowax 400 on Corasil, in both 1/9 and 1/4 CH_2Cl_2/isooctane at 120 to 160 kp/cm^2 (12 to 16 MPa) [7].

The singlet of the CH_3 protons of R^2 is displaced versus that of $C_5H_5FeC_5H_4\text{–}CH_3$ in CCl_4 by 0.06 ppm to lower field [4].

KOH in methanol forms the free acid in 97% yield [2]. $LiAlH_4/AlCl_3$ in ether gives $C_5H_5FeC_5H_3(CH_3)_2\text{-}1,3$ in 48% yield [1, 2].

Optically Active Forms. (+)(1S)-$C_5H_5FeC_5H_3(CO_2CH_3\text{-}1)CH_3\text{-}3$ is prepared by Method I (p. 346) from (+) (1S)-$C_5H_5FeC_5H_3(CO_2H\text{-}1)CH_3\text{-}3$ (p = 62%) [3, 6, 10]. It is purified by preparative thin layer chromatography on SiO_2 with 30/1 C_6H_6/C_2H_5OH; 91% yield [10]. Melting point 57 to 83 [3], 65 to 68°C [10]. Specific rotation (20°C, p = 100%): In C_2H_5OH (c = 0.1 to 0.6, values ± (0.2 to 0.9)°), $[\alpha]_{589} = +36.4°$, $[\alpha]_{570} = +44.0°$, $[\alpha]_{546} = +62.5°$ [10]. In C_6H_6 (c = 1.0), $[\alpha]_{589} = +18°$ [3]; (c = 0.1 to 0.6, values ± (0.2 to 0.9)°), $[\alpha]_{589} = +29.4°$, $[\alpha]_{570} = +30.4°$, $[\alpha]_{546} = +44.1°$ [10]. ^1H NMR spectrum ($CDCl_3$): δ = 2.0 (s, CH_3 of R^2); 3.79 (s, CH_3 of R^1); 4.15 (s, C_5H_5); 4.30 (m,1H), 4.68 (m,1H), and 4.70 (s,1H) (C_5H_3) ppm [10]. IR spectrum (CH_2Cl_2): 1713 (νC=O) cm^{-1} [10]. Electronic absorption spectrum (C_2H_5OH): $\lambda_{max}(\varepsilon)$ = 262 (3290), 306 (790), 346sh (290), 440 (207) nm [3]. Optical rotatory dispersion (20°C, c ≈ 0.1, p = 100%): In C_2H_5OH, $[M]_{589} = +94°$ [10], $[M]_{485} = +516°$, $[M]_{460} = +420°$, $[M]_{450} = +450°$, $[M]_{375} = -355°$, $[M]_{335} = +1930°$ [3]. In C_6H_6 $[M]_{589} = +76°$ [10]. Circular dichroism (c ≈ 0.1, 20°C): In C_2H_5OH, $\Delta\varepsilon_{537} = -0.04$, $\Delta\varepsilon_{485} = +0.04$, $\Delta\varepsilon_{455} = +0.01$, $\Delta\varepsilon_{419} = 0.12$. In C_6H_6, $\Delta\varepsilon_{526} = -0.03$, $\Delta\varepsilon_{490} = +0.01$, $\Delta\varepsilon_{467} = -0.01$, $\Delta\varepsilon_{416} = +0.15$ [10]. The absolute configuration is given by the correlation with the acid [10].

$LiAlH_4$ reduces the compound to (−)(1S)-$C_5H_5FeC_5H_3(CH_2OH\text{-}1)CH_3\text{-}3$ [3]. $CH_3OCHCl_2/AlCl_3$ in CH_2Cl_2 formylates it to (1S)-OHC–$C_5H_4FeC_5H_3(CO_2CH_3\text{-}1)CH_3\text{-}3$ [6]. Oxidation with active, acidic MnO_2 powder in $Cl_2HCCHCl_2$ at 110°C for 1 h leads to (−)(1S)-No. 3, which is isolated in a 39% yield by preparative thin layer chromatography on SiO_2 with 20/1 C_6H_6/C_2H_5OH [10].

References on p. 350

$C_5H_5FeC_5H_3(CO_2CH_3\text{-}1)CHO\text{-}3$ (Table **25**, No. **3**). **Optically Active Forms.** (−)(1S)-No. 3, prepared from (+)(1S)-No. 2 (see p. 349), melts at 125 to 136°C. Specific rotation (20°C, c = 0.1 to 0.6, values ±(0.2 to 0.9)°, p = 100%): In C_2H_5OH $[\alpha]_{589} = -108.5°$, $[\alpha]_{570} = -134.0°$, $[\alpha]_{546} = -175.5°$; in C_6H_6 $[\alpha]_{589} = -173.5°$, $[\alpha]_{570} = -229.0°$, $[\alpha]_{546} = -339.5°$. ^{1}H NMR spectrum ($CDCl_3$): δ = 3.87 (s, CH_3), 4.34 (s, C_5H_5); 4.98 (m, 1H), 5.16 (m, 1H), and 5.43 (m, 1H) (C_5H_3); 9.80 (m, CHO) ppm. IR spectrum (CH_2Cl_2): 1675 and 1720 (νC=O) cm^{-1}. Optical rotatory dispersion (p = 100%): $[M]^{20}_{589} = -295°$ (C_2H_5OH), −472° (C_6H_6). Circular dichroism (20°C, c ≈ 0.1, p = 100%): In C_2H_5OH $\Delta\varepsilon_{487} = -0.23$, $\Delta\varepsilon_{407} = +0.64$; in C_6H_6 $\Delta\varepsilon_{482} = -0.41$ [10].

Reaction in pyridine first with $NH_2OH \cdot HCl/H_2O$, then $CuSO_4$, $(C_2H_5)_3N$, and finally dicyclohexylcarbodiimide in CH_2Cl_2 in a single vessel, without isolation of intermediate products, gives (1S)-$C_5H_5FeC_5H_3(CO_2CH_3\text{-}1)CN\text{-}3$ in 83% yield. The attempts by oxidation with AgO in tetrahydrofuran/KOH to obtain (1S)-$C_5H_5FeC_5H_3(CO_2CH_3\text{-}1)CO_2H\text{-}3$ have in contrast remained unsuccessful [10].

$C_5H_5FeC_5H_3(CO_2C_2H_5\text{-}1)CHO\text{-}3$ (Table **25**, No. **10**). Preparation is by irradiation for 14 h of a solution of I and II in CH_2Cl_2 with a 300 W sunlamp, cooling, and then reaction with 1/1 2N $NaOH/C_2H_5OH$. The concentrated organic layer gives No. 10 as the first band on a short Al_2O_3 chromatographic column, recrystallization from ether/n-hexane follows; 80% yield [13].

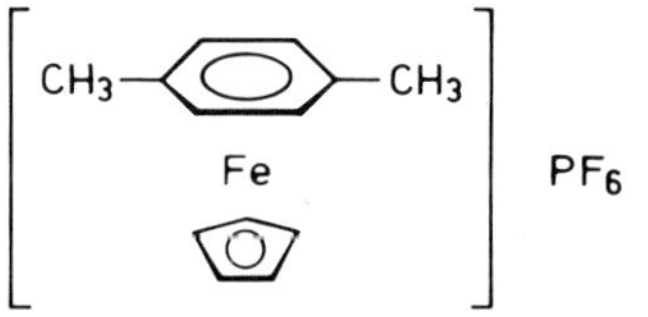

I

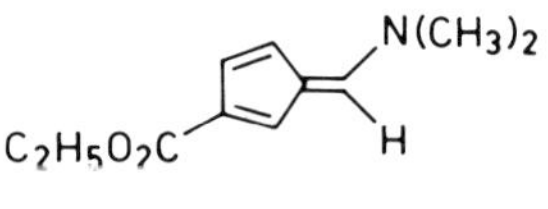

II

$C_5H_5FeC_5H_3(CO_2CH_2(C_6H_4Br\text{-}4)\text{-}1)CH_3\text{-}3$ (Table **25**, No. **11**). The mixture of No. 11 and its 1,1′- and 1,2-isomers won from $FeC_{10}H_8(COCl)CH_3$ with 4-$BrC_6H_4CH_2OH$ can be separated by chromatography on Al_2O_3 only little better than $FeC_{10}H_8(CO_2CH_3)CH_3$ (see "Further information" for No. 2, p. 349), so that it offers practically no advantage over the latter for the separation of the isomers of the free acids [2].

References:

[1] R. A. Benkeser, J. L. Bach (J. Am. Chem. Soc. **86** [1964] 890/5). – [2] H. Falk, G. Haller, K. Schlögl (Monatsh. Chem. **98** [1967] 592/602). – [3] G. Haller, K. Schlögl (Monatsh. Chem. **98** [1967] 603/18). – [4] K. Schlögl, H. Falk, G. Haller (Monatsh. Chem. **98** [1967] 82/94). – [5] K. Schlögl, M. Walser (Monatsh. Chem. **100** [1969] 840/3).

[6] H. Falk, K. Schlögl (Monatsh. Chem. **102** [1971] 33/6). – [7] R. Eberhardt, H. Lehner, K. Schlögl (Monatsh. Chem. **104** [1973] 1409/20). – [8] V. Rapić, K. Schlögl, B. Steinitz (J. Organometal. Chem. **94** [1975] 87/98). – [9] T. Akiyama, T. Kitamura, T. Kato, H. Watanabe, T. Serizawa, A. Sugimori (Bull. Chem. Soc. Japan **50** [1977] 1137/41). – [10] V. Rapić, K. Schlögl, B. Steinitz (Monatsh. Chem. **108** [1977] 767/80).

[11] M. Hisatome, O. Tachikawa, M. Sashō, K. Yamakawa (J. Organometal. Chem. **217** [1981] C17/C20). – [12] A. Kasahara, T. Izumi, Y. Yoshida, I. Shimizu (Bull. Chem. Soc. Japan **55** [1982] 1901/6). – [13] P. Bickert, B. Hildebrandt, K. Hafner (Organometallics **3** [1984] 653/7).

5.2.3.7.3.2 Carboxylic Acid Esters of the Type $C_5H_5FeC_5H_3(A{-}CO_2R'\text{-}1)R^2\text{-}3$ $(A{-}CO_2R' = R^1)$

Although compounds No. 5 to 18 in Table 26, p. 352, with $R^1 \neq R^2$ all possess a plane of chirality (cf. "Ferrocene" 7, 1980, pp. 1/4 (English) and 4/7 (German)), no optically active forms have been obtained. All data therefore refer to the respective racemic forms.

The compounds are prepared by the following methods and by those given under "Further information" at the end of the table, pp. 356/7:

Method I. Friedel-Crafts alkylation of ferrocene with $CH_3O_2C(CH_2)_nCl$ ($\triangleq$ RCl; n = 1 to 3, 5) in heptane in the presence of $AlCl_3$ as catalyst and Al powder as reductant, forming the mixture of the 1,3-compound with its two positional isomers, $FeC_{10}H_8R_2$ (m), along with $C_5H_5FeC_5H_4{-}R$. Preparative thin layer chromatography on SiO_2 with 75/5/30/8 $C_6H_6/CH_3CO_2C_4H_9$/heptane/$(CH_3)_2SO$ yields three bands (ferrocene, $C_5H_5FeC_5H_4{-}R$, and m), which after removal are extracted individually with methanol and ether. A separation of the 1,3-compound (Nos. 1 to 4) from its positional isomers does not take place.

a) Reaction for 4 h at 80 to 95°C [11].

b) Reaction first at 68, then >110°C, for 3 h, with a molar ratio of the reactants of 1/>1/(1.7 to 1.8)/>2 [12].

c) At 95 to 98°C, with a mole ratio 1/2/2.6/2.2 [12].

d) As for Method Ib, but for 6 h at 95 to 98°C [12].

Method II. Friedel-Crafts acylation of $C_5H_5FeC_5H_4{-}CH_2CO_2CH_3$ with 4-$CH_3C_6H_4COCl$ and $AlCl_3$ in a 1/1/1 mole ratio in CH_2Cl_2 at room temperature. No. 5 is isolated mixed with its 1,1'- and 1,2-isomers in a 2/5/3 ratio by dry column chromatography on SiO_2 with 2/1 hexane/$CH_3CO_2C_2H_5$ [13].

Method III. Vilsmeier-Haack formylation of $C_5H_5FeC_5H_4{-}R^1$ with $C_6H_5N(CH_3)CHO$ and $POCl_3$

a) in CH_2Cl_2 at room temperature in the dark [2],

b) at 65 to 70°C for 3 d without solvent. Subsequent column chromatography on SiO_2 with C_6H_6 gives a mixture m of No. 14 and its two positional isomers which cannot be separated by this technique. However, separation experiments by thin layer chromatography of very dilute solutions on Silufol 254 show that the isomers are separable [7, 8].

Method IV. Wittig reaction in the dark:

a) Of $C_5H_5FeC_5H_3(CHO\text{-}1)CH_3\text{-}3$ with $(C_6H_5)_3P{=}CHCO_2C_2H_5$ in a 1/5.8 mole ratio in CH_2Cl_2 for 2 d at room temperature [14].

b) Of No. 8 with $(C_6H_5)_3P{=}CHCO_2C(CH_3)_3$ in a 1/4 mole ratio in boiling benzene for 4 h [14].

c) Of No. 8 with $(C_6H_5)_3P{=}C(CH_3)CO_2C_2H_5$ in a 1/4 mole ratio in CH_2Cl_2 for 4 h at room temperature [14].

d) Of No. 8 with $(C_6H_5)_3P{=}C(CH_3)CO_2C(CH_3)_3$ in a 1/2.2 mole ratio in CH_2Cl_2 for 15 h at room temperature [14].

In each case a careful chromatographic purification of the (E)-ester on an SiO_2 column follows to remove the (Z)-isomer, for Nos. 11 to 13 in CH_2Cl_2, for No. 7 with 5/1 hexane/ether as eluent [14].

References on p. 357

Table 26

Carboxylic Acid Esters of the Type $C_5H_5FeC_5H_3(A\text{–}CO_2R'\text{-}1)R^2\text{-}3$.

Further Information for numbers preceded by an asterisk is given at the end of the table, pp. 356/7.
For abbreviations and dimensions see p. VIII.

No.	$C_5H_5FeC_5H_3(R^1\text{-}1)R^2\text{-}3$ $R^1 = A\text{–}CO_2R'$	R^2	method of preparation (yield in %; "m" indicates isomeric mixture, number following solidus is fraction present of 1,3-compound), properties, and remarks	Ref.
compounds with $R^1 = R^2$:				
1	$CH_2\text{–}CO_2CH_3$		Ia (4 m)	[11]
2	$(CH_2)_2\text{–}CO_2CH_3$		Ib (22 m) deep red oil (m) IR (KBr): 820, 824, 1005 to 1029, and 3100 (C_5H_4); 890 to 896, 900, 915, 965, and 1285 (C_5H_3); 1107, 1123, and 1140 (C_5H_5); 1450, 1585, and 2865 to 2880 (CH_2); 2935 and 2965 (CH_3); 1675 (νC=C) (m) mol. wt: 341 from hydrolysis data (m)	[12]
*3	$(CH_2)_3\text{–}CO_2CH_3$		Ic (42 m), see "Further information", p. 356 IR: 1000, 1100, 1450, 1750 (m) mass spectrum: m/e = 386 $[M]^+$ (m)	[9, 10, 12]
4	$(CH_2)_5\text{–}CO_2CH_3$		Id (18 m) deep red, viscous oil (m)	[12]
compounds with $R^1 \neq R^2$:				
*5	$\overset{6}{C}H_2\text{–}\overset{7}{C}O_2\overset{8}{C}H_3$	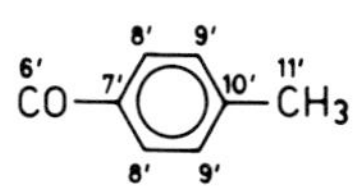	II (76 m/20%) photosensitive, deep red oil (m) ^{13}C NMR ($CDCl_3$): δ = 21.6 (C-11′), 35.1 (or 34.6?) (C-6), 51.8 (C-8), 70.9 (C-1′ to 5′ of C_5H_5), ~84.6 (or ~84.7?) (C-1); 128.2 (1.4C), 128.4 (0.6C), and 128.8 (1.4C) (C-8′,9′); 136.9 (C-7′), 142.1 (C-10′), 170.9 (or 171.5?) (C-7), ~199.5 (C-6′); C-2 to 5 of C_5H_3 are not discernible or indeterminate	[13]

			^{13}C NMR (CD_2Cl_2): δ = 34.2 (or 34.7?) (C-6), 70.6 (C-1′ to 5′ of C_5H_5), 77.1 (C-3 of C_5H_3?), 84.0 (or 84.1?) (C-1), 170.7 (or 171.4?) (C-7); C-2 and 5 of C_5H_3 are not discernible or indeterminate mass spectrum: m/e = 317 $[M-CO_2CH_3]^+$, 184, 119, 91, 65; no $[M]^+$	
*6	$(CH_2)_2$–CO_2CH_3	CHO	IIIa (65) red oil, semicarbazone: m.p. 45 to 50° (ether/petroleum ether) IR: 1000, 1105, 1685, 1741	[1,2]
*7	CH=CH–$CO_2C_2H_5$-(E)	CH_3	IVa (77) red oil ^1H NMR ($CDCl_3$): 1.35 (t, CH_3 of C_2H_5; J = 7), 2.04 (s, CH_3 of R^2) 4.11 (s, C_5H_5), 4.25 (q, CH_2 of C_2H_5; J = 8), 4.38 (m, C_5H_3); 5.99 (d, 1H; J = 15) and 7.52 (d, 1H; J = 16) (CH=CH) IR (neat): 1628, 1708 mass spectrum (CJ): m/e = 299 $[M·H]^+$	[14]
*8	CH=CH–$CO_2C_2H_5$-(E)	CHO	see No. 7, p. 356 red oil ^1H NMR ($CDCl_3$): 1.33 (t, CH_3 of C_2H_5; J = 7), 4.26 (q, CH_2 of C_2H_5; J = 7), 4.29 (s, C_5H_5); 4.88 (s, 1H), 4.99 (s, 1H), and 5.13 (s, 1H) (C_5H_3); 6.17 (d, 1H; J = 16) and 7.54 (d, 1H; J = 16) (CH=CH); 9.99 (s, CHO) IR (neat): 1630, 1678, 1708 mass spectrum (CI): m/e = 313 $[M·H]^+$	[14]
*9	CH=CH–$CO_2C_2H_5$-(E)	CH=CH–CO_2H-(E)	see No. 11, p. 356	[14]
*10	CH=CH–$CO_2C_2H_5$-(E)	$\overset{6'}{C}H$=C(CH_3)–CO_2H-(E)	see No. 13, p. 356 crystals, m.p. 300° ^1H NMR ($CDCl_3$): 1.32 (t, CH_3 of C_2H_5; J = 7), 2.14 (s, CH_3 of R^2), 4.15 (s, C_5H_5), 4.23 (q, CH_2 of C_2H_5; J = 7); 4.72 (s, 2H) and 4.84 (s, 1H)(C_5H_3); 6.12 (d, 1H; J = 16) and 7.55 (d, 1H; J = 16) (CH=CH); 7.57 (s, CH-6′) IR (KBr): 1380, 1630 br., 1710 br. mass spectrum (CI): m/e = 369 $[M·H]^+$	[14]

References on p. 357

Table 26 (continued)

No.	$C_5H_5FeC_5H_3(R^1\text{-}1)R^2\text{-}3$ $R^1 = A\text{–}CO_2R'$	R^2	method of preparation (yield in %; "m" indicates isomeric mixture, number following solidus is fraction present of 1,3-compound), properties, and remarks	Ref.
*11	$CH{=}CH\text{–}CO_2C(CH_3)_3$-(E)	$CH{=}CH\text{–}CO_2C_2H_5$-(E)	IVb (75) red oil ^{1}H NMR ($CDCl_3$): 1.32 (t, CH_3 of C_2H_5; J = 7), 1.52 (s, C_4H_9), 4.14 (s, C_5H_5), 4.22 (q, CH_2 of C_2H_5; J = 7); 4.66 (s, 2H) and 4.77 (s, 1H) (C_5H_3); 6.02 (d, 1H; J = 15), 6.07 (d, 1H; J = 15), 7.40 (d, 1H; J = 16), and 7.51 (d, 1H; J = 16) (CH=CH of R^1 and R^2) IR (neat): 1118, 1344, 1514, 1590, 1608, 1708 mass spectrum (CI, NH_3): m/e (relative intensity in %) = 411 (100) $[M \cdot H]^+$, 428 (82) $[M \cdot NH_4]^+$	[14]
*12	$CH{=}C(CH_3)\text{–}CO_2C_2H_5$-(E)	$CH{=}CH\text{–}CO_2C_2H_5$-(E)	IVc (95) red oil ^{1}H NMR ($CDCl_3$): 1.33 (t, CH_3 of C_2H_5; J = 7), 2.05 (s, CH_3), 4.14 (s, C_5H_5), 4.25 (q, CH_2 of C_2H_5; J = 7); 4.66 (s, 2H) and 4.81 (s, 1H) (C_5H_3); 6.10 (d, 1H; J = 16), 7.42 (s, 1H), and 7.56 (d, 1H; J = 16) (CH= + CH=CH) IR (neat): 1350 br., 1700 br. mass spectrum (CI): m/e = 397 $[M \cdot H]^+$	[14]
*13	$CH{=}C(CH_3)\text{–}CO_2C(CH_3)_3$-(E)	$CH{=}CH\text{–}CO_2C_2H_5$-(E)	IVd (84) red oil ^{1}H NMR ($CDCl_3$): 1.33 (t, CH_3 of C_2H_5; J = 7), 1.53 (s, C_4H_9), 2.01 (s, CH_3), 4.14 (s, C_5H_5), 4.25 (q, CH_2 of C_2H_5; J = 6); 4.66 (s, 2H) and 4.90 (s, 1H) (C_5H_3); 6.11 (d, 1H; J = 16), 7.33 (s, 1H), and 7.57 (d, 1H; J = 15) (CH= + CH=CH) IR (neat): 1630, 1695 br. mass spectrum (CI): m/e = 425 $[M \cdot H]^+$ for chemical behavior see "Further information" for No. 11, p. 356	[14]

References on p. 357

*14	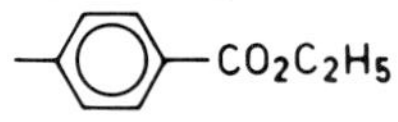	CHO	IIIb (86 m/27%) oil (m) ^{1}H NMR ($CDCl_3$): 1.37 (t, CH_3 of C_2H_5), 4.09 (s, C_5H_5); 4.96 (m, H-5), 5.12 (m, H-4), 5.30 (m, H-2) (C_5H_3); 7.0 to 8.0 (m, C_6H_4), 10.00 (s, CHO)	[7, 8]
*15	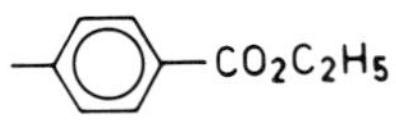	$COCH_3$	see "Further information", p. 357 ^{1}H NMR ($CDCl_3$): 1.37 (t, CH_3 of C_2H_5), 2.41 (s, $COCH_3$), 4.02 (s, C_5H_5), 4.36 (q, CH_2 of C_2H_5), 4.95 (m, H-4,5 of C_5H_3), 5.28 (m, H-2 of C_5H_3), 7.2 to 8.1 (m, C_6H_4)	[7, 8]
16	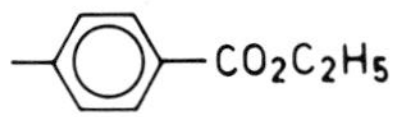	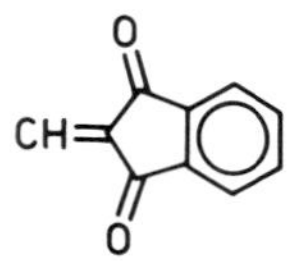	see No. 14, p. 357 m.p. 198 to 201°	[7]
*17	$CO(CH_2)_2–CO_2CH_3$	CH_3	see "Further information", p. 357 red oil, b.p. 176 to 179°/0.05 Torr (No. 17 + 1,1′-isomer) ^{1}H NMR ($CDCl_3$): 2.06 (s, CH_3 of R^2)	[5, 6]
*18	$CO(CH_2)_2–CO_2CH_3$	C_2H_5	see "Further information", p. 357 red oil	[3, 4]

References on p. 357

* Further information:

$C_5H_5FeC_5H_3([CH_2]_3{-}CO_2CH_3)_2$-1,3 (Table **26**, No. **3**). Ferrocene and $CH_3O_2C(CH_2)_3CO_2H$ (40% neutralized with KOH) are electrolyzed at 20°C in methanol between Pt electrodes with 0.023 A/cm^2, 3 V, and 0.3 F. Reduction with $Na_2S_2O_3$, extraction into benzene, and evaporation of the extract, then gives a raw product from which the mixture of the methoxycarbonylferrocenes is separated by column chromatography on SiO_2 with benzene. This mixture is resolved by thin layer chromatography on SiO_2. No. 3 together with its 1,1'-and 1,2-isomers is isolated in 1.6% yield (relative to the amount of reacted ferrocene) following $C_5H_5FeC_5H_4{-}(CH_2)_3CO_2CH_3$ (7%) and before $FeC_{10}H_7([CH_2]_3CO_2CH_3)_3$ (1.8%). Due to the similarity of the R_f values the components are not completely separable [9, 10].

$C_5H_5FeC_5H_3(CH_2{-}CO_2CH_3$-1)CO($C_6H_4CH_3$-4)-3 (Table **26**, No. **5**). Method II (p. 351) or esterification of $FeC_{10}H_8(CH_2CO_2H)CO(C_6H_4CH_3$-4) with CH_2N_2 yield a mixture of No. 5 and its two positional isomers, with the following ^{1}H NMR spectrum ($CDCl_3$): $\delta = 2.43$ (s, CH_3 of R^2), 3.24 (s, 0.5 H, CH_2 of 1,1'-isomer), 3.42 (s, 0.2 H, CH_2 of No. 5), 3.67 (s, CH_3 of R^1), 3.73 (s, 0.3 H, CH_2 of 1,2-isomer), 4.1 to 4.3 (m, 5 H, $C_5H_5 + C_5H_4{-}R^1$), 4.5 to 4.7 (m, 2 H, H-3,4 of $C_5H_4{-}R^2$), 4.8 to 5.0 (m, 1.4 H, H-2,5 of $C_5H_4{-}R^2$), 7.1 to 8.0 (m, C_6H_4) ppm. This mixture is separable by gas chromatography at 250°C on a column filled with 3% OV-225 on Chromosorb W HP. No. 5 appears ($R_f = 13.5$ min) after the 1,2- and 1,1'-isomers ($R_f = 9.3$ and 12.5 min, respectively). Thin layer chromatography on SiO_2 with 2/1 hexane/$CH_3CO_2C_2H_5$ can separate only the 1,2-isomer ($R_f = 0.47$), not the 1,1'-isomer, which possesses along with No. 5 an R_f value of 0.39. Saponification takes place with 2N NaOH in boiling C_2H_5OH [13].

$C_5H_5FeC_5H_3([CH_2]_2{-}CO_2CH_3$-1)CHO-3 (Table **26**, No. **6**) is purified by column chromatography on Al_2O_3 with benzene and subsequent bulb tube distillation at a bath temperature of 160 to 170°C and 0.5 Torr. It is reduced with $LiAlH_4$ to $C_5H_5FeC_5H_3([CH_2]_3OH$-1)$CH_2OH$-3. It undergoes a Knoevenagel condensation with malonic acid in pyridine/piperidine with simultaneous saponification during the processing to form $C_5H_5FeC_5H_3([CH_2]_2CO_2H$-1)CH=$CHCO_2H$-3 [1, 2].

(E)-$C_5H_5FeC_5H_3$(CH=CH$-CO_2C_2H_5$-1)CH_3-3 (Table **26**, No. **7**) reacts with activated MnO_2 in the dark in toluene in 4 h at 90°C to form No. 8, which is isolated in 20% yield by column chromatography on SiO_2 in CH_2Cl_2 [14].

(E)-$C_5H_5FeC_5H_3$(CH=CH$-CO_2C_2H_5$-1)CHO-3 (Table **26**, No. **8**) undergoes a Wittig reaction corresponding to Methods IVb to IVd (see p. 351) to form the compounds No. 11 to 13 in yields of 75, 95, and 84%, respectively [14].

(E,E)-$C_5H_5FeC_5H_3$(CH=CH$-CO_2C_2H_5$-1)CH=CR$-CO_2H$-3 (Table **26**, Nos. **9** and **10** with R = H, CH_3) are esterified to (E,E)-$C_5H_5FeC_5H_3$(CH=CH$-CO_2C_2H_5$-1)CH=CR$-CO_2(C_6H_4NO_2$-4)-3 with 4-$O_2NC_6H_4OH$ in CH_2Cl_2 with addition of dicyclohexylcarbodiimide and 4-dimethylaminopyridine in the dark at room temperature [14]. See also the next paragraph.

(E,E)-$C_5H_5FeC_5H_3$(CH=CR$-CO_2C(CH_3)_3$-1)CH=$CHCO_2C_2H_5$-3 (Table **26**, Nos. **11** and **13** with R = H, CH_3). Stirring in CF_3CO_2H (for 30 min for No. 11, 2 h for No. 13) selectively removes the $(CH_3)_3C$ groups with formation of Nos. 9 (R = H) or 10 (R = CH_3). No. 9 is repeatedly boiled down with benzene, while No. 10 (yield 74%) is purified by column chromatography of the evaporation residue on SiO_2 with 20/1 CH_2Cl_2/CH_3OH [14].

(E,E)-$C_5H_5FeC_5H_3$(CH=$CCH_3{-}CO_2C_2H_5$-1)CH=$CHCO_2C_2H_5$-3 (Table **26**, No. **12**) is saponified on both ester groups by 6N NaOH in C_2H_5OH at room temperature [14].

References on p. 357

$C_5H_5FeC_5H_3([C_6H_4$–$CO_2C_2H_5$-4]-1)CHO-3 (Table **26**, No. **14**) in the mixture with its 1,1'- and 1,2-isomers (fractions 27, 60 and 13%, respectively) yields on condensation with I in C_6H_6 in the presence of a few drops of $(C_2H_5)_3N$ the $FeC_{10}H_8(C_6H_4CO_2C_2H_5$-4)CH=C(CO)$_2C_6H_4$ (mixture of No. 16 and its 1,1'-isomer). Column chromatography on SiO_2 with C_6H_6 as eluent gives No. 16 ahead of the 1,1'-isomer, in 15% yield [7].

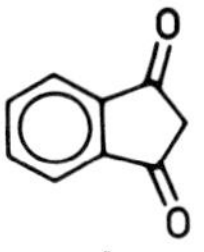

I

$C_5H_5FeC_5H_3([C_6H_4$–$CO_2C_2H_5$-4]-1)$COCH_3$-3 (Table **26**, No. **15**). $C_5H_5FeC_5H_4$–$C_6H_4CO_2C_2H_5$-4 is subjected to Friedel-Crafts acetylation at 90 to 95°C in a $(CH_3CO)_2O$ reaction medium plus 85% H_3PO_4, in a 1/20/4 molar ratio. Column chromatography on SiO_2 with benzene as eluent gives a 46% yield of an oil composed 27% of No. 15 and 73% of its 1,1'-isomer. The two are not separable chromatographically [7, 8].

$C_5H_5FeC_5H_3(CO(CH_2)_2$–$CO_2CH_3$-1)$R^2$-3 (Table **26**, Nos. **17** and **18** with $R^2 = CH_3$, C_2H_5). Friedel-Crafts acylation of $C_5H_5FeC_5H_4$–R^2 with $CH_3O_2C(CH_2)_2COCl$ and $AlCl_3$ is carried out in CH_2Cl_2 at room temperature with a mole ratio of 1/0.75/1.1 for No. 17 [6] and 1/1/1 for No. 18 [4]. Nos. 17 [6] and 18 [3, 4] are obtained mixed with their two positional isomers. The separation of excess $C_5H_5FeC_5H_4$–R^2, small quantities of diacylation products, and the 1,2-isomers proceeds smoothly by column chromatography on deactivated Al_2O_3 with benzene, petroleum ether, and their mixtures. More difficult is the separation from the 1,1'-isomers. It is effected for No. 18 by repeated column chromatography on Al_2O_3 in benzene [4]. However, No. 17 and its 1,1'-isomer are not separable chromatographically. Only after saponification of the mixture with KOH in C_2H_5OH can the free acids be separated by fractional crystallization [6]. A mixture of No. 17 and its positional isomers also forms on esterification of $FeC_{10}H_8(CO(CH_2)_2CO_2H)CH_3$ with CH_2N_2 in ether (88% yield) [5].

Nos. 17 and 18 are saponified with KOH in CH_3OH [5] or C_2H_5OH [4, 6] at room temperature [4 to 6].

References:

[1] K. Schlögl, H. Seiler (Tetrahedron Letters **1960** No. 7, pp. 4/8). – [2] K. Schlögl, M. Peterlik, H. Seiler (Monatsh. Chem. **93** [1962] 1309/27). – [3] J. Tirouflet, G. Tainturier (Tetrahedron Letters **1965** 4177/9). – [4] G. Tainturier, J. Tirouflet (Bull. Soc. Chim. France **1966** 595/600). – [5] G. Haller, K. Schlögl (Monatsh. Chem. **98** [1967] 2044/57).

[6] P. Dixneuf, R. Dabard (Bull. Soc. Chim. France **1969** 2164/71). – [7] Š. Toma, A. Maholányiová, E. Solčániová (Acta Fac. Rerum Nat. Univ. Comenianae Chim. **24** [1976] 53/80). – [8] Š. Toma, A. Maholányiová, E. Solčaniová (Collection Czech. Chem. Commun. **42** [1977] 1013/26). – [9] L. I. Denisovich, M. G. Peterleitner, S. P. Gubin (Izv. Akad. Nauk SSSR Ser. Khim. **1978** 465/8; Bull. Acad. Sci. USSR Div. Chem. Sci. **1978** 403/6). – [10] S. P. Gubin, L. I. Denisovich, N. V. Zakurin, M. G. Peterleitner (J. Organometal. Chem. **146** [1978] 267/78).

[11] N. A. Maier, S. A. Shklyar (Zh. Obshch. Khim. **48** [1978] 702/3; J. Gen. Chem. [USSR] **48** [1978] 645/6). – [12] N. A. Maier, S. A. Shklyar, Yu. A. Ol'dekop (Zh. Obshch. Khim. **49** [1979] 2518/21; J. Gen. Chem. [USSR] **49** [1979] 2223/5). – [13] B. E. Maryanoff (J. Org. Chem. **46** [1981] 70/4). – [14] R. Breslow, G. Trainor, A. Ueno (J. Am. Chem. Soc. **105** [1983] 2739/44).

5.2.3.8 Carboxylic Acid Chlorides

5.2.3.8.1 1,1′-Compounds

5.2.3.8.1.1 1,1′-Di(chlorocarbonyl)ferrocene $Fe(C_5H_4{-}COCl)_2$

Preparation of the compound starts with $Fe(C_5H_4{-}CO_2H)_2$ [1 to 4, 9, 11, 14, 17, 19, 20, 22, 31, 35]. It is obtained practically quantitatively by 24 h refluxing with excess **PCl_3**, without indications of incipient decomposition. It is isolated by column chromatography on Al_2O_3 with 50/1 C_6H_6/CH_3OH after evaporating off the excess PCl_3 and treatment first with methanol and then ether, with evaporation off of the solvents in each case [2]. Refluxing for 12 h gives a 46% yield [14]. Reaction in a 1/2 mole ratio in benzene, first for 3 h cold, then 2 h at 50 to 60°C, gives yields of 30 [1] or 46% [9]. The reaction is conducted by [31] with a molar ratio of 1/<4 in benzene with addition of pyridine, for 31 h at 50 to 60°C.

The reaction with **ClCOCOCl** in CH_2Cl_2 in the dark in the presence of traces of pyridine, first at room temperature, then at the boiling point, gives yields of 80 to 85% [3, 11, 17].

PCl_5 in a 1/2 molar ratio in benzene at room temperature [22], at 30 to 50°C, gives a 49% yield [4], see also [35].

For preparation with **$SOCl_2$** see [20].

Properties. The compound purified by recrystallization (solvent indicated in parentheses) consists of red [1, 3, 11], red-orange [2] needle-shaped [2, 11] crystals, which melt (temperatures in °C) at 92 to 93 (petroleum ether) [1], 93 to 95 [6], 95 (petroleum ether) [11], 94 to 96 [4], 97 to 99 [16], 98 to 99 [17], 98 to 100 (petroleum ether) [3], 114.0 to 114.6 (cyclohexane) [2].

The IR spectrum has C–Cl bands at 697 and 752, ferrocene bands at 1047, 1170, and 3125, and a C=O double band at 1695 and 1770 cm^{-1} (see figure in the paper) [11].

The electronic absorption spectrum displays in cyclohexane a band at $\lambda_{max}(\varepsilon_{max})$ = 462(533) nm. In this region of the A_{2g} transition an S-shaped magnetic circular dichroism band is observed with peaks at $\lambda_1 = 502$ and $\lambda_2 = 433$ nm ($\lambda_{\varepsilon=0} = 460$ nm), which is partially overlapped by a Gaussian MCD curve with $\lambda = 375$ nm [46].

$Fe(C_5H_4{-}COCl)_2$ is soluble in benzene [9] and other organic solvents [1].

Chemical Reactions

General Reference:

H.-J. Lorkowski, Ferrocen als Grundbaustein der makromolekularen Chemie, Fortschr. Chem. Forsch. **9** [1967/68] 207/53.

With Inorganic Compounds

Hydrolysis takes place relatively slowly, similar to the behavior of aromatic acid chlorides. With 20% Na_2CO_3 solution heating or long standing when cold is required [1]. Shaking with H_2O in benzene/pyridine forms $ClCO{+}C_5H_4FeC_5H_4{-}CO{-}O{-}OC{+}_xC_5H_4FeC_5H_4{-}COCl$ with x = 1 to 5, as a light brown powder moderately soluble in CH_2Cl_2 or tetrahydrofuran [5]. Passing NH_3 into the benzene solution gives $Fe(C_5H_4{-}CONH_2)_2$ practically quantitatively [1, 9, 14]. Low temperature polycondensation with H_2NNH_2 in $OP(N(CH_3)_2)_3$ leads to ${+}CO{-}NH{-}NH{-}CO{-}C_5H_4FeC_5H_4{+}_x$ [15, 16]. Attempts at high temperature polycondensation in $C_6H_5NO_2$ at 180°C failed [15]. Reaction with NaN_3 forms $Fe(C_5H_4{-}CON_3)_2$ [6, 11, 12].

References on pp. 369/70

With Alcohols

The **monoalcohols** ROH lead to esterification to form exclusively $Fe(C_5H_4{-}CO_2R)_2$. If $R = CH_3$ [11, 18, 35] or C_2H_5 [35], the alcohol also serves as the solvent. In other cases the reaction is conducted in benzene ($R = CH_2C{\equiv}CH$ [31]) with addition of pyridine ($R = (CH_2)_2N_3$ [17], the radical I (see figure) [20, 30]) or in ether with addition of K_2CO_3 ($R = CH_2C{\equiv}CH$ [17]) or Na (R = II [6, 24]).

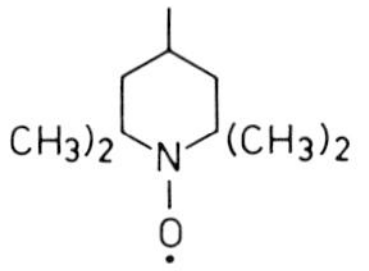
I

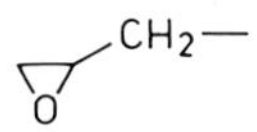
II

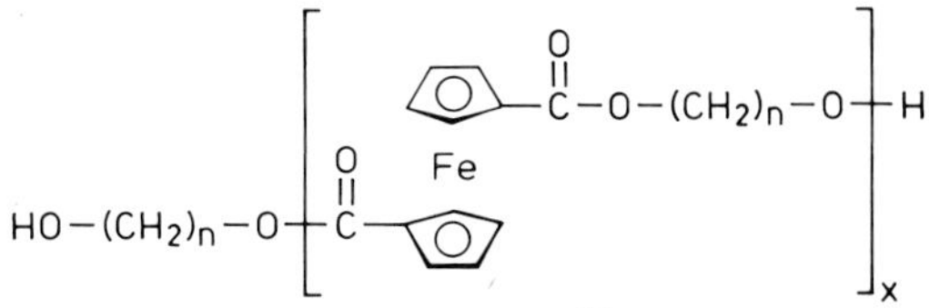
III

Dialcohols give diesters such as V or VI, polyesters such as III or IV, and mononuclear (VII, VIII, XII, XIII) and binuclear (IX, X, XI) ferrocenophane derivatives. Table 27, pp. 360/2, lists these reactions and their conditions. The abbreviations used are the following:

RT = room temperature.

b.p. = boiling point or at boiling temperatures.

hdc = high dilution condition, i.e., the solutions of the reactants are simultaneously dropped into a receiver with pure solvent, see also [43, 44].

ip = interfacial polycondensation, i.e., the aqueous phase of base + dialcohol (for Table 27, pp. 360/2) or diamine (for Table 28, pp. 364/6) and the organic phase of the $Fe(C_5H_4{-}COCl)_2$ solution are mixed in a Waring blender [3], see also [28].

$\bar{x}$ = average value of x.

When two different reaction times are given, the first is the time required for mixing of the reactants, the second the additional reaction time needed.

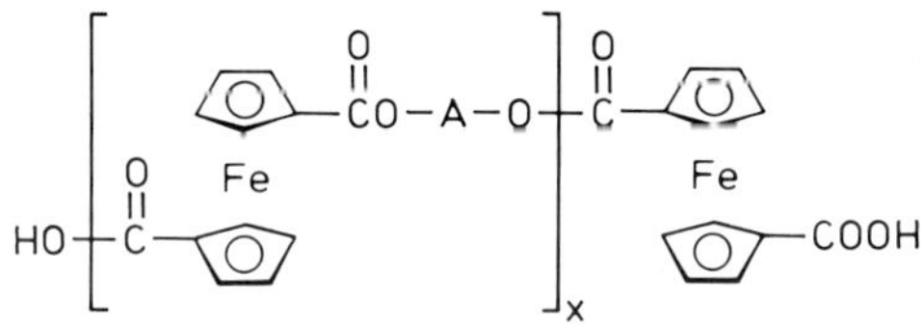
IV

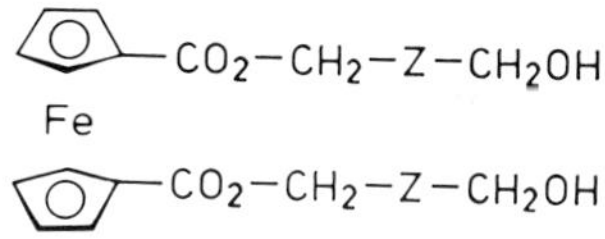

V: Z = 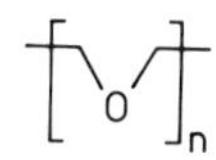

VI: Z = A

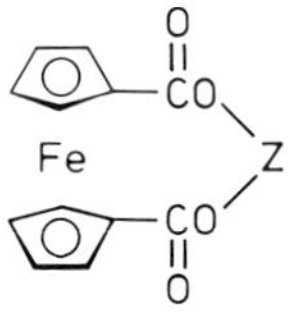

VII: Z = 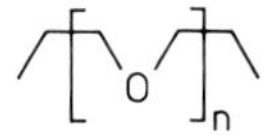

VIII: Z = $-CH_2-A-CH_2-$

XII: Z = 

XIII: Z = 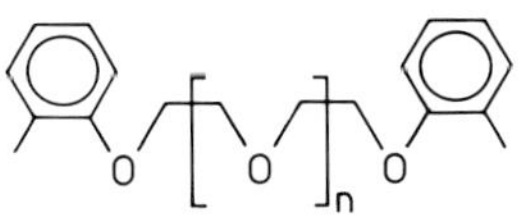

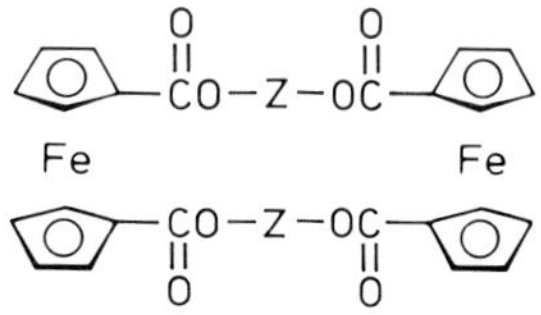

IX: Z = 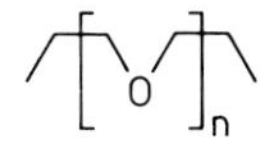

X: Z = $-CH_2-A\ CH_2$

XI: Z = 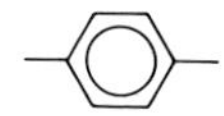

References on pp. 369/70

Table 27

Reactions of $Fe(C_5H_4{-}COCl)_2$ with Dialcohols.
For abbreviations see p. 359.

dialcohol	mole ratio $Fe(C_5H_4{-}COCl)_2$/ dialcohol/ added base	reaction conditions	reaction products	Ref.
$HO{-}(CH_2)_2{-}OH$	1/2	C_6H_6; hdc; 20 + 40 h	V (51%) + VII (28%) (n = 0)	[42]
$HO{-}(CH_2)_3{-}OH$	1/1	toluene/pyridine; 70 to 75°; 17 to 24 h	III (n = 3)	[7]
$HO{-}(CH_2)_6{-}OH$	1/1	toluene/pyridine (4/1); 70 to 75°; 17 to 24 h	III (n = 6; $\bar{x}$ = 4)	[13, 18]
$HO{-}(CH_2)_{10}{-}OH$	1/1	toluene/pyridine (4/1); 70 to 75°; 17 to 24 h	III (n = 10; $\bar{x}$ = 11)	[7, 13, 18]
	1/1	xylene/pyridine; b.p.	polyester (90%)	[47]
$(HOH_2C)_4C$	1/1	xylene/pyridine; b.p.	polyester (53%)	[47]
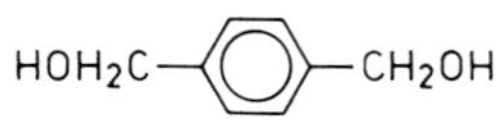	1/1	toluene/pyridine (4/1); 70 to 75°; 17 to 24 h	IV ($\bar{x}$ = 4; A = C_6H_4-1,4)	[13, 18]
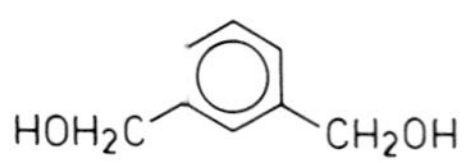	1/2/2 ($N(C_2H_5)_3$)	C_6H_6/CH_3CN (4/5); b.p.; 20 + 20 h	VI (58%) + VIII (2%) + X (4%) (A = C_6H_4-1,3)	[42]
$HOH_2C{-}CH_2{-}O{-}CH_2{-}CH_2OH$	1/1/2 (pyridine)	C_6H_6; b.p.; hdc; 8 + 0 h	VII (36%; n = 1)	[32]
	1/2	C_6H_6; b.p.; hdc; 20 + 40 h	V (59%) + VII (26%) (n = 1)	[42]
$HOH_2C{-}(CH_2{-}O{-}CH_2)_2{-}CH_2OH$	1/1	xylene/pyridine; b.p.	polyester (53%)	[47]
	1/1/2 (pyridine)	C_6H_6; b.p.; hdc; 8 + 0 h	VII (29%; n = 2)	[32]
	1/2	C_6H_6; b.p.; hdc; 20 + 40 h	V (53%) + VII (23%) (n = 2)	[42]

$HOH_2C–(CH_2–O–CH_2)_3–CH_2OH$	1/1/2 (pyridine)	C_6H_6; b.p.; hcd; 8 + 0 h	VII (29%; n = 3)	[32]
	1/2	C_6H_6; b.p.; 20 + 40 h	V (55%) + VII (22%) (n = 3)	[42]
$C_2H_5C(CH_2OH)_2CH_2OCH_2CH{=}CH_2$	1/1	xylene/pyridine; b.p.	polyester (ca. 90%)	[47]
$HOH_2C–CH_2–S–CH_2–CH_2OH$	1/1/2 ($N(C_2H_5)_3$)	C_6H_6/CH_3CN (4/5); b.p; 20 + 20 h	VI (67%) + VIII (2%) ($A = CH_2–S–CH_2$)	[42]
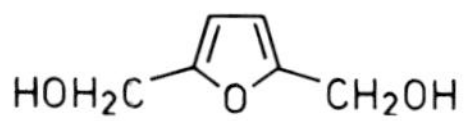	1/2/2 ($N(C_2H_5)_3$)	C_6H_6/CH_3CN (4/5); b.p.; 20 + 20 h	VIII (1%) + X (9%) (A = 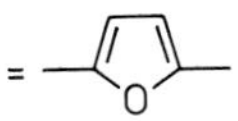)	[42]
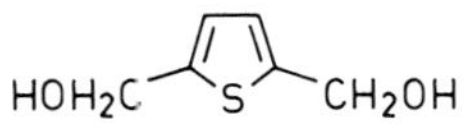	1/2/2 ($N(C_2H_5)_3$)	C_6H_6/CH_3CN (4/5); b.p.; 20 + 20 h	VI (49%) + VIII (2%) (A = 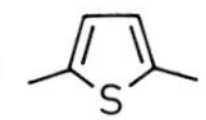)	[42]
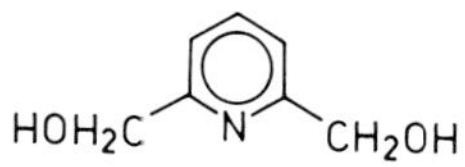	1/2/2 ($N(C_2H_5)_3$)	C_6H_6/CH_3CN (4/5); b.p.; 20 + 20 h	VI (19%) + VIII (9%) (A = 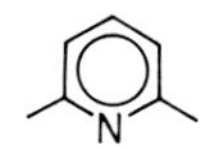)	[42]
$Fe(C_5H_4–CH_2OH)_2$	1/1	toluene/pyridine (4/1); 70 to 75°; 17 to 24 h	IV ($\bar{x} = 4$; $A = –C_5H_4FeC_5H_4–$)	[7, 13, 18, 47]
V (n = 0)	1/1	C_6H_6; b.p.; 20 + 48 h	IX (23%)	[42]
V (n = 1)	1/1	C_6H_6; b.p.; 20 + 48 h	IX (34%)	[42]
V (n = 2)	1/1	C_6H_6; b.p.; 20 + 48 h	IX (36%)	[42]
V (n = 3)	1/1	C_6H_6; b.p.; 20 + 48 h	IX (35%)	[42]
VI ($A = C_6H_4$-1,3)	1/1/1 ($N(C_2H_5)_3$)	C_6H_6; RT	X (39%)	[42]
VI ($A = –CH_2–S–CH_2–$)	1/1/1 ($N(C_2H_5)_3$)	C_6H_6; RT	X (5%)	[42]
VI (A = 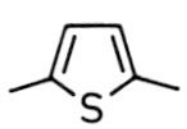)	1/1/1 ($N(C_2H_5)_3$)	C_6H_6; RT	X (5%)	[42]

References on pp. 369/70

Table 27 (continued)

dialcohol	mole ratio $Fe(C_5H_4{-}COCl)_2$/ dialcohol/ added base	reaction conditions	reaction products	Ref.
VI (A = 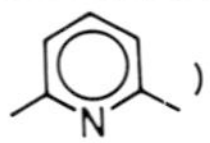)	1/1/1 ($N(C_2H_5)_3$)	C_6H_6; RT	X (5%)	[42]
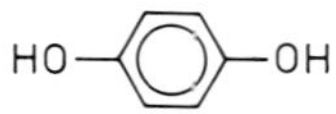	1/1/1 (NaOH)	H_2O, CH_2Cl_2; ip; 3 min	polyester (63%)	[3]
	1/3	C_6H_6/CH_3CN; b.p.; 24 + 70 h	$Fe(C_5H_4{-}CO_2C_6H_4OH\text{-}4)_2$ (38%) + XI (21%)	[39]
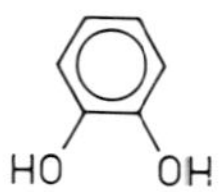	1/1/2 (pyridine)	C_6H_6; RT; hdc; (6 to 8) + 1 h	XII (33%)	[32]
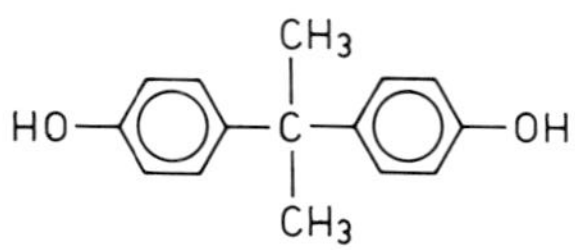	1/1/1 (NaOH)	H_2O, CH_2Cl_2; ip; 3 min	polyester (70%)	[3]
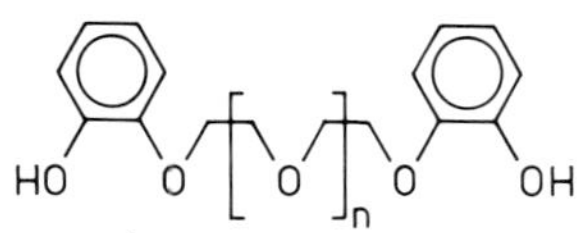				
n = 1	1/1/2 (pyridine)	C_6H_6; b.p; hdc; (6 to 8) + 1 h	XIII (37%)	[32]
n = 2	1/1/2 (pyridine)	C_6H_6; b.p.; hdc; (6 to 8) + 1 h	XIII (34%)	[32]
$Fe(C_5H_4{-}CO_2C_6H_4OH\text{-}4)_2$	1/1	C_6H_6/CH_3CN; b.p.; 24 + 70 h	XI (6%)	[39]

References on pp. 369/70

With Nitrogen Compounds

The **monoamines** RNH_2 form the $Fe(C_5H_4{-}CO{-}NHR)_2$. The reactions have been carried out with R = CH_3, C_2H_5, C_4H_9, C_6H_5, C_6H_4Cl-3, $C_6H_4OCH_3$-3, $C_6H_4OCH_3$-2 [22], XIV to XVI [40], and XVII [33]. While for the former seven [22] no HCl acceptor was added, the latter reactions [33, 40] were conducted with additions of $(C_2H_5)_3N$ (for R = XIV, XVII), $C_6H_5N(CH_3)_2$ (for R = XV, XVI), or pyridine (for R = XVII). R = XVII gives (+)-$Fe(C_5H_4{-}CO{-}NHR)_2$ [33]. Reaction with the secondary amines $R(CH_3)NH$ yields $Fe(C_5H_4{-}CO{-}N(CH_3)R)_2$ for R = CH_3 [21] or $C_6H_4NO_2$-4 [6]. In the last reaction a twice stoichiometric amount of amine was used for capture of the HCl released [6].

Diamines form mono- (XVIII to XXIII) and binuclear (XXIV) ferrocenophane derivatives as well as polyamides. The reactions and their conditions are listed in Table 28, pp. 364/6.

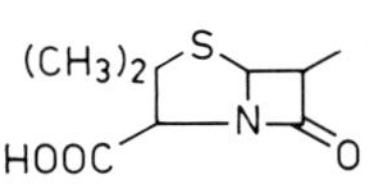

XIV

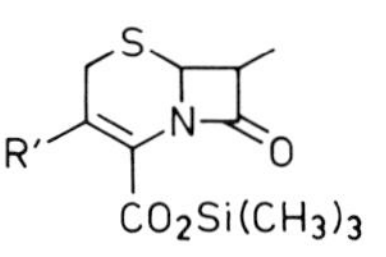

XV: R′ = $CH_2{-}OOCCH_3$

XVI: R′=

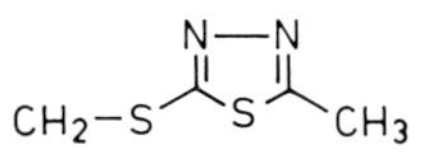

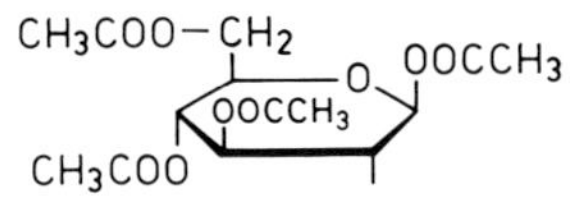

XVII

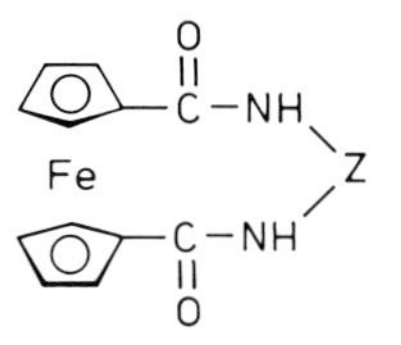

XVIII: Z = $-(CH_2)_8-$

XIX: Z =

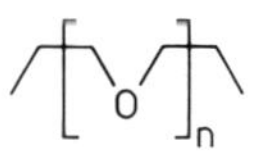

XX: Z =

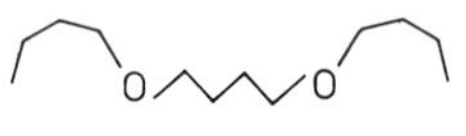

XXI: Z =

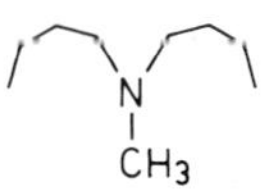

XXII: Z =

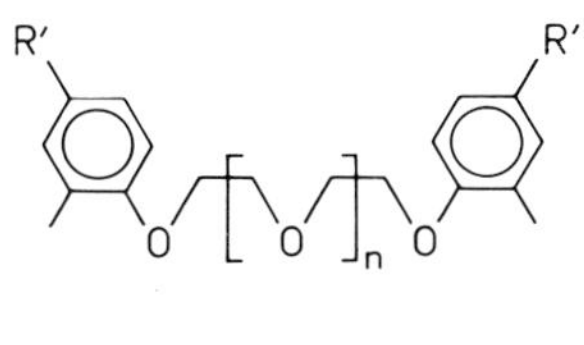

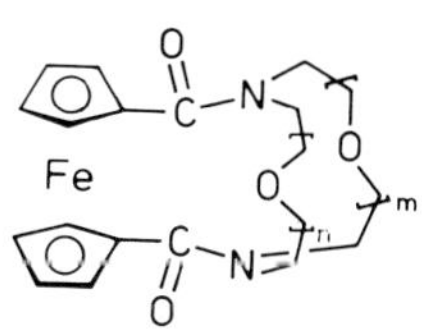

XXIII

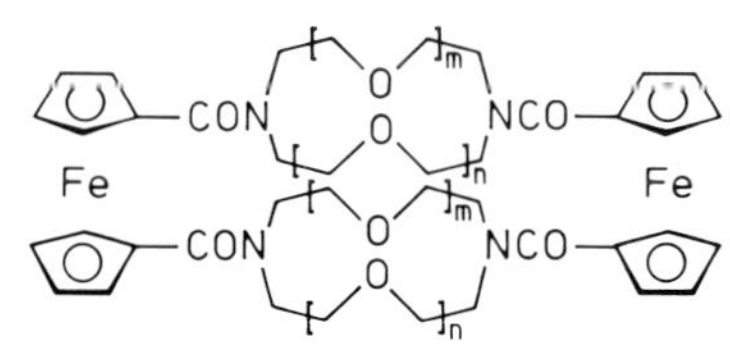

XXIV

References on pp. 369/70

Table 28

Reactions of $Fe(C_5H_4{-}COCl)_2$ with Diamines.

For abbreviations see p. 359.

diamine	mole ratio $Fe(C_5H_4{-}COCl)_2$/ diamine/ added base	reaction conditions	reaction products	Ref.
$H_2N{-}(CH_2)_2{-}NH_2$	1/>1*)	H_2O, C_6H_6; RT; ip; 2 min	polyamide (79%)	[3]
$H_2N{-}(CH_2)_4{-}NH_2$	1/1	xylene/pyridine; b.p.	polyamide (56%)	[47]
	—	toluene/pyridine	$[HN{-}CO{-}C_5H_4FeC_5H_4{-}CO{-}NH{-}(CH_2)_4]_x$	[7]
	—	ip; RT	polyamide	[4]
$H_2N{-}(CH_2)_6{-}NH_2$	1/1	xylene/pyridine; b.p.	polyamide (56%)	[47]
	—	toluene/pyridine	$[HN{-}CO{-}C_5H_4FeC_5H_4{-}CO{-}NH{-}(CH_2)_6]_x$	[7]
	1/1/2 (Na_2CO_3)	H_2O, C_6H_6; RT; ip; 2 min	polyamide (87%)	[3, 4]
$H_2N{-}(CH_2)_8{-}NH_2$	1/2*)	C_6H_6; RT; hdc; (6 to 8)+1h	XVIII (68%)	[32]
$H_2N{-}(CH_2)_{11}{-}NH_2$	1/1	xylene/pyridine; b.p.	polyamide (53%)	[47]
	—	toluene/pyridine	$[HN{-}CO{-}C_5H_4FeC_5H_4{-}CO{-}NH{-}(CH_2)_{11}]_x$	[7]
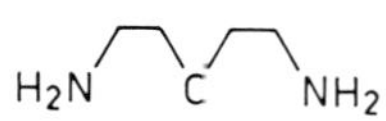	1/2*)	C_6H_6; RT; hdc; (6 to 8)+1h	XIX (72%; n=1)	[32]
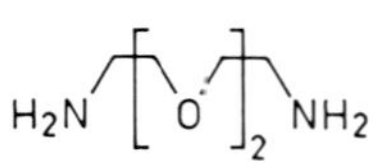	1/2*)	C_6H_6; RT; hdc; (6 to 8)+1h	XIX (57%; n=2)	[32], see also [34]
	1/1/>2($N(C_2H_5)_3$)	C_6H_6; C_2H_5OH; hdc; RT	XIX (33%; n=2)	[38]
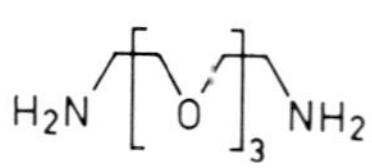	1/2*)	C_6H_6; RT; hdc; (6 to 8)+1h	XIX (54%; n=3)	[32]

$H_2N-(CH_2)_3-O-(CH_2)_4-O-(CH_2)_3-NH_2$	1/2*)	C_6H_6; RT; hdc; (6 to 8) + 1 h	XX (55%)	[32]
$H_2N-(CH_2)_3-N(CH_3)-(CH_2)_3-NH_2$	1/2*)	C_6H_6; RT; hdc; (6 to 8) + 1 h	XXI (47%)	[32]
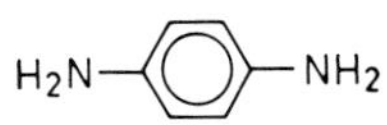	—	xylene/pyridine; b.p.	polyamide (90%)	[47]
	—	toluene/pyridine	$\text{+}HN-CO-C_5H_4FeC_5H_4-CO-NH-C_6H_4\text{+}_x$	[7]
	1/1/2 (NaOH)	H_2O, C_6H_6; ip; 3 min	polyamide (93%)	[3]
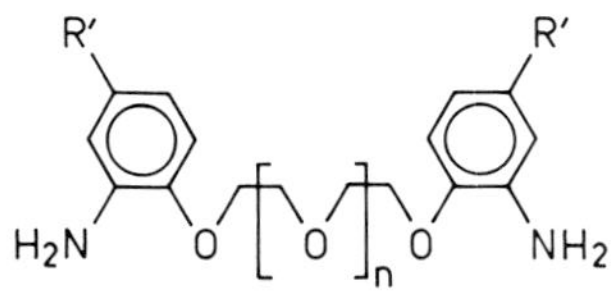				
n = 0, R′ = H	—	—	XXII	[34]
n = 1, R′ = H	—	—	XXII	[34]
n = 2, R′ = H	—	—	XXII	[34]
n = 2, R′ = CH_3 (^{15}N labelled)	—	—	XXII (^{15}N labelled)	[36]
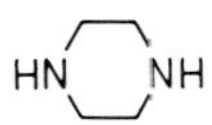	1/1	xylene/pyridine; b.p.	polyamide (54%)	[47]
	1/1/2 (Na_2CO_3)	H_2O, C_6H_6; ip; 3 min	polyamide (80%)	[3]
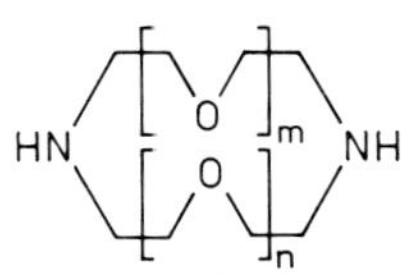				
m = n = 1	1/1/1 ($N(C_2H_5)_3$)	toluene; hdc (?)	XXIII + XXIV	[41]
m = 1, n = 2	1/2*)	C_6H_6; RT; hdc; (6 to 8) + 1 h	XXIII (31%)	[32]
	1/1/2 ($N(C_2H_5)_3$)	toluene; hdc	XXIII + XXIV	[41]
m = n = 2	—	—	XXIII	[34]
	1/2*)	C_6H_6; RT; hdc; (6 to 8) + 1 h	XXIII (37%)	[32]
	1/1/2 ($N(C_2H_5)_3$)	C_6H_6; RT; hdc; 3 + 2.5 h	XXIII (40%) + XXIV (26%)	[38]

References on pp. 369/70

Table 28 (continued)

diamine	mole ratio $Fe(C_5H_4{-}COCl)_2$/ diamine/ added base	reaction conditions	reaction products	Ref.
m = n = 2	1/1/2 ($N(C_2H_5)_3$)	toluene; hdc; 3 h; 100°**)	XXIII (63%)+XXIV (2%)	[37]
		20°	XXIII (65%)+XXIV (5%)	[37]
		−20°	XXIII (62%)+XXIV (9%)	[37]
		−40°	XXIII (17%)+XXIV (15%)	[37]
		−70°	XXIII (6%)+XXIV (20%)	[37]
m = 2, n = 3	1/1/2 ($N(C_2H_5)_3$)	toluene; hdc (?)	XXIII +XXIV	[41]
m = n = 3	1/1/2 ($N(C_2H_5)_3$)	toluene; hdc (?)	XXIII +XXIV	[41]
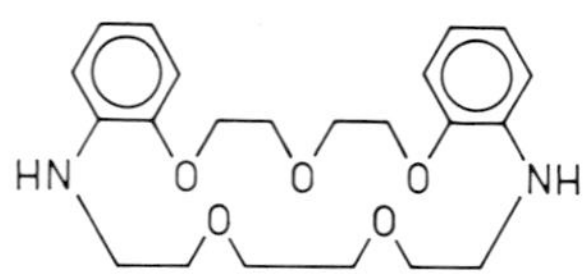	—	—	polyamide (no cryptand)	[34]

*) Half of the amine serves for HCl capture.
**) See p. 367 and for the reaction mechanism the scheme in XXVI.

The condensation process with diazo-18-crown-6 (c in XXVI), which leads via the intermediate product XXV to XXIII (m = n = 2), XXIV (m = n = 2), and various oligomers, is described by the mechanism in XXVI, in which F and C stand for an $Fe(C_5H_4–CO)_2$ or a crown unit. Each reaction step involves HCl elimination. Variation of the conditions shows that the course of the reaction is determined less by concentration than by temperature changes (cf. Table 28, pp. 365/6). At high temperatures the formation of XXIII is strongly favored over that of XXIV, thus k_2 must be $> k_3$ [f] or k_4 [XXV] or k_5 [c]. The fact that at very low temperatures more XXIV forms than XXIII can be traced to the rotation around the CO–N bond being hindered in XXV or to that fact that k_2 is $< k_3$ [f] or k_4 [XXV] or k_5 [c] [37].

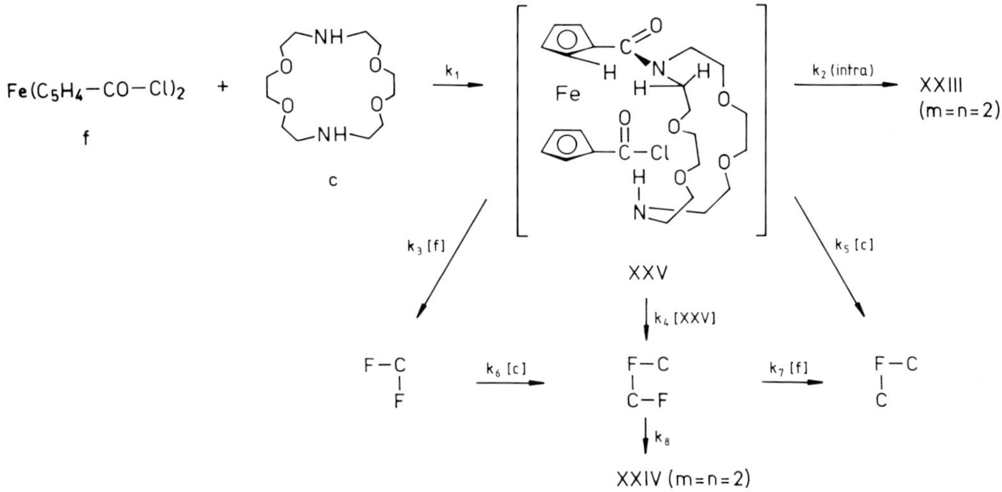

XXVI

Other N Compounds. $C_6H_5CONHNH_2$ in a 1/3 mole ratio in $OP(N(CH_3)_2)_3$, initially under ice cooling, then at room temperature, gives $Fe(C_5H_4–CONHNHCOC_6H_5)_2$. $C_5H_5FeC_5H_4–CONHNH_2$ initially at room, then at boiling temperature gives $Fe(C_5H_4–CONHNHCO–C_5H_4FeC_5H_5)_2$ [15]. Polycondensation with ice cooling with $H_2NNH–CO–A_n–CO–NHNH_2$ (n = 0 or 1; A = $(CH_2)_4$, C_6H_4-1,3, or $C_5H_4FeC_5H_4$) in $OP(N(CH_3)_2)_3$ gives $[C_5H_4FeC_5H_4–CONHNHCO–A_n–CONHNHCO]_x$ [15, 16]. Attempts failed at polycondensation at 180°C in $C_6H_5NO_2$ [15].

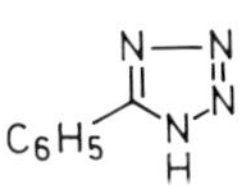

XXVII

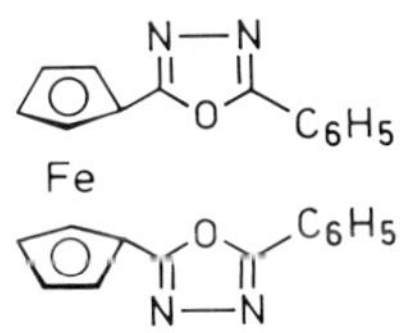

XXVIII

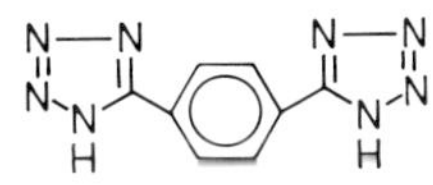

XXIX

XXVII in a 1/4 mole ratio forms XXVIII in xylene on an oil bath (bath temperature 140°C). A 1/1 mole ratio with XXIX in $C_6H_5N(CH_3)_2$ forms XXX in over 95% of the theoretical yield on warming for 25 min from 100 to 160°C [15].

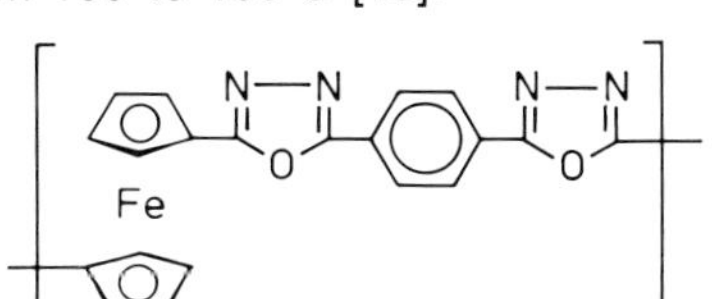

XXX

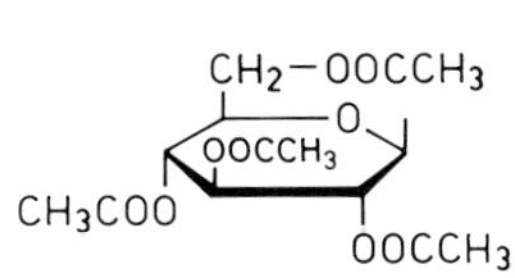

XXXI

References on pp. 369/70

With Other Compounds

Reaction with RSH, R = XXXI, yields (−)-$Fe(C_5H_4{-}COSR)_2$ [33]. The reaction with the carborane derivatives XXXIIa and b and subsequently H_2O leads to XXXIIIa and b, respectively. That with XXXIV, $ClSi(CH_3)_2OSi(CH_3)_2Cl$, and H_2O gives

$HO_2C{-}C_5H_4FeC_5H_4{-}CO{-}C_2B_{10}H_{10}{-}Si(CH_3)_2OSi(CH_3)_2{-}C_2B_{10}H_{10}{-}CO{-}C_5H_4FeC_5H_4{-}CO_2H$ [23].

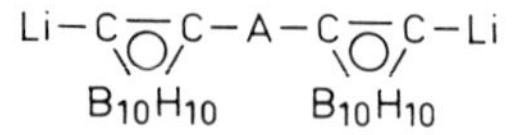

XXXII

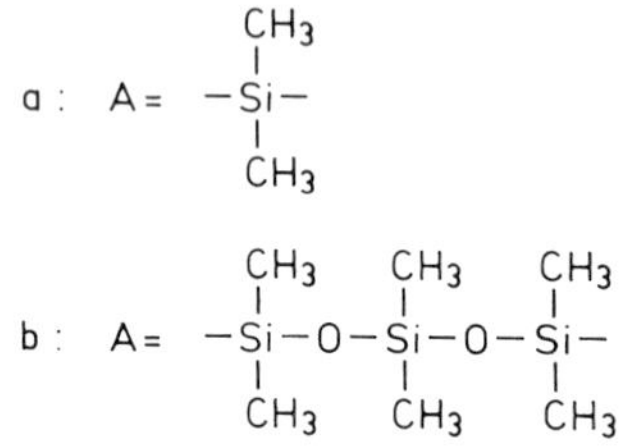

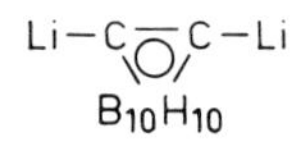

XXXIV

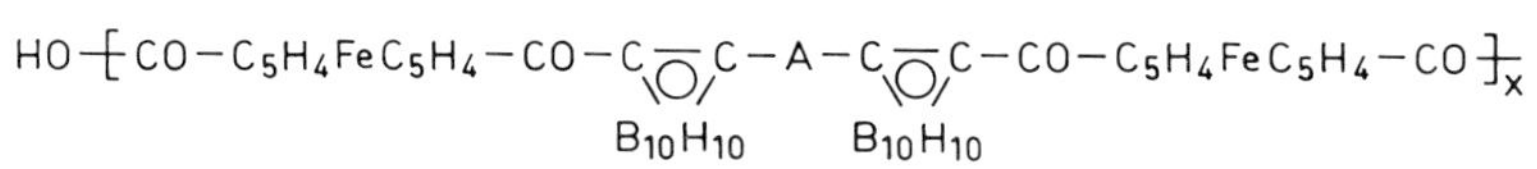

XXXIII

The action of $LiAlH(OC(CH_3)_3)_3$ in $CH_3O(CH_2)_2O(CH_2)_2OCH_3$, first at −78°C, then at room temperature, produces $Fe(C_5H_4{-}CHO)_2$ [2].

The Friedel-Crafts reaction with ferrocene and $AlCl_3$ in $CHCl_3$ yields XXXV (cf. "Ferrocene" 6, 1977, p. 152) in a low yield [8, 45].

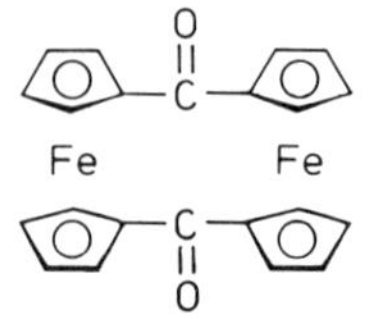

XXXV

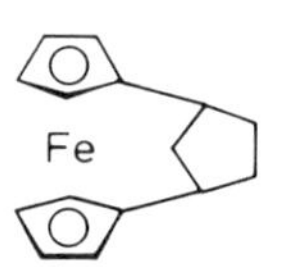

XXXVI

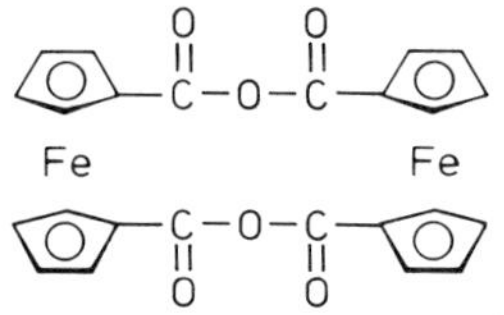

XXXVII

Polymeric products were obtained with $BF_3 \cdot O(C_2H_5)_2$ as catalyst in a molar ratio of 1/(1.0 to 1.5)/(>1 or >2) in sulfolane at 25 to 50°C. These consist essentially of $\{C_5H_4FeC_5H_4{-}CO\}_x$. Formation of side chains and networks are favored by higher temperatures and longer reaction times and similarly by a shift in favor of the acid chloride for the $Fe(C_5H_4{-}COCl)_2$/ferrocene ratio. A side reaction apparently involves a cleavage with formation of C_5H_5COCl, which is incorporated in the polymer as a chlorocarbonylcyclopentylene group. Ferrocene also undergoes a cleavage of Fe and ring, as established by the occurrence of XXXVI in the reaction product. A partial replacement of CO by CH_2 groups may be due to reduction of intermediately occurring acylium ions by ferrocene. With $ZnCl_2$ as catalyst the chain formation proceeds extraordinarily slowly and cross-linking takes place. In the absence of solvents such high temperatures are necessary for the condensation in the melt that only carbonized, insoluble products are formed [10].

The action of pyridine on $Fe(C_5H_4{-}COCl)_2$ in $CHCl_3$ forms XXXVII [1], cf. "Ferrocene" 6, 1977, p. 181.

Applications

Photocrosslinkable polyesters are obtained if instead of pure $ClCO(CH_2)_8COCl$ this with 12 to 15% $Fe(C_5H_4–COCl)_2$ is reacted with a 56/44 mixture of XXXVIII and XXXIXa or a 65/35 mixture of XXXVIII and XXXIXb in $ClCH_2CH_2Cl/Cl_2CHCHCl_2$ in the presence of $(C_2H_5)_3N$. The high molecular weight polymers are soluble in organic solvents and are used in photolithography as photoresists [27, 29].

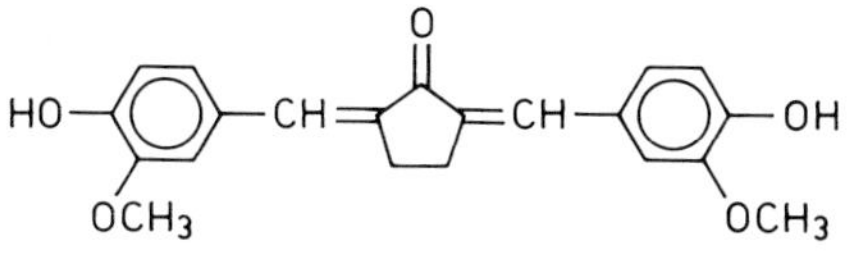

XXXVIII

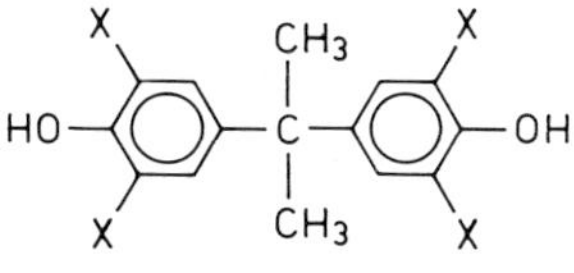

XXXIX

a: X = Cl
b: X = Br

Additions of 0.01 wt% $Fe(C_5H_4–COCl)_2$ increase the electron beam radiation sensitivity of positive working radiation resists consisting of methacrylic acid/methyl methacrylate in a 1/4 molar ratio [26] or of cellulose acetate solution coated on a Cr-coated glass support [25].

References:

[1] N. A. Nesmeyanov, O. A. Reutov (Dokl. Akad. Nauk SSSR **120** [1958] 1267/70; Proc. Acad. Sci. USSR Chem. Sect. **118/123** [1958] 503/6). – [2] S. I. Goldberg (J. Org. Chem. **25** [1960] 482/3). – [3] F. W. Knobloch, W. H. Rauscher (J. Polym. Sci. **54** [1961] 651/6). – [4] M. Okawara, Y. Takemoto, H. Kitaoka, E. Haruki, I. Imoto (Kogyo Kagaku Zasshi **65** [1962] 685/8 from C.A. **58** [1963] 577). – [5] W. A. Lee (RAE-TN-CPM-5 [1963] 1/10; X 63-13082 [1963] 1/10).

[6] R. L. Schaaf, C. T. Lenk (J. Chem. Eng. Data **9** [1964] 103). – [7] C. U. Pittman, J. A. Eikenberry (J. Alabama Acad. Sci. **36** [1965] 338/40). – [8] T. H. Barr, W. E. Watts (Chem. Soc. Meeting, Brighton 1966, Abstr. A 28 from W. E. Watts, Organometal. Chem. Rev. **2** [1967] 231/54, 246). – [9] I. A. Golubeva, T. P. Vishnyakova (Neftekhim. Pererabotka Nefti Gaza Temat. Analit. Obshch. Tsentre Nauchn. Issled. Inst. Tekhn. Eneftekhim. **1966** 36/42). – [10] E. W. Neuse, R. M. Trahe (J. Macromol. Chem. **1** [1966] 611/22).

[11] P. Petrovitch (Double Liaison No. 133 [1966] 1093/109). – [12] P. Petrovitch, H. Valot (Compt. Rend. C **263** [1966] 214/6). – [13] H. Valot (Compt. Rend. C **262** [1966] 403/5). – [14] I. A. Golubeva, T. P. Vishnyakova, N. N. Bul'on (Tr. Mosk. Inst. Neftekhim. Gazov. Prom. No. 72 [1967] 61/7; Ref. Zh. Khim. **1968** No. 13 Zh 519; C.A. **68** [1968] No. 87384). – [15] H.-J. Lorkowski, R. Pannier, A. Wende (J. Prakt. Chem. **35** [1967] 149/58).

[16] E. W. Neuse, H. Rosenberg (J. Polym. Sci. Polym. Chem. Ed. **6** [1968] 1567/82). – [17] V. G. Shershun, B. S. Sakharov, T. P. Vishnyakova, Ya. M. Paushkin (Tr. Mosk. Inst. Neftekhim. Gasov. Prom. No. 3 [1969] 48/54). – [18] H. Valot (Bull. Soc. Chim. France **1969** 1972/5). – [19] H. Imai (Kagaku [Kyoto] **25** [1970] 628/37 from C.A. **73** [1970] No. 109805). – [20] Yu. A. Krylov, A. A. Usvyatsov, V. P. Ivanov, E. G. Rozantsev (Izv. Akad. Nauk SSSR Ser. Khim. **1971** 2296/8; Bull. Acad. Sci. USSR Div. Chem. Sci. **1971** 2167/9).

[21] U. Mueller-Westerhoff (Tetrahedron Letters **1972** 4639/42). – [22] S. V. Sunthankar, S. T. Mahadik (Indian J. Chem. **11** [1973] 1207). – [23] S. P. Brown (AD-A000753-4-GA [1974] 1/23; C.A. **82** [1975] No. 171781). – [24] S. L. Sosin, V. P. Alekseeva, M. D. Litvinova, V. V. Korshak, A. F. Zhigach (Vysokomol. Soedin. B **18** [1976] 703/7). – [25] M. Miyamura, O. Tada, A. Miura, Tokyo Shibaura Electric Co., Ltd. (Japan. Kokai 77-146216 [1976/77] from C.A. **88** [1978] No. 201075).

[26] M. Miyamura, Tokyo Shibaura Electric Co., Ltd. (Japan. Kokai 77-146220 [1976/77] from C.A. **88** [1978] No. 201079). – [27] D. G. Borden (Org. Coat. Plast. Chem. **37** [1977] 72/8). – [28] C. E. Carraher (Interfacial Syn. **2** [1977] 367/416, 402/5). – [29] D. G. Borden (Organometal. Polym. Symp., New Orleans 1977 [1978], pp. 115/27). – [30] B. Nickel, A. Rassat (Tetrahedron Letters **1978** 633/6).

[31] A. M. Shevchik, Yu. P. Losev, Ya. M. Paushkin, R. V. Chibrikova, Zh. I. Zaremba (Vestsi Akad. Navuk Belarusk. SSR Ser. Khim. Navuk **1978** No. 4, pp. 114/6; C.A. **89** [1978] No. 163719). – [32] G. Oepen, F. Voegtle (Liebigs Ann. Chem. **1979** 1094/101). – [33] M. J. Adam, L. D. Hall (Can. J. Chem. **58** [1980] 1188/97). – [34] A. P. Bell, C. D. Hall (J. Chem. Soc. Chem. Commun. **1980** 163/5). – [35] V. P. Red'ko, V. K. Skubin, D. F. Kutepov, V. V. Korshak, V. G. Mochalov (Vysokomol. Soedin. B **22** [1980] 222/4).

[36] C. D. Hall, A. P. Bell, D. Z. Denney (Org. Magn. Resonance **15** [1981] 94/5). – [37] P. J. Hammond, P. D. Beer, C. D. Hall (J. Chem. Soc. Chem. Commun. **1983** 1161/3). – [38] P. J. Hammond, A. P. Bell, C. D. Hall (J. Chem. Soc. Perkin Trans. I **1983** 707/15). – [39] A. Kasahara, T. Izumi, K. Saito (Bull. Chem. Soc. Japan **56** [1983] 2865/6). – [40] C. Simionescu, T. Lixandru, L. Tătaru, I. Mazilu, M. Vâtă, S. Luca (J. Organometal. Chem. **252** [1983] C43/C46).

[41] P. D. Beer, J. Elliot, P. J. Hammond, C. Dudman, C. D. Hall (J. Organometal. Chem. **263** [1984] C37/C42). – [42] T. Izumi, T. Tezuka, S. Yusa, A. Kasahara (Bull. Chem. Soc. Japan **57** [1984] 2435/9). – [43] F. Voegtle (Chemiker-Ztg. **96** [1972] 396/403). – [44] F. Voegtle (Chem. Ind. [London] **1972** 346). – [45] T. H. Barr, H. L. Lentzner, W. E. Watts (Tetrahedron **25** [1969] 6001/13).

[46] D. Nielson, D. Boone, H. Eyring (J. Phys. Chem. **76** [1972] 511/5). – [47] C. U. Pittman (J. Polym. Sci. Polym. Chem. Ed. **6** [1968] 1687/95).

5.2.3.8.1.2 Other 1,1′-Compounds

The compounds of Table 29 are prepared from the carboxylic acids by the following methods:

Method I. Reaction with PCl_3

a) without solvent at room temperature [4, 8],

b) in benzene at an elevated temperature: 60°C [7], 75°C [5], the boiling point [6].

Method II. Reaction with ClCOCOCl in CH_2Cl_2 at room temperature [1].

Chemical Behavior

In most cases the reactive compounds are not isolated but are immediately further reacted, for example with NH_3 or RNH_2 to the acid amides. Thus, addition of No. 3 in ether solution to liquid NH_3 forms $Fe(C_5H_4{-}(CH_2)_3CONH_2)_2$ [7], reaction of No. 7 with 25% aqueous NH_3 solution gives $H_2NCO{-}C_5H_4FeC_5H_4{-}CO_2CH_3$ [2]. No. 1 yields with RNH_2 the $Fe(C_5H_4{-}CONHR)_2$ with R = I or II, No. 8 the 2-$RHNCOC_6H_4{-}C_5H_4FeC_5H_4{-}CH_2OCH_3$ with R = I [11]. For additional reactions see under "Further information" for Nos. 2, 5, and 7, pp. 371/2.

Applications

Additions at the 0.01 to 20 wt% level of the compounds No. 1 to 4 increase the sensitivity to electron beam radiation of positive working photoresists made of methacrylic acid/methyl methacrylate in a 1/4 molar ratio [10] or from cellulose acetate solutions coated on Cr-coated glass supports [9].

References on p. 372

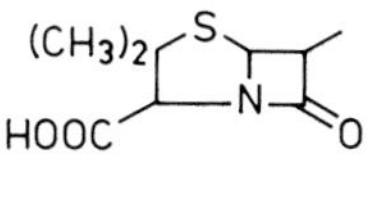

I

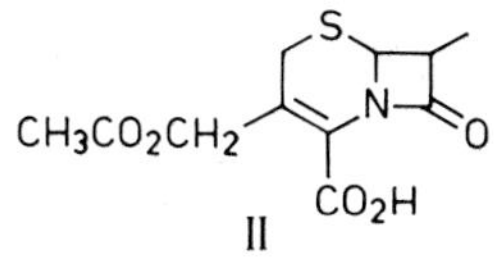

II

Table 29
Other Carboxylic Acid Chlorides of the Type $R^1–C_5H_4FeC_5H_4–R^2$.
Further information for numbers preceded by an asterisk is given at the end of the table.
For abbreviations and dimensions see p. VIII.

No.	$R^1–C_5H_4FeC_5H_4–R^2$ R^1	R^2	method of preparation, properties, and remarks	Ref.
compounds with $R^1 = R^2$:				
1	$CH_2–COCl$		I for "Chemical Behavior" and "Applications" see p. 370	[11]
*2	$(CH_2)_2–COCl$		Ia unstable brown oil IR (neat): 1800 (νC=O) for "Applications" see p. 370	[4,8]
3	$(CH_2)_3–COCl$		Ib for "Chemical Behavior" and "Applications" see p. 370	[7]
4	$(CH_2)_4–COCl$		for "Applications" see p. 370	[9, 10]
compounds with $R^1 \neq R^2$:				
*5	COCl	CH_3	Ib	[5]
6	COCl	$COCH_3$	II bright red crystals, m.p. 97.5 to 99.5° (cyclohexane/petroleum ether) IR (CCl_4): see figure in [1]	[1]
*7	COCl	CO_2CH_3	Ib	[2,5,6]
8	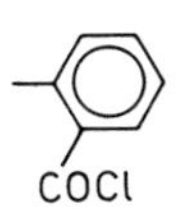	CH_2OCH_3	I for "Chemical Behavior" see p. 370	[11]

* Further information:

$Fe(C_5H_4–(CH_2)_2COCl)_2$ (Table **29**, No. **2**) undergoes a double intramolecular Friedel-Crafts reaction on dropping its CH_2Cl_2 solution into $AlCl_3$ in CH_2Cl_2 with ice cooling [3, 4, 8]. The chromatographic processing of the reaction products on SiO_2 columns with 10/1, 8/1, and 5/1 $C_6H_6/CH_3CO_2C_2H_5$ as eluent yields five bands, of which the first two yield two unknown products, the third and fourth compounds III (0.9% yield) and IV (33%), and the fifth V and/or VI (2.1%) [8]. Only IV (yield 37%) was isolated on Al_2O_3 by [3, 4]. Formation of small quantities of VII was not ruled out by [4].

References on p. 372

$ClOC{-}C_5H_4FeC_5H_4{-}CH_3$ (Table **29**, No. **5**) with $Cd(CH_3)_2$ forms $CH_3CO{-}C_5H_4FeC_5H_4{-}CH_3$ [5].

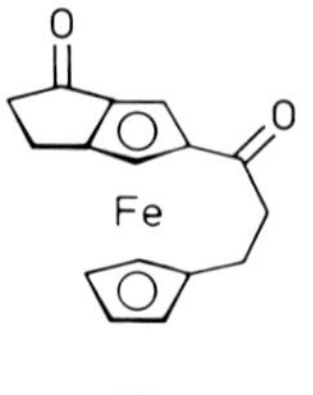

III

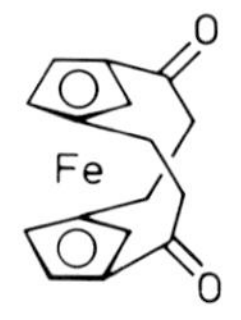

IV

$ClOC{-}C_5H_4FeC_5H_4{-}CO_2CH_3$ (Table **29**, No. **7**) yields on reduction with $LiAlH(OC(CH_3)_3)_3$ the $CH_3O_2C{-}C_5H_4FeC_5H_4{-}CHO$ [5]. The Friedel-Crafts reaction with ferrocene and $AlCl_3$ in CH_2Cl_2 gives $CH_3O_2C{-}C_5H_4FeC_5H_4{-}CO{-}C_5H_4FeC_5H_5$ [6]. For the reaction with NH_4OH see p. 370.

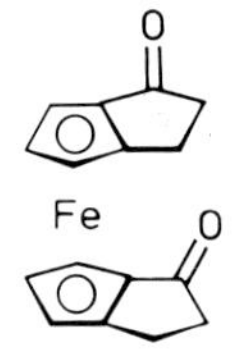

V

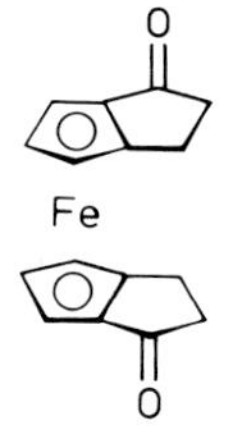

VI

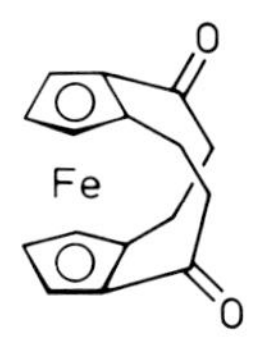

VII

References:

[1] M. Rosenblum (Diss. Harvard Univ. 1953, p. 119). – [2] N. A. Nesmeyanov, O. A. Reutov (Dokl. Akad. Nauk SSSR **120** [1958] 1267/70; Proc. Acad. Sci. USSR Chem. Sect. **118/123** [1958] 503/6). – [3] K. Schlögl, M. Perterlik (Tetrahedron Letters **1962** 573/6). – [4] K. Schlögl, M. Peterlik, H. Seiler (Monatsh. Chem. **93** [1962] 1309/27). – [5] K. Schlögl, H. Soukup (Monatsh. Chem. **99** [1968] 927/46).

[6] T. H. Barr, H. L. Lentzner, W. E. Watts (Tetrahedron **25** [1969] 6001/13). – [7] A. N. Nesmeyanov, M. I. Rybinskaya, G. B. Shul'pin, A. A. Pogrebnyak (J. Organomatel. Chem. **92** [1975] 341/67). – [8] M. Hisatome, T. Sakamoto, K. Yamakawa (J. Organometal. Chem. **107** [1976] 87/101). – [9] M. Miyamura, O. Tada, A. Miura, Tokyo Shibaura Electric Co., Ltd. (Japan. Kokai 77-146216 [1976/77] from C.A. **88** [1978] No. 201075). – [10] M. Miyamura, Tokyo Shibaura Electric Co., Ltd. (Japan. Kokai 77-146220 [1976/77] from C.A. **88** [1978] No. 201079).

[11] E. I. Edwards, R. Epton, G. Marr, G. K. Rogers, K. J. Thompson (Chem. Soc. [London] Spec. Publ. No. 28 [1977] 92/100; C.A. **88** [1978] No. 23001).

5.2.3.8.2 1,2-Compounds

$C_5H_5FeC_5H_3(COCl\text{-}1)CH_3\text{-}2$ is prepared from the acid in benzene with PCl_3 at 50 [2], 60 to 70°C [4]. Reaction with NH_3 in absolute C_6H_6 gives $C_5H_5FeC_5H_3(CONH_2\text{-}1)CH_3\text{-}2$ [2], with (+)- or (−)-$C_6H_5CH(CH_3)NH_2$ forms (1RS, 7R)- or (1RS, 7S)-$C_5H_5FeC_5H_3(CONH\overset{7}{C}H(CH_3)C_6H_5\text{-}1)CH_3\text{-}2$ [4]. Reaction with $C_5H_5FeC_5H_3(CO_2K\text{-}1)CH_3\text{-}2$ leads to the anhydride $(C_5H_5FeC_5H_3(CH_3\text{-}2){-}CO)_2O$ [2].

Optically Active Forms. (1S)-$C_5H_5FeC_5H_3(COCl\text{-}1)CH_3\text{-}2$ is obtained from (+)(1S)-$C_5H_5FeC_5H_3(CO_2H\text{-}1)CH_3\text{-}2$ in benzene with PCl_3 and a drop of pyridine [2, 3]; for the centrochiral R, S nomenclature used for the "ferrocene chirality" see "Ferrocene" 7, 1980, pp. 1/4 (English) and 4/7 (German). The compound reacts with NH_3 to (+)(1S)-$C_5H_5FeC_5H_3(CONH_2$-1)–CH_3-2 [3, 7], with NaN_3 in acetone to (1S)-$C_5H_5FeC_5H_3(CON_3$-1)CH_3-2 [7], with $Cd(CH_3)_2$ to (−)(1S)-$C_5H_5FeC_5H_3(COCH_3$-1)CH_3-2 [1, 3], and with (−)- or (+)-$C_6H_5CH(CH_3)NH_2$ to (1S, 7S)- or (1S, 7R)-$C_5H_5FeC_5H_3(CONH\overset{7}{C}H(CH_3)C_6H_5$-1)$CH_3$-2 (m.p. 122 to 126 and 150 to 155°C, respectively) [4].

$C_5H_5FeC_5H_3(CH_2COCl$-1)$(CH_2)_2C_6H_5$-2 is gotten from the acid in benzene with PCl_3 and traces of pyridine at 70 to 80°C after 90 min in a 98% yield. An oily substance remains after decanting the phosphoric acid and evaporating to dryness three times with C_6H_6. It cyclizes in CH_2Cl_2 with $AlCl_3$ catalyst to form I via an intramolecular Friedel-Crafts reaction [8].

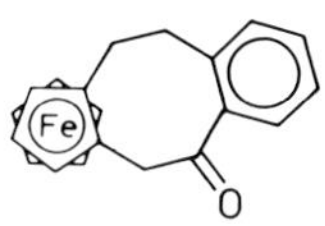

I

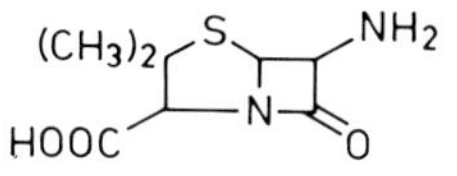

II

$C_5H_5FeC_5H_3(CH_2COCl$-1)X-2 with X = Cl, Br, or I. 6-Aminopenicillanic acid (≙ RNH_2, see Formula II) reacts with the acid chlorides won from $C_5H_5FeC_5H_3(CH_2CO_2H$-1)X-2 to form $C_5H_5FeC_5H_3(CH_2CONHR$-1)X-2 [6].

(1RS, 6SR)- and (1RS, 6RS)-$C_5H_5FeC_5H_3(\overset{6}{C}H(C_6H_5)CH_2COCl$-1)$CH_3$-2 are obtained from the corresponding acids (m.p. 148 and 188°C, respectively) by dissolution in PCl_3 at 0°C and then slowly warming to room temperature. They undergo the Arnd-Eistert reaction with CH_2N_2 in ether to form the diazoketones, which are converted by heating in collidine with $C_6H_5CH_2OH$ into the benzyl esters and are then subsequently saponified to form (1RS, 6SR)- and (1RS, 6RS)-$C_5H_5FeC_5H_3(\overset{6}{C}H(C_6H_5)CH_2CH_2CO_2H$-1)$CH_3$-2, respectively [5].

5.2.3.8.3 1,3-Compounds

$C_5H_5FeC_5H_3(COCl$-1)CH_3-3. Optically Active Forms. (1S)-$C_5H_5FeC_5H_3(COCl$-1)CH_3-3 is obtained from (+)(1S)-$C_5H_5FeC_5H_3(CO_2H$-1)CH_3-3 with PCl_3 and small amounts of pyridine in C_6H_6 [3, 7]. It gives on saturation with NH_3 of the C_6H_6 solution (1S)-$C_5H_5FeC_5H_3(CONH_2$-1)CH_3-3, with NaN_3 in acetone (1S)-$C_5H_5FeC_5H_3(CON_3$-1)CH_3-3 [7], and with $Cd(CH_3)_2$ in ether (+)(1S)-$C_5H_5FeC_5H_3(COCH_3$ 1)CH_3 3 [3].

References:

[1] H. Falk, K. Schlögl (Tetrahedron **22** [1966] 3047/53). – [2] H. Falk, K. Schlögl, W. Steyrer (Monatsh. Chem. **97** [1966] 1029/44). – [3] G. Haller, K. Schlögl (Monatsh. Chem. **98** [1967] 603/18). – [4] P. Reich-Rohrwig, K. Schlögl (Monatsh. Chem. **99** [1968] 1752/63). – [5] C. Moïse, Y. Mugnier, J. Tirouflet (J. Organometal. Chem. **51** [1973] 343/52).

[6] E. I. Edwards, R. Epton, G. Marr, G. K. Rogers, K. J. Thompson (Chem. Soc. [London] Spec. Publ. No. 28 [1977] 92/100, C.A. **88** [1978] No. 23001). – [7] V. Rapič, K. Schlögl, B. Steinitz (Monatsh. Chem. **108** [1977] 767/80). – [8] M. Benedikt, K. Schlögl (Monatsh. Chem. **109** [1978] 805/22).

5.2.3.9 Heterocyclic Compounds with Oxygen as Hetero Atom

5.2.3.9.1 1,1′-Compounds

5.2.3.9.1.1 Epoxyethyl Derivatives

All compounds in Table 30 contain the OH_3C_2 group, see Formula I.

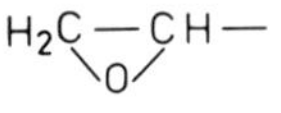

I

They are prepared by the following methods:

Method I. $Fe(C_5H_4$–$CH{=}CH_2)_2$ is oxidized with H_2O_2 [6].

Method II. $Fe(C_5H_4$–CH_2CHOH–$CH_2Cl)_2$ is epoxidized in C_2H_5OH with 30% KOH, for

a) 12 min at 0 to 10°C,

b) 30 min at 0 to 10°C, then 45 min at room temperature.

Isolation takes place by column chromatography on Al_2O_3 in benzene, with elution first of No. 2 with benzene and then No. 6 with 1/1 benzene/ether [5].

Method III. a) The product of lithiation of ferrocene with n-C_4H_9Li in ether is reacted with $OH_3C_2CH_2Cl$ (in a 1/5 molar ratio of ferrocene to $OH_3C_2CH_2Cl$) initially for 7 h at −65°C, subsequently with raising of the reaction temperature to −40°C over 16.5 h. The reaction product isolated after hydrolysis is dissolved in a little benzene and separated chromatographically on an Al_2O_3 column. Following the recovered ferrocene, C_6H_6 elutes in succession the $C_5H_5FeC_5H_4$–$C_5H_4FeC_5H_5$, $C_5H_5FeC_5H_4$–$CH_2C_2H_3O$, No. 2, and $C_5H_5FeC_5H_4$–$CH_2CHOHCH_2Cl$, then 1/1 C_6H_6/ether elutes No. 5, and finally 9/1 ether/CH_3OH the $Fe(C_5H_4$–$CH_2CHOHCH_2Cl)_2$. The respective fractions in the product are 1.1, 16.7, 10.0, 18.0, 20.0, and 32.2% [5].

b) The reaction of $Fe(C_5H_4$–$Li)_2$ with $OH_3C_2CH_2Cl$ is conducted at −40°C for ca. 100 h [10].

Method IV. $Fe(C_5H_4$–$CH_2OH)_2$ is reacted with n-C_4H_9Li in tetrahydrofuran with cooling, and the reaction mixture is refluxed for 6 h with $OH_3C_2CH_2Br$ in the presence of $(CH_3)_2SO$ [12].

Method V. $Fe(C_5H_4$–$COCl)_2$ and $OH_3C_2CH_2OH$ are reacted in ether by use of Na [10].

Method VI. $Fe(C_5H_4$–$CHOHCH_3)_2$ is reacted with Na in benzene or ether for 1.5 h at room temperature and 4 h at the boiling point and then with $OH_3C_2CH_2Cl$ for 2.5 h at boiling temperature [9].

References on p. 378

Table 30

Epoxyethyl Derivatives of the Type $OH_3C_2–A_n–C_5H_4FeC_5H_4–R^2$ ($OH_3C_2–A_n = R^1$; n = 0, 1).
Further information for numbers preceded by an asterisk is given at the end of the table.
For abbreviations and dimensions see p. VIII.

No.	$R^1–C_5H_4FeC_5H_4–R^2$ $R^1 = A_n–C_2H_3O$	 R^2	method of preparation (yield in %), properties, and remarks	Ref.
compounds with $R^1 = R^2$:				
*1	C_2H_3O		I	[6]
*2	$CH_2–C_2H_3O$		IIa (92.0), IIb (66.9), IIIa, IIIb (44), see No. 5, p. 377 dark red, slightly viscous liquid, b.p. 160 to 161°/0.09 to 0.12 Torr n_D^{20}: 1.5968 ^{1}H NMR ($CDCl_3$): 2.6 (m, $CH_2 + CH_2$ of C_2H_3O), 3.0 (m, CH of C_2H_3O), 4.04 (s, C_5H_4) IR: (neat): 845, 900 to 960, 1256 (C_2H_3O); 2970 and 3050 (νC–H of C_2H_3O) IR: 835 and 1250 (C_2H_3O); 2540 to 2870 (CH_2); 3100 (C_5H_4)	[1, 5, 10]
*3	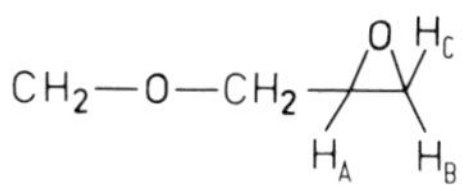		IV (80) viscous liquid ^{1}H NMR ($CDCl_3$): 2.66 (m, H_A), 3.10 (q, H_B), and 3.50 (q, H_C) (ABC pattern of C_2H_3O); 4.10 (CH_2OCH_2), 4.30 (s, C_5H_4) IR(neat): 740(m), 820(m), 840(s), 900(m), 1000(m), 1040 (m), 1090 (s), 1240 (s), 1350 (m), 1385 (m), 1470 (s), 1640 (w), 2875 (s), 2950 (s), 3100 (m)	[12]
*4	$CO_2CH_2–C_2H_3O$		V (91, 57) deep orange crystals, m.p. 38 to 40° (ether), 51.5 to 52.5°, 52 to 53° (in N_2; 1/1 hexane/toluene) ^{1}H NMR: 2.9 to 3.3 and 3.5 to 3.8 (C_2H_3O), 5.0 to 5.1 (C_5H_4) IR: 840, 920, and 1280 (C_2H_3O); 1140 (C–O), 1720 (C=O), 3100 (C–H of C_5H_4) IR (Nujol): see figure in [2] UV (cyclohexane): λ_{max} (log ε) = 219 (4.52), 255 (3.95), 450 (2.41)	[1 to 4, 10]

References on p. 378

Table 30 (continued)

No.	R^1–$C_5H_4FeC_5H_4$–R^2 $R^1 = A_n$–C_2H_3O	R^2	method of preparation (yield in %), properties, and remarks	Ref.
compounds with $R^1 \neq R^2$:				
*5	CH_2–C_2H_3O	$CH_2CHOHCH_2Cl$	IIIa dark red, medium viscous liquid n_D^{20}: 1.6014	[5]
6	CH_2–C_2H_3O	CH_2CHOH– $CH_2OC_2H_5$	IIb (6.4), see No. 2, below, and No. 5, p. 377 dark red, slightly viscous liquid n_D^{20}: 1.5711 stable in air	[5]
*7	$CH(CH_3)OCH_2$– C_2H_3O	$CHOHCH_3$	VI (<90) viscous, orange product IR: 840 and 1245 (C_2H_3O); 1110 (CH_2OCH), 3400 (OH) mol. wt: 343.3 (cryoscopically in C_6H_6)	[9]

* Further information:

$Fe(C_5H_4$–$C_2H_3O)_2$ (Table **30**, No. **1**) can be detected in small amounts by thin layer chromatography in the product of the reaction of $Fe(C_5H_4$–$CHOHCH_3)_2$ with $HOCH_2CH_2OH$ in the presence of $BF_3 \cdot O(C_2H_5)_2$ at 50°C [7]. No. 1 is a combined burning-rate catalyst and curing agent for solid composite propellants containing a CO_2H-terminated polybutadiene prepolymer, tris(methylaziridinyl)phosphine oxide, iron lineolate, NH_4ClO_4, and Al [6].

$Fe(C_5H_4$–$CH_2C_2H_3O)_2$ (Table **30**, No. **2**). Gravimetry shows a weight loss of 5% at 170°C in air or at 250°C in Ar [10]. No. 2 reacts with KOH in C_2H_5OH (mole ratio 1/~8.5/~400) to form No. 6, longer reaction times giving higher yields. Reaction for 25 min with and 40 min without cooling gives a 6% yield with 81.1% recovery of No. 2. No. 2 elutes first with benzene from an Al_2O_3 chromatographic column, then No. 6 with 1/1 benzene/ether [5]. Solventless copolymerization with equivalent quantities of the compounds II to V in inert atmospheres leads to dark red oligomers, which are well soluble in acetone and in part also in benzene or tetrahydrofuran: With II in 1 h at 50°C a product is obtained in a 30% yield that is up to ~70% crosslinked, with an average molecular weight (av. mol. wt) of 650. III gives in 6.0 h at 100°C in 96.5% yield a product (av. mol. wt 540) that contains no free epoxy groups. IV and V are especially reactive, giving after only 18 min at 20°C in 99% yields oligomers with av. mol. wts of 1250 and 720, respectively [10, 11]. No. 2 copolymerizes readily with a conventional, benzyldimethylamine-accelerated methyl nadic anhydride-crosslinked EPON 825/1031 system, thus incorporating Fe with great effectiveness into an epoxy matrix [8].

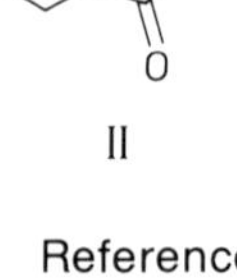

II

III

References on p. 378

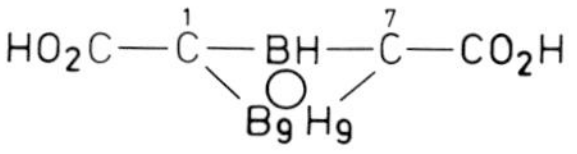

IV

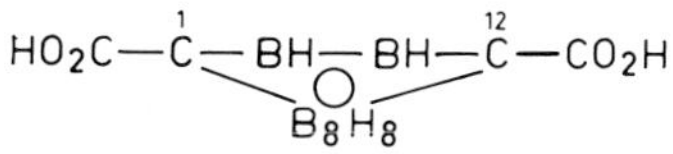

V

$Fe(C_5H_4–CH_2OCH_2C_2H_3O)_2$ (Table **30**, No. **3**) behaves as a bonding agent for composite solid propellants and also as a crosslinking agent for the binder CO_2H-terminated polybutadiene (CTPB). It is a better alternative to the usually used VI for curing the CTPB in composite propellants [12].

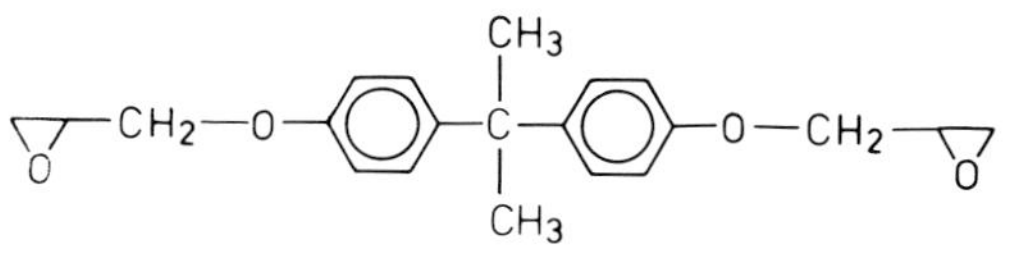

VI

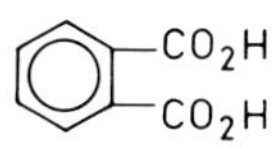

VIII

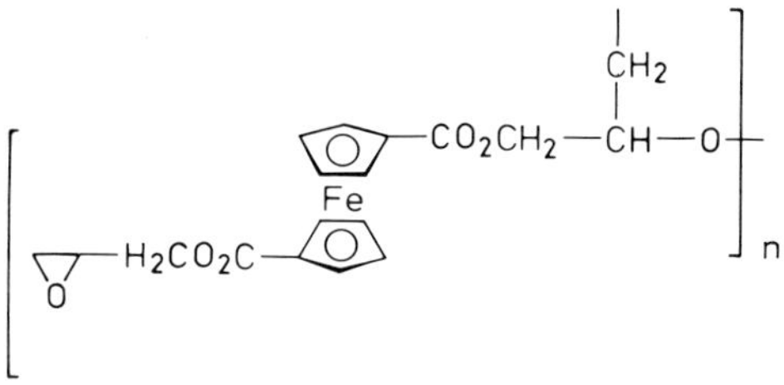

VII

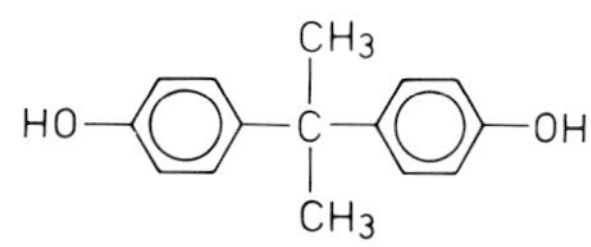

IX

$Fe(C_5H_4–CO_2CH_2C_2H_3O)_2$ (Table **30**, No. **4**) is purified chromatographically on Florisil and by careful recrystallization from 1/1 hexane/toluene [4]. It crystallizes in the monoclinic space group $P2_1/c-C_{2h}^5$ (No. 14) with a = 11.94, b = 5.96, and c = 23.88 Å, β = 102°; Z = 4. The calculated density is 1.543 g/cm³ [3].

No. 4 polymerizes in the presence of $(C_2H_5)_3N$ in 38% yield to the homopolymer VII, which is soluble in acetone and still contains 14% free epoxy groups [10]. Copolymerization without solvent with equivalent quantities of the compounds II, III, VIII, or IX in inert atmospheres with warming produces oligomers well soluble in acetone: II gives in 8 h at 80°C in 29% yield a product with av. mol. wt 2400, 70% crosslinked. III in 2 h at 130°C (in 91% yield; av. mol. wt of 16000 [11]) or in 4 h at 80°C (in 80% yield) gives a product with 10 to 20% free epoxy groups (relative to those in No. 4). VIII gives in 2 h at 100°C or 1 h at 50°C a yield of 47.5 or 37%, respectively. The latter product has an av. mol. wt of 500 [10, 11]. In 6 h at 100°C with IX is won in 100% relative yield an oligomer which contains no free OH_3C_2 groups [10].

$OH_3C_2CH_2–C_5H_4FeC_5H_4–CH_2CHOH–CH_2Cl$ (Table **30**, No. **5**) is stable in air. The epoxidation in C_2H_5OH with 30% KOH (mole ratio 1/68/9), first with cooling, then with vigorous stirring and letting the temperature rise over 45 min from 10 to 25°C, leads to No. 2 (77.3% yield) and No. 6 (11.6%) [10]. Separation is described in connection with No. 2, p. 376.

$OH_3C_2CH_2OCH(CH_3)–C_5H_4FeC_5H_4–CHOH–CH_3$ (Table **30**, No. **7**) polymerizes in C_6H_6 in the presence of 1 to 1.5% BF_3 to $\{OCH(CH_3)–C_5H_4FeC_5H_4–CH(CH_3)OCH_2CHOHCH_2\}_n$, whose low thermal stability can be raised by crosslinking with p-phenylenediamine or II. Hardening of Novolak with No. 7 gives an insoluble resin with high thermal stability [9].

References on p. 378

References:

[1] K. Hata, I. Motoyama, H. Watanabe (Bull. Chem. Soc. Japan **36** [1963] 1698/9). – [2] R. L. Schaaf, K. L. Rinehart (WADD-TR-61-108-Pt. III [1963] 1/80; AD-403929 [1963] 1/80, 50). – [3] D. W. Fischer (Acta Cryst. **17** [1964] 619). – [4] R. L. Schaaf, C. T. Lenk (J. Chem. Eng. Data **9** [1964] 103). – [5] H. Watanabe, I. Motoyama, K. Hata (Bull. Chem. Soc. Japan **39** [1966] 784/90).

[6] D. C. Sayles (U.S. 3762969 [1968/73]; C.A. **80** [1974] No. 61706). – [7] F. M. Dewey (U.S. 3598850 [1969/71]; C.A. **75** [1971] No. 152333). – [8] E. W. Neuse (J. Mater. Sci. **7** [1972] 708/9). – [9] M. D. Galkina, S. L. Sosin, V. V. Korshak (Vysokomol. Soedin. B **15** [1973] 164/7). – [10] S. L. Sosin, V. P. Alekseeva, M. D. Litvinova, V. V. Korshak, A. F. Zhigach (Vysokomol. Soedin. B **18** [1976] 703/7).

[11] V. V. Korshak, S. L. Sosin (Organomet. Polym. Symp., New Orleans 1977 [1978], pp. 25/38; C.A. **89** [1978] No. 6573). – [12] K. Kishore, V. R. P. Verneker, G. V. Dharumaraj (J. Polym. Sci. Polym. Letters **22** [1984] 607/10).

5.2.3.9.1.2 With Five-Membered Heterocycles

2-Furyl derivatives (Nos. 3, 5 to 10), the lactone No. 1, the maleic acid anhydride derivative No. 2, and the dioxolane No. 4 are collected in Table 31. The compounds are prepared by the following methods:

Method I. Catalytic hydrogenation

a) of $Fe(C_5H_4{-}CO(CH_2)_2CO_2H)_2$ in C_2H_5OH over RuO_2 at 85°C and 165 atm H_2 pressure (16.7 MPa) [1],

b) of $Fe(C_5H_4{-}CO(CH_2)_2CO_2C_2H_5)_2$ over Raney Ni at 75°C and 135 atm H_2 (13.7 MPa) [1].

Method II. Diels-Alder reaction of I with maleic acid anhydride in boiling benzene. Only the endo isomer forms [3].

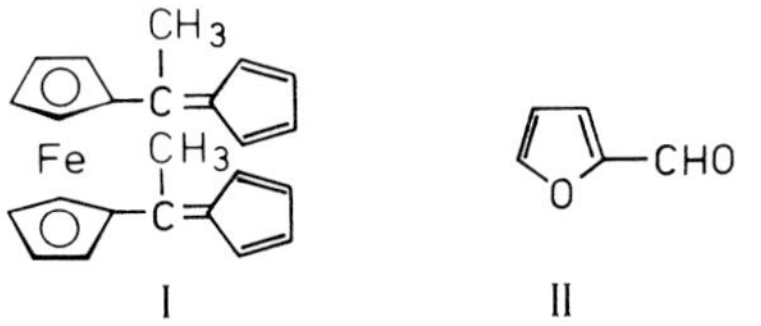

I II

Method III. Alkali-catalyzed aldol condensation with II:

a) The solution of $Fe(C_5H_4{-}COCH_3)_2$ and II in 95% C_2H_5OH is added to a solution at 0°C of NaOH in 2/1 H_2O/C_2H_5OH and the mixture is cooled for a further hour with stirring [13, 15].

b) Aqueous 30% NaOH is added dropwise to the solution of $Fe(C_5H_4{-}COCH_3)_2$ and II in CH_2Cl_2 and the mixture subsequently refluxed for 6.5 h [2].

c) The solution of $C_6H_5CO{-}C_5H_4FeC_5H_4{-}COCH_3$ and II in ethanol/water is reacted for 2.5 h with 10% NaOH and the mixture then stored for 2 h in a refrigerator [4].

References on p. 382

Method IV. $Fe(C_5H_4–COCH_3)_2$ and epichlorhydrin in CCl_4 are reacted dropwise with $SnCl_4$ in CCl_4 [9].

Method V. Friedel-Crafts reaction

a) of $C_5H_5FeC_5H_4–R^1$ with $CH_3COCl/AlCl_3$ (mole ratio 1/~1.5/3) in CH_2Cl_2 at room temperature [6, 7],

b) of $C_5H_5FeC_5H_4–CH_2–C_4H_3O$ with $(CH_3CO)_2O$ using 85% H_3PO_4 (mole ratio 1/6/3.3), in benzene first at room temperature and then boiling [7].

The compounds are isolated chromatographically on Al_2O_3 columns with 97/3 $C_6H_6/CH_3CO_2C_2H_5$ as eluent [6, 7].

Table 31

Heterocyclic Compounds of the Type $R^1–C_5H_4FeC_5H_4–R^2$.
Further information for numbers preceded by an asterisk is given at the end of the table, pp. 381/2.
For abbreviations and dimensions see p. VIII. R_f values for thin layer chromatography.

No.	$R^1–C_5H_4FeC_5H_4–R^2$ R^1 R^2	method of preparation (yield in %), properties, and remarks	Ref.
compounds with $R^1 = R^2$:			
1	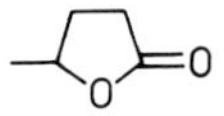	Ia (17.5), Ib golden yellow crystals ($CHCl_3 + CH_3OH$), m.p. 165° dec.	[1]
2	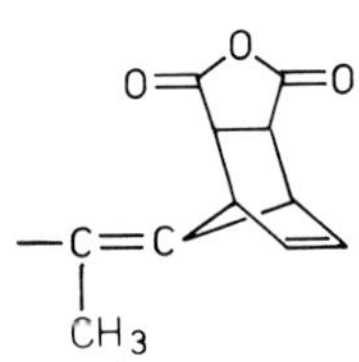 endo isomer	II (10.7) yellow crystals, m.p. 176 to 180° (1/1 dioxane/CH_3OH) IR (KBr): 1790 and 1855 (νC=O)	[3]
*3	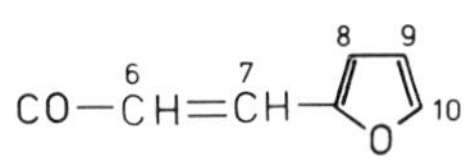	IIIa, IIIb (52.5) brick red product, m.p. 147 to 150°, 160°, 162° (CH_3CO_2H/H_2O) ^1H NMR ($CDCl_3$): 4.50 (H-3,4) and 4.92 (H-2,5) (C_5H_4); 6.42 (H-9), 6.63 (H-8), 6.87 (H-7), 7.35 (H-10), 7.45 (H-6) IR: 830 to 855, 1090, 1190, 1380, 1450, and 1545 (C_5H_4); 680, 735 to 745, 1010 to 1030, 1140, and 1240 to 1290 (C_4H_3O), 1590 (νC=C), 1650 (νC=O)	[2, 12 to 15]
4	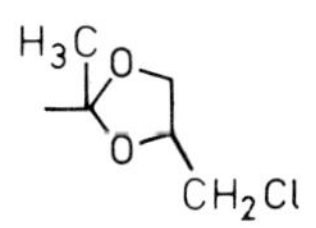	IV yellow orange crystals, m.p. 39 to 40° hydrolyzes easily with acids, very stable against saponification	[9]

References on p. 382

Table 31 (continued)

No.	$R^1–C_5H_4FeC_5H_4–R^2$ R^1	R^2	method of preparation (yield in %), properties, and remarks	Ref.
compounds with $R^1 \neq R^2$:				
5	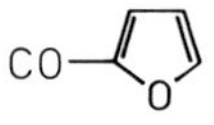	$COCH_3$	Va (65) m.p. 89° (acetone/petroleum ether) IR (paraffin oil): 1024 (δC–H of C_5H_4), 1117 (ν_sC(1′)–C_{acetyl}) UV (C_2H_5OH): λ_{max} (log ε) = 220 (4.52), 294 (4.41), 370 (3.50) R_f: 0.18 (on SiO_2 with 4/1 $C_6H_6/CH_3CO_2C_2H_5$)	[5, 7]
6	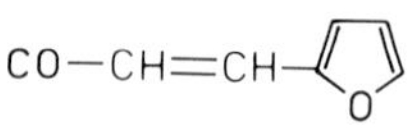	$COCH_3$	Va (1.4) m.p. 106 to 107° (acetone/petroleum ether) IR (paraffin oil): 747 (γC–H of C_4H_3O), 1008 (δC–H of C_5H_4), 1117 (ν_sC(1′)–C_{acetyl}), 1655 (νC=O) UV (C_2H_5OH): λ_{max} (log ε) = 346 (4.46) UV (n-heptane): λ_{max} (log ε) = 323 (4.77) R_f: 0.28 (on SiO_2 with 4/1 $C_6H_6/CH_3CO_2C_2H_5$) R_f: 0.17 (on Al_2O_3 with 19/1 $C_6H_6/CH_3CO_2C_2H_5$)	[5, 6]
*7	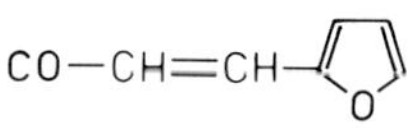	COC_6H_5	IIIc (31.7) red to red-violet compound, m.p. 81 to 83° (C_2H_5OH)	[4]
8	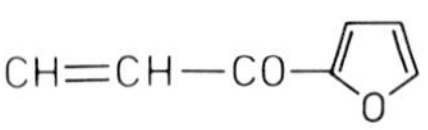	$COCH_3$	Va (29) m.p. 138 to 139° (C_6H_6) IR (paraffin oil): 773 (γC–H of C_4H_3O), 1015 (δC–H of C_5H_4), 1118 (ν_sC(1′)–C_{acetyl}); 1650 and 1663 (νC=O) UV (C_2H_5OH): λ_{max} (log ε) = 286 (4.38), 343 (4.35) UV (n-heptane): λ_{max} (log ε) = 260 (4.34), 315 (4.78) R_f: 0.23 (on SiO_2 with 4/1 $C_6H_6/CH_3CO_2C_2H_5$) R_f: 0.12 (on Al_2O_3 with 19/1 $C_6H_6/CH_3CO_2C_2H_5$)	[5, 6]
*9	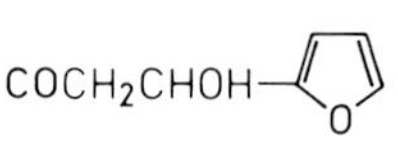	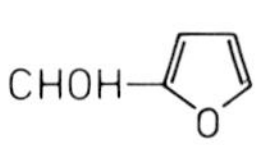	see "Further information", p. 382	[10,11]
10	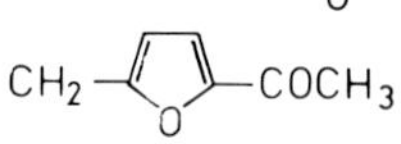	$COCH_3$	Vb (4) m.p. 58 to 59° (acetone/petroleum ether) IR (paraffin oil): 1022 (δC–H of C_5H_4), 1122 (ν_sC(1′)–C_{acetyl}) UV (C_2H_5OH): λ_{max} (log ε) = 226 (4.49), 280 (4.54), 335 (3.44) R_f: 0.12 (on SiO_2 with 4/1 $C_6H_6/CH_3CO_2C_2H_5$)	[5, 7]

References on p. 382

* Further information:

$Fe(C_5H_4\text{–}COCH{=}CHC_4H_3O)_2$ (Table **31**, No. **3**) is soluble in C_6H_6, CH_2Cl_2, $CHCl_3$, dioxane, acetone, CH_3OH, C_2H_5OH, or CH_3COOH, in contrast insoluble in water [13].

The compound exhibits a higher thermal stability than $Fe(C_5H_4\text{–}COCH_3)_2$, which might be due to the stabilization of the two C(1)–CO bonds owing to the expansion of the electron conjugation [12]. On heating in air at 10 K/min the decomposition proceeds in a single stage beginning at 230 and ending at 660°C, with a maximum rate constant at 350°C. The reaction is of 0.20 order. The weight loss amounts to 84.0%, the activation energy is 7.4 kcal/mol (31 kJ/mol) [12, 13]. The differential thermal analysis curve shows an exothermal process without weight loss at the melting point. It might be assumed that an exothermal tranfusion takes place simultaneously with and prevails over the endothermal melting [12].

No. 3, being more reactive than $Fe(C_5H_4\text{–}COCH_3)_2$, gives larger conversion yields than the latter compound in condensation reactions. Polycondensation with biuret in a 1/1 molar ratio in dioxane or $(CH_3)_2SO$ in the presence of $ZnCl_2$ as a dehydrating agent or in acetic acid, which is simultaneously a solvent and a condensation agent, leads to III. Raising the temperature and extending the reaction time favors the formation of insoluble polymeric products with three-dimensional structures, see Formulas IIIa, b, c, p. 34, with $R{=}CH{=}CHC_4H_3O$ and A = CO–NH–CO. The brown to black products possess semiconductor properties and good thermal stability up to 225°C [12, 13, 15]. Polycondensation with the aromatic diamines $H_2N\text{–}A\text{–}NH_2$ (A = 4,4′-$C_6H_4\text{–}C_6H_4$ [12, 15], 4,4′-$C_6H_4\text{–}S\text{–}C_6H_4$, 4,4′-$C_6H_3(NO_2\text{-}2)\text{–}S\text{–}S\text{–}(2\text{-}O_2N)C_6H_3$ [12, 14, 15]) in a 1/1 mole ratio with $ZnCl_2$ as catalyst leads to the paramagnetic polyazomethines. If the reaction is conducted at 150 or 200°C in the melt, only insoluble polymers with three-dimensional structures are formed (see Formulas IIIa and b, p. 34, with $R{=}CH{=}CH\text{–}C_4H_3O$). Similar results are found in dioxane at 120°C with reaction times of 20 h or more. With a reaction time of 10 h, the polymer product in contrast also contains the soluble IV [12, 14, 15]. IV with A = 4,4′-$C_6H_3(NO_2\text{-}2)\text{–}S\text{–}S\text{–}(2\text{-}O_2N)C_6H_3$ occurs on refluxing equimolar quantities of No. 3 and V in C_6H_6 with pyridine, which activates the O=S=N- group by formation of a nucleophile. The reaction mechanism corresponds to that given for $Fe(C_5H_4\text{–}COCH_3)_2$ on p. 35 [12, 15]. Prolonging the reaction time here also leads to crosslinked structures [12, 14, 15].

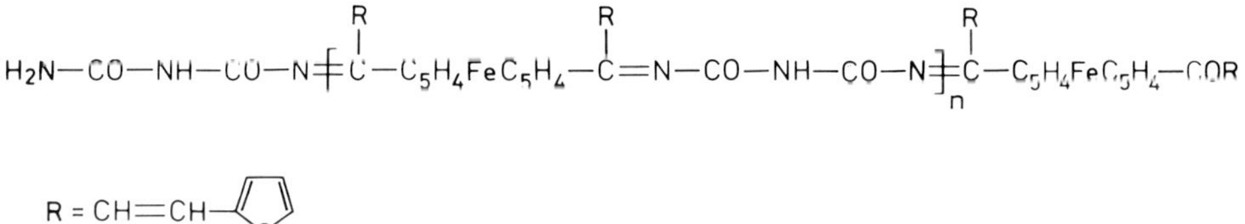

III

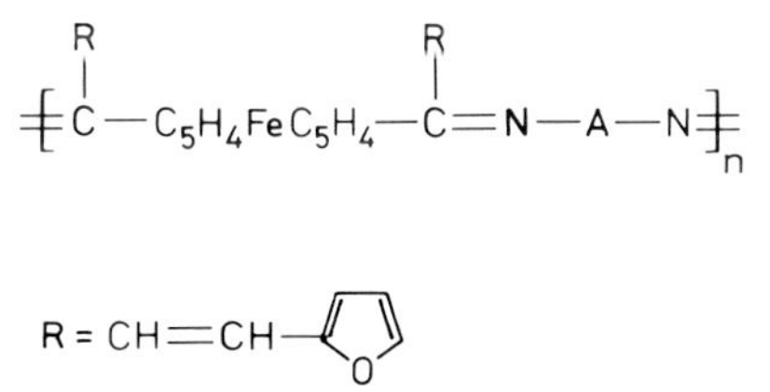

IV

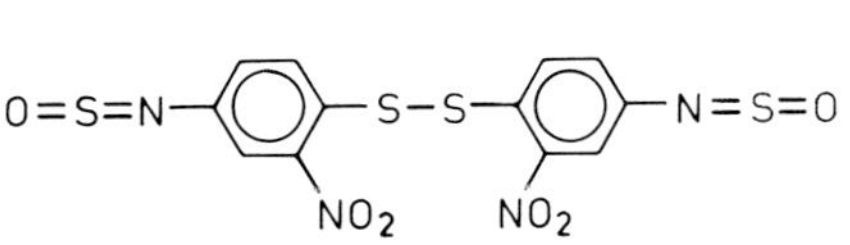

V

References on p. 382

OH_3C_4–CH=CH–CO–$C_5H_4FeC_5H_4$–COC_6H_5 (Table **31**, No. **7**) forms a hydrogen bond with phenol. The intermolecular hydrogen bonding enthalpy at room temperature amounts to −4.8 kcal/mol (−20.1 kJ/mol). The basicity of No. 7 is larger than that for the analogous compounds that contain instead of the 2-furyl the 2-thienyl or even a phenyl group, but smaller than that with a ferrocenyl group [8].

OH_3C_4–$CHOHCH_2CO$–$C_5H_4FeC_5H_4$–CHOH–C_4H_3O (Table **31**, No. **9**) probably occurs as an intermediate product during the polycondensation of $C_5H_5FeC_5H_4$–$COCH_3$ with II (p. 378) in the presence of H_2SO_4 at 45°C. It reacts with II and further with the already existing No. 9 to form VI [10, 11].

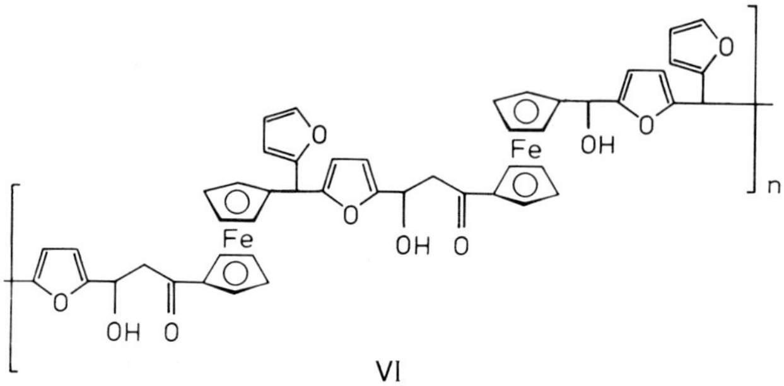

VI

References:

[1] P. J. Graham, R. V. Lindsey, G. W. Parshall, M. L. Peterson, G. M. Whitman (J. Am. Chem. Soc. **79** [1957] 3416/20). – [2] V. N. Kotrelev, S. P. Kalinina, G. I. Kuznetsova (Plasticheskie Massy **1961** No. 3, pp. 24/6; Soviet Plast. **1961** No. 3, pp. 22/3). – [3] M. Furdík, Š. Toma, J. Suchý, M. Dzurilla (Chem. Zvesti **16** [1962] 719/40). – [4] Š. Toma (Chem. Zvesti **19** [1965] 703/10). – [5] A. Perjéssy, Š. Toma (Chem. Zvesti **23** [1969] 533/9).

[6] Š. Toma (Collection Czech. Chem. Commun. **34** [1969] 2235/48). – [7] Š. Toma, E. Kalužayová (Chem. Zvesti **23** [1969] 540/52). – [8] Š. Toma, A. Perjéssy (Chem. Zvesti **23** [1969] 343/51). – [9] Ya. M. Paushkin, A. M. Shevchik (Vestsi Akad. Navuk Belarusk.SSR Ser. Khim. Navuk **1973** 95/8). – [10] M. Omori, M. Kurono, S. Yajima (Bull. Chem. Soc. Japan **48** [1975] 1291/5).

[11] S. Yajima (Zairyo Kagaku **14** [1977] 335/9; C.A. **90** [1979] No. 73945). – [12] L. Tătaru, M. Dărânga, M. Vâţă, T. Lixandru, C. Simionescu (J. Therm. Anal. **22** [1981] 259/68). – [13] L. Tătaru, I. Mazilu, M. Vâţă, T. Lixandru, C. Simionescu (J. Organometal. Chem. **214** [1981] 107/14). – [14] L. Tătaru, M. Vâţă, I. Mazilu, T. Lixandru, C. Simionescu (Angew. Makromol. Chem. **101** [1981] 19/27). – [15] C. Simionescu, T. Lixandru, L. Tătaru, I. Mazilu, M. Vâţă (J. Organometal. Chem. **238** [1982] 363/81).

5.2.3.9.2 1,2-Compounds

The Compounds of Table 32 are prepared by the following methods:

Method I. Friedel-Crafts acetylation of $C_5H_5FeC_5H_4$–R^1 with $CH_3COCl/AlCl_3$ (mole ratio 1/1.6/3) in CH_2Cl_2 at room temperature. No. 1 is isolated by column chromatography on Al_2O_3 with 97/3 $C_6H_6/CH_3CO_2C_2H_5$ following the monosubstituted ferrocene derivative and before its 1,1′-isomer [2].

References on p. 384

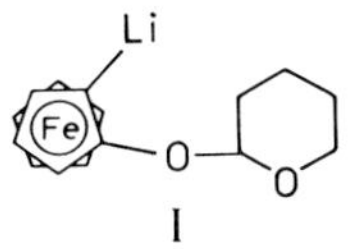

I

Method II. Reaction of I

a) with paraformaldehyde in ether [4],

b) by condensation with $HCON(CH_3)_2$ [4],

c) by slow pouring onto solid CO_2 in ether [4].

The compounds No. 3 to 6 yield on hydrolysis in C_2H_5OH with 2 M HCl the potentially pharmacologically active ferrocene derivatives $C_5H_5FeC_5H_3(OH\text{-}1)R^2\text{-}2$ [4].

Table 32

Heterocyclic Compounds of the Type $C_5H_5FeC_5H_3(R^1\text{-}1)R^2\text{-}2$.
Further information for numbers preceded by an asterisk is given at the end of the table.
For abbreviations and dimensions see p. VIII. R_f values for thin layer chromatography.

No.	$C_5H_5FeC_5H_3(R^1\text{-}1)R^2\text{-}2$ R^1	R^2	method of preparation (yield in %), properties, and remarks	Ref.
1	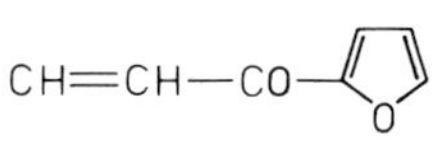	$COCH_3$	I (7.2) m.p. 113 to 114° (C_2H_5OH) IR (paraffin oil): 762 (γC–H of C_4H_3O), 1007 (δC–H of C_5H_4), 1115 (antisymmetrical C_5H_4 ring breathing); 1645 and 1663 (νC=O) UV (C_2H_5OH): λ_{max} (log ε) = 282 (4.26), 315 (4.20) UV (n-heptane): λ_{max} (log ε) = 273 (4.51), 305 (4.58) R_f: 0.49 (on SiO_2 with 4/1 $C_6H_6/CH_3CO_2C_2H_5$) R_f: 0.31 (on Al_2O_3 with 19/1 $C_6H_6/CH_3CO_2C_2H_5$)	[1,2]
*2	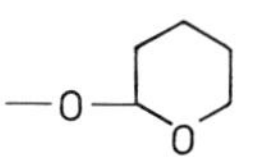	CH_2OH	IIa (61) orange crystals, m.p. 59 to 60° (ether/petroleum ether) ^{1}H NMR (CCl_4): 1.65 (s, 7H of R^1), 2.53 (s, OH), 3.65 (t, CH_2 of R^2), 3.80 (m, 1H of R^1 + 1H of C_5H_3), 4.00 (s, C_5H_5), 4.30 (d, 2H of C_5H_3), 5.10 (s, 1H of R^1)	[4]
*3	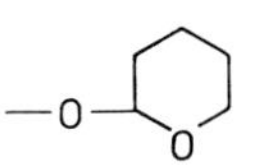	CHO	IIb, see No. 2, p. 384 IR: 1680 (νC=O)	[4]

References on p. 384

Table 32 (continued)

No.	$C_5H_5FeC_5H_3(R^1\text{-}1)R^2\text{-}2$ R^1	R^2	method of preparation (yield in %), properties, and remarks	Ref.
*4	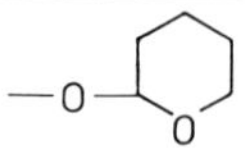	CO_2H	IIc (88) red crystals, m.p. 124° (cold ether/petroleum ether) IR: 1660 (νC=O) mass spectrum: m/e = 330 $[M]^+$	[3,4]
5	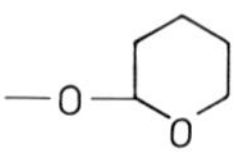	$CH{=}CH{-}CO_2H$	see No. 3, below IR: 1670 (νC=O)	[4]
6	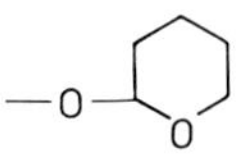	CO_2CH_3	see No. 4, below orange crystals, m.p. 91.5° (petroleum ether) IR: 1710 (νC=O) mass spectrum: m/e = 344 $[M]^+$	[4]

* Further information:

$OC_5H_9O{-}C_5H_4FeC_5H_4{-}CH_2OH$ (Table **32**, No. **2**) is oxidized to No. 3 by MnO_2 in $CHCl_3$ [4].

$OC_5H_9O{-}C_5H_4FeC_5H_4{-}CHO$ (Table **32**, No. **3**) yields No. 5 with malonic acid in pyridine [4].

$OC_5H_9O{-}C_5H_4FeC_5H_4{-}CO_2H$ (Table **32**, No. **4**) is esterified with CH_2N_2 in 5 h in ether to form No. 6, which is isolated by chromatography on an Al_2O_3 column with petroleum ether, in 87% yield [4].

References:

[1] A. Perjéssy, Š. Toma (Chem. Zvesti **23** [1969] 533/9). – [2] Š. Toma (Collection Czech. Chem. Commun. **34** [1969] 2235/48). – [3] R. Epton, G. Marr, G. K. Rogers (J. Organometal. Chem. **110** [1976] C42/C44). – [4] R. Epton, G. Marr, G. K. Rogers (J. Organometal. Chem. **150** [1978] 93/100).

Formula Index

The following index contains all the compounds described in "Eisen-Organische Verbindungen", Teil A, Ferrocen 7 (1980), and "Organoiron Compounds", Part A, Ferrocene 8 (1986). The compounds are listed in the order of increasing carbon content. Empirical formulas of ionic compounds are given in brackets; ions as well as components of solvates and adducts are separated by a dot.

No differentiation is made among stereochemical isomers.

References to "Eisen-Organische Verbindungen", Teil A, Ferrocen 7 (1980), are designated by **A 7**, and references to "Organoiron Compounds", Part A, Ferrocene 8 (1986, present volume) by **A 8**. Page references are printed in ordinary type, table numbers in bold-face, and compound numbers from the tables in italics.

Table of Conversion Factors

Following the notation in Landolt-Börnstein [7], values which have been fixed by convention are indicated by a bold-face last digit. The conversion factor between calorie and Joule that is given here is based on the thermochemical calorie, $cal_{th\,ch}$, and is defined as 4.184**0** J/cal. However, for the conversion of the "Internationale Tafelkalorie", cal_{IT}, into Joule, the factor 4.1868 J/cal is to be used [1, p. 147]. For the conversion factor for the British thermal unit, the Steam Table Btu, BTU_{ST}, is used [1, p. 95].

Force	N	dyn	kp
1 N (Newton)	1	10^5	0.1019716
1 dyn	10^{-5}	1	1.019716×10^{-6}
1 kp	9.8066**5**	9.8066**5** × 10^5	1

Pressure	Pa	bar	kp/m²	at	atm	Torr	lb/in²
1 Pa (Pascal) = 1 N/m²	1	10^{-5}	1.019716×10^{-1}	1.019716×10^{-5}	0.986923×10^{-5}	0.750062×10^{-2}	145.0378×10^{-6}
1 bar = 10^6 dyn/cm²	10^5	1	10.19716×10^3	1.019716	0.986923	750.062	14.50378
1 kp/m² = 1 mm H_2O	9.8066**5**	0.98066**5** × 10^{-4}	1	10^{-4}	0.967841×10^{-4}	0.735559×10^{-1}	1.422335×10^{-3}
1 at = 1 kp/cm²	0.98066**5** × 10^5	0.98066**5**	10^4	1	0.967841	735.559	14.22335
1 atm = 76**0** Torr	1.0132**5** × 10^5	1.0132**5**	1.033227×10^4	1.033227	1	760	14.69595
1 Torr = 1 mm Hg	133.3224	1.333224×10^{-3}	13.59510	1.359510×10^{-3}	1.315789×10^{-3}	1	19.33678×10^{-3}
1 lb/in² = 1 psi	6.89476×10^3	68.9476×10^{-3}	703.069	70.3069×10^{-3}	68.0460×10^{-3}	51.7149	1

Work, Energy, Heat	J	kWh	kcal	Btu	MeV
1 J (Joule) = 1 Ws = 1 Nm = 10^7 erg	1	2.778×10^{-7}	2.39006×10^{-4}	9.4781×10^{-4}	6.242×10^{12}
1 kWh	3.6×10^{6}	1	860.4	3412.14	2.247×10^{19}
1 kcal	4184.0	1.1622×10^{-3}	1	3.96566	2.6117×10^{16}
1 Btu (British thermal unit)	1055.06	2.93071×10^{-4}	0.25164	1	6.5858×10^{15}
1 MeV	1.602×10^{-13}	4.450×10^{-20}	3.8289×10^{-17}	1.51840×10^{-16}	1

$1\,\text{eV} \triangleq 23.0578\ \text{kcal/mol} = 96.473\ \text{kJ/mol}$

Power	kW	PS	kp·m/s	kcal/s
1 kW = 10^{10} erg/s	1	1.35962	101.972	0.239006
1 PS	0.73550	1	**75**	0.17579
1 kp·m/s	9.80665×10^{-3}	0.01333	1	2.34384×10^{-3}
1 kcal/s	4.1840	5.6886	426.650	1

References:

[1] A. Sacklowski, Die neuen SI-Einheiten, Goldmann, München 1979. (Conversion tables in an appendix.)
[2] International Union of Pure and Applied Chemistry, Manual of Symbols and Terminology for Physicochemical Quantities and Units, Pergamon, London 1979; Pure Appl. Chem. **51** [1979] 1/41.
[3] The International System of Units (SI), National Bureau of Standards Spec. Publ. 330 [1972].
[4] H. Ebert, Physikalisches Taschenbuch, 5th Ed., Vieweg, Wiesbaden 1976.
[5] Kraftwerk Union Information, Technical and Economic Data on Power Engineering, Mülheim/Ruhr 1978.
[6] E. Padelt, H. Laporte, Einheiten und Größenarten der Naturwissenschaften, 3rd Ed., VEB Fachbuchverlag, Leipzig 1976.
[7] Landolt-Börnstein, 6th Ed., Vol. II, Pt. 1, 1971, pp. 1/14.
[8] ISO Standards Handbook 2, Units of Measurement, 2nd Ed., Geneva 1982.